Mammals of Thailand

A list of the mammals of Thailand as of 2019 (2562)

Compiled by Michael Birt.

ISBN 978-1-912271-54-2

Published by Michael Birt, 2020
5/2 Moo 8,
Soi Mor Poon,
Sukhumvit Pattaya 15 Road,
Banglamung,
Chonburi,
Thailand 20150

Mammals of Thailand

A list of the mammals of Thailand as of 2019 (2562)

Compiled by Michael Birt.

Mammals of Thailand.

A list of the mammals of Thailand as of 2019 (2562)

Compiled by Michael Birt.

The last major work carried out with regards to information regarding the mammals of Thailand was the book

Lekagul, Boonsong; McNeely, Jeffrey A. (1977). *Mammals of Thailand*.

This book is an excellent publication but from 1977 until 2019, some 42 years, is a very long time, so I thought it would be a great idea due to the improved research techniques and advances in gathering information, that the time had come to update this excellent work.

Many mammals of Thailand occur throughout Asia and reference has been made to where these species occur in other parts of the world.

Index

Lesser Great Leaf-Nosed Bat
(Hipposideros turpis)
Page 330-331

Order: Pholidota (pangolins)

The order Philodota comprises the eight species of pangolin. Pangolins are anteaters and have the powerful claws, elongated snout and long tongue seen in the other unrelated anteater species.

Family: Manidae
Genus: Manis
Sunda Pangolin *(Manis javanica)*
Pages332-335

Chinese Pangolin *(Manis pentadactyla)*
Pages 336-339

Order: Cetacea (whales)

The order Cetacea includes whales, dolphins and porpoises. They are the mammals most fully adapted to aquatic life with a spindle-shaped nearly hairless body, protected by a thick layer of blubber, and forelimbs and tail modified to provide propulsion underwater.

Suborder: Mysticeti
Family: Balaenopteridae
Subfamily: Balaenopterinae
Genus: Balaenoptera
Minke Whale *Balaenoptera acutorostrata*
Pages 340-343

Sei Whale *(Balaenoptera borealis)*
Pages 344 – 349

Suborder: Odontoceti
Superfamily: Platanistoidea
Family: Phocoenidae
Genus: Neophocaena
Finless Porpoise
(Neophocaena phocaenoides)
Pages 350-352

Family: Physeteridae
Genus: Physeter
Sperm Whale *(Physeter macrocephalus)*
Pages 353-358

Family: Delphinidae (marine dolphins)
Genus: Steno
Rough-Toothed Dolphin
(Steno bredanensis)
Pages 359-361

Genus: Sousa
Indo-pacific Hump-Backed Dolphin,
(Sousa chinensis)
Pages 362-364

Genus: Tursiops
Bottlenose Dolphin *(Tursiops aduncus)*
Pages 365-368

The Indo-Pacific Bottlenose Dolphin
(Tursiops aduncus)
Pages 369- 372

Genus: Stenella
Pantropical Spotted Dolphin
(Stenella attenuata)
Pages 373-376

Spinner Dolphin
(Stenella longirostris)
Pages 377-379

Genus: Lagenodelphis
Fraser's Dolphin
(Lagenodelphis hosei)
Pages 380-382

Genus: Grampus
Risso's Dolphin
(Grampus griseus)
Pages 383- 385

Genus: Feresa
Pygmy Killer Whale *(Feresa attenuata)*
Page 386-387

Genus: Orcaella
Irrawaddy Dolphin
Orcaella brevirostris
Page 388-391

Order: Carnivora (carnivores)

The carnivores include over 260 species, the majority of which eat meat as their primary dietary item. Carnivores have a characteristic skull shape and dentition.

Suborder: Feliformia
Family: Felidae (cats)
Subfamily: Felinae
Genus: Catopuma
Asian Golden Cat (*Catopuma temminckii)*
Pages 392- 394

Genus: Felis
Jungle Cat *Felis chaus*
Pages 395-399

Genus: Pardofelis
Marbled Cat (*Pardofelis marmorata)*
Pages 400-401

Genus: Prionailurus
Leopard Cat (*Prionailurus bengalensis)*
Pages 402- 405

Flat-headed Cat (*Prionailurus planiceps)*
Pages 406-408

Fishing Cat (*Prionailurus viverrinus)*
Pages 409 -411

Subfamily: Pantherinae
Genus: Neofelis
Clouded Leopard (*Neofelis nebulosa)*
Pages 412-414

Genus: Panthera
Leopard (*Panthera pardus)*
Pages 415-421

Tiger *Panthera Tigris*
Pages 422-434

Family: Viverridae (civets, mongooses, etc.)

Subfamily: Paradoxurinae
Genus: Arctictis
Binturong *Arctictis binturong*
Pages 435-437

Genus: Arctogalidia
Small-Toothed Palm Civet
(*Arctogalidia trivirgata)*
Pages 438- 440

Genus: Paguma
Masked Palm Civet (*Paguma larvata)*
Pages 441-443

Genus: Paradoxurus
Asian Palm Civet
(*Paradoxurus hermaphroditus)*
Pages 444-446

Subfamily: Hemigalinae
Genus: Cynogale
Otter Civet *Cynogale bennettii*
Pages 447-449

Genus: Hemigalus
Banded Palm Civet
(*Hemigalus derbyanus)*
Pages 450- 451

Genus: Prionodon
Banded Linsang *(Prionodon linsang)*
Pages 452-454

Spotted Linsang (*Prionodon pardicolor)*
Page 455-457

Subfamily: Viverrinae
Genus: Viverra
Large-Spotted Civet (*Viverra megaspila)*
Pages 458- 460

Malayan Civet (*Viverra tangalunga)*
Pages 461-462

Large Indian Civet (*Viverra zibetha)*
Pages 463-464

Genus: Viverricula
Small Indian Civet (*Viverricula indica)*
Pages 465-466

Family: Tragulidae
Genus: Tragulus
Lesser Mouse Deer (*Tragulus javanicus)*
Pages 532-534

Napu (*Tragulus napu)*
Pages 535- 537

Family: Cervidae (deer)
Subfamily: Cervinae
Genus: Cervus
Eld's Deer *Cervus eldii*
Pages 538-545

Schomburgk's Deer (*Cervus schomburgki)*
Pages 546- 547

Sambar Deer (*Cervus unicolor)*
Pages 548-560

Subfamily: Muntiacinae
Genus: Muntiacus
Indian Muntjac *Muntiacus muntjak*
Pages 561-564

Family: Bovidae
(cattle, antelope, sheep, goats)
Subfamily: Bovinae
Genus: Bos
Gaur *Bos frontalis*
Pages 565-575

Banteng (*Bos javanicus)*
Pages 576-583

Kouprey *Bos sauveli*
Pages 584-586

Genus: Bubalus
Wild Asian Water Buffalo
Bubalus bubalis
Pages 587-591

Subfamily: Caprinae
Genus: Nemorhaedus
Mainland Serow
(*Nemorhaedus sumatraensis)*
Pages 592-593

Chinese Goral (*Nemorhaedus caudatus)*
Pages 594-596

This book is dedicated to the conservationists around the world who are trying against great odds to protect out fragile wildlife:

With thanks also to:

Saipin my wife, who as supported me throughout the 6 years it took to compile this book.

Along with

To Khun Wu, Khun Pui. Miss Jinta and all at Big Bee Farm, Pattaya, Thailand for their help, kindness and use of facilities, without, this book would never have been completed

Mammals of Thailand.

About Thailand

The country's official name was Siam (Thai: สยาม RTGS: Sayam, pronounced [sàjǎ:m]) until 23 June 1939, when it was changed to Thailand. It was then renamed Siam from 1945 to 11 May 1949, after which it was again renamed Thailand. Also spelled Siem, Syâm or Syâma, it has been identified with the Sanskrit Śyâma (श्याम, meaning "dark" or "brown"). The names Shan and A-hom seem to be variants of the same word, and Śyâma is possibly not its origin but a learned and artificial distortion.

The word Thai (ไทย) is not, as commonly believed, derived from the word Tai (ไท) meaning "independence" in the Thai language; it is, however, the name of an ethnic group from the central plains (the Thai people). A famous Thai scholar argued that Tai (ไท) simply means "people" or "human being" since his investigation shows that in some rural areas the word "Tai" was used instead of the usual Thai word "kon" (คน) for people.

The Thai use the phrase "land of the free" to express pride in the fact that Thailand is the only country in Southeast Asia never colonized by a European power. While the Thai people will often refer to their country using the polite form Prathet Thai (Thai: ประเทศไทย), they most commonly use the more colloquial word Mueang Thai (Thai: เมืองไทย) or simply Thai (Thai: ไทย); the word mueang (Thai: เมือง) meaning nation but most commonly used to refer to a city or town. Ratcha Anachak Thai (Thai: ราชอาณาจักรไทย) means "Kingdom of Thailand" or "Kingdom of Thai".

Etymologically, its components are: -Ratcha- (from Sanskrit raja, meaning "king, royal, realm") ; -ana- (from Pāli āṇā, "authority, command, power", itself from Sanskrit ājñā, same meaning) -chak (from Sanskrit cakra or cakraṃ meaning "wheel", a symbol of power and rule). The Thai National Anthem (Thai: เพลงชาติ), composed and written by Peter Feit during the extremely "patriotic" 1930s, refers to the Thai nation as: prathet-thai (Thai: ประเทศไทย). The first line of the national anthem is: prathet thai ruam lueat nuea chat chuea thai (Thai: ประเทศไทยรวมเลือดเนื้อชาติเชื้อไทย) and was translated in 1939 by Colonel Luang Saranuprabhandi as: "Thailand is the unity of Thai blood and body."

Totalling 513,120 square kilometers (198,120 sq. mi),[1] Thailand is the world's 51st-largest country by total area. It is slightly smaller than Yemen and slightly larger than Spain.

Thailand is home to several distinct geographic regions, partly corresponding to the provincial groups. The north of the country is the mountainous area of the Thai highlands, with the highest point being Doi Inthanon in the Thanon Thong Chai Range at 2,565 metres (8,415 ft) above sea level. The northeast, Isan, consists of the Khorat Plateau, bordered to the east by the Mekong River. The centre of the country is dominated by the predominantly flat Chao Phraya river valley, which runs into the Gulf of Thailand.

Southern Thailand consists of the narrow Kra Isthmus that widens into the Malay Peninsula. Politically, there are six geographical regions which differ from the others in population, basic resources, natural features, and level of social and economic development. The diversity of the regions is the most pronounced attribute of Thailand's physical setting.

The Chao Phraya and the Mekong River are the sustainable resource of rural Thailand. Industrial scale production of crops use both rivers and their tributaries. The Gulf of Thailand covers 320,000 square kilometer's (124,000 sq mi) and is fed by the Chao Phraya, Mae Klong, Bang Pakong and Tapi Rivers. It contributes to the tourism sector owing to its clear shallow waters along the coasts in the Southern Region and the Kra Isthmus. The Gulf of Thailand is also an industrial centre of Thailand with the kingdom's main port in Sattahip along with being the entry gates for Bangkok's Inland Seaport.

The Andaman Sea is regarded as Thailand's most precious natural resource as it hosts the most popular and luxurious resorts in Asia. Phuket, Krabi, Ranong, Phang Nga and Trang and their lush islands all lay along the coasts of the Andaman Sea and despite the 2004 Tsunami, they continue to be and ever more so, the playground of the rich and elite of Asia and the world.

Plans have resurfaced of a logistical connection of the two bodies of water which would be coined the Thai Canal, analogous to the Suez and the Panama Canal. Such an idea has been greeted with positive accounts by Thai politicians as it would cut fees charged by the Ports of Singapore, improve ties with China and India, lower shipping times and increase ship safety owing to pirate fears in the Strait of Melaka and, support the Thai government's policy of being the logistical hub for Southeast Asia.

The ports would improve economic conditions in the south of Thailand, which relies heavily on tourism income, and it would also change the structure of the Thai economy moving it closer to a services centre of Asia. The canal would be a major engineering project and has expected costs of 20–30 billion dollars.

The local climate is tropical and characterized by monsoons. There is a rainy, warm, and cloudy southwest monsoon from mid-May to September, as well as a dry, cool northeast monsoon from November to mid-March. The southern isthmus is always hot and humid.

Chiang Rai
Phayao
Chiang Mai
Nan
Mae Hong Son
Lampang
Phrae
Lamphun
Bueng Kan
Nong Khai
Uttaradit
Udon Thani
Nakhon Phanom
Loei
Nong-Bua-Lamphu
Sakon Nakhon
Sukhothai
Phitsanulok
Tak
Kalasin
Mukdahan
Khon Kaen
Kamphaeng-Phet
Phichit
Phetchabun
Maha-Sarakham
Yasothon
Chaiyaphum
Amnat-Charoen
Roi Et
Nakhon Sawan
Uthaithani
Ubon-Ratchathani
Chainat
Lopburi
Nakhon-Ratchasima
Singburi
Buriram
Surin
Sisaket
Suphanburi
Ang-Thong
Saraburi
Kanchanaburi
Ayutthaya
Nakhon-Nayok
Prachin-Buri
Nakhon-Pathom
Pathum-Thani
Sa Kaew
Ratchaburi
Chachoengsao
Chonburi
Phetchaburi
Rayong
Chanthaburi
Trat
Samut Prakan
Bangkok
Nonthaburi
Samut Sakhon
Samut Songkhram
Prachuap Khiri Khan
Chumphon
Ranong
Surat Thani
Phang Nga
Nakhon Si Thammarat
Krabi
Phuket
Phatthalung
Trang
Songkhla
Satun
Pattani
Yala
Narathiwat

Order: Proboscidea (Elephants)

Asian Elephant ***(Elephas maximus)***

Asian elephants (*Elephas maximus*) are smaller than their African savannah relatives (*Loxodonta africana*) and have many other physical features that distinguish them. The ears are smaller and the back is more rounded so that the crown of the head is the highest point of the body. One of the characteristic features of an elephant are the modified incisor teeth which are known as tusks, however, only some male Asian elephants have tusks, whilst females (cows) have 'tushes' instead, that are seldom visible. Elephants support their stocky body on stout, pillar-like legs, and the nose and upper lip are joined and elongated into a trunk. The trunk provides a wide variety of functions from feeding, vocalisation, bathing and fighting; those of the Asian elephant have only a single finger-like process on the base, whilst the African elephant has two. The thick, wrinkly skin covering the body is a greyish-brown colour and very dry.

Size:

Bull Asian elephants commonly weigh from 4,500 to 5,400 kg (9,900 to 12,000 lb) and are often around 2.7 m (8.9 ft) high at the shoulder. Females typically weigh around 2,000 to 4,000 kg (4,400 to 8,800 lb) and reach an average of 2.4 m (7.9 ft) at the shoulder. Head-and-body length is from 5.5 to 6.4 m (18 to 21 ft), with the tail measuring 1–1.5 m (3.3–4.9 ft) long. The skeleton constitutes about 15% of their body weight. One extraordinarily large bull elephant living in captivity at the Oregon Zoo, weighed 14,500 pounds (6,590 kg) in 2008.

The sizes of wild Asian elephants, especially males, have been exaggerated in the past. Record elephants may have measured as high as 3.7 m (12 ft) at the shoulder. Shoulder height is estimated using the rule of thumb of twice the forefoot circumference. The height of the adult male usually does not exceed nine feet [2.7 m], and that of the female eight feet [2.4 m]; but these dimensions are occasionally considerably exceeded. George P. Sanderson measured a male standing nine feet seven inches [2.9 m] at the shoulder, and measuring twenty-six feet two and one-half inches [8 m] from the tip of the trunk to the extremity of the tail; and he records others respectively reaching nine feet eight inches [2.9 m] and nine feet ten inches [3 m] at the shoulder. An elephant shot by General Kinloch stood upward of ten feet one inch [3.1 m]; and another measured by Sanderson ten feet seven and one-half inches [3.2 m]. These dimensions are, however, exceeded by a specimen killed by the late Sir Victor Brooke, which is reported to have reached a height of eleven feet [3.4 m]: and there is a rumour of a Ceylon elephant of twelve feet [3.7 m]. That such giants may occasionally exist is indicated by a skeleton in the Museum at Calcutta, which is believed to have belonged to an individual living between 1856 and 1860 in the neighbourhood of the Rajamahal hills, in Bengal. As now mounted this enormous skeleton stands eleven feet three inches [3.4 m] at the shoulders, but Mr. O. S. Fraser, in a letter to the Asian newspaper, states that it is made to stand too low, and that its true height was several inches more. If this be so, there can be no doubt that, when alive, this elephant must have stood fully twelve feet.

The heaviest bull elephant recorded was shot by the Maharajah of Susang in the Garo Hills of Assam, India in 1924, and weighed 8 tonnes (8.8 short tons), stood 3.35 m (11.0 ft) tall and 8.06 m (26.4 ft) long from head to tail.

Range Description:

In Thailand, the species occurs mainly in the mountains along the border with Myanmar, with smaller fragmented populations occurring in the peninsula in the south (in several forest complexes, south to the border with Malaysia); in the northeast (in the Dong Phaya Yen-Khao Yai forest complex, including Khao Yai National Park, and the Phu Khieo-Nam Nao forest complex); and in the east (in a forest complex composing the Khao Ang Runai Wildlife Sanctuary, Khao Soi Dao Wildlife Sanctuary, Khao Khitchakut National Park, and Khao Cha Mao National Park).

Population:

A recent estimate for the global population size of the Asian elephant was 41,410–52,345 animals Sukumar (2003) The estimated population size for each country was: Bangladesh 150–250; Bhutan 250–500; Cambodia 250–600; China 200–250; India 26,390–30,770; Indonesia 2,400–3,400; Lao PDR 500–1,000; Malaysia 2,100–3,100; Myanmar 4,000–5,000; Nepal 100–125; Sri Lanka 2,500–4,000; **Thailand 2,500–3,200;** and Vietnam 70–150 (Sukumar, 2003) . However, Blake and Hedges (2004) and Hedges (2006) argue that the oft-repeated global population 'estimate' of about 40,000 to 50,000 Asian elephants is no more than a crude guess, which has been accepted unchanged for a quarter of a century. They argue that with very few exceptions all we really know about the status of Asian elephants is the location of some (probably most) populations, with in some cases a crude idea of relative abundance; and for some large parts of the species range we do not even know where the populations are, or indeed if they are still extant. These difference of opinion are due in part to the difficulty in counting elephants in dense vegetation in difficult terrain, different survey techniques being used in different places, and a too-widely held belief that population monitoring is unimportant. Nevertheless, whatever the error margins, it appears almost certain that over 50% of the remaining wild Asian elephants occur in India.

The overall population trend of the Asian elephant has been downwards, probably for centuries. This remains the case in most parts of its range, but especially in most of the countries of South-east Asia. Within India, there is evidence that the large population in the Western Ghats in south of the country has been increasing in recent years due to improved conservation effectiveness.

Population Trend: Decreasing

Habitat:

_Asian elephants inhabit a wide range of grasslands and forest types, including scrub forest, rainforest and semi-cultivated forests, preferring areas that combine grass with low woody plants and trees.

Ecology and Biology:

Elephants are highly intelligent and long-lived animals; Asian elephants may live as long as 70 years, in captivity. They are extremely sociable and occur in groups of related females, led by the oldest female known as the 'matriarch'. Groups of Asian elephants average six to seven individuals, and will occasionally join with other groups to form herds; although these are more transient than those of African savannah elephants. Males leave their natal group when then reach sexual maturity at around six to seven years of age, after which time they are predominantly solitary. When males reach 20 years old they start coming into 'musth', an extreme state of arousal when levels of testosterone in the blood may increase 20 times. This state lasts about three weeks and during this time the individual will become aggressive and wander widely in search of females. Musth may cause males to fight for access to females and also increases their attractiveness to females. Cows only reach sexual maturity at ten years of age, and the interval between births may be as long as four years owing to the long gestation time and infant dependency. The single calf may suckle from other females in the group as well as their own mother.

Elephants use their dextrous trunk to pluck at grasses and pass them into their mouths; the average daily intake of food is 150 kilograms of vegetation a day. Grasses make up the mainstay of the Asian elephant's diet but scrub and bark are also eaten, and calves may eat their mothers dung to obtain nutrients. Consuming such large quantities of vegetation each day mean that Asian elephants substantially alter their ecosystem by creating new habitats for emergent vegetation. They also defecate up to 18 times per day, which has an important role in dispersing the seeds of many plant species. Where elephants occur near plantations they will readily feed on banana or rice crops. Asian elephants have had a close relationship with man over the centuries; they are still used to clear timber particularly in some of the more inaccessible forests of the continent, and play an important role in the religious and cultural history of the region. Adult females and calves may move about together as groups but adult males disperse from their mothers upon reaching adolescence. Bull elephants may be solitary or form temporary 'bachelor groups.

Cow-calf unit sizes generally tend to be small, typically consisting of 3 adult females who are most likely related, and their offspring; however, larger groups containing as many as 15 adult females may occur. There can also be seasonal aggregations containing over 100 individuals at a time, including calves and sub-adults. Until recently it was thought that Asian elephants, like African elephants, typically follow the leadership of older adult females, or matriarchs. But recently it has been shown that females can form extensive and very fluid social networks, with a lot of individual variation in the degree of gregariousness. Social ties generally tend to be weaker than in African elephants.

Elephants are crepuscular. They are mega herbivores and consume up to 150 kg (330 lb) of plant matter per day. They are generalist feeders, and both grazers and browsers, and were recorded to feed on 112 different plant species, most commonly of the order Malvales, and the legume, palm, sedge and true grass families. They browse more in the dry season with bark constituting a major part of their diet in the cool part of that season.

They drink at least once a day and are never far from a permanent source of fresh water. They need 80–200 litres of water a day and use even more for bathing. At times they scrape the soil for clay or minerals.

Elephants are able to distinguish low amplitude sounds. They use infrasound to communicate; this was first noted by the Indian naturalist M. Krishnan and later studied by Katharine Payne.

A healthy adult Asian elephant is not known to have natural predators, but there have been rare instances of tigers preying on young or weak elephants. Bulls will fight one another to get access to estrus females. Strong fights over access to females are extremely rare. Bulls reach sexual maturity around the age of 12–15. Between the age of 10 to 20 years, bulls undergo an annual phenomenon known as "musth". This is a period where the testosterone level is up to 100 times greater than non-musth periods, and they become extremely aggressive. Secretions containing pheromones occur during this period, from the paired temporal glands located on the head between the lateral edge of the eye and the base of the ear.

The gestation period is 18–22 months, and the female gives birth to one calf, or occasionally twins. The calf is fully developed by the 19th month but stays in the womb to grow so that it can reach its mother to feed. At birth, the calf weighs about 100 kg (220 lb), and is suckled for up to 2–3 years. Once a female gives birth, she usually does not breed again until the first calf is weaned, resulting in a 4–5-year birth interval. Females stay on with the herd, but mature males are chased away

Elephants' life expectancy have been exaggerated in the past; they live on average for 60 years in the wild and 80 in captivity.

Females produce sex pheromones; a principal component thereof, (Z)-7-dodecen-1-yl acetate, has also been found to be a sex pheromone in numerous species of insects.

Intelligence:

Asian elephants are highly intelligent and self-aware. They have a very large and highly convoluted neocortex, a trait also shared by humans, apes and certain dolphin species. Asian elephants have the greatest volume of cerebral cortex available for cognitive processing of all existing land animals. Elephants have a volume of cerebral cortex available for cognitive processing that exceeds that of any primate species, and extensive studies place elephants in the category of great apes in terms of cognitive abilities for tool use and tool making. Elephants are reported to go to safer ground during natural disasters like tsunamis and earthquakes, although there have been no scientific records of this.

Threats:

Numbers of Asian elephants were decimated by habitat loss and hunting throughout their historical range. Vast tracts of land have been logged or simply cleared to accommodate the growing human population in the region. Such disturbance from infrastructure development can also cause increased stress and confusion amongst elephants. Elephant populations have become increasingly isolated in the fragmented habitat that remains, often coming into conflict with local farmers. Crops are damaged and lives lost; up to 300 people a year are killed by elephants in India, leading to retaliation on local elephants. Poaching for ivory is also a threat and because only males have tusks, populations can become extremely skewed towards females, thus affecting breeding rates.

Major Threat(s):

The pre-eminent threats to the Asian elephant today are habitat loss, degradation, and fragmentation (Leimgruber et al., 2003; Sukumar, 2003; Hedges, 2006), which are driven by an expanding human population, and lead in turn to increasing conflicts between humans and elephants when elephants eat or trample crops. Hundreds of people and elephants are killed annually as a result of such conflicts. The long-term future of elephants outside protected areas, as well as in some protected areas, is therefore inextricably linked to mitigating such human–elephant conflicts, and this is one of the largest conservation challenges in Asia today (Sukumar, 1992, 2003; Hedges 2006).

Asian elephants live in the region of the world with the densest human population, growing at a rate of between 1–3% per year. Because elephants require much larger areas of natural habitat than most other terrestrial mammals in Asia, they are one of the first species to suffer the consequences of habitat fragmentation and destruction and because of its great size and large food requirements, the elephant cannot co-exist with people in areas where agriculture is the dominant form of land-use. In extreme cases, elephants have been confined as so called 'pocketed herds' in small patches of forest in landscapes dominated by man. Such 'pocketed herds' represent an extreme stage in the human–elephant conflict (Olivier, 1978). In other cases elephants have been caught and taken to so-called Elephant Training Centres where they languish, lost to the wild population (Hedges et al., 2005, 2006).

Poaching is a major threat to elephants in Asia too, although reliable estimates of the number of elephants killed and the quantities of ivory and other body parts collected and traded are scarce (Sukumar et al., 1998; Milliken, 2005). It has been argued that poaching is a relatively minor threat to Asian elephant because some males and all females lack tusks (Dawson and Blackburn, 1991). However, the reality is that elephants are poached for a variety of other products (including meat and leather) in addition to ivory, and poaching is now acknowledged as a threat to the long-term survival of some Asian elephant populations (e.g. Kemf and Santiapillai, 2000; Menon, 2002). Moreover, poaching of elephants for ivory is a serious problem in some parts of Asia (Sukumar, 1992; Menon et al., 1997). In Periyar Tiger Reserve in southern India, for example, ivory poaching has dramatically skewed adult sex ratios: over the 20-year period from 1969 to 1989 the adult male: female sex ratio changed from 1:6 to 1:122 (Chandran, 1990). Selective removal of tusked males has several implications for the affected populations: sex ratios obviously become highly female biased, genetic variation is reduced, and fecundity and recruitment may decline (Sukumar et al., 1998; Sukumar, 2003). Poaching of elephants is also a major problem in other parts of Asia. Large-scale hunting of elephants for ivory, bush meat, hides, and other products has reduced their populations significantly over a wide area from Myanmar to Indonesia (Menon et al., 1997; Duckworth and Hedges, 1998; Kemf and Santiapillai, 2000; Martin and Stiles, 2002; Menon, 2002; World Wide Fund for Nature, 2002a; Hedges et al., 2005).

Conservation Actions:

This species is listed on CITES Appendix I. The most important conservation priorities for the Asian elephant are:

1) conservation of the elephant's habitat and maintaining habitat connectivity by securing corridors; 2) the management of human–elephant conflicts as part of an integrated land-use policy that recognizes elephants as economic assets from which local people need to benefit or at least no suffer; 3) better protection to the species through improved legislation and law enforcement, improved and enhanced field patrolling, and regulating/curbing trade in ivory and other elephant products. Monitoring of conservation interventions is also needed to assess the success or failure of the interventions so that adjustments can be made as necessary (i.e. adaptive management). Reliable estimation of population size and trends will be needed as part of this monitoring and adaptive management approach.

Order: Sirenia (manatees and dugongs)

Sirenia is an order of fully aquatic, herbivorous mammals that inhabit rivers, estuaries, coastal marine waters, swamps, and marine wetlands. All four species are endangered.

Family: Dugongidae

Genus: Dugong

Dugong *(Dugong dugon)*

Description:

The dugong is the only entirely marine mammal to feed exclusively on plants, a trait that leads to its other common name of 'sea cow'. These large, rotund animals have short, paddle-like front flippers and a fluke-like tail, with a straight or concave edge, that is used for propulsion and when on the ocean floor are used for walking and not as some suggest for feeding . The thick skin is a brownish-grey colour but they are also variable in colour and there are short, coarse hairs sparsely distributed over the body but concentrated as bristles on the muzzle. All dugongs grow tusks but these only break the skin, and therefore become visible, in mature males. The large, rounded snout ends in a cleft, muscular upper lip that hangs over the down-turned mouth. As with all grazing mammals, dugongs have large complex stomachs but they also have an unique a adaptation for feeding on marine grasses. The upper lip protrudes considerably beyond the lower lip and is deeply cleft, forming a large u shaped muscular pad overhanging a small downwards opening mouth. The dental formation in immature dugongs is 2/4 0/0 0/0 5/5x2 =32 but is reduced to 1/0 0/0 /0/ 2-3/2-3 x 2 =10-14 in adults. The upper incisors are rootless and straight forming short thick tusks in males. In females the incisors do not pierce the gum suggesting that the incisors are not used for obtaining food but are used in some kind of sex linked function, and it is noted that males carry conspicuous scars which look to have been made by competing males.

Range Description:

Coastal and Island waters between East Africa and Vanuatu between latitudes of about 27° North and South of Equator. See Marsh et al. (2002).

Countries: Native:

Australia; Bahrain; Brunei Darussalam; Cambodia; China; Cocos (Keeling) Islands; Comoros; Djibouti; Egypt; Eritrea; India (Andaman Is., Laccadive Is., Nicobar Is.); Indonesia; Japan (Nansei-shoto); Jordan; Kenya; Madagascar; Malaysia; Mayotte; Mozambique; New Caledonia; Palau; Papua New Guinea (Bismarck Archipelago); Philippines; Qatar; Saudi Arabia; Seychelles; Singapore; Solomon Islands; Somalia; Sri Lanka; Sudan; United Republic of Tanzania, ; **Thailand;** Timor-Leste; United Arab Emirates; Vanuatu; Vietnam; Yemen (Socotra)

Regionally extinct:

Mauritius (Rodrigues - Native); Taiwan, Province of China

FAO Marine Fishing Areas: Native:

Indian Ocean – eastern; Indian Ocean – western; Pacific – northwest; Pacific – western central

Population:

Total population size is unknown. See Marsh et al. (2002).
Population Trend: Unknown

Biology:

Dugongs are usually seen as solitary individuals or in a group of two, although larger groups of several hundred individuals have often been recorded. The most stable and long-lasting groups appear to be mother and calf pairs. The single calf, born under water, after a gestation period of around 13 to 14 months, swims to the surface to take its first breath measures over a metre in length at birth, and suckles from the female for around 18 months. The mothers breast is under the flipper and the young reach from the back to suckle, and it is said that the mother holds the young with one flipper, thus giving rise to tales of mermaids. Very little is known about maturity and period of growth to maturity but two captives in India were 1600 mm and 1969 mm in December 1959 and 2053 mm and 2130 mm in June 1996 six and a half years later so growth must be slow and continue for relatively long periods of time. Dugongs can be extremely long-lived, reaching ages of 70 years or more. Young have been found in different stages of development in each month which suggest that breading takes place throughout the year. This is in marked contrast to most seals and other herd animals but this is quiet understandable as the live in a close environment with plentiful food on hand throughout the year, making the timing of birth non dependent on food supply. A year round food source may also reinforce pair bonding and group cohesiveness.
Both dugongs and manatees have a low metabolic rate, allowing them to exist on a herbivorous diet, and consequently they usually move relatively slowly. They have pectoral mammary glands reminiscent of human breasts. These features, and their nursing behaviour, may have caused sailors to liken them to mermaids or sirens; hence the order name of 'Sirenia'. Although most sea grass beds upon which dugongs feed occur at depths of 1 to 5 metres, they are known to feed at depths of up to 33 metres. Using the flexible upper lip to rip out whole plants, dugongs leave characteristic furrows known as 'feeding trails' on the sea floor. Dugongs are more closely related to elephants than the cows after which they are named, and have a particularly long large intestine to aid digestion. Nowadays dugongs are seen only in small groups or singularly but in the past they formed huge herds of many thousands. Pair bonds have been noted amongst and it is possible that family groups exist within larger populations. The primary defence are the largeness of the herd and they have been seen to gang up on sharks in shallow water and drive them away by butting them with their heads. Also dugongs have a non-deciduous placenta, so that there is no bloody afterbirth to attract sharks or crocodiles to the breeding area.

Threats:

The dugong has been traditionally persecuted by humans throughout much of its range for its meat, hide and oil. Its rather slow movement, large size and dependence on coastal habitats have made the dugong particularly vulnerable to human impacts, while the low reproductive rate, long generation time and high investment in each offspring mean populations can take a long time to recover from any losses. Fishing nets have also been a major cause of population decline, as dugongs are unable to hold their breath for more than about 12 minutes and therefore easily drown once entangled. In addition, the sea grass ecosystems on which this species depends are highly sensitive to human impacts, such as from mining, trawling and dredging.

Dugongs have decreased in number throughout their range to the extent that in some areas only relict populations remain, but a significant stronghold still persists in Australia . Even here, however, these animals are under threat from fishing nets, habitat loss from the silting of sea grass beds, pollution, boat traffic and illegal hunting

Order: Scandentia (tree shrews)

The tree shrews are small mammals native to the tropical forests of Southeast Asia. Although called tree shrews, they are not true shrews and are not all arboreal.

Family: Tupaiidae (tree shrews)

Genus: Dendrogale:

Northern Smooth-Tailed Tree Shrew ***(Dendrogale murina)***

Range Description:

Historical records include sightings in southern Vietnam, **South-East Thailand** and probably Cambodia. Recent records show their range extending much further north in Vietnam (at least to 17º20'N and probably to 21º27'N), into Lao PDR, where it is confined to the south-east, and confirm their presence in Cambodia, where currently localized to Mondulkiri Province (Timmins et al. 2003)

Countries: Native:

Cambodia; Lao People's Democratic Republic; **Thailand;** Vietnam.

Description:

Size. Head and Body: 115mmto 135mm. Tail: 105mm to 130mm

The pelage is short and soft with the upper parts a mixture of brownish black and ochraceous, the under parts and inner side of the legs are buffy. There is a dark band from the vibrissae through the eye to the base of the ear, with a whitish stripe above and below, the lower stripe widening behind the vibrissae and merging with the buffy neck and chest. A narrow blackish line runs from the runs from the mouth to the root of the cheek vibrissae. The male is larger than the female, and much richer in colouring with a darker tail and more tawny upper parts, richer buff underparts and brighter yellow buff on the head (Kloss 1916a). The fifth digit on the hind foot is relatively longer than in *Tupaia,*and the first digit is smaller, the claws are short and blunt. The ears are larger and hairier than in *Tupaia* but considerably smaller than in *Ptilocerus.* There are 2 mammae. The skull is similar to *Tupaia* but with a more rounded braincase and less conspicuous temporal ridges; the supraorbital foramen is well developed. The dentition is quite similar to Tupaia except the upper canine as two roots as in *Ptilocerus*. Napier and Napier(1967) say it is the smallest of the tree screws but there is considerable overlap with *T.Minor.*

Habitat and Ecology:

Most records of the species are from evergreen forest (at varying stages of degradation), but it has also been found in mixed deciduous forest, extensive secondary bamboo lacking any dicotyledonous canopy, and in streamside tangles amid rocky savannah. Records range from the plains up to 1,500 m asl (Timmins et al. 2003). The species mainly uses the under- and mid-storeys, but also enters the canopy; recent observations come primarily from under-storey tangles, especially of bamboo, almost exclusively between 30 and 300 cm above ground level (Timmins et al. 2003). In **Thailand,** has been observed on the branches of fruit trees (Timmins et al. .2003).

The diet consists mainly of insects, especially beetles but they also eat fruit. Although they are quite common in some of there habitats they are quite shy and difficult to observe or capture. The omit a shrill call when they are climbing but very little else is know of their vocalizations Very little is known about the reproductive cycle but it thought that litter sizes are small due to fact that they have less fewer mammae than *T. glis*

Systems: Terrestrial

Major Threat(s):

There are no major threats to the species, but they may be undergoing localized declines in parts of their range due to habitat loss. It is present in several national parks and protected areas. Cat Tien National Park in Vietnam, and in Lao PDR in Hin Namno National Biodiversity Conservation Area, Nam Ghong Provincial Protected Area, and Hin Namno NBCA. It is listed on CITES Appendix II.

Genus: Tupaia:

Northern Tree Shrew *(Tupaia belangeri)*

Physical Description:

Northern tree shrews have greyish, olive fur with an elongated snout (Smithsonian National Zoological Park, 2012). The dental formula of the Tupaiidae is incisors 2/3, canines 1/1, pre-molars 3/3, and molars 3/3 (Fuchs and Corbach-Sohle, 2009). Northern tree shrews are moderately sexually dimorphic (Collins and Tsang, 1987; Schehka et al., 2007). Male tree shrews have a larger body size and ring of white hair around the eye compared to females. Males also have a broader skull than the females (Collins and Tsang, 1987). (Collins and Tsang, 1987; Fuchs and Corbach-Sohle, 2009; Schehka, et al., 2007; Smithsonian National Zoological Park, 2012)Fuchs and Corbach-Sohle (2009) describe the difficulty in determining the sex of young tree shrews. As pups the external genitalia of both males and females look alike. Males have a slender and elongate penis, which is posterior to scrotal testes, which can be retracted into the abdominal cavity if the individual is stressed. In females the clitoris is greatly elongated and grooved on its ventral surface. In neonatal shrews the urethra opens together with the vagina as a single opening at the base of the clitoris; this is what looks like the males penis in young female tree shrews. However the clitoris does not have a tubular sheath like the penis. (Fuchs and Corbach-Sohle, 2009) Northern tree shrews have a mass of approximately 50 to 270 grams, Head to Body Length of 12 to 21 cm and a tail length ranging from 14 to 20 cm, usually close to the length of their body (Fuchs and Corbach-Sohle, 2009). Shrew body temperature has been described, ranging from about 35 degrees Celsius to 40 degrees Celsius; this 5 degree difference is much larger than most endothermic animals (Fuchs and Corbach-Sohle, 2009). (Fuchs and Corbach-Sohle, 2009)

Reproduction:

Northern tree shrews have a monogamous mating system (Collins and Tsang, 1987). Introductions between a mating pair are possibly the hardest part of breeding for northern tree shrews. Copulation can occur within a few hours if the female accepts the male (Fuchs and Corbach-Sohle, 2009). However, they usually use more aggressive behaviour. Fuchs and Corbach-Sohle (2009) suggest two explanations: first, that the individuals merely don't like each other, and second that to mate one animal has to enter the others territory. Northern tree shrews, male or female, will defend their territory against intruders. If the couple can overcome this they will form a stable breeding pair and continue to breed together. Females have an 8 to 12 day oestrous cycle. Ovulation is thought to be induced by copulation (Martin, 1990) (Collins and Tsang, 1987; Fuchs and Corbach-Sohle, 2009; Martin, 1990)

Mating System

Monogamous:

Under natural and artificial conditions Fuchs and Corbach-Sohle (2009) observed that breeding can occur at any time of the year; no peak periods were found. Males and females become sexually reproductive around the same time. Males will be active between four and five months and females can give birth to their first litter at approximately 4 months old. Northern tree shrews have a litter size of 1 to 5 young. The gestation period occurs for approximately 41 to 45 days and females give birth to hairless, altricial young. Pups ears open at around 10 days and their eyes open at approximately 20 days after birth. Female young tend to be heavier than males, but on average young will have a birth mass of 6 to 10 grams (Fuchs and Corbach-Sohle, 2009). Young will drink their mother's milk until they are around 35 days old when they wean off the liquid diet. Northern tree shrews reach puberty at about 2 months old and can be separated from their mother at about 50 to 60 days (Fuchs and Corbach-Sohle, 2009). Older mothers may have problems with infertility, still birth, cannibalism or abortion. These problems can also occur in females that are stressed, which is likely in tree shrews as they are highly susceptible to stressors.

Habitat and Ecology:

This widespread species is found in deciduous and evergreen primary forest and secondary forest, commonly in karst and associated natural scrub vegetation, from sea level up to upper montane areas (Molur et al. 2005; W. Duckworth pers. comm.). It is highly adaptable and may be found in oil palm plantations (K. H. Han pers. comm.), coconut plantations (Parr 2003), and regenerating scrub and ruderial mixes above abandoned dry rice fields (W. Duckworth pers. comm.). It has been seen very far from tall forest, and is probably able to live independently of tall forest (such as in far northern Lao PDR; W. Duckworth pers. comm). Immediately after birth the mother will nurse her pups and thereafter only return to feed her young once every 48 hours (Fuchs and Corbach-Sohle, 2009). They receive approximately 5 to 10 millilitres of milk, with a relatively high fat content of about 25%. Pups consume this amount of milk in approximately 2 to 10 minutes. As a result pups have less than 2 hours contact with their mother during the 30 to 35 day period. Martin (1990) described that as the lowest mother to infant contact and smallest parental investment for viviparous mammals described thus far. Males are not involved in parental care after copulation (Fuchs and Corbach-Sohle (2009). (Fuchs and Corbach-Sohle, 2009. Typically northern tree shrews live in monogamous pairs (Smithsonian National Zoological Park, 2012). One adult male and female share fights are documented between adults of the same sex (Fuchs and Corbach-Sohle, 2009). Tree shrews use scent marking, excreted from a gland on their chest, to mark their territories (Fuchs and Corbach-Sohle, 2009). The young tree shrews leave the nest around the onset of puberty and they become completely self-sufficient (Collins and Tsang, 1987). (Collins and Tsang, 1987; Fuchs and Corbach-Sohle, 2009; Smithsonian National Zoological Park, 2012) Northern tree shrews have relatively stable home ranges of about 2 acres or 0.81 hectares (Smithsonian National Zoological Park, 2012). (Smithsonian National Zoological Park, 2012) Northern tree shrews make 8 distinct sounds (Fuchs and Corbach-Sohle, 2009). Of these 4 can be associated with functional categories that are described by Binz and Zimmermann (1989): alarm, attention, contact, and defence. Tree shrew noises can range from 0.4 to 20 kHz. The structure of the sounds depends on the status and motivation of the individuals; their pitch increases with fear and decreases in pitch (or increases in frequency) with increased aggression (Kirchhof et al., 2001). Kirchhof et al. (2001) also note that tree shrews do not use ultrasonic localisations. Tree shrews use scent marking to indicate boundaries of their territories (Fuchs and Corbach-Sohle, 2009).(Binz and Zimmermann, 1989; Fuchs and Corbach-Sohle, 2009; Kirchhof, et al., 2001. It has been noted that tree shrews will react to predation risk (Schehka et al., 2007; Fuchs and Corbach-Sohle, 2009). They will display defensively to a threat, specifically baring teeth and making high pitched, loud vocalizations (Schehka, 2007). Specific predators of northern tree shrews are not specified. However, due to their size and behaviour possible predators could include large birds of prey, snakes, and potentially some carnivorous mammals. While the main food source of northern tree shrews is insects, they also eat fruit to supplement their diet. It could be inferred that they play a role in the seed dispersal of fruit bearing trees.

Systems: Terrestrial

Range:

Found in South-East Asia, North of the isthmus of Kra, including: **Thailand,** Bangladesh, Burma, far North-Eastern India and Nepal, Southern China, Cambodia, Lao PDR, Vietnam, and associated coastal islands, including Hainan. It probably also occurs on Preparis Island north of the Andaman Islands (Helgen 2005). Listed as present in Bhutan by Molur et al. (2005). Recorded to 3,000 m in China (E. Smith pers. comm).

Countries: Native:

Bangladesh; Bhutan; Cambodia; China; India; Lao People's Democratic Republic; Malaysia; Myanmar; **Thailand;** Vietnam

Major Threat(s):

There are no major known threats to this species.

Conservation Actions:

It is found in numerous protected areas throughout its range, such as **Kang Kachan National Park (Thailand)** and Mehao Wildlife Sanctuary, Namdapha National Park, and Singhalila National Park (India). It is listed on CITES Appendix II.

Common Tree Shrew *(Tupaia glis)*

Description:

The common tree shrew is one of the largest among tree shrews. Average body length is between 16–21 cm (6.3–8.3 in), and average weight is around 190g, with varying colours of reddish brown, greyish or black upper parts and whitish belly. Its long bushy tail is dark greyish brown and almost reaches the length of the body. The paws are bare with sharp nails, and there is a naked patch of skin above its long nose. Both sexes are similar. The measurements of the Tupaia glis according to 21specimens are head- to-body length: 170mm to 235mm, tail length: 170mm to 242mm and hind foot: 45mm to 56mm. The common tree shrew usually has a pale stripe on each shoulder.

There are two subspecies *T.g. longipes* and *T.g. salatana*, with *T.g. longipes* being duller in coloration than *T.g. salatana.* The under parts of longipes are dull buff to reddish buff, and the underside of the tail is greyish. The under parts and underside of the tail are dark reddish in *salatana.* Similar species are *Tupaia splendidula* and *Tupaia montana.*

Distribution and Habitat:

Common tree shrews occur south of about 10° N latitude in **Southern Thailand** through mainland Malaysia and adjacent coastal islands to Singapore. In Indonesia, they are found on the islands of Siberut, Batu, Sumatra, Java, Bangka, Riau, Lingga and Anambas.Usually they are found in primary dipterocarp forest, but are tolerant to some degree of habitat modification. They have also been recorded from secondary forest, plantations, fruit orchards and trees near housing areas.

This species is probably present throughout the lowlands and hills up to 1,100 m (3,600 ft) in the Kelabit Highlands of Borneo. The subspecies *T.g. longipes* occurs in the north of Borneo, in Sarawak, and in East Kalimantan including Sabah. The subspecies *T.g. salatana* occurs in the south of Rajang River and Kayan River in Borneo.

Common tree shrews inhabit protected areas, including Pasoh Forest Reserve on the Malay Peninsula and Krau Wildlife Reserve.

Common tree shrews are active during the day, and forage for food alone or in pairs, mainly on the ground, among shrubs and tree holes. They feed on fruits, seeds, leaves, and insects especially ants and spiders. They are also reported to catch lizards.

They are very agile in climbing both large vertical tree trunks and bushes, and occasionally jump from stems of a young tree to that of another as much as 60 cm (24 in) away. Their climbing is concentrated in lower heights. They frequently scent-mark their territories by chest and and-genital rubbing with a secretion from glands on chest and scrotum. Adult males are more secretory than females and juveniles. In the Bukit Timah Nature Reserve, mean home ranges of adult males were estimated at 10,174 m2 (109,510 sq ft), of adult females at 8,809 m2 (94,820 sq ft), of juvenile males at 7,527 m2 (81,020 sq ft), and of juvenile females at 7,255 m2 (78,090 sq ft), with partial overlaps between male and female ranges varying from 0.4% to 56.8%. Home ranges of adult residents of the same sex overlap to a lesser degree than those of opposite sexes. A male's range may include the ranges of two or three females. A high overlap between ranges of one adult male and one adult female indicates that they form a stable pair. Juvenile ranges of either sex adjoin or overlap with ranges of adults, suggesting that the juveniles are family members. Individuals of the same sex are involved in aggressive territorial chases.

Juvenile males depart from their family's territory sooner than juvenile females.

Reproduction:

Both sexes of common tree shrews are sexually mature at the age of about 3 months. In captivity, females give birth for the first time at the age of about 4.5 months, usually in February. A postpartum oestrus results in more births in April. Their oestrus cycle is 8 to 39 days, and the gestation period lasts 40 to 52 days, after which a litter of one to three individuals is born. The new born offspring weighs about 10 to 12 grams. Females suckle their young every other day, and neglect their young as long as possible. They won't even be able to identify their own young if they didn't mark them with the scent produced from glands in their sternum and abdomen. Juveniles leave the nest between 25 to 35 days of age. Longevity of a captive common tree shrew has been recorded as 12 years and 5 months.

From October to December, common tree shrews are reproductively inactive. The mating season starts at the onset of the monsoon season in December and lasts until February. Oestrus and pre-oestrus behaviour is characterized by adult males pursuing adult females. Males emit chattering, and appear to be extremely excited. They also chase each other and fight. Females do not actively choose a partner among the male participants of chases. The dominant male gains access to females.

In tropical rainforest habitats in West Malaysia, density of common tree shrews varies from about two to five animals per hectare. Their annual breeding coincides with the abundance of invertebrates after the dry season. Their main reproductive period is between February and June, and their litter size invariably two. Some females breed more than once a season, and the age at first pregnancy is seven months. The main period of emigration or mortality of young is during the breeding period or monsoon.

Threats:

Common tree shrews are threatened due to deforestation and ensuing human activities in agriculture, plantation and commercial logging. Moreover, other pressures such as hunting for food and sport can create pressure to the species.

In science:

Tupaia glis are being used by researchers as animal models for human diseases because of their close relationship to primates, and their well-developed senses of vision and hearing. Research studies have included hepatitis.

Pygmy Tree Shrew (*Tupaia minor*)

Description:

The Pygmy tree shrew (Tupaia minor) is a species of tree shrew in the family *Tupaiidae*. It is found in Indonesia, Malaysia, and **Thailand.** Tupaia minor is also known as the pygmy tree shrew and is from the order Scandentia. The generic name is derived from tupai, a Malay word for squirrels (Nowak, R., 1999). The upperparts are finely ticked with black, often with a reddish tinge, especially over the lower parts of the back. The hairs are also dark at the base. The shoulder stripe is narrow and indistinct, usually whitish or buff in colour, with the hairs the same colour throughout. The dorsal surface of the tail is also darker than the body, sometimes almost black, with the tail similar in colour, but with a longitudinal of buff or orange. It as between 4 to 6 mammae. The skull and dentition is the smallest in the genus (Lyon 1913). The rostrum is short and stubby, the braincase large and rounded. The median upper incisors are larger than *T. glis* but the lower third incisors are smaller; the lower canine is better developed than in *glis.minor* can be distinguished from other tree shrews by its appearance. It has upper body hair banded light and dark, giving a speckled olive-brown appearance. The upper parts are buffy and often have a reddish tinge towards the rear (Payne et al., 1985). The limbs are equal in length and have long claws. The maximum total length is about 450 mm, half of which is the tail (Feldhamer et al., 1999). The tail is long and thin, and its upper side is darker than the body (Payne et al., 1985).

Distribution:

T.minor is distributed in **Peninsular Thailand,** peninsular Malaysia, Sumatra, the Lingga Archipelago (Indonesia), Borneo, offshore islands of Laut (Indonesia), and Banggi and Balambangan (Malaysia). From the catalogue of mammal skins in Sarawak Museum, Kuching, Sarawak, more than 30 individuals of *Tupaia minor* have been collected from 1891 to 1991. The specimens were mostly caught in Mt Penrisen, Mt Dulit, Mt Poi, Gunung Gading, Bau, Ulu Baram, Saribas, Kuching and Forest Research.

The species has no fossil record (Jacobs, 1980).

Behaviour, diet and reproduction:

T. minor is diurnal (active in the daytime). It is often seen 3 to 8 metres above the ground, sometimes up to 20 metres, travelling along lianas or branches of small trees (Payne et al., 1985). They spend most of their time on the ground and in low bushes, nesting in tree roots and fallen timber. According to Lelevier et al., 2005, *T. minor* moves in a semiplantigrade posture which allows it to keep its centre of gravity close to the tree. The claws on its hands and feet are quite sharp and moderately curved, which is useful for climbing.

Captive studies show that both sexes maintain dominance hierarchies based on aggressive interactions with conspecifics. Males and females maintain pair bonds and live within the same territory together. The mechanisms behind the formation of this bond are not known. When random pairs of individuals were put together in captivity, 20% of the interactions resulted in violence. In 60% of the cases the pairing was amicable. Although the pair did not mate, they also did not fight. Mating only occurred in the remaining 20% of cases. (Emmons, 1991; Holst, 1990; Nowak, 1999; Sargis, 2001)

Home range data for *T. minor* is sparse within the literature, but members of the genus *Tupaia* hold territories that range in size from 500 to 8,000 m^2. Mated pairs and their offspring, which have not yet reached sexual maturity, hold these territories. Either urine or secretions from neck glands are used to designate territorial boundaries. *tupaia minor* probably is similar in these regards. Holst, 1990 The digestive tract of *T. minor* is comprised of a small, simplified stomach, a long, narrow small intestine, a narrow pouch-like cecum, and a smooth large intestine. The gut of *T. minor* lacks the plant fermentation micro-organisms needed to digest plant cell walls. The through time for digested materials ranges from 20 minutes to 45 minutes.

Tupaia minor is omnivorous. Its diet is comprised of a wide variety of invertebrates (including insects and worms) and fruit.

The method of processing fruit for ingestion is dependent on the size and rigidity of the item. Captive observations showed that small fruit is placed into the mouth and chewed. The skin and seeds are ejected from the mouth, and juice and soft pulp are ingested. Ejection of inedible items is accomplished with the help of a forepaw or the tongue. Observations of *T. minor* in the wild differ slightly. Wild individuals very rarely eject seeds. Fruit that is firm and larger than the gape is often problematic. The weak jaw and procumbent incisors of these animals inhibit their ability to bite off whole chunks from tough fruits. Thus, large or tough fruit is either chewed in the side of the mouth, or licked vigorously. Wild tupaiids tend to concentrate their frugivory on soft fruits that they can handle efficiently. (Emmons, 1991; Holst, 1990; Nowak, 1999; Shanahan and Compton, 2000)

Communication in these mammals has not been extensively documented. Members of the genus Tupaia are known to scent mark their territories with secretions from glands as well as with urine. Because they are diurnal, visual cues are likely to be important in communication. Most mammals also use vocal and tactile communication, especially when mating. (Holst, 1990; Nowak, 1999)

Tupaia minor is omnivorous; its diet includes insects and fruit. *Scandentia* has little economic significance because they do little damage to crops or plantations. However, it has been suggested that *T. minor* is a seed disperser for several Ficus fig species (Shanahan and Compton, 2000).

Litters of one to three young are born after a gestation period of 45 to 55 days. They breed throughout the year with a seasonal peak from May to July. According to Hose(1893) "It breeds in a nest in an old stump covered with creepers". But he was not sure if it makes the nest itself or uses the old nest of a bird.

The rearing strategies of *Tupaidea,* including *T. minor*, are somewhat peculiar. Prior to giving birth, a female makes a large nest, consisting of leaves, wood chips, and other assorted building materials, within a fallen tree. This nest provides the young with protection from the elements and predators. The female then gives birth to her young within the nest. The young are born naked with their ears and eyes closed. Within a few minutes of birth, the young take on approximately half their initial body weight in milk. After feeding, the bellies of the young become distended from the body. The mother then leaves the nest and only returns every 48 hours to feed the young. Each visit lasts only 5 to 10 minutes. The high fat content of Tupaiid milk (~26%) enables the young to maintain a relatively high body temperature (~37 C). The high protein content of the milk (~10%) also helps the young to maintain a rapid growth rate. After approximately 4 weeks the young are able to leave the nest. This occurs with only approximately 90 minutes of parental care from the female. Males are not known to participate in care of the young. (Holst, 1990; Nowak, 1999)

Their maximum lifespan is around nine to ten years (Lelevier et al., 2005).

Threats:

The Pygmy tree shrew is listed as Least Concern (LR/lc), lowest risk. Does not qualify for a more at risk category. Widespread and abundant taxa are included in this category, on the IUCN Red List of Threatened Species

Family: Ptilocercidae:

Genus: Ptilocercus:

Pen-Tailed Tree Shrew *(Ptilocercus lowii)*

The pen-tailed tree-shrew (*Ptilocercus lowii)* is a species of tree shrew in the *Ptilocercidae* family. It is found in Indonesia, Malaysia, and **Thailand.** It is the only species in the genus *Ptilocercus* and the family *Ptilocercidae*. All other tree-shrews are in the family *Tupaiidae*.

The pen-tailed tree-shrew is the only known wild mammal that chronically (i.e., not just occasionally) consumes alcohol. A study of the tree-shrew in Malaysia found that it spends several hours consuming the equivalent of 10 to 12 glasses of wine with an alcohol content of up to 3.8% every night drinking naturally fermented nectar of the bertam palm. Despite consuming relatively large amounts of alcohol, the pen-tailed tree-shrew does not become intoxicated. Measurements of a biomarker of ethanol breakdown suggest that they may be metabolizing it by a pathway that is not used as heavily by humans, a fellow member of the grand order of mammals *Euarchonta.* The species in **Thailand** is *P.l continentus* (Thomas1910)

Description:

Pen-Tailed Tree Shrews have a slender build and a long tail. They have well developed senses of hearing, smell and vision. The fur is short and velvety, dark greyish brown above and yellow grey below; the feet are brown, the toes whitish. There is a distinct white patch on either side of the muzzle and in some specimens the patch may almost be a stripe, extending more than halfway to the base of the ears. In contrast with other tree shrews, the ears are large and thin, standing away from the head; they are quite mobile, reflecting the greater reliance on hearing in a nocturnal animal. The whiskers are also more longer and rigid than in other tree shrews, and seem to be a more developed tactile organ. The eyes are larger and face more forward than in other tree shrews, and the retina of the eye is composed entirely of rods which is an adaptation for night vision. Locomotion on the ground is with a series of hops, with the tale inclined upwards and the tail almost vertical. In sleep the body is curled up in a ball with the tail curled around the body so that the feather of the tail covers the face. If they are disturbed they make a horse snarling hiss with the moth open, rather like the moon rat *(Echinosorex).* They are not at all shy

Habitat:

Pen-Tailed Tree Shrews can be found in the forests of Indonesia, Malaysia and **Thailand.** They are nocturnal and arboreal, inhabiting primary or secondary forest and rubber plantations, from sea level to 1000 meters. They usually seen in pairs, although as many as 5 have been seen in a single nest (Muul and Lim1971) . They sometimes nest in epiphytes in tall trees or in hollow branches 15 to 20 meters high, in a nest made of twigs, leaves and fibrous material and roughly lined with green leaves.

Diet:

Pen-Tailed Tree Shrews are omnivores and they feed on a variety of insects, small vertebrates, fruit, nectar and seeds.

They are one of only seven wild animals that are known to consume alcohol. Each night they spend up to 2 hours drinking naturally fermented nectar from the bertam palm. Despite this consumption of alcohol, the Pen-Tailed Tree Shrew does not become intoxicated which suggests that the alcohol is metabolized in a different way than that used by humans.

Breeding:

After a gestation period of approximately 50 days, a litter of 3 - 4 young are born. At birth the young are blind and hairless, but they are able to leave the nest when they are a month old.

Tree Shrews reach sexual maturity at around 4 months old and they generally breed throughout the year with no defined breeding season.

Life Cycle:

The pen-tailed shrew is rare and difficult to capture so very little is known about its reproductive habits. It is known that they have two pairs of teats or nipples which suggest a litter size of one to four young. Other species of this order have gestations or pregnancies that last forty five to fifty five days. They also nurse their young, letting the young get stuffed then leaving them for up to forty eight hours before repeating the process. The mothers have a scent the spray on the nest that drives other shrews away. The young are ready to leave the nest when they are four weeks old. Whether these habits that are prevalent in other species of the pen-tailed shrew's order are the same in the pen-tailed shrew has not been confirmed.

Subspecies:

Subspecies of the Pen-Tailed Tree Shrew are:

Ptilocercus lowii continentis
Ptilocercus lowii lowii

Interesting Facts:

Tree Shrews have the highest brain to body mass ratio of any animal, even higher than humans.

Conservation:

The pen-tailed tree shrew is classified as least concern on the IUCN (International Union of Conservation of Nature and Natural Resources) red list of threatened species. This classification is the lowest and means the species has a large widespread, population and no current threats that would likely decrease its population in the foreseeable future.

Order: Sunda colugo (colugos)

The two species of colugos make up the order Dermoptera. They are arboreal gliding mammals found in Southeast Asia.

Family: Cynocephalidae (flying lemurs)

Genus: Cynocephalus:

Sunda Flying Lemur ***(Cynocephalus variegatus)***

The Sunda flying lemur *(Galeopterus variegatus)*, also known as the Malayan flying lemur, is a species of colugo (see below for notes on the common name "flying lemur"). Until recently, it was thought to be one of only two species of flying lemur, the other being the Philippine flying lemur which is found only in the Philippines. The Sunda flying lemur is found throughout Southeast Asia in Indonesia, **Thailand,** Malaysia, and Singapore.

The Sunda flying lemur is not a lemur and does not fly. Instead, it glides as it leaps among trees. It is strictly arboreal, is active at night, and feeds on soft plant parts such as young leaves, shoots, flowers, and fruits. After a 60-day gestation period, a single offspring is carried on the mother's abdomen held by a large skin membrane. It is a forest-dependent species. The Sunda flying lemur inhabits both lowland and mountainous regions, living in primary and secondary forests, coconut plantations and rubber plantations. They are totally arboreal, seldom if ever coming to the ground, and if they did they are quite helpless. The move from tree to tree by gliding; once they arrive at a tree they can climb but only slowly. When feeding and moving in a tree they always hang upside down. They can glide for over 100 meters and are quite resistant to injury.

Head-body length of Sunda flying lemur is about 34 to 38 centimetres (13 to 15 in). Its tail length is around 24 to 25 centimetres (9.4 to 9.8 in) and weight is 0.9 to 1.3 kilograms (2.0 to 2.9 lb). The hair to the upper surface of the gliding membrane is greyish brown, resembling the bark of a tree. The underside is paler and not spotted. There seems to be 2 colour phases which some authors suggest sexual dimorphism, with the females tending to be grey whilst the males to be brighter in colour, with some shades of brown and even reddish. Grey ones seem to be more common, possibly indicating a higher radio of females. Females are also lager than males. The limbs are all the same length, with all 5 toes equipped with strong, sharp claws used for climbing. The toes are connected by webs extending as far as the base of the claws. The head is broad, with rounded and rather short ears, bright red or amber in colour. The muzzle is round and blunt. The eyes are large and hazel in colour; vision is somewhat stereoscopic, giving depth perception for a safe landing. The rims of the eye orbits are projected, making the eyes well protected.

Threats:

The Sunda flying lemur is protected by national legislation. In addition to deforestation and loss of habitat, local subsistence hunting poses a serious threat to this animal. Competition with the plantain squirrel *(Callosciurus notatus)* represents another challenge for this species. More information is needed on population declines, but at present it is believed that the rate of the decline is probably not fast enough to trigger listing in any category other than Least Concern.

Classification and evolution:

The Sunda flying lemur has two forms which are not morphologically distinct from one another, the large form occurring on the mainland of the Sunda Shelf area and the mainland of Southeast Asia while the dwarf form occurring in central Laos and some other adjacent Islands. The Sunda flying lemur from Laos specimen is smaller (about 20%) than the other known mainland population. Despite the large and dwarf form, there are four known subspecies of *G. variegatus: G. v. variegatus (Java), G. v. temminckii (Sumatra), G. v. borneanus (Borneo)*, and *G. v. peninsulae* (Peninsular Malaysia and mainland of Southeast Asia) incorporating on the genetic species concept due to geographic isolation and genetic divergence. Recent molecular and morphological data provide the evidence that the mainland, Javan and Bornean Sunda flying lemur subspecies may be recognised as three separate species in the genus Galeopterus.

Behaviour and ecology:

The Sunda flying lemur is a skillful climber but is helpless when on the ground. Its gliding membrane connects from the neck, extending along the limbs to the tips of the fingers, toes and nails. This kite-shaped is known as a patagium, which is expanded for gliding. The Sunda flying lemur can glide over a distance of 100 m with a loss of less than 10 m in elevation. The Sunda flying lemur can manoeuvre and navigate while gliding, but strong rain and wind can affect its ability to glide. Gliding usually occurs in open areas or high in the canopy, especially in dense tropical rainforest. The Sunda flying lemur needs a certain distance to glide and to land in order to avoid injury.

In general, the diet of the Sunda flying lemur consists mainly of leaves. It usually consume leaves with less potassium and nitrogen but with higher tannin. The Sunda flying lemur also feeds on buds,] shoots, coconut flowers, durian flowers, fruits, and sap[from selected tree species. The Sunda flying lemur also feeds on insects in Sarawak, Malaysian Borneo. The selected food source all depends on the localities, habitat, vegetation types and the availability of food sources. The Sunda flying lemur mainly forages in tree canopies. It may forage on several different tree species in a single night, or on a single species. The Sunda flying lemur can also be seen licking tree bark of selected tree species to obtain water, nutrients, salts and minerals.

Breeding:

After a gestation period of approximately 60 days, Malayan Flying Lemurs give birth to a single offspring. The mother carries her young in a skin membrane on her abdomen. They are weaned by the time they reach 6 months old. The young is born in a marsupial like, undeveloped state and its mother cares for the young the same as a marsupial would care for her young. the patagium can be folded into the pouch, which is soft and warm, and it is where the young is carried until quite mature.

Distributions and Habitats:

The Sunda flying lemur is widely distributed throughout Southeast Asia, ranging from the Sunda Shelf mainland to other islands – Northern Laos, Cambodia, Vietnam, **Thailand,** Malaysia (Peninsular, Sabah and Sarawak), Singapore, Brunei, Indonesia (Kalimantan, Sumatera, Bali, Java), and many adjacent Islands. Conversely, the Philippine flying lemur (*C. violins)* is confined to the southern parts of the Philippines only. The Sunda flying lemur is adapted to many different vegetation types, including gardens, primary and secondary forest, rubber and coconut plantation, fruit orchards (dusun), mangrove swamps, lowlands and upland forests, tree plantations, lowland dipterocarp forests and mountainous area. However, not all of the mentioned habitats can sustain large colugo populations.

Order: Primates (Monkeys Etc)

The order Primates contains all the species commonly related to the lemurs, monkeys, and apes, with the latter category including humans. It is divided informally into three main groupings: prosimians, monkeys of the New World, and monkeys and apes of the Old World.

Suborder: Strepsirrhini Infraorder: Lorisiformes Family: Lorisidae (lorises, bushbabies)

Genus: Nycticebus:

Bengal Slow Loris *(Nycticebus bengalensis*)

Description:

The largest of the slow Loris species, the Bengal Slow Loris was only recently recognised as a distinct species, having previously been classed as a variation of Nycticebus coucang . The Bengal Slow Loris has a round head with short ears and large, forward-facing eyes, which reflect light, giving off a brilliant orange-red "eye shine", The coat is thick and woolly, with brown-grey upperparts and white under parts, and a distinct dark stripe running up the midline of the back. Having a barely noticeable, vestigial tail, this tree-dwelling species relies on its specially adapted hands and feet for climbing, which each bear an opposable thumb widely separated from the other four digits, giving it a pincer-like grip. The Bengal slow Loris produces a variety of vocalisations such as high-pitched whistles, chitters and clicks.

Range:

The Bengal slow Loris occurs in North-Eastern India, Bangladesh, Cambodia, Lao PDR, Myanmar, Vietnam, Southern China and **Thailand**

Habitat:

Generally found in tropical and sub-tropical evergreen and semi-evergreen rainforests, the Bengal slow Loris prefers areas with dense canopy cover, as well as forest edges, where insect prey appear to be more abundant

Biology:

Active during the night, the Bengal slow Loris can be found stalking through the trees with slow, deliberate movements as it searches for food . Its diet is varied, consisting mainly of plant exudates such as gums and resins, but also nectar, fruits, insects, bark and bird eggs One of the more curious aspects of slow Loris biology is the production of a toxic substance from glands on the insides of the elbows. This toxin, secreted in sweat, is licked off the gland and mixed with saliva (activating the toxin) where it appears to be channeled up the fine comb-like teeth at the front of the mouth. When defending itself, the Loris's bite may transmit this poison, and people who have been bitten have reported it to be particularly painful, with anaphylactic shock occurring in some cases.

Little is currently known about the social behaviour or reproductive biology of the Bengal slow Loris in the wild.

Threats:

There is currently severe degradation and loss of suitable Bengal slow Loris habitat occurring throughout its range. The extent of this destruction in certain regions is decimating local populations or eradicating them entirely. In north-eastern India, one of the key causes of habitat loss is a practice known as jhum, whereby hillside forest is burnt in order to create fertile agricultural land. In addition, and as a result of development in and around Bengal slow Loris habitat, it has been recorded that numerous individuals are killed by vehicles while crossing roads.

Aside from habitat destruction, the Bengal slow Loris has been extensively hunted for its meat and for use in traditional medicine. Many have also been trapped for sale in the international pet trade, and while this is now illegal, a black market trade persists.

Sunda Loris *(Nycticebus coucang)*

Description:

The Sunda slow Loris (Nycticebus coucang) or greater slow Loris is a strepsirrhine primate and a species of slow Loris native to Indonesia, western Malaysia, **Southern Thailand** and Singapore. It measures 27 to 38 cm (11 to 15 in) from head to tail and weighs between 599 and 685 g (21.1 and 24.2 oz). Like other slow lorises, it has a wet nose (rhinarium), a round head, small ears hidden in thick fur, a flat face, large eyes, small ears and a vestigial tail.

The Sunda slow Loris is nocturnal and arboreal, typically occurring in evergreen forests. It prefers rainforests with continuous dense canopies and has an extremely low metabolic rate compared to other mammals of its size. Its diet consists of sap, floral nectar, fruit and arthropods. It will feed on exudates such as gum and sap by licking wounds in trees. The species is generally solitary; one study showed only 8% of its active time was spent near other individuals. Social behaviour makes up a very small part of the activity budget, though it has monogamous mating system with the offspring living with the parents. It sleeps during the day, rolled up in a ball in hidden parts of trees above the ground, often on branches, twigs, palm fronds, or lianas. The species is polyoestrous, usually giving birth to a single offspring after a gestation period of 192 days. The young disperses between 16 and 27 months, generally when it is sexually mature.

The species is listed as "Vulnerable" on the IUCN Red List. It is threatened with extinction due to a growing demand in the exotic pet trade, and has become one of the most abundant primate species on sale at Indonesian pet markets. Its teeth are often pulled out before being sold as pets which can result in infection and/or death, this process makes reintroduction to the wild impossible. It also suffers from habitat loss, which has been severe in the areas in which it is found.

Etymology:

The common name, Sunda slow Loris, refers to the Sunda Islands, a group of islands in the western part of the Malay archipelago where it is found. Another common name for the species is the Greater Slow Loris. The specific name, coucang, derives from kukang, its common name in Indonesia. It is commonly known as malu-malu, meaning "shy" in Indonesian, and also as bukang or Kalamasan. It is sometimes called Kuskus, because local people do not distinguish between the slow Loris and Cuscus, a group of Australasian possums. In Malaysia they are sometimes known as kongkang or kera duku; kera is Malay for monkey while duku is the fruit-bearing tree, *Lansium domesticum*. In **Thailand, it is called ling lom (ลิงลม)**, which translates as "wind monkey".

Taxonomy and phylogeny:

The Sunda Slow Loris was first described in 1785 by the Dutch physician and naturalist Pieter Boddaert under the name *Tardigradus coucang*. However, its discovery dates to 1770, when the Dutchman Arnout Vosmaer (1720–1799) described a specimen of it as a type of sloth. Vosmaer gave it the French name "le paresseuz pentadactyle du Bengale" "the five-fingered sloth of Bengal"), but Boddaert later argued that it was more closely aligned with the lorises of Ceylon (now Sri Lanka) and Bengal.

Between 1800 and 1907, several other Slow Loris species were described, but in 1952 the primatologist William Charles Osman Hill consolidated all the Slow Lorises in one species, *N. coucang,* in his influential book Primates: Comparative Anatomy and Taxonomy. In 1971 Colin Groves recognized the Pygmy Slow Loris *(N. pygmaeus)* as a separate species, and divided N. coucang into four subspecies, while in 2001 Groves opined there were three species *(N. coucang, N. pygmaeus, and N. bengalensis)*, and that *N. coucang* had three subspecies *(Nycticebus coucang coucang, N. c. menagensis, and N. c. javanicus).* These subspecies were promoted in 2010 to species status—the Javan Slow Loris *(N. javanicus)* and Bornean Slow Loris *(N. menagensis)*—leaving the Sunda Slow Loris without subspecies. Species differentiation was based largely on differences in morphology, such as size, fur colour, and head markings.

When Étienne Geoffroy Saint-Hilaire defined the genus Nycticebus in 1812, he made the Sunda Slow Loris the type species. This was questioned in 1921 by British zoologist Oldfield Thomas, who noted that there was some confusion over the specimen used as the type specimen. Instead, he suggested that the type specimen was actually the Bengal Slow Loris *(N. bengalensis).* There was also some confusion during the 1800s when Boddaert's Tardigradus coucang was mistaken for Carl Linnaeus' *Lemur tardigradus,* which was described in the 10th edition of Systema Naturæ (1758) and referred to a Slender Loris. However, this was clarified in 1902 by mammologists Witmer Stone and James A. G. Rehn.

The species has 50 chromosomes (2n=50), and it genome size is 3.58 pg.Of its chromosomes, 22 are metacentric, 26 are submetacentric, and none are acrocentric. Its X chromosome is submetacentric, and its Y chromosome is metacentric.

To help clarify species and subspecies boundaries, and to establish whether morphology-based classifications were consistent with evolutionary relationships, the phylogenetic relationships within the genus Nycticebus have been investigated using DNA sequences derived from the mitochondrial markers D-loop and cytochrome b. Although most of the recognized lineages of Nycticebus (including the pygmy slow Loris (*N. pygmaeus*), Bornean slow Loris *(N. menagensis)* and the Javan slow Loris *(N. javanicus))* were shown to be genetically distinct, the analysis suggested that DNA sequences from some individuals of Sunda slow Loris (*N. coucang)* and Bengal slow Loris (N. bengalensis) apparently share a closer evolutionary relationship with each other than with members of their own species. The authors suggest that this result may be explained by introgressive hybridization, as the tested individuals of these two taxa originated from a region of sympatry in **southern Thailand**; the precise origin of one of the *N. coucan*g Individuals was not known. This hypothesis was corroborated by a 2007 study that compared the variations in mitochondrial DNA sequences between *N. bengalensis* and *N. coucang*, and suggested that there has been gene flow between the two species.

Description:

The Sunda slow Loris has dark rings around its large eyes, a white nose with a whitish strip that extends to the forehead and a dark stripe that stretches from the back of the head along the spine. Its soft, thick, woolly fur ranges from light brown to deep reddish brown, with a lighter underside. The species is distinct from the Bengal slow Loris due to the dark inverse teardrop markings around the eyes which meet the dark dorsal stripe on the back of the head. It tends to have a much more distinct white stripe between the eyes, more distinct dark colouring around the eyes, and a browner coat than the Bengal slow Loris which is larger, greyer, and shows less contrast. The Sunda slow Loris has less white facial colouring than the much smaller pygmy slow Loris. Local colour variations are known to occur. The Sunda slow Loris generally holds branches with at least three limbs at one time. It measures between 27 and 38 centimetres (11 and 15 in) and weighs 599 and 685 grams (21.1 and 24.2 oz). Unlike the Bengal slow Loris, the Sunda slow Loris does not show sexual dimorphism by weight. The vestigial tail, hidden beneath the fur, is reduced to a stump. It has a toothcomb, six forward-facing teeth on the bottom jaw, which includes the lower Incisors and the canine teeth. The structure is generally used for grooming in other strepsirrhine primates, but lorises also use it to scrape off gum when foraging. It has a shortened second digit, and the hands have a strong grip. Like other lorises, it excretes a strong-smelling liquid from glands beneath its arms which is used in communication. One major distinguishing feature between all Loris species is locomotion: the Sunda slow Loris moves slowly through trees on all four limbs, typically with three limbs attached to a support at a time. Its movement has been described as unique; similar to crawling, or as if it was climbing in every direction, the Sunda slow Loris changes direction or moves between branches with little noise or change in speed. In captivity, about a quarter of its time is spent moving quadrupedally, a quarter suspended or hanging, a quarter climbing, and a quarter clasping multiple branches (bridging).It may hang below a branch by one or both feet for long periods of time.

Behaviour and Ecology:

Like other slow lorises, the Sunda slow Loris is an arboreal and nocturnal primate, resting by day in the forks of trees, or in thick vegetation and feeding on fruit and insects by night. Unlike other Loris species, it remains in trees most of its life: while the Bengal slow Loris will often sleep on the ground, the Sunda slow Loris sleeps in a ball in branches or foliage. It usually sleeps alone but has been observed to sleep with several conspecifics (individuals of the same species), including other adults.Adults live in overlapping ranges of 0.004 to 0.25 km2 (0.0015 to 0.097 sq mi). Despite its slow metabolism rate, the Sunda slow Loris has a high-energy diet. Its slow lifestyle may be due to the energy costs of detoxifying certain secondary plant compounds in many genera of food plants their diets. The largest amount of time is spent eating phloem sap (34.9%), floral nectar and nectar-producing plant parts (31.7%), and fruits (22.5%). It also consumes gums and arthropods such as spiders and insects. Gum is taken by licking wounds on trees. They are also known to feed on molluscs, including the giant land snail *Achatina fulica*,and birds' eggs. All slow Loris species produce a toxin in glands on the insides of their elbows. This is spread across their bodies and those of their offspring using the toothcomb while grooming. When threatened with predators, the Sunda slow Loris can bite, roll into a ball exposing its toxic saliva-covered fur, or roll up and drop from the trees. However, the primary method of predator avoidance is crypsis, whereby it hides. The Asiatic Reticulated Python, the Changeable Hawk-eagle and the Bornean Orang-utan have been recorded as predators of the Sunda slow Loris.

Social systems:

The Sunda slow Loris may fit into the monogamous single male/single female social system, though are mainly known to be solitary. One study showed only 8% of the species' active time was spent near other individuals. Where home ranges overlap, spatial groups are formed. These groups consist of one male, one female and up to three younger individuals. The interactions between these individuals are largely friendly; they include allogrooming, following, pant-growling, and click-calling, although social behaviours only make up around 3% of the activity budget. When it comes into contact with conspecifics from other home ranges there is usually no reaction as home ranges are not defended. In captivity however, it can be aggressive with other individuals. Males have shown antagonistic behaviours such as attacks, pursuits, threats, assertion, fighting, and subordination. The fighting often results in serious injury. Despite this, they are known to be generally sociable in captivity, with allogrooming being the most common social behaviour.

Communication:

There are eight distinct call types made by Sunda slow Loris adults, which can be divided into two categories: contact and contact-seeking calls such as whistles and short keckers (a social play and attention-seeking call), and aggressive and defensive calls such as long keckers, screams, snarls and grunts. Because they rely on crypsis to avoid predators, they do not make alarm calls. Infants emits clicks and squeaks when disturbed. During oestrus, females make whistle calls when in visual contact with a male. When exploring new environments and during handling, it makes ultrasonic vocalisations out of the human hearing range.

Olfactory communication is very important in this species. It has roles in alerting others of the identity, physical state and position of the individual. It is also important in reproduction. Like other slow lorises, the Sunda slow Loris has glands on its elbows that exude oils. The gland is licked to spread scent and is thought to have evolved for communication, but it is toxic to humans. If the person is allergic to the animal they can go into shock and could even die. It also has glands on its anus and can signal conspecifics by perineal rubbing which deposits urine. When it locates the scent of another slow Loris it may rub its face on the substrate where the scent was found. The Sunda slow Loris may grin or bare its teeth. When stressed, infants may grin, while adults bear their teeth to show aggression or fear, but also during play.

Reproduction:

Females hang from branches during copulation. Reproduction is one of the few times the Sunda slow Loris aggregates with conspecifics, as it is largely solitary. One study recorded that the most slow lorises ever seen together was six; this appeared to be a female in oestrus and five males following her. This may suggest a more promiscuous mating system, where females mate with more than one male. Despite this, its testis size is small compared with similar sized prom simians, which is indicative of monogamy. In the wild the mating system of the Sunda slow Loris is thought to vary between populations. Sexual maturity is reached between the ages of 18 and 24 months in females, and can be reached by 17 months in males. It is polyoestrous, having many periods of sexual receptivity during a year. In captivity however, there is a clear birth peak between March and May. It has been observed that reproductive patterns of captive prosimians in the northern hemisphere are altered. The oestrus cycle lasts 29–45 days, with most copulations occurring on the same day.

Females in oestrus are followed by males, with copulation initiated by the female. The female will hang from a branch and may vocalise. The male will hold the female and the branch and copulate with her. Urine-marking and vocalising are also used by the female to solicit mating. The male may create a mating plug following copulation. The gestation period averages 192.2 days, after which one young is born, although twinning has been observed. Both males and females will disperse in the wild, this occurs between 16 and 27 months.

Distribution and Habitat:

The Sunda slow Loris is found in continuous canopy tropical rainforests. It is adaptable and will also live in other types of habitat. It is found in Indonesia, on the islands of Sumatra, Batam and Galang in the Riau Archipelago, and Pulau Tebingtinggi and Great Natuna (Bunguran) in the Natuna Islands; in Malaysia on the Malay Peninsula and Pulau Tioman; in the **Southern Peninsular of Thailand;** and Singapore.

The Sunda slow Loris is sympatric (shares its range) with the Bengal slow Loris in **Thailand** and hybridization has occurred.

Conservation and Threats:

The species is listed as "Vulnerable" on the IUCN Red List. It is threatened with extinction due to a growing demand in the exotic pet trade, and has become one of the most abundant primate species on sale at Indonesian pet markets. Its teeth are often pulled out before being sold as pets which can result in infection and/or death, this process makes reintroduction to the wild impossible. It also suffers from habitat loss, which has been severe in the areas in which it is found. According to the 2008 International Union for Conservation of Nature (IUCN) Red List assessment, the Sunda slow Loris was evaluated as "Vulnerable". In June 2007 it was transferred from CITES Appendix II to Appendix I, indicating a true concern about the state of international trade in this species, an act now forbidden by international law. It is also protected by Indonesian law, though the law does not seem to be strictly enforced. Its population size is unknown and further studies need to be carried out to confirm their conservation status. The Sunda slow Loris is greatly threatened by the pet trade, and is sold as an exotic pet throughout southeast Asia. The slow lorises are the most commonly traded protected primates in southeast Asia.[When sold as a pet it often will have its teeth pulled out to prevent injury to the owner. This may cause dental I infections which have up to a 90% death rate. Once it has lost its teeth, reintroduction to the wild is impossible. Due to a very high mortality rate in captivity due to stress, improper nutrition and infection, the pet trade is inflated by replacements. With a greater purchasing power, the increasing human populations in the species' range could have an even more serious impact. The Sunda slow Loris is further threatened by gathering for illegal traditional medicine. The fur is reported to heal wounds, the flesh to cure epilepsy, eyes are used in love potions, and the meat is reported to cure asthma and stomach problems. It is also killed as a crop pest. Severe habitat loss and degradation over large areas of its range have caused large population declines, even though the species is more adaptable to anthropogenic habitats than other primates in the region. The resulting fragmentation has restricted species dispersal as it depends on continuous canopy cover to move from tree to tree.

Suborder: Haplorrhini:

Infraorder: Simiiformes:
Parvorder: Catarrhini:
Superfamily: Cercopithecoidea:
Family: Cercopithecidae (Old World monkeys)

Genus: Macaca:

Stump-Tailed Macaque *(Macaca arctoides)*

Physical characteristics:

Aptly named after a few distinguishing characteristics, Bear Macaques or Stump-Tailed Macaques have thick, long, dark brown fur covering their bodies and short tails which measure between 3.2 and 69 mm (.12 and 2.7 in) (Fa 1989). Stump-Tailed Macaques have bright pink or red faces which darken to brown or nearly black as they age and are exposed to sunlight. They are covered with long, shaggy fur, but their short tails and faces are hairless and they go bald with age. Infants are born white and darken with age (Fa 1989; Rowe 1996; Groves 2001). Males are much larger than females, measuring between 517 and 650 mm (20.4 to 25.6 in) and weighing between 9.9 and 10.2 kg (21.8 and 22.5 lb). Females have an average height between 485 and 585 mm (19.1 and 23.0 in) and weigh between 7.5 and 9.1 kg (16.5 and 20.1 lb) (Fa 1989). This sexual dimorphism extends to more than just body size; male Stump-Tailed Macaques have elongated canine teeth compared to females, which are important for establishing dominance within social groups. All Macaques, including Stump tails, have pouches in their cheeks to store food for short periods of time (Rowe 1996). They travel quadrupedally and usually on the ground for they are not very agile in trees (Rowe 1996; Srivastava 1999). They are not known to swim, as do other species of Macaques (Macaca) (Fooden 1990).

Habitat and Distribution:

This Old World monkey travels quadrupedally, usually on the ground, as it is not very agile in trees. It is generally found in subtropical and tropical broadleaf evergreen forests, in different elevations depending on the amount of rainfall in the area. It depends on rainforests for food and shelter, and is not found in dry forests except where it ranges in the Himalayan region of India, only spending time in secondary forests if it is bordering primary tropical forests. It is distributed from north-eastern India and southern China into the northwest tip of West Malaysia on the Malay Peninsula. It is also found in Burma, **Thailand,** Vietnam, and Far Eastern Bangladesh. A study population was introduced to Tanaxpillo, an uninhabited island in Lake Catemaco, Veracruz, Mexico in 1974, where it ranges in semi natural conditions. Most Information on the species comes from the introduced population on Tanaxpillo and other captive settings, as few long-term studies have been conducted on the stump-tailed macaque in the wild.

Ecology:

Starting the day at dawn, stump-tailed macaques spend the early morning, until midday, travelling and feeding. They are frugivore-omnivores and a significant part of their diet is devoted to fruits. They also eat seeds, flowers, leaves, roots, freshwater crabs, frogs, birds, bird eggs, and insects (Fooden 1990; Rowe 1996; Srivastava 1999). They also raid crops preferring corn and other cultivated fruits. During the middle of the day, the group stops travelling and rests in the shade, spending time on social activities such as grooming while juveniles and adolescents play (Fooden et al. 1985). In the late afternoon foraging begins again as they travel to their sleeping site, usually large trees or cliffs. The daily range of stump-tailed macaques is between two and three kilometers (1.24 to 1.86 mi), but they do not have to travel as far during the rainy season when food is more abundant. Home range is unknown but thought to be several square kilometers (Srivastava 1999). Though they spend the majority of the day travelling on the ground, usually along the banks of rivers and streams, stump-tailed macaques also forage for fruit and leaves in trees and flee to trees when in danger (Fooden 1990). Potential predators of Stump-Tailed Macaques include Clouded Leopards, Leopards, Dogs, and large Raptors. When predatory animals are near, they respond by assuming threatening postures, shaking trees and branches, and baring their canine teeth in threat. No predation event has been recorded (Srivastava 1999; Cherty et al. 2002-2003). Like some human males, Stump tail Macaques become partially bald as they age. This process of balding is similar to male-pattern baldness seen in humans because hair loss starts at the forehead and advances toward the back of the skull over time, but unlike humans, this pattern is seen in both male and female Stump tails (Uno et al. 1967). Researchers have studied balding in Stump tail Macaques and have developed treatments for baldness, namely minoxidil, or as it is commercially marketed, Rogaine (Uno 1986). Minoxidil was originally developed as a drug to treat high blood pressure, but one of the side effects identified was excessive hair growth. Testing of the drug on Stump tail Macaques revealed hair regrowth and maintenance of newly regrown areas on balding scalps (Uno 1986). By first testing its efficacy and safeness on non human primates, researchers were able to develop the drug for human use.

Reproduction:

Stump-Tailed Macaques have low reproductive rates compared to other macaque species. Females reach sexual maturity around four years of age and have an ovarian cycle lasting 30 days. Most mating occurs in October and November in the wild, during February and March in Mexico, and is not seasonal in captivity (Brereton 1994). Females begin to produce offspring between 4.5 and five years of age and will reproduce until about 17 years of age (Fooden 1990). Males reach sexual maturity around four years of age as well, but do not reach adult size until around six years. Frequency of mating correlates with dominance rank among Stump-Tailed Macaques. The highest-ranking males monopolize receptive females while the highest-ranking females are also the most likely to do the most mating (Brereton 1994). Lower-ranking male Stump-Tailed Macaques use alternative methods to gain mating opportunities. One way they do this is by lagging behind with a reproductively active female as the group travels. When the dominant males are out of sight, the lower-ranking male mates with the female and the couple then moves to rejoin the group (Brereton 1992). Both males and females solicit mating, females by presenting their rumps to males and maintaining eye contact over one of their shoulders, and males by approaching a female, sitting next to her, and giving a "teeth chattering with grimace" display (Brereton 1994). During copulation, other members of the group harass the pair (Srivastava 1999). Gestation lasts 177 days and females give birth about every two years in the wild (Fooden 1990; Srivastava 1999).

Parental care:

Macaque mothers are the primary caregivers for their offspring, though all of the females in the group direct attention to infants and will approach, play with, carry, groom and protect them, especially if they are born to a high-ranking mother. By protecting a high-ranking female's infant, a lower-ranking female may expect rewards of tolerance and reduced aggression by the high-ranking female (Estrada & Estrada 1984). High-ranking adult males also direct some attention towards and give protection to infants within the group. This may be because higher-ranking males have more chances to mate with females and because there is increased likelihood that infants in the group are their offspring, males have some interest in protecting them from danger (Bauers & Hearn 1994). Stump-Tailed Macaques are considered permissive mothers compared to other species, and early on they allow the infant to independently explore the surrounding environment (Maestripieri 1995). They may be this lenient because other group members are interested in infants but never treat them roughly or "kidnap" them as is seen in other macaques (e.g., *M. nemestrina* and *M. fascicularis)* (Bauers & Hearn 1994). Stump-Tailed Macaque infants are dependent for the first nine months of life, after which they are weaned, and become increasingly independence until adolescence, at 18 months (Srivastava 1999).

Communication:

Communication between Stump-Tailed Macaques largely takes the form of vocal or gestural signals. Frequently seen gestures or postures are used to reinforce the dominance hierarchy and reconcile after aggressive interactions. "Hindquarter presentation" is the most common gesture seen among Stump-Tailed Macaques and is displayed by subordinates to appease dominants. Other submissive signals include "bared-teeth," "lip-smack," "teeth-chatter," and "present-arm," in which one arm is put directly in front of the face of the dominant individual to be bitten (Maestripieri 1996).

Vocal communication is also important among Stump-Tailed Macaques. The most common vocalization is the "coo" heard in a variety of contexts, but especially relevant as group members maintain contact with each other while foraging and when approaching one another to initiate friendly interactions such as grooming or huddling (Rowe 1996). "Basic grunts" are another ubiquitous signal among stump-tailed macaques; they are commonly heard between animals who are greeting one another, after aggressive interactions and when one animal is interested in another (Bauers 1989). Alpha males use a "roar" when displaying against predators or threats. Infants use "trilled-whistles" as a signal of distress to their mothers when they are out of visual contact with them or if they need to be retrieved because they cannot descend a structure that they have climbed (Bauers 1989; Maestripieri 1995).

Conservation status:

Habitat disturbances affecting this species' survival include selective logging, timber and firewood collection for charcoal, building roads, dams, power lines and fisheries, deliberately set fires, fragmentation, and soil loss/erosion. These animals are hunted and traded for food, sport and traditional "medicine," and accidental mortality due to trapping occurs. There is a local trade for bones, meat for food and the live animals as pets (Molur *et al.* 2003).

In India much of its habitat is affected by Jhum cultivation (shifting or slash-and-burn) (Srivastava and Mohnot 2001). It has also been indiscriminately hunted to the brink of extinction over almost its entire distribution in this country (Srivastava and Mohnot 2001).

In Vietnam, this species is heavily targeted for use in traditional "medicine," both in country and for trade with China. Within Lao PDR, Vietnam and Cambodia, hunting levels for food is also very high. Habitat loss is a relatively lower threat compared to hunting (R. Timmins pers. comm.).

In Myanmar, logging and timber extraction are major threats. Commercial rubber plantations, hunting, and trade of animals parts with China are all major threats to this species.

In China, hunting and habitat loss have reduced in population of this species, and it is locally extinct in some places.

In **Thailand,** habitat loss is a major threat, while hunting is prevalent, but not a significant threat to the species (R. Boonratana pers. comm.).

Conservation Actions:

Internationally, this species is listed under Appendix II in CITES. Regionally, India lists it as schedule II under the Indian Wildlife (Protection) Act, 1972 (Srivastava and Mohnot 2001) amended up to 2002. The species is also protected in national wildlife acts of Lao PDR, Vietnam, **Thailand,** and Myanmar.

Stump-tailed macaques are found in a number of protected areas throughout their range, including: Balpakram National Park, Gibbon Wildlife Sanctuary, Mehao Wildlife Sanctuary, Murlen National Park (India); Huay Kha Khaeng Wildlife Sanctuary**, Phu Khieo Wildlife Sanctuary (Thailand**). They may possibly occur in Namdapha Wildlife Sanctuary and Pakhui Wildlife Sanctuary (India), and in Nam Ha National Biodiversity Conservation Area (Lao PDR).

Assam Macaque (*Macaca assamensis*)

The Assam Macaque *(Macaca assamensis)* is a Macaque of the Old World monkey family native to South and Southeast Asia. Since 2008, the species is listed as Near Threatened by IUCN, as it is experiencing significant declines due to hunting, habitat degradation and fragmentation.

Characteristics:

The Assam Macaque has a yellowish-grey to dark brown pelage. The facial skin is dark brownish to purplish. The head has a dark fringe of hair on the cheeks directed backwards to the ears. The hair on the crown is parted in the middle. The shoulders, head and arms tend to be paler than the hindquarters, which are greyish. The tail is well-haired and short. Head-to-body-length measures 51 to 73.5 cm (20 to 28.9 in), and the tail is 15 to 30 cm (5.9 to 12 in) long. Adult weight is 5 to 10 kg (11 to 22 lb).

Distribution and habitat:

The *Macaca assamensis* population is endemic to Nepal and likely in some way distinct from the two recognized subspecies, which occupy adjacent areas to the southeast and east of the range of *M. assamensis.* There is a gap in north-eastern India between the two main population pockets, specifically between central Bhutan and the south side of the Brahmaputra River; the east bank of its upper course marks the division between the two recognized subspecies

The Eastern Assamese Macaque, *M. a. assamensis,* occurs in Bhutan, Arunachal Pradesh, Assam, Manipur, Meghalaya, Mizoram, Nagaland, Sikkim, and Tripura in north-eastern India, into Northern Myanmar, southeast through the Myanmar-**Thailand** border ranges as far as Chongkrong, to the upper Mekong in Tibet, into the provinces of Guangxi, Guizhou, Tibet and Yunnan in South-Western China, in Thateng in Northern Laos, and Hoi Xuan in Northern Vietnam; The Western Assamese Macaque, *M. a. pelops, is* found from central Nepal through Uttar Pradesh, Sikkim, Assam and Northernmost West Bengal in Northern India, into central Bhutan and the Sundarbans in Bangladesh. During surveys carried out in 1976, 1978, and 1984 in Nepal, Assam macaques were found to be patchily distributed along rivers in tropical and subtropical forests at altitudes from 200 to 1,800 metres (660 to 5,900 ft). They are apparently absent from areas west of the Kaligandaki River.In India, they live in tropical and subtropical semi evergreen forests, dry deciduous and montane forests, from the sea level to altitudes of 4,000 metres (13,000 ft). They usually inhabit hill areas above 1,000 m (3,300 ft), but in the wetter east they may occur even in the lowlands, and frequent areas that only marginally reach this altitude. In Laos and Vietnam, they prefer high altitudes, usually above 500 m (1,600 ft). In forests on limestone karst, they occur in much lower elevations.

Population:

There is little information available on the global population, but it is thought to be relatively high. There is decline in the wild populations in certain countries, but not throughout the distribution. It is probably the most common macaque species in Lao PDR, where it is widespread through hill and mountain forest ranges, though no population estimate is available (Duckworth et al. 1999; R. Timmins pers. comm.). In Nepal the taxon is also restricted to less than 300 mature individuals, distributed in eight subpopulations, with no subpopulation having more than 50 mature individuals. There is no information on population estimates from **Thailand** (R. Boonratana pers. comm.). In northern Myanmar it is a very common species, but information from the eastern part of the country is not available (S. Htun pers. comm.). Due to hunting pressure, macaques in Vietnam are scarce. However, there is no precise information for this species on population and status (R. Timmins pers. comm.).

The subspecies *M. a. assamensis is* relatively widespread in its extent of occurrence (Molur et al. 2003). In Namdapha National Park, Arunachal Pradesh, India, the population was found to have a group density of 1.11/km2, and an average group size of 13.93 individuals (Chetry et al. 2003). This species is the most abundant monkey in the mountains of Arunachal Pradesh (Choudhury 2001). In China, there are estimated to be over 3,000 individuals (Zhang et al. 2002).

The subspecies *M. a. pelops* is only found in adjacent populations in India, Bhutan, and adjacent China. There are likely less than 1,000 individuals remaining (Zhang et al. 2002).

A subpopulation of *M. assamensis* is endemic to Nepal and relegated to a single population there, and is considered as a possible new subspecies (M. Chalise pers. comm.). However, further taxonomic clarification is needed.
Population Trend: Decreasing

Ecology and behaviour:

Assam macaques are diurnal, and at times both arboreal and terrestrial. They are omnivorous and feed on fruits, leaves, invertebrates and cereals. It prefers dense forest (Choudhury 2001), and does not usually occur in secondary forest. It is found in tropical and subtropical semi-evergreen forests, dry deciduous and montane forest (Srivistava and Mohnot 2001; R. Boonratana pers. comm.). Particular habitats and niches vary some depending on the subspecies. This species is found from the floodplains to the high mountains, up to 2,800 m, and sometimes 3,000 m in the summer (Choudhury 2001), and perhaps up to 4,000 m (Srivastava and Mohnot 2001; Zhang et al. 2002). It is generally an upland species, and usually associated with hill areas above 1,000 m. In the wetter east the species may also frequent areas that do not or only marginally reach 1,000 m and may occur even in the lowlands. In Lao PDR and Vietnam the species is predominantly associated with high altitudes, usually above 500 m. In forests on limestone karst the species occurs in much lower elevations (R. Timmins pers. comm.). The generation of this species is 10-12 years (Molur et al. 2003)

The gestation period lasts for 165 days. *Assemese Macaque g*ive birth to single offspring which weighs around 400 g at birth. Gestation length was on average 164.2 d (range: 158–170, *n*=10), and females had their first infant at an age of 5 yr (*n*=4). Interbirth intervals were bimodally distributed, with females giving birth on average every 13.9 or 23.2 mo. Shorter interbirth intervals were linked to early parturition within the birth season. Most females displayed subcaudal sexual swellings which, however, did not reliably indicate female reproductive status or fertility. Overall, results fall within the range of findings reported for other macaque species. These results thus add to the growing body of information available for wild macaques, facilitating comparative studies for a better understanding of interspecific differences in social and reproductive patterns. Macaques provide ideal conditions for comparative studies concerning the linkage between life histories and reproductive strategies.

All macaque species share basic patterns of social organization: They all live in MultiMate, multifamily groups and they mate promiscuously, males disperse, whereas females are philopatric (Thierry 2007). On the other hand, they vary in social and reproductive traits, e.g., the degree of seasonality, with births occurring year-round or being restricted to a few months; social structure, dominance relationships; mating patterns; and the development of sexual signals, e.g., sexual swellings, sex skin coloration, copulation calls, all of which may influence the reproductive strategies of both sexes (Dixon 1998; van Noordwijk and van Schaik.Timing of reproduction was strictly seasonal (births: April–July, 86% in April–June, 4 yr, *n*=29; conceptions: October–February, 65% in December–January, 2 yr, *n*=17). Females showed no cyclic ovarian activity outside the mating season and conceived in their first or second cycle (mean: 1.2 cycles to conception, *n*=13). Gestation length was on average 164.2 d (range: 158–170, *n*=10), and females had their first infant at an age of 5 yr (*n*=4). Interbirth intervals were bimodally distributed, with females giving birth on average every 13.9 or 23.2 mo. Shorter interbirth intervals were linked to early parturition within the birth season. Most females displayed subcaudal sexual swellings which, however, did not reliably indicate female reproductive status or fertility. Overall, our results fall within the range of findings reported for other macaque species. These results thus add to the growing body of information available for wild macaques, facilitating comparative studies for a better understanding of interspecific differences in social and reproductive patterns. 2004).

Major Threat's):

The threats to this species' habitat include selective logging and various forms of anthropogenic development and activities, alien invasives, and hunting and trapping for sport, "medicine," food, and the pet trade. Additionally, hybridization with adjacent species poses a threat to some populations (Molur et al. 2003). Although it has a wide distribution, the species is considered to be threatened in most parts of its range. Habitat destruction is the primary cause of the decline.

Habitat destruction poses the greatest risk to this species in north-eastern India (Srivistava and Mohnot 2001). However, it has been hunted in the Himalayan regions of North Bengal, Sikkim, and Arunachal Pradesh, where it frequently invades crops (Srivistava and Mohnot 2001). Locals use skulls as an "evil eye" in front of houses in north-eastern India (Das pers. comm.). There has been extensive habitat loss over the last 15 years in several states of north-eastern India (from 30-60%), with major impacts on M. a. assamensis. The Nepal population of Macaca assamensis is threatened due to its restricted distribution of less than 2,200 km2 extent of occurrence and 914 km2 area of occupancy and continuing decline in area, extent and quality of habitat, the number of locations and in the number of mature individuals—the latter two conditions being inferred from threats to habitat and population from degradation and hunting, respectively. Given its restricted extent of occurrence, threats on its population and habitat, and small numbers in fragmented patches, the Nepal population of this macaque is categorized as Endangered. In **Thailand** habitat loss is the primary threat, hunting for food less so (R. Boonratana pers. comm.). It is protected only in temples. In eastern Myanmar habitat loss is the primary threat, but hunting is presumed to occur. They are hunted to make footwear, and the skins are taken to Tibet as it is more profitable than taking it to Yunnan province of China. In northern Myanmar, hunting and habitat loss due to conversion are the major threats. There is more than 30% decline in forest cover over the last 30-35 years. The combination of habitat loss and hunting heavily impacts the subpopulations, and the threats are likely to continue in the next three decades if the demand for this continues in Tibet (S. Htun pers. comm.). In Lao PDR and Vietnam the primary threat is hunting for food and for bones to make balm and/or glue. The bones are not used within Lao PDR, but sold to Vietnamese and traders within Lao PDR. The balm is used for pain relief and other speculative "medicinal" purposes. The trade also goes into China (R. Boonratana pers. comm.). This species has declined in Vietnam and Lao PDR in the last 30-35 years by more than 30%, and is expected to continue to decline in the future.

In Tibet the habitat is good, but there is some hunting, and no detailed data exists for *M. a. pelops* from that region of China (Z. Zhou et al. pers. comm.). For *M. a. assamensis,* hunting is a major threat. Logging has ceased, but was a major threat for the last 30 years. Conversion to pastures is still ongoing, but not a significant threat. There are extensive habitats still left for this subspecies and the taxon is relatively safe.

Conservation Actions:

This species is listed in Appendix II of CITES (Chetry et al. 2003). It is legally protected in all countries of occurrence. For the populations that reside in India, the species is listed under Schedule II, part I of the Indian Wildlife (Protection) Act (amended up to 2002) (Chetry et al. 2003). In Bangladesh it is listed as Schedule III in the Bangladesh Wildlife (Preservation) (Amendment) Act, 1974. In Myanmar it is legally protected according to the 1994 Wildlife Protection Law. In **Thailand the species is protected by the Wildlife Protection Act, 1960.**

The species has been recorded from at least 41 protected areas in northeastern India, and maybe four others (Choudhury 2001), and may also be found in the following protected areas: Nam Ha National Biodiversity Conservation Area (Lao PDR); Langtang National Park, Makalu Barun National Park (Nepal); **Doi Suthep-Pui National Park, Huay Kha Khaeng Wildlife Sanctuary, Phu Khieo Wildlife Sanctuary (Thailand);** Pu Mat National Park (Vietnam).

Crab-Eating Macaque *(Macaca fascicularis)*

Etymology:

The scientific name of the Crab-Eating Macaque is *Macaca fascicularis.* Macaca comes from the Portuguese word macaco, which was picked up from makaku, a Fiot (West African language) word (kaku means 'monkey' in Fiot). Fascicularis is Latin for 'a small band or stripe'. Sir Thomas Raffles, who gave the animal its scientific name in 1821, did not specify what he meant by the use of this word, although it is presumed it had something to do with his observation of the animal's colour. This animal has several common names. It is often referred to as the long-tailed macaque because its tail is usually about the same length as its body and because its long tail distinguishes it from most other macaques. The species is also commonly known as the crab-eating macaque because it is often seen foraging beaches for crabs. Another common name for *M. fascicularis* is the cynomolgus monkey, which literally means "dog-milker" monkey; this is the name most commonly used in laboratory settings. In Indonesia, *M. fascicularis* and other macaque species are known generically as kera, possibly because of the high-pitched alarm calls they give when in danger ("krra! krra!").

Physical characteristics:

Depending on subspecies, the body length of the adult monkey is 38-55 cm (15–22 in) with comparably short arms and legs. The tail is longer than the body, typically 40–65 cm (16–26 in). Males are considerably larger than females, weighing 5-9 kilograms (11-20 lb) compared to the 3–6 kg (7-13 lb) of females.

Social structure:

M. fascicularis is a very social animal that lives in groups of 5–60 or more animals. These are multiple-male groups, normally containing two to five males and two to three times as many females. The number of juveniles is usually comparable to the number of females. Group size often depends on the level of predation and availability of food. Groups are female-centred, as the females are philopatric (i.e. remain in one place across generations), while the males move in and out of these female-based groups. Males generally first emigrate from their natal group at the age of four to six years. They will remain in a group up to four or five years and thus will emigrate several times throughout their lives. These monkeys are highly despotic and have a strict dominance hierarchy. Adult males rank higher than females. Female ranks are more stable than males, as males are defeated from time to time and then lose rank. High-ranked males are generally the most successful at reproduction, and high-ranked females generally fare best at raising surviving offspring. The females are organized into matrilines, which are the female-based families consisting of the related females and their descendents. Matrilines are ranked, and some families have greater social power than others; this difference in rank is maintained over several generations. Matrilineal overthrows rarely occur, and when they do, they have severe consequences to the reproductive success of the defeated matriline in the following year.

Reproduction:

After a gestation period of 167–193 days, the female gives birth to one infant. The infant's weight at birth is approximately 350 grams (12 oz). Infants are born with black fur which will begin to turn to a yellow-green, grey-green, or reddish-brown shade (depending on the subspecies) after about three months of age. This natal coat is suggested to indicate to others the status of the infant, and other group members treat infants with care and rush to their defence when distressed. Immigrant males will sometimes kill infants not their own, and high-ranking females sometimes kidnap the infants of lower-ranking females. These kidnappings usually result in the death of the infants, as the other female is usually not lactating. Young juveniles stay mainly with the mother and relatives, and as male juveniles get older, they become more peripheral to the group. Here they play together, forming crucial bonds that may help them when they leave their natal group. Males that emigrate with a partner seem to be more successful than those that move off alone. Young females, on the other hand, stay in the centre of the group and become incorporated into the matriline into which they were born.

Male Crab-Eating Macaques will groom females to increase the chance of mating. A female is more likely to engage in sexual activity with a male that has recently groomed her than with one that has not.

Crab-Eating Macaques typically do not consume crabs; rather, they are opportunistic omnivores, eating a variety of animals and plants. Although fruits and seeds make up 60 - 90% of the dietary intake, they also eat leaves, flowers, roots and bark. They also prey on vertebrates (including bird chicks, nesting female birds, lizards, frogs and fish), invertebrates and bird eggs. Although the species is ecologically well-adapted in its native range and poses no particular threat to the overall populations of prey species, in areas where the Crab-Eating Macaque is not native, it can pose a substantial threat to biodiversity.

The Crab-Eating Macaque can become a synanthrope, living off human resources. They are known to feed in cultivated fields on such items as young dry rice, cassava leaves, rubber fruit, taro plants, coconuts, mangos and other crops, often causing significant losses to local farmers. In villages, towns, and cities, they frequently take food from garbage cans and refuse piles. The species can become unafraid of humans in these conditions, which can lead to Macaques directly taking food from people, both passively and aggressively.

In **Thailand** and Myanmar, Long-Tailed Macaques use stone tools to open nuts, oysters and other bivalves, and various types of sea snails (nerites, muricids, trochids, etc.) along the Andaman sea coast and offshore islands.

Distribution and habitat:

The Crab-Eating Macaque is found in a wide variety of habitats, including primary lowland rainforests, disturbed and secondary rainforests, and riverine and coastal forests of nipa palm and mangrove. They also easily adjust to human settlements; they are considered sacred at some Hindu temples and on some small islands, but a pest around farms and villages. Typically, they prefer disturbed habitats and forest periphery. The native range of this species includes most of mainland Southeast Asia, and the Maritime Southeast Asia islands of Sumatra, Java, and Borneo, offshore islands, the islands of the Philippines, and the Nicobar Islands in the Bay of Bengal. **Thailand** and Myanmar

M. fascicularis is an introduced alien species in several locations, including Hong Kong, Western New Guinea, Anggaur Island in Palau, and Mauritius. Where it is not a native species, particularly on island ecosystems whose species often evolved in isolation from large predators, *M. fascicularis* is a documented threat to many native species. This has led the World Conservation Union (IUCN) to list *M. fascicularis* as one of the "100 worst invasive alien species". Insofar as it is present as an invasive alien species on several islands, it has been labelled a "weed" species and is yet another significant ecological threat to those ecosystems and the species within them. However, *M. fascicularis* is not a biodiversity threat in its native range, where other species have adapted to their presence through evolutionary time.

Relationship with humans:

M. fascicularis is used extensively in medical experiments, in particular those connected with neuroscience and disease. Due to their close physiology, they can share infections with humans. Some cases of concern have been an isolated event of Ebola-Reston virus found in a captive-bred population shipped to the US from the Philippines, which was later found to be a strain of Ebola that has no known pathological consequences in humans, unlike the African strains. Furthermore, they are a known carrier of monkey B virus (Herpes virus simiae), a virus which has produced disease in some lab workers working mainly with Rhesus Macaques *(M. mulatta)*. Nafovanny, the largest facility for the captive breeding of nonhuman primates in the world, houses 30,000 Macaques. The Crab-Eating Macaque is one of the types of monkeys that have been used as space test flight animals. It has been discovered recently that *Plasmodium knowlesi,* which causes malaria in *M. fascicularis,* can also infect humans. A few cases have been documented in human, but for how long humans have been getting infections of this malarial strain is unknown. It is, therefore, not possible to assess if this is a newly emerging health threat, or if just newly discovered due to improved malarial detection techniques. Given the long history of humans and macaques living together in SE Asia, it is likely the latter.

Conservation status:

The Crab-Eating Macaque has the third-largest range of any primate species, behind only humans and Rhesus Macaques. The IUCN Red List categorizes the species as Least Concern, and CITES lists them as Appendix II ("not necessarily threatened with extinction", in which trade must be controlled to avoid use incompatible with their survival). A recent review of their populations suggests a need for better monitoring of populations due to increased wild trade and rising levels of human-macaque conflict, which are reducing overall population levels despite the species being widely distributed.

Each subspecies faces differing levels of threats, or simply lacks sufficient information to assess their situation. The *M. f. umbrosa* subspecies is argued to be of important biological significance. It has been recommended as a candidate for protection in the Nicobar Islands, where its small, native population has been seriously fragmented, and is listed as vulnerable on the IUCN Red List. The Philippines Long-Tailed Macaque *(M.f. philippensis)* is listed as near threatened, and *M.f. condorensis* is vulnerable All other subspecies are listed as data deficient and need further study, although recent work is showing *M.f aurea* and *M.f. karimondjawae* need increased protection. One concern for conservation is, in areas where *M. fascicularis* is not native, their populations need to be monitored and managed to reduce their impact on native flora and fauna.

Northern Pigtail Macaque *(Macaca leonina)*

The Northern Pig-Tailed macaque *(Macaca leonina)* is a species of primate in the Cercopithecidae family. It is found in Bangladesh, Cambodia, China, India, Laos, Myanmar, **Thailand,** and Vietnam. Traditionally, it has been considered a subspecies of *M. nemestrina*. In India, it is found in south of the Brahmaputra River, in the north eastern part of the country. Its range in India extends from Assam and Meghalaya to eastern Aruanchal Pradesh, Nagaland, Manipur, Mizoram and Tripura. A detailed report on the ecology and behaviour of Northern Pig-Tailed Macaque has been published recently.

Description:

As its name suggests, this Macaque is characterised by its short, 'pig-like' tail, which it normally carries in an erect backward arch over the back, with the tip partially resting on the rump. This species resembles the Sunda Pig-Tailed Macaque *(Macaca nemestrina)*, but is smaller in size and has comparatively short limbs and face. The Macaque possesses a relatively long, uniformly agouti golden-brown coat, with markings confined only to the brown crown, buff-coloured cheek whiskers and the red streak extending from the outer corner of each eye. A distinct tuft of hair also exists at the end of the tail. Young are a blackish colour when born, but juveniles are rather more brightly coloured than adults.

Range:

This species occurs in eastern Bangladesh, Cambodia, southern China (southwestern Yunnan), northeastern India (Arunachal Pradesh, Assam, Manipur, Meghalaya, Mizoram, Nagaland and Tripura), Lao PDR, Myanmar (including the Mergui Archipelago), **Thailand (from about 8°N and including adjacent islands),** and central and southern Vietnam. There might be a gap in the distribution in central and northeastern Myanmar between about 20 and 25°N, where it has not been recorded except on the coast at Arakan. In India found north to the Brahmaputra River (Groves 2001).

Records from Xizang (China) are probably misidentified rhesus macaques (MacKinnon in press). It has recently been recorded from Namdapha National Park in northeastern India (Chetry et al. 2003).

The precise taxonomic boundary between *M. leonina* and *M. nemestrina is* not well defined. There are populations of the two taxa found on either side of the distribution limits in the Isthmus of Kra, but many of these populations are the result of release by humans. **The two species hybridize in a small area of southern peninsular Thailand, as well as on the islands of Phuket and Yao Yai (Groves 2001).**

In Vietnam, there are historical records from as far north as Nghe An province, but there is uncertainty as to whether the species was ever found north of this province. It is widely distributed throughout the lower elevations (below 500 m) of Lao PDR and Cambodia (R. Timmins pers. comm.). It is found over much Myanmar except in areas of human settlements. Eastern Bangladesh, Cambodia, China (Yunnan), India (Arunachal Pradesh, Assam, Manipur, Meghalaya, Mizoram, Nagaland, and Tripura), Laos, Myanmar (including the Mergui Archipelago), **Thailand,** and Southern Vietnam. There is an additional, introduced population on the Andaman Islands (India)

Habitat:

This is a predominantly terrestrial animal, although it readily climbs and forages in the canopy. It is diurnal and frugivorous. It occupies tropical evergreen and semi-evergreen forest, tropical wet evergreen forest, tropical moist deciduous forest, coastal forest, swamp forest, low elevation pine forests (in Lao PDR and China) and montane forest, including degraded forests. In China the species occupies elevations between 50-2,000 m (Molur et al. 2003; Choudhury 2003). In Lao PDR and Vietnam and **Thailand** the species is associated with lowlands, usually below 500 m. Its generation time is 10-12 years (Molur et al. 2003). According to secondary information and recent records, populations in Myanmar live between 190-400 m (S. Htun pers. comm.).

Systems: Terrestrial

Found in lowland primary and secondary forest, as well as coastal, swamp and montane forest. Dense rainforest is preferred, but agricultural lands may also be occupied. Groups often sleep in dipterocarp trees.

Biology:

This diurnal species occupies all levels of the forest canopy and also spends much of its time foraging on the ground. Fruits and seeds form the bulk of the diet, together with young leaves, buds, shoots, fungus and animal prey (including insects, river crabs and nesting birds). As an opportunistic feeder, however, this macaque also has a tendency to raid crops such as corn, papaya, oil palm and grain, earning it a reputation as a serious pest over much of its range. The Northern Pigtail Macaque lives in multi-male / multi-female groups of 5 to 40 (average 15 to 22), with around five to eight females to every male. Females remain with their natal group, which is structured by a matrilineal dominance hierarchy. By contrast, males disperse at puberty and remain solitary or peripheral to a group. Mating occurs year-round, although a reproductive peak occurs between January and May. Females have a 30 to 35-day reproductive cycle, and display an enormous, purplish-pink genital swelling at oestrous. These swellings provide a visual cue to males that the female is about to ovulate, and adult males rarely attempt to copulate otherwise. Mating is initiated by the male, whose courtship approach involves retracting the ears and pushing the lips forward. Since mates are usually familiar with each other within a group, cercopithecines (Guenons, Macaques and Baboons) typically display only minimal courtship behaviour, confined to signals that indicate an immediate readiness to mate. Single offspring are born after a gestation period of 162 to 186 days, and the young are then nursed for 8 to 12 months. Juveniles reach sexual maturity at around four years.

Threats:

Habitat disturbances that affect this species' survival include: selective logging; timber and firewood collection for making charcoal; building roads, dams, power lines; and deliberately setting fires. These threats lead to forest fragmentation and soil loss/erosion. Specifically, a decrease in habitat quality has been due to the loss of fruiting trees and sleeping sites through monocultures and plantations, selective felling, and a subsequent increase in the canopy gaps. These animals are hunted and traded for food, sport and traditional "medicine", and accidental mortality due to trapping occurs. There is a local trade for bones, meat for food and the live animals as pets (Molur et al. 2003). Habitat loss and poaching are the major threats in India and Bangladesh. There has been a reduction in forest in Assam by over 10% in two years between 2001 and 2003 (Forest Survey of India 2003).

In Lao PDR, Vietnam and Cambodia, hunting for food and trade is the primary threat, but as a predominantly lowland species habitat loss likely is also a major threat to the species. In **Thailand,** the males of this species are exploited for picking coconuts by the industry. Sometimes, a well-trained macaque is sold for 1,000USD. They are also in demand by resorts for show (R. Boonratana pers. comm.). In Myanmar, hunting, trade, habitat loss in varying degrees, shifting cultivation in the north, logging in the east and south, and rubber plantations are the major threats (S. Htun pers. comm.).

In China, hunting, habitat loss and disturbance are major threats. There is a perceptible change in habitat quality that has an impact on the species (Huang et al. pers. comm.). Macaques are used extensively in animal testing and vivisection, often being trapped in the wild or captive bred in poor conditions , and this species is no exception. Pig-Tailed Macaques are very popular for use in laboratories, being almost ideally suited for both psychological studies and HIV research. Threatened also by loss of habitat, the species is declining rapidly in many areas across its range. The Macaque's taste for agricultural crops has also deemed it a pest, and it is therefore frequently shot on sight. Sadly, as its forest habitat is destroyed, the species is likely to become ever more dependent upon such crops for food.

Conservation Actions:

This species is listed under CITES Appendix II. It is listed as Schedule III in the Bangladesh Wildlife (Preservation) (Amendment) Act, 1974, Category I under the Chinese Wildlife Protection Act (1989), and as Schedule II under the Indian Wildlife (Protection) Act, 1972 (Chetry et al. 2003) amended up to 2002.

Northern Pig-Tailed Macaques are known to occur in numerous protected areas, including Chunati Wildlife Sanctuary, Lawachara National Park, Rema-Kelanga Wildlife Sanctuary (Bangladesh); Daxueshan Nature Reserve, Nanguanhe Nature Reserve, Wuliangshan Nature Reserve, Xishuangbanna Nature Reserve (China); Balpakhram National Park, Dampa Wildlife Sanctuary, Dibru-Saikhowa Wildlife Sanctuary, Fakim Wildlife Sanctuary, Garampani Wildlife Sanctuary, Gibbon Wildlife Sanctuary, Gumti Wildlife Sanctuary, Intanki National Park, Kamlang Wildlife Sanctuary, Lengteng Wildlife Sanctuary, Mehao Wildlife Sanctuary, Murlen National Park, Namdapha National Park, Ngengpui Wildlife Sanctuary, Nongkhyllem Wildlife Sanctuary, Padumoni-Bherjan-Borajan Wildlife Sanctuary, Phawngpui Blue Mountain National Park, Sepahijala Wildlife Sanctuary, Siju WS, Trishna Wildlife Sanctuary, Yangoupokpi-Lokchao Wildlife Sanctuary (India); Pidaung Wildlife Sanctuary (Myanmar); **Doi Suthep-Pui National Park, Huai Kha Khaeng Wildlife Sanctuary (Thailand);** Cat Tien National Park, Pu Mat National Park (Vietnam). May possibly occur in Nam Ha National Biodiversity Conservation Area (Lao PDR) (M. Richardson pers. comm.; Molur et al. 2003)

Rhesus Macaque *(Macaca mulatta)*

Characteristics:

The Rhesus Macaque is brown or grey in colour and has a pink face, which is bereft of fur. Its tail is of medium length and averages between 20.7 and 22.9 cm (8.1 and 9.0 in). Adult males measure approximately 53 cm (21 in) on average and weigh about 7.7 kg (17 lb). Females are smaller, averaging 47 cm (19 in) in length and 5.3 kg (12 lb) in weight. Rhesus Macaques have on average 50 vertebrae. Their intermembral index (ratio of arms to legs) is 89%. They have dorsal scapulae and a wide rib cage. The Rhesus Macaque has 32 teeth with a dental formula of 2.1.2.3/2.1.2.3 and bilophodont molars. The upper molars have four cusps: paracone, metacone, protocone and hypocone. The lower molars also have four cusps: metaconid, protoconid, hypoconid and entoconid.

Distribution and Habitat:

Rhesus Macaques are native to northern India, Bangladesh, Pakistan, Nepal, Burma, **Thailand,** Afghanistan, Vietnam, Puerto Rico, southern China, and some neighbouring areas. They have the widest geographic ranges of any non human primate, occupying a great diversity of altitudes throughout Central, South and Southeast Asia. Inhabiting arid, open areas, Rhesus Macaques may be found in grasslands, woodlands and in mountainous regions up to 2,500 m (8,200 ft) in elevation. They are regular swimmers. Babies as young as a few days old can swim, and adults are known to swim over a half mile between islands, but are often found drowned in small groups where their drinking waters lie. Rhesus Macaques are noted for their tendency to move from rural to urban areas, coming to rely on handouts or refuse from humans.

The southern and the northern distributional limits for Rhesus and Bonnet Macaques, respectively, currently run parallel to each other in the western part of India, are separated by a large gap in the centre, and converge on the eastern coast of the peninsula to form a distribution overlap zone. This overlap region is characterized by the presence of mixed-species troops, with pure troops of both species sometimes occurring even in close proximity to one another. The range extension of Rhesus Macaque – a natural process in some areas and a direct consequence of introduction by humans in other regions – poses grave implications for the endemic and declining populations of Bonnet Macaques in southern India.

Distribution of subspecies and populations:

The name "rhesus" is reminiscent of the Greek mythological king Rhesus. However, the French naturalist Jean-Baptist Audient, who applied the name to the species, stated it had no meaning.
According to Zimmermann's first description of 1780, the Rhesus Macaque is distributed in eastern Afghanistan, Bangladesh, Bhutan, as far east as the Brahmaputra Valley in peninsular India, Nepal and northern Pakistan. Today, this is known as the *Indian rhesus macaque M. m. mulatta,* which includes the morphologically similar *M. rhesus villosus* described by True in 1894 from Kashmir and *M. m. mcmahoni* described by Pocock in 1932 from Kootai, Pakistan. Several Chinese subspecies of Rhesus Macaques have been described between 1867 and 1917. The molecular differences identified among populations, however, are alone not consistent enough to conclusively define any subspecies.

The Chinese subspecies can be divided in:
M. m. mulatta is found in western and central China, in the south of Yunnan and southwest of Guangxi;
M. m. lasiota (Gray, 1868), the west Chinese Rhesus Macaque, is distributed in the west of Sichuan, northwest of Yunnan, and southeast of Qinghai;[7] it is possibly synonymous with *M. m. sanctijohannis (Swinhoe, 1867),* if not with *M. m. mulatta.*
M. m. tcheliensis (Milne-Edwards, 1870), the North Chinese Rhesus Macaque, lives in the north of Henan, south of Shanxi and near Beijing. Some consider it as the most endangered subspecies.Others consider it possibly synonymous with *M. m. sanctijohannis,* if not with *M. m. mulatta.M. m. vestita (Milne-Edwards, 1892),* the Tibetan Rhesus Macaque, lives in the southeast of Tibet, northwest of Yunnan (Deqing, and perhaps including Yushu;it is possibly synonymous with *M. m. sanctijohannis,* if not with *M. m. mulatta.*
M. m. littoralis (Elliot, 1909), the south Chinese Rhesus Macaque, lives in Fujian, Zhejiang, Anhui, Jiangxi, Hunan, Hubei, Guizhou, northwest of Guangdong, north of Guangxi, northeast of Yunnan, east of Sichuan and south of Shaanxi; it is possibly synonymous with *M. m. sanctijohannis,* if not with *M. m. mulatta.M. m. brevicaudus,* also referred to as *Pithecus brevicaudus (Elliot, 1913),* lives on the Hainan Island and Wanshan Islands in Guangdong, and the islands near Hong Kong; it may be synonymous with *M. m. mulatta. M. m. siamica (Kloss, 1917),* the Indochinese Rhesus Macaque, is distributed in Myanmar, **in the north of Thailand** and Vietnam, in Laos and in the Chinese provinces of Anhui, northwest Guangxi, Guizhou, Hubei, Hunan, central and eastern Sichuan, and western and south-central Yunnan; possibly synonymous with *M. m. sanctijohannis,* if not with *M. m. mulatta.*
Around the spring of 1938, a colony of Rhesus Macaques was released in the Silver River State Park in Florida by a tour boat operator known locally as "Colonel Tooey" to enhance his "Jungle Cruise". A traditional story that the monkeys were released for scenery enhancement in the Tarzan movies that were filmed at that location is false, as the only Tarzan movie filmed in the area, 1939's Tarzan Finds a Son! does not contain rhesus macaques. In addition, various colonies of Rhesus and other monkey species are speculated to be the result of zoos and wildlife parks destroyed in hurricanes, most notably Hurricane Andrew. A notable colony of Rhesus Macaques on Morgan Island, one of the Sea Islands in the South Carolina Low country, was imported in the 1970s for use in local labs and are by all accounts thriving.

Ecology and behaviour:

A roadside band of Rhesus Macaque in Rishikesh, Uttarakhand, India: Although they are infamous as urban pests, which are quick to steal not only food, but also household items, it is not certain if the pair of jeans draped over the wall on the right is their handiwork.

Rhesus Macaques are diurnal animals, and both arboreal and terrestrial. They are quadrupedal and, when on the ground, they walk digitigrade and plantigrade. They are mostly herbivorous, feeding on mainly fruit, but also eating seeds, roots, buds, bark, and cereals. They are estimated to consume around 99 different plant species in 46 families. During the monsoon season, they get much of their water from ripe and succulent fruit. Macaques living far from water sources lick dewdrops from leaves and drink rainwater accumulated in tree hollows. They have also been observed eating termites, grasshoppers, ants and beetles.When food is abundant, they are distributed in patches and forage throughout the day in their home ranges. They drink water when foraging and gather around streams and rivers. Rhesus Macaques have specialized pouch-like cheeks, allowing them to temporarily hoard their food.

In psychological research, Rhesus Macaques have demonstrated a variety of complex cognitive abilities, including the ability to make same-different judgments, understand simple rules, and monitor their own mental states. They have even been shown to demonstrate self-agency, an important type of self-awareness.

Group structure:

Like other Macaques, Rhesus troops comprise a mixture of 20–200 males and females. Females may outnumber the males by a ratio of 4:1. Males and females both have separate hierarchies. Females have highly stable matrilineal hierarchies in which a female's rank is dependent on the rank of her mother. In addition, a single group may have multiple matrilineal lines existing in a hierarchy, and a female outranks any unrelated females that rank lower than her mother. Rhesus Macaques are unusual in that the youngest females tend to outrank their older sisters. This is likely because young females are more fit and fertile. Mothers seem to prevent the older daughters from forming coalitions against her. The youngest daughter is the most dependent on the mother, and would have nothing to gain from helping her siblings in overthrowing their mother. Since each daughter had a high rank in her early years, rebelling against her mother is discouraged. Juvenile male Macaques also exist in matrilineal lines, but once they reach four to five years of age, they are driven out of their natal groups by the dominant male. Thus, adult males gain dominance by age and experience. In the group, macaques position themselves based on rank. The "central male subgroup" contains the two or three oldest and most dominant males which are codominant, along with females, their infants and juveniles. This subgroup occupies the centre of the group and determines the movements, foraging and other routines. The females of this subgroup are also the most dominant of the entire group. The farther to the periphery a subgroup is, the less dominant it is. Subgroups on the periphery of the central group are run by one dominant male which ranks lower than the central males, and maintains order in the group and communicates messages between the central and peripheral males. A subgroup of subordinate, often sub adult males occupy the very edge of the groups and have the responsibility of communicating with other macaque groups and making alarm calls.

Communication:

Rhesus Macaques interact using a variety of facial expressive, vocalizations and body postures, and gestures. Perhaps the most common facial expression the Macaque makes is the "silent bared teeth" face. This is made between individuals of different social ranks with the lower ranking one giving the expression to its superior. A less dominant individual will also make a "fear grimace" accompanied by a scream to appease or redirect aggression. Another submissive behavior is the "present rump", where an individual raises its tail and exposes its genitals to the dominant one. A dominant individual will threaten another individual standing quadrupedally making a silent "open mouth stare" accompanied by the tail sticking straight. During movements, macaques will make "coos" and "grunts". These are also made during affiliative interactions and approaches before grooming. When they find rare food of high quality, Macaques will emit "warbles," "harmonic arches", or "chirps." When in threatening situations, macaques will emit a single loud, high-pitched sound called a "shrill bark" Screeches," "screams", "squeaks", " pant-threats", "growls", and "barks" are used during aggressive interactions. Infants make "geckers" during weaning conflicts.

Reproduction:

Adult male Macaques try to maximize their reproductive success by entering into consort pairs with females, both in and outside the breeding period. Females prefer to mate with males that will increase the survival of their young. Thus, a consort male provides resources for his female and protects her from predators. Larger, more dominant males are more likely to provide for the females. The breeding period can last up to 11 days, and a female usually mates with four males during that time. Male Rhesus Macaques have not been observed to fight for access to sexually receptive females, although they suffer more wounds during the mating season. Female Macaques first breed when they are four years old, and reach menopause at around 25 years of age. Male Macaques generally play no role in raising the young, but do have peaceful relationships with the offspring of their consort pairs. Mothers with one or more immature daughters in addition to their infants are in contact with their infants less than those with no older immature daughters, because the mothers may pass the parenting responsibilities to her daughters. High-ranking mothers with older immature daughters also reject their infants significantly more than those without older daughters, and tend to begin mating earlier in the mating season than expected based on their dates of parturition the preceding birth season. Infants farther from the centre of the groups are more vulnerable to infanticide from outside groups. Some mothers abuse their infants, which is believed to be the result of controlling parenting styles.his adaptable species is highly promiscuous and both males and females mate with as many members of the opposite sex as possible. They travel in groups of between 8 and 180 individuals, usually with two to four times as many females as males

Breeding takes place whenever the seasons permit, with no defined period in non-seasonal areas. Females undergo a regular oestrus cycle of 26 – 29 days, but unlike many other Macaques, the genital region swells and darkens in colour only slightly during the fertile period, and only in younger adult females. Gestation lasts around 165 days, and females give birth to a single young or, rarely, twins. The young is fed milk for a year, first clinging to the mother's belly, but riding on her back when older. After weaning, female juveniles may remain with the same group whereas males often disperse to another. Females become sexually mature between 2.5 and 4 years and males between 4.5 and 7 years. Females who reach ages of more than 25 years go through the menopause, eventually becoming infertile.

In science:

The Rhesus Macaque is well known to science. Due to its relatively easy upkeep in captivity, wide availability and closeness to humans anatomically and physiologically, it has been used extensively in medical and biological research on human and animal health-related topics. It has given its name to the rhesus factor, one of the elements of a person's blood group, by the discoverers of the factor, Karl Landsteiner and Alexander Wiener. The Rhesus Macaque was also used in the well-known experiments on maternal deprivation carried out in the 1950s by controversial comparative psychologist Harry Harlow. Other medical breakthroughs facilitated by the use of the Rhesus Macaque include: development of the rabies, smallpox, and polio vaccines creation of drugs to manage HIV/AIDS understanding of the female reproductive cycle and development of the embryo and the propagation of embryonic stem cells.

The U.S. Army, the U.S. Air Force, and NASA launched Rhesus Macaques into outer space during the 1950s and '60s, and the Soviet/Russian space program launched them into space as recently as 1997 on the Bion missions. One of these primates ("Able"), which was launched on a suborbital spaceflight in 1959, was one of the two first living beings (along with "Miss Baker" on the same mission) to travel in space and return alive.

In January 2000, the Rhesus Macaque became the first cloned primate with the birth of Tetra. January 2001 saw the birth of ANDi, the first transgenic primate; ANDi carries foreign genes originally from a jellyfish.

Though most studies of the Rhesus Macaque are from various locations in northern India, some knowledge of the natural behavior of the species comes from studies carried out on a colony established by the Caribbean Primate Research Centre of the University of Puerto Rico on the island of Cayo Santiago, off Puerto RicoThere are no predators on the island, and humans are not permitted to land except as part of the research programmes. The colony is provisioned to some extent, but about 50% of its food comes from natural foraging. In other more controlled settings, these macaques often enjoy Fig Newtons, and are particularly keen on "pouching" large quantities of marshmallows

Threats:

Whilst the rhesus macaque is threatened in the wild, a large captive population is maintained around the world for use in biological, psychological and medicinal research, especially for studies into perception, learning and behaviour. In the wild, the Rhesus Macaque is a generalist with great adaptability, allowing it to make the most of changes in land use. In India they are known for crop-raiding but their status as sacred animals in the Hindu religion prevents persecution by humans .Interspecies breeding is known to occur but appears to have no effect on the offspring's fertility, as other interspecies crosses usually do .

Southern Pig-Tailed Macaque *(Macaca nemestrina)*

The Southern Pig-Tailed Macaque *(Macaca nemestrina)* is a medium-sized Old World monkey found in the southern half of the Malay Peninsula (**only just extending into Southernmost Thailand),** Borneo, Sumatra and Bangka Island. This omnivorous macaque is mostly found in forest, but will also enter plantations and gardens. It formerly included the northern pig-tailed, Pagai Island, and Siberut macaques as subspecies.

Etymology:

The species epithet, nemestrina, is an adjective (derived from Latin Nemestrinus, the god of groves) modified to agree in gender with the feminine generic name.

Morphology:

Southern Pigtail Macaques have olive brown fur over their entire bodies, except for their undersides, which are white. The fur on the top of their heads is dark brown or black and grows in a pattern that makes them look like there is a depression in the center of the top their heads (Rowe 1996; Groves 2001). Northern Pigtail Macaques have golden brown fur and the fur on the top of their heads is brown. They have red streaks of fur extending from the outer cornier of each eye towards the ear. Pigtail Macaque infants are born black and as they mature, their pelage changes to the adult coloration (Crockett & Wilson 1980). Pigtail macaques are sexually dimorphic with males measuring 495 to 564 mm (1.62 to 1.85 ft) and weighing 6.2 to 14.5 kg (13.7 to 32.0 lb) while females measure 467 to 564 mm (1.53 to 1.85 ft) and weigh between 4.7 and 10.9 kg (10.4 and 24.0 lb) (Fa 1989). Males have much larger canine teeth than females, measuring 12 mm (.472 in), on average, which are used in aggressive interactions (Rowe 1996). Pigtail Macaques have an abbreviated tail, less than the length of the body from head to rump, which is often bare or covered only by sparse fur (Rowe 1996; Groves 2001). Pigtail Macaques get their popular name from their tails, which are short and carried half-erect so that they somewhat resemble a pig's tail (Choudhury 2003). They move quadrupedally on the ground and through trees (Rowe 1996). Pigtail Macaques have an average lifespan of about 26 years (Sponsel et al. 2002)

Range:

Pigtail Macaques have a wide range throughout Southeast Asia. Southern pigtail macaques are found in northeastern India, southern China, in Indonesia on Borneo (Kalimantan) and Sumatra, in eastern Bangladesh, as well as in Burma, **Thailand,** Laos, Cambodia, and Vietnam. They are also found in Malaysia, both on the Malay Peninsula and on Borneo (Feeroz et al. 1994; Groves 2001; Choudhury 2003). **Northern Pigtail Macaques are found in peninsular Thailand ,** through Burma and Indochina and into Bangladesh , India , and Southern China . There is some evidence of hybridization of the two species in Thailand, but interbreeding is not widespread (Groves 2001).

Habitat:

Pigtail Macaques are found in lowland and hilly primary rainforests and occasionally are found in swamp and secondary forests (Crockett & Wilson 1980). They prefer undisturbed forests and are found in the highest densities in intact rainforests (Supriatna et al. 1996). Rainforests are maintained in warm and humid climates where temperatures range between 18 and 30° C (64 and 86° F) and where there is more than 2500 mm (8.20 ft) of rainfall each year, though there may be seasonality in rainfall. Where they are found in Bangladesh, the annual precipitation is 2034 mm (6.67 ft) with the highest rainfall occurring during the monsoon season, from May to September, and the lowest in December. The coldest months of the year are December and January, which have average temperatures of 12.3° C (54.4° F) and 9.7° C (49.5° F), respectively. August is the warmest month of the year with an average maximum temperature of 33° C (91.4° F) (Feeroz et al. 1994). In neighbouring northeastern India, the climate and rainfall is quite variable, ranging from less than 1000 mm (3.28 ft) to more than 10,000 mm (32.81 ft) annually (Choudhury 2003). The climate in this part of India can also be as cool as 4°C (39.2° F) in December to early February to 30° C (86° F) from June to August. Most of the rain occurs from May to September, and snow falls in winter at the higher altitudes (Choudhury 2003). Rainfall in southern Sumatra ranges from 2000 to 3267 mm (6.56 to 10.7 ft) per year with indistinct wet or dry seasons (Supriatna et al. 1996). The climate there is also warm and humid (Lucas & Corlett 1991). In Peninsular Malaysia, rainfall averages 2100 mm (6.89 ft) per year with the least amount of rainfall during January and February and the period of highest rainfall occurring during September and October (Saiful & Nordin 2001). Pigtail Macaques range from sea level to above 2000 m (6562 ft) (Srivastava & Mohnot 2001; Choudhury 2003).

Ecology:

Pigtail Macaques are highly frugivorous, with 74% of their diet consisting of fruit, but they also consume a wide variety of foods including insects, seeds, young leaves, leaf stems, dirt, and fungus (Crockett & Wilson 1980; Caldecott 1986). Pigtail Macaques spend most of their time on the ground, but the Northern Pigtail Macaque seems to be more arboreal than the southern species (Maestripieri pers. comm.). Spending most of their time on the ground foraging, they are particularly adept at raiding agricultural fields and obtaining coconuts from oil palm plantations, papaya, corn, and cassava. They are stealthy crop raiders, sneaking silently into a garden one at a time, with one acting as a lookout and calling an alarm vocalization if humans are seen. Pigtail Macaques are especially likely to raid crops during rainstorms, when farmers are inside, away from their crops (Crockett & Wilson 1980). In some areas of the Malay Peninsula, farmers keep and train Pigtail Macaques to retrieve coconuts and fruits from cultivated trees (Crockett & Wilson 1980; Sponsel et al. 2002). Because of their large group size, between nine and 81 individuals and larger, Pigtail Macaques often split up into foraging groups to decrease direct competition for fruit at feeding sites. They travel in small subgroups, from two to six monkeys, along the ground, foraging as they move and keeping in contact with other subgroups through vocalizations (Crockett & Wilson 1980; Caldecott 1986). In addition to spreading out over the landscape as they forage, pigtail macaques cover large areas each day. They have home ranges between .6 and 8.28 km^2 (.232 and 3.20 mi^2) and in areas of high density, groups' home ranges can overlap each other by as much as 50% (Sponsel et al. 2002). The day range length varies between 825 and 2964 m (.513 and 1.84 mi), depending on weather conditions and seasonal fruit availability (Caldecott 1986).The Sunda Pig-Tailed Macaque spends more time on the forest floor and in the open than other macaques, where it forages for leaves, buds, shoots, insects and small animals. It has cheek pouches to carry food while it forages, and often returns to the safety of the trees to feed. This macaque is a social primate and lives in group sizes of 5 – 40 (average 15 – 22) individuals, though the group splits into smaller units to forage. In the group males and females live together. Females remain in their natal group, though males will disperse shortly before they reach sexual maturity. Breeding occurs all year round with one young born. Mothers care for their offspring alone. Infants are born with a black coat which lightens as they grow.Gestation: 5.7 months (170 days)

Threats:

This species faces many threats. Destruction of forests by felling, encroachment, slash and burn cultivation of the hill tribes and monoculture are all major threats to the Sunda Pig-Tailed Macaque's habitat . The rate of forest destruction is alarming, and is not only reducing this primate's habitat, but also fragmenting it. The macaque's taste for agricultural crops has also deemed it a pest, and it is therefore frequently shot on sight . Furthermore, Sunda pig-tailed macaques are very popular for use in laboratories, being almost ideally suited for both psychological studies and HIV (AIDS virus) research. Classified as Vulnerable (VU) on the IUCN Red List 2006 Further studies are needed into the distribution, abundance, and threats to this species.

Subfamily: Colobinae:

Genus: Trachypithecus:

Dusky Leaf Monkey (*Trachypithecus obscurus)*

Geographic Range:

Dusky Leaf Monkeys are found primarily on the Malay Peninsula, including southern Burma and parts of **Thailand.** They also inhabit the islands of Langkawi, Penang, and Perhentian Besar. (Lekagul and McNeely, 1977; Medway, 1969) Dusky leaf monkeys can be found in a wide range of habitats. Being arboreal forest dwellers, they prefer dense forests with tall trees. (Medway, 1969)

Biology:

Dusky Leaf Monkeys are widely variable in colour. Their upper parts may be any shade of brown, grey, or black, whereas the under parts, hind legs, and tail are paler. The face is grey and is often marked with a patch of white fur located around the eyes and mouth. The hands and feet are capable of grasping and closely resemble those of humans. The palms and soles are hairless and usually black. The fingers of dusky leaf monkeys are well developed, but are distinct because of their opposable thumb. The no prehensile tail varies in length and fur coverage from short and hairless, to long and hairy.

Head and body length ranges from 42 to 61 cm, and tail length from 50 to 85 cm. There are no significant morphological differences between males and females except that males tend to be slightly larger and heavier than females. On average, a healthy adult male weighs 7.4 kg, whereas a healthy adult female weighs approximately 6.5 kg. Newly born dusky leaf monkeys are bright yellow or orange in colour, and have a pink face; the fur changes to a greyish colour within six months. (Grzimek, 1990; Medway, 1969). The average lifespan is recorded as 15.3 years in captivity. Dusky Leaf Monkeys are diurnal. They are very active during the day, but return to their roosts in the trees by night. These monkeys are active in the tree canopy, and prefer to stay at heights of 35 meters or higher in trees. They move from tree to tree by climbing, leaping, and running quadrupedally along branches, When feeding, dusky leaf monkeys pluck leaves and shoots off by hand. They also pull down leafy branches and browse on them directly.

These monkeys travel in groups that consist of 5 to 20 individuals. Social groups usually have one or more adult males, and two or more adult females. The adult male has three main responsibilities, which include detecting predators, holding the group together, and patrolling the boundaries of the territories. The young monkeys play in groups near the vicinity of an adult female. Overall, Dusky Leaf Monkeys are quite social animals.

Dusky Leaf Monkeys have a wide range of calls that are considered to be quite complex. A variety of snorts, hoots, murmurs, and squeaks are used to communicate with other members of their social group. The motion of the tail plays a significant role in maintaining balance. (Grzimek, 1990; Lekagul and McNeely, 1977; Medway, 1969) The diet of these monkeys consists of young leaves, shoots, and seedlings. They feed from 87 different species of trees, ingesting both leaves and fruit. In general, a dusky leaf monkey eats up to 2 kg of food per day. These monkeys can be maintained, in captivity, on sweet potato shoots, lettuce, cabbage, kangkong, green beans, maize, carrots, and soft fruits. Meat was refused, but certain insects were occasionally accepted. (Lekagul and McNeely, 1977; Medway, 1969)

Reproduction:

There is no information available regarding the mating system of dusky leaf monkeys. However, the social system typically involves groups with only one or two males. In other primate species, this social organization is typically associated with polygynous breeding. It is reasonable to assume that this species is like other similar primates in this respect. (Lekagul and McNeely, 1977; Medway, 1969; Nowak, 1999)They monkeys reproduce every two years. They monkeys breed intermittently throughout the year. Range number of offspring 1 to 2.The average number of offspring is 1 with an average gestation period of 145 day with an average birth mass of 300 g (10.57 oz The age at sexual or reproductive maturity (female) 3 to 4 year. The age at sexual or reproductive maturity (male)3 to 4 years

Breeding in Dusky Leaf Monkeys is intermittent and not always seasonal. Births usually take place during the months of January, February, and March, but have been documented to occur during the summer months as well. Typically one young is born. The gestation period is, on average, 145 days. Females have a menstrual cycle lasting approximately three weeks. Oestrus is often accompanied by a swelling of the genitalia. The normal interbirth interval is about 2 years. Sexual maturity is reached between 3 and 4 years of age. (Grzimek, 1990; Lekagul and McNeely, 1977; Nowak, 1999) Research on the parental behavior of these monkeys is lacking. However, we may assume that they are like other primates in that the mother provides the bulk of the parental care. She grooms, protects, and feeds the newborn. The role of the father in parental care is not known. There is no information available regarding the weaning age of dusky leaf monkeys. It is known, however, that the newborn monkey is fully furred and active. (Grzimek, 1990)

Communication and Perception:

Details on communication in these monkeys are scant. However, we know that they use vocalizations to protect their territories from other members of the species. Like other primates, tactile communication (e.g. grooming, playing, mating, aggression) and visual communication (e.g. facial expressions and body postures) are probably both important in these monkeys. (Nowak, 1999)

Predation:

Information on predation of these monkeys is not available. Because they are arboreal, it is likely that they do not have many predators. Possible predators are large carnivores, snakes, and raptors. It is probable that Dusky Leaf Monkeys are hunted for food by the human population, since a large number of primates are hunted as sources of food throughout Asia. (Grzimek, 1990). Dusky Leaf Monkeys are generally restricted to primary forests. Therefore, it is unlikely that they contribute to any serious agricultural (or other) problems that would adversely affect humans. (Grzimek, 1990)

Ecosystem Roles:

These monkeys are likely to be important predators of foliage. They may help to disperse seeds. To the extent that they serve as prey for other species, these monkeys may affect local food webs.

Major Threat(s):

Hunting for food is a major threat, as is habitat loss and degradation due to expanding oil palm plantations, agriculture, and urbanization. In Peninsular Malaysia the animals are frequent victims of road-kill.

Conservation Actions:

This species is listed on CITES Appendix II. It is known to occur in a number of protected areas, including: Krau Wildlife Reserve, Taman Negara National Park (Malaysia); Kaeng Krachan National Park, Khao Sam Roi Yot National Park, **Khao Sok National Park, Taratau National Park (Thailand).** It is relatively common in captivity. There is a need for further survey work to determine the current population status of the insular forms. Dusky leaf monkeys are confined within a relatively small area of southeast Asia. It is probable that these forested areas are under threat of development or logging. Therefore, there is reason to believe that the species is threatened to some extent because of habitat loss, but there are no studies to support this. (Grzimek, 1990

IUCN Red List of Threatened Species

Near Threatened

Griffith's Silver Langur (*Trachypithecus villosus*)

Description:

The body mass of a male Silvered Leaf-Monkey averages 6.6 kilograms and adult female averages 5.7 kilograms (Fleagle, 1999). Subadult males are about the size of adult females but smaller than fully grown adult males (Bernstein, 1968). The pelage colour is a dark gray with the tips of the hairs being paler in colour giving it a silvered effect. The groin and the ventral side of the tail are yellowish in coloration (Rowe, 1996). The face is black in coloration and the hair on the head is pointed (Furuya, 1961-2). Adult females have irregular patches of white on the inner portion of the upper thighs (Bernstein, 1968; Furuya, 1961-2). The newborn's pelage colour is orange, with the face, hands, and feet being white (Bernstein, 1968). Infants will darken and the darkening process takes 3-5 months to change to the adult pelage coloration (Bernstein, 1968; Medway, 1970). First the head starts to fade to a yellow orange, then the head, face, hands, feet, and ears turn dark gray and then black (Bernstein, 1968). This species has a sacculated stomach to assist in the breakdown of cellulose. The dental formula of this species is 2:1:2:3 on both the upper and lower jaws (Ankel-Simons, 2000). Groves (2001) describes two subspecies for the silvered Langur, although other authors have included the species he describes as *Trachypithecus germaini*. The subspecies each have their own pelage coloration:

Range:

The Silvered Leaf-Monkey is found in the countries of Brunei, Burma, Cambodia, Indonesia, Laos, Malaysia, **Thailand,** and Vietnam (Bernstein, 1968; Fooden, 1976; Ruggeri and Timmons, 1995/1996). On the Malay Peninsula this species is found on the west coast in an area between Province Wellesley and Malacca State (Furuya, 1961-2). In Vietnam this species is found in the south from Kontum to Tay Ninh province (Le, 1996/1997). This species lives in a variety of forest types including, primary and secondary growth forests, riverine forests, mangrove forests, swamp forests, and coastal forests (Furuya, 1961-2; MacKinnon and MacKinnon, 1987). In Laos the silvered Langur has been found in lowland riverine, degraded, and mosaic habitats (Ruggeri and Timmons, 1995/1996). **At Kanchanburi, Thailand, this species lives in evergreen forests (Bennett and Davies, 1994).**

There are 4 subspecies and the four subspecies each have differing ranges:
Trachypithecus cryostats cristatus: This subspecies occurs on Sumatra, some islands in the Riau Archipelago, Bangka, Belitung, Borneo, and the western coast of peninsular Malaysia (Groves, 2001).
Trachypithecus cristatus vigilans: This subspecies occurs on the Natuna Islands (Groves, 2001).
Trachypithecus cristatus koratensis: This subspecies is found in Burma, **South-Eastern Thailand including Koh Chang** as far north as Lao, Cambodia, and southern Vietnam (Groves, 2001).
Trachypithecus cristatus caudal is: This subspecies is found in northern Vietnam (Groves, 2001).

Ecology:

The Silvered Leaf-Monkey is a folivorous species, but this species also consumes fruits, seeds, flowers, and young shoots (Bernstein, 1968; Furuya, 1961-2). Young leaves and leaf buds are selected from leaves of trees and leaves of any age are selected from ground plants and vines (Bernstein, 1968). Orchid flowers and dried wood were observed to be consumed by the silvered Langur (Bernstein, 1968). This species prefers to eat immature leaves over more mature ones because they contain less lignin and tannins. Clay is eaten by scraping the teeth along a clay outcrop (Bernstein, 1968).
Group sizes range from 9 to 51 individuals. Furuya (1961-2) found that group sizes range from 22-48 individuals. This is an arboreal and diurnal species. This species will come to the ground, but only to move between certain areas (Bernstein, 1968). Travel is sometimes initiated and coordinated by the adult male, sometimes using the kwah vocalization (Bernstein, 1968). The group will start moving shortly before sunrise (Bernstein, 1968). This species will also travel right before it becomes dark out, with all of the group members retreating to a single tree to sleep at night (Bernstein, 1968). Rest differs from sleep in that it is individual, with having a peak during the day in the late morning (Bernstein, 1968). When resting, juveniles and infants stay close to adult females (Furuya, 1961-2).The silvered leaf-monkey moves through the forest quadrupedally (Fleagle, 1988; Furuya, 1961-2).

Social Behaviour:

The Silvered Leaf-Monkey has a unimale social structure. The group is composed of a single adult male, several adult females, and immature individuals (Wolf and Fleagle, 1977). Males will live solitarily before finding a group of its own (Bennett and Davies, 1994).The groups are highly territorial and will act aggressively towards other individuals (Bernstein, 1968). Males will keep other group males out of the range of the group (Bernstein, 1968; Furuya, 1961-2). Conflicts will usually occur at the areas of home range overlap (Bernstein, 1968). Conflicts usually consist of chasing and/or fighting between the resident males (Bernstein, 1968). Adult males will move rapidly amongst the group and emit loud vocalizations when two neighbouring groups come together (Bernstein, 1968). Other members of the group will squeal and embrace each other during group conflicts (Bernstein, 1968). Fighting consists of slapping and pulling with some biting (Bernstein, 1968). Intragroup agonistic behavior is rare in the Silvered Langur (Bernstein, 1968). Males disperse from their natal group. Infanticide has been reported to occur in this species (Wolf and Fleagle, 1977; Wolf, 1980). Infanticide results from an adult male coming into the group, killing the resident adult male, and killing the infants dependent on their mothers so that the adult females will begin to ovulate again (Wolf and Fleagle, 1977). An adult male that takes over a group will also chase out the Immature males of the group (Wolf and Fleagle, 1977). Females allow other females to carry and to care for their young (allomothering) (Fleagle, 1988). Adult males have also been observed to carry and care for immature group members (Bernstein, 1968). Usually when adult males carry young, the young initiate the behavior and the adult males are passive (Bernstein, 1968). Females will carry infants ventrally (Bernstein, 1968). Infants or juveniles that fall to the ground are retrieved by juveniles or adult females (Bernstein, 1968). Social play in the Silvered Langur mostly occurs amongst juveniles of both sexes and infants, but individuals of all ages and both sexes have been found to participate (Bernstein, 1968). Wrestling is the most common form of play with loco motor play also found (Bernstein, 1968). Locomotors play includes repeated circuits along a pathway where individuals will swing from branches or drop to the ground from branches (Bernstein, 1968). Juvenile males engage in the most vigorous forms of play fighting, and they will sometimes play fight with Sub adult males (Bernstein, 1968).

Vocal Communication:

ku: This call is emitted by infants when they are in a peaceful state (Furuya, 1961-2). This call is also uttered by juveniles when addressing each other from a short distance (Furuya, 1961-2).

ho: This call is emitted by young males and females when addressing each other from a short distance (Furuya, 1961-2).

ku-ku: This call is uttered by adult females when greeting each other (Furuya, 1961-2).

kwah: This call is emitted by the adult male in order to control group travel movement (Bernstein, 1968). This call is emitted at repeated intervals (Bernstein, 1968).

squeal: This call is most often emitted by juveniles and infants (Bernstein, 1968). This call sounds like "ki" or "gyu" (Furuya, 1961-2).

territorial call: This is a two note inhale-exhale type vocalization that is used by adult males during and after territorial conflicts (Bernstein, 1968).

alarm call: This call sounds like "snick" (Bernstein, 1968). Furuya (1961-2) found that there are three different sounds associated with an alarm call: "cha-cha', "chue-chue", and "ge-ge".

fear call: This call sounds like "kie" (Furuya, 1961-2). This call expresses fear or surprise (Furuya, 1961-2).

threat call: This call sounds like "guu" or "gyu" (Furuya, 1961-2). This call is emitted by adult males (Furuya, 1961-2). This call serves to communicate threat on the part of the sender (Furuya, 1961-2).

attack call: This call sounds like "gua-gua-gua" (Furuya, 1961-2). This call is emitted by adult males when attacking an intruder (Furuya, 1961-2).

Olfactory Communication:

Visual Communication:

yawning: This behavior is used to communicate tension or general excitement on the part of the sender (Bernstein, 1968).

branch shaking: This behavior is used to communicate tension or general excitement on the part of the sender (Bernstein, 1968).

chasing: This is when one individual will chase another for a given distance. The silvered Langur will do this when there are disputes at areas of group territory overlap, with the male usually performing this behavior (Bernstein, 1968). An adult male will chase the other male out of the territory, but when they reach the other group's territory a reversal will occur where the one who was chasing is now being chased (Bernstein, 1968).

head-shaking: This is where the female shakes her head back and forth before she presents to the male (Bernstein, 1968).

presenting: This behavior is performed by the female to elicit a copulation from the male; this pattern tells the male that she is ready for copulation. Head-shaking precedes this display.

Tactile Communication:

social grooming:

This is when one individual grooms another and is used to reinforce the bonds between individuals. All group members participate in social grooming (Bernstein, 1968). Three or four individuals may groom in a chain or two individual may simultaneously groom a third (Bernstein, 1968). Grooming is often reciprocal, where two individuals groom each other, but not necessarily at the same time (Bernstein, 1968).

embracing:

This behavior is seen when there is distress amongst group members (Bernstein, 1968). This behavior can occur with squeal (Bernstein, 1968).

Reproduction:

The Silvered Leaf-Monkey gives birth to a single offspring. This species has not been found to have a birth season, with births occurring throughout the year (Medway, 1970).

A typical sexual encounter will begin with the female making rhythmic side-to-side movements and then presenting and head shaking towards the male (Bernstein, 1968). A male will not always mount a female that is performing solicitation behavior, but when a male does mount, the mounts are multiple in frequency (Bernstein, 1968). If two females solicit a male simultaneously, both females will be mounted (Bernstein, 1968). If a male moves away from a soliciting female, the female will reorient towards the male (Bernstein, 1968).

Conservation Status:

Red List Category & Criteria: Trachypithecus villosus is listed as DD ver 2.3 (1994) (IUCN Red List).

Superfamily: Hominoidea:

Family: Hylobatidae (gibbons)

Genus: Hylobates:

Agile Gibbon (*Hylobates agilis*)

Description:

Pelage colour for the Agile Gibbon ranges from buffy coloured to reddish-brown, and can even be of a brown or black colour. Males differ from the females in having both white eyebrows and cheeks while the females only have white eyebrows. This species has relatively long forearms which assist it in suspensory behavior. This species has throat sac located beneath the chin to help enhance the calls. The Agile Gibbon lacks a tail, caudal vertebrae. The average body mass for an adult male agile gibbon is around 5.8 kilograms, and for the female it is around 5.4 kilograms (Fleagle, 1988).

Range:

The Agile Gibbon is found in the countries of Indonesia, Malaysia, and **Thailand.** This species is found in semi deciduous monsoon forests and tropical evergreen forests. The Agile Gibbon prefers the upper canopy of the forest. This species occurs at highest densities in dipterocarp-dominated forests, but their known habitat ranges from swamp and lowland forests to hill, submontane, and montane forests (O'Brien et al. 2004). Additionally, populations in Bukit Barisan Selatan National Park in Sumatra do not seem to avoid forest edges near human habitations (O'Brien et al. 2004). In southern Sumatra, populations were found up to 1,400 meters (O'Brien et al. 2004). These arboreal and diurnal primates are primarily frugivorous (preferring fruits high in sugar, such as figs), but they will consume immature leaves and insects as well (Gittins 1979, 1982). An average home range size of 29 ha has been determined in a study at Sungai Dal (Gunung Bubu Forest Reserve) on the Malayan peninsula (Gittins 1979, 1982; Gittins and Raemaekers 1980).

Ecology:

The Agile Gibbon is a frugivorous species, but will also consume immature leaves and insects. The Agile Gibbon prefers to consume fruits high in sugar such as the fig (Ficus). This species forages for fruit in the middle canopy of the forest and for immature leaves in the middle and upper canopies. This an arboreal and a diurnal species. This species sleeps and rests in the emergent trees (Leighton, 1987).This species occurs at highest densities in dipterocarp -dominated forests, but their known habitat ranges from swamp and lowland forests to hill, submontane, and montane forests (O'Brien et al. 2004). Additionally, populations in Bukit Barisan Selatan National Park in Sumatra do not seem to avoid forest edges near human habitations (O'Brien et al. 2004). In southern Sumatra, populations were found up to 1,400 meters (O'Brien et al. 2004).Usually agile gibbons can reach 30-40 years in captivity; however, the oldest captive agile gibbon alive today is 49 years old!

Locomotion:

The Agile Gibbon is a true brachiator which means it moves by suspensory behavior (Fleagle, 1988). The brachiation is of a type where the Agile Gibbon throws itself from tree to tree over gaps of 10 meters or more using there arms (Fleagle, 1988). This species also climbs when moving slowly and feeding (Fleagle, 1988). This species is also able to move for short distances by bipedalism (Fleagle, 1988).

Social Behaviour:

The Agile Gibbon has a monogamous mating and social system. Females give birth to a single offspring after seven months' gestation. The young gibbon is weaned at barely 2 years of age. When fully mature, at about 8 years, it leaves its family group in order to look for a mate.

Female cycle: 24 - 26 days

Gestation length: 210 days

Weight newborn: not known

Weaning age: 2 years

Interbirth interval: 2 – 4 years

Sexual maturity: probably 7 and 9 years in the wild, but like other gibbons thought to start breeding earlier (at age 4 – 5 years) in captivity.

Each family has a distinct territory and it is defended against all

The basic group structure is the breeding pair and their offspring. Both males and females emigrate from their natal group around adolescence. This is a territorial species. Adolescent and subadult males participate in the defending of the territory against conspecific males with their fathers (Gittins, 1980). Each family has a distinct territory and it is defended against all conspecifics. The female agile gibbon leads the family group around their territory. If the family group comes across a neighbouring group, the opposing males will stare at each other, and sometimes even chase the other group back across the boundary. Meanwhile the female and her offspring keep to the background during such encounters, but will very actively sing their territorial duet with the male. These ritualized group encounters may last for 15-60 minutes in duration. Observations and playback experiments (where calls are played by scientists interested in observing their impacts on gibbon behaviour) have shown that encounters made by strange single females are 'seen off' by females, who sing, approach and defend their territorial boundary, with the help of her male mate. Whilst encounters made by strange single males stimulates the paired male to approach aggressively, but the paired females does not join in (Mitani 1986 and 1987). The social bond within the group is maintained by grooming. An infant Agile Gibbon usually sleeps clinging to the mothers' ventral area (stomach), but as it grows into a juvenile it will often sleep in the arms of its father. Subadult offspring usually keep a greater distance away from their parents and younger siblings and often sleep in a different tree.

Vocal Communication: Duetting:

These are vocalizations which occur between the breeding male and female, and is dominated by the female. This vocalization is important because it helps to maintain the pair bond between the breeding pair and also it helps to establish and maintain the territory.

Major Threat(s):

On Sumatra, this species is threatened by conversion of their forest habitats by humans and a subsequent opportunistic capture for the pet trade. These threats extend to populations within national parks and forests, including illegal agricultural development inside the parks. In Bukit Barisan Selatan National Park in southwestern Sumatra, deforestation rates are linked to the coffee market; coffee plantations serve to completely strip arboreal primates of their canopy habitats (O'Brien et al. 2004). The expansion of oil palm plantations is a major cause of forest loss on Sumatra. In nearby Java, agile gibbons are one of the most commonly seen gibbons in the wildlife markets (Nijman 2005).

The species status in West Malaysia is uncertain; in Indonesia, it was certainly affected by fires and deforestation of the 1990s. There has been a probable 50%-plus range reduction in last 10 years (C. Groves pers. comm.), and oil palm plantations are expanding rapidly in the country. In **Thailand** there is extensive conversion of forests to rubber plantations and other crops (even inside protected areas), as well as hunting for the pet trade.

Conservation Actions:

Agile gibbons are protected throughout their range by local laws as well as by listing on CITES Appendix I, although the extent to which national or international laws actually protect the species is uncertain.
The species occurs in a number of protected areas, including Bukit Barisan National Park, Kerinci Seblat National Park, Selantan National Park, and Way Kambas National Park in Indonesia; Mudah Wildlife Reserve in Malaysia; and **Hala Bala Sanctuary in Thailand.** Unfortunately, many of these are merely proposed or gazetted, and their actual protected status is uncertain. Moreover, many of the Sumatran reserves are in montane regions where the species occurs only at low densities. In Bukit Barisan Selatan National Park in southwestern Sumatra, populations are presently secure and healthy but will depend on the regaining of control by the Indonesian government over illegal deforestation of its parks for their continued survival (O'Brien et al. 2004).
As the validity of the subspecies is questionable, a research priority is to clarify this issue so that the species can be further assessed.The species is considered Endangered in light of a continued decline inferred from habitat loss, calculated to be ≥ 50% over the past 45 years (3 generations), in combination with illegal trade for the pet market. Threats are driven primarily in the Sumatran portion of the Agile Gibbon's range where the species seems to be declining rapidly and is certainly in danger of extinction. In peninsular Malaysia and **Thailand there seem to be a number of stable populations,** but the overall range for the population has contracted dramatically. The species is entirely confined to closed canopy forest, and thus, habitat conversion, road building and fragmentation are increasingly threatening the species. Furthermore on Java, agile gibbons are one of the most commonly seen gibbons in the wildlife markets (Nijman 2005). Agile Gibbons are protected throughout their range by local laws as well as by listing on CITES Appendix I, although the extent to which national or international laws actually protect the species is uncertain. The species occurs in a number of protected areas, including Bukit Barisan National Park, Kerinci Seblat National Park, Selantan National Park, and Way Kambas National Park in Indonesia; Mudah Wildlife Reserve in Malaysia; and **Hala Bala Sanctuary in Thailand.** Unfortunately, many of these are merely proposed or gazetted, and their actual protected status is uncertain. Moreover, many of the Sumatran reserves are in montane regions where the species occurs only at low densities. In Bukit Barisan Selatan National Park in south western Sumatra, populations are presently secure and healthy but will depend on the regaining of control by the Indonesian government over illegal deforestation of its parks for their continued survival (O'Brien et al. 2004).
Again as the validity of the subspecies is questionable, a research priority is to clarify this issue so that the species can be further assessed. Agile Gibbons have been documented in captivity since the early 1960s. By the early to mid 1980s breeding with many of these animals was successful and very few animals have been taken from the wild in recent years. Those few wild born animals that have come to zoos in recent years tend to have been rescue-animals rather than captured for zoos. As of April 2010, there were 8.8.1 Agile Gibbons surviving in captivity within European zoos, with an additional 2.0 thought to be hybrids. Genetic analyses of the population is being undertaken to help researchers tackle the issue of whether subspecies do indeed exist; this will influence how we manage them in captivity.

Lar Gibbon (*Hylobates lar*)

Other names: *H. albimana, H. longimana, H. variegates, H. varius,* common gibbon, lar gibbon, white-handed gibbon; withandgibbon (Dutch); gibbon à mains blanches, gibbon lar (French); gibbon de manos blancas (Spanish); lar, vanlig gibbon, vithandad gibbon (Swedish); *H. l. carpenteri*: Carpenter's lar; *H. l. entelloides*: central lar; *H. l. lar*: Malaysian lar; *H. l. vestitus*: Sumatran lar.

The sub specific status of the Yunnan lar, *H. l. yunnanensis*, is doubtful as it is likely synonymous with *H. l. carpenteri* and as such, is not considered as a subspecies here (Geissmann 1995; Brandon-Jones et al. 2004; Groves & Geissmann pers comm. cited in Geissmann 2007). Some publications still do consider *H. l. yunnanensis* a valid subspecies, however all of this may be moot as the populations which may have made up *H. l. yunnanensis* are in all likelihood extinct (Geissmann 2008). All subspecies need to be better defined and are not highly distinct (Brockelman & Geissmann 2008).

Description:

Lar Gibbons weigh on average between 5.0-7.6 kg (11.0-16.8 lb)(M) and 4.4-6.8 kg (9.7-15.0 lb) (F). Head and body lengths range from 43.5-58.5 cm (17.1-23.0 in) (M) and 42.0-58.0 cm (16.5-22.8 in) (F) (Roonwal & Mohnot 1977). Like all apes, lar gibbons have no tail and also possess small Ischia callosities as well as long arms suited for brachiation (Ankel-Simons 2007).

Pelage is creamy, brown, or black in both sexes and at all ages (Roonwal & Mohnot 1977; Mootnick 2006; Bartlett 2009). Adults have a ring around the hairless face (less apparent in lighter individuals) and have white feet and hands that contrast with the rest of the body (Marshall & Sugardjito 1986; Groves 2001; Mootnick 2006; Ankel-Simons 2007; Bartlett 2009). Individuals are either light or dark in overall colour (often called light and dark phases), with light individuals having buff or creamy pelage, and dark individuals with brown or black pelage (Groves 2001). The deciding factor whether a lar gibbon is dark or light is simple genetic inheritance from its parents, in which the dark form is the dominant allele (Brockelman 2004). The proportion of dark to light individuals varies in lar gibbons, with some populations being mostly light and others being mostly dark and with various proportions in-between, often varying greatly between populations (Marshall & Sugardjito 1986; Brockelman 2004). Mating is random with respect to pelage colour (Brockelman 2004).

H. l. carpenteri are long-haired and almost black or brownish charcoal in the dark form and are creamy white or white -buff in the light form (Groves 2001; Mootnick 2006). *H. l. entelloides* are black or black-brown in the dark form and honey-coloured in the light form with darker legs (Groves 2001; Mootnick 2006). *H. l. lar* are medium brown or dark chocolate brown in the dark form with a lighter torso and darker legs (Groves 2001; Mootnick 2006). This subspecies is creamy in the light form (Groves 2001). *H. l. vestitus* are always light with no dark form and are typically light brown, with a buff back, pale lower back and a brown head, ventrum and limbs (Marshall & Sugardjito 1986; Groves 2001). However, due to variation across their distribution, if a specific lar gibbon's place or origin is unknown, it is difficult to assign it sub specifically based on morphology (Geissmann 1995). Morphological differences between subspecies need to be further documented (Brandon-Jones et al. 2004).

Lar Gibbons are adapted for arboreal locomotion and usually move through their environment using brachiation, but also through other forms of movement (Roonwal & Mohnot 1977; Vereecke et al. 2006). These include walking (bipedal, tripedal, and quadrupedal), hopping, running (bipedal, tripedal, and quadrupedal), climbing, swinging, bridging and leaping (Baldwin & Teleki 1976; Carpenter 1976). Lar Gibbons are also capable of rapid terrestrial movement using 7 types of gait even though they are adapted for arboreal locomotion (Vereecke et al. 2006).

In captivity, Lar Gibbons have lived up to around 50 years of age (Weigl 2005).

Range:

Hylobates lar | Hylobates lar carpenteri | Hylobates lar lar | Hylobates lar vestitus | Hylobates lar entelloides | Hylobates lar yunnanensis

Lar Gibbons are mainly found in Southeast Asia and in a small portion of South Asia. They have the greatest north-south range of any of the gibbons (Bartlett 2007). The species is found in Indonesia, Laos, Malaysia, Myanmar and **Thailand** (Brandon-Jones et al. 2004). Lar gibbons are probably extinct in China but if they still exist, they would only be found in Southwest Yunnan, their former range (Brandon-Jones et al. 2004; Brockelman & Geissmann 2008; Geissmann 2008). Moving south from southwest Yunnan, **Lar Gibbons are found in the majority of Thailand excepting the north-eastern areas of the country.** The range extends south through southern and eastern Myanmar but only east of the Salween River (Brockelman & Geissmann 2008). They are found through the Malay Peninsula except for a discontinuity in their distribution between the Perak and Muda Rivers, very roughly near the Thai-Malay border (Brockelman & Geissmann 2008). Lar gibbons also exist west of the Mekong River in north-western Laos and in northern Sumatra (Brandon-Jones et al. 2004; Brockelman & Geissmann 2008). Sub specifically, *H. l. lar* are found in peninsular Malaysia and possibly **Thailand.** *H. l. carpenteri* is found in Myanmar, west Laos, and **north Thailand.** *H. l. entelloides* is found in southeast Myanmar, and in central and **South Thailand** (Brandon-Jones et al. 2004). In Indonesia, H. l. vestitus is found over the northern third of the island of Sumatra (MacKinnon 1984; Brandon-Jones et al. 2004; Brockelman & Geissmann 2008).

In Thailand, there are probably between 15,000-20,000 lar gibbons (Geissmann 2007).

Habitat:

In general, Lar Gibbons are found in lowland dipterocarp forest, hill dipterocarp forest, and upper dipterocarp forests, including primary lowland and submontane rainforest, mixed deciduous bamboo forest, and seasonal evergreen forests, but are not usually found higher than 1200 meters above sea level (Chivers 1972; Ellefson 1974; Caldecott 1980; Palombit 1992; Ungar 1996; Reichard & Sommer 1997; Brockelman et al. 1998; Yimkao & Srikosamatara 2006; Bartlett 2009). Lar gibbons are also found in peat swamp forest (Norhayati et al. 2004).

Perhaps the most important study site of Lar Gibbons is at **Khao Yai, Thailand, northeast of Bangkok, home to many studies (some long-term) since the 1970's (Brockelman et al. 1998; Bartlett 2009).** The vegetation type at this site is mostly seasonally wet evergreen rainforest (Brockelman et al. 1998; Bartlett 2009). Three seasons are experienced in this part of **Thailand;** a cool season (November to February), a hot season (March-May) and a wet season (June-October) (Bartlett 2009). Annual rainfall at this site averages between 200 and 300 cm (78.7 and 118.1 in) (mostly falling June-September) and daytime high temperatures usually average between 25°C and 30°C (77°F and 86°F) year-round. The coldest lows are usually no colder than 10°C (50°F) (Tangtham 1991 cited in Bartlett 2009; Bartlett 2009).

Lar gibbons are usually found high in the canopy and are rarely found in the understory (Ungar 1996). **At Khao Yai,** the average height of feeding trees was 23.7 meters (77.8 ft) (Bartlett 2009).

Ecology:

Lar Gibbons eat a large variety of foods, especially fruit, including figs and other small, sweet fruits, liana fruit, tree fruit and berries but also young leaves, buds & flowers, new shoots, vines, vine shoots, insects including mantids and wasps, and even birds' eggs (Carpenter 1940; Ellefson 1974; Raemaekers 1979; MacKinnon & MacKinnon 1980; Ungar 1995; Palombit 1997; Yimkao & Srikosamatara 2006; Bartlett 1999; 2009). They are known to eat parts of over 100 species of plants (Bartlett 1999; 2009). In general, when data is combined among study sites, the lar gibbon diet includes fruit (66%), leaves (24%), flowers (1%), and insects (9%) although in individual studies there can be some significant variation by locality (see review in Bartlett 2007; Bartlett 2007). Figs often make up a large proportion of the diet, on average around 27%, but sometimes up to half of the total foods eaten (Palombit 1997; Bartlett 2007). Feeding on figs increases during times that other preferred resources are scarce (Bartlett 1999). Food transfer between Lar Gibbons has been seen often accompanied by begging of one animal for food from another (Nettelbeck 1998). Most leaves that are eaten are young. One method that Lar Gibbons drink water is from tree holes by cupping their hands (Bartlett 1999). Diet can vary quite a bit with the seasons. For example, at **Khao Yai,** non-fig fruits dominated in the diet through the year excepting November and December. At this study site, flowers are most available during the cool season, while ripe fruit is most abundant during the hot and wet seasons. At this study site, ripe fruit is preferred when available, with the variability of the diet increasing when ripe fruit is less available during the cool season. However, fruit including figs never fell below 50% of the diet year-round (Bartlett 2009). Food competition may exist between Lar Gibbons and Pigtail Macaques (*Macaca nemestrina*) as the two species have been seen foraging near one another and sometimes interacting (Whitington 1992). Feeding competition between Lar Gibbons and Long-Tailed Macaques *(M. fascicularis)* and Dusky Leaf Monkeys *(Trachypithecus obscurus)* has also been seen (Ellefson 1974). Lar Gibbons compete with *sympatric siamangs (Symphalangus syndactylus),* the presence of which usually causes conflict and probably reduces the foraging success of Lar Gibbons (Palombit 1996). Diurnal Lar Gibbons in **Thailand** are active for an average of 8.7 hours per day, leaving their sleeping sites right around sunrise and entering sleeping tress an average of 3.4 hours before sunset (Bartlett 2009). On average, Lar Gibbons spend their days feeding (32.6%), resting (26.2%), travelling (24.2%), in social activities (11.3%), vocalizing (4.0%) and in inter-group encounters (1.9%), although actual proportions of activities can change significantly over the course of the year (Bartlett 2009). Daily, if the morning is clear before dawn, an adult male will solo call, usually from its sleeping tree. At sunrise, all group gibbons will awaken, then defecate and urinate while hanging from a branch. Usually, the group will then move to a feeding tree. There is normally a duet call among the adult pair before noon and the rest of the day is spent alternating between feeding and travelling to new trees to feed. With seasonal reductions in the amount of available fruit, feeding time increases and, correspondingly, less time is spent in social activities (Bartlett 2009). A group pause for about an hour of rest and social activities usually occurs each day (Bartlett 2009). During the cool season, Lar Gibbons can often be found for up to several hours a day in big fig trees (Bartlett 2009). Average home range is 0.4 km^2 (0.15 mi^2), with a daily path length of 1.4 kilometres (0.87 mi) although there is considerable variation among study sites (Bartlett 2007). Daily path length tends to increase with increased fruit availability (Bartlett 1999). Group progressions at some locations are usually led by the adult male, while at other sites females may do so, and coordinate group movements as well (Yimkao & Srikosamatara 2006; Barelli et al. 2008a). Within the home range is a territory defended from other groups averaging around 76% of the home range (Bartlett 2007). The home ranges of different groups often overlap, sometimes extensively (Ellefson 1974; Reichard & Sommer 1997; Bartlett 1999). Sleeping trees are not manipulated and group members usually sleep one to a tree nearby each other. Often the tallest trees available are chosen for sleeping and lar gibbons try to avoid notice when moving to sleeping sites, possibly as a way of reducing predation (Reichard 1998). Often, sleeping trees are located on steep slopes or on cliffs (Yimkao & Srikosamatara 2006).While the specific composition of the primate community varies across the distribution of the Lar Gibbon, the species can be found living in sympatry with several other primates and apes, including orang-utans *(Pongo pygmaeus), siamangs (S. syndactylus),* pileated gibbons (*Hylobates pileatus),* purple-faced langurs (*Trachypithecus sp.*), Thomas's Langur *(Presbytis thomasi),* slow loris *(Nycticebus coucang),* and several macaque species (Macaca sp.) (Ellefson 1974; Geissmann 1991; Palombit 1992; Ungar 1995; Palombit 1996; Bartlett 1999; Yimkao & Srikosamatara 2006).

Reproduction:

Sexually, they are similar to other gibbons. Mating occurs in every month of the year, but most conceptions occur during the dry season in March, with a peak in births during the late rainy season, in October. On average, females reproduce for the first time at about 11 years of age in the wild, much later than in captivity. Gestation is six months long on average, and pregnancies are usually of a single young. Young are nursed for approximately two years, and full maturity comes at about eight years. The life expectancy of the lar gibbons in the wild is about 25 years.

Predators and Threats:

There are a number of potential predators of adult or infant lar gibbons. These may include Clouded Leopards (*Neofelis nebulosa)*, Marbled Cats (*Pardofelis marmorata)*, Asian Golden Cats (*Catopuma temmincki)*, Leopard Cats *(Prionailurus bengalensis)* Tigers *(Panthera tigris)*, Leopards *(Panthera pardus)*, Pythons (Python sp.), Hawk Eagle species (*Spizaetus sp.)*, Black Eagles *(Ictinaetus malayensis)*, and Crested Serpent Eagles *(Spilornis cheela)* (Reichard 1998; Uhde & Sommer 2002). White-Bellied Sea eagles *(Haliaeetus leucogaster)* are also capable of taking a Lar Gibbon (Ellefson 1974). Lar Gibbons may mob and chase off potential predators (Reichard 1998; Uhude & Sommer 2002).

Major Threat(s):

The major threat to this species is hunting (having replaced even forest clearance as the top threat); they are hunted both for subsistence food use and for the pet trade. Hunting pressure varies across the range, but takes place even within protected areas. Much of the hunting is done by villagers exploiting Aquilaria spp. trees prized for their aromatic wood, and other forest products (W. Brockelman pers. comm.).

Construction of roads through protected areas (for example, the Security Highway through Nam Poyi in Lao PDR, the north-south highway in Peninsular Malaysia) may also pose a threat since it promotes forest clearance and strip development, possibly increases fragmentation, and increases access by hunters into protected areas. Ongoing localized forest loss due to shifting agriculture and commercial plantations of palm oil poses a threat. On northern Sumatra, most of the lowland forests have been logged out and the threat of Ladia Galaskar, a network to link the west and east coasts of Aceh province, means that much of the remaining forest is at risk.

Conservation Actions:

Like all gibbons, is a nationally protected species in all the countries across its range, and is listed under CITES Appendix I. In most of its range it is confined to protected conservation areas (for example in **Thailand,** where no significant populations survive outside of protected areas). However, in most countries, these areas are not well patrolled, even if they are well managed for tourism. There is an urgent need for improved protection of these areas, ideally involving local communities that should benefit from as well as participate in management. Illegal use of forest products, as well as poaching, is common in most protected areas. Inadequate management and protection, rather than forest destruction, are the main long-term threats and conservation efforts must seek to identify the hunters and incorporate them into new management priorities.

Further survey work is needed to determine current population numbers within protected areas across the range. One such priority area is southwest Yunnan, where it is unclear whether the species still survives (W. Brockelman pers. comm.).

Order: Rodentia (rodents)

Rodents make up the largest order of mammals, with over 40 percent of all mammalian species. They have two incisors in the upper and lower jaw which grow continually and must be keep short by gnawing. Most rodents are small though the capybara can weigh up to 45 kg (100 lb).

Suborder: Hystricognathi:

Family: Hystricidae (Old World porcupines)
Genus: Atherurus:

Asiatic Brush-Tailed Porcupine *(Atherurus macrourus)*

Description:

HB:381-525; T: 139-228; HF: 64-75; E: 30-36.

The upper parts are covered with flattened, brown spines interspersed with fewer slender, round pale coloured spines (Medway 1969). The underparts are whitish. The tail bears small scales for much of its length, with short spiny bristles between them. The long tail ends in a tuft of dirty coloured quills or hairs with each hair looking like a string of beads, but the beads are not globular, but flattened and looks like the hair as been modelled with plasticine, and then pinched from alternate sides along its length (Harrison 1966). the nasal bones are short, usually 30% of the greatest length of the skull. The sagittal crest is prominent , and the palate is narrow and conspicuously pitted with small cavities.

Range Description:

This species is distributed in Northeastern South Asia, through much of Central and Southern China and mainland Southeast Asia. In South Asia, this species has been recorded in Assam, India and probably also occurs in Bangladesh. It has been recorded up to 750 m asl (Molur et al. 2005). In China, it has been recorded from Hainan Island, Sichuan, Guizhou, Yunnan, Hubei, Hunnan and Guangxi (Smith and Xie 2008). In Southeast Asia, it is widely recorded in Myanmar, **Thailand (including the island of Tarutau),** Lao PDR, Vietnam, Cambodia and Peninsular Malaysia (including the islands of Aor, Pemanggil and Tioman). It is recorded from Sumatra (Indonesia) by Woods and Kilpatrick (2005), but this requires verification.

Countries: Native:

Bangladesh; (Hainan, Hubei, Sichuan, Yunnan); (Assam); Lao People's Democratic Republic; Malaysia; Myanmar; **Thailand;** Vietnam .

Ecology:

It is a nocturnal and fossorial species occurring in subtropical and tropical montane forests. It is found on the forest floor, often in areas with profuse undergrowth interspersed with cane and bamboo brakes and palms (Molur et al. 2005). It constructs burrows, which may be occupied by up to three animals. Like other porcupines it usually defecates at a certain spot near the burrow (Medway) It is mainly terrestrial but sometimes climbs. fairly well.It is found in various types of forest habitats as well as open areas near forests. It may stray into nearby agricultural areas. It digs into the ground and inhabits dens near rocky areas, where it lives in small groups. It eats roots, fruit some cultivated crops and barks off some kind of trees and also carrion especially bones. It has a gestation period of 110 days and a litter size of two or three. The species may give birth to two litters annually. Their habitat is terrestrial where they are living in the hole of tree barks or roots. It also living in a burrow, from which a network of trails penetrate into surrounding habitat. It can be found in all forest types up to 1500m.

Major Threat(s):

In its limited South Asian range, it is threatened by habitat loss due to jhum (shifting) agriculture, small-scale logging, subsistence harvesting for food, and accidental mortality (Molur et al. 2005).

Status:

The species is protected under Schedule II of the Indian Wildlife (Protection) Act, though not listed in CITES. It has been recorded from Namdapha National Park in Arunachal Pradesh, India (Molur et al. 2005). It is present in a number of protected areas in Southeast Asia. It has been recorded from Namdapha National Park in Arunachal Pradesh, India (Molur et al. 2005). It is present in a number of protected areas in Southeast Asia.

Malayan Porcupine *(Hystrix brachyura)*

Description:

The Malayan Porcupine or Himalayan porcupine (*Hystrix brachyura)* is a species of rodent in the family Hystricidae.Three subspecies are extant in South and South-East Asia.

The front half of the body is covered with covered with short, dark brown spines of about 30 to 50 mm long. The long pointed quills of about 200 to 230mm long protrude from the hindquarters. The long posterior quills are whitish, usually with a distinct blackish ring located about half to two thirds of the distance from the base. The tail is short , with 2 types of quills, long black and white, pointed quills and rattling quills. The rattling quills are hollow, open ended flattened cylinders mounted on a short, narrow stem. The tail is hidden by long quills because the hind part of the body gradually tapers towards the tail. The whiskers (vibrissae)) are very long and dark. The ears are small. The nasal bones are shorter than in *hodsoni* averaging 43.3% of the occipitonasal length (Ellerman 1940)The occipital crest is well developed, but the sagittal crest is rather low. The palette is pitted with numerous small holes

Range:

The Malayan porcupine ranges from Nepal through North-east India (Arunachal Pradesh, Sikkim, West Bengal, Manipur, Mizoram, Meghalaya and Nagaland), to central and southern China (Xizang, Hainan, Yunnan, Sichuan, Chongqing, Guizhou, Hunnan, Guangxi, Guangdong, Hong Kong, Fujian, Jianxi, Zhejiang, Shanghai, Jiangsu, Anhui, Henan, Hubei, Shaanxi, Gansu), throughout Myanmar, **Thailand,** Lao PDR, Cambodia and Vietnam, through Peninsular Malaysia, to Singapore, Sumatra (Indonesia) and throughout Borneo (Indonesia, Malaysia, Sarawak and Brunei). It is also present on the island of Penang, Malaysia. It can be found from sea level to at least 1,300 m This species and its close relatives is believed to have originated from southern Asia based on their current distribution. Their origin may lie from a common Late Pleistocene ancestor when Sumatra, Borneo and Palawan were part of Sundaland.

Habitat and ecology:

It is found in various types of forest habitats as well as open areas near forests. It may stray into nearby agricultural areas. It digs into the ground and inhabits dens near rocky areas, where it lives in small groups. It has a gestation period of 110 days and a litter size of two or three. The species may give birth to two litters annually.

Their habitat is terrestrial where they are living in the hole of tree barks or roots. It also living in a burrow, from which a network of trails penetrate into surrounding habitat. It can be found in all forest types up to 1500m.With a huge habitat range of nearly the entire continent of Asia and even further into neighbouring continents, the Malayan Porcupine is one of the most widely yet sparsely spread out rodents on Earth. What makes the habitat of these creatures unique, however, is the type of terrain that they live in. They live in heavily wooded areas and make their dens in rocky areas of the ground. Malayan Porcupines create burrows which lead into huge networks of tunnels beneath their surrounding habitats, allowing them to get around during the daytime without being exposed to sunlight. They live with other adults of their species and usually produce litters of two or three pups after a pregnancy of around 110 days.

The Malayan Porcupine is covered at birth with soft quills which sharpen into rigid, hard quills upon reaching adulthood. This is the animal's defence mechanism and is effective to most predators and hunters that come after these porcupines. Though not predators themselves, they eat fruits, roots, nuts, and insects, as well as carrion (or dead animal matter).

Threats:

In general, the Malayan Porcupine is endangered because of over-hunting and human overpopulation into their habitats. There are tribes of people who eat them and who use their quills for decorative purposes. They are also hunted by feral cats, some types of wild hogs, and other predatory animals that live in the same region as they do. In some countries like Vietnam, their paws are considered good-luck charms just as rabbits' feet are in the United States. Without a little consideration and help from mankind, this unique species will become extinct very quickly. The International Union for Conservation of Nature ranks the Malayan porcupine as a vulnerable species with a high likelihood of facing extinction in the wild. This is due to the fact that they are hunted for their meat. They are also hunted to be displayed as decoration for their unique quills and colouring. The Malayan Porcupine is listed as Vulnerable (VU), considered to be facing a high risk of extinction in the wild, on the IUCN Red List of Threatened Species It is a large and stout bodied rodent covered with quills which are sharp, rigid structures. The quills/spines are modified hair.[6] The quills or spines on their upper body parts are rough with black and white or yellowed stripe in colour. The young's soft quills become hard as they enter adulthood. It has short stocky legs covered in brown hairs which have four claws on the front and five on the hind legs. Both front and hind legs have smooth soles. The head and body measurement are around 63-72.5 cm and their tail is about 6–11 cm. Their weight is around 0.7 kg-2.4 kgIUCN has categorized this species as Least concern (LC) species. The quills of the Malayan Porcupine are used for ornamental purposes. They are also hunted for meat In Southeast Asia, it is hunted for food but this not thought to impact populations. In South Asia, it is threatened by habitat loss due to construction of dams, human settlements and other infrastructure development. It is harvested for subsistence food and medicinal purposes (Molur et al. 2005).

Conservation Actions:

This species is present in many protected areas. It is known from the following protected areas in South Asia, Namdapha National Park in Arunachal Pradesh in north-eastern India, Lang Tang National Park in Central Nepal, and Sagarmatha National Park in Eastern Nepal (Molur et al. 2005). In South Asia it is protected by Schedule II of the Indian Wildlife Protection Act.

Crestless Himalayan Porcupine (*Hystrix hodgsoni)*

Description:

HB: 650-735; T: 60-115; HF:75-93; E: 27-44.

H.hodgsoni is very similar to *H. brachyur,* though the crest is said to be lower(an aged linked character) and the quills often have a wider black band than in brachyura, (although this may be an individual variable character) The nasals are comparatively large being 55.6 of the occipitonasal length. The sagittal crest is well developed.

Range Description:

This species ranges from Nepal, through northeastern India (Arunachal Pradesh, Sikkim, West Bengal, Manipur, Mizoram, Meghalaya and Nagaland) (Molur *et al.* 2005), to central and southern China (Xizang, Hainan, Yunnan, Sichuan, Chongqing, Guizhou, Hunnan, Guangxi, Guangdong, Hong Kong, Fujian, Jianxi, Zhejiang, Shanghai, Jiangsu, Anhui, Henan, Hubei, Shaanxi, Gansu) (Smith and Xie 2008), throughout Myanmar, **Thailand,** Lao PDR, Cambodia and Vietnam, through Peninsular Malaysia, to Singapore, Sumatra (Indonesia) and throughout Borneo (Indonesia, Malaysia and Brunei). It is also present on the island of Penang, Malaysia. It can be found from sea level to at least 1,300 m asl.

Countries Native:

Bangladesh; China; India; Indonesia; Lao People's Democratic Republic; Malaysia; Myanmar; Nepal; **Thailand;** Vietnam.

Habitat and Ecology:

It can be found in various forest habitats up to about 1000 m elevation. It builds large burrows with more than one entrance in clay banks or amongst boulders on hill or mountain sides. Long , wide runways lead in many directions from the burrow., and in scrubby, Also found in open areas close to forest. It can be found in agricultural areas, but needs to have rocky outcrops or other areas in which it can create a den or dig burrows. Burrows are generally occupied by family groups. Following a gestation period of about 110 days, two or three young are born. Two litters may be produced annually. These porcupines can cause a great deal of damage in plantations. They feed on the inner and outer bark of trees, and sometimes girdles a tree causing it to die. . They also damage twigs causing loss of vigour and death to the tree.. When threatened , this species raises its long quills, rattles its cylindrical quills and also stamps its feet and snorts. In attack , it hurls itself backwards against its enemy, driving home some of the quills with great force.. The small compact, mass of short, solid quills do most of the damage. The longer quills do less damage as they are less firmly fixed. They have been known to kill large felines with their quills.

Systems:

Terrestrial

Major Threats:

In Southeast Asia, it is hunted for food but this not thought to impact populations. In South Asia, it is threatened by habitat loss due to construction of dams, human settlements and other infrastructure development. It is harvested for subsistence food and medicinal purposes (Molur *et al.* 2005).

Conservation Actions:

This species is present in many protected areas. It is known from the following protected areas in South Asia, Namdapha National Park in Arunachal Pradesh in northeastern India, Lang Tang National Park in Central Nepal, and Sagarmatha National Park in Eastern Nepal (Molur *et al.* 2005). In South Asia it is protected by Schedule II of the Indian Wildlife Protection Act.

Suborder: Sciurognathi:
Family: Sciuridae (squirrels)
Subfamily: Ratufinae:
Genus: Ratufa:

Cream-Coloured Giant Squirrel (*Ratufa affinis)*

The Cream-Coloured Giant squirrel (*Ratufa affinis)* is a large tree squirrel in the genus Ratufa found in Brunei, Indonesia, Malaysia, and **Thailand.** It is probably extinct in Singapore, as no recent sightings have been made. Reported sightings in Vietnam in 1984 are considered to be dubious.

Description:

Along with its large size, this squirrel's colourful markings, that vary widely from one individual to the next, makes it readily-identifiable and easily-spotted in the wild. The back and head of the body varies from dark brown to gray, and the belly varies from dark yellow to white. The ears are short and round. The adult squirrel's head and body is from 322 to 350 millimetres (12.7 to 14 in) in length, and the tail is 373 to 438 millimetres (14.7 to 17.2 in) long, with an overall length of 695 to 788 millimetres (27.4 to 31.0 in); the weight is between 929 to 1,575 grams (2.05 to 3.47 lb). **The subspecies found in Thailand *R.a pyrsonota* is characterizes by relatively dark pelage, light orange brown underparts, ochreous underparts and dark brown feet. The ears are also dark brown, and the tail darker than the body. The cheek area is light grey agouti. The underparts are often speckled.** The female possesses three pairs of mammae.

Habitat:

This species is the only giant squirrel in Borneo (elsewhere its distribution overlaps with that of the black giant squirrel). It is one of the mammal species found in the extensive Belum-Temengor forest preserve in the Malaysian state of Perak, on the Malay Peninsula.

The cream-coloured giant squirrel makes its home in lower montane and secondary forests, frequenting dipterocarp trees. It rarely enters plantations or settlements, preferring the forest. Although this squirrel primarily inhabits the upper canopy of the forest, it will at times come to ground in order to hunt smaller species of squirrels or to cross gaps in the trees. This is a closed canopy species that is never seen on ground, and is dependent on tall, wet evergreen forests (Giman pers. comm.). This species can tolerate plantations and secondary forest (Giman pers. comm.). This is an arboreal species (Saiful and Nordin 2004). It has been suggested that one of the reasons for low densities of this species in Malaysian tropical rain forest is competition from the great variety of other arboreal vertebrates (such as birds, and especially primates) for food, especially fruits and leaves, which are among the food items preferred by squirrels (Saiful and Nordin 2004).

Behaviour:

This species is diurnal, active during morning and evening. They live either in pairs or alone. When it is angry or shocked, it will give a loud sound that can be heard from afar. Although this squirrel often will make holes in trees for shelter, during the breeding season it constructs a large globular drey (or nest) in tree branches, roughly the size of an eagle's aerie. Little is know of the breeding habits but the young are born and raised in this nest.
The main dietary habits of Ratufa affinis are seeds, which it supplements with leaves, fruits, nuts, bark, insects, and eggs. The squirrel has a very short thumb that it uses to hold and control its food while feeding. Unlike other tree squirrels, the cream-coloured giant squirrel does not sit upright with its tail arched over its back while feeding; instead, it balances itself with its hind feet on a branch so that its hands are free to control its food. In this position the axis of the squirrels body is held at right angles to the support, with its head and forequarters on one side of the branch, and the tail as a counterweight on the other side.

Major Threat(s):

This species is threatened by logging and hunting in forest fragments, it is probably sensitive to even selective logging (Giman pers. comm.).

Conservation:

This species occurs in many protected areas (Han and Giman pers comm.). It is protected in Sarawak and Malaysia (Giman and Han pers. comm.), as well as probably in **Thailand (**Giman pers. comm.). It also protected by CITES Appendix II (Han and Giman pers. comm.). Saiful and Nordin (2004) state the need for further comparative study on this species' abundance, density and distribution and its relationship to forest structure or habitat quality, spatially and temporally, in hill dipterocarp forest of Malaysia

Black Giant Squirrel (*Ratufa bicolor*)

The beautiful Black Giant Squirrel is one of the largest squirrels in the world. It inhabits tall primary forest and remains high in the canopy, but at times may be found at lower levels when feeding. It is largely solitary in habits and extremely shy, rarely coming to ground. It can confidently make huge leaps from branch to branch in the high canopy.

Its upperparts and tail are jet black, but the cheeks, chest, front of the forelimbs and underparts are cream or orange. The tail is long and dorso-ventrally flattened. Fruits, seeds and young leaves make up its diet, supplemented by occasional insects and sometimes birds eggs. It builds a large, spherical nest of leaves and twigs. The species once ranged extensively from Nepal and Assam through Burma, Indochina and **Thailand** to Peninsular Malaysia, Sumatra, Java and Bali. Its numbers have declined with the large-scale clearance of primary forest. It is absent from Singapore.

Description:

Head and body length varies from 35 to 58 centimetres (14 to 23 in) in length, and the tail is up to 60 centimetres (24 in) long, with an overall length of up to 118 centimetres (46 in). The back, ears and bushy tail are deep brown to black with a lighter buff-coloured belly. The upperparts are black or dark brown and the underparts range from buff to orange colour. The buff throat and cheeks are separated by a thick moustache mark. The top of the head is black. There are 2 dark spots on the chin. The under fur of the dorsum is reddish brown, and the under parts especially the rump and crown are often tinged reddish brown. The hairs on the underparts have black bases. In sum subspecies there is a bluff or white patch, called a flash mark between the wrist and elbow of the upper surface of the foreleg. The rest of the upper surface of the foreleg is black or dark brown. The female as tree pairs of mammae.

Population:

The population of this species is declining in South Asia, but the rate of decline is unknown (Molur et al. 2005). In Lao PDR, Vietnam and **Thailand** the species is abundant in suitable habitat where hunting is controlled. It is easily eradicated from fragmented habitat (Duckwoth pers. comm.). In Lao PDR it is recorded from most survey areas; however, populations are now so low in some areas that it has not been recorded in recent surveys, including Sangthong District, Nam Et and Phou Louey NBCAs (Duckworth et al. 1999). It was found to be locally common during a survey in 1994-95 of the Nadi Limestone area in Lao PDR by Evans et al. (2000). It is still widespread and abundant in Cambodia, largely as hunting has not yet impacted populations in this country (R. J. Timmins pers. comm.). This species was rarely found in a survey conducted by Saiful and Nordin (2004) in Peninsular Malaysia (Weng River catchment area). This species is rare and declining on Java, where it is found only in remote areas (away from human habitation) within remaining patches of forest (primary and secondary) (Boeadi pers. comm.). This species is declining in population, in parallel with forest loss on Java (Boeadi pers. comm.).

Population Trend: Decreasing

Habitat:

Ratufa bicolor's range includes a variety of bioregions that all share the commonality of being forested. It ranges in elevation from sea level up to at least 1,400 metres (4,600 ft), in some of the most rugged land in the world. However, in recent decades, *R. bicolor's* habitat has been steadily encroached upon by human settlement, timber harvesting and agriculture, which along with over hunting by human predation in parts of its range, has resulted in a total loss of up to 30% of the population in the past ten years. However, in some places this species is protected from hunting by law or tradition.

In South Asia *R. bicolor* dwells among tropical and subtropical coniferous and broadleaf forests. In Southeast Asia *R. bicolor* lives in tropical broadleaf evergreen and semi-evergreen forests, but is rarely seen in coniferous forests. In the tropical rainforest of the Malay Peninsula and Indonesia, *R. bicolor* is not as abundant as elsewhere in its range, which is probably due to competition from other arboreal species (especially primates) for food in the upper forest canopy. Among the better places to sight the black giant squirrel is the Kaziranga National Park in the state of Assam, India.

Ecology:

R. bicolor is diurnal and arboreal, but sometimes climbs down from the forest canopy to feed on the ground. The black giant squirrel rarely enters plantations or settlements, preferring the wild forest.

Its diet consists of seeds, pine cones, fruits and leaves, but will eat small insects and birds eggs. It is primarily solitary, and has a litter of from 1 to 2 young, which it raises in a drey (or nest), often located within a hollow space of a tree. This is a shy forest species which very rarely enters the ground and remains in the canopy. It moves through the canopy making long leaps of 20 feet or more. When feeding it assumes a conspicuous posture. It hangs across a branch with its foreparts on one side and its long black tail across the other . The nest is built on thin branches on the highest part of a tall tree for protection from predators where a nest is constructed of twigs and leaves. Medway (1969) described the voice as being a loud, resonant, bubbling chatter, Chak, Chak, Chak, repeated and with a quieter note Ch'k, Ch'k. It is know to live for 11 years in captivity. It is not tolerant of habitat modification, and has a long generation time of eight to nine years, with a litter size of one or two young.

Major Threat(s):

Human induced habitat degradation due to shifting (jhum) agriculture practices, small-scale logging, clear-cutting, forest fires, expansion of human settlement, harvesting for local consumption have been observed to be major threats for this species in South Asia (Molur et al. 2005). It is threatened by hunting and habitat loss in China (Wang et al. 1989). This species is especially vulnerable to hunting and habitat loss (often through logging operations) in Lao PDR (Evans et al. 2000), Vietnam and **Thailand.**

This species was "formerly one of the most commonly sold mammals in the That Luang fresh food market in Vientiane" (Duckworth et al. 1999). Hunting pressure is predicted to increase in Cambodia, as following the over harvesting of large bodied mammals, smaller species are now being targeted (W. Duckworth pers. comm.). It is not hunted on Java, as people are generally aware of the legislation protecting this species; however, habitat loss remains a considerable threat (Boeadi pers. comm.).

Conservation Actions:

The species is included in the Schedule II (Part II) of the Indian Wildlife (Protection) Act, 1972. It is known from the following protected areas in India and Bangladesh - India: Eagle's Nest Wildlife Sanctuary, Kamlang Wildlife Sanctuary, Mehao Wildlife Sanctuary, Namdapha National Park, Pakhui Wildlife Sanctuary, Tale Valley Wildlife Sanctuary in Arunachal Pradesh, Buxa Tiger Reserve, Gorumara Wildlife Sanctuary, Jaldapara Wildlife Sanctuary, Mahananda Wildlife Sanctuary in West Bengal; Bangladesh: Lawachara National Park (Molur et al. 2005). Survey, life history, limiting factors studies and monitoring are recommended for this species in South Asia (Molur et al. 2005). In Southeast Asia, it occurs in a number of protected areas throughout its range, including Gunung Masigit Kareumbi Hunting Park, Pananjung Pangandaran Nature Reserve, and Ujong Kulon National Park in Java (Boeadi pers. comm.). Saiful and Nordin (2004) state the need for further comparative study on this species' abundance, density and distribution and its relationship to forest structure or habitat quality, spatially and temporally, in hill dipterocarp forest of Malaysia. This species is protected from hunting by legislation in Java (Boeadi pers. comm.). It is listed on CITES Appendix II regulating international trade in this species.

Hairy-Footed flying Squirrel *(Belomys pearsonii)*

Description:

The back and top of the head range in colour from speckled grey to speckled red- brown. The hairs are blackish with light or red brown tips. The hairs of the gliding membrane are very similar, but the black shows through much more. The tail is somewhat flattened below the fairly bushy top; it is plain, light reddish brown with a greyish base. The throat, breast and belly creamy white, sometimes wit a slight reddish tinge. The under parts of the gliding membrane is sparsely covered with red brown hairs, and the membrane as a white fringe. Tufts of long hair arise from the ears, and there is usually a light area behind the ear. The feet have long tactile hairs which even covers the claws to protect against the cold in the higher altitudes. The body has a length of about 22 cm; the tail is another 13 cm long. As it is closely related to the complex-toothed flying squirrel, some taxonomists add the species to the genus *Trogopterus.*

Population:

The Hairy-Footed flying squirrel *(Belomys pearsonii)* is a flying squirrel found in the mountains of the eastern Himalaya, Southeast Asia, and southern China till the island of Taiwan. It lives at altitudes between 1500 and 2400 m above sea level. This poorly known species appears to have a wide, but patchy distribution, in northeastern South Asia, East Asia and Southeast Asia. In South Asia, it has been recorded from Nepal, India (Arunachal Pradesh, Manipur, Mizoram, Nagaland, Sikkim and West Bengal) and possibly Bhutan (Molur et al. 2005). In China, it has been recorded from Yunnan, Guizhou, Guangxi, Hainan and Guangdong (Smith and Xie 2008). It is present on the island of Taiwan (Smith and Xie 2008). In Southeast Asia, there are recorded from Myanmar, **Thailand,** Lao PDR, Vietnam and possibly Cambodia although there are no records from this country (Walmart pers. comm.). This species occurs at elevations of 800 to 2,400 m asl.

The population abundance of this species is poorly known, however, it is believed to be significantly declining, or at least very rare, in some areas.

Habitat and Ecology:

It is a mostly nocturnal and arboreal species. It occurs in temperate and subtropical dry deciduous forests. It has been found to occupy tree hollows of dense broadleaved forest patches and also in rock crevices (Molur et al. 2005). It appears to be limited to primary forest habitat. The generation time is thought to be relatively long, and might be five or six years, with a litter size of one or two young.

Major Threat(s):

Major threats to this species over most of its range include general habitat loss (often through logging operations and the conversion of land to agricultural use) and significant levels of hunting for food. In South Asia, it is threatened by shifting (jhum) cultivation, forest fires, monoculture plantations and hunting for local consumption (Molur et al. 2005).

Conservation Actions:

It is listed in the Schedule II of the Indian Wildlife (Protection) Act, 1972. It is known from the following protected areas in South Asia - Pakhui Wildlife Sanctuary, Arunachal Pradesh and Nengpui Wildlife Sanctuary, Mizoram in India; Royal Chitwan National Park in Central Nepal. Further studies are needed into the distribution, abundance, ecology, and threats to this poorly-known species. Populations of this species should be monitored to record changes in abundance and distribution.

Particolored Flying Squirrel ***(Hylopetes alboniger)***

Description:

HB: 195-210; T:181-200; HF: 33-45; E: 30-34.

The back and head range in colour from grey to drab brown, depending on whether the hairs are tipped with whitish or brown. The tail is flattened underneath, but only slightly flattened underneath. The tail is dark grey to drab brown, or a mixture with the brownish colour at the base and along the sides. The underparts are creamy white, and there is a whitish mark behind the ear.

Closely resembling *H. phayrei,* but larger. The bulla is less than 20% o f the skull length. Much smaller than most other flying squirrels. They are arboreal and nocturnal and nest in hollow trees at altitudes of 1500-3400 m. The nest consists of a ball leaves and ferns and lined with fine grass. At night the squirrels presence can be detected by a high pitched trill or a repeated scream vocalization. The diet consists of fruits, nuts, leaves and buds.

Range Description:

This species is present in Northeastern South Asia, southern and Central China, and mainland Southeast Asia. In South Asia, it has been recorded from mountainous regions of Nepal, Bhutan (distribution in this country is unclear) and Northeastern India up to 4,000 m asl (Molur et al. 2005). In China it has been recorded from Hainan, Yunnan, Sichuan, Guizhou, Guangxi and Zhejiang (Smith and Xie 2008). In Southeast Asia, it has been recorded from Myanmar, **Northern Thailand**, Lao PDR (only three localities known), Vietnam and marginally in Cambodia.

Countries: Native:

Bangladesh; Bhutan; Cambodia; China; India; Lao People's Democratic Republic; Myanmar; Nepal; **Thailand;** Vietnam Listed as Least Concern because, although it is seldom recorded, it has a relatively wide distribution, has a presumed large population, and because it is unlikely to be declining fast enough to qualify for listing in a more threatened category.

Habitat and Ecology:

This is an arboreal and nocturnal species, found in tropical and subtropical montane forests, and in more temperate oak and rhododendron forests at middle to high elevations (1,500 to 3,400 m asl). Populations can be found in primary forests as well as secondary, degraded forests and scrubby habitat. The breeding season runs from April thru to mid June where 2 or 3 young are born in each litter.

Major Threat(s):

In South Asia, the species is threatened by habitat loss due to shifting (Jhum) agriculture, small wood plantations, mining activities, infrastructure development, establishment of human settlements, construction of dams and forest fires (Molur et al. 2005). In certain parts of northeastern India this species is hunted for food (S. Molur pers. comm.). The threats to the species in Southeast Asia and China are unclear.

Conservation Actions:

The species is included in the Schedule II (Part II) of the Indian Wildlife (Protection) Act, 1972 and is known from Namdapha National Park, in Arunachal Pradesh, India (Molur et al. 2005). Survey, taxonomic research and monitoring are recommended (Molur et al. 2005). Additional studies are needed into the taxonomy, distribution, abundance, general ecology and threats to this species in Southeast Asia

Indochinese Flying Squirrel ***(Hylopetes phayrei)***

The squirrel is known as Phayre's flying squirrel. The Phayre in question is Sir Arthur Purvis Phayre, an officer in the British Army who worked mainly in Burma, then a British colony, including serving several years as commissioner of Burma under Queen Victoria. He also ruled as the governor of Mauritius for a few years. Phayre wrote the first comprehensive English-language history of Burma.

Description:

The back and head range in colour from grey to drab brown, depending on whether the hairs are tipped with whitish or brown. The tail is flattened underneath, but only slightly flattened underneath. The tail is dark grey to drab brown, or a mixture with the brownish colour at the base and along the sides. The underparts are creamy white, and there is a whitish mark behind the ear. Very similar to Particolored Flying Squirrel *Hylopetes alboniger* but smaller.

Range Description:

According to Nowak (1999) the range of this species is Myanmar, **Thailand,** Lao PDR, north-western Vietnam, Guizhou, and possibly Fujian provinces in China. According to Smith and Xie (2008), it occurs on Hainan Island and in the provinces of Guizhou, Guangxi, and Fujian in China. The 1999 Status Report for Wildlife of Lao PDR indicates that no historical specimens have been confirmed and that the distribution within Lao PDR is provisional (Duckworth et al. 1999). This species occupies an elevational range of 0-1,500 m asl (Duckworth 2004).
Countries: Native:
(Guangxi, Guizhou, Hainan); Myanmar; **Thailand;** Vietnam

Population:

There are currently no data regarding the population status of this species.

Habitat and Ecology:

This species occurs in lower montane forests and mixed deciduous forests (Smith and Xie 2008). It can exist in partly cleared forests near cultivation. They are nocturnal, staying in hollow trees by day and foraging on (cultivated) fruit at night (Lekagul and McNeely 1988).

Major Threat(s):

There are no major threats to this species.

Red-Cheeked Flying Squirrel *(Hylopetes spadiceus)*

Description:

Size.

HB: 134, Tail: 116, HF: 24

The back and head are bright orange brown with black under fur. The upper side of the gliding membrane is black. The tail is distinctively flattened; it is blackish with a buffy or orange brown base and white tipped. The feet are fulvous brown and the underparts are creamy white. The cheeks are greyish, often with an orange tinge. There is a pale patch behind the ear.

Habitat and Ecology:

The attractive Red-Cheeked Flying Squirrel inhabits a variety of forest types, but is more commonly found in tall, lowland primary rainforest. Their nesting holes are generally located towards the base of medium-sized trees : the squirrels appear to excavate their own nesting holes or inhabit holes made by other species, rather than inhabit natural tree holes. Reportedly, they have also been found nesting inside coconuts. By day the entrance to the nesting hole is plugged with dry grass or other vegetation : this is removed at dusk when the squirrels emerge and quietly ascend to the canopy before feeding. The nest can be as little as one metre from the ground. The usually have two young. Adams(1958) reported one as ascending from about one metre during a six metre flight by flapping; this is a case of a flying squirrel flying as apposed to merely gliding.

Habitat:

The species is wide-ranging, occurring in Burma, **Thailand**, southern Indochina, Peninsular Malaysia, Singapore, Sumatra, Java and Borneo. These squirrels are known from primary forests, degraded secondary forest and scrub, and cultivated areas up to 1,500 m.

Scientific interest and potential value:

So little is known of these secretive creatures that the scope of scientific study on them is very great.

Major Threat(s):

The habitat of this species is being deforested for timber, firewood and conversion to agricultural land.

Genus: Petaurista:

Red Giant Flying Squirrel *(Petaurista petaurista)*

Distribution:

The Red Giant Flying Squirrel (*Petaurista petaurista)* is a species of flying squirrel, which ranges from the eastern regions of Afghanistan, into northern India and Pakistan through to Java, and Taiwan, and also Sri Lanka. It can also be found in parts of Borneo and **Thailand.** This species was recorded in Peninsular Malaysia, included Penang, Tioman Island and also Singapore. This species also recorded from many localities throughout Sabah and Sarawak, up to 900m on Mount Kinabalu, excluding the range of *P. p. nigrescens*, which is known only from the forests around Sandakan Bay north of Kinabatangan River. Giant flying squirrels *(Petuarista sp.)* have highest diversity in term of species richness and population diversity in Southeast It is often a locally common to abundant species.

Identification:

Size. HB;398-449; T; 485-570; HF;63-84

Like all other species of flying squirrels, it has a membrane of skin between its legs, which is used to glide between trees. It is characterised by its dark red colouring and large eyes. When compared to other species of squirrels, this species is large, being on average 422mm long. Entire body dark reddish except for black on nose, chin, eye-ring, behind the ears, feet and tail tip. **The squirrel from the North part of Thailand are much different from the ones from the South, being those from the South are much smaller than those from the North of the country. The ones from the North are dark brown or dark bay instead of bright Rufus and have a blackish tail instead of a black tipped Rufus tail like those in the south of the country. Also those from the north have a frosted or hoary appearance because the hairs at the back and the top of the head are tipped with white, whilst the southern variety have uniform coloured upper parts.**

Ecology and Habitat:

In the wild, it feeds primarily on conifer cones, leaves and branches, and, when in season, fruits and nuts, and occasionally insects. It is able to glide for long distances. There have been reports of distances up to 75 metres (250 ft.) or greater; glide angles are generally 40-60 degrees from the horizontal, occasionally steeper for shorter glides. Their nest holes usually at least 10m above ground. The Red Giant Flying Squirrel is nocturnal and does not hibernate, but migrates to areas with more food. *P. petaurista* also able to explore secondary conifer plantations and use this habitat as feeding and resting areas. *P. petaurista* is most active between sunset and midnight and the home range of adult females in conifer plantation was estimated to be 3.2ha. The Red Giant Flying Squirrel is believed to mate twice a year and the young are typically born in March or August in litters of two to three. Adult females tended to be heavier than adult males (1,334 and 1,260 g, respectively, $P < 0.01$), but the sexes showed no differences in other body characteristics. Some males maintained reproductive capacity in all months except February, and most were sexually active in March-June and October-November.

Most females were sexually active in May-July and November-January. About one-half of the adult females (49%) became pregnant during each breeding season, in winter (December-February) and summer (June-August). Pregnancy rates ranged from 50 to 57% in all months except February. Litter sizes ranged from one to two, with an average of 1.04 young/litter. The estimated birth weight for P. petaurista is 56 g. This species is a seasonal breeder with low reproductive rates, which may be a result of a low predation rate combined with their high degree of folivory.

Major Threat(s):

Habitat conversion due to logging, agriculture, dam construction, infrastructure development and urbanization is occurring throughout the range of the species. It has been harvested for the pet and fur trade in South Asia (Molur et al. 2005).

Conservation Actions:

This species is present in many protected areas (eg. Namdapha National Park, Arunachal Pradesh, India). This is considered to be a species complex for which taxonomic revision is needed. Surveys and monitoring are recommended for this species in South Asia (Molur et al. 2005).

Indian Giant Flying Squirrel *(Petaurista philippensis)*

Taxonomic Notes:

Petaurista philippensis possibly represents a complex of several similar species. Further studies are needed to clarify the taxonomic status of populations currently allocated to this species. Listed as Least Concern in view of its wide distribution, tolerance of a degree of habitat modification, presumed large population, and because it is unlikely to be declining fast enough to qualify for listing in a more threatened category.

Description:

Very similar to the Red Giant Flying Squirrel *(Petaurista petaurista)*

HB;398-449; T; 485-570; HF;63-84

Like all other species of flying squirrels, it has a membrane of skin between its legs, which is used to glide between trees. It is characterised by its dark red colouring and large eyes. When compared to other species of squirrels, this species is large, being on average 422mm long. Entire body dark reddish except for black on nose, chin, eye-ring, behind the ears,

Habitat and Ecology:

This is an arboreal and nocturnal species. In South Asia it occurs in dry deciduous forests and evergreen forests. In addition to natural forest, it has been recorded from plantations. It is found to occupy tree canopies and holes (Molur et al. 2005). On Hainan Island it has been found only in large patches of forest, where it was considered abundant (Smith and Xie 2008). On Taiwan they are most abundant in hardwood compared with coniferous forest (Smith and Xie 2008). *Castanopsis cuspidata* is eaten most often (24.95% of annual diet), followed by *Ficus superba* (14.67%) and *Glochidion acuminatum* (12.18%). *P. philippensis* fed mainly on young leaves (27.7% of annual diet) and mature leaves (24.0%). Leaf parts (including buds, petioles, young leaves, and mature leaves) constituted 74.0% of the annual diet. Monthly and annual diet variation was obvious, and this was related to availability of species-specific plant parts. *P. philippensis* fed on mature leaves and petioles in winter; buds, young leaves, and fruit in spring; and young leaves and fruit in summer and autumn. They preferred buds, young leaves, flowers, and fruit to mature leaves and petioles. However, mature leaves and petioles seemed to be relied on when more favourable food items were scarce. The Red Giant Flying Squirrel is believed to mate twice a year and the young are typically born in March or August in litters of two to three.

Population:

It is locally abundant in Southeast Asia where suitable habitat protection is in place (Duckworth et al. 1994; Duckworth 1998; Evans et al. 2000). It could be in local decline in a few areas in India where it is under heavy hunting pressure (Rajamani pers. comm.). It appears to be in decline in the Western Ghats and also in northeastern India (Molur et al. 2005).

Population Trend: Decreasing

Major Threat(s):

There are no major threats to this species overall. In Southeast Asia, the species seems to be very resilient to threats, but it could decline with increasing habitat loss; although there is no serious evidence of this except where forest is being completely replaced. In South Asia, habitat loss and degradation resulting from logging, shifting cultivation, expansion of human settlements and forest fires are considered to be threats for this species (Molur et al. 2005). It is hunted for local consumption and medicinal purposes in South Asia and China.

Conservation Actions:

It occurs in numerous protected areas in Southeast Asia. In India it is present in Eturnagaram Wildlife Sanctuary in Andhra Pradesh; Valmiki Tiger Reserve, Kaimur Wildlife Sanctuary in Bihar; Bandipur National Park, Nagarhole National Park in Karnataka; Chinnar Wildlife Sanctuary, Parambikulam WS, Peechi-Vazhani Wildlife Sanctuary, Periyar Tiger Reserve, Thathekad Bird Sanctuary in Kerala; Bori Wildlife Sanctuary, Kanha National Park in Madhya Pradesh; Phulwari Wildlife Sanctuary, Sitamata Wildlife Sanctuary in Rajasthan; Kalakkad-Mundunthurai Tiger Reserve in Tamil Nadu. In Sri Lanka it is known from Horton Plains National Park, Knuckles Forest Reserve in Central Province and Sinharaja Forest Reserve, Sabaragamuwa (Molur et al. 2005). The species is included in the Schedule II (Part II) of the Indian Wildlife (Protection) Act, 1972. A taxonomic review of populations currently allocated to this species is needed.

Genus: Petinomys:

Temminck's Flying Squirrel *(Petinomys setosus)*

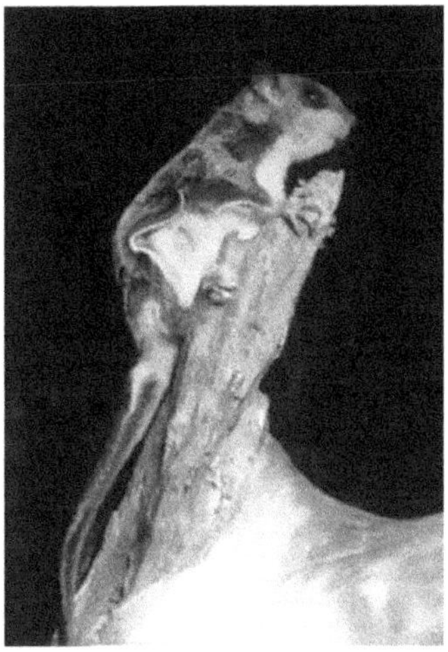

Taxonomic Notes:

Northern populations are distinct subspecies and may be a distinct species, owing to a large geographic gap.

Description:

HB: 128; T: 118; HF: 21

The upper parts are slaty grey with buffy tips to the hairs resulting in a lustrous grey apperance.The underparts are creamy white, the feet are brownish, and the tail is blackish with a cream coloured base. There are often white marks behind the ears which form a partial collar. The gliding membrane is blackish above, and sparsely covered with buff hairs below. It as a white fringe. There is a black ring around the eyes, and a black line running from the eye ring to the nose.

Range Description:

This species is patchily distributed in Myanmar, **Thailand,** Malaysia, Sumatra, and Northern Borneo (Thorington and Hoffmann 2005).

Countries: Native:

Brunei Darussalam; Indonesia; Malaysia; Myanmar; **Thailand**

Habitat and Ecology:

This is nocturnal and arboreal, specialized to deciduous forest in Northern **Thailand** and possibly elsewhere. This species is found in higher elevations in the north and lowland areas in the south.

Systems: Terrestrial

Major Threat(s):

In the southern part of its range, it is threatened by habitat loss due to logging and habitat conversion

Conservation Actions:

It occurs in several protected areas across its range. Further studies are needed into the taxonomy, distribution and abundance of this species.Temminck's Flying Squirrel or *Petinomys setosus is* listed on the IUCN Red list (1996) as Lower Risk/Least Concern

Genus: Pteromyscus:

Smoky Flying Squirrel *(Pteromyscus pulverulentus)*

Description:

The body length ranges from 22 to 29 cm; tail length from 18-23 cm; Hind foot from 38-45 mm; ear length from 17 to 23mm and weight from 134-252 grams. The dental formula is 1/1 0/0 2/2 3/3. The upperparts are dark brown to blackish, the basal hairs are grey with buffy or whitish sub terminal bands, giving a frosted apperance.The underparts are buffy white, the feet are light brown, and the cheeks are grey. The tail exhibits longer hairs on sides than the top of bottom of the tail, giving it a slightly flattened profile, but not as pronounced as in *Gyaucomys* species. The bushy tail is greyish brown with blackish hairs at the tips.

Range Description:

This species has been recorded from **Southern Thailand**, Peninsular Malaysia (from Selangor and Pahang, to Johore), to Sumatra and Borneo. The potential range on Borneo and in Malay peninsula is probably more extensive than currently mapped. It is also found on the island of Penang (Medway 1983). but its range may be more extensive than currently mapped. Its natural habitat is subtropical or tropical dry forests. According to records of sightings from Borneo and Malaysia the species is only rarely sighted even in primary/pristine forest, although it lives mainly in these conditions.

Countries: Native:

Brunei Darussalam; (Kalimantan, Sumatera); (Peninsular Malaysia, Sabah, Sarawak); **Thailand**

Population:

According to Bornean and Malay records it is probably a rare species, as it is rarely seen even in primary forest. So, even in optimum habitat it is not abundant. There are probably two relatively secure populations (with not large numbers) in Mt. Kinabalu National Park and in a nearby protected area.

Population Trend: Decreasing

Habitat and Ecology:

It is nocturnal, lives in tree hollows in tall, undisturbed primary forest. It eats mostly fruits, nuts and fungi found on nocturnal forays through the trees. they breed in all months of the year, producing a litter of 1 to 3 babies. Litters and pregnant females are found throughout the year but only in small numbers. This species feeds on plant material including leaves, buds, blossoms and young shoots. They are nocturnal animals are mostly solitary and roost in trees at about 3 to 4 metres from the ground

Systems: Terrestrial

Major Threat(s):

This species is affected by habitat degradation because of low elevational range. A threat to the population is habitat loss due to lower elevational range, restricting or preventing long-distance gliding needed to maintain a large range. It eats mostly fruits, nuts and fungi found on nocturnal forays through the trees.

Subfamily: Callosciurinae:

Genus: Callosciurus:

Gray-Bellied Squirrel ***(Callosciurus caniceps)***

Description:

An adaptable species, the Grey-Bellied Squirrel occurs in a variety of habitats including primary and secondary forests, and mature gardens. It is not too shy and may approach close to human habitation at the forest edge where it feeds on fruits, seeds, flowers and sometimes insects. The upper side of the body is greenish brown, and the belly light grey. The long tail is thick with fur, has grey bands and a black tip. During the dry season the upper side becomes more orange-brown in colour, thus the species is sometimes called the Golden-backed Squirrel. The size and shape of the skulls of these squirrels Indicated the differences between N group and S group from t-test and U-test. These results may be influenced by the two transitions of the phytogeography around the southernmost locality in N group and the northernmost locality in S group in the peninsular **Thailand** and Malay Peninsula. Localities which are located between N and S groups were called the Middle (M) group. From the PCA among N, S groups and each locality of M group, the plots of localities such as **Prachuap Khiri Khan, Chumphon, Krabi, Nakhon Si Thammarat and Trang** in both sexes of M group could not be separated from those of N and S groups. We suggest that the sympatric distribution of N and S groups and the hybrid of N and S populations may be seen in these localities of M group.

These are medium sized squirrels which usually have speckled grey or olive brown upperparts and silver grey underparts,The upperparts or sides, and, in a few geographical areas the underparts may have a reddish wash and some in the north of Thailand have a rich Rufus back with contrasts sharply with the grey legs and tail, and always as a black tip to the tail.

Range

The species occurs in southern Burma, **Thailand** and Peninsular Malaysia. It has also been introduced to Japan. Population: This species was rarely found in a survey conducted by Saiful and Nordin (2004) in Peninsular Malaysia (Weng River sub-catchment). It is common in Kuala Lumpur Parks (W. Duckworth pers. comm.) and in **Khao Yai National Park in Thailand (W. Duckworth pers. comm.)** The geographical variation of the gray-bellied squirrel *(Callosciurus caniceps)* was examined using optometry of skull in Southeast Asia. From the principal component analysis (PCA), the plots of the northern localities from **Nan to Kanchanaburi** and those of the southern localities from Narathiwat to Kuala Lumpur in male were completely separated. In female, the plots of the locality from **Uttradit to Kanchanaburi and those of the locality from Pattani to Negri Sembilan were completely separated.** We call these northern localities and southern localities which are distinguished by the PCA as N group and S group.

Habitat and Ecology:

This squirrel is well adapted to the presence of people. It may be found in plantations, cultivated areas, second growth and gardens, as well as forest. In natural habitats it seems to prefer dense dipterocarp forests with thick brushy vegetation. It may be found up to 2,500 m, but is usually found at lower elevations (Smith et al. 2008). The Gray-bellied Squirrel is normally diurnal and arboreal (Saiful and Nordin 2004), although it sometimes descends to the ground to pick up food, which it then carries into a tree and eats. The diet consists of fruit and some insects. The spherical nest is built on the upper branches of a bush or small tree. The home range is small compared with other arboreal squirrels, and does not change in size seasonally.

It has been suggested that one of the reasons for low densities of this species in Malaysian tropical rain forest is competition from the great variety of other arboreal vertebrates (such as birds, and especially primates) for food, especially fruits and leaves, which are among the food items preferred by squirrels (Saiful and Nordin 2004). The nest of course twigs or leaves, lined with shredded fibres is usually found in in the upper branches of a small scrub or tree, nest have also been found amongst the ground in leaf litter at the base of bamboo. There are 3 to 4 young in a litter. The longest recorded life span was one in captivity at nine years and six months (Medway1969).

Systems: Terrestrial

Major Threat(s):

The are no major threats to this species.

Conservation Actions:

Saiful and Nordin (2004) state the need for further comparative study on this species' abundance, density and distribution and its relationship to forest structure or habitat quality, spatially and temporally, in hill dipterocarp forest of Malaysia.

Pallas's Squirrel (*Callosciurus erythraeus)*

Taxonomic Notes:

A major structured program of collecting is required to clarify the complex taxonomy of this species in Lao PDR (Evans et al, 2000). Often occurs in reports under the name *C. flavimanus* even in lists that also include *C. erythraeus*. This is symptomatic over the general confusion of squirrel identification in the region (Duckworth et al. pers. comm.). There are many named and still unnamed forms in **Thailand,** Lao PDR and Vietnam some of which have small ranges. Whether any merit species rank is unclear but there is no persuasive reason for any as yet detected form to be considered at risk (Duckworth et al. pers. comm.).

Description:

Head-Body Length : approx 20 cm
Tail Length : approx 20 cm
Weight : approx 250 grams

Pallas's squirrel is a medium-sized tree squirrel, with a head-body length of 16 to 28 cm (6.3 to 11.0 in), and a tail 11 to 26 cm (4.3 to 10.2 in) in length. Both sexes are of similar size and appearance, and weigh between 310 and 460 g (11 and 16 oz). The colour of the pelt varies considerably between the many different subspecies, but is generally brownish on the upper body with a more reddish tint on the belly, and often with some black on the tail. The precise pattern and shades of the fur are often used to distinguish subspecies from one another, but make it difficult to distinguish the species as a whole from other, similarly variable, tree squirrel species.

The species is identified by its medium size, its olive-brown upperparts and reddish underparts. The absence of black and buff stripes along the flank distinguishes the species from the similar Plantain Squirrel *Callosciurus notatus,* which preferslowland areas.

Range Description:

This species is widely distributed in northeastern South Asia, much of central and southern China, and mainland Southeast Asia. In South Asia, the species is known to occur in Bangladesh and northeastern India (Molur et al. 2005). It is widely distributed in the region. In China, it has been recorded from Sichuan, Yunnan, Guandong, Guangxi, Hunan, Guizhou, Hainan Island, Xizang, Zhejiang, Fujian, Jiangsu, Anhui, Henan and Hubei (Smith and Xie 2008). It has been recorded from the island of Taiwan. In Southeast Asia, it is present in Myanmar, **Northern Thailand**, Lao PDR, Vietnam, Eastern Cambodia and Peninsular Malaysia .Tree squirrels were considered to make particularly successful invasive species due to their high reproductive potential, ability to disperse effectively, diverse food habits, ability to build nests, and adaptability to human-impacted landscapes (Palmer *et al.,* 2007). Bertolino (2009)

Native:

Bangladesh; Cambodia; China; India; Lao People's Democratic Republic; Malaysia; Myanmar; Taiwan, Province of China; **Thailand;** Vietnam

Habitat and Ecology:

It is diurnal and arboreal species typically occurring in subtropical montane evergreen and broadleaved forests, although in China it is also present in sub alpine coniferous forests or in a mix of conifers and broadleaf trees at altitudes above 3,000 m asl (Smith and Xie 2008). It has been found to occupy tree hollows in mid high canopy. Very flexible in terms of habitat; Duckworth and Robichaud (2005) found in heavily degraded scrub landscapes with small degraded forest patches in far northern Lao PDR, It has a generation time of two to three years. Three types of ectoparasites are found: sucking louse *Neohaematopinus callosciuri* (Anoplura: Haematopinidae), 26 fleas *Ceratophyllus (Monopsyllus) anisus Rothschild (Siphonaptera: Ceratophyllidae)* and four nymphs of the tick *Haemaphysalis flava* Neumann *(Acari: Ixodidae)* Pallas's squirrel (C. erythraeus) is 1 tree squirrel species with body mass of approximately 250–400 g. They are distributed in Southern China and Southeast Asia and are also introduced in Japan and France (Xu and Ran 2004). Pallas's squirrels are found as important pests by gnawing barks and young shoots in many conifer plantations (e.g., Kuo and Liao 1986; Xu and Ran 2004; Tamura and Ohara 2005). Previous studies showed that Pallas's squirrels can hoard plant seeds during fruiting seasons in autumn (Chou et al. 1985; but see Setoguchi 1990). According to recent and ongoing studies in Qingcheng Mt, this tree squirrel is a key scatter-hoarder for the dispersal of large seeds from Fagaceae species and other large-seeded species (e.g., Camellia spp.). The most exciting finding is that Pallas's squirrels remove acorn embryos from several Fagaceae species (e.g., *Castanea spp., Quercus spp.,* and *Cyclobalanopsis spp.),* like Sciurus squirrels to white oak acorns reported in North America.

Like all tree squirrels, Pallas's squirrels are primarily herbivorous. They eat a wide range of different foods, and have differing diets in different parts of their broad range. However, primary foodstuffs include leaves, flowers, seeds, and fruit. They also eat small quantities of insects, as well as occasional bird eggs. The squirrels breed throughout the year, and may mate again as soon as they have weaned a previous litter. Pregnancy lasts 47 to 49 days, and results in the birth of up to four young, with two being typical. The young leave the nest at 40 to 50 old, and are sexually mature at one year of age. They have lived for up to seventeen years in captivity. Pallas's squirrels are diurnal, and inhabit much of the forest canopy, and construct both leaf nests 7 to 18 m (23 to 59 ft) above the ground, and, less commonly, underground burrows. Females occupy home ranges of just 0.5 to 0.8 hectares (1.2 to 2.0 acres), which usually do not overlap, while males occupy much larger ranges of 1.3 to 3.8 ha (3.2 to 9.4 acres), which overlap with those of both nearby males and females. Like many other squirrels, they have been observed to cache acorns in the autumn. The squirrels make calls to warn neighbours of predators, and have been observed to mob tree-climbing snakes, with females protecting young being particularly likely to join in.Males also make loud calls prior to, and after, mating.

Major Threat(s):

There are no major threats to this species. Hunting for consumption has depleted some South Asian populations (Molur et **al. 2005).**

Conservation Actions:

It is known from the following protected areas in India: Eagle's Nest Wildlife Sanctuary, Kamlang Wildlife Sanctuary, Namdapha National Park, Pakhui Wildlife Sanctuary, Sessa Orchid Sanctuary and Tale Valley Wildlife Sanctuary in Arunachal Pradesh (Molur et al. 2005). It Is presumably present in many protected areas in China and Southeast Asia.

Finlayson's Squirrel *Callosciurus finlaysonii)*

Description:

HB;212-218; T;225-240; HF;46-49; E;19-23

Occurring in diverse habitats from primary and secondary forests to open woodland and plantations, the medium-sized Variable Squirrel is, as its name suggests, extremely variable in colour and patterning. Different subspecies and populations can be partly black, brown, orange-red, cream or white with diverse patterning. Typically, however, the upper side is of darker colour than the underside. Colour variation can also occur within subspecies or populations, and it is possible that seasonal colour variations occur too. Around 13 to 17 subspecies are generally recognised. These attractive squirrels are arboreal, diurnal and feed on a variety of seeds and fruits.

Range Description:

This species occurs only in central Indochina from central Myanmar southeast through much of **Thailand** and Lao PDR and Cambodia to the Mekong Delta of Vietnam
(Moore and Tate 1965; Corbet and Hill 1992; Oshida et al. 2001).
An apparently undescribed form of this species was found in the limestone mountains of central Lao PDR (Evans et al. 2000). A small introduced population exists in Singapore (B. Lee pers. comm.).

It is common across at least its Lao range living at very high densities in large blocks of habitat. Small numbers are found in fragmented and degraded patches (Evans et al. 2000; W. Duckworth et al. pers. comm.). The introduced population on Singapore is less than 20 individuals (B. Lee pers. comm.).

Countries: Native:

Cambodia; Lao People's Democratic Republic; Myanmar; **Thailand;** Vietnam

Introduced:

Singapore

Habitat and Ecology:

This is an arboreal species, found in many habitats including open woods, coconut plantations, and dense forest (Lekagul and McNeely 1988). It is very tolerant of degradation and fragmentation (W. Duckworth et al. pers. comm.). Ecological overlap with *C. erythraeus* seems to be extremely restricted but is know from parts of Khammouane Limestone National Biodiversity Conservation Area in Lao PDR (Duckworth et al. 1999). Finlayson's squirrel is extremely variable in body size and fur colour, and even within subspecies there is a great deal of variation (Lekagul and McNeely, 1988). Squirrels spent the majority of active time foraging, switching between resources according to availability. In winter, they fed mainly on buds or stripped bark, shifting to flowers when plants bloomed in spring, and to mature seeds and fruits bark-stripping became progressively more important. This activity accounted for 36.5% of the yearly feeding time. The flexibility to exploit a wide range of foods, the ability to change food habits throughout the year, and hoarding behavior are factors that could indicate a proclivity to use different habitats successfully. The female as two pairs of functioning mammae.Most squirrels in Callosciurus live in tropical rain forests, but some individuals live in parks and gardens in cities. In the trees, they build their nests out of plant material. They are solitary, and give birth to one to five young. Their food consists of nuts, fruits, and seeds, and also of insects and bird eggs.

Systems: Terrestrial

Major Threat(s):

There are no major threats to this species. This species is listed as Least Concern in because of its wide distribution, occurrence in a number of protected areas, tolerance to some degree of habitat modification, and because it is unlikely to be declining at nearly the rate required to qualify for listing in a threatened category.

Conservation Actions:

A major structured program of collecting is required to clarify the complex taxonomy of this species in Lao PDR (Evans et al. 2000)

Black-Striped Squirrel *(Callosciurus nigrovittatus)*

Description:

HB: 170-240: T; 145-230: HF:40-50.

The under parts and tail are olive brown, the underpants are grey, and there is a black stripe on the side bordered above by a parallel buff stripe. The female has 3 pairs of mammae. Medway describes the voice as a rattling chatter or chuckle. Also a repeated bird like chirp and now and again a bird like shrill call "twist"

Range Description:

This species is found in the southern part of Malaysia, on Sumatra and Java, and on the islands of Tioman and Tambelan (Oshida et al. 2001). It is also found in **Southern Thailand** (Lekagul and McNeely 1988).

Countries: Native:

Indonesia (Jawa, Kalimantan, Sumatera); Malaysia; **Thailand**

Population:

The population of this species is stable. This species was rarely found in a survey conducted by Saiful and Nordin (2004) in Peninsular Malaysia (Weng River sub-catchment).

Population Trend: Decreasing

Habitat and Ecology:

It is an arboreal squirrel, living in primary and secondary forest, from lowlands to montane areas; occurring also in mixed areas with cropland and tree cover. This is a diurnal and arboreal species (Saiful and Nordin 2004). It has been suggested that one of the reasons for low densities of this species in Malaysian tropical rain forest is competition from the great variety of other arboreal vertebrates (such as birds, and especially primates) for food, especially fruits and leaves, which are among the food items preferred by squirrels (Saiful and Nordin 2004).

Systems: Terrestrial

Major Threat(s):

There are no major threats to this species.

Conservation Actions:

Saiful and Nordin (2004) state the need for further comparative study on this species' abundance, density and distribution and its relationship to forest structure or habitat quality, spatially and temporally, in hill dipterocarp forest of Malaysia.

Plantain Squirrel *(Callosciurus notatus)*

Range Description:

This species is found from Peninsular Malaysia and **Thailand** to Java, Bali, Lombok, and Borneo (Lekagul and McNeely 1988; Thorington and Hoffmann 2005). It is also widespread on smaller islands. The population on Salajar island, south of Sulawesi, has been introduced possibly from Java (Musser 1987). This species is found from lowland up to 1,500 m.

Countries: Native:

Brunei Darussalam; Indonesia (Bali, Jawa, Kalimantan, Lesser Sunda Is., Sumatera); Malaysia (Peninsular Malaysia); Singapore; **Thailand.**

Description:

Main features: Small; head and body 20-25cm, tail about the same length.

These agile squirrels are a beautiful greyish brown, with a chestnut belly and a black-and-white line between the two. In fact, their genus name Callosciurus means Beautiful Squirrels and the genus includes some of the most colourful mammals. This is a medium sized squirrel with olive brown upperparts and solid brown underparts. The tail is olive brown, tipped with red found in the subspecies of **Thailand.** On the flank there is a narrow there is a narrow black strip which borders the red brown colour of the under parts.Immediatly above this black stripe there is a narrow buff stripe. The female as 2 or 3 pairs of mammae.

Ecology:

Plantain Squirrels eat mainly fruits and nuts but also snack on insects and other titbits that they come across including birds eggs. They have been known to break open ant plants to eat the ant larvae.

C. notatus is a very adaptable species, as it is found in secondary forests, plantations, and all kinds of habitats, though not so common in pristine forest (Han pers. comm.). Ancestral habitats appear to be mangroves and major natural disturbances along big rivers (Payne et al. 2005). Now it is very common in towns, scrub, secondary forest, forest edge and urban parks (W. Duckworth et al pers. comm.).

This is a diurnal and arboreal species (Saiful and Nordin 2004), though it does descend to the ground to forage (Han pers. comm.). The diet of this species consists of fruit (42%) and bark 40% (A. Saiful pers. comm.), as well as insects and rubber tree latex (Lekagul and McNeely 1988). The Plantain squirrel *(Callosciurus notatus),* is also one of several species called the Oriental squirrel or Tricoloured squirrel. It is found in Indonesia, Malaysia, Singapore and **Thailand,** in a wide range of habitats: forests, mangroves, parks, gardens, agricultural areas.

Fruit farmers consider them to be pests.

Its diet consists mostly of leaves and fruits, but also may include insects and bird eggs. Plantain squirrels are known to break open twigs that contain ant larvae to eat them, and they can eat fruits much bigger than them, such as mangoes, jackfruits or coconuts. They are very quick and agile in trees, able to jump a few meters between trees and rarely wander on the ground. Plantain Squirrels forage mostly in trees and undergrowth and rarely come to the ground. They are active during the day, more so in the morning and evening. They rest in hollow trees and some construct a nest out of leaves and twigs, in tree branches or large bushes. Like other squirrels, they scamper energetically among the undergrowth and trees. To do this, they have good eyesight, sharp claws and a long counterbalancing tail. Plantain Squirrels are very quiet, and often the first sign of their presence is a quick movement seen from the corner of your eye! Role in the habitat: Plantain Squirrels disperse plants by eating their fruits. But they are considered a pest by fruit farmers and coconut plantation owners. They nest throughout the year with the nest being a spherical structure made out an outer wall of twigs and broad leaves lined with fibrous material such a shredded palm spathe, and with a lateral entrance. Litter sizes vary from 1 to 4. The maximum recorded life span in captivity is 9 years and 7 months.

Status and threats:

Plantain Squirrels are fairly common because they have adapted to humans. They are often found in gardens, parks and agricultural areas.. There are no major threats to this species.

Conservation Actions:

It occurs in a number of protected areas across its range and there are no conservation measures needed for this species (W. Duckworth et al. pers. comm.).

Prevost's Squirrel *(Callosciurus prevostii)*

Population:

Prevost's squirrel *(Callosciurus prevostii),* or Asian tri-coloured squirrel, is found in forest in the **Thai-Malay** Peninsula, Sumatra, Borneo and nearby smaller islands, with an introduced population in northern Sulawesi. They eat fruits, nuts, seeds, buds, flowers, insects and bird eggs. These squirrels carry the fruits far from the tree and drop the seeds when finished with their meal. Spread far from the parent tree the seeds have a reduced likelihood to be eaten by other animals and greater opportunity to produce a new generation of plants.

The "typical" subspecies (for example *C. p. prevostii* from the **Thai-Malay Peninsula) o**f Prevost's Squirrel are among the most colourful mammals in the world with their black upperparts and tail, reddish-orange underparts, and whitish thighs and flanks. The markings in some subspecies are duller, and *C. prevostii pluto* from northeastern Borneo is reddish-orange below and black above (no whitish thighs or flanks)., or Asian tri-coloured squirrel, is found in forest in the **Thai-Malay Peninsula,** Sumatra, Borneo and nearby smaller islands, with an introduced population in northern Sulawesi. This species was the second most abundant of its genus found in a survey conducted by Saiful and Nordin (2004) in Peninsular Malaysia (Weng River sub-catchment), with a density of 3.29 ± 1.54 individuals/km2. In general, this species is found at low densities in unlogged forest in Malayasia: in Danum Valley, Sabah, Norhayati (2001) found 7.3 individuals/km2, while Zainuddin et al. (1996) found 2.98 individuals/km2 in Nanga Gaat, Sarawak. In general, throughout its range it is fairly common (Han pers. comm.) and stable.

Population Trend:

Decreasing

Description:

HB: 252; T: 270; HF; 59

The ears back and top of the head are black. The tail is usually uniform black, but some have brown hairs in the tail.. The underparts and legs are chestnut red, except for the upper thigh of the hind leg, which is white. A wide white band divides the red under surface from the black upper surface. It extends from the thigh of the hind leg to above the foreleg. A dark grey extension of this band covers the side of the face to the nose. Some red suffuses into the white band above the foreleg and the shoulder is brown red in some animals. The female as 3 pairs of mammae.

Habitat and Ecology:

This is a diurnal and arboreal species (Saiful and Nordin 2004), descending only occasionally to ground to forage (Han pers. comm.). This species can tolerate secondary forest, and often enters fruit orchards (Han pers. comm.). Davies (1962) found that they inhabit smaller trees of the middle story, though they may feed on the crowns of fig trees. Stomach contents consist of mainly pulpy fruit, though sometime arthropods are present such as ants termites and beetles. Food is usually eaten sitting along and not across a branch Banks (1931). They are antagonistic towards other squirrels and tree shrews, suggesting they are territorial. The nest is very large, located high up in a tree, made up of a thick outer layer of large twigs which they gnaw off and lined with shredded bark and a few pieces of grass; the entrance is low down on one side of the nest (Banks 1931). 2 to 3 young are born usually early in the year; the longest recorded life span in captivity is 5 years and 8 months (Medway1969)

This species is found in the dense rainforests of the **Southern Thai peninsula,** eating mostly fruit, but also some arthropods, and is considered a pest of oil palm and coconut plantations (Lekagul and McNeely 1988).
Systems: Terrestrial

Major Threat(s):

The remaining habitat of this lowland species is very small. A substantial part of its native range has been replaced by plantations. In parts of Sarawak it is very heavily hunted for pets (Giman and Han pers. comm.).

Conservation Actions:

It occurs in many protected areas. Saiful and Nordin (2004) state the need for further comparative study on this species' abundance, density and distribution and its relationship to forest structure or habitat quality, spatially and temporally, in hill dipterocarp forest of Malaysia.

Genus: Dremomys:

Asian Red-Cheeked Squirrel *(Dremomys rufigenis)*

Range Description:

This species is widely distributed, from Northeastern South Asia and southern China, into much of mainland Southeast Asia. It has been recorded in only two locations in India, once each in Arunachal Pradesh and Nagaland at around 1,500 m asl (Molur et al. 2005). In southern China, it has been recorded from Yunnan and southwestern Guangxi (Smith and Xie 2008). In Southeast Asia, it is distributed from Myanmar in the west, through **Western Thailand,** much of Lao PDR and Vietnam, and northern areas of Peninsular Malaysia. It is generally found at less than 1,500 m asl.

Description:

HB: 185; T: 158; HF:42. The pelage of the upper parts is dark grey agouti. The tail is darker than the back, and the hairs of the tail are tipped with white. The underparts are cream coloured or greyish whit. The under sides of the tail and the anal region, as well as the cheeks and muzzle are reddish brown. There is a small white spot immediately behind the ear. The nose is longer and more tapered than in most squirrels, but not as much as in *Rhinosciurus*. According the Medway(19690, the ventral pelage and upper hindquarters are suffused with buff in some geographical areas. The female as 3 pairs of mammae

Habitat and Ecology:

A Mountainous species found at elevations above 900 metres. It is diurnal and largely arboreal. It occurs in subtropical montane evergreen and broadleaved forests. It is predominantly found in thick shrub layer vegetation but does go into the forest canopy to feed and forages on the floor, where it feeds on insects, root's and other plant matter. It persists well in degraded and fragmented habitats (Duckworth et al. 1999). It has been found to occupy tree hollows in mid and high canopy of dense oak, bamboo, fir and pine forest patches (Molur et al. 2005).

Systems:

Terrestrial

Major Threat(s):

There are no major threats to this species in Southeast Asia (W. Duckworth pers. comm.). Hunting for food is a major concern in parts of northeastern India (Molur et al. 2005).

Conservation Actions:

It occurs in numerous protected areas in Southeast Asia, with no direct conservation measures needed (J.W. Duckworth pers. comm.). In South Asia, it is known from Namdapha National Park in Arunachal Pradesh, India (Molur et al. 2005). Survey, taxonomic research and monitoring are recommended for this species in South Asia (Molur et al. 2005).

Genus: Lariscus:

Three-Striped Ground Squirrel (*Lariscus insignis)*

Description:

HB:165-184; T:96-105; HF:42-46; E :14-18

The upper parts are deep brown agouti, and there are tree cleat black stripes on the back running from the shoulder nearly to the base of the tail. The central black stripe runs along the mid dorsal line. The tail is darker than the back, and is fairly short. The underparts are pail buff. The female as 3 pairs of mammae

Range Description:

This lowland species is found on Borneo, Peninsular Malaysia, and **Southern Thailand,** probably below 500 m (Giman and Han pers. comm.), as well as Sumatra and Java (Lekagul and McNeely 1988) and adjacent islands. This species has been recorded up to 1,500 m.

Countries: Native:

Brunei Darussalam; Indonesia (Jawa, Kalimantan, Sumatera); Malaysia (Peninsular Malaysia, Sabah, Sarawak); **Thailand** This species is listed as Least Concern as although uncommon, it has a wide distribution, occurs in a number of protected areas, it is tolerant to some degree of habitat modification, and because it is unlikely to be declining at nearly the rate required to qualify for listing in a threatened category.

Population:

This species is not common (Han pers. comm.). This species was rarely found in a survey conducted by Saiful et al. (2001) at the Field Studies Centre of the University of Malaya, Ulu Gombak, Selangor Darul Ehsan, Malaysia. The vegetation at this site is lowland dipterocarp forest that has been logged twice thus it is a mixture of primary and secondary forest.

Population Trend: Decreasing

Habitat and Ecology:

This is a diurnal and terrestrial species (Saiful and Nordin 2004; Saiful et al. 2001). It is found mostly in tall, primary forest, but can tolerate secondary habitat (Han and Steinmetz pers. comm.). In Gunung Gading it probably feeds on the buds of *Raffleisa flower* (Han pers. comm.). It is found in limestone hills in Kanthan Ipoh in Peninsular Malaysia (Han pers. comm.). There is a recent record from lowland evergreen forest in Kuiburi National Park at about 300 m (Steinmetz pers. comm.). It is believed that they have a specialized diet of fruits and roots and run into burrows in the ground when frightened. They nest in the trunks of trees, preferring cool areas. (Han pers. comm.).The nests are located at the base of rotted trees.

Systems: Terrestrial

Conservation Actions:

It is found in several protected areas, including Gunung Gading National Park, Pasoh Forest Reserve, and Krau Wildlife Reserve (Han pers. comm.), as well as Kuiburi National Park (Steinmetz pers. comm.). It is protected in Indonesia.

Genus: Menetes:

Berdmore's Ground Squirrel *(Menetes berdmorei)*

Distribution:

The Berdmore's Ground Squirrel *(Menetes berdmorei)* is a ground squirrel found in Southeast Asia, from the east of Myanmar to Vietnam. It is however absent on the Malay peninsula, as well as the islands. In **Thailand** it is found everywhere including the peninsular.

Description:

HB:184; T:140; HF:40

This rarely known squirrel has a grey brown back and a white belly. Most striking are stripes on the side - on each side one beige and below a black stripe. The head is pointy, so this squirrel resembles a mouse or a tree shrew. Its length is 20 cm (7.9 in), not including the 15 cm (5.9 in) long tail. The pelage of the upperparts and tail is grey agouti (or reddish brown agouti on the back); the hairs on the upperparts have alternating dark and light bands. The underparts are very light buff. The females have 3 sets of mammae.

Ecology:

As a ground squirrel it is rarely found on trees, but spends most of the time in the thick under woods of the rainforests. However it is also found in fields or villages, especially in rice fields it is sometimes omnipresent. Despite it being quite common only very little is known about the life of this squirrel. Despite its name, it is not a close relative of the ground squirrels of the tribe Marmotini

Red List Category:

Least Concern

Justification:

Listed as Least Concern because it is a widespread species that is abundant in suitable habitat. It can tolerate habitat degradation, and the only current major threat to this species is total habitat conversion.

Genus: Rhinosciurus:
Shrew-Faced Squirrel *(Rhinosciurus laticaudatus)*

Description:

HB: 195-215; T: 115-140; HF: 44-49; E:19-20
Though a true squirrel of the family Sciuridae, this species superficially resembles the Common Tree shrew *Tupaia glis* (family : Tupaiidae) in appearance. In both species the snout is long and pointed, but the gape of the Shrew-faced Ground Squirrel is much shorter in comparison to the Common Tree shrew.
In addition, the tail of the Shrew-faced Ground Squirrel is characteristically short and bushy, unlike the Common Tree shrew's which is nearly twice as long.
The fur of the Shrew-Faced Ground Squirrel is dark brown above, and pale buff to white below. The upper incisors are much reduced in comparison to other squirrel species, and the tongue is very long : these are adaptations to its insectivorous diet which comprises ants, termites, earthworms and beetles. The tail is speckled with grey and is often carried fluffed out. The shrew-faced squirrel, *Rhinosciurus laticaudatus* (Müller), is a medium-sized and monotypic squirrel with a head-body length of 19–23 cm and a tail length of 11 –17 cm (Payne & Francis, 1985; Francis, 2008). *Rhinosciurus laticaudatus* has uniform dark brown upperparts and buff or white underparts (Lekagul & McNeely, 1988). The short and bushy tail is often held upwards when the squirrel is active. While the shrew-faced squirrel is similar in size to the plantain squirrel (*Callosciurus notatus*), the former possesses a pointed snout that is unique. Rhinosciurus superficially resembles tree shrews (order Scandentia), but the latter have relatively more pointed snouts . In comparison, *Rhinosciurus laticaudatus* is distinguished by its shorter gape, and shorter and bushier tail. It differs from the diurnal squirrels in Singapore in that it is primarily terrestrial in habit, and its diet comprises mainly insects (Lekagul & McNeely, 1998).

Distribution:

The species occurs in **Southern Thailand,** Peninsular Malaysia, Sumatra and parts of Borneo. In Singapore it is considered rare. *Rhinosciurus laticaudatus* occurs in the tropical humid lowland forests of Thailand, Peninsular Malaysia, Singapore, Sumatra, Borneo, and small adjacent islands (Duckworth et al., 2008)

Population:

This species has been rarely encountered.
Population Trend: Decreasing

Habitat and Ecology:

The Shrew-faced Ground Squirrel is a forest floor inhabitant of undisturbed primary and mature secondary forests : it is secretive in habits and rarely seen. It is medium-sized, feeds primarily on forest floor insects, and probably nests in hollow logs. They eat ants, termites beetles, earthworms and fruit. The teeth of old individuals are worn down probability because of the grit which is taken in from ants and termites. This is a diurnal and terrestrial species (Saiful and Nordin 2004), occurring in lowland primary rainforest, logged forests, and 40 year old regrowth. A young baby of a shrew faced squirrel is called a 'pup, kit or kitten'. The females are called 'doe' and males 'buck'. A shrew faced squirrel group is called a 'dray or scurry'. The female as 2 pairs of mammae. They nest in hollow gaps logs and have a litter of 1 or 2 offspring. According to Medway(1969)the voice is a typical squirrel like, churring chatter. It differs from the diurnal squirrels in Singapore in that it is primarily terrestrial in habit, and its diet comprises mainly insects (Lekagul & McNeely, 1998).

Status:

The species is categorised as ‗Near Threatened' in the IUCN Red List due to the high forest conversion rates throughout its range, and it almost qualifies as threatened under criterion A2c' (Duckworth et al., 2008). Perhaps owing to the extremely low encounter rate and poor trap success for *Rhinosciurus laticaudatus* (Payne & Francis, 1985; Charles & Ang, 2010), very little is known about the shrew-faced squirrel. When performing a search in scientific databases, the term *Rhinosciurus laticaudatus*' appeared predominantly in articles on biodiversity surveys (e.g., Saiful & Nordin, 2004).

Systems: Terrestrial

Major Threat(s):

It is threatened by habitat loss due to logging and agricultural conversion.

Conservation Actions:

It occurs in several protected areas.

Horse-Tailed Squirrel *(Sundasciurus hippurus)*

Habitat:

Range elevation
1500 (high) m
(4921.26 (high) ft)

Sundasciurus hippurus is most often seen in lowland primary forest but has been found in re-grown logged forests and secondary forests. (Medway,1978; Payne,1985). (Medway, 1978; Payne, J, et al., 1985)

Physical Description:

Range mass
260 to 420 g
(9.16 to 14.80 oz)

Range length:

21.5 to 25 cm
(8.46 to 9.84 in)

Sundasciurus hippurus is the largest and most colourful of the Sunda tree squirrels, Sundasciurus, with considerable geographical colour variation in Borneo. This medium-sized squirrel always has a grey head, shoulders, and fore feet. This grey pelage may be more or less grizzled. The upperparts are reddish brown to chestnut. Subspecies differ, the hind legs may be grey or reddish brown and the underside is whitish, dull orange, or reddish brown. The tail is glossy black or grey and black banded. Despite the common name, horse-tailed squirrel, the tail is not very similar to that of a horse. There are some individuals resembling *Callosciurus erythraeus*. But in northern Malaysia, where both species occur, horse-tailed squirrels have uniform red undersides, not agouti, and darker and more bushy tails. (Payne, 1985, Lekagul & McNeely, 1977 ; Medway, 1978 ; Corbet & Hill, 1992)

Head and body length ranges from 21.5 cm to 25 cm and tail length from 24 to 29 cm. Hind foot length measures from 54 to 64 mm. They weight from 260 to 420 g. (Lekagul & McNeely, 1977; Medway, 1978)

The dental formula is 1/1 0/0 2/1 3/3 = 22. (Payne, 1985) (Corbet, G. B. and J. E. Hill, 1992; Lekagul B. and McNeely J. A., 1977; Medway, 1978; Payne, J, et al., 1985)

Reproduction:

Very little is known about mating behavior and systems in Sundasciurus species.

Breeding season

These squirrels probably breed throughout the year.

Females have two or three pairs of mammae. Little is known of reproduction in these squirrels but perhaps, as in other diurnal squirrels in that region, they produce young throughout the year. Two close relatives, *Sundasciurus lowii* and *S. tenuis,* have litter sizes of 2 to 4.

Key Reproductive Features:

Iteroparous ; year-round breeding ; gonochoric/gonochoristic/dioecious (sexes separate); viviparous

Little is known of parental investment in these squirrels. Females care for and nurse their offspring until independence in a tree nest.

Lifespan/Longevity:

Status: captivity

14 (high) years

Lifespan is unknown in these squirrels. Most squirrel species have lifespan averaging 3 to 7 years.

Behaviour:

Horse-tailed squirrels are diurnal. They are solitary or occur in pairs (Medway, 1978) They feed and move mainly in the lower and middle part of the understory, but nest in the upper canopy. Sometimes they came to the ground. In Malaysia they share their habitat with *Callosciurus notatus* and *Callosciurus nigrovittatus* in the same understory (8-18 m). Larger diurnal squirrels tend to live in the upper canopy and smaller species from the ground level through the lower canopy. (McKinnon, in McDonald, 2001). (McDonald David, 2001)

Key Behaviours

arboreal ; diurnal ; sedentary ; solitary

Communication and Perception:

The most commonly heard call is "CHEEK!.....CHEEK!.......chekchekchekchek....." (Payne,1985)

Horse-tailed squirrels probably also communicate through visual, chemical, and tactile cues. Diurnal tree squirrels typically have exceptional vision and vibrissae on the chin and limbs that aids in the perception of surfaces, making these animals quite agile climbers. (Payne, J, et al., 1985)

Communication Channels

visual ; tactile ; acoustic ; chemical

Perception Channels

visual ; acoustic

Food Habits:

Horse-tailed squirrels feed on seeds, fruits, and arthropods. In Malaysia they are reported to feed on bark and sap, while sympatric beautiful squirrels (Callosciurus species) feed more opportunistically on different plant material and insects (McKinnon in McDonald, 2001). Like other squirrels, they may include a diversity of foods in their diet, such as eggs, young vertebrates, and fungi. (McDonald David, 2001; Nowak, 1997)

Animal Foods:

insects; terrestrial non-insect arthropods

Plant Foods:

leaves; wood, bark, or stems; seeds, grains, and nuts; fruit; sap or other plant fluids

Predation:

Horse-tailed squirrels avoid predation primarily through their agility and vigilance in the trees. Few predators can chase and capture adults the forest canopy. Young squirrels are vulnerable to predation in the nest by small, arboreal predators such as snakes, cats, or other squirrels. Their coloration may make them cryptic in the forest canopy.

Ecosystem Roles:

Horse-tailed squirrels are important seed dispersers in primary and secondary lowland forests throughout their range. They may also serve as an important prey base for large predators, such as raptors.

Ecosystem Impact:

disperses seeds

Economic Importance for Humans: Positive:

Horse-tailed squirrels may help to disperse the seeds of important lowland tree species.
Economic Importance for Humans: Negative
There are no negative impacts of horse-tailed squirrels on humans.

Conservation Status:

IUCN Red List of Threatened Species
Near Threatened
More Information
Horse-tailed squirrels have no special status, although they may be threatened by habitat destruction throughout their range.

Low's Squirrel *(Sundasciurus lowii)*

Description:

HB:130-136; T: 80-90; HF: 30-31; E:11-12;W;55-60g

Low's Squirrel (*Sundasciurus lowii)* is a species of rodent in the Sciuridae family. It is found in Indonesia, Malaysia, and **Thailand.** Named after Sir Hugh Low, the well-respected British colonial civil servant of Malaya, this small squirrel inhabits primary forest, both lowland and montane. The body measures up to 15 cm, and the short, bushy tail 10 cm. The upperparts are brown, and the underparts cream to buff. The nose is straight and pointed. Head-Body Length : 13-15cm Tail Length : 8-10cm Weight : 60-70 grams Maximum longevity11.1 years (captivity) *S. lowii* is a small, brown squirrel. The dark brown pelage of the upperparts and tail forms a sharp clear border with the creamy underparts. The tail is short and bushy. The head as a straight Roman nose. The female as 3 pairs of mammae.

Population:

It is very common in suitable habitat (Han pers. comm.). This species was the most abundant of its genus found in a survey conducted by Saiful and Nordin (2004) in Peninsular Malaysia (Weng River sub-catchment), with a density of 13.00 ± 1.85 individuals/km2. In general, this species is found at low densities in unlogged forest in Malaysia; in Danum Valley, Sabah, Norhayati (2001) found 14.4 individuals/km2, while Zainuddin et al. (1996) found 1.28 individuals/km2 in Nanga Gaat, Sarawak. The species ranges from **Southern Thailand** and Peninsular Malaysia to Sumatra, Borneo and other smaller Indonesia islands including Natuna. The subspecies *S .l robinsoni* (Bonhole 1903) is known only from the southernmost provinces.

Population Trend: Stable

Habitat and Ecology:

It prefers degraded and secondary tall forest habitats in both lowland and mountain. It usually forages very near to the ground but T J Marshall reports reported them eating Lichen 12 feet from the ground in the bole of a large tree. In captivity it prefers to eat soft fruit. Three researchers have recorded that it increases in number after shifting cultivation. This is a diurnal and arboreal species (Saiful and Nordin 2004). It has been suggested that one of the reasons for low densities of this species in Malaysian tropical rain forest is competition from the great variety of other arboreal vertebrates (such as birds, and especially primates) for food, especially fruits and leaves, which are among the food items preferred by squirrels (Saiful and Nordin 2004). It is occasionally found along river banks, as seen at Temerloh, Pahang (Han pers. comm.). Active mainly in early morning and late afternoon, it feeds on insects, such as ants and termites, and small fruits. Litter size 2 or 3.

Systems: Terrestrial

Major Threat(s):

There are no major threats to this species (Han pers. comm.).

Conservation Actions:

It occurs in several protected areas across its range (Han pers. comm.). Saiful and Nordin (2004) state the need for further comparative study on this species' abundance, density and distribution and its relationship to forest structure or habitat quality, spatially and temporally, in hill dipterocarp forest of Malaysia

Slender Squirrel *(Sundasciurus tenuis)*

Description:

HB: 212-218; T:217-224; HF:48-40; E:19-22

The Slender Squirrel *(Sundasciurus tenuis)* is a species of rodent in the Sciuridae family. The body is brown on the upper parts and light grey on the underparts. The body measures about 13-16 cm, with a slightly shorter slender tail. *S. tenuis* is a nondescript squirrel with an olive brown back, darker tail and underparts which are buff grey or brownish grey. The demarcation between dark upperparts and lighter underparts is indistinct. The tail is long, nearly as long as the head and body and also slender. The head as a turned up nose rather than a straight Roman nose as found in *S. Lowii* (Harrison 1962) The female as 3 sets of mammae

Range Description:

This species is found in Peninsular Malaysia and **Thailand,** Sumatra, Borneo and several small adjacent islands.

Countries: Native:

Brunei Darussalam; Indonesia (Kalimantan, Sumatera); Malaysia (Peninsular Malaysia, Sabah, Sarawak); Singapore; **Thailand**

Population:

This species is common in suitable habitat. This species was rarely found in a survey conducted by Saiful and Nordin (2004) in Peninsular Malaysia (Weng River sub-catchment).

Population Trend: Decreasing

Habitat and Ecology:

It occurs in primary and disturbed forests, in both lowlands and mountains (Lekagul and McNeely 1988). It seems to cope well in forests that have been affected by logging. This is a diurnal and arboreal species (Saiful and Nordin 2004). It feeds on soft bark, fruits and insects.

Systems: Terrestrial

Major Threat(s):

There are no major threats to this species.

Conservation Actions:

It occurs in numerous protected areas across its range. Saiful and Nordin (2004) state the need for further comparative study on this species' abundance, density and distribution and its relationship to forest structure or habitat quality, spatially and temporally, in hill dipterocarp forest of Malaysia.

Genus: Tamiops:

Himalayan Striped Squirrel *(Tamiops macclellandi)*

Description:

HB:111; T:111;HF:27

The Himalayan, or Burmese, Striped Squirrel is largely a species of montane, forested areas above 700 metres elevation, though in some parts of Burma it is reported as being found at much lower altitudes. It is almost exclusively arboreal in habits, preferring tall forests, fruit trees and coconut palms.
The species is small in size, and quick in habits : it can easily be overlooked as it frantically moves from tree to tree searching for their food source - mainly insects. The striped pattern serves as an effective camouflage on fissured tree trunks.
The thickness and length of the alternating cream and dark brown dorsal stripes is the key identifying feature, and is diagnostic in distinguishing it from other, larger striped squirrels. In this species the stripes are thick and bold and run from near the eye to the base of the tail, which is narrow and short-haired.

There are 3 sub-species in Thailand.

Tamiops mcclellandi collinus; Tamiops mcclellandi konggensis; Tamiops mcclellandi leucotis;

Range Description:

This widespread squirrel has been recorded from northeastern South Asia, southern China and much of mainland Southeast Asia. In South Asia, this species has been recorded from Bhutan, India (Arunachal Pradesh, Assam, Manipur, Mizoram, Nagaland and Sikkim [Molur et al. 2005]) and Nepal where it widely distributed up to 1,500 m asl (Molur et al. 2005). In China, it has been recorded from Yunnan and Xizang (Smith and Xie 2008). In Southeast Asia, it ranges from Myanmar, through much of **Thailand,** Western Cambodia, Northern Vietnam, Western Lao PDR and Northern Peninsular Malaysia. There is a recent published record from Sumatra (Indonesia), but this is considered erroneous.

Countries: Native:

Bhutan; Cambodia; China; India; Lao People's Democratic Republic; Malaysia; Myanmar; Nepal; **Thailand;** Vietnam

Habitat and Ecology:

This species is found in a wide variety of habitats with trees, including secondary growth forest, scrub forest, and gardens. It is highly adaptable to habitat degradation. They are also found in association with humans in fruit trees and coconut palm plantations (Smith and Xie 2008). They live in the upper storeys of tall forests, fruit tree and coconut palms n Thai villages, especially in the Chiangmai area. The dorsal stripes serve as a camouflage when they on the bark of a tree; when frightened they often spread themselves out against the bark to heighten the effect, which makes then very hard to detect. They use holes in trees to shelter and often moves through the trees using long leaps. The diet consists of insects, fruit and vegetable matter. There are 2 calls: a bird like chirp monotonously repeated at intervals of about 1 second; and a long descending trill of shrill chirps which lasts about 3 seconds. They are seen alone or in family parties.

Systems: Terrestrial

Major Threat(s):

There are no known major threats to this species. In South Asia, habitat loss due to forest fire, encroachments, fragmentation, jhuming (slash and burn agriculture) and hunting are threats for this species (Molur et al. 2005).

Conservation Actions:

The species occurs in numerous protected areas (eg. Namdapha National Park, Arunachal Pradesh, India). In South Asia, surveys, taxonomic research and monitoring are recommended for this species (Molur et al. 2005).

Cambodian Striped Squirrel *(Tamiops rodolphei)*

Range Description:

This species occurs in southern Vietnam, Southern Lao PDR, Cambodia and **Eastern Thailand** (Lekagul and McNeely 1988).

This species appears to become rare or absent northwest of a line from 18°00'N, 104°10'E to 18°30'N, 105°00'E (Evans et al. 200)

Description:

Size: HB: 123; T: 122; HF;26. Skull length 32.4. The tail is narrower and shorter than the body. There are usually 5 dark stripes , alternating with white stripes on the back and sides, somewhat like the pattern of a chipmunk *Utami-as* The general dorsal background colour is yellowish brown, and the underparts are white, pale brown or bright red. The ears are usually tufted with white, and the tail is rather slender and thinly furred. There are 2 moults per year for winter and summer and the coats are somewhat different in these periods. Females have 6 mammae.

There are 2 sub species in Thailand.
***Tamiops rodolphei rodolphei* (Milne-Edwards 1867) and *Tamiops rodolphei elbeli* (Moore 1958)**

A young / baby of a Cambodian striped squirrel is called a 'pup, kit or kitten'. The females are called 'doe' and males 'buck'. A Cambodian striped squirrel group is called a 'dray or scurry'.

Habitat and Ecology:

This species is found in nearly every habitat with trees, including secondary growth forest, scrub forest, and gardens. Like *T. mcclellandii,* this species uses holes in trees for shelter and travels by jumping long distances between trees (Lekagul and McNeely 1988) his species is abundant in suitable habitat. This species is common in evergreen or semi-evergreen forest, but less common where the forest is degraded or includes mixed deciduous components (Evans et al. 2000).

Population Trend: Stable

Major Threat(s):

There are no significant threats to this species.

Conservation Actions:

It occurs in numerous protected areas.

Genus: Cannomys:

Lesser Bamboo Rat *(Cannomys badius)*

Description:

Size.

HB;147-265; T: 60-75; E: 7-10;

Dental formula. is 1/1 0/0 0/0 3/3x2=16. The incisors are thick and relatively blunt; and grow thought-out the animals life. The incisors are used for both eating and digging and protrude beyond the lips. The root of the lower incisor continues to the hind edge of the mandible, forming a knob on the outer side of the ramus; the root of the upper incisor extends back to above the molars, but is difficult to observe than in the lower. The molars are flat crowned, with a pattern of folds which become isolated as islands as the teeth wear. The surface of the upper molars is slanted markedly outwards, with the surface, with the surface of the lower molars slanted correspondingly inward, with the last molar the smallest.

The fur is soft and fine and as season variations. They are adapted to burrowing with a short, stout, rounded body, small eyes and ears, and a short smooth tail, short legs and short stout claws. There are no cheek pouches.

The skull is rather flat with a marked postorbital constriction and a triangular braincase The footpads are smooth.

Range:

This species ranges from eastern Nepal (up to 2,000 m asl [Molur et al. 2005]), through northeast India (Assam, Manipur, Meghalaya, Mizoram, Nagaland and West Bengal), Bhutan, south-eastern Bangladesh, Myanmar, South China, Northwest Vietnam, **Thailand** and Cambodia (Musser and Carleton 2005). Species within this complex have generally been recorded from roughly sea level to around 4,000 m asl. This is an unresolved species complex, with some taxa restricted to certain elevations, the altitudinal range is not constant throughout the known range.

Countries: Native:

Cambodia; India; Lao People's Democratic Republic; Myanmar; Nepal; **Thailand;** Vietnam

Habitat and Ecology:

This species complex has been recorded from a wide variety of habitats, ranging from bamboo forest to cultivated land, and other disturbed areas (although it is not present in rice paddies). In South Asia it occurs in montane temperate forest and bamboo forests in subtropical forest tracts; with animals found under bamboo clumps (Molur et al. 2005). Diet consists of young grass, leaves and bamboo roots. The do not normally drink water, obtaining liquid from their food or licking dew from leaves. They usually spend the day in their burrows, coming out in the evening and roaming widely through bamboo groves. Their major predators are owls and small carnivores, though hill tribes go to great lengths to catch them, because of their excellent flesh.Dispite hunting and predation pressures they seem to be secure because so much of the country is covered with dense bamboo stands which is the preferred habitat. They are long-lived species that have only one or two young in a litter. The gestation period is 42 days and the young are weaned at 61 days. The young are 7 g at birth. Can be locally very abundant in some areas. It is said to be vicious when cornered, charging the enemy and biting with the incisors.

Systems: Terrestrial

Major Threat(s):

In general the species is quite heavily exploited in some areas for food, and isolated populations in particular may be declining through over harvesting. It is also killed as a pest of rubber plantations in parts of its range (such as Myanmar), and may occur in densities of up to 600 animals per hectare (Ken Aplin pers. comm.). In South Asia, it is locally threatened by habitat loss due to jhum (shifting) cultivation, forest fires and harvesting for subsistence use (Molur et al. 2005).

Conservation Actions:

In general, animals are presumably found in a number of protected areas. It is known from the following protected areas in India and Nepal - India: Dampa Wildlife Sanctuary, Mizoram; Nepal: Royal Chitwan National Park, Central Nepal and Makalu Barun National Park, Eastern Nepal (Molur et al. 2005). The species is included in the Schedule V (considered as vermin) of the Indian Wildlife (Protection) Act, 1972. Further studies are needed into the distribution, abundance, ecology, and threats to these poorly-known taxa.

Genus: Rhizomys:

Hoary Bamboo Rat *(Rhizomys pruinosus)*

Description:

Size. HB: 256-350; T: 100-124; E: 18-26; HF: 45-61; W: 1-3 kg

A stocky rodent with long greyish brown fur, with many white tipped hairs intermingled; the posterior footpads are granular and separate. The upper parts are slightly darker than the underparts. Many hairs are grey at the base, brown in the middle and white tips, giving the pelage a frosted effect.. The feet are brown, and have granular pads on the soles; the 2 posterior pads are separate. The fur is thick and lustrous in the northern part of the range, but is thinner and harsher in the southern range. The young are snowy white, with black eyes and reddish feet and tail (Thomas 1927) There is no sexual dimorphism in this species. There are one or 2 pairs of pectoral mammae and 3 pairs of abdominal mammae. The skull is broad and flattened with prominent occipital condyles; the sagittal ridge is much lower than in *sumatrenis. F*rom the dorsal view the zygomatic arch joins the frontal bones at an obtuse angle. The ridges on each side of the palette are less developed than in *sumatrenis.* The first upper molar is slightly smaller than the second molar. The incisors do not project as much as in sumatrenis, and the diastema is less than 60% of the occipitonasal length.

Range:

This widely distributed species is found in Southern China (Yunnan, Ghizhou, Sichuan, Jiangxi, Hunan, Guangxi, Guangdong, and Fujian), Northern and North-Eastern India (Meghalaya, Nagaland, and Manipur up to 1,500 m asl [Molur et al. 2005]), Eastern Myanmar, **Thailand,** Lao PDR, Cambodia, Vietnam, south to Perak on the Malay Peninsula, at altitudes between 100 m and 4,000 m asl (Musser and Carleton, 2005).In general, this species can be locally abundant. It was found to be rare during market surveys in the lowlands of Lao PDR, but commonly traded in markets in mountainous areas (Francis 1999). There is no information available on the population abundance of this species in South Asia (Molur et al. 2005).

Countries: Native:

Cambodia; China; India; Lao People's Democratic Republic; Malaysia; Myanmar; **Thailand;** Vietnam

Habitat and Ecology:

It is found in bamboo hill and montane forest in **Thailand (**Lekagul and McNeely 1977). Xu (1984) reports records of this species from grassland intermixed with secondary forests, shrub forests, bamboo groves, with a smaller number of specimens from pine, fir, and other artificial forests. In South Asia, it occurs in tropical and subtropical deciduous forests. It is found to occupy bamboo thickets (Molur et al. 2005). There seems to be consensus that they are not found in agricultural fields and rarely in deep forest, with some affiliation with bamboo. It is a long-lived, relatively slow-maturing, with a small litter size (one or two young).

Systems: Terrestrial

Major Threat(s):

The only threat to this species is that it is hunted for food, since they are easy to find and conspicuous. In South Asia, the species is threatened by habitat loss and degradation due to shifting cultivation, bamboo extraction and hunting for local consumption (Molur et al. 2005).

Conservation Actions:

It is present in several protected areas across its range. In South Asia, it is known from the following protected areas in India: Dampa Wildlife Sanctuary, Mizoram and Namdapha National Park, Arunachal Pradesh (Molur et al. 2005). The species is included in the Schedule V (considered as vermin) of the Indian Wildlife (Protection) Act, 1972. In South Asia, surveys, life history studies, and population monitoring are recommended for this species (Molur et al. 2005).The Hoary bamboo rat is listed as Least Concern , lowest risk. Does not qualify for a more at risk category. Widespread and abundant taxa are included in this category, on the IUCN Red List of Threatened Species

Family: Cricetidae:
Subfamily: Arvicolinae:
Genus: Eothenomys:
Père David's Vole ***(Eothenomys melanogaster*** **)**

Description:

Size: HB: 101; T 40; HF: 18.7;E:11; W 27g.

A soft shaggy, soft brown mouse, whose blunt nose, short legs, short ears and short tail give it a stubby appearance, unlike that of any murine rodent. The mammae are)+ 2. The tail is well covered with short hairs so as almost conceal the ring of scales. The short broad skull has a zygomatic plate whose anterior border is concave. The tall molar crowns bear a complex zigzag enamel pattern consisting of a series of closed triangles The soles of the feet are said to be hairy behind the pads. The young are blackish and the female has 4 mammae.

Range:

This species is present in northeastern South Asia, it is widespread in South China, and ranges into Northern Southeast Asia. In India it has been recorded from Mishmi hills (Agrawal 2000) and upper catchment area of Noa Dihing river (S.S. Saha pers. comm.) in Arunachal Pradesh at over 300 m asl. In China, it has been recorded from Sichuan, Anhui, Guangdong, western Yunnan, Southeastern Xizang, southern Gansu and southwestern Shaanxi (Smith and Xie 2008). It has been recorded from the island of Taiwan (Smith and Xie 2008). In Southeast Asia, it is present in northern Myanmar, extreme north-western Vietnam; **and there is an isolated locality in Thailand that possibly represents a relictual population (Musser and Carleton 2005),** there may be additional isolated populations of this species that have not been recorded. The species has a known altitudinal range of 700 to 3,000 m asl (Musser and Carleton 2005).It is a common species in suitable habitat.

Countries: Native:

China; India; Myanmar; Taiwan, Province of China; **Thailand;**

Habitat and Ecology:

In China it is found in temperate montane evergreen forest (pine and rhododendron), in southern parts of the range it is more likely to be associated with open or cultivated areas close to the forest edge. In South Asia found in tropical and subtropical montane, temperate forest, where it leads a subterranean life, found in the leaf litter on the forest floor. Frequents wooded jungles and grassy meadows (Agrawal, 2000) and transition zones between temperate broad leaved and subtropical forests (Molur et al. 2005). They burrow in soft earths of banks. It is a grazer like other member of the sub family. Harsh silicone in grass wears down teeth hence most voles have tall, permanently growing molars. The ones that live in meadows make runways and store cut grass and they also adaptable to forest life.

Systems: Terrestrial.

Major Threat(s):

There are no major threat to this species as a whole. In South Asia this species is locally threatened by habitat loss and degradation due to expansion of agriculture, small-scale logging, human settlements (Molur et al. 2005).

Conservation Actions:

It is present within many protected areas. **The isolated population in Thailand is present in Doi Inthanon National Park.** In South Asia, it has been recorded in Namdapha National Park in Arunachal Pradesh, India (Molur et al. 2005).

Family: Muridae (mice, rats, voles, gerbils, hamsters, etc.)
Subfamily: Murinae:
Genus: Bandicota:
Greater Bandicoot Rat ***(Bandicota indica)***

Description:

Size: large-sized rat.
Fur: dorsum brown-black, ventrum dark grey, not sharply demarcated; guard hair developed on the back.
Tail: shorter than head and body (about 10% shorter); uniformly dark with a white ring at its basis.
Ears: short and thick,
Feet: black with long claws,
Teeth: yellow-orange broad incisors,
Hind foot: long, over 40 mm for adults, dark hairs above, stronger than that of Bandicota savilei,
Mammae: 1 + 2 + 3

It can grow to about 13 inches without including the tail which can grow to be just as long. The greater bandicoot rat *(Bandicota indica)* is the largest, weighing 0.5 to 1 kg (1.1 to 2.2 pounds). The shaggy, blackish brown body is 19 to 33 cm (7.5 to 13 inches) long, not including a scantily haired tail of about the same length.

Range Description:

This species is found throughout most of India, Sri Lanka, Bangladesh, the lowlands of Nepal through Myanmar, southern China (recorded from Yunnan, Sichuan, Fujian and Guangdong) (Smith and Xie 2008), Taiwan, **Thailand,** Lao PDR, Cambodia, and Vietnam; also on Cat Ba Island, off the coast of northern Vietnam. Introduced into Kedah and Perlis regions of Malay Peninsula as well as Java (Musser and Carleton 2005). Its spotty distribution may reflect other geographic introductions (Taiwan for example) (Musser and Carleton 2005). Sea level to about 1,500 m asl.
Countries: Native:

Bangladesh; Cambodia; China; Hong Kong; India; Lao People's Democratic Republic; Malaysia; Myanmar; Nepal; Sri Lanka; **Thailand;** Vietnam
Introduced:
Indonesia

Ecology:

The description of the habitat preference for the species we present thereafter is based on data gathered using the standard protocol defined during the two projects Ceropath and ANR roboviroses.
Animals were identified either using Aplin's criteria or molecular methods.
Animals caught in trap lines (sampling effort: 5600 nights/traps/habitat). Forest Flood plain Dry land Total
Bandicota savilei 0 53 56 109
Bandicota indica 0 11 6 17
Total 0 64 62 126

In females, body, uterine and preputial gland weights, occurrences of pregnancies and placental scars, and in males, testicular weights and histology, and sizes of accessory sex glands, were recorded. Pregnancies occurred predominantly, but not exclusively, in the wet season, with a higher incidence pregnancies in the second, than in the first, dry season. Uterine and preputial gland weights tended to be lower in the first, but not the second dry season, with placental scars occurring at all times of year. Males tended to have heavier testes in the wet season but some somniferous tubules contained sperm even in the dry season. Seminal vesicles, but not prostates and preputial glands, tended to be heavier in animals in the wet season. The greater bandicoot rat in **Southern Thailand** shows maximal reproductive activity in the wet season with some reproductive activity, albeit variable from year to year, occurring in the dry season depending upon environmental conditions. Studies have also shown that females, as well as males, have large preputial glands, and that males invariably have small testes regardless of the time of year. These observations suggest a similar timing of reproduction, but a different breeding biology and perhaps social organisation, from that of the sympatric rice field rat, *Rattus argentiventer* Greater Bandicoot Rat-borne is a carrier of the diseases leptospirosis

Habitat:

Bandicota indica live in paddy fields during the growing season, when the rice is regularly flooded. When paddy fields are dry, Bandicota indica moves near water ways or in dry cultivated lands where the food is more abundant. During this period, they live in tall grasslands, corn, pineapple fields and mixed orchards.
Burrow: Large and complex burrows with many rooms used for nesting and food storage, at the edge of fields, dikes, streams banks, and even of city streets.
Behaviour: ferocious nature, good swimmer and diver. When trapped in a cage, *Bandicota indica* can be recognized by its noisy growling and its threatening posture.
Food habit: omnivorous in diet, can feed largely on products of cultivation, such as rice, grains, sugar cane.

Population:

Common, and even extremely abundant; it has increased in abundance with agricultural expansion.
Population Trend: Increasing

Major Threat(s):

There are no major threats to the species.

Conservation Actions:

It is present in several protected areas; it has been reported from more than 15 protected areas in India (Molur et al. 2005). It is listed in the Schedule V (considered as vermin) of the Indian Wildlife (Protection) Act, 1972.

Savile's Bandicoot Rat *(Bandicota savilei)*

Description:

Size. HB:196; T: 159; HF:37; E:24; W: 199g

Savile's bandicoot rat *(B. savilei)* have dark brown or brownish grey body fur, weigh up to 350 grams, and measure up to 40 cm long including their brown tails. A chunky medium sized rat. The colour effect on the back is produced by a mixture of black and buff hairs, similar to *Rattus argentiventer* . the black colour of the bandicoot is more variable than that of the rat. The unicoloured tail, feet and underparts are of various shads of grey, which distinguish this species from the Norway Rat,*Rattus norviegicus* which has white feet.

Range Description:

Any of five Asiatic species of rodents closely associated with human populations. This species occurs in mainland South East Asia: central Myanmar, **Thailand,** Vietnam (Dang et al. 1994), and Cambodia (Aplin et al. 2003). It is probably present in lowland areas of Lao PDR, within the valley of the Mekong River (Aplin et al. 2003). This species is sympatric in much of its range with B. indica.

Countries: Native:

Cambodia; Myanmar; **Thailand**; Vietnam

Taxonomic Notes:

Aplin et al. (2003) described external morphological criteria for distinguishing *B. indica* from *B. savilei,* based upon large samples collected in sympatry in Cambodia and Vietnam. Populations in Myanmar, attributable to *savilei,* appear to be genetically distinct from those in the neighbouring part of the range for which the available name would be curtata (K. Aplin pers. comm.).

Population:

B. savilei is locally abundant across its range.
Population Trend: Stable

Habitat and Ecology:

This is an adaptable species, present primarily in degraded habitats and regarded as a major agricultural pest in certain areas, being commonly found in rain fed paddy fields and corn fields. They are excluded from southerly inundated parts of mainland South East Asia, and appear to be absent from forested environments.

Systems: Terrestrial:

The lesser bandicoot rat is found on the Indian subcontinent, Sri Lanka (formerly Ceylon), and Myanmar (Burma) and has been introduced on Pinang Island off the western coast of the Malay Peninsula, northern Sumatra, eastern Java, Saudi Arabia, and Patta Island in Kenya. Savile's bandicoot rat, on the other hand, occurs only on the mainland of Southeast Asia. These three terrestrial species are nocturnal or active at twilight, constructing burrows where they nest and where they bear their litters, which number from 2 to 18. They subsist on grains, fruit, and invertebrates and are destructive to cultivated crops. The lesser bandicoot rat, an especially aggressive burrower, has been reported to make tunnels in the concrete cellars of rice warehouses in Calcutta.

Major Threat(s):

There are no major threats to the species. However, it is hunted and sold for food, and they are also poisoned extensively as an agricultural pest.

Conservation Actions:

It is present in several protected areas across its range.

Genus: Berylmys:

Small White-Toothed Rat *(Berylmys berdmorei)*

Description:

Size: medium-sized rat,
Fur: dorsum silver grey, ventrum purely white,
Tail: about 20% shorter than head and body length,
Teeth: white incisors,
Hind foot: white-grey above,
Mammae: 1 + 2 + 2 A medium sized iron grey rat with crisp fur, white feet, and dark tail that is paler on the under surface than on top. It is host to the flea, *Xenopsylla vexabilis*. *Berylmys* species have rarely been reported from **Thailand.** Thai farmers usually call them the handsome rat, referring to their pure grey dorsal fur. Unlike *Bandicota indica*, *Berylmys* species moan discretely while trapped in cages. Females can be easily differentiated from *Bandicota* species with the different number of pair of mammae.

Range Description:

This species ranges from Southern Myanmar, and Southern China (Southern Yunnan) (Smith et al. 2008), through **Thailand** (Robinson et al. 1995), Lao PDR and Cambodia (Aplin et al. 2003), to southern Vietnam (Dang et al. 1994; Kuznetsov 2000). It roughly occurs from around 200 to 2,000 m. Although the species occurs at low densities, it can be locally abundant. It is absent from southern tropical peninsula of **Thailand**.

Population Trend: Stable

Countries: Native:

Cambodia; China; Lao People's Democratic Republic; Myanmar; **Thailand**; Vietnam

Habitat and Ecology:

This species inhabits forests from near sea level to at least 1,400 m, but most common in upland areas where it is reported to favour swampy areas. Occasionally an agricultural pest, but generally avoids human habitations. It is nocturnal, spending the day in their burrows. Occurs in bushy upland areas or dry cultivated land as cassava, corn, grassland, banana's orchard. This species also inhabits various forested areas and can occur at proximity of rice fields.

Behaviour: very gentle and quiet.

Systems: Terrestrial

Major Threat(s):

There appear to be no major threats to this species. It is considered to be a pest species in parts of its range.

Conservation Actions:

It is presumably present in a number of protected areas within its wide range. Further studies are needed into the taxonomy of this species.

Bower's White-Toothed Rat (*Berylmys bowersi*)

Description:

Size: HB: 245; T: 256; HF:55; E:30; W:420g

A large sized rat, with white teeth, recurved incisors, small bullae, third upper molar small compared to the larger one of *R.mulleri*. 2+2 mammae, feet white with a broad black band on the top, tail long.

Range:

This species is recorded from China (Yunnan and Sichuan to Zhejiang and Fujian [Smith and Xie 2008]), north-east India (Agrawal, 2000), north and central Myanmar, Northern and Peninsular **Thailand,** Northern Lao PDR, Vietnam, Malay Peninsula, and Medan in North-West Sumatra, Indonesia (see Musser and Carleton 2005). Appears to have a wide elevational range 600-1,800 m (Molur et al. 2005).

Countries: Native:

China; India; Indonesia; Lao People's Democratic Republic; Malaysia; Myanmar; **Thailand**; Vietnam.

Common and locally abundant, albeit patchily distributed.

Population Trend: Decreasing

Habitat and Ecology:

In general this species is present in a wide variety of habitats where there is suitable continuous tree cover, from forest to plantations. In South Asia, it occurs in temperate and subtropical montane forests, and is found in burrows in cultivated fields, disturbed primary and secondary forests, abandoned jhum (shifting agriculture), scrub, moist deciduous and evergreen patches (Molur et al. 2005).

Systems: Terrestrial

Major Threat(s):

It might possibly be threatened in some parts of its range where there has been extensive forest loss.

Conservation Actions:

It is presumably present in several protected areas. It is not known from any of the protected areas within its range in South Asia (Molur et al. 2005). Further taxonomic work is required to determine whether populations on the Malay peninsula south of the Isthmus of Kra may be a separate species. It is listed in the Schedule V (considered as vermin) of the Indian Wildlife (Protection) Act, 1972.

Fea's Tree Rat *(Chiromyscus chiropus)*

Description:

Size: HB:138; T:233; HF:30; E:20.

A medium sized rat with fairly spiny fur. 2+2 mammae, white belly, long dark tail, simple teeth as in the Niviventer group, and supraorbital shelf. Resembles *R.cremoriventer* very much but differs in colour as follows: brighter golden brown peppered with black hairs, blackish mask through the eye, cinnamon colour on the face and black rimmed molars.

Range Description:

This species is recorded from South West China (south Yunnan), Eastern Myanmar, **Northern Thailand**, Vietnam, and Lao PDR (Musser 1981; Musser and Carleton 2005; Smith et al. 2008).

Countries: Native:

China; Lao People's Democratic Republic; Myanmar; **Thailand;** Vietnam

Population:

This species is common and can be locally abundant.

Population Trend:

Stable

Habitat and Ecology:

This is an arboreal rat, it inhabits moist deciduous and evergreen forest (Corbet and Hill 1992), using mosaic scrub areas as extensions of forest. It may also occur in regrowth and other degraded habitats. It is probably present at higher elevations above 700-800 m.

Systems: Terrestrial

Major Threat(s):

There are no major threats to the species.

Conservation Actions:

It is present in several protected areas across its range.

Genus: Chiropodomys:

Pencil-Tailed Tree Mouse *(Chiropodomys gliroides)*

Chiropodomys (or pencil-tailed tree mice) is a genus of Old World rats and mice native to Southeast

Asia and Indonesia. They are tree-dwelling, very small mice, mostly found in tropical rainforest. In total six extant species have been identified, but only one of these, *Chiropodomys gliroides*, is common and widely distributed, and has been extensively studied.

Range Description:

This widespread species is present in North Eastern India, South China (including Hainan island), Myanmar, **Thailand**, Lao PDR, eastern Cambodia, Vietnam, Peninsular Malaysia and Indonesia (Java, Southern Sumatra, the Mentawai Islands and other islands) (Musser and Carleton 2005). It occurs between sea level to 1,600 m (Molur et al. 2005).

Countries: Native:

Cambodia; China; India; Indonesia; Lao People's Democratic Republic; Malaysia; Myanmar; **Thailand**; Vietnam

Arboreal, found in primary and secondary forest; it does not appear to be restricted to a particular forest type and specimens have been obtained from tropical moist deciduous forest and evergreen and semi-evergreen rain forests on lowlands, coastal plains, and hills, and from the montane evergreen rain forests on mountains and plateaus (Musser 1979). It can be found in very degraded habitat with some forest. In South Asia, it is found in tropical and subtropical dry deciduous forest where it is seen among primary and secondary forests with abundance of bamboo and also in coastal habitats and hills. Bamboo species with moderate girths are favoured nesting sites between 1 to 3 m high from the ground (Molur et al. 2005).

Systems: Terrestrial

Description:

Species of Chiropodomys have a body length of 7 to 12 cm, plus a tail of 9 to 17 cm.Head-Body Length : Up to 10 cm

Tail Length : Up to 14 cm Weight : Up to 35 gm They are generally grey or brown on the back and white underneath. The tail is only sparsely covered with hair, This small mouse is characterised by its relatively long tail, which is equivalent to 135% of the head-body length. The soft fur on the upperparts varies from pale fawn to pale grey, and the underparts are white. The tail is generally hairy, ending in a brush-like tip. The head is short, the eyes relatively large, the ears rounded, and the whiskers long. but has somewhat more at the end, giving the appearance of a pencil, thus the genus name.

Ecology:

Chiropodomys gliroides is particularly common in bamboo forest. It is active at night, sleeps during the day in a nest in the bamboo, padded with leaves. It eats exclusively plants. It lives in the bamboo internodes reached by way of a neat round hole, 22mm in diameter chiselled by the mouse. They have been found in the daytime high up in a flowering tree in **Chantaburi, Thailand.**

A close connection between Chiropodomys and the genus Hapalomys (marmoset rats) is accepted. The Haeromys (pygmy tree mice) are also thought to be closely related. On the other hand, an earlier-posited connection with *Crateromys (cloudrunners)* is no longer considered probable.

Population:

It can be locally abundant, although it is patchily distributed.
Population Trend: Stable

Major Threat(s):

Overall, there appear to be no major threats to this species other than heavy deforestation. The major threats of the species in South Asia include, habitat loss due to jhum (shifting) cultivation, small-scale logging, forest fires, clearing of bamboo patches, expansion of human settlements and dam constructions. It is also in local trade where it is harvested for consumption (Molur et al. 2005).

Conservation Actions:

It is present in many protected areas (eg. Namdapha National Park, Arunachal Pradesh, India). In India it is listed in the Schedule V (considered as vermin) of the Indian Wildlife (Protection) Act, 1972.

Marmoset Rat *(Hapalomys longicaudatus)*

Description:

Size. HB: 155-168;T: 179; HF: 28; E: 13.

A chubby, blunt nosed rat with tufts of long whiskers, extending from the ear and face. The fur is soft and brown. The molars have even rows of cusps in a checkerboard pattern of threes both crosswise and lengthwise.. Otherwise the rat and its skull look like a larger edition of *Chiropodomys*. The mammae are 2+2.

Range Description:

The species has a patchy distribution, it has been recorded from Myanmar and **Western Thailand,** however, it may be extirpated from these countries as little natural vegetation remains. An extant population occurs in Peninsular Malaysia (Musser and Carleton 2005). Specimens of *H. longicaudatus* reported from Yunnan by Wang (2003) probably represent *H. delacouri*.

Countries: Native:

Malaysia

Possibly extinct:

Myanmar; **Thailand**

Population:

It was formerly recorded as being common in Myanmar and **Thailand** ,but all of the natural habitat within the known range has now been cleared.

Population Trend: Decreasing

Habitat and Ecology:

This species is known only from natural stands of bamboo in evergreen lowland rainforest. It is a lowland species in the northern part of its range; it is recorded from around 500 m in Malaysia. The diet of this species is restricted to various parts of bamboo.

Systems: Terrestrial

Major Threat(s):

This species appears to be associated with natural stands of bamboo, a habitat of which very little remains within the known range of the species *(D. Lunde and K. Aplin* . Remaining stands of this bamboo are being cut for building materials and other uses. **It is possibly extirpated in Myanmar and Thailand.**

Conservation Actions:

It is present in the Cameron Highlands (Malaysia) and is likely to be present in the Cameron Highlands Wildlife Sanctuary, however, this needs to be confirmed. There is a need to protect remaining stands of natural undisturbed bamboo where the species has been recorded, and to conduct additional surveys to locate any additional remaining populations in habitat patches.

Genus: Leopoldamys:

Edwards's Long-Tailed Giant Rat *(Leopoldamys edwardsi)*

The black colour is uniform and pale brown rather than the slightly buff brown of sabanus, which is also darker down the middle of the back than on the flanks. The tail is rather patchy dark, with white hairs on the underside and tip.

Range Description:

This widespread species ranges from northeastern South Asia, into Central and Southern China, and parts of Southeast Asia. In South Asia, it has been recorded from Arunachal Pradesh, Meghalaya, Nagaland and West Bengal in India (Molur et al. 2005). In China, the species has been recorded from Guizhou to Zhejiang and Guangdong, Sichuan, south Gansu, south Shaanxi, Hubei and Hainan (Smith and Xie 2008). In Southeast Asia, it has been recorded from northern Myanmar, northern Lao PDR, Northern and Central Vietnam, with an isolated population present in **Northern Thailand (Musser and Carleton 2005). It has been recorded from close to sea level to around 1,400 m asl.**

Countries: Native:

China; India; Lao People's Democratic Republic; Malaysia; Myanmar; **Thailand;** Vietnam

Habitat and Ecology:

This mainly terrestrial large species is found in both primary and secondary closed canopy moist lowland evergreen and montane forest. It has been recorded at the forest edge and can be found in disturbed habitats close to forested areas. It generally occurs at higher altitudes than *Leopoldamys sabanus* where they are sympatric (Marshall 1977). Lives in burrows on the forest floor and has been recorded inside hollow logs. *Leopoldamys edwardsi* uses its long tail to climb easily and is semi-arboreal, nocturnal, non-commensal.
Systems: Terrestrial

Major Threat(s):

There appear to be no major threats to this species. It is hunted, but this seems unlikely to be causing any population declines. In South Asia, the species may be affected by habitat loss and degradation due to small-scale logging, non-farmland management and harvest for local consumption (Molur et al. 2005).

Conservation Actions:

It is presumably present in many protected areas over its wide range (eg. Namdapha National Park, Arunachal Pradesh, India). It is listed in the Schedule V (considered as vermin) of the Indian Wildlife (Protection) Act, 1972. Surveys and monitoring are recommended for this species in South Asia (Molur et al. 2005).

Neill's Long-Tailed Giant Rat *(Leopoldamys neilli)*

Description:

Size. HB:217; T: 270; HF: 42; E: 27; W: 219g

This rats upperparts are course variegated black and buffy brown. The short tail well under 290mm is dark, basally and along the top for about two thirds of its length, without vivid contrast from the white areas. There is a dusky line along the top of the foot. Juveniles are plain dark above. Mammae formula is 2+2. The skull is under 54 mm and tooth roes less than 8.9 mm.

Range Description:

The species has been recorded from a few locations in limestone areas of **Northern and South Western Thailand, north of the peninsular region.** The range may extend into adjacent areas of Myanmar and Vietnam, however, these populations might represent a separate species (Musser and Carleton 2005). It has a rough elevational range of 100 to 800 m.

Countries: Native:

Thailand

Population:

The population abundance is not well known. According to Wiles (1981) this species is rare in lowland bamboo forest in **Salak Phra Wildlife Sanctuary in South Western Thailand.**

Population Trend: Unknown

Habitat and Ecology:

Very little is known about this species of rat and of the few specimens collected, some have been taken from wooded limestone cliffs and others from lowland bamboo forest (Musser and Carleton 2005).

Systems: Terrestrial

Major Threat(s):

There are possibly no major threats to the limestone areas where this species is found. There may be some trapping, but probably not a level that would threaten the species.

Conservation Actions:

It is present in **Salak Phra Wildlife Sanctuary and may be present in Phu Hin Rong Kla National Park.** Further studies are needed into the taxonomy, distribution, abundance, ecology, and threats to this poorly-known species.

Long-Tailed Giant Rat *(Leopoldamys sabanus)*

Description:

Size. HB: 231; T353; HF: 49; E 30; W 343 g

This large nocturnal murid species lives semi-arboreal in tropical forested habitats and reaches an average body mass of 368 g feeding on plant material and arthropods. A slender sleek, buff, brown, giant rat, with a very long tail which is distinctively patterned and black and white, with the black extending continually, or sometimes in patches, along the upper surfaces except for their dorsal third, which is white all round

Range:

This species is present in North Eastern South Asia, and is widespread in Southeast Asia. In South Asia, it has been reported from only two locations, one each from Bangladesh (exact location unknown) and India (Gandhigram in Changlang district, Arunachal Pradesh) at about 300 m elevation (Molur et al. 2005). In Southeast Asia, the species has been recorded from **Thailand,** Northern Vietnam, Lao PDR, Southern and South-Western Cambodia, Southern Myanmar, Peninsular Malaysia, Indonesia (Sumatra, Java, and smaller islands off the Sunda Shelf [except Bali) and the island of Borneo (Brunei, Indonesia and Malaysia) (Musser and Carleton, 2005). It can be found up to 3,100 m asl on the slopes of Mount Kinabalu in Sabah, Malaysia.

Countries: Native:

Bangladesh; Cambodia; India; Lao People's Democratic Republic; Malaysia; **Thailand;** Vietnam

Population:

This is generally a common species over most of its range. Md Nor et al. (2001) recorded 103 specimens between 500 and 1,350 m on Mount Nuang, Hulu Langat, Selangor, Malaysia, making it by far the most abundant species found during the survey.

Population Trend: Stable

Habitat and Ecology:

This species is generally found in lowland forest habitats (infrequently in montane habitats), and probably at lower altitudes than *L. edwardsi* (Corbet and Hill 1992). It is similar to *L. edwardsi* in ecology, being semi-arboreal (Gorog et al. 2004), and foraging on the ground as well as in the canopy (Wells et al. 2004). The diet consists of insects, fruit, and other vegetable matter (as well as snails) (Lim, 1970). In South Asia, it is a nocturnal and subterranean species, occurs in tropical, subtropical, wet montane temperate forests (Molur et al. 2005).
Systems: Terrestrial

Major Threat(s):

In Southeast Asia, There are no major threats, although it may be susceptible to forest loss in some parts of its range. It is also hunted, and is one of the most common murids in markets in Lao PDR. In South Asia, the species could be affected by habitat loss and degradation due to shifting agriculture, small-scale logging and harvest for local consumption (Molur et al. 2005).

Conservation Actions:

This species is found in several protected areas across its range. There is a need for further research to elucidate the taxonomic status of this species complex. It is listed in the Schedule V (considered as vermin) of the Indian Wildlife (Protection) Act, 1972. In South Asia, it is not known from any of the protected areas and surveys and monitoring are recommended for this species (Molur et al. 2005).

Genus: Maxomys:
Rajah Spiny Rat *(Maxomys rajah)*

The Rajah Spiny Rat *(Maxomys rajah)* also known as the Brown Spiny Rat is endemic to **Thailand** and Malaysia, Sumatra, Borneo and the adjacent islands (Payne et al., 1985). Corbet and Hill (1992) mention that Maxomys rats are often the most common rodent in the Southeast Asian tropical forest, from most of the Malay Archipelago to Sulawesi, Palawan and Borneo. This species can be found in primary forest and logged-over forest. According to Payne et al.,(1985) this species lives in primary or secondary forest and tends to favour sandy and lowland sites. This terrestrial species is mostly active on the ground but occasionally climbs into the upper canopy. Its tend to live separately from other rats. *M. rajah is* medium in size where the upperparts are brown, darker in the midline, with numerous stiff gray-brown spines. The underparts are white with many short, white spines, and usually with a dark brown streak along the middle in adults, but never with an orange throat patch. The white colour of the underparts extend down in a narrow line to the feet. Above the tail is brown, pale below and thinly haired. Usually, the range of the measurement of *M. rajah* are HB 138.1-218, T 142-210, T/HB = 102.9-109.3%, HF 33.8-43, E 21.9-22.3, Wt 71-218 g, D 1003/1003=16, M 2+2=8. Skull: gl 40.9-48.6, iob 6.6-7.4, mt 6.9-8.1.The immature Red Spiny Rats, *M. surifer is* similar *M. rajah* and quite difficult to distinguish. (Payne et al., 1985). *M.rajah* is a common species but Yasuda et al., (2000) reported that very little is known about the ecological features of Maxomys rats. The smooth pads and long hind feet suggest a scampering gait along the forest floor.

Taxonomic Notes:

Once regarded as conspecific with *M. surifer* (Ellerman and Morrison-Scott 1951; Misonne 1969), but occurs sympatrically with that species, and although samples of each are regularly misidentified, the two differ in a suite of morphological, ecological, behavioural, and biochemical traits, as well as albumin immunology (Chan et al. 1979, Corbet and Hill 1992, Musser et al. 1979, Yong 1972, Watts and Baverstock 1994).

Range Description:

This species is found in **Peninsular Thailand** south of the Isthmus of Kra, Peninsular Malaysia, Riau Archipelago, Sumatra, and the island of Borneo (Brunei, Malaysia and Indonesia). It has been recorded from sea level to 1,100 m.

Countries: Native:

Brunei Darussalam; Indonesia; Malaysia; **Thailand**

Population:

It is common in appropriate habitat. Md Nor (2001) recorded 34 specimens between 500 and 1,100 m on Mt. Nuang, Hulu Langat, Selangor province (Malaysia).

Population Trend: Decreasing

Major Threat(s):

It is threatened by habitat loss, largely through commercial logging and the conversion of land to agricultural use.

Conservation Actions:

It is present in many protected areas.

Red Spiny Rat *(Maxomys surifer)*

Description:

This handsome species of rat occurs in primary and mature secondary forest, and adjacent cleared or cultivated areas. It does not occur in heavily degraded habitats or in large-scale oil palm plantations. It is nocturnal and mainly terrestrial in habits. The species is known to be omnivorous, feeding on fallen fruits and other plant matter, as well as insects and other invertebrates. Reaching up to 21 cm in head-body length, the Red Spiny Rat can be identified by its intense orange-brown or red-brown dorsal fur, which contrasts with a pale yellowish to white belly. The demarcation between the dorsal and ventral fur is sharp. Short, stiff spines occur amongst both the dorsal and ventral fur. There is often an orange-brown collar under the neck. The tail is dark above and light below, and lacks significant fur. The snout is long and pointed, and the limbs short. The hind feet are long and slender.

Range Description:

This species is widespread throughout Southeast Asia including, Southern Myanmar, **Thailand,** Lao PDR, Southern and Southwestern Cambodia, throughout Vietnam (including Thom and Phu Qoc Islands off the south coast), extreme southern Yunnan, China, the Malay Peninsula, Borneo, Sumatra, Java, and many smaller islands (Musser and Carleton 2005). It is a lowland species that ranges from around sea level to 1,680 m asl (Mount Kinabalu) (Nor 2001).

Countries: Native:

Brunei Darussalam; Cambodia; China (Yunnan); Indonesia (Jawa, Kalimantan, Sumatera); Lao People's Democratic Republic; Malaysia (Peninsular Malaysia, Sabah, Sarawak); Myanmar; **Thailand;** Vietnam

Population:

The abundance of this species ranges from locally common to quite rare depending on the population surveyed (Aplin et al. 2006). Nor (2001) recorded 17 specimens between 700 and 1,200 m asl on Mount Kinabalu, Borneo, Sabah, Malaysia. Wells et al. (2004) recorded nine specimens in Kinabalu National Park.

Population Trend: Decreasing

Habitat and Ecology:

This is generally a lowland, exclusively terrestrial species. It is present in primary forest and forest edge habitats (including adjacent gardens (Marshall, 1977)), but does not appear to be present in secondary forest (Aplin et al. 2006). It lives in burrows (Smith and Xie 2008). It is a nocturnal species that feeds on roots, fallen fruit, insects, and small vertebrates (Smith and Xie 2008). Litter size ranges from 2-5 (Smith and Xie 2008).

Systems: Terrestrial

Major Threat(s):

There are no major threats to this widespread species.

Conservation Actions:

This species is present in many protected areas. Further studies are needed into the taxonomy of this species. In China, this species has been regionally Red Listed as Least Concern (Wang and Xie 2004).

Whitehead's Spiny Rat *(Maxomys whiteheadi)*

Description:

Named after the British explorer John Whitehead who first documented its existence,Whitehead's spiny rat *(Maxomys whiteheadi)* has reddish-brown upperparts with grey under fur.. As its name suggests, Whitehead's spiny rat also has numerous inflexible grey spines with black tips on its upperparts, which are thought to be for protection. The underparts are orange-buff with grey under fur and numerous soft, pale spines The tail, which is shorter than the head and body length combined, is bicoloured, with a brown-black upper side separated from a white underside by a sharp line. Male Whitehead's spiny rats generally have longer hind feet than females. This species of rat has the largest teeth of all Maxomys species. Six subspecies of Whitehead's spiny rat are currently recognised, all of which differ slightly in appearance. Size Head-body length: 10 - 15 cm
Tail length: 9 - 13 cm
Weight 50 g

Range:

Whitehead's spiny rat is native to Southeast Asia, ranging from peninsular **Thailand** South to Malaysia, Indonesia and Brunei. This includes the islands of Borneo and Sumatra, as well as various neighbouring islands.

Habitat:

Belonging to a genus of rodents which are the most common in Southeast Asian tropical forests, Whitehead's spiny rat is found in old undisturbed areas of forest, as well as areas that have re-grown following a disturbance (secondary forest) . This ground-dwelling spiny rat also inhabits rice paddies and gardens adjacent to forests . Although generally a lowland species, Whitehead's spiny rat has been recorded at elevations as high as 2,100 metres

Biology:

Believed to be monogamous, with each breeding pair occupying a home range that is defended against other spiny rats . It is believed to breed throughout the year, at least in peninsular Malaysia, and has one to six young per litter.

A nocturnal animal , Whitehead's spiny rat spends its nights foraging on the forest floor for a wide range of plant material, such as fallen fruits and seeds, with a preference for oil palm fruit . However, as an omnivorous species, its diet also includes arthropods. Like other spiny rats, it is a hoarder, storing seeds in deep underground burrows that are dug by other forest animals. Whitehead's spiny rat is eaten by a number of predators, including the leopard cat *(Prionailurus bengalensis)*.

Threats:

The main threat to Whitehead's spiny rat is the widespread loss of habitat on Sumatra and Borneo, as a result of commercial logging and agriculture, particularly oil palm plantations . Lowland forests have been particularly hard hit, with some predicting that if this rate of deforestation continues, lowland forest in Sumatra and Kalimantan (Indonesian Borneo) may soon disappear completely

Conservation Actions:

It is recorded from several protected areas

Genus: Mus

Ryukyu Mouse (*Mus caroli*)

Range Description:

This species is widespread throughout Southern China and South-East Asia. It is known from Okinawa Island, Japan (it is not known whether this population was introduced (Abe et al. 2005)) to Taiwan, Province of China, Southern China (Yunnan, Guizhou, Guangxi, Guangdong, Fujian, Hainan Island, and Hong Kong provinces) (Smith and Xie 2008), Vietnam, Lao PDR, Cambodia, and **Thailand** (Musser and Carleton 2005). It has also been recorded from the Malay Peninsula, and in Indonesia in Sumatra, Java, Madura and Flores Islands in Nusa Tenggara; all places where they may possibly have been inadvertently introduced (Musser and Newcomb 1983), but they may also represent relictual populations (although those on Flores are definitely introductions). The spotty distribution of this species on the Sunda Shelf is best explained by inadvertent human introductions (Musser and Newcomb 1983).

Countries: Native:

Cambodia; China (Fujian, Guangdong, Guangxi, Guizhou, Hainan, Yunnan); Hong Kong; Lao People's Democratic Republic; Myanmar; Taiwan, Province of China; **Thailand**; Vietnam

Present - origin uncertain:

Indonesia (Jawa, Lesser Sunda Is., Sumatera); Japan (Nansei-shoto); Malaysia (Peninsular Malaysia)

Description:

Size. HB: 76; T:78; HF:18; E: 14; W: 12 g.

A cute tiny mouse; the pelage is brownish grey to brown above, and white to creamy white below; the head is small with large ears; projecting upper incisors are tan or brownish; the long tail is longer than or equal to the head-to-body length, dark above and pale below; foot pads are small and oval in shape.

Habitat and Ecology:

This species inhabits grasslands, scrub and secondary growth as well as other grassy agricultural fields such as rice paddies where it feeds on seeds and invertebrates (Smith and Xie 2008). They are mainly active at night, although they sometimes also are active for brief periods during the day (Smith and Xie 2008). It digs tunnels in the clay sod of the terrarium by biting off mouthfuls and raises a litter of 5 or 6 inside a ball of dry grass deep in the sod.

Systems: Terrestrial

Major Threat(s):

There are no major threats to the species throughout its range.

Conservation Actions:

This species probably occurs in several protected areas throughout its range. In China, it occurs in Changshanerhai and Tongbiguan Nature Reserves (CSIS 2008). Research should be conducted to determine the status of a potentially distinct species in central Myanmar. In China, this species has been regionally Red Listed as Least Concern (Wang and Xie 2004).

Fawn-Coloured Mouse (Mouse *Mus cervicolor)*

Description:

Size. HB: 83; T: 58; HF: 16; E: 13; W: 15g.

A short haired. soft furry, long nosed mouse, coloured brownish grey above, whitish underneath. Differs from *Mus caroli* in its shorter tail, coloured brownish grey above instead of blackish, long nasals, smaller feet which are always white, paler incisors, incisive foramina cutting back between the anterior molars, and posterior palatal foramina located exactly in the middle of the palatal bridge as in all Mus except *caroli.*

Range Description:

This species is patchily distributed from Pakistan and northern India, through Nepal, Myanmar, **Thailand,** Lao PDR, Cambodia, Vietnam and Southern China (Anning and Ruili, both localities in Yunnan [Smith and Xie 2008]) (Musser and Carleton 2005). It has also been recorded from northern Sumatra and Central and Eastern Java, both to Indonesia, where it may represent either introduced or natural, relictual populations (Ken Aplin pers. comm.). It is found from sea level to roughly around 2,000 m asl. it is a common species, and can be locally very abundant.

Countries: Native:

Cambodia; India; Lao People's Democratic Republic; Myanmar; Nepal; Pakistan; Sri Lanka; **Thailand;** Vietnam

Habitat and Ecology:

In South Asia, it is a nocturnal, fossorial and sometimes terrestrial species. It occurs in all types of habitats except deserts. It is often found in fields and tall grassland habitat close to rivers. It has been found to occupy irrigated and cultivated fields (Molur et al. 2005). It is a paddy field mouse and is also found in grass raised for fodder. A litter of 5 or 6 with a gestation period of 3 weeks with 3 weeks nursing. In China, it occurs in secondary growth, grass, brush, rice fields and other agricultural areas (Smith and Xie 2008). In Southeast Asia, this species is found in open areas in deciduous dipterocarp forest, grasslands and paddy fields (Marshall 1977). This species is often sympatric with Mus caroli (Marshall 1977).

Systems: Terrestrial

Major Threat(s):

There appear to be no major threats to this species.

Conservation Actions:

It is presumably present in many protected areas (eg. Jaldapara Wildlife Sanctuary, West Bengal in India). No direct conservation measures are currently needed for this widespread and adaptable species. It is listed in the Schedule V (considered as vermin) of the Indian Wildlife (Protection) Act, 1972. Further studies are needed to clarify the taxonomic status of populations currently allocated to this species.

Cook's Mouse *(Mus cookii)*

Description:

Size.HB:96.,T: 19.5., E15,. W: 23g

A larger edition of *M. Caroli* with longer and stiffer fur. But the elongated skull, with its long nasal bones, recurved incisors that are pale, and big molars is not similar to caroli except for the incisive foramina and s shaped anterior border of the zygomatic plate. Baby *Mus cookii* are fairly dark beneath , appearing somewhat like a house mouse, from which they are distinguished by larger molars.

Range Description:

This species is found in two distinct populations, one centred on north-eastern South Asia (Bangladesh, Bhutan, North-eastern India and Nepal [Molur et al. 2005]) and North Western Myanmar, and a second ranging through central and eastern Myanmar, Southern China (Yunnan, West of the Salween River [Smith and Xie 2008]), **Thailand,** Lao PDR and Vietnam. Both populations are widespread but patchily distributed. The North Western population ranges from around 50 to 2,500 m asl (Molur et al. 2005), while the South Eastern population occurs from about 200 to 1,500 m asl.

Countries: Native:

Bangladesh; Bhutan; China; India; Lao People's Democratic Republic; Myanmar; Nepal; **Thailand;** Vietnam

Population:

The species is abundant.

Population Trend: Stable

Habitat and Ecology:

It is present in a wide variety of primary and secondary forest types. In South Asia it is found in subtropical dry deciduous forests, shola grasslands, temperate coniferous and broadleaved forests, and has been found to occupy arable land near Lantana bushes (Molur et al. 2005). In Southeast Asia it is only found in forested areas and occasionally in moderately disturbed areas such as upland gardens in forests. In China, it has been reported from upland rice fields and other disturbed habitats (Smith and Xie 2008).

Systems: Terrestrial

Major Threat(s):

There are no major threats to the species.

Conservation Actions:

It is presumed to be present in many protected areas. It is possibly present in some protected areas in north-eastern India (eg. Namdhapha National Park, Arunachal Pradesh), and is almost certainly present in some protected areas in Southeast Asia. It is categorised as a vermin (Schedule V) of the Indian Wildlife (Protection) Act. No direct conservation measures are currently needed for this widespread and adaptable species. In South Asia, general taxonomic research, field surveys, and monitoring of populations are recommended for this species (Molur et al. 2005).

Gairdner's Shrewmouse ***(Mus pahari)***

Description

HB:96., T:86., HF: 22,. E;14., W28g.

A large spiny mouse, dark bluish grey on the back. The underparts are silvery with grey bases. The feet are white. The tail is long and the screw like appearance of the head is enhanced by the small eyes and ears and long nose. The smoothly rounded and tapered skull is further distinguished, as in other *Coelomys*, by a narrower zygomatic plate whose front border slants backward. The frontal bones are rounded off along the edge of the orbit.

Range Description:

This species is widespread from North-Eastern India (Arunachal Pradesh, Assam, Meghalaya, Mizoram, Nagaland, Sikkim and West Bengal), Bhutan (Putlibir), Myanmar, Southern China (Southeast Xizang, Yunnan, South Sichuan, Guizhou, Guangxi), **Thailand,** Southwest Cambodia, Lao PDR, and Central and Northern Vietnam (Molur et al. 2005; Musser and Carleton 2005; Francis 2008; Smith and Xie 2008). It is found between 200 and around 2,000 m asl.

Countries: Native:

Bhutan; Cambodia; China; India; Lao People's Democratic Republic; Myanmar; **Thailand;** Vietnam

Population

: In general, this species seems to be relatively uncommon.

Population Trend: Stable

Habitat and Ecology:

It is a nocturnal, semi-arboreal and terrestrial species (Molur et al. 2005). This species occurs mostly in montane forests (Marshall 1977; Musser and Newcomb 1983), it occurs in both primary and secondary forest. It can be found at the forest edge, but is closely associated with forest habitats. It builds globular nests in dry grass (Molur et al. 2005; Smith 2008).

Systems: Terrestrial

Major Threat(s):

There appear to be no major threats to this species. It is present in secondary and modified areas of forest.

Conservation Actions:

Globally, it is presumably present in many protected areas. Further taxonomic studies are needed for this species complex. It is listed in the Schedule V (considered as vermin) of the Indian Wildlife (Protection) Act, 1972. It is known from Dampa Wildlife Sanctuary, Mizoram in India. Survey, life history studies and monitoring are recommended for the South Asian populations of the species (Molur et al. 2005).

Shortridge's Mouse *(Mus shortridgei)*

Description:

HB:130,. T:71; HF:20.5; E; 18; W:34g

The largest mouse in **Thailand,** spiny, pale brownish grey above, white beneath with grey bases and white feet. The tail is lead covered or bi coloured , shorter than the head plus body. It as 46 chromosomes including a metacentric pair. You will be in for a surprise if you try to pick this mouse up by the tail as it is the common procedure as the entire tail integument slips off as id specifically greased. A constriction at its base marks the point of release. This is a protective device to foil predators. Another structural peculiarity is the large skin, lose and big enough to hold 2 bodies as can be seen when this gentle animal allows itself to be picked up by the scruff of the neck. The young are NOT spiny, but they have much larger ears than Mus cervicolor in the same habitat.

Range Description:

This species has a genuinely fragmented distribution in mainland Southeast Asia and is recorded in Myanmar, **Thailand,** South West Cambodia, central Lao PDR, and North West Vietnam (Musser and Carleton 2005).

Countries: Native:

Cambodia; Lao People's Democratic Republic; Myanmar; **Thailand;** Vietnam

Population:

Although patchily distributed, it can be locally abundant.
Population Trend: Stable

Habitat and Ecology:

It is found in dry grass and pygmy bamboo in dry dipterocarp forest (Marshall 1977). It is tolerant of habitat disturbance, although they require forest cover. It shares this habitat with *Mus cervicolor popacus* and *Rattus surifer.* In captivity they do not burrow in the sod and have instead built a nest and raised their young in a bamboo tube
Systems: Terrestrial

Threat(s):

There are currently no major threats.

Conservation Actions:

It is present in several protected areas across its range.

Genus: Niviventer

Chinese White-Bellied Rat *(Niviventer confucianus)*

Taxonomic Notes:

This species is sometimes included in *N. niviventer,* but that allocation is not supported by present evidence (see Musser and Carleton 2005). Specimens reported from Hainan are actually *N. tenaster,* and those from Taiwan are *N. culturatus,* which in the past was treated as a subspecies of *N. confucianus* (Smith and Xie 2008). There exists considerable geographic variation between populations of *N. confucianus* but the significance of this is still difficult to determine (Smith and Xie 2008). Deng et al. (2000) recognized ten subspecies, but one of these (lotipes) represents *N. tenaster* and another, culturatus, is now recognized as a valid species (Musser and Carleton 2005). Wang (2003) recognized eight subspecies from China, but the status of these needs to be reassessed, and Musser and Carleton (2005) consider them synonyms.

Description:

A small long tailed white bellied rat with 2+2 mammae. The skull has relatively smaller bullae and shorter palate than *Rattus rattus.* The short palate figured as a percentage of total skull length is an illusory consequence of the nasal bones projecting far ahead of the incisors, whereas the palate extends aft of the molars in the subgenus *Rattus* (to give a long palate). A characteristic though rare louse *Hoplopleura sicita Polyplax pricie* s also been found on this mammal. The 46 chromosomes are mostly telocentric but the smallest 6 are metacentric. In **Thailand** there are 3 species, this one, Chinese White-bellied Rat *(Niviventer confucianus),* Dark-tailed Tree Rat *(Niviventer cremoriventer) and* Chestnut White-bellied Rat *(Niviventer fulvescens)* all are very closely related. Please see the other 2 species for more information

Range Description:

This species is widely distributed in China (but not on islands off the coast), extending to Northern Myanmar, **Northwest Thailand** (summit Doi Inthanon, Chiengmai Province), and extreme Northwest Vietnam (summit Mount Fan Si Pan west of the Red River); it may also occur on summits of mountains in northern Lao PDR, although there is as yet no confirmed record (Musser and Carleton 2005). It has an elevational range of between 150-4,000 m asl (Corbet and Hill 1992).

Countries: Native:

China (Anhui, Beijing, Fujian, Gansu, Guangdong, Guangxi, Guizhou, Hebei, Henan, Hubei, Hunan, Jiangsu, Jiangxi, Jilin, Liaoning, Nei Mongol, Ningxia, Qinghai, Shaanxi, Shandong, Shanghai, Shanxi, Sichuan, Tianjin, Tibet or Xizang], Yunnan, Zhejiang); Myanmar; **Thailand;** Vietnam

Population:

Large museum collections suggest that this species is common in the highlands of Northern Myanmar and China (Musser 1981). In a survey by Wu et al. (1996) in Xishuangbanna Nature Reserve, Southern China this species was by far the most abundant in both primary and secondary forest. It is known from only a handful of specimens in **Thailand** and Vietnam.

Population Trend: Stable

Habitat and Ecology:

It occurs in montane mossy forest in **Thailand,** and is present in a wide range of habitats from forest to cultivated land in China (Corbet and Hill 1992). In China, the species is found in both primary and secondary forest, though it is about twice as common in primary forest (Wu et al. 1996).

Systems: Terrestrial

Major Threat(s):

There are no major threats to this species.

Conservation Actions:

It is present in **Doi Inthanon National Park in Thailand.** In China, it is present in Xishuangbanna (Wu et al. 2006) and Jiuzhaigou National Nature Reserves (Liu et al. 2005) and many other protected areas see CSIS (2008) for a complete list. Taxonomic revision of the subspecies is needed. In China, this species has been regionally Red Listed as Least Concern (Wang and Xie 2004).

Dark-Tailed Tree Rat ***(Niviventer cremoriventer)***

Range Description:

This widespread species is found in peninsular Thailand and Malaysia, as some offshore islands; including the Mergui Archipelago, Anambas islands, Sumatra, Nias, Billiton, Bangka, Borneo, Java, Bali, and small islands off the northern tip of Sabah (Musser and Carleton 2005). This generally lowland species reaches 1,530 m on the slopes of Mt. Kinabalu in Sabah (Musser and Carleton 2005). In Indochina this species occurs only south of the Isthmus of Kra (Musser and Carleton 2005).

Native

Indonesia; Malaysia; Singapore; **Thailand**

Habitat and Ecology

This species is frequently trapped in both the canopy and on the ground, showing no preference for either (Wells et *al*. 2004). It prefers primary forest habitats but it is relativley tolerant of disturbance and may be found in secondary wooded habitats. It is not found outside of forested areas.

Systems:

Terrestrial

Major Threats

The habitat of this species, especially in lowland areas, is being deforested for timber, firewood and conversion to agricultural land (such as oil palms. Listed as Vulnerable because of a population decline, estimated to be more than 30% over the last ten years, inferred shrinkage in distribution, and habitat destruction and degradation.

Conservation Actions:

It is present in many protected areas throughout its range.

Chestnut White-Bellied Rat *(Niviventer fulvescens)*

Description:

A small long tailed white bellied rat with 2+2 mammae. The skull has relatively smaller bullae and shorter palate than *Rattus rattus.* The short palate figured as a percentage of total skull length is an illusory consequence of the nasal bones projecting far ahead of the incisors, whereas the palate extends aft of the molars in the subgenus *Rattus* (to give a long palate). A characteristic though rare louse *Hoplopleura sicita Polyplax pricie* s also been found on this mammal. The 46 chromosomes are mostly telocentric but the smallest 6 are metacentric. In **Thailand** there are 3 species, this one, Chinese White-Bellied Rat *(Niviventer confucianus),* Dark-Tailed Tree Rat *(Niviventer cremoriventer) and* all are very closely related. Please see the other 2 species for more information

Range Description:

This species is found in Nepal, possibly Pakistan, Northern India (Arunchal Pradesh, Assam Himachal Pradesh, Manipur, Meghalaya, Sikkim, Uttarachal and West Bengal), possibly Bangladesh, Southern China (Xizang, Yunnan, Guizhou, Hunan, Guangxi, Guangdong, Hainan, Jiangxi, Fujian, Zhejiang, Anhui, Henan, Shaanxi, Gansu, Sichuan, Hong Kong, Macao), Lao PDR, Vietnam (including a number of offshore islands such as Con Son), **Thailand**, Malaysia and Indonesia (Sumatra, Java and Bali) (Molur et al. 2005 Musser and Carleton 2005; Francis 2008; Smith and Xie 2008). It is absent from Borneo and other islands on the Sunda Shelf (Musser and Carleton 2005). It occurs from around sea level to over 2,200 m asl.

Countries: Native:

China; Hong Kong; India; Indonesia; Lao People's Democratic Republic; Macao; Malaysia; Myanmar; Nepal; **Thailand.**

Population:

This species is very common in Southeast Asia. There is no information available on the population abundance of this species in South Asia (Molur et al. 2005).

Population Trend: Decreasing

Habitat and Ecology:

It is found in a wide variety of forest types throughout its range. It has been trapped in gardens and other forest edge habitats. This species is found in subtropical evergreen broadleaved forest in Dujiangyan Region of Sichuan Province, China. In Lao PDR, it is found in evergreen, pine, deciduous, and secondary forests (Marshall 1977). In South Asia, it occurs in tropical evergreen, temperate broadleaf, grass and bushy land, riverbeds in hilly forest. It has been found to occupy evergreen broadleaved forests, shrubs, rocks, also found near water (Molur et al. 2005). This species is sympatric with N. confucianus over an extensive part of its range (Allen, 1940).

Systems: Terrestrial

Major Threat(s):

There appear to be no major threats to the species other than total loss of forest cover. In South Asia it is locally threatened by habitat loss and degradation (Molur et al. 2005).

Conservation Actions:

It is presumably present in many protected areas. It is not known from any protected area in South Asia. It is listed in the Schedule V (considered as vermin) of the Indian Wildlife (Protection) Act, 1972. In South Asia, general field surveys, research into the natural history and monitoring of populations are recommended for this species (Molur et al. 2005).

Limestone Rat *(Niviventer hinpoon)*

Range Description:

This species is currently known only from the Korat Plateau in South East Thailand (Musser and Carleton 2005). It is likely to be confined to this karst area of Thailand, but is currently known only around the immediate vicinity of the type locality. Six *N. hinpoon* samples were trapped from two locations. One adult male of *N. hinpoon* was captured in the cave at Wat Tham Prathat in Lop Buri province (14°48′ 15″ N 100°49′ 30″ E), and five *N. hinpoon* samples (two adult males and three adult females) were captured from the mixed deciduous forest at the foot hill of Wat Khudkham (Dondung) in Lop Buri province (15°08′ 35″ N 100°36′ 44″ E). In these areas, limestone rat has never been reported of its occurrence before. Therefore, Wat Khud Kham (Dondung) is a new distributional range of limestone rats and becomes the largest population site in this study. During the field survey, the number of *N. hinpoon* was extremely small. After the physical measurements of 4 samples were performed on *N. hinpoon*, Only one adult female rat was instantly released to minimize biological and physiological impacts that might be arising in the rats. *N. hinpoon* was first trapped by William Neil in 1973 from the areas close to the entrance of the bat cave at Wat Tham Prapothisat, Saraburi province (14°34′ 31″ N 101°08′ 43″ E) (Lekagul and McNeely, 1988). The second location of *N. hinpoon* was discovered in 1976 at Wat Tham Prathat, Lop Buri province (14°48′ 15″ N 100°49′ 30″ E), 43 kilometers northwest of Wat Tham Prapothisat. The specimen was captured by Mr. Songsakdi Yenbutra, a former mammalogists of the TISTR, from inside the cave. The survey results revealed that the current distributional ranges of *N. hinpoon* were geographically restricted. *N. hinpoon* samples were trapped from two sites in Lop Buri province, at Wat Tham Prathat, and Wat Khudkham (Dondung) (15°08′ 35″ N 100°36′ 44″ E), the new distributional range of *N. hinpoon* (Figure 4). However, it was unsuccessful to trap *N. hinpoon* from its type locality at Wat Tham Prapothisat, Saraburi province. The results indicated very small ecological ranges of the species . The geographical distances of these three habitat sites were considerably adjacent. Therefore, it was expected that new locations of *N. hinpoon* on the limestone hill were situated along the ranges of the current and former distributional sites in Saraburi and Lop Buri provinces. At Wat Tham Prathat, Saraburi province, one adult male was captured inside the cave on the last night of 4-night trapping with 0.5% of the successful rate. Previously, 50 live traps were placed around the cave entrance and adjacent areas for three consecutive nights where none of *N. hinpoon* was found. This indicated an extremely low number of its population and restricted local distribution. At Wat Khudkham (Dondung), five rats were captured with 2.5% of the successful rate. Moreover, some pieces of its skull were found inside the bat cave near the temple. In addition, *N. hinpoon* was not found in places away from Wat Khudkham (Dondung). The results indicated that the local geographical ranges of *N. hinpoon* were very narrow and relatively small as its type locality (Lekagul and McNeely, 1988). It was likely that they lived on the foot hill near the temple. Many factors such as the availability of food may influence the home range size of the rodents (Boutin, 1990). In Panama, the resource availability was related to the home range sizes of the spiny rat (*Proechimys semispionosus*), an endemic rodent (Endries and Adler, 2005).

Nonetheless, it was uncertain about the biological and ecological factors that might influence the distribution ranges of *N. hinpoon* in the study sites. *N. hinpoon* lived in the temple areas which appeared to have a good protection from illegal poaching. At Wat Tham Prapothisat, Saraburi province, a type locality of *N. hinpoon*, total of 400 live traps were placed around the areas of its habitats but none of them was caught. It was possible that the rats might be extinct from the area or its population was too small to detect. The study was not focused on identifying the causes of vanishing. However, it was observed that there were about 80 domestic dogs and 30 cats living in the temple. Some cats returned to the wild and became free-living animals. Both dogs and cats were high potential predators, especially for the rats. Therefore, *N. hinpoon* rats in the area were relatively under the pressure of the predators and under the serious situation of local extinction.

Countries: Native:

Thailand

Population:

It is patchily distributed, being tied to karst habitats, but there is no current information on population status.

Population Trend: Unknown

Description:

The external measurements of *Niviventer hinpoon* captured from Watt Khud Kham (Dondung), **Lop Buri province,** a new distributional range, compared to two specimens in CTNRC. *Niviventer hinpoon* (n=4) External measurements Head and body length (mm.) 129.25 122.80-142.00 8.67 144 138 Tail length (mm.) 136.05 132.40-139.40* 3.51 141 131 Tail length (% of head and body) 104.98 98.17-111.03* 6.47 97.9 94.9 Ear length (mm.) 18.23 17.80-18.60 0.39 28 18 Hind foot length (mm.) 23.83 23.00-24.80 0.75 20 24 Weight (g) 65.00 59.00-65.00 7.79 54.5 60 The limestone rat *N. hinpoon* is a small rat with the spiny upper-part, dark belly and Bicolored tail. Its tail length is relatively equal to its head and body length. Belly dull duff with grey bases; back very spiny, coloured dull Buffy grey like the dullest of *R.bukit* . Tail relatively short, sharply bi-coloured for its entire length. Skull undistinqshable from that of *R.bukit.* Feet smaller than that of *R.bukit.* There are 46 chromosomes of which the 6 smallest are metacentric, 2 of the largest are subtelocentric and the rest are telocentric.

Habitat and Ecology:

This species is closely associated with limestone cliff habitat. It was collected halfway up the face of a forested limestone cliff outside the entrance to a cave, while another record is from "high above the valley floor, in scrubby vegetation at the base of vertical limestone cliffs" (Lekagul and McNeely 1988). It has a very narrow ecological range, shared with a very endemic species of bird, the limestone babbler, *Napothera crispirons calcicola* and with *Rattus neilli*

Systems: Terrestrial

Major Threat(s):

The major threat is likely to be habitat loss and degradation, particularly of the limestone forest, although it may be shown to occur in the nearby **Khao Yai National Park** where there is improved habitat protection.

Conservation Actions:

It may occur in **Khao Yai National Park**, but as far as is known it has not yet been recorded. This species is in need of further survey work to establish whether it really is endemic to, and more widespread on, the Korat Plateau.

Lang Bian White-Bellied Rat (*Niviventer langbianis)*

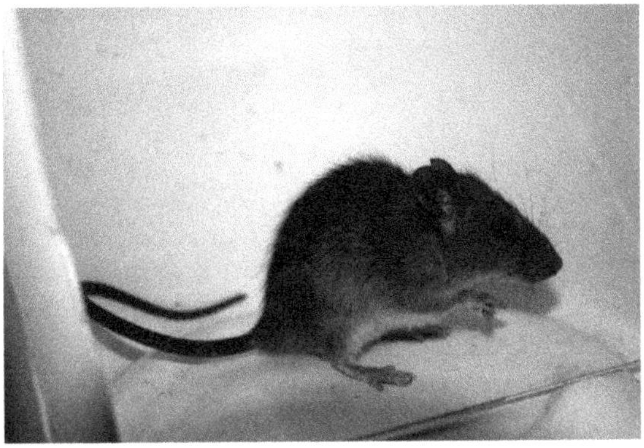

Range Description:

This species is found in Myanmar, **Thailand** (North of the Isthmus of Kra), Southwestern Cambodia, Lao PDR, Vietnam, and northeastern India (Margherita in Assam [Molur et al. 2005]). It is present in southern China (a single site in Yunnan) (Smith and Xie 2008). It has an altitudinal range of roughly 200 to 1,600 m asl.
Countries: Native:
China; India; Lao People's Democratic Republic; Myanmar; **Thailand**; Vietnam

Population:

In Southeast Asia, it is widespread but not particularly abundant at any site. There is no information available on the population abundance of this species in South Asia (Molur et al. 2005).
Population Trend: Decreasing

Habitat and Ecology:

This species is arboreal, but commonly captured on the ground. It is primarily found in tropical evergreen rainforest formations, but there are records from mixed deciduous seasonal forest (Marshall 1977). Throughout its range, this species is found with N. fulvescens in the same habitats (Musser and Carleton, 2005). It can be found in lightly disturbed, but not secondary forest (Guy Musser pers. comm.). In South Asia, it can be found in both primary and secondary habitat (Molur et al. 2005)
Systems: Terrestrial

Major Threat(s):

The habitat of this species is generally being deforested over its range for timber, firewood and conversion to agricultural land.

Conservation Actions:

It is presumably present in several protected areas. It is not known from any protected area in India. It is listed in the Schedule V (considered as vermin) of the Indian Wildlife (Protection) Act, 1972. General field surveys, research into the natural history and monitoring of populations are recommended for this species (Molur et al. 2005).

Genus: Rattus

Rice-field Rat (*Rattus argentiventer*)

Range:

Rice field Rat can be found throughout Southeast Asia consists of Indochina region, **Thailand,** Peninsular Malaysia, Indonesia, Philippine, and New Guinea as major rodent pest in rice field area.

Description:

The Rice-field Rat is a medium-sized rat with a grizzled yellow-brown and black pelage. Its belly is gray in the midline with whiter flanks. The tail is uniformly medium brown. They have chisel-like incisor. The Rice-field Rat is between 304-400mm long with a tail length of 140-200mm and a skull length of 37-41mm. The average weight of *Rattus argentiventer* is around 97 to 219 g. Female have 12 mammae. Young have an orange-coloured tuft in front of each ear.

Behaviour:

Rice-field rat lives in large groups which consist of a dominant male and high ranking female. When attacked or disturbed they will make squeals and whistles sound. *Rattus argentiventer's* main diet includes termites, insects, grasshopper, snails, seeds, nuts, rice, vegetables and fruits. They feed at night and actively moving at dusk and dawn. During daytime, they can be seen among vegetation, weeds or maturing field. Rice-field rat undergoes 3 week gestation giving birth about 5 to 10 young per litters. Parasites of Rice-field Rat include: *Schistosoma spindale*

Reproduction:

Rattus argentiventer is polyoestrous with a 4 to 5 day estrus and a continuous breeding season. Gestation lasts 3 weeks, with 3 to 8 young per litter and 1 to 12 litters a year. Rice field rats have 12 mammae. Female rats build a nest 3 to 5 days before parturition in which the young are born. They are born naked and blind but fully furred. After 15 days, their eyes open. Weaning occurs and the young leave the nest after 3 weeks. Young *Rattus argentiventer* reach sexual maturity at 3 months. All young experience maternal care and are reared with their litter mates. The male rat plays little part in the care of the young. (Ansell 1960, Nowak 1991, Hamilton 1939)

Food Habits:

Rattus argentiventer is omnivorous, with a diet that includes termites, grasshoppers, snails, insects, rice, seeds, nuts, vegetables, and fruit. (Nowak 1991, Grzmick 1990)

Economic Importance for Humans: Negative

Rattus argentiventer is often responsible for depredations on rice fields and gardens. It is the fourth most damaging rodent to rice crops. (Grzmick 1990, Nowak 1991)

Habitat:

Rice-field Rats primarily reside in cultivated areas such as rice paddies and grasslands. It is largely dependent on human rice fields and plantations. Rice field rats shelter in burrows in soil, under rocks, and in logs.

Major Threat(s):

There are no major threats to this species

Conservation Actions:

It is present in several protected areas across its range.

Polynesian Rat (*Rattus exulans*)

Description:

The Polynesian rat, or Pacific rat *(Rattus exulans)*, known to the Māori as kiore, is the third most widespread species of rat in the world behind the brown rat and black rat. The Polynesian rat originates in Southeast Asia, but like its cousins, has become well travelled – infiltrating Fiji and most Polynesian islands, including New Zealand, Easter Island and Hawaii. It shares the ability to easily adapt to many different types of environments, from grasslands to forests. Its habits are also similar, becoming closely associated with humans because of the easy access to food. As a result, it has become a major pest in almost all areas within its distribution. The Polynesian rat is similar in appearance to the black, or ship, rat (*Rattus rattus)*, but smaller. Its fur is brown-grey on the back, with the underside a light grey. The tail is uniformly dark. Females have four pairs of teats. Body plus head length: 110-130mm. Tail length: 120 to 150mm. Weight: usually 60-80g, but animals of up to 180g have been found. The animals on small islands, such as the Hawaiian islands, are smaller, weighing only 37 to 39g.he Polynesian rat is s similar in appearance to other rats, such as the black rat and the brown rat. It has large, round ears, a pointed snout, black/brown hair with a lighter belly, and comparatively small feet. It has a thin, long body, reaching up to 6 inches (15 cm) in length from the nose to the base of the tail, making it slightly smaller than other human-associative rats. Where it exists on smaller islands, it tends to be smaller still [e.g. 4.5 inches (11 cm)]. It is commonly distinguished by a dark upper edge of the hind foot near the ankle. The rest of its foot is pale

Distribution:

The Polynesian rat is widespread throughout the Pacific and Southeast Asia. It cannot swim over long distances, so is therefore considered to be a significant marker of the human migrations across the Pacific, as the Polynesians accidentally or deliberately introduced it to the islands they settled. The species has been implicated in many of the extinctions that occurred in the Pacific amongst the native birds and insects; these species had evolved in the absence of mammals and were unable to cope with the predation pressure posed by the rat. This rat also may have played a role in the complete deforestation of Easter Island by eating the nuts of the local palm tree, thus preventing regrowth of the forest. Although remains of the Polynesian rat in New Zealand were dated to over 2000 years old during the 1990s, which was much earlier than the accepted dates for Polynesian migrations to New Zealand, this finding has been overturned by later research showing the rat was introduced to both of the country's main islands around 1280 AD. They are found from Bangladesh to Vietnam, throughout the East Indies and on many Pacific islands including Easter Island, Hawaii and New Zealand. Their distribution is so wide because they have been introduced both accidentally and deliberately by people to many of these places. Native range: The Pacific rat is native to Southeast Asia.

Present in India, Indonesia, Japan, Malaysia, Myanmar/Burma, **Thailand,** Philippines, Taiwan, USA, Chile, Australia, New Zealand, Papua New Guinea, American Samoa, Cook Islands, Fiji, French Polynesia, Guam, Kiribati, Marshall Islands, Federated States of Micronesia, Nauru, New Caledonia, Niue, Northern Mariana Islands, Palau, Pitcairn, Samoa, Solomon Islands, Tokelau, Tonga, Tuvalu, Vanuata, Wallis and Futuna, and US minor outlying islands. It has dispersed with humans across the Western and central Pacific. They occur from the Asiatic mainland south to New Guinea and New Zealand, and east to the Hawaiian Islands and Easter Island. R. exulans may have been present in the Mariana Islands for at least 3500 years, introduced by the Chamorro people. There appear to be no island groups that were reached by the Polynesians that did not receive Rattus exulans. However, a number of islands within groups, particularly those uninhabited, are still free of it (Atkinson and Atkinson, 2000).

Introduced:

American Samoa (American Samoa); Brunei Darussalam; Christmas Island; Cocos (Keeling) Islands; Cook Islands; Fiji; French Polynesia; Guam; Kiribati; Marshall Islands; Nauru; New Caledonia; New Zealand; Niue; Norfolk Island; Northern Mariana Islands; Palau; Papua New Guinea; Philippines; Samoa; Singapore; Solomon Islands; Timor-Leste; Tokelau; Tonga; Tuvalu; United States (Hawaiian Is.); United States Minor Outlying Islands; Vanuatu; Wallis and Futuna

Habitat:

They are found in a wide variety of habitats, including grasslands, scrublands, scrub and various types of forest. In Southeast Asia the Polynesian rat has a very close association with people, inhabiting houses and rice fields.

Diet:

Polynesian rats are omnivorous, eating insects, centipedes, spiders, worms, snails, fruit, seeds, leaves and roots. Sometimes they eat lizards and birds or strip bark from saplings. bark, insects, earthworms, spiders, lizards, avian eggs and hatchlings. Polynesian rats have been observed to often take pieces of food back to a safe place to properly shell a seed or otherwise prepare certain foods. This not only protects them from predators, but also from rain and other rats. These "husking stations" are often found among trees, near the roots, in fissures of the trunk, and even in the top branches. In New Zealand, for instance, such stations are found under rock piles and fronds shed by nikau palms. They are considered pests in many places, as they feed on sugar cane, coconuts, cocoa and other crops. Polynesian rats are nocturnal like most rodents, and are adept climbers, often nesting in trees. In winter, when food is scarce, they commonly strip bark for consumption and satisfy themselves with plant stems.

Behaviour:

Polynesian rats are nocturnal, starting to become active around dusk or just after. They are agile climbers and sometimes nest and feed in trees. Each rat has a small home range of up to 280m diameter. Females avoid each other in the breeding season.

Reproduction:

Females are capable of having up to 13 litters per year, but 1 to 3 litters is normal for wild populations. Gestation period is 19 to 30 days. Litters are normally born in the summer. There are 2 to 5 offspring per litter, and these are weaned when 2 to 3 weeks old. They have common rat characteristics regarding reproduction: polyoestrous, , litter size affected by food and other resources (6–11 pups), weaning takes around another month at 28 days. They diverge only in that they do not breed year round, instead being restricted to spring and summer. Placental, sexual. Females are polyoestrous and ovulate spontaneously. Breeding largely determined by food availability. Litter size normally 6 - 11, gestation is 21-24 days, young weaned at about 28 days. Females can be sexually active in the season of their birth, and can have up to six litters a year (Wittenberg, R. (ed.) 2005). In New Zealand, the Norway rat has been observed breeding throughout the year (Innes, 2001).

Lifecycle stages:

Gestation 19-21 days. Weaning 2-4 weeks. Sexual maturity 8-12 months, though maturity can be achieved during the same season as birth (Atkinson and Moller 1990). In captivity: Gestation minimum 23 days. Weaning 2-3 weeks. Sexual maturity 60-70 days (Tobin 1994). Total life is estimated between 12 - 15 months. Pacific rats have been observed to mature earlier and survive better, with some adults surviving to a second breeding season, where high quality food and year-round shelter were available (Atkinson and Towns, 2001). Strong seasonal fluctuations occur in the density of kiore populations on New Zealand's northern offshore islands. Breeding is restricted to spring and summer, so densities reach a peak in autumn, then decline to low levels in spring (Campbell et al., 1984).

Major Threat(s):

There are no major threats to this species.

Conservation Actions:

It is presumably present in many protected areas.

Conservation status

Polynesian rats are not listed by the IUCN.

Notes

Polynesian rats carry various diseases that are transmittable to humans, including plague and leptospirosis.

Lesser Rice-Field Rat (*Rattus losea)*

Description:

HB: 146; T: 133; HF: 30; E:17; W 77g

A small dark, woolly or sleek furred rat, only slightly bigger than *Rattus exulans* from which it differs with larger feet, teeth and with a 3+3 mammae. It as a blunt nose, short ears, and tail which is considerably shorter than the head and body. No ectoparasites have been found on this species. The colours vary from dull, dusky or plain brown upperparts and feet with belly grey or buff grey, depending on the locality. The tall squared off cranium and short rostrum (with short nasals) are very distinctive. The parietal ridge are high atop the narrow atop the narrow cranium so that the parietal ridge is separated by a considerable vertical space about half the height of the cranium, from the rear zygomatic root. The sooty population of Trad are covered like the frequent individuals of *Rattus rattus*

Range Description:

This species occurs in South East China (Fujian, Guangdong, Guangxi, Jiangxi, Guizhou, Chongqing, East Sichuan, South Shaanxi,, Hainan) and Taiwan, extending into central and Southern Lao PDR, South and South West Cambodia, **Thailand**, and the Pescadores islands (Dang et al. 1994; Marshall 1977; Musser and Carleton 2005; Robinson et al. 1995; Smith et al. 2008).

Wang (2003) recognized three subspecies in China: *R. l. exiguus,* Howell, 1927, found in continental China; *R. l. losea,* (Swinhoe, 1871), occurring in Taiwan; and *R. l. sakeratensis,* Gyldenstolpe, 1917, from Hainan island.

Countries: Native:

Cambodia; China; Lao People's Democratic Republic; Malaysia; Taiwan, Province of China; **Thailand;** Vietnam

Habitat and Ecology:

It inhabits grass, scrub, mangroves, cultivated fields and other human modified areas from sea level up to about 1,000 m. Studies in Vietnam suggest that population sizes may fluctuate in relation to the availability of field crops (Smith et al. 2008). Males performed more locomotion behaviour in frequencies and aggression in time and frequencies when they encountered unfamiliar females than that when they encountered familiar ones. However, amicable behaviour of both females and males were not significant. The factors of stimulus rats were not significant. All rats involved in encounters performed no aggression to familiar opposite sex and no sexual behaviour to unfamiliar ones

Systems: Terrestrial

Major Threat(s):

There are no major threats to the species.

Conservation Actions:

It is present in several protected areas across its range.

Himalayan Field Rat (*Rattus nitidus*)

Description:

HB: 177; T:168; HF: 37; E: 21; W: 122g

A medium sized, brown greyish rat with a dark tail, dull grey underparts and pearly white feet.; mammae 3+3, sort short fur, long rostrum, and broad, flat cranial roof. There are 8 pairs of small metacentric chromosomes. No ectoparasites have been found.

Range Description:

This very widespread Asian species, ranges from North-Eastern South Asia (Bhutan, India and Nepal [Molur et al. 2005]; probably present in Bangladesh), much of Central, Southern and Eastern China (including the island of Hainan) (Smith and Xie 2008), ranging into Southeast Asia, where it has been recorded from Myanmar, **Thailand,** Lao PDR, Vietnam and probably occurs more widely. It has been introduced to Benguet Province, Luzon in the Philippines; Palau; central Sulawesi (Indonesia); Seram (Indonesia); and the Vogelkop Peninsula of Papua Province (Indonesia) (Corbet and Hill 1992; Flannery 1995; Musser and Carleton 2005). In South Asia the species occurs from 686 to 2,740 m asl (Molur et al. 2005). It is presumably found as low as sea level in parts of its range.

Countries: Native:

Bhutan; China; India; Myanmar; Nepal; **Thailand**; Vietnam

Introduced:

Indonesia; Palau; Philippines

Habitat and Ecology:

This very adaptable species occurs in various forest types, cropland and human settlements, and lives exclusively inside houses of hill tribe villages. It is a really good climber. They have litters of around 6.

Systems: Terrestrial

Major Threat(s):

There are no major threats to this species.

Conservation Actions:

It is present in many protected areas. It is listed in the Schedule V (considered as vermin) of the Indian Wildlife (Protection) Act, 1972. Additional taxonomic studies are needed for this species.

Sikkim Rat (*Rattus sikkimensis)*

Description:

HB:173; T: 209, HF: 33.5; E: 22.5; W: 129g

Dorsal colour like that of the roof rat but overlaid with long black hairs down the middle of the back, giving it a darker affect; ventral colour creamy white sharply demarcated along the flanks and never with a dark chest streak. Feet and tail dark. the fur is long, thick and lustrous. The paw pads are large, projected and laminated for climbing. Mammae 3+3 with 2 postaxial pairs at least a centimetres apart,(longitudinally). The molars are usually black rimmed. Compared to the slender nosed *Rattus rattus* the rostrum is wider and the sides of the cranium are not vertical, they slant outwards from the parietal ridges.

Range Description:

This widespread species is known from southern China, Vietnam, Lao PDR, Cambodia, **Thailand, C**entral and Northern Myanmar, North Eastern India, Bhutan, and Eastern Nepal (Musser and Carleton 2005). In China, it is known from the provinces of Tibet, Guizhou, Hainan (Guangdong), and Hong Kong (Smith and Xie 2008). It could possibly be present in the remnant Northern and North-Eastern forests of Bangladesh (S. Molur pers. comm.). **It is not known from mainland peninsular Thailand South of the Isthmus of Kra (10E, 30N), but it occurs on four islands off the coast (Koh Tau, Koh Phangan, Koh Samui, and Koh Kra) (Musser and Carleton 2005). It is also found on the Andaman Islands (islands of North Andaman, South Andaman, and Little Andaman) and Car Nicobar (Musser and Carleton 2005).** It roughly occurs from sea level (Aplin and Frost 2006) up to around 2,000 m asl (Sarker et al. 2005).

Countries: Native:

Bhutan; Cambodia; China (Guizhou, Hainan, Tibet [or Xizang]); Hong Kong; India (Andaman Is., Arunachal Pradesh, Assam, Manipur, Meghalaya, Nagaland, Nicobar Is., Sikkim, West Bengal); Lao People's Democratic Republic; Myanmar; Nepal; **Thailand;** Vietnam

Population:

It can be moderately abundant in forested areas.

Population Trend: Stable

Habitat and Ecology:

This is a highly arboreal species that is present in a variety of forest and forest edge habitats. It can occur in secondary and modified habitats with wooded areas. It is often found in agriculture lands, scrubland, and around houses (Smith and Xie 2008).

Systems: Terrestrial

Major Threat(s):

There are no major threats to this species.

Conservation Actions:

This species is listed in the Schedule V (considered vermin) of the Indian Wildlife (Protection) Act, 1972 (Sarker et al. 2005). In China, it occurs in Fanjingshan (Guizhou) and Jianfengling Nature Reserves (CSIS 2008) and it is presumably present in many other protected areas. Further studies into the taxonomy of this species, particularly a review of genetic material, is needed. In China, it has been regionally Red Listed as Least Concern (Wang and Xie 2004).

Tanezumi Rat (*Rattus tanezumi*)

Range Description:

Both the native and introduced geographic distribution of this very widespread species remains somewhat unclear. We follow Musser and Carleton (2005) who consider the species native distribution to range from Eastern Afghanistan (Niethammer and Martens 1975) through Central and Southern Nepal (below about 2,000 m asl), Bhutan, Northern India, Northern Bangladesh and North-Eastern India (distributional relationships between *R. rattus and R. tanezumi* along the Northern portion of the Indian subcontinent still requires resolution) into Southern and Central China (including Hainan Island), the Korean Peninsula, and mainland Southeast Asia (including offshore islands) south to the Isthmus of Kra, being also probably native to Mergui Archipelago. Musser and Carleton (2005) report that it is presently unclear whether the species is native or introduced to Taiwan and Japan. But is most likely introduced to Peninsular Malaysia and islands on the Sunda Shelf and nearby archipelagos, including the Mentawai Islands (Musser and Califia 1982; Musser and Newcomb 1983). It has been introduced to Little Nicobar and Great Nicobar, in the Andaman and Nicobar Islands of India. It has been widely introduced throughout the Philippines (Heaney et al. 1998; Musser and Carleton 2005; Larry Heaney pers. comm.). It appears to have been widely introduced throughout insular Southeast Asia, including the Greater and Lesser Sunda Islands and the Molucca Islands, to Western New Guinea (range map very provisional for this island) (Flannery 1995), and from here into Micronesia to the islands of Eniwetok and Fiji (Musser and Carleton 2005).

Countries: Native:

Afghanistan; Bangladesh; Bhutan; Cambodia; China; India; Japan; Korea, Democratic People's Republic of; Korea, Republic of; Lao People's Democratic Republic; Malaysia; Nepal; Taiwan, Province of China; **Thailand;** Vietnam

Introduced:

Fiji; Indonesia; Papua New Guinea; Philippines

Description:

The fur on the dorsal side is olive-brown, and the ventral side generally lighter. The tail is very dark grey and nearly naked. The ears are large and the eyes jet black. Juveniles have a relatively larger head and smaller body. Head-Body Length : Up to 22cm Tail Length : Up to 22cm Weight : Up to 200 grams. *Rattus tanezumi* is the oldest name for the 2n = 42 group of Asian house rats that is distinguished from the 2n = 38/40 *Rattus rattus* not only by chromosomal characters but also morphological and biochemical traits (Wilson, 2005) Many endoparasites infecting the different viscera of *R. tanezumi* belong to taxonomic groups Nematoda (roundworms), Cestoda (tapeworms), and Trematoda (flukes). The more important ones are those that are transmissible to humans. In contrast, there have been few reports on the ectoparasites that infest these rats.

These are the mites, ticks, and fleas, some of which may serve as vectors of microbial infections to humans and domestic animals. *Rattus tanezumi* is an alternative meat source for the rural folk. Rat meat is a favourite accompaniment to alcoholic beverages during drinking sprees. Unknowingly, some zoonotic infections can be transmitted to humans through improperly cooked rat meat and viscera, accidental wound contamination with rat urine, or rat bites during handling and precooking preparations

This species is regularly misidentify with *Rattus rattus* and treated as *Rattus rattus* species group and often confused in field with *Rattus argentiventer, Rattus andamanensis* and *Rattus tiomanicus* .

he species is closely related to the European House Rat *Rattus rattus,* but recent studies suggest it is a separate species. It probably occurs throughout Southeast Asia.

Population:

It is a common to abundant species.

Population Trend: Increasing

Habitat and Ecology:

This highly adaptable species is commonly found in and around villages and agricultural areas. In the Philippines it is common in disturbed lowland and montane forest up to 1,800 m (Danielsen et al. 1994; Heaney et al. 1989. A supremely adaptable species, is to be found in many man-made habitats including agricultural and wholly urban areas. It is omnivorous, feeding on all manner of farmyard waste and food scraps. It is a fast runner, can climb well and can jump up to 50 cm.

Systems: Terrestrial

Major Threat(s):

There are no major threats to this species.

Conservation Actions:

It is presumably present in numerous protected areas. No direct conservation measures are needed for this common and adaptable species. There is a need to further study the taxonomic relationships between populations of Rattus currently allocated to this species.

Malayan Field Rat (*Rattus tiomanicus)*

Description:

Mainly nocturnal. Occurs in a variety of habitats including coastal forests, especially mangroves, secondary forests and grassland. It as also adapted to plantations such as rubber and sometimes oil palm. The upper parts are brown, and the undersides light grey to white, the latter being an identification feature. the fur is soft, and the tail is dark in colour. The feet are broad and the upper side is devoid of hair.

Range Description:

This species is found on peninsular **Thailand (south of the Isthmus of Kra),** the Malaysian Peninsula, and Indonesia (Sumatra, Borneo). In the Philippines, it occurs in the Palawan Faunal Region only being recorded from Arena, Bancalan, Busuanga, Calauit, and Palawan (Musser and Carleton 2005).
It is also found on many smaller islands, including Tioman island, Enggano island (South-West of Sumatra) and the Maratua Archipelago (east of Borneo), as well as the smaller islands of Pandjang, Rabu Rabu, Sangka Laki, Bilang Bilangan, Eraban, and Miang Besar off the east coast of Borneo (Musser and Califia 1982). It is also found on Busuanga (Musser 1977).
Countries: Native:
Brunei Darussalam; Indonesia; Malaysia; Philippines; **Thailand**

Population:

This species is common and widespread.
Population Trend: Increasing

Habitat and Ecology:

The species is apparently found only in a variety of habitats in the lowland areas: from secondary forest, agricultural areas, plantations, gardens, scrub, to grassland (Sanborn 1952; Payne et al. 1985), while Esselstyn et al. (2004) found it in grassland-forest mosaic, selectively logged forest, rice fields, mossy forest, and the transition zone between mossy and montane forest.
Systems: Terrestrial

Major Threat(s):

There are no major threats to the species.

Conservation Actions:

It is present in several protected areas across the range.

Black Rat (*Rattus rattus)*

Description:

HB: 182; T: 188; HF: 33; E:23; W: 139gm average. A slender and graceful mammal, with a pointed snout, large waving ears, tail about a long as its head and body, basically 2 + 3 mammae (with a tendency for twinning of the second pair from the front., and laminated plantar pads. Coloration of most **Thailand** examples are light brown, reddish brown, or buff brown on the back with pure white under parts and often with a little dark streak on the chest.

Range:

Native to Asia and spread around the world by man. In **Thailand** they are found in all provinces and all the islands and are the most abundant mammal in the country. Very similar in appearance to the common rat but with larger eyes and ears and longer, thinner, unicoloured tail. When in hand the ears are almost hairless compared to the common rats furry ears. Smaller than the common rat with head/body length: about 240mm and tail up to 260mm. Weight: average approx 150 to 200 grammas with the males generally larger than to females. Pelage is grey brown above and pale grey beneath but some individuals are completely black which gives rise to the name of the black rat

General Ecology:

Ship rats are mainly nocturnal and rarely seen during the day. They live in large groups each dominated by a male and unlike the common rates do not live out of doors but favours buildings with cavity walls, paneling and false ceilings. It climbs very well and can sometimes be seen running along pipes and even wires. It can swim but unlike the common rats does not enter water voluntarily . Ship rats rates are Omnivores but are very often vegetarian and are very partial to fruit, most agricultural crops and in particular cereals. They live in very dense populations but vary greatly with the environment. Females sexually mature at about 90g and have an average litter size of 3 to 5 babies. Most breeding is done between March to November. Ship Rats were once widespread in all Parts of the United Kingdom where they spread from India to Egypt in the 4th Century and then into the British Isles during the 3rd century. Replaced by the common rat by the beginning of the 18th century and were confined to major ports, a few inland towns and small islands by the 1956 Now very rare indeed and is now probably one of the rarest mammals in Britain but in the tropics is very well established

Mortality is high among young rats who are fair game to most predatory animals.

Ship rats will forever by associated in peoples minds with the Black Death (bubonic plague) which caused catastrophic outbreaks of the disease in the Middle Ages

Conservation:

Ship rats are offered no protected by any laws but in theory are one of the most deserving cases for conservation action. It is indeed our rarest rodent, rapidly declining and now reduced to a few isolated populations. Nevertheless, legislation for an alien species appears inappropriate and attempts to conserve it would be widely opposed. In Lundy where a small population exists and do little harm, its English Nature who are seeking to remove them so that Puffins can be re-introduced. Its strange that with the Puffin being one of Britain's commonest sea birds and the Ship Rat being probably our most endangered British mammal that this should be the case.

Brown Rat *(Rattus Norvegicus)*

Description:

Large size: pointed muzzle: Long scaly tail
Head/body length: about 280mm, tail 80 to 100 of body
Weight: average approx 500grammes

The fur is coarse and usually brown or dark grey, while the underparts are lighter grey or brown. The Brown Rat is a rather large true murid and can weigh twice as much as a Black Rat and many times more than a House Mouse. The length is commonly in the range of 20 to 25 cm (8 to 10 in), with the tail a further 18 to 25 cm (7 to 10 in), thus being roughly the same length as the body. Adult body weight averages 350 g (12 oz) in males and about 250 g (9 oz) in females. Exceptionally large individuals can reportedly reach 900 to 1,000 g (32 to 35 oz) but are not expected outside of domestic specimens. Stories of rats attaining sizes as big as cats are exaggerations, or misidentifications of other rodents, such as the coypu and muskrat. In fact it is common for breeding wild Brown Rats to weigh (sometimes considerably) less than 300 g (11 oz).

Brown rats have acute hearing, are sensitive to ultrasound, and possess a very highly developed olfactory sense. Their average heart rate is 300 to 400 beats per minute, with a respiratory rate of around 100 per minute. The vision of a pigmented rat is poor, around 20/600, while a non-pigmented (albino) with no melanin in its eyes has both around 20/1200 vision and a terrible scattering of light within its vision. Brown rats are dichromates which perceive colors rather like a human with red-green colorblindness, and their colour saturation may be quite faint. Their blue perception, however, also has UV receptors, allowing them to see ultraviolet lights that some species cannot.

General Ecology:

Common rats are mainly nocturnal and rarely seen during the day. The have a nest usually below ground but nests can be amongst sacks or straw bales. Common rates are opportunist feeders and prefer starch rich or protein rich foods but they will eat almost anything that is left lying around Common Rates live in very dense populations which compose of mostly young individuals. The females in these dense populations are pregnant throughout the year with litter sizes of between 6 and 11 babies. Common Rats are widespread in all Parts of the United Kingdom where they spread from eastern USSR into the British Isles during the middle of the18th century where they mostly replaced the Black Rat (Rattus rattus) Mortality is high among young rats who are fair game to most predatory animals. Larger rats may deter weasels and owls due to their aggressive behaviour. The maximum life expectancy of a Common Rat is about 1 year though very few survive so long. Common rats in terms of success are in a league of there own due to the fact that they have developed the ability to live in such close association with man.

Reproduction and life cycle:

The brown rat can breed throughout the year if conditions are suitable, with a female producing up to five litters a year. The gestation period is only 21 days, and litters can number up to 14, although seven is common. They reach sexual maturity in about five weeks. The maximum life span is up to three years, although most barely manage one. A yearly mortality rate of 95% is estimated, with predators and interspecies conflict as major causes. When lactating, female rats display a 24-hour rhythm of maternal behavior, and will usually spend more time attending to smaller litters than large ones.

Brown rats live in large, hierarchical groups, either in burrows or subsurface places, such as sewers and cellars. When food is in short supply, the rats lower in social order are the first to die. If a large fraction of a rat population is exterminated, the remaining rats will increase their reproductive rate, and quickly restore the old population level.

Communication:

Brown rats are capable of producing ultrasonic vocalizations. As pups, young rats use different types of ultrasonic cries to elicit and direct maternal search behavior, as well as to regulate their mother's movements in the nest. Although pups will produce ultrasounds around any other rats at 7 days old, by 14 days old they significantly reduce ultrasound production around male rats as a defensive response. Adult rats will emit ultrasonic vocalizations in response to predators or perceived danger; the frequency and duration of such cries depends on the sex and reproductive status of the rat. The female rat will also emit ultrasonic vocalizations during mating.

Chirping:

Rats may also emit short, high frequency, ultrasonic, socially induced vocalization during rough and tumble play, before receiving morphine, or mating, and when tickled. The vocalization, described as a distinct "chirping", has been likened to laughter, and is interpreted as an expectation of something rewarding. Like most rat vocalizations, the chirping is too high in pitch for humans to hear without special equipment. Bat detectors are often used by pet owners for this purpose. In clinical studies, the chirping is associated with positive emotional feelings, and social bonding occurs with the tickler, resulting in the rats becoming conditioned to seek the tickling. However, as the rats age, the tendency to chirp appears to decline. Rat chirp also can be used for mosquito control. Other ultrasonic vocalizations, including a lower-frequency 'boom' or 'whoom' noise can be produced by bucks in a calm state, when grooming or settling down to sleep.

Audible communication:

Brown rats also produce communicative noises capable of being heard by humans. The most commonly heard in domestic rats is bruxing, or teeth-grinding, which is most usually triggered by happiness, but can also be 'self-comforting' in stressful situations, such as a visit to the vet. The noise is best described as either a quick clicking or 'burring' sound, varying from animal to animal. In addition, they commonly squeak along a range of tones from high, abrupt pain squeaks to soft, persistent 'singing' sounds during confrontations.

Social behavior:

Rats commonly groom each other and sleep together. As with dogs, rats create a social hierarchy, and each rat has its own place in the pack. Rats are said to establish an order of hierarchy, so one rat will be dominant over another one. Groups of rats tend to "play fight", which can involve any combination of jumping, chasing, tumbling, and "boxing". Play fighting involves rats going for each other's necks, while serious fighting involves strikes at the others' back ends. If living space become limited, rats may turn to aggressive behavior, which may result in the death of some animals, reducing the burden over the living space.

Rats like most mammals also form family groups, a mother and her young. This applies to both groups of males and females. However, rats are territorial animals, meaning that they usually act aggressively or scared of strange rats. Rats will fluff up their hair, hiss, squeal, and move their tails around when defending their territory. Rats will chase each other, groom each other, sleep in group nests, wrestle with each other, have dominance squabbles, communicate, and play in various other ways with each other. Huddling is an additional important part of rat socialization. Huddling is often supposed to have a heat-conserving function. Nestling rats especially depend on heat from their mother, since they cannot regulate their own temperature. Huddling is an extreme form of herding. Other forms of interaction include, crawling under, which is literally the act of crawling underneath one another, walking over, also explained in the name, then there is allo-grooming, so-called to distinguish it from self-grooming. And lastly there is another type of contact called nosing, where a rat gently pushes with its nose at another rat near the neck.

Burrowing:

Rats are known to burrow extensively, both in the wild and in captivity, if given access to a suitable substrate. Rats generally begin a new burrow adjacent to an object or structure, as this provides a sturdy "roof" for the section of the burrow nearest to the ground's surface. Burrows usually develop to eventually include multiple levels of tunnels, as well as a secondary entrance.Older male rats will generally not burrow, while young males and females will burrow vigorously.

Burrows provide rats with shelter and food storage, as well as safe, thermo-regulated nest sites. Rats use their burrows to escape from perceived threats in the surrounding environment; for example, rats will retreat to their burrows following a sudden, loud noise or while fleeing an intruder. Burrowing can therefore be described as a "pre-encounter defensive behavior", as opposed to a "post-encounter defensive behavior", such as flight, freezing, or avoidance of a threatening stimulus.

Distribution and Habitat:

Likely originating from the plains of Asia, Northern China and Mongolia, the brown rat spread to other parts of the world sometime in the Middle Ages The question of when brown rats became commensal with humans remains unsettled, but as a species, they have spread and established themselves along routes of human migration and now live almost everywhere humans are, including being widespread throughout **Thailand.**

The brown rat may have been present in Europe as early as 1553, a conclusion drawn from an illustration and description by Swiss naturalist Conrad Gesner in his book Historiae animalium, published 1551–1558. Though Gesner's description could apply to the black rat, his mention of a large percentage of albino specimens—not uncommon among wild populations of brown rats—adds credibility to this conclusion. Reliable reports dating to the 18th century document the presence of the brown rat in Ireland in 1722, England in 1730, France in 1735, Germany in 1750, and Spain in 1800,becoming widespread during the Industrial Revolution.It did not reach North America until around 1750–1755.

As it spread from Asia, the brown rat generally displaced the black rat in areas where humans lived. In addition to being larger and more aggressive, the change from wooden structures and thatched roofs to bricked and tiled buildings favored the burrowing brown rats over the arboreal black rats. In addition, brown rats eat a wider variety of foods, and are more resistant to weather extremes.

In the absence of humans, brown rats prefer damp environments, such as river banks. However, the great majority are now linked to man-made environments, such as sewage systems.

It is often said that there are as many rats in cities as people, but this varies from area to area depending on climate, living conditions, etc. Brown rats in cities tend not to wander extensively, often staying within 20 m (66 ft) of their nest if a suitable concentrated food supply is available, but they will range more widely where food availability is lower. There is great debate over the size of the population of rats in New York City, with estimates from almost 100 million rats to as few as 250,000. Experts suggest New York is a particularly attractive place for rats because of its aging infrastructure, high moisture, and high poverty rates. In addition to sewers, rats are very comfortable living in alleyways and residential buildings, as there is usually a large and continuous food source in those areas.

In the United Kingdom, some figures show the rat population has been rising, with estimations that 81 million rats reside in the UK.Those figures would mean there are 1.3 rats per person in the country. High rat populations in the UK are often attributed to the mild climate, which allow them higher survival rates during the winter months.

The only brown rat-free zones in the world are the Arctic, the Antarctic, some especially isolated islands, such as Iceland, the province of Alberta in Canada, and certain conservation areas in New Zealand.

Antarctica is almost completely covered by ice and has no permanent human inhabitants, making it uninhabitable by rats. The Arctic has extremely cold winters that rats cannot survive outdoors, and the human population density is extremely low, making it difficult for rats to travel from one habitation to another. When the occasional rat infestation is noticed and eliminated, the rats are unable to re-infest it from an adjacent one. Isolated islands are also able to eliminate rat populations because of low human population density and geographic distance from other rat populations.

Diseases:

Similar to other rodents, brown rats may carry a number of pathogens, which can result in disease, including Weil's disease, rat bite fever, cryptosporidiosis, viral hemorrhagic fever, Q fever and hantavirus pulmonary syndrome. In the United Kingdom, brown rats are an important reservoir for *Coxiella burnetii,* the bacterium that causes Q fever, with seroprevalence for the bacteria found to be as high as 53% in some wild populations.

This species can also serve as a reservoir for *Toxoplasma gondii,* the parasite that causes toxoplasmosis, though the disease usually spreads from rats to humans when domestic cats feed on infected brown rats. The parasite has a long history with the brown rat, and there are indications that the parasite has evolved to alter an infected rat's perception to cat predation, making it more susceptible to predation and increasing the likelihood of transmission. Surveys and specimens of brown rat populations throughout the world have shown this species is often associated with outbreaks of trichinosis,but the extent to which the brown rat is responsible in transmitting Trichinella larvae to humans and other synanthropic animals is at least somewhat debatable. *Trichinella pseudospiralis,* a parasite previously not considered to be a potential pathogen in humans or domestic animals, has been found to be pathogenic in humans and carried by brown rats.

Brown rats are sometimes mistakenly thought to be a major reservoir of bubonic plague, a possible cause of the Black Death. However, the bacterium responsible, *Yersinia pestis,* is commonly endemic in only a few rodent species and is usually transmitted zoonotically by rat fleas—common carrier rodents today include ground squirrels and wood rats. However, brown rats may suffer from plague, as can many non rodent species, including dogs, cats, and humans. The original carrier for the plague-infected fleas thought to be responsible for the Black Death was the black rat, and it has been hypothesized that the displacement of black rats by brown rats led to the decline of bubonic plague. This theory has, however, been deprecated, as the dates of these displacements do not match the increases and decreases in plague outbreaks.

In captivity:

Uses in science:

Selective breeding of albino brown rats rescued from being killed in a now-outlawed sport called rat baiting has produced the albino laboratory rat. Like mice, these rats are frequently subjects of medical, psychological and other biological experiments, and constitute an important model organism. This is because they grow quickly to sexual maturity and are easy to keep and to breed in captivity. When modern biologists refer to "rats", they almost always mean *Rattus norvegicus.*

As pets:

The brown rat is kept as a pet in many parts of the world. Australia, the United Kingdom, and the United States are just a few of the countries that have formed fancy rat associations similar in nature to the American Kennel Club, establishing standards, orchestrating events, and promoting responsible pet ownership.

The many different types of domesticated brown rats include variations in coat patterns, as well as the style of the coat, such as Hairless or Rex, and more recently developed variations in body size and structure, including dwarf and tailless fancy rats.

Conservation:

Common Rats are offered no protected by any laws and are listed on the Pest Act of 1949 which requires notification of local authorities of substantial numbers And due to the contamination of foodstuffs have become a major problem throughout the world. They are also a vector of some human diseases particularly leptospirosis among rice workers in Spain and Italy.

On the credit side has served man for decades as a laboratorial animal

Genus: Sundamys:

Müller's Giant Sunda Rat (*Sundamys muelleri)*

Description:

HB: 250; T: 269; HF: 50; E: 24; W: 379g

A giant shaggy rat which is blackish brown above, buff underneath, not sharply demarcated along the flanks. feet larger than 45mm in length and a black tail. Mammae 2+2. The skull is like Rattus but has larger teeth, smaller bullae, and no post dental palate

Range Description:

This species is known from south western peninsular Myanmar, **Peninsular Thailand,** Malay Peninsula, Sumatra, Borneo, and Palawan, as well as on the smaller islands of Siantan (Anamba Island), many of the Riau Islands, Tuangku and Bangkuru (Banjak Islands), Mansalar (Western Sumatra), Pinie, Tanahmasa and Tanahbala (Batu Islands), Banka, Bunguan and Serasan (Natuna Islands), Karimata Island (south western Borneo), Sebuku (South Eastern Borneo), Balembangan and Banggi (Northern Borneo), and Balabac, Culion, and Busuanga (Palawan Faunal Region) (Musser and Carleton 2005). It does not occur on Java.

Countries: Native:

Brunei Darussalam; Indonesia; Malaysia; Myanmar; Philippines; **Thailand**

Population:

This species is presumably common, as there are large series from several collecting localities, but population numbers are likely to be declining.

Population Trend: Decreasing

Habitat and Ecology:

This species is primary forest and secondary growth (Sanborn 1952; Md Nor 2001). It is generally a lowland species, reaching 1,650 m on slopes of Mt. Kinabalu in Sabah (Md Nor 2001). It is often found near streams and prefers moist habitats (Lim 1970; Esselstyn et al. 2004). It is primarily terrestrial, though it is found on branches of trees (Lim 1970). The diet consists of insects, fruit, leaves and shoots, other vegetable matter, crabs, and land snails (Medway 1969; Lim 1966, 1970).

Systems: Terrestrial

Major Threat(s):

It is likely to be affected by ongoing habitat loss across much of its range. It is probably hunted in non-Muslim parts of its range.

Conservation Actions:

It is present in several protected areas across its range.

Genus: Vandeleuria:

Asiatic Long-Tailed Climbing Mouse *(Vandeleuria oleracea)*

Description:

HB: 68; T: 105; HF: 17; E:13; W: 10g

Adapted to a life high in the canopy, the long-tailed climbing mouse *(Vandeleuria nilagirica)* has a long prehensile tail that acts as a balancing aid, and an opposable digit on the hands and feet that is used to grasp to branches whilst climbing. In common with other Vandeleuria climbing mice, the fur is soft and silky, and highly variable in colour, ranging from pale, dull brown to dark, reddish brown. The long-tailed climbing mouse is very similar in appearance to the Asiatic long-tailed climbing mouse *(V. oleracea),* and indeed was once considered a subspecies of its more widespread congener, but is distinguished by a substantially longer tail and greyer underparts

Range Description:

This species is widespread in South Asia, Southern China and mainland Southeast Asia. In South Asia, this species has a wide distribution in Bangladesh, Bhutan, India, Nepal and Sri Lanka, occurring in elevation from 200 to 1,500 m asl (Molur et al. 2005). In China, it is limited to Western and Southern Yunnan (Smith and Xie 2008). In Southeast Asia, the species ranges from Myanmar in the west into **Thailand** (north of the Isthmus of Kra), south-western Cambodia, possibly Lao PDR, and Northern and Southern Vietnam (Musser and Carleton, 2005).

Countries: Native:

Bangladesh; Bhutan; Cambodia; China; India; Myanmar; Nepal; Sri Lanka; **Thailand;** Vietnam

Population:

It is widespread, but never very abundant.

Population Trend: Stable

Biology:

A highly arboreal and active species, at night the long-tailed climbing mouse runs along branches and twigs, even climbing vertical shoots, to forage for a variety of fruits and buds Whilst feeding, this diminutive species may loosely wind its long tail around a branch as a balancing aid, or hang from it to reach food sources on fragile twigs. During the day the long-tailed climbing mouse seeks shelter in tree holes or nests high up in the canopy. Pairs nest between October and February and will descend to the ground to collect grass and leaves to construct a shallow, oval shaped nest in the fork of a tree As is typical of many small mice species, three or four young are born after a gestation period of 20 to 30 days, with a life expectancy of little more than one year.

Threats:

Occupying an area no more than 500 square kilometres, the long-tailed climbing mouse is threatened by the loss and degradation of its habitat Changes in forestry management, with the increasing use of pesticides and native tree canopy species being replaced with exotics, has resulted in forest fragmentation and gaps in the canopy. Coffee plantations have also been felled and replaced with ginger cultivation. Consequently, many populations of the long-tailed climbing mouse have been left isolated, greatly increasing the species extinction risk.

Conservation Actions:

It is present in several protected areas across its Southeast Asian range. In South Asia, it has been recorded from the Indian protected areas of Eturnagaram Wildlife Sanctuary, Gundla Brahmeshwaram Metta Wildlife Sanctuary, Nagarjunasagar Srisailam Tiger Reserve, Pocharam Wildlife Sanctuary in Andhra Pradesh; Banneraghatta National Park in Karnataka; Eravikulam National Park in Kerala and Sanjay Gandhi National Park in Maharshtra. In Sri Lanka it is known from Knuckles Forest Reserve in Central Province (Molur et al. 2005). It is listed in the Schedule V (considered as vermin) of the Indian Wildlife (Protection) Act, 1972. There is a need for further research to elucidate the taxonomic status of this species complex.

Order: Lagomorpha (lagomorphs)

The lagomorphs comprise two families, Leporidae (hares and rabbits), and Ochotonidae (pikas). Though they can resemble rodents, and were classified as a superfamily in that order until the early 20th century, they have since been considered a separate order. They differ from rodents in a number of physical characteristics, such as having four incisors in the upper jaw rather than two.

Family: Leporidae (rabbits, hares)

Genus: Lepus:

Burmese Hare ***(Lepus peguensis)***

Description:

HB: 440-500;T: 65-80; HF:95-105;

The upper parts are grizzled, with the individual hairs white or greyish at the base,, followed by a dark brown band, ten tipped with fulvous. The under parts are pure white, except for the upper breast, which is light Rufus; the chin is white. There is a poorly defined wide light band from the nostril through the eye, continuing to the base of the ear in some individuals. The ears are brown with lighter tips. The back of the neck is Rufus, as are the front legs and feet. The hind legs are fulvous. The tail is white, with a thick dark brown band down the dorsal surface. In the skull, the palate is short, not extending beyond the premolars; the postorbital processes are broad and triangular, forming a nearly complete bony orbit. the nasal cavity is greatly enlarged; and the auditory bullae are small. The dental formula is 2/1 0/0 3/2 3/3x2 =28. The anterior surfaces of the first upper incisors are distinctly grooved; the grooves are enlarged and deepened, rectangular in shape, and their internal walls are fluted and channelled in a direction parallel with the main groove; the entire groove may be filled with a soft cement in some individuals. The cheek teeth are high, crowned and rootless; in chewing, the lower jaws are moved from side to side (usually backwards and forwards in rodents). The upper tooth rows are farther apart than the lower tooth row, so that only one side occludes at a time.

Range Description:

This species occurs in central and Southern Myanmar from the Chindwin River valley east through **Thailand,** Cambodia, Southern Lao PDR, Southern Vietnam, and south into the upper Malay Peninsula (Myanmar, **Thailand**) (Hoffmann and Smith 2005). This distribution may include the northern and central regions of Lao PDR, as signs of its presence in the wild and fresh kills in markets have been identified, but only to the genus level (Duckworth et al. 1994; Duckworth 1996). Distribution is enigmatic in the north and east, with few confirmed records. **Thai** distribution would suggest that the species has spread with human-related forest loss in the northern highlands of Lao PDR, particularly one might surmise along the lowland valleys. The mapped range in Myanmar is largely speculative, further work is likely to show that the species is more restricted within this country. **A collection expedition in Thailand (pre-1950's) recorded this species as occurring at 4,300 feet on Doi Ang Ka (modern name Doi Inthanon) (Allen and Coolidge 1940).** Further work is needed to determine the upper limit of this species.

Countries: Native:

Cambodia; Lao People's Democratic Republic; Myanmar; **Thailand;** Vietnam

Population:

This species is locally abundant.

Population Trend: Stable

Habitat and Ecology:

This species occupies mostly in low altitude dipterocarp forest, and plain non-forest habitats with numbers especially high in the grass and shrub vegetation of seasonally exposed large river channels. It can be found in rain/flood-fed low intensity rice fields, but avoids irrigated multi-crop rice fields which cover much of **Thailand** (Duckworth pers. comm.). Species appears to be absent from Nakai plateau, Lao PDR, despite what appears to be suitable habitat (possibly due to high altitude) (Evans et al. 2000). **Also absent from Thung Yai Wildlife Sanctuary in Thailand,** again despite suitable habitat at a relatively high altitude (Steinmetz pers. comm.). For these reasons and the fact that there have been numerous field surveys for the species above 700 m without detecting the species (in **Thailand** and Lao PDR) the 1,200 m upper elevation limit may be too high, though a comprehensive examination of specimens is lacking (Duckworth, Steinmitz, Pattanavibool pers. comm.). There are insufficient data regarding the home range and population density of Lepus peguensis (Flux and Angermann 1990). Total length of this species ranges from 36.0-50.0 cm (Corbet and Hill 1992). *L. peguensis* may have several litters per year with litter size ranging from one to seven (three to four average) (Lekagul and McNeely 1977). Gestation lasts approximately 35-40 days (Lekagul and McNeely 1977). with an average litter of 3-4 young. It is estimated that the longevity of this species is six years (Lekagul and McNeely 1977). This species is crepuscular and nocturnal (Duckworth pers. comm.) L. peguensis actively feeds at night on grass, bark and twigs (Lekagul and McNeely 1977). Hares do not dig or occupy burrows, but spend the day hiding in a particular spot in bushes or tall grass that is known as a "form". The come out in the evening or at night to feed on grass, bark and twigs. Their main defence is their speed. The are usually solitary and territorial and in the breeding season they become very aggressive and you will often see males fighting with the fore feet and kicking with the hind leg which we call boxing.. When mating females can become seriously mauled by over energetic males. who bite and kick them. In **Thailand** they are found where the original forest has been cleared, especially in lalang grass or around hill tribe villages, thus indicating that human disturbance as occurred in the area.

Systems: Terrestrial

Major Threat(s):

Extension of irrigated rice fields destroys habitat in localized areas (Duckworth pers. comm.). The species is heavily hunted, but this does not seem to constitute a major threat (Duckworth, Steinmitz, Pattanavibool pers. comm.). Habitat in Lao PDR and Viet Nam is regularly burnt during the dry season (February-May), posing a threat to young that may be present (Duckworth pers. comm.)

Conservation Actions:

Lepus peguensis is found within protected areas throughout its range. Research is needed in the following areas: taxonomy, distribution, and behaviour (Flux and Angermann 1990). The evident lack of hares within the

Order: Erinaceomorpha (hedgehogs and gymnures)

The order Erinaceomorpha contains a single family, Erinaceidae, which comprise the hedgehogs and gymnures. The hedgehogs are easily recognised by their spines while gymnures look more like large rats.

Family: Erinaceidae (hedgehogs)
Subfamily: Galericinae:
Genus: Echinosorex :
Gymnure or Moonrats *(Echinosorex gymnura)*

The Gymnure, also called a hairy hedgehog or moon rat, is a type of mammal belonging to the subfamily *Galericinae,* in the family *Erinaceidae* and the order *Erinaceomorpha*. Although more closely related to hedgehogs, their resemblance is akin to rats. Gymnures are thought to have appeared in Eastern Asia before their closest relatives, and changed little from the original ancestor, who was supposedly as well common with shrews.

Distribution:

Gymnures inhabit moist jungle terrain in various locales of Southeast Asia, including Vietnam, **Thailand,** Sumatra, China and the Malay Peninsula. Range Description: The Moon rat occurs in the Sundaic region of Indonesia, Malaysia, Brunei, Southern Thailand and southern Myanmar. It can be found throughout Borneo (Brunei, Kalimantan, Sabah and Sarawak), Labuan island (off Sabah), Sumatra, Tebingtinggi island (off the east coast of Sumatra), Peninsular Malaysia, Peninsular Thailand south of 12ºN, and extreme southern Myanmar (Corbet 1988). It has not been reported from Singapore, though it might once have occurred there. It occurs from sea-level up to at least 1,000 m asl.Population: The population size of this species is not known, but it is believed to be relatively common in suitable locations.

Population Trend: Unknown

Countries: Native:

Brunei Darussalam; Indonesia; Malaysia; Myanmar; **Thailand**

Description and Ecology:

HB: 260-450; T: 200-240; HF: 68-75; E: 20-30; W: 0.5 - 1.4gm.

The body is usually black with a white mask and shoulders and black spots or strips to the eyes. but some albinos do occur. There are 2 kinds of fur, a thick short under wool and an overcoat of course shaggy bristles. The tail is scaly, scantly hired, and white on the distal third. The feet are nearly hairless; the hind feet are digitigrade, with the heel held well clear of the ground. The body is very narrow, an adaptation to finding food in narrow crevices. The ears are flesh coloured, rather small and rounded. The mouth is very large, and the pointed and mobile snout is grooved almost to the incisors. The lower stout muscles are less developed than in moles or shrews.. They are not adapted for burrowing and do not use the braincase for pushing up soil, so the skull as not become flattened, retaining more of a generalised shape. Their are 4 mammae, 2 inguinal and 2 pectoral. Although its closest relative is the hedgehog, full grown specimens more closely resemble large rats, or the North American Virginia Opossum *(Didelphis virginianis)*, with which it shares similar habits and ecological niches (an example of parallel evolution). The gymnure's body plan is believed to resemble that of the earliest mammals, with a large toothy head about 1/3 the length of the total body, a naked furless tail for balance and thermoregulatory purposes, and a plantigrade stance. They also have an outstanding sense of smell, and tactile response in the snout region.Gymnures are primarily carnivorous. They are nocturnal or crepuscular: they come out to forage at twilight or in the night to search the forest floor, using smell to find the animals that they eat. Gymnures eat various arthropods, mice, small reptiles and amphibians, with occasional fruit and fungi. An animal of the night, also eat worms, insects, crabs, and other invertebrates found in moist areas. They will also eat fruit, and occasionally frogs or fish.
Gymnures keep territories, and individuals are solitary except when breeding. Gymnures have a very strong scent, typically described as a rancid garlic or onion smell, which is produced by its territory-marking scent glands. Several creatures similar in form and niche, such as the opossum and solenodon smell similar to the gymnure.The Moon rat prefers primary and secondary lowland forests, including very moist areas such as mangroves and swamp forests. It is also found in hilly forests. The species can also tolerate a certain degree of habitat modification, and can be found in rubber plantations and other cultivated areas. According to Lekagul and McNeely (1977), Moonrats prefers wet areas and often enters water. When they are not looking for food at night, Malayan Moonrats rest in hiding places among tree roots, inside hollow logs, or in other tight spaces. Adults live alone. They release strong odours to mark the edges of their territories and warn other Moonrats to stay away with threatening hisses. They also release doors to ward off predators. When they are preparing to have young, they will make nests mostly from leaves. Females usually have two babies at a time, either once or twice a year. Scientists know little more about moon rat adults or young. When they are not looking for food at night, they rest in hiding places among tree roots, inside hollow logs, or in other tight spaces. Adults live alone. They release strong odours to mark the edges of their territories and warn other Moonrats to stay away with threatening hisses. They also release doors to ward off predators. When they are preparing to have young, they will make nests mostly from leaves. Females usually have two babies at a time, either once or twice a year. Scientists know little more about moon rat adults or young.
Systems: Terrestrial

Major Threat(s):

Serious forest loss is taking place through much of the range of this species, especially at lower elevations. In particular, forest in being cleared for oil palm plantations, and forests are being burnt in many places. Although these large-scale changes must be having an impact on the species, it is relatively adaptable, and so is probably not seriously threatened at present.

Conservation Actions:

This species is present in many protected areas. Further studies are needed into the taxonomy, distribution, and abundance of this species. If it is shown to comprise several species, it is possible that some of these might prove to be threatened. Research should be carried out to determine whether or not this species can survive in large-scale oil-palm plantations.

Genus: Hylomys:

Short-Tailed Gymnure *(Hylomys suillus)*

Description:

HB: 105-145; T:12-10; HF: 23-26; E:15-17; W: 15-20 g

Hylomys is the smallest of the gymnures and has the shortest tail. The coat is soft; the hairs are grey at the base and yellow at the tip, or black throughout, giving the coat a drab brown colour; the underparts are buffy. An indistinct black nape or dorsal strip may be present. The ears are rather large, naked and dark brown in colour. The hands and feet are long and slim, and only scantly haired; there are 5 digits, but the first and fifth reach only the bases of the three middle toes. A strong odour is present. though less apparent than in *Echinosorex.* There are 4 mammae, 1 pair inguinal and one pair thoracic. The teeth are slightly reduced in size, with corresponding reduction of the facial parts of the skull.

Range Description:

The species ranges widely in southeast Asia. It ranges from southern Yunnan (Smith et al. 2008), through Eastern and Southern Myanmar, most of Lao PDR and Vietnam, northern Cambodia and most of **Thailand,** south to Penininsular Malaysia, Tioman (Malaysia), Sumatra (mainly in the mountain chain), Java, and Borneo (Brunei, Northern Kalimantan, Sabah and Northern Sarawak) (Ruedi et al. 1994; Ruedi and Fumagalli 1996). It is generally, though not always, found in hilly areas. A survey conducted by Nor (2001) on Mount Kinabalu, Malaysia, found I individuals of this species at 1,700 m asl and higher. In other areas, such as in Sumatra, the species can be found as high as 3,000 m asl, while on the Asian mainland it has been found as low as 90 m asl.

Countries: Native:

Brunei Darussalam; Cambodia; China; Indonesia; Lao People's Democratic Republic; Malaysia; Myanmar; **Thailand;** Vietnam

Population:

This is a common species in many places. Its population is probably stable, as it is adaptable to anthropogenic habitats.

Population Trend: Stable

Habitat and Ecology:

This species prefers humid montane habitats, but it is also found in lowland forests. Individuals found on the Bolaven Plateau in southern Lao PDR were living in degraded scrubland, and the species appears to be adaptable to a number of anthropogenic habitats. These animals find shelter in nests of dead leaves located in hollow structures on the ground and under rocks. The diet is predominately composed of invertebrates, including insects and earthworms, but it can also feed on fruits. The species breeds throughout the year, giving birth to up to three young. The lifespan does not usually exceed two years.
Systems: Terrestrial

Major Threat(s):

There are no major threats to this widespread and adaptable species.

Conservation Actions:

This species is found in many protected areas.

Order: Soricomorpha (shrews, moles, and soledons)

The "shrew-forms" are insectivorous mammals. The shrews and soledons closely resemble mice while the moles are stout bodied burrowers.

Family: Soricidae (shrews)

Subfamily: Crocidurina:

Genus: Crocidura:

Grey Shrew (*Crocidura attenuata*)

Range Description:

This widespread Asian species ranges from India and perhaps Pakistan in the west of its range, through much of Eastern China and mainland Southeast Asia. It is present on the islands of Hainan and Taiwan. In the Philippines, the species was believed to be restricted to Batan Island (Heaney and Ruedi 1994), however, a 2004 survey recorded the species from Calayan Island and the Babuyan Islands (Oliveros et al. 2004). It is unclear if the species is present on the island of Sumatra, Indonesia. It is found from sea level up to 3,000 m asl.This widespread Asian species ranges from India and perhaps Pakistan in the west of its range, through much of eastern China and mainland Southeast Asia. It is present on the islands of Hainan and Taiwan. In the Philippines, the species was believed to be restricted to Batan Island (Heaney and Ruedi 1994), however, a 2004 survey recorded the species from Calayan Island and the Babuyan Islands (Oliveros et al. 2004). It is unclear if the species is present on the island of Sumatra, Indonesia. It is found from sea level up to 3,000 m asl.

Countries: Native:

Cambodia; China; India; Lao People's Democratic Republic; Malaysia; Myanmar; Philippines; Taiwan, Province of China; **Thailand;** Vietnam

Population:

This species is generally abundant and stable, although there have been some localised population declines. It is often the most common shrew present in any given habitat throughout its range.

Population Trend: Unknown

Description:

HB:90-100; T:48-52; HF: 14-15; E:10-11;

A small-sized shrew; the pelage is light brownish-grey above, and paler grey to faintly tinged brown below; eyes are very small with prominent ears, and the snout is shrew-characterized tapered and pronounced, which projects beyond the lower lip; the tail is short, which is about 50% of the head-to-body length, and it is bi-coloured which is dark brown above, light below; both feet bear five digits with claws, thinly covered by short pale hairs on ventral side, and characterized by enlarged fore claws.

Habitat and Ecology:

The species occurs in a wide variety of habitats including lowland and montane tropical and subtropical moist forest, bamboo forest, herbaceous vegetation, scrubland, foothills of Terai and Babbar regions (South Asia). It can be found in secondary forest close to rivers (Molur et al. 2005).
Systems: Terrestrial

Major Threat(s):

There are no major threats to this species as a whole. In South Asia it is locally threatened by habitat loss and invasive species (Molur et al. 2005).

Conservation Actions:

This species is present in several protected areas (eg. Jaldapara Wildlife Sanctuary, West Bengal, India). General taxonomic research, field surveys, and monitoring of populations are recommended for this species in South Asia (Molur et al. 2005).

Southeast Asian Shrew (*Crocidura fuliginosa)*

Description:

HB:70-84; T:44-55; HF:11-13; E:9-11;

A grey shrew with a silver gloss. The tail is 60% to 65% of head and body length. *C.fuliginosa* has grey to blackish fur with a dull silvery gloss; the underparts are lighter. The tail is thin, with a few faint white hairs; the basal portion of the tail is thicker during the breeding season (Blanford 1888)The ears are naked and prominent, they eyes small. The feet are covered with a few short white hairs.

Range Description:

This species is widely distributed in central and southern China and most of mainland Southeast Asia. The species has been recorded from India, however, the distribution in this country is unclear (Sanjay Molur pers. comm.) and it is not mapped here.

Countries: Native:

Cambodia; China; India; Lao People's Democratic Republic; Malaysia; Myanmar; **Thailand;** Vietnam

Population:

There is little information available on the abundance of this species.
Population Trend: Unknown

Habitat and Ecology:

This species is found in a wide variety of habitats, and has been recorded from primary and degraded tropical moist forest, open disturbed areas, and possibly from drier habitats (Francis 2008; Smith and Xie 2008). This shrew is rather common, found in various habitats from montane to lowland forests, cultivated areas and even caves. Medway (1969) states that it is easier to easier to trap in open areas of close ground cover or grass. There does not seem to be any regular breeding season; the litter size being 1-2 (Medway)
Systems: Terrestrial

Major Threat(s):

There are no major threats to this species as a whole.

Conservation Actions:

It is presumably present in many protected areas.

Genus: Suncus:
Etruscan Shrew (*Suncus etruscus*)

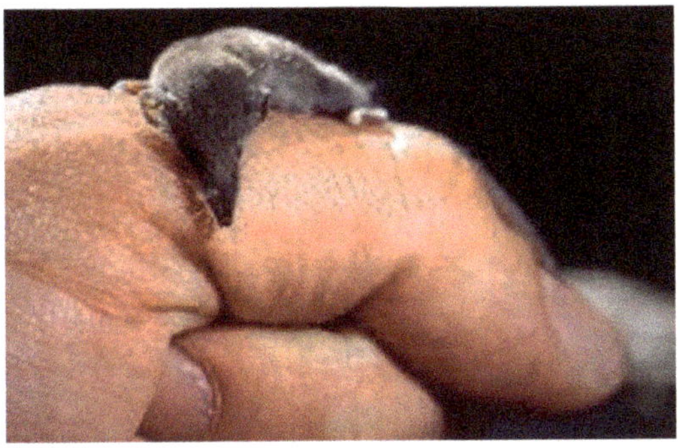

The Etruscan shrew (*Suncus etruscus*), also known as the Etruscan pygmy shrew or the white-toothed pygmy shrew is the smallest known mammal by mass, weighing only about 1.8 grams (0.063 oz) on average The bumblebee bat *Craseonycteris thonglongyai* is regarded as the smallest mammal by skull size.

The Etruscan shrew has a body length of about 4 centimetres (1.6 in) excluding the tail. It is characterized by very rapid movements and a fast metabolism, eating about 1.5–2 times its own body weight per day. It feeds on various small vertebrates and invertebrates, mostly insects, and can hunt individuals of the same size as itself. These shrews prefer warm and damp climates and are widely distributed in the belt between 10° and 30°N latitude stretching from Europe and North Africa up to Malaysia. They are relatively rare and are endangered in some countries.

Description:

The Etruscan shrew has a slender (not truncated) body, with a length between 3 and 5.2 cm (1.2 and 2.0 in) excluding the tail, which adds another 2.4 to 3.2 cm (0.94 to 1.3 in). The body mass varies between 1.3 g (0.046 oz) and 2.5 g (0.088 oz)and is usually about 1.8 g (0.063 oz).In comparison, the related Greater White-toothed Shrew can be twice as long and weighs four to five times more. The head is relatively large, with a long, mobile proboscis, and the hind limbs are relatively small. The ears are relatively large and protuberant. The Etruscan shrew has a very fast heart beating rate, up to 1511 beats/min (25 beats/s) and a relatively large heart muscle mass, 1.2% of body weight The fur colour on the back and sides is pale brown, but is light gray on the stomach. The fur becomes denser and thicker from fall through the winter. The shrew usually has 30 teeth, but the 4th upper intermediate tooth is very small (rudimentary), and is absent in some individuals. Near the mouth growth a dense array of short whiskers, which the shrew actively uses in searching for prey, especially in the night. Dimorphism in body features between males and females is absent.

Activity:

Etruscan shrews live alone, except during mating periods. Their lifespan is estimated at typically around two years, but with a large uncertainty. They protect their territories by making chirping noises and signs of aggressiveness. They tend to groom themselves constantly when not eating, and are always moving when awake and not hiding. The hiding periods are short, and typically last less than half an hour. Clicking sounds are heard when these shrews are moving, which cease when they rest. The shrews are more active during the night when they make long trips; during the day, they stay near the nest or in a hiding place. They reach their maximum level of activity at dawn.

The movements of the Etruscan shrew are rapid, with a rate of about 780 min−1 (13 s−1). In cold seasons and during shortages of food, the shrews lower their body temperatures down to about 12 °C (54 °F) and enter a state of temporary hibernation to reduce energy consumption. Recovery from this state is accompanied by shivering with the frequency of about 3500 min−1 (58 s−1). This induces heating, with the rate up to 0.83 °C/min, which is among the highest values recorded in mammals; the heart rate increases exponentially with time from 100 to 800–1200 beats/min, and the respiratory rate rises linearly from 50 to 600–800 beats/min.

Etruscan shrews mate primarily from March to October, though they can be pregnant at any time of the year. Pairs usually form in the spring and may tolerate each other and their young for some time at the nest. The gestation period is 27-28 days, and they have 2–6 cubs per litter. Cubs are born naked and blind, weighing only 0.2 g (0.0071 oz). After their eyes open at 14 to 16 days old, they mature quickly. The mother usually moves the young when they are 9 to 10 days old and if disturbed leads them by caravanning them to a new location. The young Etruscan shrews are weaned at 20 days old. By three to four weeks of age, the young are independent and are soon sexual mature.

Distribution

The Etruscan shrew inhabits a belt extending between 10° and 40°N latitude across Eurasia. In Southern Europe, it has been found in Albania, Bosnia and Herzegovina, Bulgaria, Croatia, Cyprus, France, Macedonia, Malta, Montenegro, Greece, Italy, Portugal, Slovenia, Spain, and Turkey, with unconfirmed reports in Andorra, Gibraltar and Monaco; it has been introduced by humans to some European islands, such as Canary Islands.

The shrew also occurs in North Africa (Algeria, Egypt, Libya, Morocco, Tunisia) and around Arabian Peninsula (Bahrain, Israel, Jordan, Lebanon, Oman, Syria, and Yemen including Socotra). In Asia, it was observed in Afghanistan, Azerbaijan, Borneo, Bhutan, China (Gengma County only), Burma, Georgia, Guinea, India, Iran, Iraq, Kazakhstan, Laos, Malaysia (Malaysian part of Borneo island), Nepal, Pakistan, Philippines, Sri Lanka, Tajikistan, **Thailand,** Turkey, Turkmenistan and Vietnam. There are unconfirmed reports of the Etruscan shrew in West and East Africa (Guinea, Nigeria, Ethiopia) and in Armenia, Brunei, Indonesia, Kuwait and Uzbekistan. The shrew is relatively rare, especially in Azerbaijan, Georgia (included into the Red Book), Jordan and Kazakhstan (Red Book). Even where not endangered, its density is always lower than of the other shrews living in the area.

Habitat

The Etruscan shrew favours warm and damp habitats covered with shrubs, which it uses to hide from predators. Areas where open terrain such as grasslands and scrub meet deciduous forests are usually inhabited. It can be found at sea level but is usually confined to the foothills and lower belts of mountain ranges, though has been found up to 3,000 m (9,800 ft) above sea level. It colonizes riparian thickets along the banks of lakes and rivers, as well as human-cultivated areas (abandoned gardens, orchards, vineyards, olive groves and edges of fields). The shrew, however, avoids intensively cultivated areas, as well as dense forests and sand dunes. It is poorly adapted to digging burrows, so arranges its nests in various natural shelters, crevices and others uninhabited burrows. They frequent rocks, boulders, stone walls and ruins, darting quickly in and out between them.

Hunting and feeding

Because of its high ratio of surface area to body volume, the Etruscan shrew has an extremely fast metabolism and must eat 1.5–2.0 times its body weight in food per day. It feeds up to 25 times per day, mostly on various invertebrates (insects, their larvae, earthworms, etc.), as well as small vertebrates (the young of frogs, lizards and rodents), and can hunt prey of nearly the same body size as itself. It prefers species with a soft, thin exoskeleton, so avoids ants when given a choice. Grasshoppers, where common, are often regular prey. It kills large prey by a bite to the head and eating it immediately, but takes small insects back to its nest. When hunting, the Etruscan shrew mostly relies on its sense of touch rather than vision, and may even run into its food at night.

Predators and threats

The largest threat to Etruscan shrews originates from human activities, particularly destruction of their nesting grounds and habitats as a result of farming. Etruscan shrews are also sensitive to weather changes, such as cold winters and dry periods. Major predators are birds of prey, especially owls.

Conservation Actions:

This species is present in many protected areas. It is listed on Appendix III of the Bern Convention

Malayan Pygmy Shrew (*Suncus malayanus)*

The species is known from three specimens collected in three widely separate localities in the Peninsular **Thailand** and Peninsular Malaysia (Hutterer 2005). It is presumably occurs more widely within its mapped range (the distribution map is, in any case, highly generalised).
Countries: Native:
Malaysia; **Thailand**

Population:

There is almost no information. The species has a very uneven occurrence through its range.
Population Trend: Unknown

Habitat and Ecology:

It is a lowland forest species. It is not known whether or not it can adapt to anthropogenic habitats.
Systems: Terrestrial

Major Threat(s):

The threats to this species are not known. If it is dependent on forest, then threats are likely to include forest clearance for plantations (especially oil palm), and logging.

Conservation Actions:

It is not known if this species is present in any protected areas. Further studies are needed into its distribution, abundance, natural history and threats.

Subfamily: Soricinae:
Tribe: Anourosoricini:
Genus: Anourosorex:
Mole Shrew (*Anourosorex squamipes)*

Description:

HB: 90-95; T: 14-17; HF: 13-16; E: 0.

Adapted for burrowing with minute eyes, short tail and limbs, and a relatively short and pointed snout. The ears are valvular and can be closed at will to prevent anything entering; they are not visible externally. The feet are scaly, with well developed claws; in contrast to with most fossorial mammals, the hind foot is larger than the front. The backs of the feet are dusky, with white digits and claws. The summer coat is grey, with a bluish tinge, while the winter coat has a greenish tinge.. The hairs of the rump form an elevated tuft, almost like a brush and the hairs are usually glistened with a mucilaginous secretion. The skull is more robust than in Crocidura, with a low, strong sagittal ridge and low occipital ridges that project slightly behind forming a continuous crescent. Dentition is usually heavy, with prominent cusps; the first upper ante molar is the largest.

Range Description:

This species is found in Northeastern South Asia, Central and southern China and Northern Southeast Asia. In South Asia, the species is now known only from Mizoram in India (Molur et al. 2005). In China, it has been recorded from Gansu, Shaanxi and Hubei to Yunnan. In Southeast Asia, it has been recorded from northern Myanmar, **Northern Thailand,** Northern Lao PDR and Northern Vietnam.

Countries: Native:

China; India; Lao People's Democratic Republic; Myanmar; **Thailand;** Vietnam

Population:

There is little information available on the abundance of this species.

Population Trend: Unknown

Habitat and Ecology:

This is a terrestrial and fossorial species, mainly associated with montane forests (Molur et al. 2005; Francis 2008; Smith and Xie 2008). In 2006, the Centre for Disease Control reported that this species can carry a version of the hantavirus which is genetically distinct from rodent hantavirus and new to science.

Systems: Terrestrial

Major Threat(s):

There appear to be no major threats to the species as a whole.

Conservation Actions:

In India, it has been recorded from Murlen National Park, and is presumably present in several more protected areas throughout its range. General field surveys, research into the natural history and monitoring of populations are recommended for this species (Molur et al. 2005).

Tribe: Nectogalini:
Genus: Soriculus:
Lowe's Shrew (*Soriculus parca*)

Description:

The species' colour has been characterised as "dark mouse gray" by Allen (1923), with feet distinctly brown with touches of white. This shrew measures approximately 70 mm, compared to 80 mm in the larger species, *C. smithii.*

Range Description:

This species ranges from South Western China (Sichuan, Guizhou and Yunnan) to Northern and Eastern Myanmar, **Northern Thailand** and Northern Vietnam (Hutterer 2005; Smith et al. 2008). It presumably also occurs in Lao PDR, though there have not so far been any records. Its elevational range is 1,500 to 3,000 m asl.

Countries: Native:
China; Myanmar; **Thailand;** Vietnam

Population:

The abundance and population size of this species are not known.
Population Trend: Unknown

Habitat and Ecology:

Specimens collected from near the Chinese border in Vietnam were from montane bamboo forests between 1,500-2,000 m asl, and in Western Yunnan they have been recorded up to 3,000 m (Smith et al. 2008). It is not known whether or not it can adapt to disturbed habitats. The species can be mistaken for *C. smithii* due to similar external dimensions, but in *C. parca,* the skull, ears and feet are relatively smaller than those of smithii (Allen 1923). The species' colour has been characterised as "dark mouse gray" by Allen (1923), with feet distinctly brown with touches of white. This shrew measures approximately 70 mm, compared to 80 mm in the larger species, *C. smithii.*
Systems: Terrestrial

Major Threat(s):

The threats to this species are not known. If it is dependent on forest, then it might be impacted by forest loss in some parts of its range.

Conservation Actions:

It presumably occurs in several protected areas. Further studies are needed on the distribution, abundance, natural history, threats and conservation needs of this species.

Family: Talpidae (moles)
Subfamily: Talpinae
Tribe: Talpini:
Genus: Euroscaptor:
Kloss's Mole (*Euroscaptor klossi)*

Range Description:

This species is found in Southern China (Southern Yunnan (Smith et al. 2008)), extending to Eastern Myanmar, the highlands of **Northern and Western Thailand**, and most of the mountainous regions of Lao PDR and Vietnam. There is an isolated population in the mountains of Peninsular Malaysia.

Countries: Native:
China; Lao People's Democratic Republic; Malaysia; Myanmar; **Thailand;** Vietnam

Population:

This species is locally abundant throughout its range.
Population Trend: Unknown

Habitat and Ecology:

This species inhabits forest habitat, and areas with loose sandy soil, especially near streams. It is not known whether or not it can adapt to anthropogenic habitats.
Systems: Terrestrial

Major Threat(s):

The threats to this species are not known. If it is dependent on forest, then threats are likely to include forest clearance for agriculture, plantations, logging and human settlement.

Conservation Actions:

Given its wide range, this species is likely to be present in a number of protected areas. Further studies are needed on the distribution, abundance, natural history and threats to this species

Order: Chiroptera (bats)

Bats are mammals of the order Chiroptera karptərə; from the Greek χείρ - cheir, "hand" and πτερόν - pteron, "wing") whose forelimbs form webbed wings, making them the only mammals naturally capable of true and sustained flight. By contrast, other mammals said to fly, such as flying squirrels, gliding possums, and colugos, glide rather than fly, and can only glide for short distances. Bats do not flap their entire forelimbs, as birds do, but instead flap their spread-out digits, which are very long and covered with a thin membrane or patagium. Bats represent about 20% of all classified mammal species worldwide, with about 1,240 bat species divided into two suborders: the less specialized and largely fruit-eating 'megachiroptera', or flying foxes, and the more highly specialized and echolocating 'microchiroptera'. About 70% of bats are insectivores. Most of the rest are frugivores, or fruit eaters. A few species, such as the fish-eating bat, feed from animals other than insects, with the vampire bats being the only parasitic mammalian species. Bats are present throughout most of the world, performing vital ecological roles of pollinating flowers and dispersing fruit seeds. Many tropical plant species depend entirely on bats for the distribution of their seeds. Bats are important in eating insect pests, reducing the need for pesticides. The smallest bat is the Kitti's hog-nosed bat, measuring 29–34 mm (1.14–1.34 in) in length, 15 cm (5.91 in) across the wings and 2–2.6 g (0.07–0.09 oz) in mass. Found only in **Thailand**, it is also arguably the smallest extant species of mammal, with the Etruscan shrew being the other contender. The largest species of bat is the giant golden-crowned flying fox, which is 336–343 mm (13.23–13.50 in) long, has a wingspan of 1.6 m (5 ft 3 in) and weighs approximately 1.1–1.2 kg). There are few fossilized remains of bats, as they are terrestrial and light-boned. An estimated only 12% of the bat fossil record is complete at the genus level. An Eocene bat, Onychonycteris finneyi, was found in the 52-million-year-old Green River Formation in Wyoming, United States, in 2003. It had characteristics indicating it could fly, yet the well-preserved skeleton showed the cochlea of the inner ear lacked development needed to support the greater hearing abilities of modern bats. This provided evidence flight in bats developed well before echolocation. The team that found the remains of this species, named Onychonycteris finneyi, recognized it lacked ear and throat features present not only in echolocating bats today, but also in other known prehistoric species. Fossil remains of another Eocene bat, Icaronycteris, were found in 1960. The appearance and flight movement of bats 52.5 million years ago were different from those of bats today. Onychonycteris had claws on all five of its fingers, whereas modern bats have at most two claws appearing on two digits of each hand. It also had longer hind legs and shorter forearms, similar to climbing mammals that hang under branches such as sloth's and gibbons. This palm-sized bat had broad, short wings, suggesting it could not fly as fast or as far as later bat species. Instead of flapping its wings continuously while flying, Onychonycteris likely alternated between flaps and glides while in the air. Such physical characteristics suggest this bat did not fly as much as modern bats do, rather flying from tree to tree and spending most of its waking day climbing or hanging on the branches of trees.

Family: Pteropodidae (flying foxes, Old World fruit bats)
Subfamily: Pteropodinae:
Genus: Balionycteris:
Spotted-Winged Fruit Bat (*Balionycteris maculata)*

Description:

The spotted-winged fruit bat, (*Balionycteris maculata)* is the smallest megabat in the world ,and the only species in the genus Balionycteris. It inhabits forests in Indonesia and Malays and **Thailand.**

Spotted-winged fruit bats are unusually small megabits, with a head-body length of 5 to 7 centimetres (2.0 to 2.8 in), a wingspan of 28 centimetres (11 in), and an adult body weight of around 13 grams (0.46 oz). Most of the head and body are covered in thick blackish-brown fur, while the under parts are a pale grey-brown. Their wings are dark brown, with distinctive pale cream spots, which are most noticeable on the joints, but are also found in a variable pattern across the wing membranes. The wings have been reported to have an aspect ratio of 5.8 and a wing loading of 9.51 N/m2.

There are also pale markings in front of the large eyes, and on the ears, which are simple in shape, compared with those of microbars. The nostrils are elongated, and splayed to the side of the snout. The species can be distinguished from other small fruit bats by the presence of the spots on its wings, but also by its unusual dental formula of 2.1.3.2 over1.1.3.2

Distribution and habitat:

Spotted-winged fruit bats inhabit primary rain forest up to 1,500 metres (4,900 ft) elevation, and coastal mangrove forests. They are found across Malaysia and Brunei, and in the northern and western parts of the island of Borneo in Indonesia. At the edges of their range, they have also been reported from **Trang Province in southern Thailand,** from a small area in Sumatra, and from the Riau Archipelago. No subspecies are known.

Behaviour and Ecology:

Spotted-winged fruit bats forage in the under story of dense forest where they feed on a range of generally small, inconspicuous fruit, such as figs and persimmon, and on some insects and spiders. They spend the day roosting alone or in small groups consisting of a male and up to nine females and their young. Rather than remaining with a single male, females may visit up to three different males. Males return regularly to their roosts during the night, suggesting that they gain access to females by controlling and defending prime roosting sites. They generally roost in small groups, huddled together : roosts typically occur beneath palm leaves, or amongst epiphytes. The absence of such plants in recently logged forest may encourage these bats to roost under man-made structures. They have also been recorded roosting in caves.

They forage as individuals, rather than in a group, and rarely travel more than 1 kilometre (0.6 mi) from their roost site. They have been observed to make high-pitched "peep" noises when foraging, as well as more complex series of sounds when socialising. Roosts are of a consistent shape and size, and may be partially excavated by the bats, a behaviour that is otherwise unknown in this group of animals. They are generally located in the root masses of epiphytic plants, such as ferns, and also in the nests of ants and termites.

Females give birth to a single young up to twice a year, typically between June and January. The young are born blind, and weighing around 3.5 grams (0.12 oz), after a gestation of 135 days. They are weaned by 40-80 days, and are able to fly by the time they have reached 5 grams (0.18 oz) in weight. Females are sexually mature at ten months of age.

Major Threat(s):

This species is locally threatened throughout much of its distribution by deforestation for agriculture, logging and plantations, and as a result of fire. Scattered populations in **Southern Thailand,** Sumatra, and the Riau Archipelago may be particularly vulnerable (Hodgkison and Kunz 2006).

Conservation Actions:

On the Malay Peninsula and Borneo (Sabah, Sarawak, Brunei, and Western Kalimantan), this species occurs within several reserves and protected areas, including production forests (Hodgkison and Kunz 2006).

Genus: Chironax:

Black-Capped Fruit Bat (*Chironax melanocephalus)*

Description:

Head and body length about65-70 mm. No external tail is present. Forearm 40-50mm. Weight 12 to 17 grams. The head is black. The back is grey brown to reddish brown. The undersides are pale buff. No spots on the wings.

Distribution:

Usually found at elevations above 600 metres, with groups of between 2 and 8 individuals being found during the day resting on the underside of tree ferns several metres from the ground. Seven specimens of *C. melanocephalus* were collected from the lowland forest at Kubah and Lambir, and beach forest at Samunsam, Sarawak, Borneo. The specimen that was mist-netted in Kubah National Park on 12 July 1995 was a first record for Sarawak (Abdullah et al. 1997). According to Payne et al. (1985) the distribution of *C. melanocephalus* in Borneo is only known from Sepilok in Sabah and Temburong in Brunei. The range includes **Thailand,** Peninsular Malaysia, Sumatra, Java and Sulawesi (Payne et al. 1985).

Biology and Ecology:

The habitat where it was caught was lowland mixed dipterocarp forest. Two specimens were caught in the under story of primary dipterocarp forest in Kubah National Park; one form the under story of mixed beach forest habitat in Samunsam Wildlife Sanctuary; and four from the canopy (between 15 to 30 m) of the primary dipterocarp forest in Lambir Hill National Park. These sites are new distributional records for *C. melanocephalus* in Sarawak and have extended the range of the species to the western part of Borneo. The bat normally roosts in small groups in tree ferns and in shallow caves (Payne et al. 1985).

All *C. melanocephalus* caught were adults, three males and four females. Two were in non-reproductive condition and others were in various reproductive stages. A female (MTA96041) caught on 21 May 1996 from Kubah was pregnant. In November 1996, two females caught in Lambir Hills National Park were in post-lactating condition, suggesting recent detachment or loss of juveniles. A male caught at the same site was observed to have enlarged testes. Medway (1978) recorded pregnant females in February and April in the upland area of Cameron highlands in Peninsular Malaysia. It occurs in lowland, hill and montane forest. The Sabah specimen was taken in the under story of dipterocarp forest. It has been found roosting in small groups in tree ferns and in a shallow cave (Payne et al. 1985). It is probably not dependent on water. They have been seen feeding on wild Fig *(Ficus)*

Systems: Terrestrial

Major Threat(s):

Habitat loss is a threat for this species, it is usually not found in disturbed forest. Deforestation is occurring throughout its range for logging, agriculture and plantations, and as a result of fires.

Conservation Actions:

The species occurs in protected areas.

Genus: Cynopterus

Lessser Short Nosed Fruit Bat (*Cynopterus brachyotis*)

Description:

With an appearance typical of a fruit bat, the lesser short-nosed fruit bat is a beautiful example. Its dog-like face with large, appealing eyes and white edging on the ears give this bat a magical quality. When roosting, the bat wraps its black wings tightly around its body like a cloak, leaving only the head visible. The white finger bones stand out against the black wing membranes, adding to the striking effect. The fur is short and greyish brown to yellowish brown on the back and paler on the underside. Adult males have a dark orange-red collar and females a more yellow-orange collar. Juveniles lack this collar and tend to be uniformly grey.

Distribution:

This widespread species ranges from South Asia, through parts of Southern China to parts of Southeast Asia. In South Asia, this species is presently known from Bangladesh (Sylhet division) (Sarker and Sarker 2005; Srinivasulu and Srinivasulu 2005), India (Andaman and Nicobar Islands, Gujarat, Karnataka, Kerala, Madhya Pradesh, Maharashtra, Meghalaya, Orissa, Uttar Pradesh and West Bengal) and Sri Lanka (North Central, Uva and Western provinces) (Srinivasulu et al. in press; Molur et al. 2002). In southern China, it has been recorded from Guangdong, with possible records from Xizang (Medog) and southern Yunnan (Wang 2002; Smith and Xie 2008). In continental Southeast Asia, it is known from southern Myanmar, **Thailand,** Lao PDR, Viet Nam (identity of records from northern Vietnam need verification), Cambodia (known only from Phnom Phen [G. Csorba pers. comm.), and Peninsular Malaysia. In Insular Southeast Asia, it is known from the islands of Sumatra and Java (Indonesia), Borneo (Indonesia and Malaysia only), the island of Sulawesi (Indonesia), the island of Timor (East Timor and Indonesia), the Talaud Islands (Indonesia) and Ternate Island (Indonesia). It might be present on the island of Palawan in the Philippines, but this requires confirmation. This species of fruit bat is found across Southern and Southeast Asia, from Sri Lanka to Indonesia and the Philippines. It has many subspecies that vary in size and colouration: *Cynopterus brachyotis altitudinis* (Cameron Highlands of Peninsular Malaysia); *Cynopterus brachyotis brachyotis* (Borneo, Lombok, Peninsular Malaysia, the Philippines, Sulawesi); *Cynopterus brachyotis brachysoma* (Andaman Islands); *Cynopterus brachyotis ceylonensis (*Sri Lanka); *Cynopterus brachyotis concolor* (Enganno Island); *Cynopterus brachyotis hoffeti* (Vietnam); *Cynopterus brachyotis insularum* (Kangean Island) (3*); Cynopterus brachyotis javanicus* (Java) ; *Cynopterus brachyotis minutus (*Nias Islands).

Countries: Native:

Cambodia; China; India (Andhra Pradesh, Bihar, Goa, Karnataka, Maharashtra, Nagaland, Tamil Nadu); Indonesia (Sulawesi, Sumatera); Lao People's Democratic Republic; Malaysia; Myanmar; Singapore; Sri Lanka; **Thailand**; Timor-Leste; Vietnam

Habitat.

The lesser short-nosed fruit bat occurs in many different habitats, from pristine primary rainforests, to oil palm plantations, gardens and mangroves.

Ecology

Under favourable conditions, female lesser short-nosed fruit bats give birth to one pup twice each year, once between mid January and mid April, and again between mid June and early October. Pregnancy lasts between five and six months and the birth of the pups does not necessarily occur in time with flowering or fruiting . Females carry their pup in flight for the first few months of life, until it has learnt to fly with confidence. The young become sexually mature at seven months, and females will give birth to their first pup at just over 12 months old .

Lesser short-nosed fruit bats become active shortly after sunset and fly directly to fruiting trees up to 2 km away to feed on small fruits, including mangoes and figs, as well as on nectar. They fly around the trees several times before settling on the fruit , where they use claws on the first and second digits of the hands, as well as their strong feet, to cling on to bunches of fruit whilst feeding. As fruit bats do not echolocate, they must find their food using their large eyes and strong sense of smell. During the day, they return to their roosts under shaded trees, tree-ferns and near the entrances of caves . This species is a particularly important seed-disperser; it is a seasonal specialist, and over an annual fruiting cycle can consume the fruits of 54 species, the leaves of 14 species and the flower parts of four species. This species can be found from habitats ranging from orchards, gardens to forested tracts. It roosts in palms especially seed clusters of palms either solitary or in small groups of a few individuals in rural and urban landscapes and in forested areas. Bears a single young after a gestation period of 105-120 days (Bates and Harrison 1997). In South Asia, the species is believed to be more restricted to higher elevations when compared to C. sphinx, making it specifically a hill forest species.
Systems: Terrestrial

Major Threat(s):

There are no major threats to this species as a whole. In South Asia, this species is locally threatened by deforestation, generally resulting from logging operations and the conversion of land to agricultural and other uses (Molur et al. 2002).

Conservation Actions:

Other than further taxonomic studies, no conservation actions are currently needed for the species as a whole. It is present in many protected areas throughout its range. In South Asia, this species like most other fruit bats in India is considered a vermin under Schedule V of the Indian Wildlife (Protection) Act. The species has been recorded from protected areas in India like Nagarhole National Park in Karnataka and Kalakkad-Mundunthurai Tiger Reserve in Tamil Nadu, and in Hakgalla National Park in Sri Lanka.

Horsefield's Fruit Bat *Cynopterus horsfieldi*

Description:

Horsfield's fruit bat is a medium-sized megabat, intermediate in size between flying foxes and pygmy fruit bats. Adults weigh around 55 to 60 grams (1.9 to 2.1 oz), and have light grey to brown fur, with a reddish brown or orange mantle around the shoulders. In some males the mantle extends across the chest, and the fur is often brighter in colour than in females. The rim of the ears and the skin overlying the metacarpals and phalanges within the wing are white. Juveniles have a more bland coat pattern, with uniformly dull buff or grey fur. The bats have a short, broad snout, ending in a pair of almost tubular nostrils. Both the eyes and ears are large, although the latter have a simpler structure than in most other bats, and lack a tragus. The wings have a low aspect ratio and high wing loading, typical of many mega bats, and indicating a relatively slow flight speed and moderate manoeuvrability. It is named for Thomas Horsfield, an American naturalist who presented the type specimen to the British Museum.

Distribution and Habitat:

Horsfield's fruit bat is found in Thailand, Indonesia, Malaysia, and Brunei. Within this region, it inhabits a broad range of lowland habitats, from dense primary rain forest to agricultural land and suburban gardens. Range Description: This species is found in **Thailand**, Cambodia, Malaysia, Borneo, Java, Sumatra, Lesser Sunda Islands, and adjacent small islands in Indonesia. It has also been recorded from Bali, Lombok, Simelue and Nias (I. Maryanto pers. comm.).

Countries: Native:

Indonesia; Malaysia; **Thailand**

There are four recognised subspecies:

Cynopterus horsfieldii horsfieldii - Java and islands East to Sumbawa

Cynopterus *horsfieldii harpax* - Thailand, peninsular Malaysia, Sumatra

Cynopterus horsfieldii persimilis - Borneo

Cynopterus horsfieldii princeps - Nias

Population:

This species is common in Indonesia (I. Maryanto pers. comm.), but locally rare in northern and **Western Thailand** (S. Bumrungsri pers. comm.).

Population Trend: Unknown

Biology and Behaviour:

Horsfield's fruit bats eat the fruit of strangler figs, Elaeocarpus, and Payena, and the flowers of bitter beans. They have been reported to pluck fruit from trees and carry it to roosts elsewhere in order to feed. During the dry season, hen fruit is in short supply, they instead feeds on pollen, which they take from a wide variety of different plants. It is found in many habitats in Indonesia but only pristine forests in **Thailand.** In Malaysia it has been found in agricultural areas, suburban parks, fruit orchards and secondary forest as well as good forest (Campbell and Kunst, 2006). They have a harem social structure, and are highly gregarious when roosting in caves, but are found in smaller groups when roosting in forests. It makes tent roosts.

Systems: Terrestrial

They live in small groups, consisting of a single adult male and up to five females and their young. Although these groups are maintained year round, individual females often move between different groups, and may spend some time nesting alone between leaving one group and joining another. They roost in trees and cave mouths, reportedly favouring banana trees. They often modify their roosting sites by constructing tents from the leaves, partly cutting through them to make an inverted "V" shape. They breed throughout the year, but most commonly give birth at two times of the year: between February and March and between July and August. They have been reported to live for at least 31 months.

Major Threat(s):

There are no major threats to this species throughout its range. Deforestation is a threat in **Thailand,** but elsewhere this species adapts to disturbed habitats.

Conservation Actions:

Isolated populations from Thailand need to be examined taxonomically (S. Bumrungsri pers. comm.). It is found in a number of protected areas throughout its range.

Greater Short-Nosed Fruit Bat (*Cynopterus sphinx)*

Description:

These bats have a relatively long snout. Their upper parts are brown to grey-brown with paler under parts. The fur is very fine and silky. The ears and wing bones of *C. sphinx* are edged in white. Lower cheek teeth rounded without accessory cusps. The wing span of the adult is about 48 cm. Juveniles are lighter than adults. Average forearm length 70.2mm (64-79mm).This species is found in a wide variety of habitats from rural areas, primary and secondary forested habitats to urban landscapes. It is found as small colonies consisting of 3-7 individuals, sometimes more; it roosts underside leaves, in flower and fruit clusters of Kitul Palm, on Saraka asoka trees and is known to build tents in the roosting trees. It feeds on a variety of fruits both wild and cultivated. It has a low but fast flight. It breeds twice a year, and in some areas is known to breed throughout the year and bears a single young (Bates and Harrison 1997).

Systems: Terrestrial

Habitat:

It is found in Pakistan, Bangladesh, Bhutan, Cambodia, China, India, Indonesia, Laos, Malaysia, Myanmar, Philippines, Sri Lanka, **Thailand,** and Vietnam. *C. sphinx* is common in tropical forests and areas where fruit crops are cultivated. They can also be found in grassland and mangrove forests. They typically nest high in palm trees. The bats chew the fronds of the palms to construct fairly simple tents. These bats are also known to construct tents by closely interweaving the leaves and twigs of creeping vines which cover buildings, but such nests are constructed only when palms are not available. They are common in tropical forests, where fruit crop acreage. They can also be found in the grass and mangroves. They usually nest in the height of palm tree. Bat bite the hand of the leaf structure is very simple tent. These bats are also known as the structure closely intertwined with branches, leaves and vines climbing tents, covers the construction, but construction of the nest only palm is not available. Generally, they are found at lower altitudes. The habitats are including agricultural areas, forested regions and parks in cities. They can also be found from Pakistan, India and Sri Lanka to southern China, Malaysia, Java, Lesser Sunda Island, Borneo and Sulawesi which is ranging wide.

Native:

Bangladesh; Bhutan; Cambodia; China; Hong Kong; India; Indonesia; Lao People's Democratic Republic; Malaysia; Myanmar; Nepal; Pakistan; Sri Lanka; **Thailand;** Vietnam

Behaviour and breeding:

The species is gregarious, and typically roosts in same sex groups of 8-9 individuals. The sexes remain separate until the mating season, when group size increases. They are polygynous and it is usual for 6-10 males and 10-15 females to share palm frond tents during the breeding season. *C. sphinx* is the only non-primate species to show fellatio, which enhances copulation time in the species. Copulation by males is dorsoventral and the females lick the shaft or the base of the male's penis, but not the gland which has already penetrated the vagina. While the females do this, the penis is not withdrawn and research has shown a positive relationship between length of the time that the penis is licked and the duration of copulation. Post copulation genital grooming has also been observed. Males stay with females for some time after mating, but later return to same sex groups. Female bats perform fellatio to increase copulation time. This species is the only non-primate to exhibit this behaviour The adult sex ratio is very female biased. Researchers attribute this to the relatively rapid maturation of females compared to males. In Central India, *C. sphinx* breeds twice per year. Females produce a single young at a time. Each half of the bicornate uterus functions during alternate breeding cycles. The first pregnancy cycle occurs from October through February/March. Mating occurs immediately postpartum, and a second offspring is born in July. Gestation period is about 3–5 months. In 72% of bats, the first pregnancy occurs in the right horn of the uterus. The corpus luteum in the right ovary persists for some time after the pregnancy and prevents ovulation from occurring in the right ovary during the second breeding cycle. This creates the pattern of alternate functioning of the two horns of the uterus. However, the corpus luteum in the left ovary does not persist until the beginning of the next breeding cycle. As yet, no reason has been found for the dominance of the right horn during the first breeding cycle. Newborn bats weigh about 13.5 g and have a wingspan of 24 cm. By the time of weaning at 4 weeks of age, young bats weigh 25 g and have wings spanning 36 cm. Female short-nosed fruit bats reach sexual maturity at 5–6 months of age, but males are not capable of breeding until they are a year old.

These bats are frugivorous, locate their preferred food items by scent. They have been described as voracious feeders, eating more than their body weight in food in one sitting. Some preferred fruits include ripe guava, banana, chikoo, dates and lychees.

Short-nosed fruit bats inflict serious damage on many fruit crops, and are considered pests. In addition, these bats are possible vectors for Japanese encephalitis, which is serious disease in humans. These bats are important dispersers of date palm seeds, and pollinate many night blooming flowers.

Major Threat(s):

There are no major threats to this species as a whole. In South Asia, it is locally threatened in parts of its range through deforestation, generally resulting from logging operations and the conversion of land to agriculture and for construction of dams and other developmental activities. It is also under threat due to hunting for medicinal purposes (Molur *et al.* 2002). In Southeast Asia, it is a pest species of orchards and is hunted for food in some parts of the range. In China, there is some habitat loss and the species is hunted for medicinal purpouses

Conservation Actions.

As it is present in many protected areas and is very adaptable, no direct conservation measures are currently needed for this species as a whole. In South Asia, this species like most other fruit bats in India is considered a vermin under Schedule V of the Indian Wildlife (Protection) Act. It has been recorded from protected areas in India like Point Calimere Wildlife Sanctuary in Tamil Nadu, Coringa Wildlife Sanctuary, Kawal Wildlife Sanctuary, Nagarjunasagar Srisailam Tiger Reserve in Andhra Pradesh, Kanha National Park in Madhya Pradesh, Indravati National Park in Chattisgarh. Taxonomic studies, population monitoring and lobbying with the government to accord it a non-vermin status are recommendations (Molur *et al.* 2002). Molecular taxonomic studies to know the variations in the population are recommended (C. Srinivasulu pers. comm. 10 October, 2007).

Genus: Megaerops:
Tailless Fruit Bat (*Megaerops ecaudatus*)

Description:

HB: 90-100; T: None; FA: 54-59; E: 13; HF: 12; W; 20-38g

The body is yellowish brown to reddish above, with a pale grey nape contrasting with the brown of the back; the chest a belly are silvery grey, with a yellowish spot under the chin. The individual hairs are grey at the base. The ears are large and hairless. The mussel is short and broad with the nostrils separated by a deep groove and the inner margins of the nostrils distinctively projected, almost tubular; the nasal bones extend forwards to a point well in front of the canines and the nasal cavity is quite large. This structure of the nose suggest specialisation adaptation for keeping the mouth clear whilst eating liquid, over ripe fruit. The dentition is very similar to *Cynopterus* but without surface cusps on the canines , the lower canines are also large and close together and occlude with the lower incisors. There is only one pair of lower incisors and the lower incisors are reduced, supporting the hypothesis of a softer, more liquid diet.

Range Description:

This species occurs in Borneo, Sumatra, Malaysia, and **Thailand.** It probably occurs also in North-eastern India (Srinivasulu et al. in prep.).

Countries: Native:

Brunei Darussalam; Indonesia; Malaysia; **Thailand**

Population: This species is widespread but uncommon (P. Bates and S. Bumrungsri pers. comm.) throughout its range.

Population Trend: Unknown

Habitat and Ecology:

It is a forest associated species found in a variety of habitats from lowland evergreen forest to disturbed secondary forest. In Peninsula Malaysia the species moves roosting sites to follow food availability (S. Bumrungsri pers. comm.). Two males and two females have been captured in various habitats. An adult female was collected in April, 1995, and adult male in April, 1996. A sub adult female was taken from the Balui secondary forest in June, 1994, while an immature was collected from the primary forest in Taleban on 28 March 1997. In Cameron Highlands, Peninsula Malaysia, females were pregnant in February and extending to March and June. Females with suckling juveniles were also observed in November. A specimen MTA96004 was collected in the canopy net approximately 30 m above ground level. The specimen BD013 was collected at 1,100 m elevation in Bario upland. According to Lekagul and McNeely (1977), *M. ecaudatus* has been recorded both in the lowlands and on mountains up to 3,000 m in **Thailand.**

Systems: Terrestrial

Major Threat(s):

There are no major threats to this species, although deforestation is occurring in some parts of its range.

Conservation Actions:

This species occurs in a number of protected areas throughout its range.

Ratanaworabhan's Fruit Bat (*Megaerops niphanae)*

Range Description:

This species has been recorded from India (Arunachal Pradesh, Manipur, Mizoram, Nagaland and West Bengal) at an altitude range of 100 to 2,100 m asl (Molur et al. 2002), from **Thailand (Nakhon Ratchasima Province, Amphoe Pak Thong Chai, Sakaerat Environmental Research Station),** Lao PDR, Cambodia and Vietnam. Some previous records of Megaerops ecaudatus may actually represent *M. niphanae.*

Countries: Native:

Cambodia; India; Lao People's Democratic Republic; **Thailand;** Vietnam

Population:

In **Thailand** the species is locally rare and is possibly nomadic (S. Bumrungsri pers. comm.). The population size and trends for this species are not known in India (Molur et al. 2002).

Population Trend: Unknown

Habitat and Ecology:

In **Thailand** it is found from 140 to 240 m asl in a variety of habitats including pristine forest and farmland adjacent to forest (S. Bumrungsri pers. comm.). In South Asia, little is known about the habitat and ecology of this species. It is presumably it is found in deciduous forests, coniferous forests, bamboo forests and subtropical mixed forests across the foothills of Himalaya.

Systems: Terrestrial

Major Threat(s):

There are no major threats to this widespread species as a whole. It may be locally affected by deforestation in **Thailand** (S. Bumrungsri pers. comm.). Threats to this species not known in India.

Conservation Actions:

This species is present in many protected areas. It is classified as a vermin under Schedule V of the Indian Wildlife (Protection) Act. There is a debate over the taxonomic status of two species of the genus Megaerops, with additional taxonomy, survey, ecological studies needed (Molur et al. 2002).

Genus: Pteropus:

Island Flying-fox (*Pteropus hypomelanus)*

Description.

HB:170-200; T: None; FA: 130-144; E:23-25; HF: 45-50; W: 45g

Its fur colour is variable but is generally dark grey-brown on the lower back to orange-brown in the underparts and around the chest and face. The populations towards the west of the species range is generally darker in colour. The hairs around the eyes are greyish. Postorbital processes are variable, sometimes reaching halfway around the orbit. The premaxillaries are open , with quite large palatal foramina. Dentition is heavy, and the large premolars have a distinct ridge on the posterior surface, lacking or very faint in other Thai species. The canines have long, deep longitudinal grooves, causing the tooth to split. The upper incisors are in a semi circle , well separated from the canine. The first upper premolar is very tiny.

Range Description:

This species is widespread, ranging from the Maldives and Andaman and Nicobar Islands (India) in the West, to Melanesia in the east. In South Asia, this species is restricted to only four locations in the Andaman and Nicobar Islands in India and Addu Atoll in Maldives (Molur et al. 2002). The species is found on many offshore islands and in coastal lowlands in the Southeast Asian region, including the Philippines and Indonesia (Sulawesi). In the Philippines it is found throughout the country with the possible except of the Palawan faunal region; records are from: Bohol, Cagayan Sulu, Camotes (Paguntulan pers. comm. 2006) Camiguin, Cebu (Paguntulan pers. comm. 2006) Cuyo, Dinagat, Guimaras, Leyte, Luzon (Camarines Sur, Ilocos Norte, and Nueva Ecija) Mactan, Marinduque, Maripipi, Masbat, Mindanao (Gunther 1897) Negros, Panay (including Boracay and Batbatan), Polillo, Romblon (Timm and Birney 1980), Samar, Siargao, Sibuyan, Siquijor (Heaney et al. 1998) Tablas (Paguntulan pers. comm. 2006). There are few records from mainland Papua New Guinea; the species is also found in the D'Entrecastreaux Archipelago of Papua New Guinea, on Manus in the Admiralty Islands. Records from New Britain and Tabar are questionable (probably *P. admiraltatum)*. It is confirmed only from Mbanika and the Russel Islands in the Solomons. In South Asia, this species has been recorded from sea level to an elevation of 100 m asl. It has been recorded from sea level up to 900 m asl in the Philippines; it is primarily found in low elevation areas in Melanesia (under 500 m asl).

Countries: Native:

India; Indonesia; Malaysia; Maldives; Myanmar; Papua New Guinea; Philippines; Solomon Islands; **Thailand;** Vietnam

Population:

This species is common to abundant throughout its range, and can form colonies of up to 5,000 individuals. The abundance, population size and trends for this species are not known in South Asia (Molur et al. 2002).

Habitat and Ecology:

In South Asia, this species roosts in large colonies of several individuals and is found in forests, orchards, coconut palm groves (Molur et al. 2002). It feeds both on wild and cultivated fruits (Bates and Harrison 1997). In most parts of its range in the Philippines, this species roosts on small offshore islands and near coastlines, but forages on the mainland where it is common in agricultural areas and absent from primary forest (Heideman and Heaney 1992, Rickart et al. 1993, Utzurrum 1992). In Melanesia, it is generally an insular species found on small offshore islands. Animals may commute infrequently to larger islands for foraging. It is known to roost in trees, often in large groups at the coastline. It can be found foraging for food in both primary and secondary tropical forest habitats, rural gardens and plantations. The females give birth to a single young which take about one year to reach maturity (Flannery 1995; Bonaccorso 1998).

Systems: Terrestrial

Major Threat(s):

There are no major threats to this species as a whole. In South Asia, this species is threatened by deforestation, generally resulting from logging operations and the conversion of land to agricultural and other uses. It is also threatened due to tourism related activities (Molur et al. 2002). The species is under heavy hunting pressure in the Philippines, although it may be able to withstand this, at least in the short term, as populations have remained stable. Animals are sold locally for 30-50 Philippine Pesos each Cariño pers. comm. 2006). In Melanesia, the species is locally vulnerable to hunting. Island populations of this species, especially on the coastal islands of northern New Guinea, are vulnerable to overexploitation by hunting for food (Bonaccorso 1998). Threats to this species are unknown in Indonesia.

Conservation Actions:

This is accorded vermin classification under Schedule V of the Indian Wildlife (Protection) Act. This species has been recorded from Barren Island Wildlife Sanctuary, Andaman and Nicobar Island in India. Taxonomic studies, ecology, population monitoring and habitat management are recommended. Awareness needs to be created to mitigate threats to this species (Molur et al. 2002). This species is listed on Appendix II of CITES. This species apparently does well in captive-breeding programmes (Bonaccorso 1998).

Lyle's Flying Fox (*Pteropus lylei)*

Description:

HB: 200-250; T: None; FA: 145-160; E: 35-39; HF: 40-45; W: 390-480g

Lyle's flying fox (*Pteropus lylei)* is a medium-sized flying fox which forms large colonies high up in trees. Lyle's flying fox has a long dark muzzle, large eyes and a ring of orange fur around the neck, and bears some resemblance to a fox, hence the common name 'flying fox'. Lyle's flying fox (*Pteropus lylei*) is medium-sized bat species, weighting between 390-480 g. It distributes in Cambodia, **Thailand** and Vietnam.[9] It has the nocturnal lifestyle like other bat species, forages at night and returns to the colony before sunrise. Via satellite, Hondo *et al.* (2010) showed high mobility of Lyle's flying fox, speeding up to 49.83 km/h for foraging on fruit, nectar and pollen.[8] Along the way, it usually distributes seeds in its droppings.[5] This has secured its position as keystone species-animal that is disproportionately important in determining the ecological diversity, distribution and abundance of plants and other animals.[12] However, Lyle's flying fox is considered as a pest by fruit growers in agricultural areas of **Thailand**. Owing to this attitude, it is threatened by humans.[2] Consequently, it was classified as vulnerable species in IUCN red list. Killing is not only a factor resulted in the population size declining, reduction in roosting area due to habitat change by humans and effects of continual use by bats may have an impact on population size of these bats in future. The effective management and conservation for this bat species are essential. The wings and back of Lyle's flying fox are dark brown or black, which strongly contrast against the bright fur around the head and neck. Its lower body varies from a deep dark-brown to a brighter yellow-brown. Its breast and belly are black-brown, which is similar to the large flying fox (*Pteropus vampyrus).*

Range{

Lyle's flying fox is native to Southeast Asia and is found in **Thailand,** Vietnam, Cambodia, Malaysia, and Yunnan in China. This species has been mostly documented in **Thailand,** where at least 11 colonies have been identified, the largest containing around 3,000 individuals .

Habitat:

Lyles's flying fox can be found in some of Southeast Asia's most secluded jungles, as well as in the middle of villages or even large urban metropolises such as **Bangkok**. The species usually roosts in temples in the middle of urban areas. Colonies inhabiting the forest roost in the tops of tall canopy trees, which they revisit year after year. Lyle's flying fox is a vulnerable species of **Thailand.** Large colonies are established in the central Plain of **Thailand,** but the ecology and behaviour of this bat remain largely unknown.

Systems: Terrestrial

Ecology:

The diet of Lyle's flying fox consists mainly of ripe fruit. However, this species will also feed on nectar, pollen and blossoms to ensure it gets enough energy. Fruit is very low in protein and sodium, so the salivary gland of Lyle's flying fox has become specially adapted to ensure this species can extract the required nutrients .

Lyle's flying fox's primary sense when foraging is vision, as it lacks the echolocation abilities of insectivorous bats. It has well developed teeth which are used to chew fruit while spitting out most of the seeds and pulp. Some seeds are ingested, which is important for the local ecosystem as it allows for the dispersal of seeds into other areas.

Although it is a nocturnal species, Lyle's flying fox is very sociable and noisy during the day, as this is when females suckle their young . Large noisy colonies are very conspicuous, but they have few natural predators and so can hang safely up in the trees all day .

Unlike other bat families, fruit bats do not hibernate. Instead, Lyle's flying fox produces heat by shivering, which keeps its body temperature between 33 and 37 degrees Celsius.

Lyle's flying fox, and other species of bat, have been found to harbour large reservoirs of the Nipah virus. Although the virus does not harm the bat, it can be deadly to humans, and there is substantial data suggesting that recent outbreaks were caused by bat to person transmission of the virus .Nothing is known from China about the ecology of this species (Smith et al. 2008). In other areas, it is known to form large colonies in trees that can become stripped of leaves by the bats? activity. It feeds in orchards and is regarded as a serious pest in **Thailand** (S. Bumrungsri pers. comm.). It occurs in mangrove forest in Vietnam (Son Nguyen Truong pers. comm.). This species travels up to 50 km between colonies (S. Bumrugsri pers. comm.).

The most common behaviours performing by bats were sleeping, grooming and wing flapping. Sleeping, grooming, wing spreading, movement, wing flapping, aggression and mating/courtship were varied significantly throughout the day, while nursing/maternal was the only behaviour that did not change between times of day. Sleeping, grooming and mating/courtship occurred mostly in morning and late afternoon, while aggression and wing flapping occurred mostly in morning and noon respectively. Seasonal variation was mostly found in grooming, wing spreading, movement and wing flapping. Grooming, wing spreading and movement were more frequency found in wet season than in dry season, while wing flapping was less frequent in wet season than dry season. This study can conclude that the season and the times of day have influenced on daytime behaviours of Lyle's flying foxes.

Mating and Courtship:

There were significant differences in number of bats performing this behaviour between wet and dry seasons in some durations of the day ($p<0.05$. The percentages of bat mating and courtship in dry season were significantly higher than in wet season between 0600 to 0800 hours, and between 1600 to 1800 hours. There were also significant differences in the percentage of bats mating and courtship during the day both in wet and dry seasons ($p<0.05$). The highest percentage of mating and courtship was recorded in morning (0600 to 0900 hours) and in evening (1700 to 1800 hours). Increasing in mating and courtship activity levels was suggested to be the result from the elevations of testosterone. Testosterone secretion has a circadian rhythm, highest in the morning and lower in the afternoon.

Nursing:

There were differences in number of bats performing this behaviour between wet and dry season between 0700 to 1100 hours, but there was no significant difference in the percentage of bats performing this behaviour among time of day both in wet and dry seasons (*p*>0.05, figure. 8). This result can be explained by the number of lactating females with juvenile attached. Similar numbers of lactating females with juvenile attached were observed in two sampled periods (wet season and dry season). Lactating females usually take care their pups throughout the day, so the time of day has no influence on this behaviour. However, Boonneung (1977) found the negative relationship between precipitation and juvenile number. In the middle of wet season, number of juveniles was low because rainfall has negative impact on giving birth and maternal care.1 During field observations, maternal care was exhibited by females only. Environmental factors such as temperature and precipitation would be the cause of seasonal variation in daytime behaviour of Lyle's flying foxes because they are exposed to open environment while roosting on tree. Internal factors such as hormones can also affect on Lyle's flying foxes behaviour. Owing to the short observation periods, we are not able to explain all behaviours that we observed in Lyle's flying foxes. We need further study for intensive knowledge on Lyle's flying foxes behaviour.

Major Threat(s):

In **Thailand** the species is threatened through loss of roosting habitat, as existing trees die and are not replaced, it is also subject to hunting (S. Bumrungsri pers. comm.). It is also threatened in Cambodia by hunting (P. Bates pers. comm.).

Conservation Actions:

This species is listed on Appendix II of CITES. There are no known populations within protected areas in Vietnam (Son Nguyen Truong pers. comm.) or Cambodia. The species is protected by monks in **Thailand** (S. Bumrungsri pers. comm.).

Large Flying Fox (*Pteropus vampyrus)*

The Large Flying Fox *(Pteropus vampyrus)*, also known as the Greater Flying Fox, Malaysian Flying Fox, Kalang or Kalong, is a species of megabat in the family Pteropodidae. Like the other members of the genus Pteropus, or the Old World fruit bats, it feeds exclusively on fruits. It is noted for being the largest member of the bat family by wingspan. It, as with all other Old World fruit bats, lacks the ability to echolocate. It should not be confused with *Acerodon jubatus* which is also large. Large flying foxes inhabit tropical forests and swamps. They occur primarily in secondary forests and use agricultural areas during forging bouts. Populations also occur on oceanic islands. During the day, groups often roost in large trees. Roost sites are often used for many years and trees become stripped of bark and foliage by the bats' sharp claws. Roosting trees are often found in mangrove forests, coconut groves, and mixed fruit orchards. Studies in Subic Bay, Philippines have shown that foraging locations range between 0.4 and 12 km from the roost. They prefer undisturbed forests in lowlands, beaches, and mangroves, for roosting and select against disturbed and agricultural areas. Large flying foxes are commonly found in riparian areas. (Tampa's Lowry Park Zoo, 2010; Bates, et al., 2011; Mildenstein, et al., 2005; Tampa's Lowry Park Zoo, 2010)

Description:

HB: 300-340; T: None; FA: 190-210; E:40; HF:50; W: 645-1100g

The large flying fox is among the largest species of bats. It weighs 0.6 (1.3 lbs) to 1.1 kg (2.4 lbs) and has a wingspan of 1.5 m (4.9 ft) on average. As with all mega bats, it has a fox-like face, hence its name. It lacks a tail and has pointed ears. The hairs on much of its body is long and woolly but are shorter and more erect on the upper back. The mantle hairs tend to be the longest. The colour and texture of the coat differ between sexes and age classes. Males tend to have slightly stiffer and thicker coats than females. Immature individuals are almost all dull gray-brown. Young have a dark-coloured mantle that becomes lighter in males when they mature. The head has hairs that range in colour mahogany-red and orange-ochreous to blackish. The ventral areas are brown or blackish, tingled with chocolate, gray or silver. The mantle can vary from pale dirty-buff to orange yellow, while the chest is usually dark-golden brown or dark russet. The large flying fox has a large and robust skull. The dental formula is2.2.1.1over 2.3.2.3 . It has a total of 34 teeth. The large flying fox's wing are short and somewhat rounded at the tips. This allows them to fly slowly but with great manoeuvrability. Wing membranes are only haired near the body.Pteropus vampyrus is one of the largest bats in the world. Forearm length ranges from 180 to 220 mm, mean wingspan is 1.5 m, and body mass ranges from 0.6 to 1.1 kg. It has long pointed ears and a dog-like or fox-like face and head. Pelage varies in colour and texture with age and sex .

Upper dorsal fur is short and stiff, with longer, woollier fur on the Venter. Head and upper body are covered with a dark mantle ranging in colour from mahogany-red to black, and the Venter is often darker than the rest of the body. Its wings have short rounded tips. Except for parts close to the body and the edge of the wing membrane, wing membranes lack fur. Young are born with dark skin and fur but become paler as they develop. Males have thicker and stiffer pelage than females and glandular neck tufts with dark bases. *Pteropus vampyrus* is different from most Pteropus in that it has darker underparts and a dark mantle. For example, *P. giganteus* and *P. lylei* have pale underparts that contrast with the darker dorsal pelage. Pelage colour occasionally varies as a few *P. vampyrus* specimens have lighter mantles, and some have a gray or silver Venter. (Hollister, 1913; Kunz and Jones, 2000;

Large flying foxes have robust skulls, with a nearly complete orbit and a thick, wide zygomatic arch. The postorbital processes reaches more than halfway to the zygomatic arch. The dental formula is 2/2, 1/1, 3/3, 2/3 for a total of 34 teeth. Upper canines have a prominent anterior groove and a smaller groove on the inner surface. (Hollister, 1913; Kunz and Jones, 2000.

Ecology:

The large flying fox ranges from **Thailand** to the Malay Peninsula, stretching to the Philippines in the east and Sumatra, Java, Borneo and Timor in the south. In certain areas, the bat prefers coastal regions, but it can also be found at elevation up to 1,370 m. Flying foxes inhabit as primary, mangrove forest, coconut groves, mixed fruit orchards and a number of other habitats. During the day, trees in mangrove forests and coconut groves may be used as roosts. In Malaysia, flying foxes prefer in lowland habitats below 365m. In Borneo, flying foxes inhabit the coastal areas by move to nearby islands to feed on fruit. Flying fox roost in the thousands (maximum). One colony was recorded numbering around 2,000 individuals in a mangrove forest in Timor. In general, mangrove roosts have lower numbers of resting bats compared to lowland roost sites which could mean that mangroves forests are only used temporarily. This species primarily feeds on flower, nectar and fruit. When all three food item are available, flower and nectar are preferred. The pollen, nectar and flower of coconut and durian trees as well as the fruits of rambutan, fig and langsat trees are consumed. Flying foxes will also eat mangoes and bananas. With fruit, the flying fox prefers the pulp, and slices open the rind to get it. With durian tree flowers, the flying fox can lick up the nectar without doing apparent damage to the flower.Behavior Large flying foxes are highly social and vocal animals that live in groups sometimes numbering in the thousands. They prefer to roost in tall trees that rise above the forest canopy. Roost sites are often loud and may include several species. Large flying foxes are nocturnal, leaving the roost at around sunset and returning at dawn. Some individuals fly up to 50 km each night to reach their feeding grounds. They often fly the same route to a feeding ground, returning until all resources are exhausted. They form groups ranging from 2 to 50 at feeding grounds. They usually land on the tips of the branches and fall into a position with their head down to feed. (Kunz and Jones, 2000; Oakland Zoo, 2011; Tampa's Lowry Park Zoo, 2010)

Large flying foxes are highly territorial and communicate ownership by spreading their wings, growling, or making other vocalizations. The presence of flowers on trees appears to encourage territorial behaviour. They are often met with hostile vocalizations and aggressive behaviour that promotes spacing between roosting individuals. Large flying foxes rest by hanging upside down with wings wrapped around their bodies. During the warmest periods of the day, they sometimes cool themselves by fanning their wings, licking their bodies, or by panting. Roost activity increases as the day progresses and may include short flights around the roost. (Kunz and Jones, 2000;

Behaviour and life history:

Colonies of large flying foxes fly in a scattered stream. They may fly to their feeding grounds for up to 50 km in one night. Vocalization are not made during flight. Large flocks fission into family or feeding groups upon arrival at feeding grounds. Flying foxes may circle a fruit tree before landing, and usually land on the tips of branches in an upright position, then falling into a head-down position from which they feed. Feeding aggregations tend to be very noisy.

Flowers on trees are the basis territories in this species. Territorial behaviour includes growling and the spreading of wings. During antagonistic behaviour, individuals maintain spacing with wrists/thumbs sparing, bites, and loud vocalizations. When moving to a suitable resting place after landing, an individual may fight with conspecifics along the way. A roosting flying fox is positioned upside down with its wings wrapped up. When it gets too warm, a flying fox fans itself with its wings. Roosting bats are restless until mid morning.

Female large flying fox gestations are at their highest between November to January in Peninsular Malaysia, but some births occur in other months. In Thailand, gestations may take place during the same period with young being born in March or early April. Females apparently enter partus during April and May in the Philippines. This species usually gives birth to only one young. For the first days, the mothers carry their young but leave them at the roost when they go on their foraging trips. The young are weaned by 2–3 months. Unlike other pteropods, which have fused horns on the baculum, Pteropus vampyrus has a saddle-shaped baculum. It ranges from 4.5 to 8.2 mm and is wider than it is long. Females typically give birth to only one offspring per year. Synchronous birthing occurs within each population, and timing depends on local geography and seasonality. In peninsular Malaysia, mating peaks from November to January. In **Thailand,** birthing peaks during March and April and in the Philippines it peaks during April and May. In captive populations, birthing peaks during May and June. Mean body mass newborns is 133g (20-30% of maternal body mass) and their forearm length is around 79.5 mm. Mothers carry their young during the first few days after parturition, then leave them at the roost during foraging bouts. Young are weaned by 2 to 3 months after birth. (Kunz and Jones, All parental care is provided by the mother, though males often help protect and defend their harem. 2000)

Status:

A recent update by the IUCN has listed the species as Near Threatened and mentioned its near-vulnerable status with the following reasons:
Listed as Near Threatened because this species is in significant decline (but at a rate of probably less than 30% over ten years or three generations) because it is being over-harvested for food over much of its range, and because of ongoing degradation of its primary forest habitat, making the species close to qualifying for Vulnerable under criterion A.

One threat to the large flying fox is habitat destruction. Flying fox are sometimes hunted for food and the controls on hunting seem to be unenforceable. In some areas, farmers consider flying foxes pests as they sometimes feed on their orchards.

Genus: Rousettus:

Geoffroy's Rousette (*Rousettus amplexicaudatus*)

Distribution:

The Geoffroy's Rousette occurs throughout the whole of Southeast Asia: Myanmar, **Thailand,** Cambodia, Laos, Vietnam, Singapore, Indonesia, the island of Borneo, East Timor, the Philippines, the Solomon Islands and Bismarck Archipelago, and Papua New Guinea. This species is known from Yunnan in China (Smith et al. 2008) extending to Cambodia, **Thailand,** Myanmar, Vietnam (Borissenko and Kruskop 2003), Lao PDR (Duckworth et al. 1999), Peninsular Malaysia through Indonesia (including Java and Bali), and the Philippines. It is present on many of the Molucca Islands, including the islands of Halmahera, Morotai, Buru and Seram. It is widespread throughout much of New Guinea (Indonesia and Papua New Guinea), the Bismarck Archipelago and adjacent islands. It has been recorded from the island of Bougainville and Buka (S. Hamilton pers. comm.) and many of the Solomon Islands as far south as the island of San Cristobal.

It is found throughout the Philippines, with records from Balabac, Babuyan Group (Garcia pers. comm.), Barit, Biliran, Bohol, Boracay, Busuanga, Caluya, Camotes (L. Paguntalan pers. comm.), Carabao, Catanduanes, Cebu, Dalupiri, Dinagat, Fuga, Ilin (J. C. Gonzales pers. comm.), Jolo, Leyte, Lubang, Luzon (Abra, Cagayan, Ilocos Norte, Isabela, Laguna, Quezon, Rizal, Sorsogon, Zambales provinces), Maripipi, Marinduque, Mindanao (Bukidnon, Davao del Norte, Davao del Sur, Lanao del Norte, Lanao del Sur, Misamis Oriental, Misamis Occidental (Ramayla pers. comm.), South Cotabato, Surigao del Sur, Zamboanga del Sur), Mindoro, Negros, Palawan , Panay, Polillo, Samal, Samar (J. C. Gonzales pers. comm.), Semirara, Siargao, Sibay, Sibuyan, Siquijor, Tablas, Ticao (L. Paguntalan pers. comm.), Tincansan (Alcala and Alviola 1970; Heaney et al. 1998). It has been recorded from sea level up to 2,200 m asl.

Countries: Native:

Brunei Darussalam; Cambodia; China; Indonesia (Bali, Jawa, Lesser Sunda Is., Maluku, Sulawesi, Sumatera); Lao People's Democratic Republic; Malaysia; Myanmar; Papua New Guinea (Bismarck Archipelago); Philippines; Singapore; Solomon Islands; **Thailand;** Timor-Leste; Vietnam. Population: This is a locally abundant species in Southeast Asia and the Philippines (Utzurrum 1992). In Southeast Asia it has probably experienced declines due to hunting as a large cave roosting species (C. Francis pers. comm.).

Population Trend: Unknown

Description:

HB: 105-115; T:15-17; FA: 79-87; E: 18.5-19.7; HF: 20-22.8; W: 54-75g

Like other fruit bats, *R. amplexicaudatus* have sensitive hearing and sense of smell and good eyesight which help them to manoeuvre well during flight, specifically at night. What makes them different from other fruit bats is their echolocating ability. It can be distinguished by their grey-brown to brown upperparts which is darker on top of their

They have long pale hairs on the chin and neck despite having short and sparse fur. They sometimes have pale yellow tufts of hair on the side of the neck which occur in adult for this species, especially males. Most males are substantially larger than females. The most distinguishable figure of this bat besides producing a distinctive, audible clicking call is their wings. It is attached to the sides of the back and separated by a broad band of fur. The lower incisors are bifid, the canines have a longitudinal groove on the outer surface which is slightly medial to centre and the first premolars are smaller than second premolars, especially on the upper jaw.

Biology and Ecology:

This is a colonial species which forms cave roosts of several thousand animals. Roosts are known from caves, rock crevices, and old tombs. It can be found in a wide variety of habitat types including secondary forest, agricultural areas, and other disturbed habitats like rural gardens, fruit orchards and at the forest edge (Heaney et al. 1991, 1998; Heideman and Heaney 1989; Lepiten 1995; Rickart et al. 1993). It is present, but less common, in primary tropical moist forest.

They may travel long distances each night, in search of appropriate fruit. Rousettus are known to use a primitive form of echolocation while foraging. Gestation is thought to be about 15 weeks, and lactation about three months. The species may have two litters of a single young annually in New Guinea (Bonaccorso 1998).

Systems: Terrestrial

Specimens in the Sabah Museum were collected from coconut plantations on Mantani Island and the highland of Crocker Range while the one from Sarawak was from Niah Cave. This medium-sized bat normally roosts in caves; feeds on fruit, nectar and pollen (Payne et al. 1985). It roosts dark caves, rock crevices and old tombs (Lekagul and McNeely 1977).

The Monfort Bat Cave in the Southern Philippines has the largest gathering of these bats.

Major Threat(s):

Overall there are no major threats to this species. It is regarded as a pest in some parts of its range. In the Philippines and Indochina it is subject to intense hunting at some cave roosts (Utzurrum 1992).

Conservation Actions:

It is present in several protected areas. Identification and protection of important roosting sites would benefit the conservation of this species across its range. In the Philippines, caves are protected under the Cave Management Act, although this is not fully enforced. Elsewhere in southeast Asia the species requires protection from hunting and education is needed on the actual impact of the species on orchards (C. Francis pers. comm. 2006).

Genus: Sphaerias:

Blanford's Fruit Bat (*Sphaerias blanfordi)*

Description:

FA = 52.14, TL = none, EL = 18.58, HD = 27.56, TB = 20.25, WT = 31.00, HB = 78.63, HF = 11.23, D5 = 69.36; 1 sample. An immature (B0777) had a forearm length exceeding the range recorded by Lekagul and McNeely (1977). All other adult, sub adult and immature were within range of external measurements given by Lekagul and McNeely (1977).

The basic colour is dull, greyish brown above and light paler, underneath. The inter femoral membrane is the least developed amongst *cynopterine* fruit bats, extending along the femur and upper parts of the tibia, with no calcar; the membrane is covered with soft, fine, dense hair. The ears are dark and hairless, with the anterior margin edged in white, as in *Cynopterus* The nostrils are situated well back from the mouth, with the lower jaw slightly underslung.The skull is less heavily built than in *Cynopterus,* with very slight post orbital processes and a weak zygomatic arch; there is no postorbital foramina. The incisors are close together, robust, and have triangular pointed crowns; the incisors tooth rows are slightly curved, with both upper and lower incisors situated entirely in front of a line connecting the canines. The canines have no secondary cusps and are slanted outwards. Anderson (19120 suggests that because of the unique shape and position of the incisors. these teeth must be used differently than in other *Cynopterus* forms.

Range Description:

This species ranges from northern South Asia, into southern China and Northern parts of Southeast Asia. In South Asia, this species is presently known from Bhutan (Ganghlakha), India (Arunachal Pradesh, Mizoram, Uttar Pradesh and West Bengal) and Nepal (Eastern Nepal) (Molur et al. 2002). In China it is present in southern Xixang and eastern Yunnan. In Southeast Asia, there are records from Myanmar, **Northern Thailand** and Vietnam. It has been recorded from 308 to 2,710 m asl.

Countries: Native:

Bhutan; China; India; Myanmar; Nepal; **Thailand**; Vietnam

Population:

In South Asia, although it is widespread, it is not a common species (Molur et al. 2002). In China, the species is little known, but is believed to be most common in lower montane forests. It is uncommonly encountered in Southeast Asia.
Population Trend: Unknown

Blanford's Fruit Bat *(Sphaerias blanfordi)* is a mountain species of megabat. The only recording of it was from the **Doi Suthep-Pui National Park in Chiengmai, Thailand.** According to Lekagul and McNeely (1977), the species is confined to the Thai-Burmese border, Uttar Pradesh in India and Eastern Nepal.

Biology and ecology:

A male and three female *S. blanfordi* were netted in a secondary habitat at 1020 m elevation. In May 1997, an adult female was lactating, two immature female fledglings and a sub adult were observed at Doi Suthep. The bats were found in the nets the morning, presumably caught between the last net check at midnight and dawn. All *S. blanfordi* were caught approximately 0.8 m to 1.5 m above the ground. One site had about 50% canopy cover, relatively dense undergrowth with saplings and the net was set across a fire break on a slope. The other catching site had about 90% canopy cover, relatively sparse undergrowth with some seedlings, ginger and Pandanus plants. The net was set between two clumps of banana plants on a slope. During the survey, the night temperature was cool (approximately 20oC) and misty around daybreak. There is little ecological information recorded for this species (Lekagul and McNeely 1977).Little has been recorded about the habitat or ecology of this species other that it is known to inhabit bamboo forests (Molur et al. 2002).
Systems: Terrestrial

Major Threat(s):

In South Asia, this species is threatened by habitat loss largely through commercial harvest of bamboo (Molur et al. 2002). There appear to be no major threats to the species in Southeast Asia. In China, the threats to this species are unclear although it may be threatened by ongoing habitat degradation.

Conservation Actions:

In South Asia, it is considered to be vermin under Schedule V of the Indian Wildlife (Protection) Act. It has been recorded from protected areas in India like Namdapha National Park in Assam, Phambong Lho Wildlife Sanctuary in Sikkim. In Southeast Asia it has been recorded from a number of protected areas. In South Asia, field surveys, population monitoring, habitat management and public awareness initiatives are immediate recommendations (Molur et al. 2002).]

Subfamily: Macroglossinae:

Genus: Eonycteris:

Lesser Dawn Bat (*Eonycteris spelaea*)

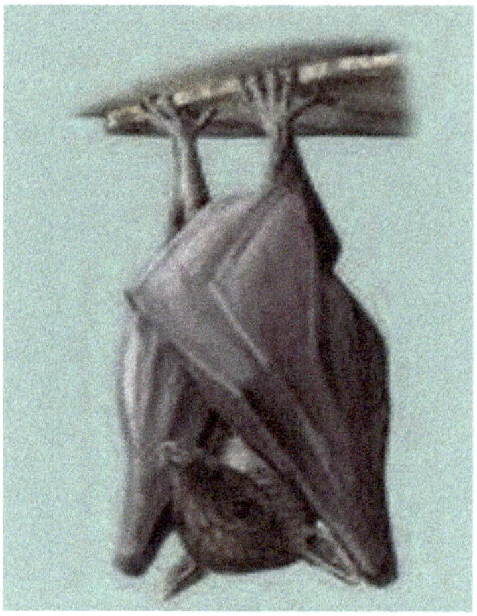

Description:

HB: 80-100; T:15-20; FA: 60-75; E: 18-24; HF: 20-24; W;37-82g

The fur is short and thin, with the upperparts dark to light brown and the underparts greyish; the head is darker, with a more thinly haired area at the back of the neck which is often tinged yellow. Males on average are larger than females, and adult males have a well marked ruff on the sides of the neck, which is darker than the back and chest. On each side of the anal orifice is a small kidney shaped subcutaneous gland, the function is yet unknown. The mandible is modified for nectar eating, with the front portion formed into a spout shaped depression. Tate (1942b) says this is for the attachment of the greatly developed muscles whose contraction causes the tongue to protrude. The first lower incisor as a shallow groove on the occlusal surface. The canines are long, strong, well separated and grooved on the front. The first premolars are quite small, with the second premolars dominant and the following teeth progressively lower crowned. The last molars are very small, and the lower ones may be absent.

Range Description:

This species ranges widely from Northern South Asia, into Southern China, and much of Southeast Asia. In South Asia it is known from India (Andaman and Nicobar Islands, Andhra Pradesh, Assam, Karnataka, Manipur, Meghalaya, Mizoram, Nagaland, Sikkim, Tamil Nadu (Vanitharani et al. 2005) and Uttaranchal) and Nepal (Far Western Nepal) (Molur et al. 2002). In China it is found in south-western Guangxi and Yunnan (Smith and Xie 2008). In mainland Southeast Asia, it ranges from Myanmar in the west, through **Thailand,** Lao PDR, and parts of Vietnam and Cambodia to Peninsular Malaysia. In insular Southeast Asia, it ranges from Indonesia (Sumatra, Java, Bali, Lombok, Sumba, Sulawesi, Muna, Sanana, Halmahera, Batjan and Tidore), much of the island of Borneo (Brunei, Indonesia and Malaysia), the island of Timor (East Timor and Indonesia) to the Philippines where is found throughout the country, except the Batanes/Babuyan region. It has been recorded from Biliran, Bohol, Camotes (Paguntalan pers. comm. 2006), Carabro (Alcala and Alviola 1970), Catanduanes, Cebu, Danjugan (Carino 2004), Ilin (Gonzalez pers. comm. 2006), Leyte, Luzon (Abra, Cavite, Ilocos Norte, Laguna, Pampanga, Rizal, and Sorsogon provinces), Marinduque, Maripipi, Masbate, Mindanao (Agusan del Norte, Davao del Sur, Davao Oriental, Lanao del Norte, Misamis Oriental, Misamis Occidental (Ramayla pers. comm.), South Cotabato, Surigao del Norte, and Zamboanga del Sur provinces),

Mindoro, Negros, Palawan, Polillo, Samar (Gonzalez pers. comm. 2006), Sanga-sanga, Sabtang, Siargao, Sibuyan, Siquijor, Tablas, and Ticao (Paguntalan pers. comm. 2006, Heaney et al. 1998). It occurs from sea level to 1,000 m asl.

Countries: Native:

Brunei Darussalam; Cambodia; China; India (Andaman Is., Andhra Pradesh, Assam, Karnataka, Manipur, Meghalaya, Nagaland, Nicobar Is., Sikkim, Tamil Nadu, Uttaranchal); Indonesia; Lao People's Democratic Republic; Malaysia; Myanmar; Philippines; Singapore; **Thailand;** Timor-Leste; Vietnam

Population:

More than 4,000 individuals were recorded from Batu caves in Malaysia (Bates and Harrison 1997). The abundance, population size and trends for this species are not known in South Asia although a recent discovery of this species in Kalakkad-Mundanthurai Tiger Reserve in Tamil Nadu suggests a comparatively rare occurrence compared to the *Rousettus leschenaulti* (Vanitharani et al. 2005). In the Philippines, this species is often found in colonies of thousands in caves, even in agricultural areas (Heaney et al. 1989; Heideman and Heaney 1989; Lepiten 1995; Rickart et al. 1993). On Palawan Island two populations were found, one exceeding 2,000 individuals and another that probably exceeded 50,000 individuals (Esselstyn et al. 2004).

Population Trend: Unknown

Ecology:

The Dawn Bat is a small Southeast Asian rainforest bat which lives in the mangrove and lowland forests of Malaysia, Indonesia and Borneo. Its unique and special place in the rainforest ecosystem has only been recently discovered. This is a cave roosting bat forming compact clusters and cohabits with other bats. It roosts in large groups consisting of thousands of individuals in caves in forested areas. They are more often found in disturbed and agricultural areas, although they do occur in primary forest as well. It is a nectar eating bat with a slow flight, and has adapted to using the flowers of many important agricultural and orchard crops (Smith and Xie 2008). It has been also reported living in small groups from the attics of village huts in northeast India (Tarapada Bhattacharyya pers. comm. June 2005) and Myanmar (Khim Maung Swe pers. comm. January 2000).

Systems: Terrestrial Dawn Bats have long and slender snouts which give them a dog-like appearance. Their ears are small and pointed and their eyes are round and small. The fur on the dawn bat's back is dark brown and lighter on their underside. Males have a ruff, or fur collar around their neck, which is a little darker in colour then their head and chest hair. Females only have sparse hairs around their necks. The tongue of a Dawn Bat is long and can be thrust forward into flowers, while brush-like projections allow it to pick up pollen and nectar.

The male dawn bat is larger than the female. They weigh from 55 to 82 grams while females weigh 35 to 78 grams. Dawn Bats are 8.5 to 12.5 cm long from head to tail. They either have a very small tail or none at all. Their forearm length is 6 to 8 cm long. The genus Eonycteris doesn't have a claw on their index finger like most other bats.

Female Dawn Bats reach sexual maturity after one year while males become sexually mature in two years. Females can have babies at any time of the year. Gestation is a little longer than 6 months, sometimes as long as 200 days. Usually one pup is born at a time. It attaches itself to a nipple and holds on for 4 to 6 weeks. After that time it starts to take little practice flights of its own. They aren't completely weaned until after 3 months.

Dawn Bats roost in limestone caves found in Malaysia and the Indonesian archipelago. Small groups of several dozen will roost in small shelters, while tens of thousands can inhabit larger caves. Dawn Bats are nocturnal and emerge from their caves at dusk to feed on pollen and nectar of night blooming plants.

Their favourite flowers are from the Mangrove Apple (Sonneratia alba)and the Durian (Durio zibethinus). The flowers of the durian tree are large and waxy and only bloom at night. As the Dawn Bat hangs onto the flower and pushes its nose deep inside to lick the nectar, pollen covers its face and chest. They are the main pollinators of these two trees and very important to their survival.

Peaches, bananas, avocados, kapok, hemp and latex are some other plants which depend on bats for pollination. In Southeast Asia some of the largest caves with bat colonies are being destroyed by limestone quarrying. Mangrove swamps, home to the Mangrove Apple, are being filled in for development, destroying one of the Dawn Bat's main food supplies. Bats have been accused of spreading disease and destroying commercial fruit crops. However, Dawn Bats prefer ripe and strong-smelling fruit. Commercial fruit is picked green for shipping and is therefore rarely damaged by these bats.

Very little is known about the importance of bats and how forests depend on them for their survival. But it is known that they are one of the main pollinators and seed dispersers for many tropical plants. Bats often make up more than half of the mammal species in a rainforest ecosystem and most species of bats have not been studied.

Although Dawn Bats are being killed by the thousands, they are not on the CITES or IUCN list as a vulnerable or protected species. Malaysia has a Wildlife Protection Ordinance (Amendment 76). However, Dawn Bats can be killed if they pose a threat to crops or property. Since many still think they are fruit pests, the ordinance does little to protect them. People need to become more educated about the Dawn Bat to understand how important they and other bats are to the forests and crops.

Major Threat(s):

There are no major threats to this species as a whole. In parts of South Asia it is locally threatened by deforestation, generally resulting from logging operations and the conversion of land to agricultural and other uses (Molur et al. 2002). Cave tourism and lighting is an emerging threat in some caves, as at Borra Caves, Andhra Pradesh (Srinivasulu and Srinivasulu 2003). In China they are often under heavy hunting pressure (Smith and Xie 2008).

Conservation Actions:

This species is adaptable, and is present in many protected areas throughout its range. No direct conservation measures are currently needed for the species as a whole. In South Asia, this species like most other fruit bats in India is considered a vermin under Schedule V of the Indian Wildlife (Protection) Act. The species has been recorded from protected areas like Kalakkad-Mundunthurai Tiger Reserve in Tamil Nadu (Vanitharani et al. 2005). Populations of this species should be monitored to record changes in abundance and distribution, and lobbying to remove it from vermin category are recommended (Molur et al. 2002). Awareness programmes on ill effects of cave/roost disturbance is recommended (C. Srinivasulu and Bhargavi Srinivasulu pers. comm. September, 2007).

Genus: Macroglossus:

Lesser Long Toothed Fruit Bat *(Macroglossus minimus)*

Description:

HB: 60-65; T:3-4; FA; 38-42; E:13-18; HF: 13-16; W; 12-18g

Having a clawed index finger, very short tail, long, feathery tongue, pale bark pelage without marks, and light brown membranes. It as a shorter rostrum than *sobrinus*. It as a total skull length of 26-28 mm as opposed to *sobrinus* 30 mm. Also a shorter muzzle and nostrils pointing more forward. There is a deep internarial groove that extends to the margin of the upper lip; there are shallower lateral grooves about 1mm on each side of the median groove. The underparts are lighter than in *sobrinus*

Range Description:

This widespread species ranges from Vietnam, Cambodia, **Thailand** and Peninsular Malaysia, to the Natuna Islands, Borneo (Brunei, Indonesia and Malaysia), the Philippines, the island of Sulawesi to the island of Timor (East Timor and Indonesia), and through much of the Moluccan Islands (Indonesia) including the islands of Halmahera, Buru, Seram and Ambon. It is present on the Kai Islands and the Aru Islands and throughout much of the island of New Guinea (Indonesia and Papua New Guinea), the Bismarck Archipelago (Papua New Guinea) and Fergusson Island (Papua New Guinea). It is present on the island of Bougainville (Papua New Guinea) and is found throughout much of the Solomon Islands. It ranges across much of Northern Australia (Corbet and Hill 1992; Flannery 1995; Strahan 1995; Bonaccorso 1998).

n the Philippines this species was recorded from Batu-bato, Biliran, Bohol, Boracay, Busuanga, Cagayan de Sulu, Calauit, Caluya, Camiguin, Carabao, Catanduanes, Cebu, Dinagat, Leyte, Luzon (Albay, Aurora, Cagayan, Camarines Norte, Camarines Sur, Isabela, Kalinga, La Guna, Quezon, Rizal, Sorsogon, Tarlac provinces), Marinduque, Maripipi , Masbate, Mindanao (Bukidnon, Davao del Norte, Davao del Sur, Davao Oriental, Lanao del Norte, Lanao del Sur, Misamis Oriental, South Cotabato, and Zamboanga del Sur provinces), Mindoro, Negros, Palawan, Panay, Polillo, Reinard, Sanga-sanga, Seimirara, Siargao, Sibay, Sibutu, Sibuyan, Simunul, and Siquijor (Alcala and Alviola 1970; Catibog-Sinha 1987; Heaney et al. 1998, Heaney et al. 2004).

Countries: Native:

Australia; Brunei Darussalam; Cambodia; Indonesia; Malaysia; Papua New Guinea; Philippines; Singapore; Solomon Islands; **Thailand**; Timor-Leste; Vietnam

Population:

It is a common species (Strahan 1995; Bonaccorso 1998). It is abundant in disturbed areas but rare in old growth forest. In the Philippines, it is abundant and widespread, with populations stable or increasing (Utzurrum 1992; Heaney et al. 1998).

Population Trend: Stable

Habitat and Ecology:

It is found in both primary and secondary tropical moist forest, it has also been reported from paper bark woodlands, mangroves, swamp forest, plantations, rural gardens and urban areas. In the Philippines, it occurs in virtually every habitat from sea level up to at least 2,250 m asl (Heaney et al. 1998), preferring disturbed habitats. It usually roosts as single animals, or in small groups, under large leaves (such as palm fronds), under branches and loose bark, in bamboo or in abandoned buildings. The females give birth to a single young three times per year (Flannery 1995; Strahan 1995; Bonaccorso 1998). Gestation time of that species in Malaysia ranges from 110 to 130 days, followed by a lactation period of 60 to 70 days. On average, the length between pregnancies ranges from 140 to 160 days, so that *Macroglossus minimus* likely produces 2 to 2.5 offspring per year. Like most other bats, they give birth to a single young. The degree of seasonality present in reproduction of M. minimus varies with location. For example, reproduction on the island of Maripipi was not seasonal, while 9 km away on Biliran reproduction was more synchronous. (Bates and Harrison, 1997; Lekagul and McNeely, 1988) During the day, when inactive, *Macroglossus minimus* of Papua New Guinea roost individually. They have home ranges of around 5.8 hectare with very little overlap. This, along with prominent sternal scent glands, suggests territoriality, although this has not been definitively shown. In addition, *Macroglossus minimus* engages in "probing forays," potentially to scout new habitat. When temperatures reach 11-29 degrees Celsius, *Macroglossus minimus* enter a torpid state, maintaining a body temperature of 2-5 degrees Celsius above the ambient temperature.
(Bonaccorso and McNab, 1997; Winkelmann, et al., 2003)

Systems: Terrestrial

Major Threat(s):

There are no major threats to this species.

Conservation Actions:

The species occurs in a number of protected areas throughout its range.

Greater Long Toothed Fruit *Bat* *(Macroglossus sobrinus)*

Distribution:

M. sobrinus were mist-netted from Doi Suthep, Sungai Enam, Kuala Gandah, Wang Pinang Malaysia to Taleban on the Asia Mainland. This species is distributed in Burma, **Thailand,** Vietnam and Peninsular, Sumatra, Java and Bali (Lekagul and McNeely 1977, Medway 1978). This species is commonly found in hilly areas in Malaysia (Lim 1966).Range Description: This species ranges from north-eastern South Asia and Southern China, to much of Southeast Asia. In South Asia it is found in India (Arunachal Pradesh, Meghalaya, Tripura and West Bengal) (Molur et al. 2002). In Southern China, it is found in Mengla, Xishuangbanna and Yunnan (Smith and Xie 2008). In Southeast Asia, it ranges from Myanmar in the West, through **Thailand,** Lao PDR, Vietnam, possibly Cambodia (reports of this species from Cambodia cannot be confirmed [Kock 2000]) and Peninsular Malaysia, and from here occurs in Indonesia (the Mentawi Islands, Sumatra and Java). It has been recorded from elevations of up to 2,000 m asl.greater long-tongued fruit bats occupy a variety of habitats, including mangrove swamps, montane forests, and lowland forests. When range overlaps occur between *Macroglossus sobrinus* and *Macroglossus minimus,* it appears that *M. minimus* is restricted to coastal mangrove swamps while *M. sobrinus* is found in montane and lowland forests. A separate study in **Thailand f**ound the species in evergreen forests below 1918 m, where pollen is available year round. (Bates and Harrison, 1997; Lekagul and McNeely, 1988)

Population:

This is generally a locally common species. Nothing much is known of its population in South Asia (Molur et al. 2002).

Population Trend: Stable

Biology and ecology and Description

This species is found to prefer lowland and montane forest. It occurs in secondary habitats and also in old fruit orchards, villages, and banana plantations, rarely recorded in mangroves also common in association with humans, particularly under roofs of houses. Roosts singly or as a few animals, commonly in rolled up banana leaves. Not recorded roosting in caves (Bates and Harrison 1997).

Systems: Terrestrial

Five adult males and five adult females were recorded. On 25 March 1997, a full term foetus was aborted by a full term pregnant female (MTA97347) taken from Wang Pinang. Males with enlarged testes suggested that they were of sexually active state at Taleban in March 1997, Kuala Gandah in July 1997 and Doi Suthep in May 1997.

M. sobrinus that were netted were all associated with banana plants in most sites and with flowering trees in Doi Suthep.

External measurements

FA = 46.12±1.25, TL = 3.81±1.07 (6), EL = 15.38±1.00, HD = 30.22±1.39, TB = 18.29±0.61, WT = 20.22±3.23, HB = 73.92±4.45, HF = 9.29±0.89, D5 = 58.50±1.73.

M. sobrinus can be isolated from *M. minimus* by using external morphological measurements. Comparatively, *M. sobranus* is than *M. minimus* in the forearm length (43.35 to 46.95 mm vs 40 to 43 mm) and head length (30.05 to 30.54 mm vs 26 to 28 mm) and heavier body weight (17 to 23 g vs 13 to19 g) (Medway 1978). However, the tail length (0-2 mm versus 0-5 mm) and hind foot (8 - 9 mm versus 15-18 mm) disagreed to that of Lekagul and McNeely (1977).Greater long-tongued fruit bats have slender, delicate skulls with long rostra. Length of skull is at least 30 mm. The skull has a low occipital crest, a palate ending beyond the last molar, and weakly developed postorbital processes. The braincase is deflected downward. Males have larger skulls than females. The mandible has low processes and is long and slender with a groove for the tongue at the front. The highly protrusible tongue has a feathery tip made of filiform papillae and is used in feeding on pollen. (Bates and Harrison, 1997; Lekagul and McNeely, 1988)

Greater long-tongued fruit bats have reduced dentition: cheek teeth are narrow and low-crowned. All teeth are well spaced linearly along the jaw. The incisor tooth row is elliptical with a large gap between both the upper and lower incisors through which the tongue protrudes. The gap between the lower incisors is larger relative to the upper gap. The incisors are forward-sloping and small, and the canines are short and sharp.
(Bates and Harrison, 1997; Lekagul and McNeely, 1988)

The long pelage of *Macroglossus sobrinus* lacks markings. It is soft, fine, and light reddish-brown, with the chest coloured lighter than the head. On the dorsal surface, the pelage is uniformly clay-brown. Hairs are present the inner wing membranes, forearms, and upper surface of the tibiae. Members of this species exhibit no sexual dimorphism in fur colour. The ears are medium-sized with small antitragal lobes. Both ear and wing skin is dark brown. Short hairs thickly cover the upper and lower surfaces of the interfemoral membrane. These bats lack an external tail and have reduced calcars. (Bates and Harrison, 1997; Lekagul and McNeely, 1988)

Basal metabolic rate has not been documented, but in their close relative *Macroglossus minimus,* it falls around 0.88 and 1.48 cm^3 of oxygen/g/hr for lower range bats and highland bats, respectively. *Macroglossus sobrinus* closely resemble *Macroglossus minimus,* but the paler under parts and longer muzzle are distinguishing characteristics of the former species. (Bates and Harrison, 1997; Bonaccorso and McNab, 1997; (Lekagul and McNeely, 1988) Little or no information exists on the mating systems of the species. However, male *Macroglossus sobrinus* have smaller testes relative to other mega chiropteran bats. Across species, testes size has a positive relationship with body size, but also correlates positively to group size, suggesting that members of this species do not live in large groups. Lower intraspecific sexual competition resulting from small or no group-living may be the cause of their reduced testes. (Hosken, 1998)

Breeding interval

Breeding interval is not known.

Breeding season

Greater long-tongued fruit bats may breed throughout the year.

Average number of offspring

One.

Greater long-tongued fruit bats probably breed year round. Other information on the reproduction of *Macroglossus sobrinus* is either limited or not available. The closely related *Macroglossus minimus* is better known. Gestation time of that species in Malaysia ranges from 110 to 130 days, followed by a lactation period of 60 to 70 days. On average, the length between pregnancies ranges from 140 to 160 days, so that *Macroglossus minimus* likely produces 2 to 2.5 offspring per year. Like most other bats, they give birth to a single young. The degree of seasonality present in reproduction of *M. minimus* varies with location. For example, reproduction on the island of Maripipi was not seasonal, while 9 km away on Biliran reproduction was more synchronous.

(Bates and Harrison, 1997; Lekagul and McNeely, 1988)

Behaviour:

Range territory size

1 to 2 km^2

Greater long-tongued fruit bats typically forage alone, with pairs found rarely. Individuals tend to roost alone, although small groups are also found. In Malaysia, these groups ranged from 5 to 10 individuals, most commonly found in palm trees, roofs, and banana plants. Macroglossus sobrinus, along with *Macroglossus minimus,* are responsible for the first peak feeding activity by Chiroptera each night. (Bates and Harrison, 1997; Gould, 1978)

Information on the behaviour of *Macroglossus sobrinus* is scarce, but more research has been done on the behaviour of its close relative, *Macroglossus minimus.* During the day, when inactive, *Macroglossus minimus* of Papua New Guinea roost individually. They have home ranges of around 5.8 hectare with very little overlap. This, along with prominent sternal scent glands, suggests territoriality, although this has not been definitively shown. In addition, *Macroglossus minimus* engages in "probing forays," potentially to scout new habitat. When temperatures reach 11-29 degrees Celsius, *Macroglossus minimus* enter a torpid state, maintaining a body temperature of 2-5 degrees Celsius above the ambient temperature. (Bonaccorso and McNab, 1997; Winkelmann, et al., 2003)

Communication and Perception:

Like most members of Pteropodidae, greater long-tongued fruit bats do not use echolocation. These bats emit audible calls of less than 9 kHz with a duration of about 15 milliseconds and inter-call intervals around 128 milliseconds. Macroglossus sobrinus occasionally emits double pulses or clicks. The purpose of these double pulses is not yet known. In stressful disturbance situations, these bats emit harsh broadband signals as distress calls. The call is a large series of clicks with duration greater than 100 milliseconds. (Funakoshi, et al., 1995)

Food Habits:

The diet of the greater long-tongued fruit bats varies among localities, but it appears to consist primarily of pollen and nectar, although soft fruit may be taken as well. In Malaysia, *Macroglossus sobrinus* was found to feed on the pollen and nectar of three species of wild banana, *Dubanga grandiflora* flowers, and soft fruit. Start (1974) estimated that an adult Macroglossus sobrinus could survive solely off two to three wild banana plants; due to the flowering habits of the banana, food would be available year-long. Generally, *Macroglossus sobrinus* is considered a banana specialist, although secondary food sources are also used. (Bates and Harrison, 1997; Lekagul and McNeely, 1988; Winkelmann, et al., 2003)

Greater long-tongued fruit bats move around 1 to 2 km per night while feeding, although this range may be significantly smaller in areas with higher densities of wild banana plants. It has been hypothesized (Gould, 1978) that individuals may follow set routes as they travel from plant to plant each night. (Bates and Harrison, 1997; Gould, 1978; Lekagul and McNeely, 1988)

Plant Foods

fruit; nectar; pollen; flowers

Major Threat(s):

There are no major threats to this widespread and adaptable species as a whole. In South Asia, this species is inferred to be locally threatened by habitat loss, largely through commercial extraction of bamboo and the conversion of land to agricultural use (Molur et al. 2002).

Ecosystem Roles:

One of the major roles that *Macroglossus sobrinus* fills is that of a pollinator to a variety of ecologically important plant species. A study in 2002 found that Macroglossus sobrinus serves as an effective pollinator for *Musa itinerans*, an important pioneer species in the rain forests of south-western China. A separate study found that *Macroglossus sobrinus* serves an important role in the pollination of a wild banana, *Musa acuminata*. (Gould, 1978; Itino, et al., 1991)

Macroglossus sobrinus populations in Cambodia serves as a potential host for a variety of viruses in the Lyssavirus family. This is evidenced by the presence of anti-lyssavirus neutralizing antibodies in serum samples. (Reynes, et al., 2004)

Economic Importance for Humans: Negative:

Greater long-tongued fruit bats have antigens for the Lyssavirus – a family of viruses containing rabies. It indicates that they may be carriers for the Lyssavirus, and could infect humans with rabies. Their large localized population density results in their being classified as vermin in India, although they do not harm people and are beneficial through pollination services. (Reynes, et al., 2004)

Predation:

Predators specific to *Macroglossus sobrinus* are unknown, but Old World fruit bats generally have few predators. Snakes, birds of prey, and carnivorous mammals occasional prey on them. Of these, snakes are most common predators of bats in roosts. While predation influences behaviour and feeding habits, it does not significantly affect population size. (Mickleburg, et al., 1992)

Conservation Actions:

This species is present in many protected areas, and no direct conservation measures are needed as a whole. Within India, it is considered to be vermin under Schedule V of the Indian Wildlife (Protection) Act. The species has been recorded from protected area in India namely Namdapha Wildlife Sanctuary in Arunachal Pradesh. Survey, population monitoring, habitat management and awareness to policy makers are critical recommendations for India (Molur et al. 2002).

Conservation Status:

IUCN Red List of Threatened Species Least Concern
More Information

US Federal List
No special status

CITES
No special status

State of Michigan List
No special status

Due to the large range of *Macroglossus sobrinus,* the species is considered stable. The lack of endemic subspecies coupled with the locally common nature of *Macroglossus sobrinus* contributes to this assessment. In addition, the species is present in multiple protected areas across South Asia, minimizing the need for conservation measures. Although the species is generally "safe," it may be locally threatened as a result of habitat loss due to bamboo extraction and the conversion of land for agricultural use. Molur et. al. (2008) recommend survey, population monitoring, and habitat management to the Indian government. (Bates and Harrison, 1997; Hutson, et al., 2008)

Other Comments:

Macroglossus sobrinus was considered to be a subspecies of *Macroglossus minimus,* from the works of E. Geoffrey in 1810 until the more recent research done by Hill in 1983. Currently, a variety of different authors consider *M. sobrinus* to be a distinct species, including Corbet and Hill (1992), Koopman (1993), Bates and Harrison (1997) and Simmons (2005). All Indian specimens are referred to by the subspecies name of *Macroglossus sobrinus sobrinus.* The species name *Macroglossus sobrinus* is also synonymous with *Macroglossus fraternus.* The common name is variable, although most sources use either "hill long-tongued fruit bats" or "greater long-tongued fruit bats". (Bates and Harrison, 1997; Lekagul and McNeely, 1988; Hutson, et al., 2008; Wilson and Reeder, 1993)

Family: Vespertilionidae:

Subfamily: Kerivoulinae:
Genus: Kerivoula:
Hardwicke's Woolly Bat (*Kerivoula hardwickii)*

Range Description:

This widespread species ranges from Southern and Northern South Asia, Southern China, and throughout continental and insular Southeast Asia. In South Asia, this species is presently known from Bangladesh (no exact location) (Das, 2003, Srinivasulu and Srinivasulu, 2005), India (Assam, Jammu and Kashmir, Karnataka, Meghalaya, Mizoram, Nagaland and West Bengal), Pakistan (Punjab) and Sri Lanka (Central Province) (Srinivasulu et al. in press). In China, it is widely distributed in the Southeast of the country, including the island of Hainan. In Southeast Asia, it is found throughout the mainland, in insular Southeast Asia it ranges from Indonesia (the Mentawi Islands Siberut and Sipura, Sumatra, Java, Bali, Kangean, Nusa Penida, Lombok, Sumba, Sulawesi, Pelang, Banggi, Karakelang), to the island of Borneo (Brunei, Indonesia and Malaysia) to the Philippines, where it has been recorded from the islands of Biliran, Leyte, Luzon (Camarines Sur Province), Mindanao (Bukidnon Province), Palawan, and Samar (Taylor 1934; Heaney et al. 1998). Over its range, it has been recorded between 60 and 2,100 m asl.

Countries: Native:

Brunei Darussalam; Cambodia; China; Indonesia (Jawa, Kalimantan, Lesser Sunda Is., Sulawesi, Sumatera); Lao People's Democratic Republic; Malaysia (Sabah, Sarawak); Myanmar; Philippines; **Thailand;** Vietnam

Description:

HB: 44-54; T;36-48; FA: 31-34; E:11-12; HF:7-10

Woolly Bats are small insectivorous bats which inhabit primary and secondary forest. The ears are somewhat rounded and barely extend above the top of the fur on top of the domed skull. The distinctive tragus is narrow, pointed and projects upwards inside the ear (see lower photo) The orange-brown fur is thick, long and fluffy and covers much of the face. The eyes are tiny and the mouth small. The tail is fully enclosed in the flight membrane.

Morphological Description:

·A very small bat: dorsally the fur is smoky brown, the ventral surface is greyish-ochre.

·The size difference between the first two upper premolars P2 and P3, and the last upper premolar P4 is more conspicuous than in Kerivoula picta (Smith & Xie 2008).

· Forearm length 31-36 mm, ear length 11-15 mm (Smith & Xie 2008).

Population:

In Southeast Asia it is a moderately common species in primary forest (L. Heaney pers. comm.). The abundance, population size and trends for this species in South Asia are not known (Molur et al. 2002).
Population Trend: Stable

Habitat and Ecology:

In South Asia, little is known about the habitat or ecology of this species except that it is found in warm valleys in Northeast India and near paddy fields in Sri Lanka and known to roost in caves and buildings in forests (Molur et al. 2002). In China, it is found in subtropical and tropical areas, and has been collected in both forested and agricultural localities. Animals sometimes forage around houses and villages, and roost in buildings and under tile roofs on occasion (Smith and Xie 2008). In Southeast Asia, there are records from primary forest in Cambodia (G. Csorba pers. comm. 2006), dry forest including disturbed sites in Lao PDR, and hill forest in Peninsular Malaysia (Francis pers. comm. 2006). On Banggi Island (Indonesia) the species was found roosting in the axial of a rattan vine leaf. In the Philippines, it has been recorded in lowland, montane, and ridge top mossy forest (Rickart et al. 1993). On Palawan Island, there are records from a bamboo thicket at 60 m and primary lowland forest at 650 m (Esselstyn et al. 2004). Hardwicke's Woolly Bat *(Kerivoula hardwickii)* lives in the forest under storey (Francis 2008). where roosts have been found in hollow trees and among clusters of dead leaves (Francis 2008). It is low flying, and very manoeuvrable. The echolocation calls are likely to be frequency modulated, brief and to cover a wide range of frequencies if they are similar to other Kerivoula species described by Kingston et al. (1999).
Systems: Terrestrial

Additional Information:

Tiny Borneo bats roost in carnivorous pitcher plants

Tiny bats, no bigger than a car key, have been discovered roosting in carnivorous pitcher plants in Borneo -- with their droppings a vital nutrient for the plants.

"It's totally unexpected," said Ulmar Grafe, an associate professor at the University Brunei Darussalam who led the study.

"There's a lot of animal-plant mutualisms, but this one is where the animal gives a nutrient to a plant. Usually it's the other way around."

The study, published in Biology Letters, began by looking at how the pitcher plant -- a vine growing up to 6-10 meters (20 ft to 33 ft) long with 25 cm (10 inch) pitchers -- managed to gain the nitrogen it needed in the nutrient-poor peat swamps and heath forest on the island of Borneo.

Grafe's team was surprised to find that the roughly 4 gram (0.14 oz) Hardwicke's woolly bat *(Kerivoula hardwickii)* consistently chose the pitchers to sleep in during the day, despite a wealth of other possible roosts in the nearby forest.
Not only single bats but male and female pairs, and mother-juvenile pairs, can fit inside comfortably. At night, they fly out to hunt insects.

"The pitcher is a very nice roost for them," Grafe said. "It's dry in there and there's no build-up of blood-sucking ectoparasites that often accumulate in other cavities."

Theoretically, there is some danger to the bat should it fall into the digestive fluid at the bottom of the pitcher. But the plant has adaptations to prevent this, including an unusually low amount of fluid and a tapering pitcher.

Instead of getting nitrogen by consuming the bats, the plants get it from their feces.

"There's no reason why the bat couldn't fly outside. But they probably defecate in there because they usually do that when they roost," Grafe said.

The find is an example of why diversity matters, he added, noting that much of the forest in Borneo is under threat.

"There's so much extinction of animals and reduction of populations and removal that this again highlights how important it is to save every individual, every creature out there."

Major Threat(s):

There appear to be no major threats to this somewhat adaptable species.

Conservation Actions:

In South Asia, although there are no direct conservation measures in place, the species has been recorded from protected areas in India like Siju Wildlife Sanctuary in Meghalaya. It has been recorded from protected areas in parts of Southeast Asia. Further studies are needed into the taxonomy of this species.

Least Woolly Bat (*Kerivoula minuta)*

Range Description:

The species is found in Borneo, Peninsular Malaysia, and **Southern Thailand. The record from Thailand requires confirmation (S. Bumrungsri pers. comm.)** as it may not be this species. The species is easily confused with *K. intermedia,* with which it is probably conspecifics.

Countries: Native:

Indonesia; Malaysia

Population:

This species occurs at low densities. As with *K. intermedia,* the species is difficult to catch unless harp traps are used.

Population Trend: Decreasing

Description:

HB: 31; T: 39; FA: 27; E:8; HF:4.4.

Least Woolly Bats are small insectivorous bats which inhabit primary and secondary forest. The ears are somewhat rounded and barely extend above the top of the fur on top of the domed skull. The distinctive tragus is narrow, pointed and projects upwards inside the ear . The orange-brown fur is thick, long and fluffy and covers much of the face. The eyes are tiny and the mouth small. The tail is fully enclosed in the flight membrane.

Habitat and Ecology:

This species inhabits the under storey of dipterocarp forest (Payne et al. 1985). It is found also in lowland and secondary forest (A. Suyanto pers. comm.).

Systems: Terrestrial

Major Threat(s):

Habitat loss due to logging, agriculture, plantations, forest fires and coal and limestone mining are threats to this species.

Conservation Actions:

This species needs taxonomic revision, however, it appears to be distinct according to local scientists (A. Suyanto pers. comm.). It does occur in a few protected areas within its range.

Whitehead's Woolly Bat *(Kerivoula whiteheadi)*

Description:

HB: 35-39; T: 36-39; FA: 28-29; E: 12.5-13.5; HF: 7-8

The upper parts are rufus orange, with slaty bases to the hairs showing through; the underparts are greyish with lighter tips. Very similar to *hardwickei* but differs quite a lot in distribution of the hair, according to Thomas(1894). There are scattered short hairs along the posterior margin of the interfemoral membrane, but not the definite fringe found in *picta.*The ears are bluntly pointed, with the anterior margin of the tragus slightly convex, the posterior margin with a small projection at the widest part of the tragus, above which there is a shallow depression. Hill(1965). The skull is small, with a greatly inflated brain case rising abruptly from rostrum. The inner, upper incisors are slender with a distinct posterior secondary cusp, the outer upper incisor is much shorter, reaching only to the cusp of the inner incisor. The second and third premolars are elongated. The anterior parts of the tooth row are markedly convergent.

Range Description:

This species occurs in **Southern Thailand,** Borneo, and the Philippines. In Malaysian Borneo, the species is known only from two localities in Sarawak (Mulu and Usun Apau) and from Sabah (Kinabatangan). In the Philippines, records are from Luzon Mindanao (Davao del Norte, Davao del Sur, Davao Oriental, and Zamboanga del Norte Provinces), Palawan, and Panay (Heaney et al. 1998; Taylor 1934) and Polillo (Alviola 1999).Disturbed forest and agricultural land (near sea level only

Countries: Native:

Indonesia (Kalimantan); Malaysia (Sabah, Sarawak); Philippines; **Thailand**

Population:

Originally this species was considered to be extremely rare but this was probably due to inappropriate survey methodology and an inadequate number of surveys. Surveys using harp traps in montane forest on mount Isarog (Camarines Sur Province, Luzon) have found the species to be common at high elevations (1,400 m asl) (J. Sedlock pers. comm. 2006). It is known only from isolated records in Malaysia, and it is not certain whether this reflects genuine rarity for this species or if it is a sign of the sampling difficulties associated with it.

Population Trend: Unknown

Habitat and Ecology:

In the Philippines, the species has been found in disturbed forest and agricultural areas (Sanborn 1952). On Palawan island, it has been reported from cogon grassland at around 60 m asl (Esselstyn et al. 2004) and in secondary forest beside a creek at 450 m asl on Mt. Makiling (Laguna, Luzon) (Heaney et al. 1998; Sedlock 2001).

Systems: Terrestrial

Major Threat(s):

There are no major threats to this species throughout its range. In the Philippines, the species is locally threatened at higher elevations in montane forest on ultramafic soils through activities associated with mining (D. Balete pers. comm. 2006).

Conservation Actions:

It occurs in protected areas throughout its range. Taxonomic research is needed to clarify the status of some populations.

Groove-toothed Bat *(Phoniscus atrox)*

Description:

HB: 43-46; T:37-38; FA:32-35; E: 13-13.6; HF: 7.4-7.6; W: 5g

A highly recognisable bat, due to its small size combined with exquisitely coloured fur. The hairs are long, soft and fluffy, and on the back and head, they are tipped with a shimmering golden colour, giving a gilded appearance. The base of each hair is light brown to black, and on the underside of the body the fur is paler. Golden hairs are also found along the forearm, the wing edges, the legs, tail and feet. The nose of the gilded groove-toothed bat is simple, with no nose leaf, and the ears are small and brown, but contain a pointed, whitish tragus (inner ear), which has a characteristic notch at the base. A tiny groove in each canine tooth of this species contributes to its common name. A membrane that stretches between the hind legs, known as the interfemoral membrane, encloses the tail. Fur: The most distinctive feature of Phoniscus atrox is the beautiful golden brown tips to the dark brown fur. The soft fluffy hairs actually have four bands of colours: light grey or brown near the base, then a buffy, light brown colour, followed by a dark brown band and then the golden tips. Teeth: Although you can't see them, the long canines have a groove running from top to bottom, and this helps us tell the Kerivoula and Phoniscus apart, because the canines in the Kerivoula are smooth.

Nose: The nose is simple.

Tail: The tail is very long and the tail membrane large. The bat uses this tail membrane partly to help it brake during flight, but it also acts as a sort of scoop that helps the bat to catch insects. After successfully catching an insect, he will duck his head down and grab it from the tail membrane scoop with his mouth.

Wings: The wings are a dark brown, and broadly rounded, making this bat very manoeuvrable when flying through the forest under story.

Size: Phoniscus atrox is a small bat, with an average forearm of about 34 mm and an average weight of just under 5 g, a little bit less than a nickel.

The forearm, legs, tail and feet are also thinly covered with fine hairs of gold.

Ears: The ears are funnel shaped like the Kerivoula to which Phoniscus is closely related, and there is also a long pointed tragus, that is typically without pigment so it appears a whitish pink, in contrast to the brown ears. At the base of the tragus is a small notch; the Kerivoula don't have this, so it helps us distinguish the Phoniscus from the Kerivoula.

Listening for the returning echo of their shouts, each of which lasts just 3 milliseconds, the bat is able to distinguish an insect from its surroundings, using such detail as the movement of its tiny, beating wings. As it approaches the insect the speed of its echolocation pulses quickens, to give pinpoint precision for the capture of its prey.

Range Description:

This species is known to occur in Peninsular Malaysia, southern **Thailand,** Sumatra (Indonesia), and Sabah in Malaysian Borneo. It is found below 400 m asl. This species may be more widespread and common than is currently known.

Countries: Native:

Indonesia (Kalimantan, Sumatera); Malaysia (Sabah, Sarawak); **Thailand**

Biology and Ecology:

This is a forest dependent species. Lekagul and McNeely (1977) reported that the type specimens were found roosting in an abandoned bird's nest (broadbill) in heavy forest on the bank of a river. A recent collection was found in Thailand in disturbed forest at 150 m asl which was adjacent to a large patch of pristine evergreen forest (Thong et al. 2006).

Systems: Terrestrial

With a small body and short, rounded wings, the gilded groove-toothed bat has a high degree of flight control, enabling it to pass nimbly amongst the leaves and branches of the forest. It feeds on flying insects, detecting their presence with ultrasonic shouts and listening for the returning echo. As it approaches the insect the speed of its echolocation pulses quickens, to give pinpoint precision for the capture of its prey.

Little is known about the breeding habits of this species, but pregnancy has been recorded in April, August and October, and lactation has been noted in January, May, June and October

Major Threat(s):

Deforestation due to logging, agriculture, plantations and forest fires represents a major threat to this species.

Conservation Actions:

This species is known from protected areas (Thong et al. 2006). It also appears on a checklist for the Bukit Barisan National Park in South Sumatra (O'Brien and Kinnaird 1996), but this record is questionable.

Subfamily: Myotinae:
Genus: Myotis:
Szechwan Myotis (*Myotis altarium)*

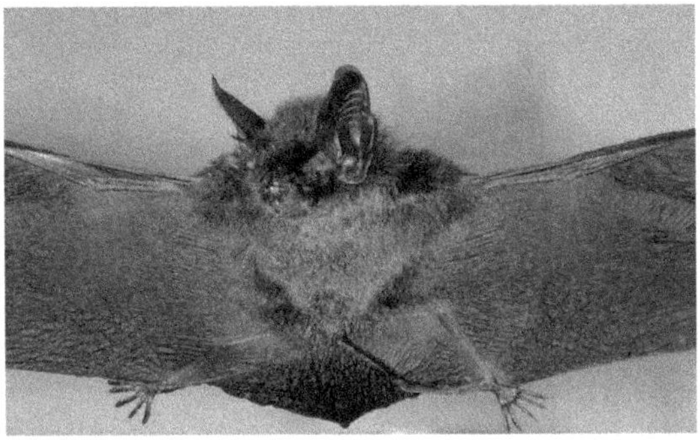

Description:

· Fur dark brown, paler ventrally.

A medium-sized Myotis with relatively large ears (22-24 mm). Forearm length is 42-46 mm (Smith & Xie 2008).

· Length of calcar more than half length of tibia. .

Range Description:

This species is found in southeast China (Sichuan, Yunnan, Guangxi, Jiangxi, Fujian, Anhui and Guizhou provinces), extending to **Thailand (**Smith et al. 2008). The only location in Thailand is around 1,000 m asl

Countries: Native:

China (Sichuan); **Thailand**

Habitat and Ecology:

All of the known specimens of *M. altarium* have been collected from caves. Other than this roosting habitat, nothing is known of the natural history of this species.

Systems: Terrestrial

Major Threat(s):

The threats to this species are not known.

Conservation Actions:

It is not known if the species is present in any protected areas. Further studies are needed into the distribution, abundance, threats and ecology of this species.

Hairy-Faced Bat (*Myotis annectans)*

Description:

HB:55; T: 48; FA: 43-46; E:16

The pelage is long and woolly. The lower individual hairs of the upperparts are dark brown at the base and pale brown at the tip. The hairs of the underparts are also dark at the base but the hairs at the base are greyish white giving a frosted effect. The hairs on the belly have orange tips. The ears are translucent and of moderate size, with small rounded tips and long, tapering tragi. the upper and second incisor is separated from the canine by diastema;it has a larger anterior cusp and a lower posterior cusp. The middle premolars are lacking, as in *M rosseti;* the canines are quite small, scarcely exceeding the height of the posterior premolar.
(Hill and Kitty Thonglongya 1972)

Range Description:

This species ranges from North-eastern South Asia, into southern China, and mainland Southeast Asia. In South Asia, this species is presently known from India (Nagaland and West Bengal) (Das 2003, Molur et al. 2002); it has been recorded from around 1,100 m asl. In China it is known only from a single record from Yunnan (Wang 2002; Smith and Xie 2008). In Southeast Asia, it ranges from northern **Thailand** (Corbet and Hill 1992), probably through much of Lao PDR (Francis et al. 1999) and in southern Cambodia (Hendrichsen et al. 2001). It is unclear if this species is present in Vietnam.

Countries: Native:

Cambodia; China; India; Lao People's Democratic Republic; **Thailand**

Population:

In South Asia the abundance, population size and trends for this species are not known (Molur et al. 2002). In Southeast Asia it is widespread but uncommon to moderately common.
Population Trend: Unknown

Habitat and Ecology:

In South Asia, little is known about the habitat or ecology of this species except that this species is found in montane forests and valleys (Molur et al. 2002). In Southeast Asia it has been recorded from wet evergreen forest. In Vietnam, one specimen has been taken in second growth along a river valley (Smith and Xie 2008).
Systems: Terrestrial

Major Threat(s):

The threats to this species remain unknown (Molur et al. 2002).

Conservation Actions:

In South Asia, there are no conservation measures in place, and this species is not recorded from any protected areas. Surveys, ecological and populations studies are recommended (Molur et al. 2002). The species has been recorded from protected areas in Cambodia and Lao PDR (C. Francis pers. comm.).

Large Myotis (*Myotis chinensis)*

Description:

HB: 91-97; T:53-58; FA: 64-69; HF: 16-18; E:20-23

Dorsal fur is grey- brown. Ventral fur is grey or white-grey.
This is a relatively large *Myotis,* being the largest in **Thailand.** The ears are rather long, extending to the tip of the nose when laid forwards. The calcar is very long and slender. The feet are noticeably smaller than those of *M. ricketti* (length of foot is under half the length of the tibia (Bates et al. 1999). Bates et al. (2005) give the forearm length of bats from Myanmar as 65.1-68 mm. Borissenko & Kruskop (2003) give forearm lengths of 65-69 mm for bats in Vietnam.
The species is closely related to the large mouse-eared bats *M. myotis* and *M. blythii* of the Palaearctic. The skull is typical of the genus, though considerably large than other **Thai** *Myotis.*The upper inner incisors are thick and blunt, with a blunt internal cusp, they are shorter than the lower incisors, which are slightly curved outwards (Dobson1876) The first upper premolar is actually pointed and the second is very small, slightly intruded from the tooth row

Range Description:

This species occurs in central and Southeast China (Jiangsu, Jiangxi, Guangdong, Hong Kong, Guangxi, Fujian, Hainan, Hunan, Zhejiang, Sichuan, Guizhou, and Yunnan provinces), extending to **Thailand**, Myanmar and Vietnam (Smith et al. 2008). It can be expected in northern Lao PDR, but this has not yet been confirmed. In Myanmar it is found between 50-1,000 m asl in elevation (P. Bates pers. comm.).

Countries: Native:

China (Yunnan); Hong Kong; Myanmar; **Thailand;** Vietnam

Population:

This is a relatively common species in at least part of its range, for example Myanmar (P. Bates pers. comm.).
Population Trend: Unknown.

Habitat and Ecology:

It is found in a wide range of habitats from lowlands through the hill country. It is known to hibernate in caves (Smith et al. 2008). Most of the recent records in Southeast Asia (Bates et al. 1999, 2001) report it being netted near cave entrances in or adjacent to limestone areas with rivers and streams (Bates et al. 1999, 2001).

Roost sites and roosting behaviour :
Roosts in caves.

Emergence and flight pattern
M. chinensis has relatively broad wings and is probably a manoeuvrable species. Its close relatives *M. myotis* and *M. blythii* hunt by gleaning (Arlettaz et al. 2001).

Foraging behaviour:
Not known.

Echolocation calls:
Not known.
Systems: Terrestrial

Major Threat(s):

There are no major threats to this species.

Conservation Actions:

This species is known to occur in protected areas throughout its wide range.

Lesser Large-Footed Bat (*Myotis hasseltii*)

Description:

HB: 53-58; T:37-44; FA:38-44, E:16-17; HF:10-12; W:7.5-10g

The pelage is short and velvety, with upperparts grey to pale brown and the underparts somewhat lighter; the individual hairs are dark at the base, with pale silver tips. The ears and wings are pale greyish brown. The interfemoral membrane is large, acutely angular behind; the wings attach at the ankle. The face and muzzle are covered with sparse hairs, with the exposed skin pink in colour. The ears are rather short, extending half the distance eve and the end of the muzzle when laid forward; the tragus is bluntly pointed, with a triangular lobe at its base. The second premolar is displaced inwards so that the first and third premolars are nearly in contact (actually touching in some specimens) (Hill1972)

Range Description:

This species has a patchily recorded distribution in South Asia and Southeast Asia. In South Asia, this species is presently known from India (West Bengal) and Sri Lanka (Eastern, Northern, North Central and Southern provinces) and has been recorded up to an elevation of 1,000 m asl (Molur et al. 2002). In Southeast Asia, the species is known from Myanmar, **Thailand,** Cambodia, Vietnam (with a record from Co-Loa [Bates et al. 1999]), Peninsular Malaysia, Indonesia (the Mentawai Islands [Pagai Islands], Riau Archipelago, Sumatra [Bukit Barisan Selatan protected area (Opo pers. comm.)] Java and Sumbawa [Maryanto, pers comm. 2006]) and the island of Borneo (records from Kalimantan [Indonesia] and Sarawak [Malaysia]).

Countries: Native:

Cambodia; India; Indonesia (Jawa, Kalimantan, Sumatera); Malaysia; Myanmar; Sri Lanka; **Thailand**; Vietnam

Population:

This is a reasonably common, but locally distributed species. In South Asia, although it is a fairly common bat a declining trend in the population has been observed (Molur et al. 2002).

Population Trend: Unknown

Habitat and Ecology:

In South Asia, this species prefers dry forests but is also seen in mangrove forests. It roosts either solitary or in groups of few individuals among bamboo, cracks of tree trunks, and in old and ruined buildings. It is a low flyer, hunting over water surfaces even the sea. It feeds on small insects like the mosquitoes, gnats, flies and moths (Bates and Harrison 1997). In Southeast Asia, this species is known to feed over open seas, and probably roosts in mangrove forests; it has also been found more inland, where its roosting and feeding habits are poorly known.

In view of the species possible association with coastal habitats, it may be the species might be more widespread on Borneo around the islands coasts. Payne et al. (1985) suggested that bats feeding over open water near Sandakan were probably this species. Borissenko and Kruskop (2003) recorded the species from the Bassac River in Cambodia and from Pnom-Penh, always near large water surfaces. The species has also been recorded from large cities, such as Rangoon and Bangkok, and there is also a locality from near Hanoi (Bates et al. 1999). In Rangoon and **Bangkok,** it is seen hawking for insects and fish over small ponds and lakes within the city. The species was found to roost in caves at Bukit Barisan Selatan (Opo pers. comm. 2006), and has been found roosting in limestone caves in Langkawi Island (Lim Boo Liat pers. comm. 2006).

Systems: Terrestrial

Major Threat(s):

There appear to be no major threats to this species as a whole. In South Asia, the habitat of this species is being deforested for timber, firewood and conversion for agricultural use and human settlements (Molur et al. 2002).

Conservation Actions:

This species has been recorded from protected areas in Southeast Asia (eg. Co-Loa, Vietnam). In South Asia, there are no conservation measures in place and it has not been recorded from any protected areas. Ecological studies and population monitoring recommended (Molur et al. 2002).

Horsfield's Bat *(Myotis horsfieldii)*

Description:

HB:51; T:35; FA: 34.6-36.2; HF: 8.5-8.8; W: 7.5-8.5g

There are two kinds of fur; the shorter coat has , individual hairs with 3 bands, dark brown or black at the base, followed by a pale buff band, then a narrow dark brown band, with flaxen tips; the longer courser coat extending beyond the short hairs is black with flaxen tips. The intermingling of the 2 types of hair gives the back a grizzled grey appearance. The underparts are lighter brown, with individual hairs having dark brown bases with pronounced buffy tips. The wings and interfemoral membrane are dark grey; the tip of the tail projects slightly beyond the membrane. The second lower premolar is half the height of the first, but within the toothrow; the second upper premolar is within the toothrow, but so small that the first and third upper premolars are nearly in contrast (Medway 1969)

Range Description:

This wide ranging species has been recorded from South Asia, Southern China and much of Southeast Asia. In South Asia, this species is presently known from India (Andaman and Nicobar Islands [Great Nicobar, Little Nicobar and Car Nicobar], Goa, Karnataka, Kerala, Madhya Pradesh, Maharashtra and Tamil Nadu) (Aul and Vijaykumar 2003, Molur et al. 2002, Vanitharani 2006). It has been recorded from sea level to around 800 m asl (Molur et al. 2002). In China, it is found in Guangdong, Hainan and Hong Kong (Smith and Xie 2008). In mainland Southeast Asia, it ranges from Myanmar in the west, through **Thailand**, Lao PDR, Vietnam, Cambodia, Peninsular Malaysia and possibly Singapore. Within insular Southeast Asia, it has been recorded from Indonesia (the islands of Java, Lombok and Sulawesi), Borneo (Brunei, Indonesia and Malaysia) and the Philippines, where it has been recorded from sea level to 800 m asl (Heaney et al. 1998) although there is a record from 1,450 m asl on mount Isarog Camarines Sur Province, in the South Eastern portion of Luzon Island.There are records from the Philippine islands of Bohol, Catanduanes, Luzon (Camarines Sur, Cagayan, Laguna, Pampanga, Quezon, and Rizal provinces), Mindanao (Lanao del Norte Province, Misamis Occidental), Negros, and Palawan (Heaney et al. 1998) and Polillo (Gonzales pers. comm. 2006).

Countries: Native:

Brunei Darussalam; Cambodia; China; Hong Kong; India (Andaman Is.); Indonesia (Bali, Jawa, Kalimantan, Sulawesi); Lao People's Democratic Republic; Malaysia; Myanmar; Philippines; **Thailand;** Vietnam

Population:

This is a moderately common species over much of its range. The abundance, population size and trends for this species in South Asia are not known (Molur et al. 2002).
Population Trend: Stable

Habitat and Ecology:

In South Asia, this species is found in prime forests and tea estates near to water source. It roosts in tunnels, caves, bridges, palm fronds, crevices in old buildings, cracks and hollows between wooden beams (Aul and Vijaykumar 2003) either singly or in small groups of a few individuals (Bates and Harrison 1997). In Myanmar the species has been recorded in lowland forests and agricultural areas adjacent to limestone karst, it has also been collected in a limestone cave in disturbed forest (Bates pers. comm. 2006). In Vietnam (Borissenko and Kruskop 2003), **Thailand** (S. Bumrungsri pers. comm.) and the Philippines, it has been recorded near to streams in lowland forest, disturbed forest and agricultural areas. In the Philippines, it sometimes roosts in caves and in tunnels and has been reported roosting beneath a large rock over a stream (Taylor, 1934).
Systems: Terrestrial

Major Threat(s):

There appear to be no major threats to this species as a whole. In South Asia, the habitat of this species is being deforested for timber, firewood and conversion for agricultural use. It is also threatened through disturbance of roosting sites (Molur et al. 2002).

Conservation Actions:

The species has been recorded from many protected areas. Within India, the species has been recorded from the Silent Valley National Park in Kerala, Agasthiyamalai Biosphere Reserve in Tamil Nadu (Vanitharani 2006) and Kanha National Park in Madhya Pradesh. Further studies on distribution, abundance, breeding biology, general ecology and population monitoring are recommended (Molur et al. 2002).

Burmese Whiskered Bat (*Myotis montivagus)*

Description:

The greater mouse-eared bat is one of the larger bats Its fur is a medium-brown colour on the upper body, and greyish-white underneath. It has large ears with a very prominent tragus, the organ which is part of the bat's echolocation system.

Greater mouse-eared bats are usually found around human settlements. They probably used caves as roosting sites, and today they hibernate in both caves and mines. They hunt in forests and adjoining cultivated areas

This bat preys on larger insects, mainly beetles, which they hunt for about four to five hours after emerging late in the evening. They are known to forage on the ground for some of their insect prey. Male greater mouse-eared bats are polygamous, and may have a harem of up to five females. The females form large maternity roosts and give birth to one offspring, usually in June. When they leave to feed, females leave their babies in a crèche and there are often several females left to guard the roost. The young bats can fly after three weeks, and become sexually mature at three months.

Range Description:

This widespread species has been recorded from South Asia, Southern and Eastern China, and both mainland and insular Southeast Asia. In South Asia this species is presently known only from India (Andhra Pradesh, Karnataka, Kerala, Maharashtra, Mizoram and Tamil Nadu) (Vanitharani 2006, Molur et al. 2002). In China, it ranges through much of Southern and Eastern, parts of the country (Smith and Xie 2008). In mainland Southeast Asia, it has been recorded from Myanmar, **Southern Thailand** (although there are no published records (Bumrungsri et al., 2006)), Northern Vietnam, Lao PDR and Peninsular Malaysia (it has been recorded at the Batu caves close to Kuala Lumpur). It is also present on the island of Borneo, where it has been reported from Sabah (Kinabatangan area) (Malaysia) and Kalimantan (Indonesia). It has been recorded from sea level to an elevation of 1,100 m asl.: Within India, it has been recorded roosting in caves, rocky crevices and subterranean habitats in forested areas (Molur et al. 2002). In Lao PDR, the species has been collected in hill forests at 1,000 m asl and in open forest at 500 m asl (Francis et al. 1996; in Duckworth et al. 1999). In Vietnam, the species has been recorded in heavily disturbed agricultural landscapes, at about 200 m asl. Animals have been found roosting in small colonies in a crevice of a bridge (Yasuma and Andau, 2000). In southern **Thailand** the species roosts in small colonies in lowland evergreen forest (S. Bumringsri pers. comm.). In Myanmar it is known from around 1,850 m asl (Bates et al. 2005). In Malaysia it has been recorded in highly disturbed secondary habitat

Countries: Native:

China; India; Lao People's Democratic Republic; Malaysia; Myanmar; Vietnam

Population:

It is believed to be an uncommon species. In South Asia, the population abundance of this species is not known. There are only a relatively few individuals in each colony and the colonies and localities are scattered (Molur et al. 2002).

Population Trend: Unknown

Habitat and Ecology:

Within India, it has been recorded roosting in caves, rocky crevices and subterranean habitats in forested areas (Molur et al. 2002). In Lao PDR, the species has been collected in hill forests at 1,000 m asl and in open forest at 500 m asl (Francis et al. 1996; in Duckworth et al. 1999). In Viet Nam, the species has been recorded in heavily disturbed agricultural landscapes, at about 200 m asl. Animals have been found roosting in small colonies in a crevice of a bridge (Yasuma and Andau, 2000). **In Southern Thailand** the species roosts in small colonies in lowland evergreen forest (S. Bumringsri pers. comm.). In Myanmar it is known from around 1,850 m asl (Bates et al. 2005). In Malaysia it has been recorded in highly disturbed secondary habitat.

Systems: Terrestrial

Major Threat(s):

In South Asia, this species is threatened due to disturbance to roosting sites by increase in tourism and habitat loss due to tourism related developmental activities (Molur et al. 2002). The threats to the species over the rest of its range are not known, although it has been reported from degraded forest in parts of its range.

Conservation Actions:

In South Asia, although there are no direct conservation measures in place, the species has been recorded from Agasthiyamalai Biosphere Reserve in Tamil Nadu (Vanitharani 2006). Surveys, ecological and population studies are recommended (Molur et al. 2002). In Southeast Asia, it has been recorded from some protected areas (eg. Vu Quang National Park, Vietnam [Borissenko and Kruskop 2003]).

Whiskered Myotis (*Myotis muricola*)

Description:

The Whiskered Myotis is a small, insectivorous occurring in forested and cultivated areas. It is well known for roosting by day in young, rolled-up banana leaves. In Thailand large colonies are also reported as roosting in caves.

The fur is dark brown to dark grey on the upperside, and medium to pale grey underneath. The wing membrane extends to the base of the toes, and almost fully encloses the tail. The ears are moderately long and pointed, the eyes small, and the snout blunt.

Range Description:

This widespread species has been recorded throughout much of northern South Asia, central and southern China and most of Southeast Asia. In South Asia, the species is presently known from Afghanistan (Balkh, Faryab, Kabul, Konarha and Kunduz provinces), Bhutan (no exact location, Das 2003), India (Assam, Himachal Pradesh, Jammu and Kashmir, Jharkhand, Meghalaya, Mizoram, Sikkim, Uttaranchal and West Bengal), Nepal (Central Nepal) and Pakistan (North West Frontier Province and Punjab), and has been recorded between 1,230 and 2,700 m asl (Molur et al. 2002). It has been found in Xizang, Sichuan, Yunnan in mainland China, and has been reported from the island of Taiwan (Smith and Xie 2008). In Southeast Asia, it appears to be present throughout the mainland and widespread in insular Southeast Asia being recorded from Indonesia (the Mentawi Islands, Sumatra, Java, Bali, Lombok, Sumbawa, Sumba, Flores, Sulawesi, Ambon and Bunguran), the island of Borneo (Brunei, Indonesia and Malaysia), and throughout the Philippines, with records from Biliran, Busuanga, Leyte, Luzon (Cagayan, Laguna, Rizal and Kalinga Provinces [Heaney et al. 2004]), Maripipi, Negros (Heaney et al. 1998). Specimens from Culion Island and Camarines Sur Province (Luzon Island) previously referred to this species are now considered to represent Myotis ater (Heaney 2005). The specimen previously reported from Bukidnon, Mindanao is of uncertain identity (Heaney 2005). In the Philippines, it has been recorded from near sea level to 1,600 m asl (Rickart et al. 1993; Heaney et al. 2004).

Countries: Native:

Afghanistan; Bhutan; Brunei Darussalam; Cambodia; China; India; Indonesia; Lao People's Democratic Republic; Malaysia; Myanmar; Nepal; Pakistan; Philippines; Singapore; Taiwan, Province of China; **Thailand;** Vietnam

Population:

This is a common bat in Southeast Asia, and is particularly abundant at higher elevations (Sedlock pers. comm. 2006). In South Asia, the localities and the colonies are scattered with small populations per colony. A declining trend in the population is being observed (Molur et al. 2002). There is only a single record of this species from the island of Ambon, Indonesia.

Population Trend: Stable

Habitat and Ecology:

Total length 69-81 mm; tail 30-38 mm; hind foot 6-8 mm; ear 11-14 mm; forearm 30-34 mm; weight 3-4.5 g. Like all Myotis, the ears are relatively long; the tragus and long and slender, coming to a blunt point; the nostrils are simple (not tubular); and the muzzle is not swollen. The dorsal pelage is dark brown, sometimes with slightly paler tips; ventral pelage is similar but grayer and paler overall. The wing membrane attaches to the side of the foot at the base of the toes. This species is found in primary and secondary montane and lowland forests, scrub, secondary growth and gardens. It roosts either singly or in small groups of few individuals among tightly rolled leaves of the broad-leaved trees especially banana, also in caves and tree hollows. It is a fast and early flyer, with bats often encountered in the forest understory and in gaps along streams
(Rickart et al. 1993; Molur et al. 2002; Heaney et al. 2004; P. Bates and G. Csorba pers. comm. 2006; Sedlock pers. comm. 2006; Smith and Xie 2008). This species of bats is often found roosting in curled up leaves of banana plants.

Size: About 3.7 cm without tail.

Diet: Consists of insects.

Activity: Nocturnal.

Systems: Terrestrial

Major Threat(s):

In Southeast Asia, there are no major threats to this adaptable and widespread species as a whole, although some populations are locally threatened by habitat degradation (largely from mining and logging operations, and ongoing human settlement). In South Asia, the habitat of this species is being deforested for timber, firewood and conversion to agricultural use. This species is also considered to be locally threatened at some localities due to scientific collection for research purposes (Molur et al. 2002).

Conservation Actions:

This species is presumably present in a number of Southeast Asian protected areas. In South Asia, although there are no direct conservation measures in place, the species has been recorded from protected areas like the Murree National Park in Pakistan and the Langtung National Park in Nepal (Molur et al. 2002). Further studies are needed on distribution, abundance, breeding biology, general ecology and population monitoring.

Thick-Thumbed Myotis (*Myotis rosseti)*

Description:

HB: 36-38; T: 37-38; FA: 27-30; HF:7; E:11-12.

The upper parts are dark greyish brown, with the underparts lighter. *M.rosseti* can easily be recognised by the by the fleshy. wrinkled pad at the base of the thumb and the broad, concave pad on the sole of the foot. The ears are funnel shaped and quite long, reaching the end of the muzzle when laid forwards; the tragus is about half the length of the ear, with a blunt, interiorly directed point. The interfemoral membrane is attached just below the ankle, with the wing attached at the base of the first digit. (Hill 1969)

Range Description:

This species occurs in Cambodia, **Thailand,** and possibly Vietnam (Borissenko and Kruskop 2003), and is known from one locality in Lao PDR (C. Francis pers. comm.).

Countries: Native:

Cambodia; Lao People's Democratic Republic; **Thailand**

Population:

This species is known from only a small number of individuals in any country.
Population Trend: Unknown

Habitat and Ecology:

This species uses disturbed habitats, in **Thailand** it is found in tree hollows and house roofs, and bamboo.
Systems: Terrestrial

Major Threat(s):

There are no major threats to this adaptable species.

Conservation Actions:

This species occurs in some protected areas throughout its range. Further studies are needed into the distribution, abundance, and ecology of this species.

Subfamily: Vespertilioninae:

- **Genus: Arielulus:**

Collared Pipistrelle (*Arielulus aureocollaris)*

Range Description:

This species is known from Lao PDR, Vietnam and **Thailand (Chiang Mai Province, Amphoe (District) Mae Ai, Doi (Mount) Pha Hom Pok, at 1,500 m asl)** and possibly neighbouring China although there are no records from there yet.

Countries: Native:

Lao People's Democratic Republic; **Thailand**; Vietnam

Population:

Its population status is unknown.
Population Trend: Unknown

Habitat and Ecology:

The mapped range probably does not represent the real range of the species, which could be due to lack of surveys. It is probably associated with hills; early records are known from mountainous regions (G. Csorba pers. comm.). In Vietnam it is found in limestone and non limestone forest (N. Furey pers. comm.), and has been recorded from primary and mildly disturbed forest and also flying around a stream there at 1,200 m asl (Son pers. comm.). In **Thailand** it is found from 2,000 m asl in hill evergreen forest (S. Bumrungsri pers. comm.).
Systems: Terrestrial

Major Threat(s):

No major threats are known for this species.

Conservation Actions:

This species occurs in a number of protected areas.

Genus: Eptesicus:

Surat Serotine (*Eptesicus dimissus)*

Description:

HB:57-59; T:39-41; FA: 42; E:14-15.5;HF10.

The fur is rather short, chestnut above, lighter below, with some hairs tipped buffy brown. The fur extends the fur extends on the interformeral membrane as a triangle at the base of the tail. The ears are short and narrow, with rounded tips; the antitragus is low, the tragus is short and broad. The skull as a well marked occipital helmet. The zygomatic arch is also very broad, and there are deep basial pits between the bullae.The upper incisors are more similar in size than in other *Eptesicus,*with the outer upper incisors 3/4the length of the inner upper incisor. The third upper molar is complete

Range Description:

This species is known from a single specimen collected in Southern **Thailand (**Bandon, Kao Nawg), and from southern Nepal (Myers et al. 2000) where it has been recorded in and near Royal Chitwan National Park.

Countries: Native:

Nepal; **Thailand**

Habitat and Ecology:

Little is known about the habitat or ecology of this species.

Systems: Terrestrial

Major Threat(s):

The threats to this species are not known.

Conservation Actions:

This species is known to occur in Royal Chitwan National Park, Nepal (Myers et al. 2000). Further studies are needed into the distribution, abundance, natural history and threats to this species.

Thick-Eared Bat (*Eptesicus pachyotis)*

Description:

HB:55-56; T:40-41; FA: 38-39; E:13-14;HF; 8-9.
The pelage is dark brown above, slightly lighter below. The head is rather flat, the muzzle broad. The ears are triangular, with rounded tips: the tragus is short, expanding above and curved inwards. The ear from below the tip of the tragus to the termination of the outer margin near the angle of the mouth is very thick and fleshy. The teeth are very small, with the outer incisor half or less the height of the inner incisor. The minute upper premolar is lacking.

Range Description:

This species ranges from north-eastern South Asia, through much of Central China, with some records from northern parts of Southeast Asia. In South Asia, this species is presently known from Bangladesh (Sylhet division) and India (Meghalaya and Mizoram) in South Asia (Molur et al. 2002). In China, it ranges throughout much of the central part of the country, being recorded as far north as Gansu and Ningxia. In Southeast Asia, it is limited to records from northern Myanmar and **Northern Thailand.**

Countries: Native:

Bangladesh; China; India; Myanmar; **Thailand**

Population:

The abundance, population size and trends for this species are not known (Molur et al. 2002).
Population Trend: Unknown

Habitat and Ecology:

In South Asia, little is known about the habitat or ecology of this species (Molur et al. 2002) except that it is reported from Khasi Hills in Meghalaya where the vegetation is tropical moist deciduous forest (Bates and Harrison 1997). There appears to be nothing recorded on the natural history of this species from other regions of its range.
Systems: Terrestrial

Major Threat(s):

The threats to this species remain unknown.

Conservation Actions:

In South Asia, there are no conservation measures in place. This species has not been recorded from any protected areas. It is not known if it is present in protected areas in other parts of the species range. Further studies are needed into the taxonomy, distribution, abundance, reproduction and ecology of this species. Populations should be monitored to record changes in abundance and distribution (Molur et al. 2002).

Genus: Glischropus:

Common Thick-thumbed Bat (*Glischropus tylopus)*

Description:

HB:39; T:37; FA::31-32;E: 12.5; HF: 6.5

The upper parts are dark grey or blackish, the under parts are paler; the ears and the wings are dark grey. There are thickened, kidney shaped pads at the base of the thumb and the sole of the foot; they are white of pink, contrasting with the dark shade of the surrounding skin. The muzzle is broad, evenly rounded in front. The ears are triangular with rounded tips; the tragus is faintly concave along the inner margin. The tip of the tail is free of interformeral membrane. The upper inner incisors are long and bifid; the outer upper incisors are much shorter, with only one cusp, but are equal in cross section to the inner ones. The lower incisors are trifid but not crowded.

Ecology:

Frequently associated with bamboo, roosting in dead or damaged stalks; they also roost in crevices or rocks and in new bamboo leaves. They also tend to fly high.

Range Description:

The species occurs in Myanmar, through **Thailand,** Cambodia, Lao PDR, Vietnam, Malaysia, Moluccas Islands (Indonesia), Borneo, Sumatra, Java and the Philippines. In the Philippines there are records from Palawan Island (Heaney et al. 1998).

Countries: Native:

Cambodia; Indonesia (Kalimantan, Maluku, Sumatera); Lao People's Democratic Republic; Malaysia (Sabah, Sarawak); Myanmar; Philippines; **Thailand;** Vietnam

Range Map: Click here to open the map viewer and explore range.

Population:

The species is uncommon throughout most of its range although it can sometimes be locally common.

Population Trend: Unknown

Habitat and Ecology:

This species roosts in rock crevices, in hollow bamboo, and in banana leaves in Peninsular Malaysia (Payne et al. 1985). Habitat and ecological requirements are unknown in the Philippines (Heaney et al. 1998).

Systems: Terrestrial

Major Threat(s):

There are no known major threats to this species.

Conservation Actions:

This species is found in a number of protected areas throughout its range.

Genus: Hesperoptenus:
Blanford's Bat (*Hesperoptenus blanfordi*)

Description:
HB:45-53; T: 28-32; FA: 26-29; E:5-10; HF:7-11.

The pelage is dark reddish brown or dark brown, with a glossy sheen to the fur; the under parts are slightly lighter. The fur over the shoulders is longer but thinner and somewhat lighter brown. The wing is brown. The head is flat and broad, the nostrils wide apart. The ears are short and triangular

Range Description:
This species is found from East of Moulmein, Tenasserim (Myanmar), extending to **Thailand** and Peninsular Malaysia. There are records also from Cambodia (P. Banks pers. comm.), Lao PDR, Vietnam, and Sabah where it is only known from two localities (the Witti Range and Sepilok), and east Kalimantan (one specimen, A. Suyanto pers. comm.) in Borneo. In Lao PDR the species is reported, by Robinson and Webber (1998), from Khammouane Limestone National Biodiversity Conservation Area, and from Xe Pian in southern Lao PDR by Robinson (1998). In Viet Nam the species is known from Don Nai province. In Thailand they have been found in the mountains Nakon Si Thammarat and Suray Thani Provinces as well as Changmaai.

Countries: Native:

Cambodia; Indonesia (Kalimantan); Lao People's Democratic Republic; Malaysia (Peninsular Malaysia, Sabah, Sarawak); Myanmar; **Thailand;** Vietnam

Range Map: Click here to open the map viewer and explore range.

Population:

This species is uncommon throughout its scattered distribution through mainland Southeast Asia and Borneo.

Population Trend: Unknown

Habitat and Ecology:
This species is found in a wide range of habitats, from lowland wet forest to dry dipterocarp forest and hill forest, and also limestone caves. It is usually associated with small rivers and streams. It usually lives in small colonies, with less than 10 individuals.

Systems: Terrestrial

Major Threat(s):
There are no major threats to this species. This species is apparently tolerant to a broad range of habitats.

Conservation Actions:
This species is found in a number of protected areas throughout its range.

Tickell's Bat (*Hesperoptenus tickelli)*

Description:

HB: 66-67; T: 50-55; FA:55-58; E: 16; HF:12.

The pelage is greyish yellow to golden brown; the black wings are marked with white. The head is grey, with a tuft of whitish hairs at the base of the ears. The head is furred to the nostrils; the ears are covered with short hairs, and the interformeral membrane has rather thick hair at the base, thinning towards the distal end. There are very small friction pads on the thumbs. The head is broad and flat; the muzzle is blunt, swollen at the sides, and slightly depressed above in the centre. The wings attach at the base of the toes. The ears are oval. The second upper incisor is located behind the first,; the upper incisor is unicuspid (bisuspid in most other *pipistrelle* like genera. The small second incisor may appear to be a basal cusp of the first incisor.

Range Description:

This species is widespread in South Asia and mainland Southeast Asia. In South Asia, it is widely distributed species and is presently known from Bangladesh (Dhaka Division), Bhutan (no exact location) (Koopman 1993; Simmons 2005), India (Andaman Islands, Andhra Pradesh, Assam, Goa, Jharkhand, Karnataka, Madhya Pradesh, Maharashtra, Orissa, Rajasthan, Tamil Nadu and West Bengal), Nepal (Mid Western Nepal) and Sri Lanka (Anuradhapura, Northern, North-western, Sabaragamuva, Southern, Central and Uva provinces) (Molur et al. 2002). In South Asia, it has been recorded up to 1,000 m asl (Molur et al. 2002). In Southeast Asia, it has been recorded from Myanmar, **Thailand (including Sunate Karnphum),** Lao PDR (Duckworth et al. 1999), Vietnam (Hendrichsen et al. 2001) and Cambodia (Hendrichsen et al. 2001), and is probably more widespread than is currently known.

Countries: Native:

Bangladesh; Bhutan; Cambodia; India; Lao People's Democratic Republic; Myanmar; Nepal; Sri Lanka; **Thailand**; Vietnam

Population:

In South Asia, although it is a widely distributed species the abundance, population size and trends for this species are presently not known (Molur et al. 2002). It is widespread and relatively common in Southeast Asia, although it often flies at between nine and twelve metres above the ground and so is not regularly captured in surveys (Bates and Harrison 1997).

Population Trend: Unknown

Habitat and Ecology:

In South Asia, this species roosts solitary or in small groups of a few individuals among dense canopied trees. It is found in lowlands, hills and near seashores (Molur et al. 2002).The species forages in open areas among paddy fields, grasslands, with a steady and slow flight, and mostly feeds on beetles, termites and other insects. A single young is born in May. Bates and Harrison 1997). In Southeast Asia, it is considered to be a forest edge species that can be found close to degraded areas (such as agricultural land). They usually appear early, flying low and steadily, wheeling in large circles. Their feeding territory is strongly defended against conspecifics ,though and male female pair may share the same territory. It is not known how permanent these pairs are

Systems: Terrestrial

Major Threat(s):

There are no major threats to this species as a whole. In South Asia, this species is locally threatened by habitat loss, largely through commercial logging and the conversion of land to agricultural use and human settlements. It is also threatened in some areas by hunting for local consumption and for medicinal purposes (Molur et al. 2002).

Conservation Actions:

In South Asia, there are no direct conservation measures in place for this species and the species has not been recorded from any protected areas. Further studies are needed into the distribution, abundance, reproduction and ecology of this species. Populations of this species should be monitored to record changes in abundance and distribution. General habitat maintenance, conservation and restoration are needed. Public awareness activities are recommended (Molur et al. 2002). In Southeast Asia, the species has been recorded from a number of protected areas and no direct conservation measures are currently needed.

Genus: Hypsugo:

Cadorna's Pipistrelle (*Hypsugo cadornae)*

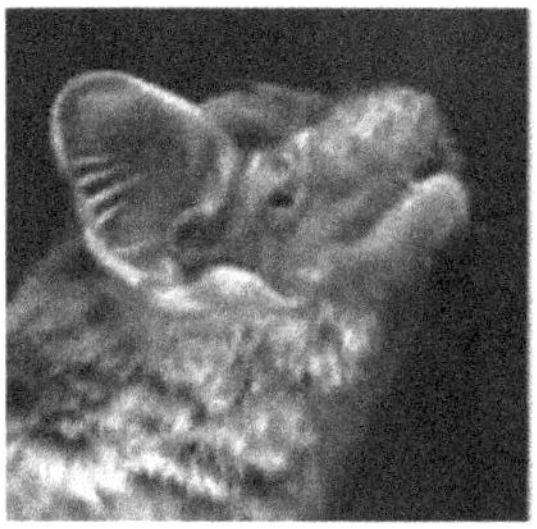

Description:

HB: 48; T: 49; FA: 32-33; E: 15.
The upper parts are dark brown, the underparts lighter. The individual hairs are slaty for their basal halves, with only the tips brown. The ears are broad and triangular, with a short, very broad tragus. The skull is long, with a full braincase, deepened xygomata, and distinct post orbital processes; the rostrum and palate are short. The first upper premolar is very tiny.

Range Description:

This species is widespread in north-eastern South Asia and mainland Southeast Asia. In South Asia, this species is presently known from India (West Bengal) and has been recorded from an elevation of 1,000 m asl (Molur et al. 2002). Lekagul and McNeely (1977) erroneously reported Assam instead of West Bengal as the range of this species range (Srinivauslu et al. in press). In Mainland Southeast Asia, it has been recorded from northern Myanmar, **Northern Thailand,** Lao PDR and Vietnam. The species is probably more widespread in Southeast Asia than is currently known (C. Francis pers. comm.).

Countries: Native:

India; Lao People's Democratic Republic; Myanmar; **Thailand;** Vietnam

Population:

This species is widespread but not common.
Population Trend: Unknown

Habitat and Ecology:

In South Asia, little is known about the habitat or ecology of this species except that this species is found in dry bamboo forests and mixed forests (Bates and Harrison 1997). In Vietnam the species was collected in a disturbed area close to the Cuc Phuong National Park headquarters (Bates et. al. 1997). Remains of a single specimen have been collected from a cave in **North-eastern Thailand** (Robinson and Smith 1997).
Systems: Terrestrial

Major Threat(s):

In Southeast Asia, there are no major threats to this species as a whole. In South Asia, the threats to this species remain unknown. However, disturbance to roosting sites by humans could affect the populations (Molur et al. 2002).

Conservation Actions:

In South Asia there are no direct conservation measures in place for this species, and it has not been recorded from any protected areas. Survey, ecological studies and impacts of roost disturbance need to be conducted (Molur et al).

Chinese Pipistrelle (*Hypsugo pulveratus)*

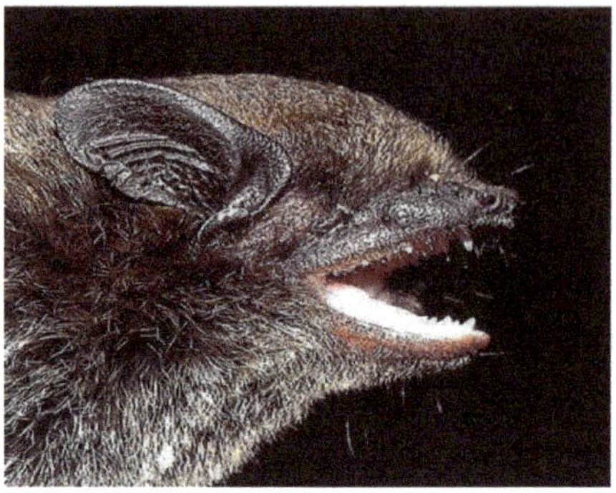

Description:

HB: 44-47; T:37-38; FA: 35-36; E:12-14; HF:7-8.

The upperparts are dark brown with a scattering of pale tipped hairs; on the underparts, the pale tips are longer and more abundant, evenly frosting the lower surface. The fur of the body does not extend to the membranes. The ears are rather large, the tragus is short, broad and bluntly rounded. In the skull, the upper incisors' are about equal in length, but the inner one is distinctly bifid, with the secondary cusp on the our side and directed slightly backwards. The second upper premolar is internal to the tooth row but visible between the first and third upper premolars, which is nearly in contact.

Range Description:

This species is found in China (Anhui, Shanghai, Fujian, Guangdong, Hong Kong, Hainan, Yunnan, Sichuan, Shaanxi, Hunan, Guizhou, and Jiangsu), extending to Vietnam, Lao PDR, and **Thailand (**Smith et al. 2008). It is also found in Myanmar (Bates et al. 2005).

Countries: Native:

China; Lao People's Democratic Republic; Myanmar; **Thailand;** Vietnam

Population: This species is widespread, never abundant but not rare (P. Bates pers. comm.) and can be locally abundant (C. Francis pers. comm.).

Population Trend: Unknown

Habitat and Ecology:

In Myanmar and **Thailand** the species is found in limestone outcrops surrounded by semi-evergreen rainforest and disturbed remnants of forests. It roosts in caves, but has also been found roosting in houses. Also found in Myanmar in dry deciduous forest and known from sea level up to 1,080 m asl (Bates et al. 2005). In Vietnam it also inhabits limestone areas. Has been found roosting in houses (Smith & Xie 2008).

Echolocation calls

· Broadband sweeps ending in a narrowband tail around 32-35 kHz (unpublished).

Major Threat(s):

There are no major threats to this species.

Conservation Actions:

This species occurs in protected areas throughout its range.

Genus: Ia:

Great Evening Bat (*Ia io)*

Description:

HB: 89-104; T: 61-83; FA: 71.5-80; E: 22-29; HF: 13-18

The largest Thai pipistrelle with a forearm of over 70. The upperparts are a uniform sooty brown, while the underparts are dark greyish brown with the individual hairs dark brown at the base , becoming greyish brown at the tips. The face is relatively hairless. The eras are rather short, not extending beyond the muzzle when laid forward; the insides of the ears are densely haired near the tips. The tragus is rather short, with a bluntly rounded tip. The tip of the tail extends slightly beyond the interformeral membrane. The skull is large and robust, with a flattened rostrum and an occuput which is distinctly elevated above the forepart of the braincase. The sagittal crest is prominent and continues forward as far as the interorbital region. The upper inner incisor has a faint indication of a cusp near the outer tip, while the lower incisors are marked trifid. The upper canine is strongly in contact with the large second upper premolar, with the tiny first premolar hidden between the canine and second premolar. · The species is phylogenetically close to Scotomanes, and is placed in the tribe Nycticeiini, not the Pipistrellini as previously thought (Thabah et al. 2007). .

Range Description:

This widespread species has been recorded from Northern South Asia, much of Southern and Central China, and Northern Southeast Asia. In South Asia this species is presently known from India (Assam and Meghalaya) and Nepal (Western Nepal) (Molur et al. 2002). In China, it has been recorded from Sichuan, Guizhou, Anhui, Jiangxi, Hunan, Guangxi, Jiangsu, Yunnan, Shaanxi and Hubei (Smith and Xie 2008). In Southeast Asia, it ranges from Myanmar in the west, eastwards into **Thailand,** Lao PDR and Vietnam. It has been recorded at elevations of 200 m to 1,700 m asl.

Countries: Native:

China; India; Lao People's Democratic Republic; Myanmar; Nepal; **Thailand;** Vietnam

Range Map: Click here to open the map viewer and explore range.

Population:

The largest known population is of close to a hundred animals in Nepal (G. Csorba pers. comm.). In **Thailand** the known populations are up to about 20 individuals (S. Bumrungsri pers. comm.).

Population Trend: Unknown

Habitat and Ecology:

This species is associated with tropical moist forest, where it is exclusively a cave roosting species. It has been found in areas of pristine mixed and deciduous forest **(Thailand),** wet evergreen forest (Vietnam). It has often been reported from limestone karst areas, however, it is found in non-limestone areas in Vietnam (Thong pers. comm.). Roosts in caves. Emergence and flight pattern

· Ia io has a high wing loading (15.4 Nm -2), average aspect ratio (6.9) and a high tip shape index (1.1), features associated with fast and efficient flight (Thabah et al. 2007).
Foraging behaviour

· This is a carnivorous bat and feds extensively on birds. It evolved this remarkable feeding habit independently of *Nyctalus lasiopterus,* which seems to specialise in capturing migrating birds in the air (Ibanez et al. 2001). In India (Meghalaya) the diet comprises mainly *Coleoptera* and birds, although *Lepidoptera* and traces of *Diptera, Orthoptera* and *Hemiptera* were also found (Thabah et al. 2007). In China, (Guizhou Province) bats were captured in early November, and fresh droppings were also collected from underneath the roost. Bird feathers comprised 82% of the bats' droppings by volume (Thabah et al. 2007).

Systems: Terrestrial

Echolocation calls:

Ia io emits relatively low frequency echolocation calls and sometimes produces two-toned calls, a characteristic of species that echolocate distant targets. A waveform and spectrogram from a bat recorded in India below (Thabah et al. 2007).

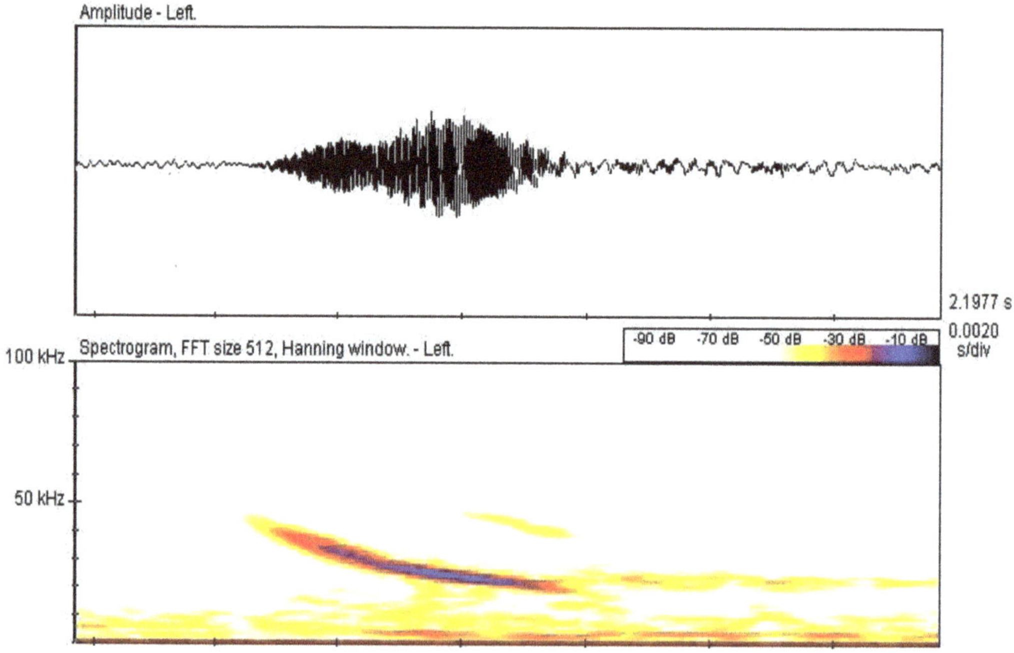

The table below shows call characteristics from two-toned calls recorded from hand-released bats in India (Thabah et al. 2007).
Call parameters

Higher Tone
Lower tone Highest energy (kHz)
27.6 ± 1.5
24.8 ± 1.5
Start Frequency (kHz)
34.9 ± 5.9
38.6 ± 2.8
End Frequency (kHz)
21.9 ± 1.8
18.0 ± 1.1
Duration (ms)
2.4 ± 0.5
3.8 ± 0.6
Pulse Interval (ms)
65 ± 20.5
93 ± 9.3

Major Threat(s):

In Southeast Asia there are no major threats to this species. It may be locally threatened by general habitat degradation. In South Asia, this species is threatened by deforestation, generally resulting from logging operations and the conversion of land to agricultural and other uses. It is also threatened by disturbance to roosting sites by humans (Molur et al. 2002).

Conservation Actions:

In South Asia there are no direct conservation measures in place for this species, and the species has not been recorded from any protected areas. In Southeast Asia, it has been recorded from some protected areas. In South Asia, further studies are needed into the distribution, abundance, reproduction and ecology of this species. Populations of this species should be monitored to record changes in abundance and distribution (Molur et al. 2002).

Genus: Pipistrellus:
Indian Pipistrelle (*Pipistrellus coromandra*)

Description:

HB: 8-9; FA: 3; WS: 19.22; W: 9-13.

The female is larger than the male. They are generally blackish brown to reddish brown above. The underside is brown with a greyish tinge. Wing membrane, face and ears are blackish brown. Wing membrane as a pale margin. Possesses fairly dense and short fur. Muzzle is blunt. Ears small, triangular shaped, with blunt tips. Tragus prominent and one third the length of the pinnae.

Range Description:

This widely distributed species is found throughout most of South Asia, parts of Southern China and much of mainland Southeast Asia. In South Asia this species is presently known from Afghanistan (Nangarhar Province), Bangladesh (no exact location), Bhutan (no exact location), India (Andaman and Nicobar Islands, Andhra Pradesh, Arunachal Pradesh, Assam, Bihar, Goa, Gujarat, Jammu and Kashmir, Jharkhand, Karnataka, Madhya Pradesh, Maharashtra, Manipur, Meghalaya, Nagaland, Orissa, Sikkim, Tamil Nadu, Tripura, Uttarakhand, Uttar Pradesh and West Bengal), Nepal (Central), Pakistan (North West Frontier Province and Punjab) and Sri Lanka (Central, North Central, North Western Northern, Southern and Uva provinces) (Das 2003; Khan 2001; Korad et al. 2007; Molur et al. 2002; Simmons 2005; Srinivasulu and Srinivasulu 2005; Vanitharani 2006). In South Asia, it has been recorded from 100 to 2,769 m asl (Molur et al. 2002). In China it has been recorded from Xizang (Smith and Xie 2008). In Southeast Asia, it is present in Myanmar, **Thailand,** Lao PDR, Vietnam, Cambodia and Peninsular Malaysia.

Countries: Native:

Afghanistan; Bangladesh; Bhutan; Cambodia; China; India; Lao People's Democratic Republic; Myanmar; Nepal; Pakistan; Sri Lanka; Vietnam

Population:

In South Asia, it is a widely distributed and common species and the population seems to be doing well in its range (Molur et al. 2002). It is fairly common in Vietnam, even in cities and similar urban habitats.

Population Trend: Unknown

Habitat and Ecology:

This species is found in varied habitat types from forested regions, agricultural landscapes to urban areas. It roosts in trees, crevices and cracks in walls and ceilings of houses, tiles of huts, old buildings, temples, under bark and in holes of large trees, signboards, tree hollows in small groups of few individuals. It is an early flyer with a slow fluttering flight and hunts on flies, ants and other small insects. There are three breeding seasons and two young ones are born (Bates and Harrison 1997).

Systems: Terrestrial

Major Threat(s):

There are no major threats to this species as a whole.

Conservation Actions:

In South Asia, although there are no direct conservation measures in place, the species has been recorded from several protected areas including Kanha National Park in Madhya Pradesh and Nagarjunasagar-Srisailam Tiger Reserve in Andhra Pradesh (C. Srinivasulu pers. comm. 6 March 2008). In Southeast Asia it has been recorded from some protected areas.

Java Pipistrelle (*Pipistrellus javanicus)*

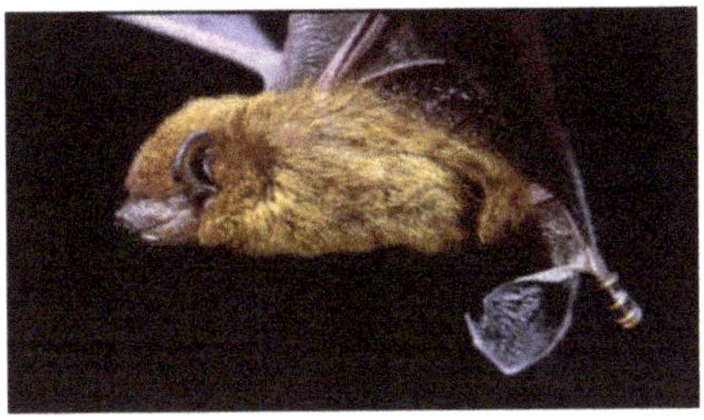

Description:

HB: 45-56; T: 31-38; FA: 31-35, E:11-13; HF:7-8.

The fur as frosting due to the individual hairs being dark at their base and pale at their tips; the underparts are paler in colour. The ears and wings are dark brown. The first upper premolar is displaced inward, but still prevents the canine from being in direct contact with the second upper premolar.

Range Description:

This widespread species has been recorded from Northern and Central South Asia, Southern China, and much of Southeast Asia. In South Asia this species is presently known from Afghanistan (Balkh, Faryab, Kabul, Konarha and Kunduz provinces), Bangladesh (Chittagong and Sylhet divisions,), India (Assam, Himachal Pradesh, Madhya Pradesh, Maharashtra, Manipur, Meghalaya, Nagaland, Nicobar Islands, Sikkim, Uttarakhand and West Bengal), Nepal (Central) and Pakistan (North West Frontier Province and Punjab) (Das 2003; Korad et al. 2007; Molur et al. 2002; Srinivasulu and Srinivasulu 2005). In South Asia, it has been recorded up to 2,380 m asl. In China, it has been recorded from Xizang and Yunnan (Smith and Xie 2008). In Southeast Asia, it ranges widely, from Myanmar in the west, through **Thailand,** Lao PDR, Vietnam, Cambodia and Peninsular Malaysia to Singapore. With insular Southeast Asia, the species is distributed in Indonesia (Sumatra, Bangka, Java, Flores, Karakelang), the island of Timor (East Timor and Indonesia), on the island of Borneo (Brunei, Indonesia and Malaysia) to the Philippines, where it has been recorded from the islands of Camiguin, Luzon (Benguet , Cagayan, Camarines Sur, Isabela, Laguna, Quezon, and Rizal provinces), Mindanao (Bukidnon, and Davao del Sur provinces), Mindoro, Negros, Palawan, Panay, Reinhard, and Sibuyan (Heaney et al., 1998). In the Philippines, it has been recorded from sea level to around 2,250 m asl (Mount Kitanglad).

Countries: Native:

Afghanistan; Bangladesh; Brunei Darussalam; Cambodia; China; India; Indonesia; Lao People's Democratic Republic; Malaysia; Myanmar; Nepal; Pakistan; Philippines; Singapore; **Thailand;** Vietnam

Range Map: Click here to open the map viewer and explore range.

Population:

This species is widely distributed, common and the population seems stable and doing well (Molur et al. 2002; C. Francis pers. comm.; Sedlock pers. comm. 2006).

Population Trend: Stable

Habitat and Ecology:

This species is found in varied habitat types from primary and secondary forested regions, agricultural landscapes (including rubber plantations) to urban areas. It roosts in trees, crevices and cracks in walls and ceilings of houses, tiles of huts, old buildings, temples, under bark and in holes of large trees, signboards, tree hollows in small groups of few individuals. It is an early flyer with a slow fluttering flight and hunts on flies, ants and other small insects. There are three breeding seasons and two young ones are born (Sanborn et al., 1952; Bates and Harrison 1997; Heaney et al., 1998; S. Bumrungsri pers. comm.; P. Bates pers. comm.).

Systems: Terrestrial

Major Threat(s):

Overall there are no major threats to this adaptable species. In South Asia, it is locally threatened by deforestation for timber, firewood and conversion to agricultural use. It is also threatened by disturbance to roosting sites by humans (Molur et al. 2002).

Conservation Actions:

This species occurs in a number of protected areas in Southeast Asia. In South Asia, although there are no direct conservation measures in place, the species has been recorded from several protected areas including Kanha National Park in Madhya Pradesh and Nagarjunasagar-Srisailam Tiger Reserve in Andhra Pradesh (C. Srinivasulu pers. comm. 6 March 2008). Further studies are needed into the distribution, abundance, breeding biology and general ecology of this species. Populations of this species should be monitored to record changes in abundance and distribution (Molur et al. 2002).

Mount Popa Pipistrelle (*Pipistrellus paterculus)*

Range Description:

This species is distributed in northern South Asia, southern China and parts of mainland Southeast Asia. In South Asia, this species is presently known from India (Assam, Bihar, Jammu and Kashmir, Manipur and Nagaland), where it has been recorded from sea level to an elevation of 1,500 m asl (Molur et al. 2002). In China, it is restricted to Yunnan (Smith and Xie 2008). In Southeast Asia, it has been recorded from Myanmar, **Northern Thailand (Bumrungsri, et al., 2006),** Lao PDR and Vietnam.

Countries: Native:

China; India; Lao People's Democratic Republic; Myanmar; Vietnam

Population:

This species is relatively common in many areas in Myanmar (P. Bates pers. comm.). In South Asia, although this is a widely distributed species a declining trend in its population is being observed (Molur et al. 2002).

Population Trend: Unknown

Habitat and Ecology:

In South Asia, this species is found in primary and secondary forests. It roosts among banana and bamboo trees, deep holes in trees and roofs of thatched huts in forested areas (Molur et al. 2002). In Myanmar, it is predominantly found in limestone caves, although it has also been recorded from forest areas without karst habitats. It can be found in disturbed areas including agricultural land (P. Bates pers. comm.).

Systems: Terrestrial

Major Threat(s):

In Southeast Asia there are no believed to be any major threats to this species as a whole (P. Bates pers. comm.). In South Asia, this species is threatened by deforestation, generally resulting from logging operations and the conversion of land to agricultural and other uses (Molur et al. 2002)

Conservation Actions:

It has been recorded from some protected areas in Southeast Asia. In South Asia, there are no direct conservation measures in place for this species and it has not been recorded from any protected areas. Further studies are needed into the distribution, abundance, reproduction and ecology of this species. Populations of this species should be monitored to record changes in abundance and distribution (Molur et al. 2002).

Genus: Scotomanes:

Harlequin Bat (*Scotomanes ornatus)*

Description:

HB: 72-78; T:50-62; FA: 55-59; E: 19-24; HF: 12.

The upper parts are russet brown, with the underparts dark brown along the midline and whitish along the sides. There is a white strip down the middle of the back, and a white patch on the crown of the head, and white spots above the wing behind each shoulder; other white spots may be scattered over the body in some individuals. Individual hairs are black at the base, then whitish with rusty tips. Dobson (1878) says that males are always rusty red, and females are darker and may lack the white stripe down the back. The tail is long, free of the interformeral membrane at the tip, the ears are moderate, with the tragus about 1/3 the length of the ear. In the skull, the anterior palatal emargination is relatively smaller than in other members of the family. The upper incisor are long and sharp, extending from well separated premaxillae. The premolars are much smaller than the molars.

Ecology and Behaviour:

They roost mainly in Trees, where the coloration may resemble fruit or leaves, giving them camouflage in their relatively exposed roosting sites. The roost is around 2 to 4 metres above the ground. Individual bats have been found roosting in folded banana leaves.

Range Description:

This species is distributed from north-eastern South Asia, into Central and Southern China, and Northern parts of Southeast Asia. In South Asia this species is presently known from Bangladesh (Sylhet Division), India (Arunachal Pradesh, Assam, Manipur, Meghalaya, Nagaland, Sikkim and West Bengal) and Nepal (Eastern Nepal) (Khan 2001; Das 2003; Molur et al. 2002; Srinivasulu and Srinivasulu 2005), and has been recorded from sea level to an elevation of 1,400 m asl (Molur et al. 2002). In China, it has been recorded from Sichuan, Yunnan, Guizhou, Anhui, Fujian, Hunan, Guangxi, Guangdong and the island of Hainan (Wang 2002; Smith and Xie 2008). In Southeast Asia, it ranges through northern and western Myanmar, **Northern Thailand** (Lekagul and McNeely 1977; S. Bumrungsri pers. comm.), Lao PDR and Viet Nam. It has been recorded up to 2,200 m asl in Thailand (S. Bumgrunsri pers. comm.).

Genus: Tylonycteris:

Greater Bamboo Bat (*Tylonycteris robustula*)

Description:

· Dorsal and ventral fur is brown, and the throat and belly are paler than the back.

· Forearm length 25.0 -29.0 mm , average 27.2 mm . Average wingspan is 195.5 mm. Normally larger than *T. pachypus.*

· Skull is extremely flat, there are pads on digit plantar (foot) and elbow joints, all these features being adaptations to roosting in bamboo stems.

· Average weight 4.8g. (all the measurement data as given by Zhang et al., 2005).
Life history

· Birth dates between late May and early June in Guangxi. Gives birth to twins. Pups can fly when 24-27 days old (Zhang et al. 2005).This bat is common where bamboo grows. It is believed that it has a habit of roosting inside the stems of the bamboo. The bats gain access via vertical slits in the stem wall; these slits can created by beetles. Size: About 4-5 cm without tail. Bamboo Bats (or 'Club-footed Bats') are so-called because they roost by day inside the stems (or 'culms') of various bamboo species.

Many such roosting sites are created by the activities of a species of leaf beetle Lasiochila goryi which, after emergence from pupation, cuts a narrow vertical slit in order to escape confinement within the bamboo stem. Larger roosting sites may naturally occur in dead or dying bamboo stems. Occasionally both species may roost in the same bamboo clump, but rarely in the same cavity.

There are two species of bamboo bat : the Greater Bamboo Bat *Tylonycteris robustula* and the Lesser Bamboo Bat *T. pachypus*. Both species have a markedly flattened skull to allow them access to their roost. The thumbs and soles of the feet have fleshy pads which give a firm foothold inside the bamboo.

Remarkably these bats are able to squeeze through slits as narrow as 4-8 mm. Roosting sites are thus largely inaccessible to predators, though an instance of predation by a Paradise Tree Snake *Chrysopelea paradisi* was documented in Singapore by Chan Kwok Wai.

Reportedly the bats can fly straight through the slit-shaped opening without pausing, though whether this is true or not is unclear.

Fur colour is variable : the Greater Bamboo Bat is medium to dark brownish overall, and the Lesser Bamboo Bat tends to be more pale brown on top, and orange underneath. The species are most easily distinguished on the basis of size.

Bamboo bats inhabit forested areas, and forage at night for flying insects, particularly termite swarms.

Range:

This species ranges from North-eastern India, through parts of Southern China, to much of mainland and insular Southeast Asia. In South Asia, this species is presently only known from India (Mizoram) (Srinivauslu et al. in press), where it has been recorded from an elevation of 800 to 1,000 m asl (Molur et al. 2002). In southern China, it has been reported from Yunnan and Guangxi (Smith and Xie 2008). In Southeast Asia, it ranges from Myanmar in the west, through **Thailand,** Lao PDR, Vietnam, Cambodia, Peninsular Malaysia and Singapore, to Indonesia (Sumatra, Java, Bali, Lombok, Sulawesi, Peleng), the island of Timor (East Timor and Indonesia), western Borneo (Brunei, Indonesia and Malaysia) and the Philippines, where it has been recorded from the islands of Luzon (Rizal and Zambales provinces), Calauit and Palawan (Heaney et al. 1998) and Negros (Carino et al. 2006).

Countries: Native:

Brunei Darussalam; Cambodia; China; India; Indonesia; Lao People's Democratic Republic; Malaysia; Myanmar; Philippines; Singapore; **Thailand;** Timor-Leste; Vietnam

Threats:

There are no major threats to the species as a whole. In South Asia the threats to this species remain unknown (Molur et al. 2002), however, it is presumed that this species may be threatened by deforestation, generally resulting from commercial logging operations and the conversion of land to agricultural and other uses. It is also possibly threatened due to disturbance to roosting sites (C. Srinivasulu pers. comm. 24 February 2008).

Conservation Actions:

This species has been recorded from nay protected areas. In South Asia, there are no direct conservation measures in place for this species and it has not been recorded from any protected areas. In South Asia, further studies are needed into the taxonomy, distribution, abundance, reproduction and ecology of this species. Populations should be monitored to record changes in abundance and distribution (C. Srinivasulu pers. comm. 24 February 2008).

Subfamily: Murininae:

Genus: Harpiocephalus:
Greater Hairy-Winged Bat (*Harpiocephalus mordax)*

Description:

HB: 55-57; T: 37-55; FA:40-54.
The colouration above is chestnut, orange red or rusty, usually mixed with grey, the underparts are greyish buffy. The fur is thick and woolly. The legs, tail membrane and occasional parts of the wings are covered in hair.

Ecology:

Very little is known about the habitats and history of these bats, but they have been observed in hilly countryside and roost most probably in vegetation. Beetle remains have been found in the stomach contents which indicate they feed on chitinous insects.

Range Description:

This species appears to range from South Asia, to mainland Southeast Asia and the island of Borneo. In South Asia, some specimens earlier identified as *H. mordax* were reassigned as *H. harpia,* and it is not currently possible to clearly identify the distribution range of this species in India (Molur et al. 2002). In Southeast Asia, the species has been patchily recorded from Myanmar, **Thailand,** Lao PDR, Vietnam, Peninsular Malaysia (reported from Taman Negara [Aik and Akbar 2002] and from Temengor Forest Reserve, Perak by [Francis 1995]) and Borneo (Sabah [Malaysia).

Countries: Native:

India; Lao People's Democratic Republic; Malaysia; Myanmar; **Thailand;** Vietnam
Range Map: Click here to open the map viewer and explore range.

Population:

There is little information available on the population of this species.
Population Trend: Unknown

Habitat and Ecology:

There appears to be little information available on the natural history of this species. It appears to be a species of mature forest.
Systems: Terrestrial

Major Threat(s):

The threats to this species are not known, but probably include degradation of forested areas.

Conservation Actions:

Taxonomy and field surveys are needed for this species throughout its range. It has been recorded from some protected areas in Malaysia.

Genus: Murina:

Hutton's Tube-Nosed Bat (*Murina huttoni)*

Description:

HB:47-50; T:31-39;

FA:29-35; HF:8-10; E: 16-18;W6-9 g

The upper parts are greyish brown, with the under parts whitish; the pelage is also rather long, and the interformeral membrane and backs of the toes are covered with long hairs. (VanPeenen 1969). The ears are oval with convex margins; there is a small process projecting from the base of the inner margin. The tragus is long and pointed and curved slightly inwards. In the skull, the outer upper incisors are larger than the inner ones. The first upper premolar is smaller than the second, and is located slightly inside the tooth row. The molar teeth have more pronounced cusps than in M. cyclotis(Hill1972)

Range Description:

This widely recorded species largely ranges from northern South Asia, through Southern China and northern mainland Southeast Asia. In South Asia, this species is presently known from India (Arunachal Pradesh, Assam, Jammu and Kashmir, Meghalaya, Uttarakhand and West Bengal), Nepal (Central) and Pakistan (North West Frontier Province) (Das 2003, Molur et al. 2002). In China, it has been reported from Fujian, Jiangxi and Guangxi (Smith and Xie 2008). In Southeast Asia, it is found in northern Myanmar, **Northern Thailand,** much of Lao PDR, parts of Vietnam, and seemingly disjunctly in Peninsular Malaysia. In South Asia it has been recorded from 1,450 up to 2,500 m asl (Molur et al. 2002).

Countries: Native:

China; India; Lao People's Democratic Republic; Malaysia; Myanmar; Nepal; Pakistan; **Thailand;** Vietnam

Range Map: Click here to open the map viewer and explore range.

Population:

In South Asia, the abundance, population size and trends for this species are not known (Molur et al. 2002). In Southeast Asia it is considered to be uncommon to rare.

Population Trend: Unknown

Habitat and Ecology:

In South Asia, this species is found in montane forests, tropical broadleaved forests and banana plantations. It roosts among banana leaves and bases of banana plants, and under tree bark (Molur et al. 2002). In China, it is found at mid-elevations, and appears to be adapted to a variety of habitats (Smith and Xie 2008). In Southeast Asia, the natural history is not well known, however, it has been recorded from pristine hill evergreen forest in **Thailand** and Lao PDR (S. Bumrungri and C. Francis pers. comm.).

Systems: Terrestrial

Major Threat(s):

In South Asia, the habitat of this species is being deforested for timber, firewood and converted for agricultural use and disturbance to roosting sites (Molur et al. 2002). In Southeast Asia and China nothing known about the threats to this species (P. Bates pers. comm.).

Conservation Actions:

In South Asia, there are no conservation measures in place and this species has not been recorded from any protected areas. In Southeast Asia, it has been recorded from some protected areas. Further studies into the natural history of this species are needed.

Scully's Tube-Nosed Bat (*Murina tubinaris)*

Description:

The name "ashy-gray" comes from the colour of their dorsal fur, while the ventral fur is dark gray, and there is some white fur on the breast area. In some areas the tips of the hair are dark. It lacks the golden guard hairs so common in other members of the genus Murina.This bat is small enough to fit in a person's hand, weighing 4.0 and 5.5 grams (0.14 and 0.19 oz). As of 2013, there are still few details known about them and their ecology and it is suspected there are many more species of bats yet to be discovered in the region.

Range Description:

This species is present in northern South Asia and mainland Southeast Asia. It does not appear to be present in China. In South Asia this widely distributed species is presently known from India (Arunachal Pradesh, Jammu and Kashmir, Meghalaya, Mizoram, Sikkim and West Bengal) and Pakistan
(North West Frontier Province) (Das 2003, Molur et al. 2002), and has been recorded from 1,200 m to 2,650 m asl. In Southeast Asia it is present in northern and eastern Myanmar, **Northern Thailand,** Lao PDR and Vietnam.
Countries: Native:
Cambodia; India (Jammu-Kashmir); Lao People's Democratic Republic; Myanmar; Pakistan; **Thailand;** Vietnam
Range Map: Click here to open the map viewer and explore range.
Population: It has only been caught in small numbers in all parts of the range (Francis pers. comm.).
Population Trend: Unknown

Habitat and Ecology:

In South Asia, little is known about the habitat or ecology of this species except that this species is found in montane forests on the mountain slopes and roosts among trees (Molur et al. 2002). One or two young are born (Bates and Harrison 1997). In northern Myanmar, it has been found altitudes ranging from 615 to 1,230 m asl (Bates and Harrison 1997). In Southeast Asia, it appears to have been recorded from somewhat disturbed areas.
Systems: Terrestrial

Major Threat(s):

In South Asia, this species is threatened due to habitat loss and disturbance to roosting sites mainly due to human unrest in the areas where it occurs (Molur et al. 2002). In Southeast Asia, the threats to this species are not known (P. Bates pers. comm.).

Conservation Actions:

In South Asia, there are no conservation measures in place and this species has not been recorded form any protected areas. Ecological and population studies recommended (Molur et al. 2002). In Southeast Asia, it has been recorded from some protected areas.

Tube Nosed Bat (*Murina cyclotis)*

Description:

HB:42-47; T: 38-42; FA:31-35;E: 12; HF:9-10

There seems to be 2 colour phases:

1.Upper parts warm Rufus brown, with the hairs dark brown at the base and lighter at the tips;underparts greyish white, with a brownish tinge.

Under parts greyer, lacking the Rufus tinge; the underparts lack the brownish tinge.

Long hairs cover the dorsal surface of the interformeral membrane, the backs of the toes and the edge of the membrane. The ears are broad and rounded, with a small lobe at the base of the anterior margin. The tragus is triangular, about half the length of the ear. The incisors are complex, with subsidiary cusps; the lower cheek teeth are equal in crown height, with the upper molars having lower crowns. The braincase is inflated interiorly, rising in a curve from the rostrum.

Round-eared tube-nosed bats range in weight from 9 to 12 g, with total length between 80 and 89 mm. Only one wingspan measurement (229 mm) has been reported for this species. Females tend to be slightly larger than males. Round-eared tube-nosed bats are diagnosed by tubular nostrils and short, round ears. They have a triangular tragus, which are long, narrow, and tapering to a point. The anterior braincase is ballooned and the upper lip possesses a protruding fringe of hairs. *Murina cyclotis* has small eyes which are closer to the ears than the nostrils. Two different color patterns have been noted. In one color variant, the dorsal fur is a reddish-brown, with the hairs increasing in lightness from the base to the tip; the ventral fur is greyish white, with a brownish tinge. In the other color variant, the dorsal fur is greyer, without the reddish tinge; the ventral fur is greyish white without any brownish tinge. This species has a semi transparent wing membrane, relatively long thumbs, and a tail shorter than the length of its head and body. The feet are hairy, small, and the claws are relatively long and sharp.

Habitat:

Murina cyclotis occurs at elevations as low as 250 m in the foothills to 1,500 m in the mon-tane forests. Round-eared tube-nosed bats are found primarily in the tropical forests of south-east Asia. Coastal areas in-habited by *M. cyclotis* are bordered by the South China Sea and Philippine Sea, which suggests reliance on humid regions. (Bates and Harrison, 1997; Heaney, 2005; Medway, 1983; Phillips, 1980) Three sub-species of M. cyclotis have been identified. *M.c. cyclotis* is found from northeastern India to Vietnam, including **Thailand.** *M. c. eileenae* is in Sri Lanka and has slightly darker and duller fur, and *M. c. peninsularis* is found on the Malay Peninsula and is identified by a relatively wider anterior rostrum and more massive teeth.

Ecology:

There is little information reported for mating systems in *M. cyclotis.* (Phillips, 1980)

There is little information reported for general reproductive behavior in round eared tube nosed bats. Pregnant females in Pahang, Malay Peninsula were reported to carry two fetuses in the months of February and May. (Medway, 1983)

Breeding interval
Nothing known about the breeding interval of *M. cyclotis*.
Breeding season
Nothing known about the breeding season of *M. cyclotis*.
Range number of offspring
1 to 2

There is no information reported for lifes-pan and longevity in *M. cyclotis*.

Behaviour:

Murina cyclotis individuals feed in open forest areas, and fly approximately 1.8 m above the ground. Their flight is slow but maneuverable. They roost in large dead and dry leaves of the cardamon plant. Several individuals often roost together on a single leaf or bunch of leaves, which serves to camouflage the bats. A closely related species *Murina florium* has been found roost-ing in enclosed plant-material nests and hang-ing nests of bird species, presumably to provide protection where other roosting options are not available. Another closely related species, *Murina aurata,* is thought to hibernate in trees during the winter. This suggests possible hibernation of *M. cyclotis,* but this is not known for certain. (Bates and Harrison, 1997; Kingston, et al., 1999; Schulz, 1998) These bats make use of echolocation to avoid obstacles and capture airborne prey. Calls are frequency-modulated (FM), with low intensity and short duration. The calls are at high frequency, starting at 152 to 180 kHz and sweeping down to an end frequency of 43 to 86 kHz. This large band-width allows these bats to accurately locate targets in the cluttered forests they in-habit. Furthermore, the use of an extremely high echolocation frequency may reduce the need for both visual and other auditory signals. (Bates and Harrison, 1997; Kingston, et al., 1999; Schulz, 1998). Tube-nosed bats eat insects in the damp forests they in-habit. A closely re-lated species, *M. florium* (Flores tube nosed bats) has been ob-served to eat in a manner possibly similar to *M. cyclotis*. While eat-ing a moth, *M. florium* individuals were observed perching by both feet and thumb claws. They placed separated pieces of the insect into the uropatagium. After feeding on the edible parts, they released the waste to the ground below. This posture was also used by *M. florium* to ingest fecal matter (coprophagy). (Bates and Harrison, 1997; Bonaccorso, 1998)

Predation:

The reddish-brown color of *M. cyclotis* makes it difficult for predators to detect as they roost among dead leaves.

There are no known predators of *M. cyclotis,* how-ever the skulls of a closely re-lated species, *M. florium*, have been recovered from owl pellets. Most bats are preyed on by owls as they fly and by snakes at roosts. (Bonaccorso, 1998; Phillips, 1980)

Threats:

The rapid increase in oil palm plantations has resulted in extensive loss of primary forest. Together, Malaysia and Indonesia export 88% of the world's palm oil, for use in products such as margarine, lipstick and detergent. Deforestation continues at a steady rate for conversion to agricultural land and building communities. Despite the contribution of many bats in the control of insect crop pests, persecution of bats is also a threat.

Conservation:

Deforestation of primary forest for oil palm plantations, including within protected areas, is an issue of major concern and one that relies on both governmental and consumer action. Some large retailers have agreed, in collaboration with the WWF, to source products containing palm oil from plantations that are not on deforested land. Many scientific and charitable groups contribute to bat monitoring and local education programmes that can help to reduce persecution and raise awareness of the natural assets of the land

Subfamily: Miniopterinae:
Genus: Miniopterus:
Western Bent-Winged Bat (*Miniopterus magnater)*

Description:

Bent-winged or Long-fingered Bats are so called because of the relatively long terminal bone of the middle finger, compared to the middle bone. As a result, the end part of the wing is folded back and tucked underneath the rest of the wing when roosting.

The true number of species of *Miniopterus* bats is unclear, but at least 11 are generally agreed to exist. There are great similarities between all species, which are often distinguished only on the basis of forearm length and weight. Identification of species on the basis of photographs alone is not possible.

In all species the fur is thick and generally brown-black to brown, or sometimes reddish. The ears are relatively small and rounded, and the eyes small.

Many bent-winged bat species roost in large colonies in caves : these bats may become spectacularly abundant in tropical rainforests where karst limestone habitats provide cave systems for roosting. For example in parts of Borneo there are colonies of over 100,000 bats. These colonies are well organised : it is known that thousands of juveniles may be left together by night under the care of a few females while their parents are out foraging for food. They feed on flying insects, particularly moths, high in the forest canopy.

Some species are wide-ranging across Southeast Asia, New Guinea and beyond. The Common Bent-winged Bat *Miniopterus schreibersi* has a range that extends from parts of Africa, across southern Europe and Asia to Australia and Oceania. The range of elevations is also large : in New Guinea some species occur up to 3200 metres above sea level.

Range Description:

This species appears to be widely distributed from Myanmar in the west (recorded from Hpoan and Mawlamyine; Bates et al. 2000), through Southern China (including Hainan Island) and much of Southeast Asia, as far east as the island of New Guinea. Within this range it has been recorded from many island localities including Sumatra (a single locality; Indonesia), Western Java (Indonesia), Northern Borneo (Sabah, Malaysia and Kalimantan, Indonesia), Timor (East Timor and Indonesia), and Seram (Indonesia). **Thailand** and Vietnam A number of records of *Miniopterus schreibersii* have been re-identified as *Miniopterus magnater (*T. Reardon pers. comm.). It has been recorded from sea level to 2,100 m asl, but this is not clear given identification problems.

Countries: Native:

Indonesia; Papua New Guinea

Population:

The population abundance of this species is not known.

Population Trend: Unknown

Habitat and Ecology:

This is a cave roosting species, often found roosting with other bat species. It possibly forms large colonies. It has been recorded foraging generally above the canopy in primary and secondary tropical forest, and also close to human settlements. Bonaccorso (1998) and Payne et al. (1985) noted that they may be seen hawking insects around street lights, but this is not certain given identification problems.
Systems: Terrestrial

Major Threat(s):

There are no known major threats to this species. It is presumed to be threatened by cave disturbance and degradation in some parts of its range.

Conservation Actions:

This species is present in a number of protected areas. Hendrichsen et al. (2001) recorded it in Vietnam at Pu Mat National Park, Cuc Phuong National Park, and at Toong Ching. The urgent priority with this species is to resolve Miniopterus taxonomy and identification in order to understand the distribution, abundance, habitat requirements, ecology, and threats to it. There is a need to protect important roosting and foraging sites of this species.

Intermediate Long-fingered Bat (*Miniopterus medius)*

Description:

HB: 50-56; T: 45-49; FA: 41-44; HF: 8-10; E: 11-12;W: 8-11gm

Very similar to *M. schreibersi* but with a shorter forearm. There are 2 colour phases, with upperparts dark reddish brown or black, with or without irregular patches of dark rusty read. The skull closely resembles that of *M. schreibersi* but is narrower in palate and braincase.

Range Description:

This species ranges from **Southern Thailand** and the Malay Peninsula, to Java, with scattered localities on the islands of Borneo and New Guinea (Papua New Guinea only), as far east as Buka Island, Papua New Guinea. It is perhaps also present in the Solomon Islands. On Borneo, there are three records from Sabah (Tepadong caves, Batu pute and Kinabatangan Wildlife Sanctuary; Lackman-Ancrenaz and Ancrenaz 1997), and at least two records from Kalimantan - Betung Kerihun National Park (Soedjito 1999) and from Gunung Niut Nature Reserve, West Kalimantan (Simons 1987). On Java, it has been recorded from the Cibodas Botanical Gardens, Mount Gede Pangrango National Park. It has been recorded from sea level to 2,600 m. However, these data are unclear, given identification problems (T. Reardon pers. comm.).

Countries: Native:

Indonesia (Jawa, Kalimantan, Sulawesi); Malaysia; **Thailand**

Population.

It is locally abundant in the highlands of the western half of Papua New Guinea, and in lowland areas with caves on the islands of New Britain and New Ireland (Bonaccorso 1998) However, these data are unclear, given identification problems (T. Reardon pers. comm.).

Population Trend: Unknown

Habitat and Ecology:

This species roosts in limestone caves, tunnels, and rock crevices. It is frequently found as tightly packed groups in clusters of mixed bat species. Bonaccorso (1998) mentions a mixed colony of 10,000 bats, containing this species, on the island of New Britain. It is an aerial insectivore that forages in lowland tropical dipterocarp forest. One of the five adult females captured at the Ulatawa Plantation in January 1997 was in early term pregnancy (Bonaccorso 1998). However, these data are unclear, given identification problems (T. Reardon pers. comm.).

Systems: Terrestrial

Major Threat(s):

There appear to be no major threats to this species. Limestone extraction and human disturbance of roosting caves may adversely affect the species in some areas.

Conservation Actions:

This species has been recorded from a number of protected areas. The urgent priority for this species is to resolve Miniopterus taxonomy and identification in order to understand the distribution, abundance, habitat requirements, ecology, and threats to it.

Small Bent-Winged Bat (*Miniopterus pusillus)*

Description:

· Dark brown, paler below. ·Forearm length 39 - 43 mm (Shek 2006). · Each wing has a highly developed second phalanx of the third finger. · Body mass 7-10g. · Smaller than *M. fuliginosus*. Very little is known of the life history of this bat and more research need to be carried out

Distribution:

This widespread species is distributed from patchy records in South Asia, into much of mainland Southeast Asia, and part of Insular Southeast Asia. In South Asia, it has been recorded from India (Karnataka, Tamil Nadu and Nicobar Islands) and Nepal (Western Nepal) (Aul and Vijayakumar 2003, Korad et al. 2007, Molur et al. 2002, Vanitharani 2006). In China it has been recorded from Hong Kong, Guangdong, Hainan and Yunnan (Smith and Xie 2008). In mainland Southeast Asia, it ranges from southern Myanmar, through **Thailand** (including the island of Terutau), Lao PDR, Vietnam and northern Cambodia. In Insular Southeast Asia the species has been recorded from Indonesia (Java Sulawesi, West Timor, Ambon, Seram and Batjan) and possibly from East Timor. In South Asia it has been recorded up to an elevation of 1,200 m asl (Molur et al. 2002).

Habitat:

In South Asia, little is known about the habitat or ecology of this species except that this species roosts in limestone caves, under culverts, in crevices in trees in colonies comprising up to 700 individuals (Molur et al. 2002). On the Nicobar Islands, it is observed to roosts in forest caves in large numbers (Aul and Vijaykumar 2003). In Myanmar it has been recorded roosting in limestone caves, and in both degraded forest and agricultural areas (P. Bates pers. comm.).

Systems: Terrestrial

Emergence and flight pattern:

Narrow wings and a high wing loading and aspect ratio mean that this species flies rapidly and is not very manoeuvrable.

Foraging behaviour:

Little known.

Echolocation calls:

Not known.

Major Threat(s):

In Southeast Asia there are no major threats to this species. In South Asia, this species is threatened by habitat loss, largely through commercial logging and the conversion of land to agricultural use. It is also threatened by disturbance and loss of roosting sites by humans (Molur et al. 2002).

Conservation Actions:

In South Asia, although there are no direct conservation measures in place, the species has been recorded from protected areas in India like Kalakkad-Mundunthurai Tiger Reserve in Tamil Nadu (Vanitharani 2006). It has been recorded from a number of protected areas in Southeast Asia. In South Asia, studies are needed into the distribution, abundance, reproduction and ecology of this species. Populations of this species should be monitored to record changes in abundance and distribution (Molur et al. 2002).

Family: Rhinopomatidae:

Genus: Rhinopoma:
Lesser Mouse-Tailed Bat (*Rhinopoma hardwickei*)

Description:

HB:55-75; T: 60-75; FA; 57-66; E22; HF: 15

The lesser mouse-tailed bat is, as its name suggests, a fairly small bat with a very long, mouse-like tail. In fact, it has one of the longest tails of all bats, often as long as the head and body combined. The fur on the body of the lesser mouse-tailed bat is usually brown-grey in colour, tending towards darker brown on the back and lighter grey on the underside, short and soft. . The face is furless, with beady black eyes, a flat, thin snout and a blunt nose, and the ears are large in proportion to the head. A thickened portion of skin at the end of the snout forms a nose leaf, which the bat uses as an amplifier for its echolocation calls . The long, thin tail is only partially enclosed within a flap of skin, known as the tail membrane, which aids the bat when flying; this is in contrast to many other bat species in which the tail membrane fully encloses the tail .

Biology and Ecology:

The lesser mouse-tailed bat feeds primarily on insects, which are caught in flight. Its flight is rather unique amongst bats as it initially flutters its wings and then partially glides . Like most bats, the lesser mouse-tailed bat uses echolocation to locate prey and avoid obstacles, allowing it to fly in complete darkness. The ultrasound calls, which are emitted through the nostrils, consist of long, high frequency chirps (of around 32 kilohertz). Feeding as much as possible during the summer months, the lesser mouse-tailed bat builds up fat in its abdominal region in preparation for the winter, when food is scarce. During the winter season, the lesser mouse-tailed bat enters a period of dormancy (hibernation) and lives off its fat reserves .After awakening from its winter dormancy in late February, the lesser mouse-tailed bat enters its breeding season, which lasts until the middle of April . The gestation period lasts from 95 to 100 days, after which the female gives birth to a single young in June or July. The lesser mouse-tailed bat is well adapted to its arid habitat. The slit-shaped nostrils can be closed to keep out dust and sand and it is able to survive without a source of fresh water, as it obtains most of the water it requires from its food. The kidneys are also able to produce highly concentrated urine, in order to conserve precious water. Inhabits arid and semi-desert vegetation zones where suitable roosts and food are available. Recorded in semi-desert grassland with areas of Acacia scrub in oases with gardens and orchards surrounded by sandy desert and Hamada, in gorges of wades with some Tamarix and Oleanders *(Nerium oleander).* Roosts in dry caves, ruins, underground tunnels (including catacombs), mosques and old buildings. In summer sometimes roosts in fissures, small crevices and among boulders. The species is sedentary and it stores fat in autumn for the winter months.

Systems: Terrestrial

Range Description:

Occurs across central and northern Africa through Arabia and Southern Asia; from Morocco to India north to Israel, Palestine, Jordan Iraq and Afghanistan and south to Kenya. Presence in Myanmar based on a very old reference with no detail of location; there is doubt about its current presence. Occurs up 1,100 m asl in Morocco and Algeria.

Countries: Native:

Afghanistan; Algeria; Bangladesh; Burkina Faso; Cameroon; Chad; Djibouti; Egypt; Eritrea; Ethiopia; India; Iran, Islamic Republic of; Iraq; Israel; Jordan; Kenya; Kuwait; Libya; Mali; Mauritania; Morocco; Nepal; Niger; Nigeria; Oman; Pakistan; Saudi Arabia; Somalia; Sudan; Syrian Arab Republic; **Thailand;** Tunisia; Western Sahara; Yemen (Socotra)

Population:

Appears to be particularly abundant near oases. However, both distribution and abundance are undoubtedly insufficiently investigated because the roosts and suitable habitats are often unreachable. Colonies range in size from a few individuals up to several hundred. Up to 500 individuals have been reported in colonies in Jordan (Amr 2000). Occurs with other species in the genus, in Iran it is normally found in low numbers and low densities and it feeds on coleoptera (M. Sharifi pers. comm. 2005). Assumed stable throughout the southwest Asia region (D. Kock pers. comm. 2005). Population information remains unknown for its African distribution.

Population Trend: Stable

Major Threat(s):

Human disturbance in roost sites and pesticide use against locusts are the main threats. In arid areas of Iran which can not support high numbers of colonies, they aggregate in a few large groups which increases their vulnerability (M. Sharifi pers. comm. 2005). These are not thought to be major threats to the species as a whole at present.

Conservation Actions:

No specific measures are known or are in place, but presumably occurs in protected areas across the range. A study on the impacts of pesticides is required, especially ways in which the impact might be minimised.

Greater Mouse-Tailed Bat (*Rhinopoma microphyllum)*

Description:

One of the few bats in the world to have a tail as long as the head and body combined, the greater mouse-tailed bat has soft, greyish-brown or dark brown fur on the body, with a paler abdomen. The face, the rump and the rear part of the abdomen are bare . It has large ears, which are connected across the forehead by a flap of skin, and slit-like nostrils. This species also has the ability to hang by its thumbs as well as its feet

Ecology:

The greater mouse-tailed bat is a desert species, occupying areas of little vegetation with an annual rainfall of less than 300 millimetres. It roosts in caves, mines, tunnels, old monuments and buildings, and is even known to have lived in the ancient pyramids of Egypt for over 3,000 years .The greater mouse-tailed bat lives in large colonies, sometimes containing over a thousand individuals. Its diet varies depending on the time of year and the location; for example, a population of bats in Iran was found to feed almost exclusively on beetles, while in Israel, during the summer months, the greater mouse-tailed bat is known to feast on carpenter ants during their massive nuptial flights, in which male and female ants emerge from nests to mate . Over autumn, the greater mouse-tailed bat accumulates fat and almost doubles in weight, allowing it to survive for several weeks without any food or water during the harsh winter months when insect prey is scarce. As a result, the greater mouse-tailed bat does not need to hibernate and instead remains active throughout the year. Female greater mouse-tailed bats mate in spring, around March, giving birth to a single young in June or July after a gestation period of around 18 weeks. The young are weaned after six to eight weeks and become sexually mature in their second year of life. Inhabits arid and semi-desert vegetation zones where suitable roosts and food are available. Recorded in semi-desert grassland with areas of Acacia scrub in oases with gardens and orchards surrounded by sandy desert and hamada, in gorges of wadis with some Tamarix and Oleanders (Nerium oleander). Roosts in dry caves, ruins, underground tunnels (including catacombs), mosques and old buildings. In summer sometimes roosts in fissures, small crevices and among boulders. The species is sendentary and it stores fat in autumn for the winter months.

Systems: Terrestrial

Range Description:

Occurs across central and Northern Africa through Arabia and southern Asia; from Morocco to India north to Israel, Palestine, Jordan Iraq and Afghanistan and south to Kenya. Presence in Myanmar and **Thailand** based on a very old reference with no detail of location; there is doubt about its current presence. Occurs up 1,100 m asl in Morocco and Algeria.

Countries: Native:

Afghanistan; Algeria; Bangladesh; Burkina Faso; Cameroon; Chad; Djibouti; Egypt; Eritrea; Ethiopia; India; Iran, Islamic Republic of; Iraq; Israel; Jordan; Kenya; Kuwait; Libya; Mali; Mauritania; Morocco; Nepal; Niger; Nigeria; Oman; Pakistan; Saudi Arabia; Somalia; Sudan; Syrian Arab Republic; **Thailand;** Tunisia; Western Sahara; Yemen (Socotra)

Population:

Appears to be particularly abundant near oases. However, both distribution and abundance are undoubtedly insufficiently investigated because the roosts and suitable habitats are often unreachable. Colonies range in size from a few individuals up to several hundred. Up to 500 individuals have been reported in colonies in Jordan (Amr 2000). Occurs with other species in the genus, in Iran it is normally found in low numbers and low densities and it feeds on coleoptera (M. Sharifi pers. comm. 2005). Assumed stable throughout the southwest Asia region (D. Kock pers. comm. 2005). Population information remains unknown for its African distribution.

Population Trend: Stable

Major Threat(s):

Human disturbance in roost sites and pesticide use against locusts are the main threats. In arid areas of Iran which can not support high numbers of colonies, they aggregate in a few large groups which increases their vulnerability (M. Sharifi pers. comm. 2005). These are not thought to be major threats to the species as a whole at present.

Conservation Actions:

No specific measures are known or are in place, but presumably occurs in protected areas across the range. A study on the impacts of pesticides is required, especially ways in which the impact might be minimised.

Bumblebee Bat *(Craseonycteris thonglongyai)*

Description:

Kitti's hog-nosed bat *(Craseonycteris thonglongyai)* is not only the smallest bat in the world, but also the smallest mammal in existence, and the sole living species of the family Craseonycteridae. Its extinction would not only be the loss of an incredibly unique species, but an entire branch of the evolutionary tree would vanish from our planet. Kitti's hog-nosed bat has long greyish-brown fur on its back, and shorter, paler fur on its underparts. Its common name arises from its flat, fleshy, pig-like muzzle, situated between extremely small eyes concealed by facial hair. Their large and membranous ears, with a long and well-developed tragus, enhance their ability to pinpoint the echoes with which they navigate. Male Kitti's hog-nosed bats possess a rounded, glandular swelling on the lower portion of the throat; this is either less prominent or completely absent in females.

Range Description:

This species occurs in **Thailand** and Myanmar. **The type locality in Thailand is Kanchanaburi, Ban Sai Yoke (= Yok), cave near the Forestry Station (14"26'N, 8"51'E).** It potentially occurs further south in Myanmar, a genetic study is being undertaken to determine if the species exists in two isolated subpopulations. There are additional limestone caves near the Thai border in Myanmar although these have not been surveyed. Its elevational range is from 0-500 m asl.

Countries: Native:

Myanmar; **Thailand**

Population:

The total population is restricted to eight caves in Myanmar (Pereira et al. 2006) and 35 in **Thailand** (Yokubo et al. 2005; Redfield, 2006). The population in **Thailand** is estimated to be 5,100 individuals and the population has decreased. From 1983 to 1997, there has been a 10% decrease while from 1998 to the present a 14% decrease has been estimated. These figures are not based on the same cave samples, as the species disappeared from some caves and appeared in others and it is not known whether this simply reflects colonies moving from one cave to another (S. Bumrungsri pers. comm.) The population in Myanmar is over 1,500 individuals (P. Bates pers. comm.), all age classes are included although this is likely to be represented by mature individuals as the species matures quickly.

Population Trend: Decreasing

Habitat and Ecology:

This species is always associated with limestone outcrops near rivers and can survive in degraded areas. It always roosts in caves. Weighing around two grams, this tiny bat is reputedly the world's smallest mammal. It is insectivorous and normal foraging range appears to be limited to an area of around 1 km from the roost site (Hutson et al. 2001). This apparently wary bat can be found deep inside small caves, hanging high on the ceiling, suspended by their toes and strong claws. A 'tendon-locking mechanism' keeps their claws bent with very little muscular effort, and hanging upside down allows the bat to swiftly take flight from the resting position. Many caves in which Kitti's hog-nosed bats have been found contain only 10 to 15 individuals, but the average group size is 100, and the maximum is 500. Females give birth to a single young in late April, the dry season, and leave their offspring in the roost whilst they venture out to forage. Kitti's hog-nosed bats emerge from their caves shortly after sunset, and again just before dawn, when they hunt for brief periods. They search for prey around the tops of teak trees and bamboo clumps, gleaning insects from foliage and seizing small flying insects from the air. Like other bats, the Kitti's hog-nosed bat can locate prey and navigate through the trees by using echolocation. They emit ultrasonic squeaks that bounce off their surroundings, and the echoes are used to create a mental map of the area, and determine the location of potential prey.

Systems: Terrestrial

Major Threat(s):

The species is affected by disturbance of roosts by religious visits, fertilizer collection and tourism - there has been development of 'show caves' (S. Bumrungsri pers. comm.). There is also the potential for the extraction of limestone which would cause habitat destruction. The smallest mammal in the world, this tiny bat weighs less than 2 grams. Its body is about the size of a large bumblebee, hence the common name "bumblebee bat". Since it was first described in 1974 this tiny mammal has been disturbed by collectors and tourists wanting to see the world's smallest mammal. Today the main threats are from burning of the forest areas near the limestone caves in which it lives.

Conservation Actions:

There is a need for the protection of the cave roosts. A survey of the limestone caves just across the **Thai border** inside Myanmar is required to determine if the species occurs there. Education of monks to prevent burning of incense is also required (P. Bates pers. comm.) as well as improvement of trans boundary collaboration on conservation of the species along with trans boundary information exchange, and identification and establishment of protected areas (P. Bates pers. comm.).

Family: Molossidae:

Genus: Chaerephon:
Wrinkle-Lipped Free-tailed Bat (*Tadarida plicata)*

Description:

HB: 65-75; T:30-40; FA: 40-40; E: 21-24; HF:8-9; W: 17=31G.

The upper lip is markedly wrinkled. The pelage is sort and dense, with the upperparts dark brown, slightly darker towards the tail; the area over the shoulders is more thinly haired. The underparts are slightly lighter, with some of the hairs on the belly having grey or whitish frosting. The face is covered with stiff, short, black bristles; the nostrils protrude, making a "button nose". The ears are thick, round and broad, joined on the front of the muzzle; the tragus id quite small, separated posteriorly from the larger antitragus by a deep notch. The body is heavy, supported by rather narrow wings. The tail is free for over half the lenght. The first and fifth toes are thickened, with numerous tactile hairs protruding from the sides. The braincase is slightly inflated, with low sagittal supraorbital crests.

These bats which roost in Rakang cave consume 30-40 million insects per night.

Range Description:

This very widely distributed species is found throughout much of South Asia, southern and Central China, most of mainland Southeast Asia, and much of insular Southeast Asia. In South Asia, this species is very widely distributed and is presently known from India (Andhra Pradesh, Goa, Karnataka, Madhya Pradesh, Maharashtra, Meghalaya, Punjab, Rajasthan, Tamil Nadu, Uttar Pradesh and West Bengal) and Sri Lanka (Central and Western Provinces) (Molur et al. 2002). It has been recorded from sea level to an elevation of 950 m asl. The species is likely to be found in Bangladesh (Srinivasulu and Srinivasulu 2005). In China it has been recorded from Gansu, Yunnan, Hainan, Guangxi, Guangdong, Hong Kong, and Guizhou (Smith and Xie 2008). It is present in Myanmar, **Thailand,** western Cambodia and Peninsular Malaysia in mainland Southeast Asia. Within insular Southeast Asia it is wide ranging, and is found on the islands of Sumatra, Java, Bali and Lombok (all to Indonesia), Borneo (Brunei, Indonesia and Malaysia), and the Philippines. In the Philippines there are records from Leyte, Luzon (Cagayan, Isabela, Pampanga, and Rizal provinces), Mindanao (Cotabato) (Taylor 1934), Cebu and Negros (Heaney et al. 1998).

Countries: Native:

Cambodia; China; Hong Kong; India; Lao People's Democratic Republic; **Thailand;** Malaysia; Philippines; Sri Lanka; Vietnam

Population:

This is a widespread but localized species which occurs in large colonies usually in caves. In South Asia, this species is widespread in its range in India, but is found in few locations. The Sri Lankan population is stable (W. Yapa pers. comm.) (Molur et al. 2002). In China, colonies ranging from a few hundred to more than 200,000 have been reported (Smith and Xie 2008). In Peninsula Malaysia the species is common and widespread although there are very few recent records. **The total population in Thailand is around eight million, with a single population of over two million individuals (S. Bumrungsri pers. comm. 2006);** it is a very abundant species in Myanmar (P. Bates pers. comm. 2006).

Population Trend: Unknown

This species can form large colonies of thousands of bats that typically roost in caves, but can also be found in crevices in rocks, old disused buildings and temples. Populations generally forage close to roost sites, and have been recorded hunting in forested areas and over rice fields. It is a high and fast flyer that feeds on insects and other invertebrates.

Systems: Terrestrial

Major Threat(s):

In South Asia, the habitat of this species is being deforested for timber, firewood and conversion to agricultural use. It is also threatened due to extraction and mining activities (Molur et al. 2002). In mainland Southeast Asia, a colony of 300,000 bats was destroyed in northern Myanmar as a result of limestone extraction for cement manufacture (P. Bates pers. comm.), and a colony of hundreds of thousands of bats was eradicated as pests in Phnom Pehn, Cambodia (P. Bates pers. comm. 2006). In the Philippines, this was formerly among the most abundant bats in some large caves. However, virtually all of the known large colonies have now been extirpated, including the loss of a very large colony from Leyte in 1984 (L. Heaney pers. comm. 2006). The remaining population is probably declining due to forest loss, disturbance in caves (guano mining, hunting), and tourism (Heaney et al., 1998). The only recent reports since 1980 are from northern Luzon (Danielsen et al. 1994; Heaney et al. 1998), and two recent records from Cebu (L. Heaney pers. comm. 2006). It parts of its range, such as Lao PDR and Borneo, over harvesting of this species for food is leading to significant population declines.

Conservation Actions:

In South Asia, there are no direct conservation measures in place for this species and it has not been recorded from any protected areas in South Asia. Protection of key roost sites, surveys and habitat management are urgently recommended along with public awareness (Molur et al. 2002). In Southeast Asia, while many large colonies are protected in countries such as Myanmar and Thailand by villagers and monks (P. Bates and S. Bumrungsri pers. comm. 2006), and the species has been recorded from a number of protected areas, there is a need to protect key roosting sites and to undertake population surveys in areas of previous decline (eg. Philippines).

Genus: Tadarida:

La Touche's Free-tailed Bat (*Tadarida latouchei)*

Very little information is available on this species

Range Description:

This species is known from north-eastern coastal China (in the provinces of Hebei, Beijing, Nei Mongol, Liaoning and Shandong), north and central Lao PDR, and just across the border into north central **Thailand.** In Japan, there are only three records, all of vagrants, on Amami-oshima, Kuchinoerabu Island, and Yoron Island (Kagoshima Prefecture) (Abe et al. 2005). Records from Lao PDR were originally identified as *T. teniotis* and are assumed to be *T. latouchei* based on range.

Countries: Native:

China; Lao People's Democratic Republic; **Thailand**

Vagrant:

Japan

Population:

It is probably locally rare, as usually only a single specimen is ever collected. One large colony is known from Lao PDR. The record from central Lao PDR was purchased in a market (C. Francis pers. comm.). It presumably originated nearby. Three specimens have been found in Japan, after the type was recorded in China 1917.

Population Trend: Decreasing

Habitat and Ecology:

This species roosts in caves, and forages in the surrounding forest. In **Thailand** it has been recorded from dry evergreen forest at 1,500 m asl, on top of a mountain (S. Bumrungsri pers. comm.).

Systems: Terrestrial

Major Threat(s):

Disturbance of the cave habitat of the species is a potential threat, including mining. In some parts of its range, such as northern Lao PDR, it is known to be trapped for food and this may be occurring in other parts of its range.

Family: Emballonuridae:

Genus: Emballonura:
Lesser Sheath-Tailed Bat (*Emballonura monticola)*

Description:

HB: 40-45; T:10-11; FA:43-45; HF:6-7; E: 10-11; W: 30-40g.

The pelage is dark brown or dark copper, with the upper surface of the interformeral membrane slightly thinly furred; hair also extends up the basal third of the ears. The eyes are black, triangular in shape, and erectile, becoming limp when the bat is at rest; the tragus is slightly narrower at the midpoint than at either end. The tail is shorter than the interfemoral membrane, with the last 2 caudal vertebrae free of the membrane. The dentition differs from Taphozous in having 2 pairs of upper incisors instead of the 1 pair. The first premolar is quite small, with the upper one tightly wedges between the canine and second upper premolar. The skull is thin and papery, with a deep central longitudinal groove and long slender postorbital process. The braincase is strongly inflated anteriorly.

Range Description:

This species occurs in Myanmar and **Thailand** to Western Malaysia, Borneo (all of the island) and Indonesia (Sumatra, Riau Archipelago, Bangka, Belitung, Enggano, Babi Islands, Batu Islands, Nias Island, Mentawai Islands, Java, Sulawesi) (Simmons 2005). The distribution in Sulawesi is uncertain. There is a new record from southern **Thailand, near Phanom,** suggesting that the Myanmar population is linked to that of Southern **Thailand** and Malaysia (S. Bumrungsri and P. Bates pers. comm.). This species may also have been recorded from Buton (T. Kingston pers. comm.). It is found at an altitude of about 50 m in southern Myanmar and 330 m asl in Sumatra (A. Suyanto pers. comm.).

Countries: Native:

Indonesia; Malaysia; Myanmar; **Thailand**

Range Map: Click here to open the map viewer and explore range.

Population:

There is very little population information for this species. It is relatively common on the islands of southern Myanmar (P. Banks pers. comm.).

Population Trend: Decreasing

Habitat and Ecology:

This species is found in limestone caves and small crevices throughout secondary forest in southern Myanmar where colony size reaches 50 individuals. In Southern **Thailand** a maximum of 100 individuals are found in man made caves while smaller colonies are found under hanging rock in tropical lowland forest (S. Bumrungsri pers. comm.). In Peninsular Malaysia E. monticola occurs in lowland rainforest, roosts under fallen trees, buttresses and tables (C. Francis pers. comm.).
Systems: Terrestrial

Major Threat(s):

There are no major threats to this species throughout its range. Extraction of limestone may lead to the destruction of caves where this species is found. Deforestation for agriculture, plantations, logging and as a result of fire is a major threat in some parts of its range.

Conservation Actions:

This species occurs in protected areas throughout its range.

Genus: Saccolaimus:

Naked-Rumped Pouched Bat (*Saccolaimus saccolaimus)*

Description:

S. saccolaimus or *T. saccolaimus* (Payne et al., 1985) have dark-reddish brown or blackish brown upperparts which are irregularly marked with white patches. Their underparts are usually white, but in one colour phase it can be dark brown. They have no wing pouch or in other words, a poorly developed radio-metacarpal pouch. They have a distinct glandular pouch on the throat. The ear is short and broadly rounded with ribbing on the interior of the pinna with a short tragus which has a semicircular margin. It has long and narrow wings with black skin and translucent whitish portions. It is the largest species with the whitest wings.

Range Description:

This widespread species ranges from South Asia, through parts of continental and insular Southeast Asia, Melanesia and Australia. In South Asia, this species is presently known from Bangladesh (Sylhet division) (Sarker and Sarker 2005; Srinivasulu and Srinivasulu 2005), India (Andaman and Nicobar Islands, Gujarat, Karnataka, Kerala, Madhya Pradesh, Maharashtra, Meghalaya, Orissa, Uttar Pradesh and West Bengal) and Sri Lanka (North Central, Uva and Western provinces) (Molur et al. 2002; Srinivasulu et al. in press). In South Asia, it has been recorded from sea level to 1,200 m asl (Molur et al. 2002). It has been recorded from continental Southeast Asia, in Myanmar, **Southern Thailand,** Cambodia (known only from a collection in Phnom Phen G. Csorba pers. comm.), southwestern Vietnam, Peninsular Malaysia and possibly Singapore. Within insular Southeast Asia, the species has been recorded from the islands of Sumatra and Java (Indonesia), Borneo (Indonesia and Malaysia only), Sulawesi (Indonesia), the island of Timor (East Timor and Indonesia), Halmahera (Indonesia), the Talaud Islands (Indonesia) and Ternate Island (Indonesia), and the Philippines. In the Philippines it has been the species has been recorded from sea level to 800 m asl (Heaney et al. 1998) from Catanduanes, Luzon and Mindoro (Corbet and Hill 1992) Mindanao (Misamis Oriental, Zamboanga del Sur provinces), and Negros (Heaney et al. 1998) although it is likely to occur throughout the country except for the Batanes/Babuyan region (L. Heaney pers. comm. 2006). It has been recorded from scattered localities on the island of New Guinea (Indonesia and Papua New Guinea), on Yapen Island (Papua Province, Indonesia), on the Bismarck Archipelago and the Trobriand Islands (Papua New Guinea), on Bougainville Island (Papua New Guinea). It has also been recorded from the island of Guadacanal in the Solomon Islands and from coastal northern and North-eastern Australia (Queensland and Northern Territory) (Corbet and Hill 1992; Flannery 1995; Strahan 1995; Bonaccorso 1998). In the Northern Territory of Australia, there have been relatively few records, with the only confirmed records from the Kakadu lowlands (Woinarski and Milne 2005). This is partly due to the lack of a diagnostic call that can be assigned to the species and can be used for its detection (Woinarkski and Milne 2005).

Countries: Native:

Australia; Bangladesh; Brunei Darussalam; Cambodia; India; Indonesia; Malaysia; Myanmar; Papua New Guinea; Philippines; Solomon Islands; Sri Lanka; **Thailand;** Vietnam

Biology and Ecology:

This medium sized bat roosts in hollow trees and rock crevices and sometimes houses in colonies varying from a few individuals to a few hundred. Roosting bats maintain individual spacing. They are alert at the roost and scurry all over the roost substrate if disturbed. Echolocation clicks produced by this bat in flight are audible. females give birth to a single young per litter.

n the Nicobar Islands in India, it is found in dense forests near to the seashore. In Sri Lanka, the species has been recorded from dense forests, swampy areas and plantations. The diurnal roosts include hollows of old and decaying trees including Kitul Palm and Arecanut Palm, old buildings and rocky crevices. No sexual segregation is observed while roosting. It roosts in small colonies of five or six. This species is known to feed on termites, beetles and other insects and forages close to the ground. Its flight is very fast and high, initially flying 300 to 400 m from ground. It is recorded to emerge very early in the evening from its roost for its foraging bouts (Bates and Harrison 1997). In Southeast Asia, the species is strongly associated with modified habitats, including agricultural areas. It is a canopy feeder which roosts in buildings and shallow caves sometimes occurring in large groups. In the Philippines, records have mainly been from hollow coconut trees (L. Heaney pers. comm. 2006). In Melanesia and Australia, this species has been recorded from wooded areas ranging from open dry sclerophyll woodland to dense tropical moist forest. It roosts in shallow caves, buildings and tree hollows. Roosts may range from a few individuals to several hundred animals. The female gives birth to a single young (Flannery 1995; Strahan 1995; Bonaccorso 1998). In Australia, it has been found in open Pandanus woodland, as well as open eucalypt tall forests and coastal lowlands. It roosts in tree hollows, as well as caves (Woinarski and Milne 2005).
Systems: Terrestrial.

Major Threat(s):

There are no major threats to this widespread and adaptable species as a whole. In South Asia, it is locally threatened by deforestation, generally resulting from logging operations and from conversion of land to agricultural and other uses (Molur et al. 2002). In Australia, it is significantly threatened in coastal Queensland by clearance of coastal tropical woodland and changes to the fire regime at the northern and southern range limits (Duncan et al. 1999). Vegetation change due to saltwater intrusion and invasion by exotic species (such as Mimosa pigra) may affect habitat suitability (Woinarski and Milne 2005).

Conservation Actions:

This species has been recorded from many protected areas, and over much of the species range no direct conservation actions are needed. In South Asia, the species has been recorded from protected areas such as Campbell Bay National Park (Andaman and Nicobar Islands) and Kanha National Park (Madhya Pradesh). With, further studies needed on distribution, abundance, breeding biology and general ecology of this species. In South Asia, populations of this species should be monitored to record changes in abundance and distribution. There is also a need to identify populations that are being threatened by human induced habitat alterations in order to develop mitigation measures (C. Srinivasulu pers. comm.). It has been recorded from several protected areas in Australia (eg. Kakadu National Park), however, there is a need to identify and protect important roosting and foraging sites for the species. Further studies are needed in Australia into the distribution, abundance, natural history and threats to this species. The eastern Australian form has been described as a subspecies, and there is a there is a need for additional taxonomic work (T. Reardon pers. comm.). There is also a need to resolve the taxonomic status of the Northern Territory population relative to that in north-eastern Queensland, as well as a study to determine its habitat, distribution, population size, and status (Woinarski and Milne 2005).

Genus: Taphozous:

Long-Winged Tomb Bat (*Taphozous longimanus)*

Description:

HB: 65-75; T: 24-33; FA: 57-63; E:20; HF:11.

This is a sexually dimorphic species, with males usually cinnamon brown and females usually dark grey; both sexes may be speckled with white. Only the males have a larger gular sac; females have a naked patch of skin at the corresponding area (female melanopogon) has fur under the chin. The fur on the head continues to below the eyes. The wings are long and narrow, and attached to the ankles .

Ecology:

They do not seen to require a totally dark roost and they bhang up in rows under the eaves of larger houses which are more exposed to the light, merely shifting their position if the sun shines directly upon them. They also roost in a wide variety of biotypes, including hollow trees, caves and wells, they are gregarious, roosting in groups of between 2-20 individuals. They hunt alone, starting about 15 minutes after dusk and returning to the same territory each evening. Each female becomes pregnant more that once a year and most probably there is continual breeding , with pregnancies in quick succession. This reproductive behaviour is unique amongst old world bats. Many of the bats seen flying over Bangkok at about 30 meters above ground about half an hour after dusk are of this species.

Range Description:

This species is widespread in South Asia and Southeast Asia. In South Asia, it has been recorded from Bangladesh (location unknown) (Khan 2001, Srinivasulu and Srinivasulu 2005), India (Andhra Pradesh, Assam, Bihar, Gujarat, Jharkhand, Karnataka, Kerala, Madhya Pradesh, Maharashtra, Nagaland, Orissa, Rajasthan, Tamil Nadu, Tripura, Uttar Pradesh and West Bengal), Nepal (location unknown [Bates and Harrison 1997]) and Sri Lanka (Eastern, Northern, Uva and Western provinces) (Srinivasulu et al. in press; Molur et al. 2002). In Southeast Asia, it ranges from Myanmar in the west, into northern **Thailand** and Cambodia. It is present in Indonesia on the islands of Sumatra, Java, Bali, Lombok, Sumbawa and Flores and is also present on the island of Borneo (Brunei, Indonesia and Malaysia). The eastern limit of the distribution in Southeast Asia is uncertain, however, the species has yet to be recorded from either Lao PDR or Viet Nam (P. Bates pers. comm.). It has been recorded to occur up to 1,200 m asl.

Countries: Native:

Bangladesh; Cambodia; India; Indonesia; Malaysia; Myanmar; Nepal; Sri Lanka; **Thailand**

Range Map: Click here to open the map viewer and explore range.

Population:

In South Asia, it is widespread living in large colonies in the hundreds. In many places the populations are considered to be stable (C. Srinivasulu pers. comm.). In Southeast Asia, the species is locally common, with colonies consisting of hundreds of bats in Cambodia (G. Csorba pers. comm.).
Population Trend: Stable

Habitat and Ecology:

Throughout its range this species is found in varied habitats from arid areas to humid zones. It roosts in caves, old tunnels, caves created due to mud excavation, old forts, dungeons (C. Srinivauslu pers. comm.), large wells, hollows and crowns of trees, eaves of houses. It roosts in colonies from single animals to hundreds of bats. It is an early and fast flyer and feeds on cockroaches and beetles. There are two breeding seasons-one in mid January and the other in mid May (Bates and Harrison 1997).
Systems: Terrestrial

Major Threat(s):

Overall there appear to be no major threats to this species. In South Asia, some populations may be threatened due to disturbance to roosting sites by humans (Molur et al. 2002), while in Southeast Asia, disturbance from guano mining is a localised threat to some populations.

Conservation Actions:

In South Asia, although there are no direct conservation measures in place, the species is reported from Hazaribagh National Park in Jharkhand and Kanha National Park in Madhya Pradesh (Molur et al. 2002), and it is likely to be present in most of the protected areas in peninsular India (C. Srinivasulu pers. comm.). In Southeast Asia, in view of its wide range it probably occurs in some protected areas. Further studies are needed into the taxonomy, distribution, abundance.

Black-Bearded Tomb Bat (*Taphozous melanopogon)*

Description:

HB:70-80;T: 20-24; FA: 55-65; E: 21-23; HF: 13-15.

The colour is widely variable, depending on the sex, age and season; the individual hairs are white, tipped with pale brown to reddish-buff on the upperparts. The underparts, ears and flight membrane are buff; juveniles are darker. Only the males have a black beard, but this is by no means universal, and the beard may only be seasonal. Females have short hair under the . There is no throat sacs but instead small pores which open into the throat. A radio-metacarpal pouch is present. The tail is rather thick, with a slight swelling at the tip. The wings attach above the ankles, and fur extends to the interfemoral membrane and the wings as far as the radius.

Ecology:

Usually found in hilly forest countryside, near water, where they form colonies of between 150 to 4000 individuals. They are often found with *T. theobaldu*. They roost in caves or large vertical faults in cliffs, with each individual occupying a defined territory determined by social hierarchy. At the roost, adult males are dispersed around a nucleus of females; if the shape of the colony is elongated the males are seen occupying the both extremities in a linear disposition. Some colonies consist of only males, usually situation close to the colony of females. The bas come out of the roost in waves of 3 to 12 individuals 25 to 30 minutes after sunset. Flight is swift and pitching and the prey is usually flying insects. In the mating season (January -February small glands under the chin of the male excrete a thick sticky substance which runs over the beard; the single young is born blind and naked in April or may and are nursed for 2 months , then leaves the mother forever. The young though stay in the colony. Juvenile mortality is very high.

This widespread insectivorous bat inhabits various forest types, including scrub and other disturbed areas. It generally roosts in caves and rock crevices, sometimes in the company of other bat species. The pair of bats illustrated here were found clinging to the roof a large limestone cave at **Krabi, Southern Thailand,** in the company of around 50 Round leaf bats. Colonies of up to 4000 individuals have been reported from other parts of **Thailand.** Its fur is pale in colour, ranging from buff to greyish-brown. Adult males of the species often possess a dark beard under the throat The wings are pale and pinkish, and mainly devoid of fur. The free tail is slender and slightly enlarged at the tip.

Range Description:

This species is distributed in much of South Asia, Southern China and Southeast Asia. In South Asia, this species is presently known from Bangladesh, India (Andaman and Nicobar Islands, Andhra Pradesh, Bihar, Chhattisgarh, Goa, Gujarat, Karnataka, Kerala, Madhya Pradesh, Maharashtra, Orissa, Rajasthan and Tamil Nadu) and Sri Lanka (Central, Northern Central, North Western, Sabaragamuwa and Western provinces) (Molur et al. 2002). It has been recorded up to 800 m asl. In southern China, it has been recorded from southern Yunnan, southern Guangxi, Guangdong and Hainan Island (Smith and Xie 2008). In addition, the type series of T. solifer Hollister, 1913, is purportedly from Beijing (Smith and Xie 2008). In Southeast Asia, it is widely distributed from Myanmar in the west, through **Thailand,** Lao PDR, Vietnam, Cambodia, Peninsular Malaysia and Singapore, to Indonesia (Sumatra, Java, Bali, Nusa Penida, Lombok, Sumbawa, Sulawesi, Waleabahi, Sanana, Buru, Halmahera and Bunguran), the island of Timor (East Timor and Indonesia), the island of Borneo (Brunei, Indonesia and Malaysia), and the Philippines. In the Philippines it has been recorded from the islands of Biliran, Cebu, Gigante, Leyte, Luzon (Ilocos Norte, Nueva Viscaya, Pangasinan, Rizal provinces), Maripipi, Mindanao (Davao del Sur), Negros, Palawan, Sibuyan, Tincasan (= Tingkasan). Also reported from Lubang, Luzon (Abra Province) and Mindoro (Lawrence 1939) (Heaney et al. 1998). It is found from sea level to 150 m asl in the Philippines (Heaney et al. 1998).

Countries: Native:

Brunei Darussalam; Cambodia; China; India; Indonesia (Jawa, Lesser Sunda Is., Sulawesi, Sumatera); Lao People's Democratic Republic; Malaysia (Peninsular Malaysia, Sabah, Sarawak); Myanmar; Philippines; Singapore; Sri Lanka; **Thailand;** Timor-Leste; Vietnam

Population:

This species is locally abundant or common and populations seem stable throughout its range.
Population Trend: Unknown

Habitat and Ecology:

This species is known from a wide variety of forested habitats in tropical regions, and has additionally been recorded from urban areas. It is found in hilly areas and roosts in caves, old dilapidated buildings, dungeons of old forts, temples, old disused mines, tunnels (Lawrence 1939; Sanborn 1952; Taylor 1934; Rickart et al. 1993; Heaney et al. 1998; Molur et al. 2002; Smith and Xie 2008). This species is colonial and roosts in colonies of a few to thousands of individuals. It is alert, agile and flies with a fast, straight pitching flight (Bates and Harrison 1997). A single young is born after a gestation period of 120-125 days (Bates and Harrison 1997).
Systems: Terrestrial

Major Threat(s):

There are no major threats to this species as a whole. In South Asia, the species is locally threatened by habitat loss, largely through commercial logging, conversion of land to agricultural use, hunting for local consumption and disturbance to roosting sites by humans (Molur et al. 2002). In the some countries, including Lao PDR and the Philippines, there is likely some localized hunting of this cave roosting species for food (Heaney and Balete pers. comm.).

Conservation Actions:

This species has been recorded from many protected areas. Within India, it has been recorded from Kanha National Park and Satpura National Park in Madhya Pradesh, Nagarjunasagar Srisailama Tiger Reserve, Coringa Wildlife Sanctuary, Kawal Wildlife Sanctuary in Andhra Pradesh, Borivili National Park in Maharashtra, and may occur in many more protected areas (C. Srinivasulu pers. comm.). In South Asia, further studies are needed into the distribution, abundance, breeding biology and general ecology of this species. Populations of this species should be monitored to record changes in abundance and distribution (Molur et al. 2002). Public awareness activities need to be taken up to highlight the importance of this species in agricultural ecosystem as controller of insect pests (C. Srinivasulu pers. comm.).

Theobald's Tomb Bat (*Taphozous theobaldi)*

Description:

Theobald's Tomb Bat is a widespread, though relatively uncommon species. It inhabits forested areas and roosts mainly in caves. Large roosts numbering thousands of bats can occur. Dorsally its fur is brown to dark brown, ventrally it is more pale. The chin is covered in fur, and some males may have a narrow 'beard' of darker fur. The limbs and flight membrane are dark brown. The ears are large and rounded, the eyes of moderate size, and the tail around 25 mm in length. The wings are long and narrow.

Range Description: This species is widely, but patchily, recorded in South Asia, mainland Southeast Asia and insular Southeast Asia with a single report from China. In South Asia, this species is presently known only from India (Karnataka, Kerala, Madhya Pradesh and Maharashtra) in South Asia and has been recorded from sea level to 1,200 m asl (Molur et al. 2002). On the Southeast Asian mainland, this species occurs in southern Myanmar, **Thailand,** Vietnam, Cambodia and Lao PDR. Within insular Southeast Asia, it has been recorded from Indonesia (Java and Sulawesi) and the island of Borneo (Kalimantan [Indonesia]), but is presumably more widespread in this region. In China, it is reportedly present in Yunnan although this needs confirmation (Wang 2002; Smith and Xie 2008). A record from Malaysia appears to be in error (see Medway 1969).

Countries: Native:

Cambodia; India; Myanmar; **Thailand;** Vietnam

Population:

In Myanmar, there are some large colonies of up to several thousand bats (P. Bates pers. comm.). In India, the species is reportedly present as colonies of several hundred bats in a few caves (D.S. Joshi pers. comm.). In **Thailand** colonies are smaller, generally consisting of between 10 and 40 bats (S. Bumrungsri pers. comm.).

Population Trend: Unknown

Habitat and Ecology:

The species tends to be associated with forest habitats, with caves and deep crevices of large caves used as roosting sites (Bates and Harrison 1997; Smith and Xie 2008).). It has been collected from limestone caves in Myanmar (P. Bates pers. comm.).

Systems: Terrestrial

Major Threat(s):

This species is threatened by disturbance and destruction of roosting sites and persecution by local people. D.S. Joshi (pers. comm.) recounts 3,000 bats killed in six caves in Maharashtra, India, earlier this decade. Guano from this insectivorous species is mined and used as fertilizer in some areas (Smith and Xie 2008). It is used for subsistence food and medicinal purposes in parts of its range.

Conservation Actions:

Within South Asia, there are no direct conservation measures in place, however, the species has been recorded from a number of protected areas in India such as Silent Valley National Park in Kerala and Bhimashanker Wildlife Sanctuary in Maharashtra (Molur et al. 2002). There is a need to protect important roosting sites for this species throughout its range. Additional field surveys, studies into distribution, abundance, breeding biology, general ecology and population monitoring of this species are needed

Family: Nycteridae:

Genus: Nycteris:
Malayan Slit-Faced Bat (*Nycteris tragata)*

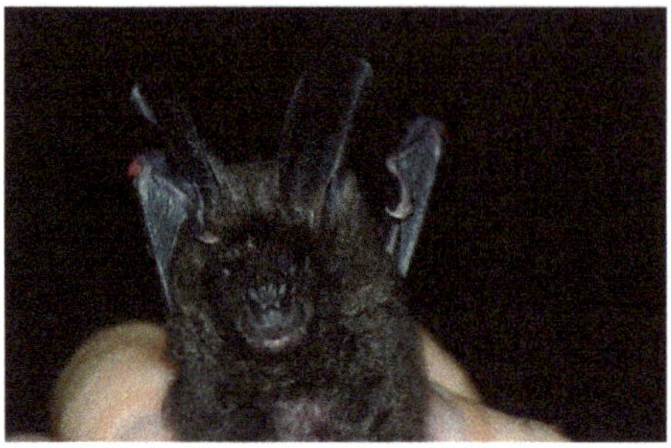

Description:

This species is named after the deep slit running down the face, from a central point between the eyes down to the nostrils. This is, in fact, part of the nose leaf, which the hollow-faced bat needs for echolocation. Folds of skin extend from either side of the slit to direct echolocation calls with great accuracy, leaving little space for the minute eyes. The ears are brown and particularly long, with a short, rounded tragus (inner ear). The tail is commonly equal in length to the head and body, and is enclosed in a membrane that stretches between the hind legs (the interfemoral membrane). A tiny triangular cut in the membrane where the tail reaches the edge helps to identify the species. The hollow-faced bat is very furry, with a greyish-brown back, paler underside, and even brown fur on the forearms. The wings are very broad and short, with a large area of membrane above the forearm (known as the propatagium). This gives the bat great agility, but reduces its flight speed.

Range Description:

This species' distribution includes Myanmar, **Thailand,** Peninsular Malaysia, Sumatra (Indonesia) and throughout Borneo (Simmons 2005).The type locality for this species is Bidi caves, Sarawak, Borneo Malaysia.
Countries: Native:
Brunei Darussalam; Indonesia; Malaysia; Myanmar; Singapore; **Thailand**

Population:

This species is uncommon.
Population Trend: Decreasing

Biology and Ecology:

Although this insect-eating species is capable of complex echolocation calls involving a rapid, low intensity sweep of the frequency range, it is also thought to hunt simply by listening for sounds made by the insects themselves. Its large wing area and comparatively low body weight allows it to take off nearly vertically from the forest floor when hunting, possibly with heavy prey . It tends to hover over its prey before snatching it up into the air . There are thought to be two breeding seasons per year, with females giving birth to a single pup. At first the pup is carried in foraging flights, which may well limit the diet of the hollow-faced bat, as it can only take lighter prey. The pup learns quickly to fly and forage alone, and at one year old will be sexually mature .It has been collected from primary Dipterocarp forest and peat swamp forest (I. Maryanto pers. comm.). It roosts in small groups in fallen tree hollows and similar man-made hollows such as culverts as well as crevices of large boulders and in limestone caves.
Systems: Terrestrial

Major Threat(s):

The major threat for this species is habitat loss due to deforestation for logging, plantations, agriculture, as well forest fire.

Conservation Actions:

The species range includes several protected areas.

Genus: Megaderma:

The Greater False Vampire Bat *(Megaderma lyra)*

The Greater False Vampire Bat (Megaderma lyra) is a carnivorous bat, just as the South American Spectral Bat. However, it lives in Asia, along with other bats of the genus Megaderma, which are also known as "false vampires".

Description:

It has a relatively large body size 65–95 mm and its weight ranges between 40-60 g. Average forearm length 66.4 mm (56.0-71.5mm). Like the Spectral Bat, it has no tail. Fur colour is bluish grey except on the underside which is brownish grey. It is smaller than the Spectral Bat. Its ears are big. The nose leaf is erect and 10mm in length. The rather long fur is uniform greyish brown above and whitish grey below. The nose leaf consists of a small fold of skin truncated at the top; the dorsal lobe is rectangular, with parallel sides; but its junction with median leaf forms an obtuse angle on each side. The scull as a very broad rostrum, with the width between the orbits (lachrymal width) greater than the lenght from the front of the orbit to the tip of the nasals. The supraorbital ridges are prominent, increasing the width of the frontal region. The postorbital process as small but present.

Range Description:

This very widely recorded species ranges through much of South Asia, southern and Central China, and throughout the Southeast Asian mainland. In South Asia it is known from Afghanistan (Qachcar and Darunta [Habibi 2003]), Bangladesh (Dhaka, Chittagong, Sylhet, Khulna and Rajsahi divisions), India (Andhra Pradesh, Arunachal Pradesh, Assam, Bihar, Chhattisgarh, Gujarat, Himachal Pradesh, Jammu and Kashmir, Jharkhand, Karnataka, Kerala, Madhya Pradesh, Maharashtra, Meghalaya, Nagaland, Orissa, Rajasthan, Tamil Nadu, Uttar Pradesh and West Bengal), Nepal (Central Nepal), Pakistan (Baluchistan and Punjab) and Sri Lanka (Central, North Central, Northern, Southern and Western provinces) (Molur et al. 2002). In China, it has been recorded in Fujian, Sichuan, Guangdong, Hainan, Xizang, Yunnan, Guizhou, Guangxi and Hunan (Smith and Xie 2008). In Southeast Asia the species occurs in Myanmar, **Thailand,** Cambodia, Lao PDR, Viet Nam and Peninsular Malaysia. In South Asia, it has been recorded from sea level to an elevation of 1,000 m asl.

Countries: Native:

Afghanistan; Bangladesh; Cambodia; China; India; Lao People's Democratic Republic; Malaysia; Myanmar; Nepal; Pakistan; Sri Lanka; **Thailand;** Vietnam

Population:

Overall this is a common species. Its population status is stable in Sri Lanka (W. Yapa pers. comm.), increasing in Bihar (Y.P. Sinha pers. comm.), decreasing in certain parts of northern India such as Rajasthan (colonies recorded in 1970s missing in the 1990s [as observed by I. Prakash pers. comm. and by Senacha pers. comm.]) (Molur et al 2002).

Population Trend:

Unknown It hunts at and after dusk. It roosts in caves, abandoned mines and attics of houses. Other species of bats avoid caves where this species roosts, probably because of fear of being eaten by them.

Ecology:

Wings are modified pectoral appendages that have been adapted for flight. Bat wings are membranous wings that attach to the side of the body and include the small hind legs and sometimes the tail as well. The last four digits have been elongated to support the wing skin connects each of the digits covering the entire wing structure. The first digit has been modified into a grasping hook that allows the bat limited ability to grasp onto objects. This hook remains free of the wing. All bat wings are shaped differently based upon the different selective pressures for each habitat and lifestyle. There are seven key measurements used to classify bat wings: Wing Span The distance between the wing tips of a bat with wings extended so that the leading edge of the wing is straight. With units of B/m Wing Area Is the combined area of the two wings, the entire tail membrane and the portion of the body between the wings. With units of S/m2 Aspect Ratio Is the square wingspan divided by the wing area; a higher aspect ratio usually corresponds with greater aerodynamic efficiency and lower energy losses in flight. Aspect ration correlates well with the outline of wing shape, because wings with rounded tips naturally have low aspect ratio. However not all shape variation in bat wings is expressed by the aspect ratio alone. Wing Loading Is the weight of the bat divided by the wing area, and it is related to the mean pressure on the wings. Therefore, flight speed is proportional to the square root of wing loading. Tip Length Ratio Is the ratio of the length of the hand wing to the length of the arm wing. Tip Area Ratio Is the ratio of the hand wing area to the arm wing area. Tip Shape Index This measurement is particularly valuable because it is independent of the overall size and shape of the arm and hand wings, it is determined simply by their relative size. A value of 1 corresponds to triangular wingtips. Higher index values indicate rounded or nearly square wing tips and with values below 1 the wingtip becomes more acute and the wing thins considerably as the tip is approached Environmental pressure put on the bat by the habitat causes selective adaptation of the wings. This permits a species to improve its capacity to use certain food sources in certain ways. Mass-carrying ability, for example, is most closely linked with wing loading. As wing loading increases, the bat must fly faster and therefore expend more energy. Because they habitually fly with substantial loads, we expect carnivores to have relatively low wing loading. Large wing area and large wingtips ensure sufficient thrust and weight support when loaded, without risk of stall. Large wing area also permits a slow controlled approach to prey and facilitates easy take-off under load. Flight in cluttered areas requires slow speeds and high manoeuvrability; this constrains a bat to a short wing, because long wings can be a physical hindrance. The optimal wing size and shape for any bat is a compromise between numerous of different, and sometimes conflicting, selection forces. *Megaderma lyra* bats are carnivorous. They hunt, using a method called gleaning, in a variety of different habitats. The wing measurements for a typical adult are: Total mass 0.0375 M/kg, Wing Span 0.440 B/m, Wing Area 0.0312 S/m2, Aspect Ratio 6.2, Wing Loading 11.8 Mg/S/N*m-2, Tip Length ratio 1.70, Tip area ratio 0.96 and Tip shape index 1.30. These measurements show that Megaderma lyra have a low wing loading and aspect ratio. Meaning that Megaderma lyra aren't very fast fliers but they have good manoeuvrability. They have wings designed to give them power and control which they need for capturing and lifting prey. A low wing loading also provides the capacity for increased life which is essential if prey is equal to 50% your body mass. *Megaderma lyra* are Gleaning bats. Gleaning bats prefer to capture prey from ground and water surfaces. They consume large rthropods and small vertebrates such as frogs, geckoes, lizards, fish, mice, birds, and even smaller bat species. All of the bats in the roost will hunt from dusk to dawn with the exception of the pups. Some bats will commute 4 km to their hunting area. Once they reach this spot they will remain in an area of approximately 0.1 km2. All foraging areas take advantage of at least two different types of habitats offering them a more diverse selection of prey. *Megaderma lyra* use a combination of hunting strategies. 85% of prey is captured during searching flights in which the bats fly 0.5 meters above the ground. These flights last for about 6 minutes. It has been observed that some will sustain flight for 45 minute and based on this observation it has been speculated that Megaderma lyra can eat while in flight. Megaderma lyra also utilize a "sit and wait" strategy. They will perch 2 meters above ground and wait. Foraging behaviours of individuals may vary with energetic restraints associated with different reproductive states. Reduces flight activity (nightly flight times) and longer perching bouts of lactating females coincided with the time spent at the juveniles night roost. Bats have the same basic auditory system as all mammals only theirs has evolved to be highly sensitive so that sound can be utilized as a method of seeing.

Bats utilize ultrasonic sound (greater than 20 kHz), which most mammals cannot hear, during echolocation. A large ossified larynx allows the bat to build up a lot of tension on the vocal cords. The bat can then emit sound waves, through either its nose or mouth, which can reach up to 150 kHz. These high frequency sound waves are far closer together than low frequency sound waves. This gives the bat more information about its surroundings when the information is processed by the brain. These waves can be emitted 100 times per second as the bat stalks its prey. Interestingly, in the Microchiroptera family, successful foraging, particularly in clutter, is sometimes hindered by echolocation. Some species use calls which are clutter-resistant, whereas others forsake echolocation while hunting. In some cases, foraging success may be more limited by echolocation than by flight performance. In any case, it has been observed that in captivity, Greater False Vampire Bats can successfully detect and catch mice and frogs in complete darkness without echo locating. Not much is known about Megaderma lyra reproduction. We do know that mating takes place from November through to January. After which females segregate from males. Gestation usually lasts 150–160 days. April to June is when child birth happens where one or sometimes two pups are born per mother. The birth disruption between males and female is equal. Males reach sexual maturity by 15 months, females by 19 months. Females will carry young with them during foraging until the pups are between one and twenty-three days old, at which point they "park" them at either a day or a special night roost. Young are nursed for 2 to 3 months.

Major Threat(s):

There appear to be no major threats to this species as a whole. It is locally threatened in parts of its range due to disturbance and loss of roosting sites due to renovation of old temples, buildings and old forts. Populations are also threatened by mining activities and hunting for local consumption (medicine and food) in India and Vietnam.

Conservation Actions:

In South Asia, although there are no direct conservation measures in place, the species has been recorded from a number of protected areas in India including Tadoba-Andhari Tiger Reserve and Radhanagari Wildlife Sanctuary in Maharashtra, Orang National Park in Assam, Kawal Wildlife Sanctuary and Nagarjunasagar Srisailam Tiger Reserve in Andhra Pradesh, Kanha National Park in Madhya Pradesh and Indravati National Park in Chattisgarh. In Southeast Asia it has been recorded from several protected areas. In South Asia populations of this species should be monitored. Captive breeding techniques are known for this species and captive stocks exists in Germany. Public awareness to mitigate threats to this species is recommended (Molur et al. 2002).

Lesser False Vampire Bat (*Megaderma spasma)*

Description:

The lesser false vampire bat, *Megaderma spasma,* is a bat found in South Asia and Southeast Asia from Sri Lanka and India in the west to Indonesia and the Philippines in the east. They live in caves and tree hollows. They are insectivorous. The Lesser False Vampire Bat has a wingspan of up to 30 centimetres (12 in) and have a head andbody length of around 10 centimetres (3.9 in). Their forearms are normally around 7 centimetres (2.8 in). Sometimes called the yellow-winged bat, the lesser vampire bat has yellowish veins through the wing, and when the wings are spread with light behind, they are given a prominent yellow/orange tinge. Their body colour ranges from grey-brown to blue-brown. Lesser False Vampire Bats live in rock crevices, caves, foliage and hollow trees, depending on availability. *Megaderma spasma* is from the order Chiroptera and family Megadermatidae which comprises four genera and five species. *M. spasma* also known as lesser false vampire. Its type locality was in Indonesia, Molucca Islands and Ternate. There are two specimens of *M. spasma* collected in Sarawak Museum Unimas, one from Niah and another one from batu 16, Ulu Gombak. *M. spasma* was distributed in India to Indochina and Malay peninsula, Sri Lanka, Andaman Islands, Sumatra, Borneo, Mollucas, Philippines, Sulawesi and other Indonesian islands. In Borneo, it can found in Sepilok, Darvel Bay area in Sabah, Niah and Kuching in Sarawak, upper S. Kapuas in West and upper S. Tengah in South Kalimantan It is most closely related to the Greater False Vampire Bat, which is the only other species in the genus Megaderma. *M. spasma* has fur pale grey to grey-brown in colour. Its nose leaf has long dorsal lobe with stiffened central ridge and broad convex flaps on the sides. Its ears are very large, joined at the base and it has no visible tail. Its echolocation pulses are short, low in density and broadband and its large ears are sensitive to echoes returning from their pulses and also sensitive to the sounds that prey generates.M. spasma usually roosts in groups in caves, pits, building, and hollow trees. *M. spasma* favours grasshoppers and moths but sometimes they eat small vertebrates including other bats. They have well developed, forward-pointing eyes and can locate prey visually. *M. spasma* usually roosts in a group of 3–30 individuals. Their ecological importance may be both positive and negative to humans. They eat some insects which harm human crops, but they may carry and transmit certain diseases. in the field, false vampires can be identified primarily by their large, rounded ears, which are joined at the base. Closer inspection reveals a tragus which is bifurcate (the tragus is the pointed structure inside the ear), and a nose leaf comprising a long lobe stiffened by a central ridge. The fur is grey to brown, and the short tail does not extend outside the interfemoral membrane.

Habitat and Ecology:

In South Asia, this species is found in humid areas and dense tropical moist forest. It roosts in small colonies in caves, old and disused buildings, temples, lofts of thatched huts, tiled roofs, hollows in large trees and disused mines (Molur et al. 2002). It has a low and fast flight and feeds on Lepidopterans, coleopterans, hymenopterans and other insects but does not feed on vertebrates. One young is born between April to the month of June (Bates and Harrision 1997). Rarely occurs in the same location with M. lyra. Individuals hang separately until disturbed, when they huddle together (S. Molur pers. obs.). In the Philippines, it is found in lowland primary and secondary forest (Heaney et al. 1991). Known to roost in caves, tree-hollows, and hollow logs (Taylor 1934;Lawrence 1939; Rabor 1986; Ingle 1992; Rickart et al. 1993; Lepiten 1995). On Palawan, the species occurs in bamboo thickets, secondary forest, and primary forest (Esselstyn et al. 2004).

Systems: Terrestrial

Range Description:

This species is very widely distributed over much of South Asia and Southeast Asia. In South Asia, it is predominantly distributed in the Western Ghats and northeastern parts of India, and in Sri Lanka. It is known from Bangladesh (Khulna division), India (Andaman and Nicobar Islands, Andhra Pradesh, Assam, Goa, Karnataka, Kerala, Maharashtra, Meghalaya, Mizoram, Tamil Nadu and West Bengal) and Sri Lanka (Eastern, Southern and Western provinces) (Molur et al. 2002) In Southeast Asia, it ranges throughout virtually the entire mainland, and most of insular Southeast Asia, including the major islands of Sumatra, Java, Sulawesi and Halmahera (all to Indonesia), Borneo (Brunei, Indonesia and Malaysia), and throughout Philippines except for the Batanes/Babuyan region. There are records in the Philippines for Biliran, Bohol, Busuanga, Catanduanes, Cebu, Dinagat, Leyte, Luzon(Abra [Lawrence 1939)], Aurora, Camarines Sur, Isabela, Laguna Province [Taylor 1934; Ingle 1992], Rizal provinces), Mindanao (Davao del Norte, Davao Oriental [Sanborn 1952], Lanao del Norte, Misamis Occidental, South Cotabato, Zamboanga del Sur provinces), Mindoro, Negros, Palawan, Panay, Polillo, Siquijor (Heaney et al. 1998). In South Asia this species has been recorded up to an elevation of 1,600 m asl.

Countries: Native:

Bangladesh; Brunei Darussalam; Cambodia; India; Indonesia; Lao People's Democratic Republic; Malaysia; Myanmar; Philippines; Singapore; Sri Lanka; **Thailand;** Vietnam

Population:

It is a common species in South Asia. In the Philippines, the species is widespread and locally common to uncommon in primary and secondary forest (Heaney et al. 1998). It is always encountered in small groups, though sometimes cumulative numbers are in the hundreds (L. Heaney, Esselstyn, and Balete pers. comm. 2006).

Population Trend: Unknown

Major Threat(s):

Over much of the species range there are no major threats to populations as a whole. In Southeast Asia, there are localized threats to caves, especially due to guano mining and limestone quarrying. In South Asia, the primary forest habitats of this species are under threat, usually deforested for timber, firewood and agricultural purposes (Molur et al. 2002). Roosts in large hollow trees and old buildings are under threat due to human interference from being cut down or broken down or cleaned up. Populations in some areas are also under threat from being exploited for medicinal purposes where the bat oil is used for massaging ailing newborns and infants (S. Molur pers. obs.).

Conservation Actions:

In South Asia, there are no direct conservation measures in place, however, the species may occur in many protected areas, but the few protected areas it has currently been recorded from include Tadoba Tiger Reserve, Melghat Tiger Reserve, Pench National Park in Maharashtra and Sunderbans National Park in West Bengal. It is presumably present within a number of protected areas in Southeast Asia. In South Asia, further studies are needed into the distribution, abundance, reproduction and ecology of this species. Populations should be monitored to record changes in abundance and distribution. Habitat maintenance, conservation and restoration are needed. Public awareness activities are needed to mitigate threats to this taxon (Molur et al. 2002).

Family: Rhinolophidae:

Subfamily: Rhinolophinae:
Genus: Rhinolophus:
Acuminate Horseshoe Bat (*Rhinolophus acuminatus)*

Description:

HB: 55-60; T: 25; FA: 47-51; E: 18-19; HF:11-12.

There are 2 colour phases. The dark or grey phase has a dorsal pelage greyish brown with hairs tipped with silver, with the underparts pale grey. The red or russet phase has the upperparts reddish brown, with the underparts somewhat paler. The sella is parallel sided; the lancet is constricted in the centre. The connecting process is tall, triangular in outline, with the straight sides converging to a point. There is a distinct expansion on the nasal region of the skull, forming a lump against which the nose leaf rests. The middle premolar is outside the tooth row, and the first and third lower premolars may be in contact. The canines are sturdy and wide apart.

Range Description:

This species occurs from **Thailand** east to Lombok (Indonesia), Borneo (Sabah, Malaysia) and the Philippines, where it is known only from the Palawan Faunal Region: recorded from Balabac, Busuanga and Palawan (Heaney et al. 1998).

Countries: Native:

Cambodia; Indonesia; Lao People's Democratic Republic; Malaysia; Myanmar; **Thailand;** Vietnam

Population:

The population status of the species is uncertain, though apparently it is locally common and occurs in small colonies (Esselstyn et al. 2004, Heaney et al. 1998).

Population Trend: Unknown

Habitat and Ecology:

The ecology and habitat of this species are unknown in the Philippines (Heaney et al. 1998). The only known roosts in the Philippines are in caves, and the species has been found from 60-250 m asl (Esselstyn et al. 2004). It occurs in lowland dipterocarp forest on Borneo (Payne et al. 1985). The species forages in primary and secondary forest, as well as bamboo thickets (Esselstyn et al. 2004). In Myanmar the species was found in a small colony roosting under a building (P. Bates pers. comm.). It is tolerant of urban areas.

Systems: Terrestrial

Major Threat(s):

There are no major threats to this species although disturbance of caves is a localized threat.

Conservation Actions:

The species occurs in protected areas within its range.

Intermediate Horseshoe Bat *(Rhinolophus affinis)*

Description:

HB:58-63; T:20-30; FA:46-50; E: 16-20; HF: 11-13.

The upper parts are greyish to yellowish brown with the hairs darkest at the tips. The wings are dark brown, the ears are medium brown and shorter than the head. The anterior nose leaf is broad, but does not conceal the muzzle when viewed from above. The anterior margin bears a notch at the midpoint over the upper lip, behind which a very slight groove extends backwards to the base of the sella. The sella is pandurate with concave sides; it as no lappets. The connecting process is semicircular and the lancet is wedge shaped. The lower lip as 3 grooves.

Range Description:

This very widespread species is present throughout much of South Asia, southern and Central China and Southeast Asia. In South Asia, this species is presently known from Bangladesh (Sylhet division) (Srinivasulu and Srinivasulu 2005), Bhutan (Gedu), India (Andaman and Nicobar Islands, Arunachal Pradesh, Assam, Sikkim, Tamil Nadu, Uttaranchal and West Bengal) and Nepal (Central and Western Nepal) (Molur et al.2002). In China, it has been reported from Hunan, Shanxii, Hubei, Guizhou, Sichuan, Yunnan, Zhejiang, Fujian, Jiangxi, Guangdong, Hong Kong, Guangxi, Jiangsu, Anhui and Hainan island. In Southeast Asia, it ranges from Myanmar in the west, through **Thailand,** Lao PDR and Vietnam, into Peninsular Malaysia, Indonesia (Sumatra, Java and the Lesser Sunda Islands), to southern parts of the island of Borneo (Indonesia and Malaysia). Reports of this species from Cambodia cannot currently be confirmed (Kock 2000). It has been recorded from 290 to at least 2,000 m asl (China).

Countries: Native:

Bangladesh; Bhutan; Cambodia; China; Hong Kong; India (Andaman Is.); Indonesia (Kalimantan, Lesser Sunda Is., Sumatera); Lao People's Democratic Republic; Malaysia (Sabah, Sarawak); Myanmar; Nepal; Singapore; **Thailand;** Vietnam

Population:

It is a highly adaptable and common species.

Population Trend: Stable

Habitat and Ecology:

In South Asia, this is a highly adaptable species. It roosts in caves, and is found commonly in man-made habitats such as orchards, degraded habitats and agriculture areas (M. Muni pers. comm. February 2002, Molur et al. 2002). In Southeast Asia, it has been recorded from primary and secondary forest, occasionally in cultivated areas, but is not found in urban areas. It has a tendency to roost in caves, and colonies can be large, up to thousands of Individuals. Species forages in under story of forest, and is not thought to be dependent on water. In China it is considered to be a cave roosting species, found both in the wet western highlands and in the more tropical eastern lowlands.

Systems: Terrestrial

Major Threat(s):

There appear to be no major threats to this widespread and somewhat adaptable species. However, limestone extraction may be a threat locally in South Asia (Molur et al. 2002).

Conservation Actions:

This species has been recorded from a number of protected areas. Other than general research activities, no direct conservation measures are needed for this species as a whole.

Croslet Horseshoe Bat *(Rhinolophus coelophyllus)*

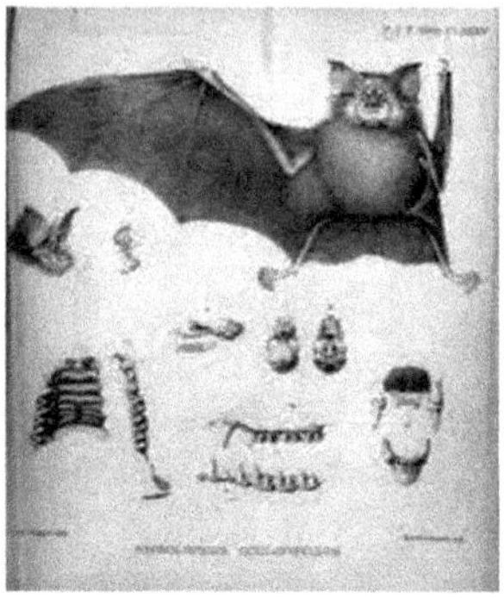

Description:

HB: 52-54; T: 20-23; FA: 41-45; E: 18-20; HF: 8.5.

There are a wide range of colour variations with intermediate forms; the upper parts may be light brown to bright orange, with the underparts much lighter. The sella is parallel sides, with a triangular tip; the connecting process begins at the upper edge of the sella and rounds gently backwards into the cavity formed by the large and hood shaped lancet; the lancet projects well forward over the connecting process and is thickly covered with hair (Shamel 1942). In the skull, the region behind the nasal swelling is scooped out; the two arms formed by the anterior divisions of the sagittal crest are beaded.

Range Description:

This species is recorded from Myanmar, Salaween River, **Thailand,** Western Malaysia, and Lao PDR. The records from Eastern Thailand may be *R. shameli* (S. Bumrungsri pers. comm.).

Countries: Native:

Lao People's Democratic Republic; Malaysia; Myanmar; **Thailand**

Population:

The species is locally common in **Thailand,** a population there may have contained 1,000 individuals (S. Bumrungsri pers. comm.). Its population status in other countries is not known but it is not thought to be common anywhere.

Population Trend: Unknown

Habitat and Ecology:

It is found in broad leaved mixed deciduous to evergreen forest in a wide altitudinal range in **Thailand** (S. Bumrungsri pers. comm.).

Systems: Terrestrial

Major Threat(s):

There are no major threats to this widespread species at present.

Conservation Actions:

It occurs in a number of protected areas.

Blyth's Horseshoe Bat *(Rhinolophus lepidus)*

Description:

HB: 51-54; T: 15-20; FA: 37-42; E: 15-16; HF: 8.3-8.7.

The wings are short, broad and nearly transparent. The nose leaves and skulls are similar to acuminatus but with the connecting process pointed. The lower second premolar is external, and the first and third premolars may be in contact. This is a small bat species with two colour phases - one reddish-brown, the other grey. In all forms the fur is quite glossy, and the underparts are pale. The nose leaf is elaborate and comprises a typically shaped anterior horseshoe structure, a pointed bifurcated Stella, and a high, pointed lancet *Rhinolophus refulgens* (Glossy Horseshoe Bat), was considered a separate species to *Rhinolophus lepidus* (Blyth's Horseshoe Bat), the latter ranging from the Indian Subcontinent and Burma to southern China and **Northern Thailand.** *R. refulgens* is now relegated to the subspecies *R. lepidus refulgens,* which has a complimentary geographical range from Southern **Thailand,** Peninsular Malaysia and Singapore to Sumatra

Range Description:

This species is very widespread in South Asia and Southeast Asia. In South Asia the species is known from Afghanistan (Faryab, Kabul, Nangarhar, Parwan and Zabol provinces), Bangladesh (Chittagong, Dhaka, Khulna, Sylhet and Rajsahi divisions), India (Andhra Pradesh, Assam, Bihar, Delhi, Karnataka, Kerala, Gujarat (Senacha, under review), Himachal Pradesh, Madhya Pradesh, Maharashtra, Meghalaya, Mizoram, Nagaland, Orissa, Rajasthan, Tamil Nadu, Uttaranchal, Uttar Pradesh and West Bengal), Nepal (Central and Eastern Nepal) and Pakistan (Punjab) (Molur et al. 2002). In Southeast Asia, it has been recorded from Myanmar, **Thailand,** Cambodia, Vietnam, Peninsular Malaysia, and (Sumatra). In South Asia, it has been recorded up to an elevation of 2,330 m asl (Molur et al. 2002).

Countries: Native:

Afghanistan; Bangladesh; Cambodia; China (Yunnan); India; Indonesia (Sumatera); Malaysia (Peninsular Malaysia); Myanmar; Nepal; Pakistan; **Thailand;** Vietnam

Population:

It is locally common throughout much of its range.

Habitat and Ecology:

This species can be found in both dry and moist forest and fringe areas (C. Srinivaulu pers. comm. September 2007). In Malaysia and **Thailand** it appears to be associated with intact lowland tropical moist forest. Roosts include caves, unused tunnels, old and ruined buildings, old temples (Molur et al. 2002). Its flight is slow and low and feeds on lepidopterans, coleopterans, dipterans, hymenopterans (Bates and Harrison 1997). Blyth's Horseshoe Bat, or Glossy Horseshoe Bat, roosts in caves and houses near forest. In semi-urban areas it is known to have adapted to roost in man-made tunnels and drain culverts.

Systems: Terrestrial

Major Threat(s):

In general there are no major threats to this species as a whole. In parts of its Indian range, this species is threatened due to roost disturbance from conversion of old forts and hovels to hotels as a part of tourist related development activities (K.R. Senacha pers. comm. January 2002, Molur et al. 2002).

Conservation Actions:

There are no direct conservation measures in place in South Asia. It is reported from Ranthambore National Park in Rajasthan, Karnala Bird Sanctuary in Maharashtra, Satpura National Park in Madhya Pradesh (Molur et al. 2002), and Nagarjunasagar Srisailam Tiger Reserve in Andhra Pradesh (C. Srinivasulu pers. comm. 10 October 2007). Further studies are needed into the distribution, abundance, breeding biology and general ecology of this species (Molur et al. 2002). In Southeast Asia it occurs in a number of protected areas.

Woolly Horseshoe Bat (*Rhinolophus luctus)*

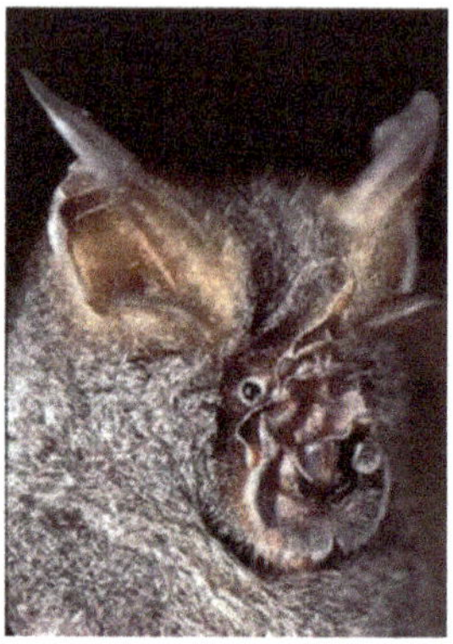

Range Description:

Description:

HB: 75-95; T: 48-56; FA: 72-75; E: 35-42;

The fur is rather woolly, dark grey approaching black; there is a grey iridescence on the hairs of the upperparts. Older individuals are browner. The nose leaf is dull black; the anterior leaf is large, projecting over the lip. It is deeply grooved in the centre. The sella has a lappet on each side. The dorsal connecting process is very low, xtending on both sides. The lancet is long, terminating in a blunt point. The ears are long and pointed, with the outer margin concave below the tip. The second premolar is usually with the tooth row; the upper first premolar is always in the tooth row, though its cusp is rudimentary or lacking. This species is widespread in South Asia, Southern China and South East Asia. In South Asia it is presently known from Bangladesh (Chittagong and Sylhet divisions), India (Andhra Pradesh, Assam, Madhya Pradesh, Meghalaya, Nagaland, Uttaranchal, Sikkim and West Bengal) and Nepal (Central and Eastern Nepal) in South Asia (Molur et al. 2002). In south-eastern China, the species has been recorded from Zhejiang, Jiangxi, Guangdong, Guizhou, Sichuan, Fujian, Guangxi, Anhui, Yunnan and Hainan Island. It has been recorded throughout most of continental Southeast Asia, and ranges into Indonesia (Sumatra, Java and Bali) and the island of Borneo (Brunei, Indonesia and Malaysia). It has been recorded from sea level to an elevation of 1,600 m asl.

Countries: Native:

Bangladesh; Cambodia; China; India; Indonesia (Bali, Jawa, Kalimantan, Sumatera); Lao People's Democratic Republic; Malaysia (Peninsular Malaysia, Sabah, Sarawak); Myanmar; Nepal; Singapore; **Thailand;** Vietnam

Population:

The species appears to be found at naturally low densities.

Population Trend: Unknown

Habitat and Ecology:

In South Asia, this species is a forest dweller, it roosts solitary or in pairs in small to large caves, rocky outcrops and overhanging ledges and large trees with hollow (Molur et al. 2002). Its flight is low just above the ground and feeds on coleopterans, termites and other insects. In Southeast Asia it is considered to be forest dependent, although it is present in degraded forest (and is apparently somewhat resistant to human disturbance). Roost in small groups in rocky outcrops, rock crevices, niches in cliffs, shallow holes in earth banks, roosts of trees, and hollow trees. In China, most have been collected in forested areas and have been collected from tunnels, old mine shafts, hollow trees and under thick bark.

Systems: Terrestrial

Major Threat(s):

In view of the species wide range, it seems probable that there are no overall major threats to the species. In South Asia, the habitat of this species is being deforested for timber, firewood and conversion to agricultural use. It is also threatened by hunting for medicinal purposes (Debojit Pukhan pers. comm. January 2002, Molur et al. 2002).

Conservation Actions:

There are no direct conservation measures in place. In South Asia, it is recorded from protected areas in India like Satpura National Park in Madhya Pradesh. Taxonomy, distribution, ecology, habitat and population monitoring are recommended for this species (Molur et al. 2002). In Southeast Asia, it has been collected in a number of protected areas, including Vu Quang National Park, Vietnam (Borissenko and Kruskop, 2003).

Big-eared Horseshoe Bat *(Rhinolophus macrotis)*

Description:

HB: 38; T:14;

FA: 36-41; E: 19-22; HF: 8.2-9.2.

Very similar to R. pearsoni but smaller. The fur is short, not woolly; the upperparts are light brown, the underparts are buffy. The anterior nose leaf is large, covering the upper lip. The sella is broad, rounded in front, with rather long and dense on its front face. The lancet is triangular. The dorsal connecting process is rounded off. The lower lip as 3 grooves The upper first premolar is very small, but as a distinctive cusp.

Range Description:

This species ranges from northern South Asia, into South Eastern China and Southeast Asia. In South Asia, this species is presently known from Bangladesh (Chittagong division), India (Arunachal Pradesh, Meghalaya, Uttaranchal and West Bengal), Nepal (Central and Western Nepal) and Pakistan (Punjab) (Molur et al. 2002). In China it has been recorded from Sichuan, Shaanxi, Zhejiang, Jiangxi, Guangdong, Guizhou and Guangxi. In Southeast Asia, it has generally been recorded from northern Myanmar and Thailand, northern Lao PDR and Vietnam, Peninsular Malaysia (and the island of Tioman), Sumatra (Indonesia), and the Philippines where it has been recorded from the islands of Guimaras, Luzon [Abra and Pampanga provinces], Mindanao [Bukidnon Province], Negros, Palawan (Esselstyn et al. 2004) and Samar (Gonzales unpublished data). It has been recorded from 200 up to 1,692 m asl (South Asia) (Molur et al. 2002).

Countries: Native:

Bangladesh; China; India; Indonesia (Sumatera); Lao People's Democratic Republic; Malaysia (Peninsular Malaysia); Myanmar; Nepal; Pakistan; Philippines; **Thailand;** Vietnam

Population:

Though this species is widely distributed in its range in South Asia it is known from a few localities and has a small colony size. The population size of this species is low and a declining trend in the population is inferred (Molur et al. 2002). In Peninsular Malaysia and **Southern Thailand** it is rare in tall lowland forest and also hill forest in Peninsular Malaysia (Bumrungsri and Francis pers. comm. 2006) and widespread but seemingly uncommon in the Philippines (Heaney et al. 1998).

Population Trend: Stable

2004).

Systems: Terrestrial

Habitat and Ecology:

In South Asia, this species roosts in abandoned mines and caves in forests (Molur et al. 2002). Its flight is fast and high and feeds on coleopterans and dipterans (Bates and Harrison 1997). Other than being a cave roosting species, there are few details available on the natural history of this species in China. In Myanmar and Vietnam it is associated with limestone caves at an altitude of around 1,000 m asl, and has been found at a large cave in disturbed secondary growth forest. It has been recorded from in caves in secondary forest in Lao PDR and lowland tropical moist forest in Peninsular Malaysia (C. Francis pers. comm.). In the Philippines, it has been recorded in lowland tropical moist forest, but is otherwise poorly known (Heaney et al., 1998). There are some records from forest caves. On Palawan Island, it has been reported from caves in disturbed lowland forest at 50 to 250 m asl (Esselstyn et al. 2004).

Systems: Terrestrial

Major Threat(s):

In South Asia, the habitat of this species is being deforested for timber, firewood and conversion for agricultural use. In Nepal it is threatened due to increase in tourism leading to disturbance to roosting sites; fumigation and chemical pesticides to rid the caves of roosts (T.K. Shreshta pers. comm. January 2002; Molur et al. 2002). In Southeast Asia, it is probably threatened in parts of its range (such as Malaysia) by ongoing habitat loss.

Conservation Actions:

In South Asia, there are no direct conservation measures in place for this species. The species has not been recorded from any protected areas. Additional studies are needed into the distribution, abundance, breeding biology, general ecology and threats to this little-known species. Populations should be monitored to record changes in abundance and distribution. Habitat maintenance, conservation and restoration are needed. Public awareness activities need to be taken up to mitigate any further threats to this taxon (Molur et al. 2002). In view of its wide range in Southeast Asia, it seems probable that the species has been recorded from some protected areas, although this needs to be confirmed.

Malayan Horseshoe Bat (*Rhinolophus malayanus)*

Description:

HB: 40-50; T: 18-25; FA: 41-43; E: 16-19; HF: 9-10; W: 3.8 - 6.2 g.
The hair is moderately long, with the hairs uniformly coloured to their bases; the is 2 colour phases, with a marked contrast between upper and lower parts in both the phases. The light phase is bright cinnamon above, buffy below, and the dark phase is brown above, pale brown below. There is a well marked light area on the chest where the hairs are dull whitish without any dark tips, withy the tendency to form a white area in the interorbital region just behind the lancet. The ears are sharply pointed, and when laid forward reach the end of the muzzle; the outer margin is concave immediately below the tip, then slightly convex and separated from the large antitragus by a deep notch. The lower lip as 3 vertical grooves. In the nose leaf , the anterior leaf is moderately broad, sharply cleft in front; the sella is moderately wide, with the sides nearly parallel, the connecting process is rounded off , rising slightly above the tip of the sella; the lancet is of moderate size, the tip elongated, the margins slightly concave.. The scull is slender, with a narrow braincase, short palate, with rather well developed nasal swellings, the anterior swellings extend laterally down the sides of the rostrum, thereby reducing the area of the lateral swellings. The dentition as the upper front premolar in the tooth row, but much reduced with a tiny cusp; the second lower premolar is slightly extruded or is internal to the tooth row.

Range Description:

This species is found in Myanmar, **Thailand,** Cambodia, Lao PDR, Viet Nam, and Peninsular Malaysia.
Countries: Native:
Cambodia; Lao People's Democratic Republic; Malaysia (Peninsular Malaysia); Myanmar; **Thailand;** Vietnam

Population:

The species is common in Lao PDR, locally common in eastern Myanmar, Vietnam, and possibly in **Thailand.**
Population Trend: Stable

Habitat and Ecology:

In eastern Myanmar it is associated with limestone caves, and found in agricultural areas and secondary forest (P. Bates pers. comm.). In Cambodia it is found in caves (not necessarily limestone) in secondary forest and degraded habitat (G. Csorba pers. comm.). This species is tolerant to some degree of habitat disturbance.
Systems: Terrestrial

Major Threat(s):

There are no major threats to this species.

Conservation Actions:

The species is found in a number of protected areas.

Marshall's Horseshoe Bat *(Rhinolophus marshalli)*

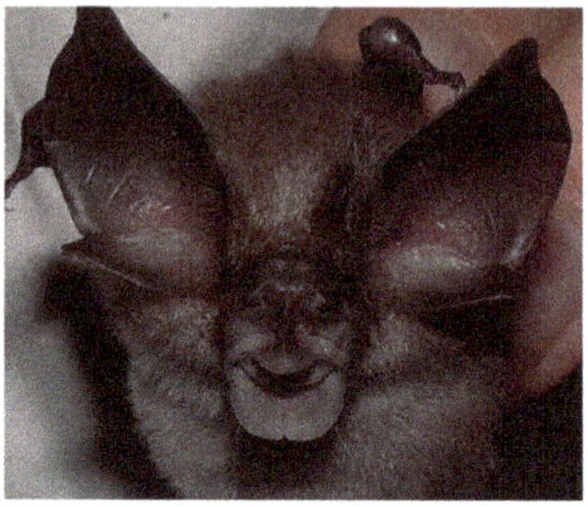

Description:

HB: 42; T: 20; FA; 45.5; E:27.

The pelage of the back is about 8mm long, dark brown with pale bases on the individual hairs above, smoky grey below. Some of the hairs on the back are tipped with Rufus. The ears are very large and funnel shaped with a large antitragus half the length of the ear. The anterior leaf is large, covering most of the upper lip and extending a few mm beyond the muzzle; it as a narrow, deep median anterior cleft. The internarial septum is cup shaped with the sides expanding into wing like structures enclosing the base of the sella and behind the sella to merge with the base of the connecting process, which is low and poorly developed. The lancet is hidden behind the sella when viewed from the front; the tip bluntly rounded . The skull is rather long, with the mastoid width greater than the zygomatic width. The nasal swelling are high and oblong, with a deep cavity behind them. The sagittal crest is low, the postorbital process obsolescent. The upper incisors are minute, widely separated and bifid, the lower incisors are trifid. The first upper and second lower premolars are small, but both are within the tooth row.

Range Description:

This species occurs in Myanmar, **Thailand (Chantaburi, Amphoe Pong Nam Ron, foothills of Khao Soi Dao Thai),** Vietnam, Lao PDR, and Peninsular Malaysia.

Countries: Native:

Lao People's Democratic Republic; Malaysia; Myanmar; **Thailand;** Vietnam

Population:

It is always found as individuals, never in large numbers (P. Bates pers. comm.; S. Bumrungsri pers. comm.). It is not clear whether it has declined significantly historically (C. Francis pers. comm.).

Population Trend: Unknown

Habitat and Ecology:

In Myanmar and Vietnam it is associated with limestone caves and is tolerant of very disturbed habitat (P. Bates pers. comm.). In **Thailand** it is also found in rock crevices (S. Bumrungsri pers. comm.). No significant differences both in body mass and in forearm length between males and females at birth. On average, young bats were not agile fliers until 31 days of age. Body mass and forearm length of pups followed a linear pattern of growth until day 13, with mean growth rates of 0.17 g/day and 1.22 mm/day, respectively, and thereafter growth rates decreased. Length of the total epiphyseal gap of the fourth metacarpal-phalangeal joint showed a linear increase up to 10 days followed by a linear decrease until day 40 with a mean rate of 0.09 mm/day. Together, two equations permitted estimation of the age of R. marshalli pups between 1 and 40 days of age. Of the three nonlinear growth models (logistic, Gompertz, and von Bertalanffy), the logistic equation provided the best fit to the empirical curves for body mass and forearm length. The echolocation call is a long constant frequency signal, with a brief frequency-modulated start and tail. Frequencies with most energy recorded from hand-held bats ranged between 39.2-41.6 kHz

Systems: Terrestrial

Major Threat(s):

There are no major threats to this species.

Conservation Actions:

This species occurs in a number of protected areas. All known populations are in protected areas in **Thailand** (S. Bumrungsri pers. comm.)

Smaller Horseshoe Bat *(Rhinolophus megaphyllus)*

Range Description:

This species ranges from the island of New Guinea (Papua New Guinea only), the Bismarck Archipelago (islands of New Britain and New Ireland) (Papua New Guinea); the D'Entrecasteaux Islands (Papua New Guinea), the Louisade Archipelago (Papua New Guinea), and on the Eastern coast of Australia from Cape York to Victoria (Flannery 1995a,b; Bonaccorso 1998; Pavey and Young 2008). It occurs from sea level up to 1,600 m asl. It as been noted in **Thailand** but more information is needed to confirm

Countries: Native:

Australia (New South Wales, Queensland, Victoria); Indonesia (Irian Jaya, Lesser Sunda Is., Maluku); Papua New Guinea (Bismarck Archipelago)

Population:

It is locally common in suitable habitats within Australia, common in New Guinea, and relatively abundant in New Ireland and New Britain, and islands of Milne Bay.

Population Trend: Unknown

Habitat and Ecology:

This is a cave (including lava tunnel) roosting species that is generally associated with closed tropical forest habitats and hot humid roosting sites. It has been recorded from rural gardens and plantations. Colonies usually consist of less than 50 individuals, but they sometimes (rarely) exceed 2,000 (Pavey and Young 2008). Females give birth to a single young. Its range has expanded in the south due to mine shafts in the last 100 years (L. Lumsden pers. comm.).

Systems: Terrestrial

Major Threat(s):

There appear to be no major threats to this species. It is generally sensitive to disturbance of cave roosting sites.

Conservation Actions:

It is present in a number of protected areas.

Bourret's Horseshoe Bat (*Rhinolophus paradoxolophus)*

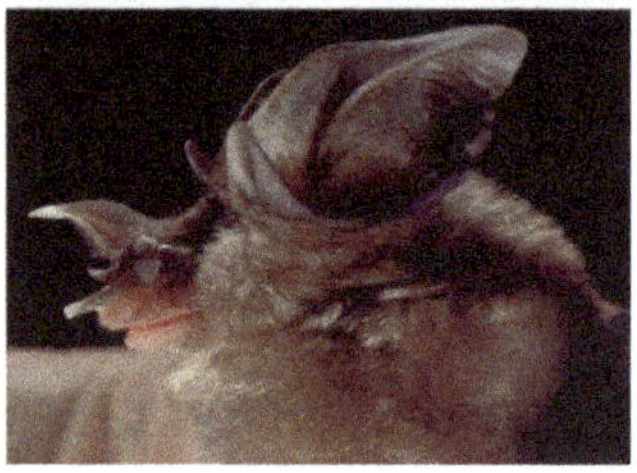

Description:

HB: 48; T: 23; FA: 54; E: 27

The fur is rather long, with dark brown upper parts, the individual hairs are pale at the base and some are tipped with rufescence. The chin and upper breast are greyish, with the remaining under parts smoky grey. The eras are large and funnel like; the antitragus is high but narrow; with nearly parallel sides and a rounded top. The anterior leaf is large, but less developed that in *marshalli,* covering the upper part of the muzzle and extending a little beyond, with a narrow, deep median anterior cleft. The edges of the internarial septum are raised to form a large cup, but do not expand into the wing like structures as in *marshalli,* instead they surround the base of the sella, merging with the base of the dorsal connecting process. The sella is very high , but narrow and tongue shaped , rounded at the tip. The connecting process is low and poorly developed. The lancet is completely hidden behind the sella when viewed from the front, it is shaped like an isosceles triangle with its tip bluntly rounded. The skull as high nasal swellings. Postorbital processes can be distinctly seen on the xygomata. The lower middle premolar is slightly extruded from the tooth row on the left mandible and lost in the right mandible, leaving a small space between the remaining premolars.

Range Description:

This species is known from northern Vietnam, adjacent Guangxi Province in China (Zhao et al. 2002), central Vietnam (Hendrichsen et al. 2001; Timmins et al. 1999), **central Thailand (Thonglongya 1973),** and northern and central Lao PDR (Francis et al. 1999).

Countries: Native:

China; Lao People's Democratic Republic; **Thailand;** Vietnam

Population:

Once considered extremely rare, colonies of up to 50 individuals were found in caves in central Lao PDR (Francis et al. 1999) and northern Vietnam (Hill and Kemp 1996). Single individuals or small groups were found in other areas of China, Lao PDR, and Vietnam (Eger and Fenton 2003). It is not difficult to find during surveys.
Population Trend: Unknown

Habitat and Ecology:

This species roosts in caves. The specimen from **Thailand** was taken in a rather dry pine forest (Thonglongya 1973), and the specimen from southwestern China was found torpid in a limestone cave in late November 1999. This species hunts flying insects around trees and bushes.
Systems: Terrestrial

Major Threat(s):

Deforestation due to logging and agricultural expansion, represents a major threat to this species in some parts of its range.

Conservation Actions

It has been reported from protected areas.

Pearson's Horseshoe Bat (*Rhinolophus pearsoni*)

Description:

HB: 50-57; T: 17-20; FA: 48-52; HF: 9-11.

The pelage is long, dense and soft, there are 2 colour phases, brown and bright yellow, with the under parts lighter. The anterior nose leaf is large, concealing the upper lip when viewed from above. The sides of the sella are parallel, with a raised longitudinal rib in the middle; the front of the sella is covered with long dense hair. The connecting process is markedly projecting rounded. The middle lower premolar is outside to tooth row, in the first and third lower premolar nearly in contact with the side. The upper first premolar is within the tooth row, but the cusp but the cusp is less distinct than in macrotis; the small lower premolar is slightly external to the tooth row so that the larger two premolars are almost in contact.

Range Description:

This species is widely distributed in northern South Asia, southeast and Central China, and much of continental Southeast Asia. In South Asia it has been recorded from Bangladesh (Khan 2001, Srinivasulu and Srinivasulu 2005), Bhutan (Phuntsholing), India (Assam, Meghalaya, Mizoram, Sikkim, Uttarakhand and West Bengal) and Nepal (Central and Eastern Nepal) (Molur et al. 2002). In China, the species has been reported from Xizang, Sichuan, Guizhou, Yunnan, Shaanxi, Hubei, Zhejiang, Anhui, Fujian, Jiangxi, Guangdong, Guangxi and Hunan (Smith and Xie 2008). In Southeast Asia it has been recorded in Myanmar, Vietnam, Lao PDR, **Thailand** and Peninsular Malaysia. It has been recorded from an altitudinal range of 610 to 3,077 m asl (Molur et al. 2002).

Countries: Native:

Bangladesh; Bhutan; China; India; Lao People's Democratic Republic; Malaysia (Peninsular Malaysia); Myanmar; Nepal; **Thailand;** Vietnam

Population:

In South Asia, while the species is still common and widely distributed, a declining trend in the population of this species is suspected because of ongoing habitat degradation in the Himalayas (Molur et al. 2002). It is common in Southeast Asia.

Population Trend: Unknown

Habitat and Ecology:

In South Asia, this species roosts in caves in hilly areas, and is found in montane forests, bamboo forests and cultivated areas (Molur et al. 2002; Bates and Harrison 1997). In Myanmar the species has been recorded from agricultural areas; in Thailand it is found in good forest which is surrounded by agricultural land. The species appears to be associated with limestone caves (P. Bates and S. Bumrungsri pers. comm.). In Vietnam, it is predominantly found in limestone areas but does not appear to be dependent on this substrate (Furey pers. comm.). In China, it has been recorded hibernating in caves and bomb shelters.

Systems: Terrestrial

Major Threat(s):

In general there appear to be no major threats to this species as a whole. In South Asia, the habitat of this species is being deforested for timber, firewood and converted for agricultural use in the Himalayas (Molur et al. 2002).

Conservation Actions:

There are no direct conservation measures in place for this species. The species has not been recorded from any protected areas in South Asia. Additional studies are needed into the distribution, abundance, breeding biology and general ecology of this species (Molur et al. 2002). It is not known if the species is present in protected areas in China or Southeast Asia.

Least Horseshoe Bat (*Rhinolophus pusillus)*

Description:

HB: 38-42; T:16-20; FA:36-40; E: 13-16; HF: 7-8.

The pelage is rather furry, with the upper parts pale brown and the underparts lighter. The ears are quite large, extending beyond the muzzle when laid flat. The anterior nose leaf is broad; the sella is widest at its base; the l ancet is taller than it is wide. The canines and the upper premolars are unmodified; the middle lower incisor is variable, sometimes outside the tooth row, or half in the tooth row, or even completely in the row.

Range Description:

This species has a very wide range from South Asia eastwards to Japan, occurring also in southern and southwestern China, including Taiwan, southwards through mainland Southeast Asia to Indonesia and Borneo. The distribution of the geographically isolated subspecies is given below.

Subspecies *pusillus* is widely distributed in South Asia, southern and Southwestern China and much of Southeast Asia. In South Asia it is widely distributed and is presently known from India (Andhra Pradesh, Arunachal Pradesh, Assam, Karnataka, Kerala, Meghalaya, Tamil Nadu, Meghalaya, Sikkim, Uttarakhand and West Bengal) and Nepal (Central and Western Nepal) (Molur et al. 2002). Recently collected from Kothaiyar Dam site in Kalakad-Mundanthurai Tiger Reserve, Tamil Nadu (Vanitharani et al. 2005). In China, it has been recorded from Sichuan, Guizhou, Hubei, Fujian, Guangxi, Guangdong, Hainan Island and Yunnan (Smith and Xie 2008). It appears to be present throughout most of continental Southeast Asia, ranging from Myanmar in the west, to Vietnam in the east and as far south as Peninsular Malaysia. It has also been recorded from Indonesia (including the Mentawi Islands and the islands of Sumatra, Java and Bali), and from Kalimantan (Indonesia) and Sabah (Malaysia). In South Asia it has been recorded from 200 to 1,370 m asl (Molur et al. 2002).

Subspecies is endemic to Taiwan, Province of China (Smith and Xie 2008). It occupies low elevations on the island (Chen et al. 2006).

Subspecies *cornutus* is endemic to Japan. It ranges from Hokkaido south to Okinoerabu on Honshu, Shikoku, Kyushu, Sado Islands, Izu-Oshima, Niijima, Miyake, Mikura, Hachijo, Tsushima, Iki, Fukue, Yaku, Tane, Kuchinoerabu, Amami-Oshima, Kakeroma, and Tokunoshima (Abe, 2005). It includes as a synonym perditus which is found only on the Yaeyama Islands (Iriomote, Ishigaki, Taketomi, and Kohama), and imaizumii (described from Iriomote) (Abe, et al., 2005).

Subspecies pumilus is endemic to Japan, where it is found on three islands in the Ryukyus: Okinawa, Iheya, and Kume. The species is believed to have become extinct on Miyako (30 years ago) and Irabu (probably around 5 years ago).

Countries: Native:

Cambodia; China; India; Indonesia (Jawa, Lesser Sunda Is.); Japan; Lao People's Democratic Republic; Malaysia (Peninsular Malaysia); Myanmar; Nepal; Taiwan, Province of China; **Thailand;** Vietnam

Range Map: Click here to open the map viewer and explore range.

Population:

It appears to be widespread and common throughout its range.

Subspecies pumilus breeding colonies are on Okinawa, where the subpopulation is estimated at 5,000 or more. Only three major breeding colonies of 200, 1500 and 2000 individuals are currently known (Ministry of the Environment, 2002). On the smaller two islands, the total subpopulation is less than 300 individuals.

Population Trend: Unknown

Habitat and Ecology:

It has been recorded from both primary and secondary tropical moist forest, roosting in caves (up to 1,500 animals in China) and houses (usually a smaller colony size). Animals have been recorded foraging low over bamboo clumps in limestone areas (Molur et al. 2002; Smith and Xie 2008). This species is characterized as a forest-interior specialist that is not readily adaptable to open habitats (Chen et al. 2006).

Systems: Terrestrial

Major Threat(s):

There appear to be no major threats to this species.

Conservation Actions:

The species has been recorded from Phamong Lho Wildlife Sanctuary in Sikkim and Kalakkad-Mundanthurai Tiger Reserve, Tamil Nadu in India. Ecological studies and population monitoring are needed (Molur et al. 2002). It is present in numerous protected areas in Southeast Asia. Subspecies is listed as Endangered (EN) in the Japanese Red List (2007)

Shamel's Horseshoe Bat *(Rhinolophus shameli)*

Description:

HB: 52-55; T: 17-19; FA: 46-48; E: 18-20; HF:9

There are two colour phases, brownish and reddish with underparts lighter; the face, rostrum and lower back of the neck are usually light brown This colour extends to the middle of the back, forming a distinct triangle The head is dark brown to the anterior base of the ears, The nose leaf is as in coelophyllusith a rudimentary supplementary nose leaf which is lacking in coelophyllusith. The skull behind the nasal swelling is flat, depressed below the level of the swelling, the sides not ridged or beaded.

Range Description:

The species is known from northern Thailand as well as islands on the south east coast of **Thailand (Koh Chang Islands),** and from eastern Myanmar (Struebig et al. 2005), Lao PDR, Cambodia, and Central Vietnam.

Countries: Native:

Cambodia; Lao People's Democratic Republic; Myanmar; **Thailand;** Vietnam

Population:

It is locally common in eastern Myanmar and **Thailand,** but locally rare in Viet Nam.

Population Trend: Unknown

Habitat and Ecology:

It inhabits undisturbed lowland evergreen forest (Thong et al. 2006) and secondary forest. In **Thailand** it occurs in secondary mixed deciduous forest (S. Bumrungsri pers. comm.); in Myanmar it is associated with limestone caves around at least 1,000 m asl (P. Bates pers. comm.) within very disturbed areas (P. Bates pers. comm.).

Systems: Terrestrial

Major Threat(s):

There are no major threats to this species.

Conservation Actions:

It is not reported from any protected areas.

Lesser Brown Horseshoe Bat (*Rhinolophus stheno)*

Description:

HB: 54-58; T:16-18;

FA: 45-48; E:16-17; HF: 8.5-9.2.

As a medium-sized member of the horseshoe bat family, this species is characterised by its complex nose leaf. The nose leaf consists of a broad, flat horseshoe-shaped area over the nostrils, a horn-like projection above, and a shorter, rounded projection in between. The function of the nose leaf is to emit echolocation calls for orientation and hunting. The nose leaf is relatively small, and the lancet is low and bluntly pointed rather than very long and acutely pointed as in *malayan*us. The ears are large and black, and the eyes extremely small, dwarfed by the nose leaf. The natural fur colour is dark brown, but this bat's habit of roosting in caves can lead to chemical bleaching, leaving the fur orange. The braincase is rather long and slender, and the skull shows an extraordinary abrupt projection of the anterior nasal swellings; the posterior swellings are much reduced. The first and third lower premolars are in contact, with the lower premolars outside the tooth row; the upper front premolar is in the tooth row and as a very large cusp.

Range Description:

This species occurs in Myanmar (Struebig et al. 2005), Vietnam, **Thailand,** Lao PDR, Peninsular Malaysia, Sumatra and Java (Indonesia). It has not yet been recorded but probably also occurs in Cambodia. It inhabits degraded forest to primary deciduous forest in **Thailand** (S. Bumrungsri pers. comm.). It is associated with limestone in Myanmar in degraded habitats including agricultural areas (P. Bates pers. comm.).

Systems: Terrestrial

Countries: Native:

Indonesia (Jawa, Sumatera); Lao People's Democratic Republic; Malaysia (Peninsular Malaysia); Myanmar; **Thailand;** Vietnam

Population:

It tends to be found in small groups in Myanmar and Viet Nam (P. Bates and G. Csorba pers. comm.). It is locally common in **Thailand,** one population of 1,000 individuals was found there, and also common in Lao PDR (C. Francis pers. comm.).

Population Trend: Stable

Biology and Ecology:

This species feeds on insects, hunting at night using echolocation to locate prey items and move amongst the dense jungle vegetation. It calls at a constant ultrasonic frequency of 86 kHz, enabling it to detect the flutter of an insect's wings. In Peninsular Malaysia, the majority of pregnant females of the lesser brown horseshoe bat have been found between February and April, and are with young pups between May and July. A single pup is born to each female, who cares for it for up to a year. At birth it weighs a massive one quarter of its mother's weight, and she may continue to carry it on feeding flights for two months.

Threats:

There are no major threats to this species, although the species is hunted in northeast **Thailand** but this is not at a level that is considered a threat to the species (S. Bumrungsri pers. comm.).

Conservation Actions:

It is known from a number of protected areas in **Thailand** and throughout its range.

Thomas's Horseshoe Bat (*Rhinolophus thomasi)*

Description:

HB: 48-50; T: 19-20; FA: 44-46; E:16-18; HF: 8-8.8.

A small pale brown bat, with individual hairs light yellow at the base and medium brown at the tips. It resembles *R. acuminatus i*n many of its external characters, but is smaller and as a more narrow horseshoe with a lancet which is broad and short. The sella is parallel sided, and rather wide; the connecting process is indistinctly notched, with the upper part rounded and bearing long hairs. The skull is short and broad similar to *R. rouxi,* but smaller; the nasal swellings are low, giving a rather smooth profile. The lower middle premolar is external to the tooth row and the first and third lower premolars are in contact. The front upper premolar is external, and the upper canine and posterior premolar are in contact.

Range Description:

This species occurs in South West China, extending down the Indochina Peninsula in Myanmar, Vietnam, **Thailand,** and Lao PDR.

Countries: Native:

China; Lao People's Democratic Republic; Myanmar; **Thailand;** Vietnam

Population:

There is no information available on the population and abundance of this species. However, it is known to be locally uncommon to common. It is also known to roost in caves.

Population Trend: Unknown

Habitat and Ecology:

This species is associated with limestone areas and found in caves from 400 up to 1,100 m asl. It can also occur in degraded habitats. Nothing is known from China, but this species is known to roost in caves with other species of Rhinolophus in other areas.

Systems: Terrestrial

Major Threat(s):

There are no major threats, although the species appears to be dependent on caves and their destruction in some parts of the range may cause localised population declines (C. Francis pers. comm.).

Conservation Actions:

It is known from a number of protected areas throughout its range. More research need to be carried out so as to accumulate more information on this species

Trefoil Horseshoe Bat (*Rhinolophus trifoliatus)*

Description:

HB: 60-70; T: 30-38; FA: 47-55; E: 25-27; HF: 11-12.

This striking and charismatic bat is identifiable by its yellow nose leaf and long, soft, fluffy, greyish-brown fur. The elbows, knees and ears are also yellow, deepening in colour with age. The wings are a light, almost orangey, brown, and the membrane also stretches between the hind legs, enclosing most of the tail. As a member of the horseshoe bat family, the nose leaf of this species is shaped like a horseshoe, with a projection that looks somewhat like a horn above it called the connecting process. This probably serves to focus the ultrasonic pulses during echolocation. The ears are large to receive the echoes. The Trefoil Horseshoe Bat is a small to medium-sized member of the Horseshoe Bat family or Rhinolophidae, which includes over 30 species in Southeast Asia. These bats are masters of echolocation, swooping down on moths and other insects in the under story of Primary Forest. The complicated nose leaf, used for detecting echoes bouncing from its prey, and distinctive ears of this species are yellow. The fur is thick and woolly and can vary from buff to grey. The fur is thick and furry, with the individual hairs buff at the base, lightly tipped with dull brown, the under parts are lighter. The nose leaf is similar to *luctus* but smaller; the erect transverse process of the sella is narrower above. The margin of the interfemoral membrane is straight, narrowly bordered with yellow between the tip of the tale and the calcar; the extreme tip of the tai projects. The skull as a triangular pit just behind the 4 nasal swellings, the 2 middle swellings are situated close together. The sagittal crest is prominent in the frontal region, less prominent towards the rear. The lower second premolar is variable, and can be in the tooth row. The first and third premolars may be in contact.

Range Description:

This species is widely distributed in Southeast Asia, with additional records from South Asia and China. In Southeast Asia, it has been recorded from southern Myanmar and **Thailand,** into Peninsular Malaysia, and from here into Indonesia (including the Mentawi Islands [Nias], Sumatra, Bangka, Billiton, Java and Banta), ranging to the island of Borneo (Brunei, Indonesia and Malaysia). In South Asia, it is presently known from Assam and West Bengal in India, recorded up to an elevation of 1,800 m asl (Molur et al. 2002). In China, it is known only from a single specimen recorded from Guizhou (Jinsha) (Wang 2002; Smith and Xie 2008).

Countries: Native:

China; India; Indonesia (Jawa, Kalimantan, Sumatera); Malaysia (Peninsular Malaysia, Sabah, Sarawak); Myanmar; **Thailand**

Population:

In Southeast Asia, this species is generally considered to be common in intact forest habitats. It is considered to be uncommon on the island of Sumatra (Indonesia). The abundance, population size and trends for this species in South Asia are not known (Molur et al. 2002).

Population Trend: Unknown

Habitat and Ecology:

In Southeast Asia, it has been recorded from primary and secondary tropical moist forest. It is generally considered a lowland species on Borneo and elsewhere in the region. Animals roost singly under leaves or palm leaves in the forest under story, and are believed to breed once a year. In South Asia, little is known about the habitat or ecology of this species except that it is found in dense evergreen forests. (Molur et al. 2002). An insectivore, the trefoil horseshoe bat catches its prey by 'perch hunting'; it hangs from a branch at a height of around three metres above the forest floor, with a clearing beneath, and waits for insects to pass. Once it has detected them using echolocation, emitted from the nose leaf, it drops from the perch and pursues the insect, returning to a hanging position to eat it. This species uses long pulses of a constant frequency to detect the flutter of an insect's wings. The trefoil horseshoe bat gives birth to a single pup which it suckles for several months before the pup is able to fly and catch insects alone. The pup begins life weighing two to five grams; a large proportion of its mother's weight. By one year of age it is fully grown and ready to have a pup of its own.
Systems: Terrestrial.

Major Threat(s):

This species is threatened over much of its range by deforestation, generally resulting from logging operations and the conversion of land to agricultural use. In South Asia, the development of tourism related activities is also considered to be a major threat (Molur et al. 2002).

Conservation Actions:

It has been recorded from a number of protected areas in Southeast Asia. In South Asia, there are no direct conservation measures in place for this species. Survey, ecological and population monitoring are recommended (Molur et al. 2002)

Dobson's Horseshoe Bat (*Rhinolophus yunanensis)*

Description:

HB: 55-60; T: 25-26; FA: 47-51; E:18-19; HF: 11-12'

There are 2 colour phases. The dark grey phase has the dorsal pelage greyish brown with the hairs silver tipped, with the under parts pale grey. The red of russet phase as the upper parts reddish brown and the underparts paler. The sella is parallel sided, the lancet is constricted in the centre. The connecting process is tall, triangular in outline, with straight sides converging to a point. There is a distinct expansion of the nasal region of the skull, forming a lump against which the nose leaf rests. The middle lower premolar is outside the tooth row, and the first and third lower premolars may be in contact. The canines are sturdy and wide apart.

Range Description:

This species is distributed in Arunachal Pradesh and Mizoram in India, Yunnan in China, Northern and Eastern Myanmar and Northern, Western and Southern **Thailand** (including the island of Terutau). It might be present in Lao, however, this needs confirmation. In India, it has been recorded at elevations up to of 1,231 m asl (Molur et al. 2002).

Countries: Native:

China; India; Myanmar; **Thailand**

Population:

It seems to be uncommon in Southeast Asia, and in **Thailand it** is mostly found in groups of 100 or fewer animals (S. Bumrungsri pers. comm.). In South Asia, the abundance, population size and trends for this species are not known (Molur et al. 2002).

Population Trend: Unknown

Habitat and Ecology:

In South Asia, little is known about the habitat or ecology of this species except that it is commonly encountered in dense hilly forests among bamboo clumps (Molur et al. 2002). In Southeast Asia all known records are from limestone caves. In China, this species has been collected from bamboo thickets, and also from thatched roofs.

Systems: Terrestrial

Major Threat(s):

In South Asia, this species is threatened by habitat loss, largely through commercial logging and the conversion of land to agricultural use and human settlements (Molur et al. 2002). In Southeast Asia, there appear to be no major threats but may be locally threatened by deforestation in some areas.

Conservation Actions:

In South Asia, there are no direct conservation measures in place for this species. The species has not been recorded from any protected areas. Further studies are needed into the taxonomy, distribution, abundance, reproduction and ecology of this species. Populations should be monitored to record changes in abundance and distribution. Habitat maintenance, conservation and restoration are needed (Y.P. Sinha pers. comm. January 2002, Molur et al. 2002). In Southeast Asia, most known populations are in protected areas (S. Bumrungsri pers. comm.).

Subfamily: Hipposiderinae:
Genus: Aselliscus:
Stoliczka's Trident Bat (*Aselliscus stoliczkanus)*

Description:

HB: 40-50; T: 30-40; FA: 39-44; E: 10-14; HF: 9-9.5.

This bat as 2 colour phases, Grey or brownish grey , with bright yellow or red. I guess if more specimens were found there would be a wider variation than this.. The coat is long, with individual hairs white at the base, coloured only at the tips. The outer margins of the anterior leaf is divided into 3 points, the middle one separated from those on the side by deep grooves. There are 2 accessory lateral leaflets margining the horse shoe. Neither sex as a frontal sac. The skull as a low brain case, deep rostrum, and sagittal crest greatly developed in the interorbital area. there is no postorbital process. The first premolar is within the tooth row.

Range Description:

This species occurs from Myanmar (Struebig et al. 2005) and southern China (Yunnan, Guangxi and Guizhou) (Smith et al. 2008), south to the Malay Peninsula.

Countries: Native:

China; Lao People's Democratic Republic; Malaysia; Myanmar; **Thailand;** Vietnam

Population:

It is locally common in limestone areas.

Population Trend: Unknown

Habitat and Ecology:

The species occurs in limestone areas and is known to roost in caves (C. Francis and P. Bates pers. comm.). The species is not dependent on primary habitat and can survive in degraded areas and extends into agricultural zones and shares roosts with other species (P. Bates pers. comm.). A colony in **Thailand** roosted in loose association in limestone caves, with individuals at least 30 cm from one another. Reproductively active females have been collected in May and June in Lao PDR and Vietnam.

Systems: Terrestrial

Major Threat(s):

The species is vulnerable to disturbance of cave roosts by tourism and also the destruction of caves through limestone extraction (P. Bates pers. comm.).

Conservation Actions:

The species occurs in a number of protected areas in **Thailand,** Lao PDR and Vietnam

Greater Roundleaf Bat *(Hipposideros armiger)*

Description:

HB:80-110; T: 48-70; FA; 88-98; E: 30-35; HF: 17-21.

The fur is soft and thick, with 2 colour phases;

1. Hairs on back tipped with warm brown, followed by a broad zone of much lighter brown, and the basal band nearly as dark as the tip.

2. Darker tips of hairs darker brown, with a broad sub terminal zone dirty whitish, the base as dark as the tip. The hairs of the mantle area are medium brown at the roots, pastel brown for the terminal two thirds .

The ears and wings are dark brown. The tip of the tail is free from the interformeral membrane. The anterior nose leaf is much wider than the posterior leaf, there are four lateral supplementary leaflets on each side. The accessory transverse fleshy structures which arise behind the posterior leaf are small in females and young males and much larger in old males. The skull has a prominent sagittal crest in old individuals. The first upper premolar is outside the tooth row.

The Great Roundleaf Bat is amongst the largest of the Hipposideros genus, and is the largest example in Southeast Asia. Typically the species roosts in caves, but is also found in abandoned buildings. Roundleaf bats are insectivorous bats characterised by a horseshoe-shaped base to the nose leaf but, unlike the Horseshoe Bats, lack a complicated 'lancet' or projection from the top of the nose leaf. Similar in appearance to the closely related, but smaller, Intermediate Roundleaf Bat *Hipposideros larvatus,* the Great Roundleaf Bat possesses four, not three, lateral accessory leaflets on each side of the main nose leaf. Males also possess a fleshy, swollen area above and behind the nose leaf. Its thick and woolly fur is medium brown, and the ears dark brown.

Range Description:

This widespread species has been recorded from India and Nepal, eastwards into central and South Eastern China, and from much of peninsular Southeast Asia. In South Asia, it has been reported from India (Assam, Manipur, Meghalaya, Sikkim, Uttaranchal and West Bengal) and Nepal (Central, Eastern and Western Nepal) in South Asia (Molur et al. 2002). In China, it is distributed south of the Yangtze river, and has been recorded in Jiangxi, Zhejiang, Guangdong, Hong Kong, Macao, Guangxi, Hainan, Hunan, Jiangsu, Anhui, Yunnan, Sichuan, Shaanxi, Guizhou, Fujian and the island of Taiwan. In Southeast Asia, it ranges from Myanmar in the west, through to **Thailand** and the Malay Peninsula. In South Asia, it has been recorded at elevations of 1,000 to 2,031 m asl (Molur et al. 2002).

Countries: Native:

Cambodia; China; Hong Kong; India; Lao People's Democratic Republic; Malaysia; Myanmar; Nepal; Taiwan, Province of China; **Thailand;** Vietnam

Population:

It appears to be fairly common throughout its range.
Population Trend: Unknown

Habitat and Ecology:

In South Asia, this usually low flying species is generally a high altitude species found in montane and bamboo forests (Mistry 1995). It has been recorded roosting either singly or in colonies of several individuals and shares its roosts with other species of bats in subterranean caves, lofts of houses, verandas of old houses, old temples. It breeds once a year and gives birth to two young (Bates and Harrison 1997). In China, it is considered to be a species found in a variety of habitats. They are known to inhabit caves and a variety of man-made structures. Colonies can number in the hundreds of individuals, and they co-occur with species of Rhinolophus and other bats. In Southeast Asia, the species has been primarily recorded near caves but occurs in quite distant areas from the roosts when foraging.
Systems: Terrestrial

Major Threat(s):

In South Asia, this species is threatened by deforestation, generally resulting from logging operations and the conversion of land for agricultural purposes, from mining activities, and disturbance to roosting sites in caves (Molur et al. 2002). In Southeast Asia, cave disturbance is occurring throughout the species' range and it is hunted (presumably for food) in Lao PDR, Vietnam and **Thailand.**

Conservation Actions:

In South Asia, there are no direct conservation measures in place. The species has been recorded from protected areas in India like Mahanandi Wildlife Sanctuary in West Bengal. In Southeast Asia, it occurs in protected areas throughout its range. In parts of its range further studies are needed into the distribution, abundance, breeding biology and general ecology of this species. Populations of this species should be monitored to record changes in abundance and distribution (Molur et al. 2002).

Dusky Roundleaf Bat *(Hipposideros ater)*

Description:

Head and body length is 6-7cm. Forearms 4cm. Wingspan 23-25cm.

Females are slightly larger than males. Reddish brown above and lighter brown to yellowish brown below. Two thirds of each hair is white, only the tip is coloured, usually bicoloured in nature. Wing membrane translucent and brown. Large ears greyish brown and ridged, feet and thumbs flesh coloured. Young are dark brown in colour. Three small nose-leaf flaps present. Frontal sac behind posterior leaf is prominent in males. Long tail is enclosed by a membrane, only the tip protrudes.

Range Description:

This very widespread species ranges from India, through much of Southeast Asia, to the islands of New Guinea and Australia. In South Asia this species is very widely distributed. It is presently known from India (Andhra Pradesh, Karnataka, Kerala, Madhya Pradesh, Maharashtra, Meghalaya, Nicobar Islands, Orissa and Tamil Nadu) and Sri Lanka (Eastern, North Central, Southern and Western provinces) (Molur et al. 2002, Aul and Vijaykumar 2003, Srinivasulu and Srinivasulu 2006). In Southeast Asia it is distributed from Myanmar, into Thailand and Peninsular Malaysia, and from here ranges to Sumatra (Indonesia), northern parts of Borneo (Brunei [possibly], Indonesia and Malaysia), the Philippines (found throughout with records from Balabac [Hill, 1963], Bohol, Catanduanes, Cebu, Leyte, Luzon [Abra, Benguet (Taylor, 1934), Cagayan, Camarines Sur, Laguna, Pampanga, Rizal, Tarlac provinces], Marinduque [Lawrence, 1939], Maripipi, Mindanao, Mindoro, Negros, Palawan [Heaney et al. 1998].), Sulawesi (Indonesia), the Moluccan Islands (Indonesia), the Aru Islands (Indonesia), the islands of Biak-Supiori and Yapen (Indonesia), New Guinea (Indonesia and Papua New Guinea), the islands of New Britain and New Ireland (Papua New Guinea), the Louisade Archipelago (Papua New Guinea). It has been recorded from northern and central parts of Australia. It has been recorded up to 1,700 m asl in (New Guinea).

Countries: Native:

Australia; India; Indonesia; Malaysia; Myanmar; Papua New Guinea; Philippines; Sri Lanka; **Thailand**

Population:

This species is common in its range in South Asia and has been recorded in small colonies of few individuals. Although there are no long-term studies on this species, informal observations reveal that the populations are stable in many localities (C. Srinivasulu pers. comm.). In the Philippines, the species has not been found to be common (Heaney and Balete pers. comm.). It is widespread, but never especially common in the remainder of Southeast Asia (Bates pers. comm.). In Australia, it appears to be a generally uncommon species.

Population Trend: Unknown

Habitat and Ecology:

In South Asia, this species roosts in small colonies in lofts of old thatched houses, old disused buildings, disused areas of buildings, mines, tunnels, culverts, wells, hollows of large trees in forested areas, large crevices in walls, caves on sea shores. It is a late flyer with a low, fast and fluttering flight and feeds on small sized coleopterans and mosquitoes. A single young is born after a gestation period of 150-160 days (Bates and Harrison 1997). In Southeast Asia, the species has been recorded from lowland and montane primary and secondary forest, over or associated with limestone (L. Heaney and Balete pers. comm.), where it usually roosts in caves (L. Heaney et al. 1991; Rickart et al. 1993; Esselstyn and L. Heaney pers. comm. 2006) although there is a record from a human-constructed tunnel in lowland secondary forest (Sedlock, 2001). The species is only very rarely found in agricultural areas near to forest (L. Heaney pers. comm.). In New Guinea and Australia, it forages in a wide variety of habitats including rainforest, dry woodland, mangroves, dry scrub, euclypt woodland and secondary growth. It roosts in caves and abandoned tunnels. Animals largely roost in small groups but can sometimes be found in groups of several hundred individuals (Flannery 1995; Strahan 1995; Bonaccorso 1998). The female gives birth to a single young.
Systems: Terrestrial

Major Threat(s):

Overall there appear to be no major threats to this widespread species as a whole. In South Asia and Australia the species is locally threatened by the disturbance of maternity caves. In the Philippines, the species has probably declined as a result of destruction of lowland forest and disturbance of caves (Heaney et al. 1998). In the Philippines, as with other cave dwelling bats, there is likely to be some localized hunting for food (Heaney and Balete pers. comm.), however, it is unlikely to be taken for food throughout most of its southeast Asian range (Francis pers. comm.).

Conservation Actions:

In view of the species wide range, it seems probable that it is present in a number of protected areas. There is a need to identify and protect important roosting sites for this species. In South Asia, additional studies into the threats to this species are needed.

Bicolored Roundleaf Bat *(Hipposideros bicolor)*

Description:

HB: 44-46; T: 30-38; FA:38-47; E: 16-18; HF:7-9.

The fur is long with upper parts smoky grey or brownish or reddish at the tip, with pale and white bases; the under parts are much paler, with a brown tinge. Some are even golden brown. The ears and wings are brown, the nose leaf brownish pink. The ears are rounded, with a distinct antitragus; they extend to the end of the muzzle when laid forwards and are haired for almost half their lenght. A band of skin connects the base of the ears across the crown of the head. The nose leaf is simple, the anterior leaf is almost square, without notches or indentations.. The posterior leaf as four poorly defined compartments with the upper margin semi circular. The skull is long and slender, with an enlarged braincase and narrow zygomata; the anterior half of the zygomata is massive. There is a low sagittal crest. The first upper premolar is very small and outside the too throw.

Range Description:

This species is found from **Thailand** to Timor. In the Philippines, the few individuals referred to this species are from Luzon (Camarines Sur province) and Mindoro (Heaney et al. 1998). A specimen from Palawan reported by Allen (1922) was recently re-identified as Hipposideros ater, which is the only report of the species from the island (Esselstyn et al. 2004).

Countries: Native:

Indonesia; Malaysia; Philippines; **Thailand;** Timor-Leste

Population:

This species is common in the southeast Asia region and probably widespread in Borneo, although there are no records (C. Francis pers. comm.). The status of the Philippines population is unknown (Heaney et al. 1998).

Population Trend: Stable

Habitat and Ecology:

A forest species preferring primary forest, it is not common in disturbed forest areas. Found in a wide range of roosts including houses, caves and burrows of foxes and porcupines. The group size varies fro 10 to 12 right up to 200 individuals, there is no sexual segregation at the roost, but males are twice as numerous as females. The diet is soft incest such as termites. The young are born in April and are carried by the mother attached to a false nipple until they reach adult size. It roosts in caves in Peninsular Malaysia (Payne et al. 1985).

Systems: Terrestrial

Major Threat(s):

There are no major threats to this species, although it is sensitive to deforestation.

Conservation Actions:

The species occurs in a number of protected areas.

Ashy Roundleaf Bat *(Hipposideros cineraceus)*

Description:

HB: 37-42; T: 25;28; FA: 33-37; E:16; HF: 5-6.

The smallest of the Thai *Hipposideros* is quite similar to H. bicolor but is rather smaller with relatively smaller eras, which do not reach the end of the muzzle when laid forwards. As in many Hipposideros, there are two colour phases, being grey and brown; the wings are medium brown. The ears are large, broad and blunt, covered with hair for half their lenght. The nose leaf is simple closely resembling that of H. fulvus. The internarial septum is markedly swollen into two kidney shaped lobes, but the lateral narial lappets are rather low. The intermediate nose leaf as 4 dark glands. As in H. fulvus, the posterior nose leaf is indistinctly divided by three vertical septa. The scull is rather small, with narrow zygomata, the nasal region is inflated. The lower first premolar is larger in Bicolor, over half the height of the lower second premolar. The upper median incisors are slightly longer than the outer upper incisors; all upper incisors are sharply pointed. The lower incisors are low and trifid. The upper first premolar is small, compressed between the canine and second upper premolar, but located within the tooth row.

Range Description:

This species is widely distributed from South Asia into much of Southeast Asia. In South Asia this species occurs in India (Assam, Arunachal Pradesh, Haryana, Meghalaya, Mizoram, Tamil Nadu, Uttaranchal and West Bengal), Nepal and Pakistan (Punjab) at elevations ranging from 62 to 1,480 m asl (Molur et al. 2002). Although it has only been recorded from only a few locations in the region, it is quite widespread (Molur et al. 2002). In Southeast Asia, the species is found from Myanmar in the west, through parts of **Thailand,** Lao PDR, Vietnam and possibly Cambodia (presence here needs to be confirmed, G. Csorba pers. comm.) to Peninsular Malaysia, and from here has been recorded on the islands of Sumatra (Indonesia) and parts of Borneo (Indonesia and Malaysia). Although there are reports of this species from Philippines, there appears to be no conclusive evidence of the species presence here L. Heaney pers. comm.).

Countries: Native:

India; Indonesia; Lao People's Democratic Republic; Malaysia; Myanmar; Pakistan; **Thailand;** Vietnam

Population:

The abundance, population size and trends for this species are not known in South Asia (Molur et al. 2002). In Thailand has been found roosting in groups of up to 30 individuals (S. Bumrungsri pers. comm.). It is usually captured in small numbers in surveys in Lao PDR (C. Francis pers. comm.).

Population Trend: Unknown

Habitat and Ecology:

There is little information available on the natural history of this species. In Southeast Asia, it seems to be a largely lowland species that has been recorded roosting in caves (S. Bumrungsri pers. comm.). In South Asia this species roosts in hollows of trees in forests (Molur et al. 2002). The young are born after a gestation period of 180 days (Bates and Harrison 1997).

Systems: Terrestrial

Major Threat(s):

This species is threatened over much of its range by deforestation, generally resulting from logging operations for conversion of land to agriculture and as human settlements. It is also facing threat due to tourism related developmental activities (Molur et al. 2002).

Conservation Actions

In South Asia, although there are no direct conservation measures in place, the species has been recorded from several protected areas in India such as the Kalakkad Mundunthurai Tiger Reserve in Tamil Nadu (Vanitharani et al. 2006) and from Nagarjunasagar Srisailam Tiger Reserve in Andhra Pradesh (C. Srinivasulu pers. comm.). Detailed studies on taxonomy, distribution, abundance, reproduction, ecology, threats and population monitoring are recommended (Molur et al. 2002). In Southeast Asia, although it is present in a number of protected areas further protection of suitable forest habitat is needed.

Diadem Roundleaf Bat *(Hipposideros diadema)*

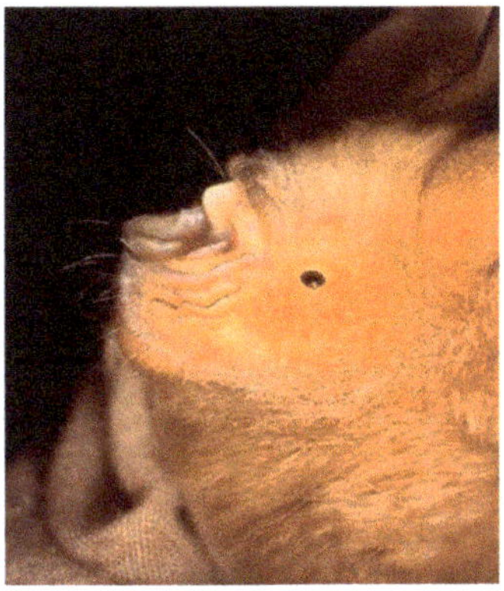

Description:

HB: 85-92; T: 50-60; FA: 80-96; E: 30-32; HF: 16-20

An impressive sight amongst the trees of the rainforest, this large bat is a formidable hunter with a wingspan of up to half a metre. Readily identifiable by its pig-like face, the Diadem Roundleaf bat also possesses distinctive white flashes on the shoulders, contrasting with the reddish-brown back and light-brown underside. The colour varies quite strongly between individuals, and females are generally more orange. The nose leaf is pink and highly convoluted, but the ears are brown. With a heavy body and long, narrow wings, the Diadem Roundleaf bat is adept at fast flight but has relatively poor maneuvering ability. It has adapted to foraging in gaps in forests, such as around tree falls or above rivers. This bat species is not restricted to rainforest and in outback Australia it forages within eucalypt woodland and open forest, deciduous vine thicket and within towns. Individuals are known to forage up to two and a half kilometres from the roost during the course of the night . During the day it roosts in small groups in caves, old mines and sheds, hollow trees and tree branches. There are 2 colour stages

1 Pale yellow or buff, with reddish tips with the underparts lighter.

2. Pale buff, with brown tips on the upper parts and the under parts paler brown with lighter tips. Both the phase have a the conspicuous buffy shoulder spot. The nose leaf is pinkish, the eras and wings brown. The ears are triangular, sharply pointed, and haired for 1/3 of the lenght; there is no antitragus. The posterior nose leaf is rounded and larger than the anterial leaf, which has a 3-4 supplementary lateral leaflets on each side, the other may be rudimentary. The posterior leaf is divided into 4 sections(only 2 in subspecies *masoni)*. The skull is large and heavily built, with a broad facial portion. The upper first premolar is outside the tooth row, so the canine and upper second premolar are in contact or nearly so.

Range Description:

This widespread species ranges from the Nicobar Islands of India through Southeast Asia to Australia. On the Nicobar Islands, the endemic subspecies *H. d. nicobarensis* is present on the islands of Bompuka, Katchal, Tillangchong, Tressa and Trinket (Aul and Vijaykumar 2003). In Southeast Asia, it ranges from southern Myanmar, into Thailand, Cambodia, Vietnam and southern Lao PDR, to Peninsular Malaysia, and from here to much of Indonesia (including the islands of Sumatra, Java, Bali and Sulawesi), East Timor, the island of Borneo (Brunei, Indonesia and Malaysia), and the Philippines. In the Philippines it is found in all regions (sea level to 900 m asl) except the Batanes/Babuyan region. Specimens have been collected from the Philippines islands of Bohol, Busuanga, Calauit, Catanduanes, Cebu (Lawrence, 1939), Dinagat, Guimaras, Leyte, Luzon [Benguet (Taylor, 1934), Bulacan, Cagayan, Camarines Sur, Ilocos Norte, Isabela, Laguna, Nueva Vizcaya, Pampanga, Quezon, Rizal (Taylor, 1934) provinces], Mindanao [Agusan del Norte, Bukidnon, Davao Oriental, Lanao del Norte, Maguindanao, South Cotabato (Sanborn, 1952), Zamboanga del Norte, Zamboanga del Sur (Taylor, 1934)], Mindoro (Lawrence, 1939), Negros, Palawan, Panay, Polillo, Samar, Siquijor (Heaney et al., 1998). There are some records from the island of Tawi-tawi (K. Helgen pers. comm.). On the island of New Guinea (Indonesia and Papua New Guinea) it is found in scattered localities and it is present throughout much of the Bismarck Archipelago (Papua New Guinea). The species has been recorded from the islands of Bougainville and Buka (Papua New Guinea), and from many of the Solomon Islands. It is present in Australia where it is largely restricted to the rainforests of northern Queensland (Corbet and Hill 1992; Flannery 1995; Strahan 1995; Bonaccorso 1998). It is found from sea level to 1,300 m asl.

Countries: Native:

Australia; Cambodia; India; Indonesia (Irian Jaya, Jawa, Kalimantan, Lesser Sunda Is., Maluku, Sulawesi, Sumatera); Lao People's Democratic Republic; Malaysia (Peninsular Malaysia, Sabah, Sarawak); Myanmar; Papua New Guinea; Philippines; Solomon Islands; **Thailand;** Timor-Leste; Vietnam

Population:

The Nicobar subspecies, *H.d. nicobarensis,* is very common and has been found to roost in large colonies of over 500 bats to small colonies of over 30 bats on four islands of Nicobar Islands (Aul and Vijaykumar 2003). It was first recorded from a cave on Trinket Island, but it does not occur on that island anymore (Aul and Vijaykumar 2003), the reason for this disappearance is unknown. In the Philippines this species is considered to be locally common (Heaney et al., 1998). There can be large roosts or thousands or more individuals, though the species often roost in small numbers (K. Helgen and L. Heaney pers. comm.). In New Guinea it is a relatively common species (Flannery 1995).

Population Trend: Unknown

Biology and Ecology

Diadem Roundleaf bats are predominantly insectivorous, feeding mostly on large insects, favouring beetles, grasshoppers and locusts, and moths . This species is also carnivorous having been recorded feeding on birds at two sites in Australia. The main foraging method is perch hunting which is a low-energy strategy. Its prey commonly move with direct and predictable flight paths, making it possible for the bats to hang from a tree branch, up to ten metres above the ground, waiting for a suitable insect to fly past. Whilst at the perch, the bat scans the area using echolocation at a constant frequency of 58 to 60 kilohertz. Once it has detected an insect, it drops from its perch and flies fast and straight to snatch its quarry from the air. The size of the pup relative to the mother in insectivorous bats is remarkable. This species can give birth to a single pup weighing 13 grams – a quarter of its mother's weight. The mother must carry the pup on foraging trips until it is developed enough to fly and feed alone. By one year the young diadem Roundleaf bat will be ready to breed. This is amongst the largest of Roundleaf bats in Southeast Asia, and is easily identified by the presence of well-defined white or pale orange patches of fur on the sides of its body and shoulders. The rest of the upper body fur is a rich, dark brown colour, and the face and underparts pale. When hunting, this bat typically hangs from a perch waiting to ambush large insects, such as moths, as they fly past. Other insects consumed may include those with a thick exoskeleton such as beetles, weevils and katydids : the soft parts of such insects are consumed and the inedible parts, such as wings, carapace and legs are discarded below the perch.

The species makes use of a variety of roosts including tree holes, caves or the shelter of large-leaved palms. Large caves may shelter maternity roosts numbering in their thousands.

Major Threats

Habitat loss and degradation is the biggest problem facing the Diadem Roundleaf bat. Deforestation continues at a steady rate for conversion to agricultural land and building communities. Of particular concern is the extensive loss of primary forest due to the rapid increase in land devoted to growing oil palm. Together, Malaysia and Indonesia export 88 percent of the world's palm oil, for use in products such as margarine, lipstick and detergent . Disturbance of roost sites is also an issue , and despite the contribution of many bats in the control of insect crop pests, persecution of bats is also a threat

Fulvus Roundleaf Bat *(Hipposideros fulvus)*

Description:

HB: 41-43; T31; FA: 38-44.

There are 2 colour phase,chesnut brown above and below and dark brown above with underparts paler. The skull is elongated, rather slender, with a low sagittal crest, the zygomata are broader than in *ater.* The first lower premolar is smaller than in *bicolor*, less than half the height of the second lower premolar. The first upper premolar is minute and opposite the tooth row so that the canine and upper premolar are in contact. the nose leaf is simple. The anterior leaf is small, well separated from the upper lip. The internarial septum is markedly swollen, and there are low lateral narial lappets. The intermediate nose leaf as not special characters, and the posterior nose leaf is indistinctly divided by three vertical septa.

Range Description:

This species is very widely distributed in South Asia in Afghanistan (Laghman and Nangarhar provinces), India (Andhra Pradesh, Bihar, Gujarat, Jharkhand, Karnataka, Kerala, Madhya Pradesh, Maharashtra, Orissa, Punjab, Rajasthan, Tamil Nadu and Uttar Pradesh), Nepal (Central Nepal), Pakistan (Baluchistan, Punjab and Sind) and Sri Lanka (Central, Sabaragamuwa and Western provinces) (Molur et al. 2002). It has been recorded from sea level to an elevation of 2,600 m asl. It has been recorded from western Yunnan in China (Smith and Xie 2008). **Records of this species from Vietnam and Thailand are probably erroneous, based on misidentified Hipposideros pomona (Borissenko and Kruskop 2003).**

Countries: Native:

Afghanistan; Bangladesh; China; India; Pakistan; Sri Lanka

Population:

This species is widespread and common throughout its range and the population seems to be stable and doing well.

Population Trend: Stable

Habitat and Ecology:

This species can be found in a variety of habitats from arid regions to thick forests. It roosts in colonies of a few to large number of individuals in old dilapidated buildings, temples, cellars, caves and old wells. It is a late flyer with a slow, fluttering and low flight and feeds on cockroaches and coleopterans. A single young is born after a gestation period of 150-160 days (Bates and Harrison 1997).

Systems: Terrestrial

Major Threat(s):

Some populations of this species are affected by habitat loss largely due to tourism related developmental activities such as fumigation and cave sealing in the Ajanta-Ellora Cave Complex, and from stone mining and quarrying (Molur et al. 2002).

Conservation Actions:

There are no direct conservation measures in place. The species has been recorded from protected areas in India including the Bhimashakar Wildlife Sanctuary in Maharashtra and is very likely to be found in other protected areas too in the range. Further studies are needed into the distribution, abundance, breeding biology and general ecology of this species. Populations of this species should be monitored to record changes in abundance and distribution.

Cantor's Roundleaf Bat *(Hipposideros galeritus)*

Description:

Slightly larger than *bicolor.* The colour is variable from grey to bright red to dark brown with juveniles very dark grey or black. The under parts are greyish brown and grey in juveniles. The nose leaf, ears and wings are brown. The eras are short, broad and pointed and continually move when the bat is at rest; they are haired for 3/4 of their lenght. The anterior nose leaf is well developed, extending almost to the end of the muzzle with 2 lateral supplementary leaflets on each side. The posterior leaf is evenly rounded behind, the surface divided into 3 cells by vertical septa. The skull is short and broad, with the inflated braincase and lower sagittal crest; the supraorbital ridges are sharply defined. The upper first premolar is minute, displaced from the tooth row so that the canine and second upper premolar are in contact or nearly so

Range Description:

This species is widely distributed in South Asia and Southeast Asia. In South Asia it is presently known from Bangladesh (Chittagong, Dhaka, Khulna, Sylhet and Rajshahi divisions), India (Andhra Pradesh, Bihar, Gujarat, Karnataka, Madhya Pradesh and Maharashtra) and Sri Lanka (Central, North Central, North Western, Sabaragamuwa, Uva and Western provinces) (Molur et al. 2002, Srinivasulu et al. in press). It is very widely distributed and has been recorded from sea level to an elevation of 1,100 m asl (Molur et al. 2002). On mainland Southeast Asia, the species has been recorded from **Western and Southern Thailand (including Terutau Island),** Cambodia (four specimens have been recorded but no exact localities are available [Borissenko and Kruskop 2003]), southern Lao PDR (Francis pers. comm.), Southern Vietnam, and Peninsular Malaysia (including Penang Island). Within insular Southeast Asia, the species has been recorded from the islands of Sumatra, Bangka and Java (all to Indonesia), the island of Borneo (Brunei, Indonesia and Malaysia), and from the island of Sanana in the Sula Islands (Indonesia) (Flannery 1995). There is a record from the island of Bali (Indonesia) but it may be erroneous (Kock and Dobat 2000).

Countries: Native:

Bangladesh; Brunei Darussalam; Cambodia; India; Indonesia (Kalimantan); Lao People's Democratic Republic; Malaysia (Peninsular Malaysia, Sabah, Sarawak); Sri Lanka; **Thailand;** Vietnam

Population:

This species, though widespread throughout its range in South Asia, is not very common as the localities are scattered and the colony size is small. The abundance, population size and trends for this species are not known (Molur et al. 2002). Bates and Harrison (1997) call this a comparatively rare bat as only isolated individuals have been recorded. In Southeast Asia this species is never abundant, but is widespread and not rare.

Population Trend: Unknown

Habitat and Ecology:

This species is found in dry to wet lowland forests, but have also been recorded from rubber estates in Southeast Asia. This species roosts in small colonies or family units in old mines and deep caves., cracks, culverts and crevices in old buildings, caves, among large boulders, overhanging ledges, tunnels, dungeons, forts, temples and churches. On Borneo it has been found roosting in caves in small groups with *H. cervinus,* as well as in colonies of hundreds of animals (Payne et al. 1998). On Sanana Island it has been found in a colony of several hundred individuals sharing a large cave with six other bat species (Flannnery 1995). It is a low flier and feeds on beetles and other insects (Bates and Harrison 1997). In Sri Lanka, the species is observed to change roost sites frequently (W. Yapa and P.C.M.B. Digana pers. comm.). These bats are very territorial and old males may even be solitary and are found in regularly used roosting sites where the daily scratching of there claws keeps the face of the rock clean and rough. But for there evening flights out of the cave, these bats form large social gatherings. Every day in the early afternoon they begin to move from there roost site deep within the cave. This gross movement is an unstoppable stream, but each individual bat in fact makes only a short flight before with unfailing precision, it somersaults abruptly in mid air and in the same movement reaches out and clings with its sharply clawed feet to the rock. By a succession of leap fogging flights the bats move down to the final forming up point. which beyond this point there is no halt.

Systems: Terrestrial

Major Threat(s):

There are no major threats to this species as a whole. In South Asia, the foraging grounds of this species are threatened by deforestation, generally resulting from logging operations and conversion of land for agriculture. It is also threatened due to disturbance to roosting sites and disturbance to caves (Molur et al. 2002).

Conservation Actions:

This has been recorded from a number of protected areas. In India, the species has been recorded from the Melghat Tiger Reserve, Borivili National Park in Maharashtra, Kanha National Park in Madhya Pradesh and Nagarjunasagar Srisailam Tiger Reserve in Andhra Pradesh (Srinivasulu, 2004). In South Asia, further studies are needed into the distribution, abundance, breeding biology and general ecology of this species. Populations of this species should be monitored to record changes in abundance and distribution. Public awareness is needed to mitigate future threats to this species (Molur et al. 2002).

Thailand Roundleaf Bat *(Hipposideros halophyllus)*

Description:

The great Roundleaf bat is similar in appearance to the closely related intermediate Roundleaf bat (*H. larvatus*). However, the great Roundleaf bat is larger and possesses four, not three, lateral accessory leaflets on each side of the main nose leaf. This bat has a forearm length up to 9.8 cm, and weighs up to 47 g.

Range Description:

This species is known from a few localities in **Central and Southern Thailand (Lop Buri, Tha Woong, Khao Sa Moa Khan),** further surveys are required to determine if it is in more locations. The two southern localities on the range map may be historical as the entrances to the caves have been blocked, however, it is not known whether the species currently occurs in nearby caves (S. Bumrungsri pers. comm.).

Countries: Native:
Thailand

Population:

The current total population estimate for the species for the known colonies is 1,000-1,400 individuals (Waengsothorn et al. 2006).

Population Trend: Decreasing

Habitat and Ecology:

The species requires a very specific roosting habitat of caves with small entrances which are underground, and this type of cave is rare (Waengsothorn et al. 2006). The species does not travel more than 2 km from the cave to forage. It is unlikely that there is any mix between the known populations as the areas between them are rice fields which are unsuitable for the species. The closest populations are more than 150 km apart (Waengsothorn et al. 2006).

Systems: Terrestrial

Major Threat(s):

There has been a loss of limestone habitat in the area due to limestone mining (S. Bumrungsri pers. comm.). In addition most known populations are outside protected areas and the forest habitat is highly disturbed due to livestock and deforestation (S. Bumrungsri pers. comm.). Known localities are surrounded by rice fields where pesticide use is high. The roosts of the species are disturbed due to harvesting of guano and the species is sensitive to this. It is hunted in some parts of its range for food.

Conservation Actions:

Surveys are required to determine if the species occurs in additional locations, and areas of suitable habitat require protection (S Bumrungsri pers. comm.).

Intermediate Roundleaf Bat (*Hipposideros larvatus)*

Description:

HB:60-80; T: 30-45;FA: 52-65; E: 20-23; HF: 10-15.

There are two colour phases, dark brown and reddish brown; the underparts are smokey grey with brown tips. The mantle is pale brown,; the ears, nose leaf, and wings are brown. The throat is sparsely haired, and there is a well marked frontal sac in males, well less marked in females. The ears are large and broad and triangular in shape. The nose leaf is similar to *bicolor,* but with 3 lateral supplementary leaflets on each side of the anterior leaf; the anterior leaf is simple, with a median notch. It does not coyer the end of the muzzle. The supraorbital ridges are barely developed and the zygomata are slender. The first upper premolar is tightly compressed between the canine and the second upper premolar.

Range Description:

This species complex is widely distributed, ranging from northeastern South Asia, throughout much of Southern China and mainland Southeast Asia, into several islands within insular Southeast Asia. In South Asia this species is found in Bangladesh (Chittagong and Sylhet divisions) and India (Arunachal Pradesh, Assam and Meghalaya) in South Asia (Molur et al. 2002). In China, it has been recorded from Hainan island, Guangdong, Guangxi and Guizhou (Smith and Xie 2008). On mainland Southeast Asia, it is distributed widely across Myanmar (K.M. Swe pers. comm., Bates and Harrison 1997, Molur et al. 2002), ranging into **Thailand, C**ambodia, Lao PDR, Vietnam and Peninsular Malaysia. In insular Southeast Asia, it has been recorded from numerous islands including the Mentawi Islands, Sumatra, Java, Bali, the Kangean Islands (all to Indonesia) and the southern part of the island of Borneo (Kalimantan [Indonesia] and possibly Sarawak [Malaysia]). This species has been taken from sea level up to around 2,000 m asl.

Countries: Native:

Bangladesh; Cambodia; China; India; Indonesia (Bali, Jawa, Kalimantan, Sumatera); Lao People's Democratic Republic; Malaysia (Peninsular Malaysia, Sabah, Sarawak); Myanmar; **Thailand;** Vietnam

Population:

As currently defined this is a common bat throughout much of its range (Bates and Francis pers. comm.). In South Asia, this species is distributed across three northeastern states of India, but the sightings are sparse and few between. Only two colonies of less than 100 individuals have been recorded in the last decade (Molur et al. 2002). In northern Myanmar, it is one of the more common bats living in big colonies. However, there is a decline in their numbers due to mining and it is estimated to have declined by 10% over the last 10-15 years (K.M. Swe pers. comm.).

Population Trend: Unknown

Habitat and Ecology:

In Myanmar it is found in a variety of habitats from highly disturbed agricultural land to secondary forest, in dry zones and humid areas (P. Bates pers. comm.; K.M. Swe pers. comm.; Bates and Harrison 1997). It is often associated with limestone caves in Myanmar and Indonesia (A. Suyanto pers. comm.). On Peninsular Malaysia, the species has been recorded from primary lowland tropical moist forest (Frances pers. comm.). In South Asia, this species roosts in caves, mine shafts, pagodas and buildings (Molur et al. 2002). It is also found roosting in human habitations (P. Bates pers. comm.).
Systems: Terrestrial

Major Threat(s):

There are no major threats to this species as a whole. In South Asia, there currently appear to be no major threats to this species in northeastern India. The situation in Bangladesh is not known. However, disturbance to roosting sites such as caves by humans may pose a threat to this species as observed in Meghalaya (A. Tabah pers. comm.). In neighbouring Myanmar mining activities in limestone caves for manufacture of cement is a threat to some populations of this species (K.M. Swe pers. comm., Molur et al. 2002). It may be affected by hunting in some parts of its range (Francis pers. comm.).

Conservation Actions:

This species is present in a number of protected areas. In South Asia, it has been recorded from protected areas in India like Orang National Park in Assam. Surveys, habitat management, population monitoring are important recommendations. In areas where this taxon is facing threat due to mining activities public awareness is needed (Molur et al. 2002). Further studies are needed to clarify the taxonomic status of populations currently allocated to this species.

Large Asian Roundleaf Bat *(Hipposideros lekaguli)*

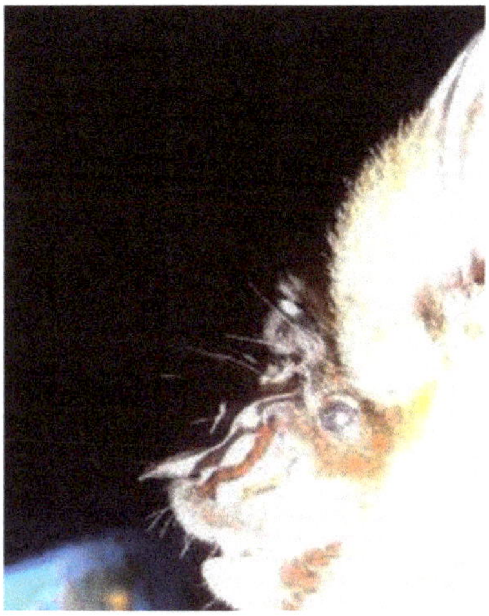

Description:

HB: 66-79; T: 43.7-53.6; FA:61.5-79.3; E: 26.5-33.3; HF: 12.1-13.8.

The upper parts are greyish white with a brown tinge, with the lower back more brownish, the interface between the shades forming a poorly defined V shape.. The under parts and throat are a dull creamy white some slightly darker, with those from Rayong and Pattani being slightly darker... The ears are triangular, large and broad, sparsely covered with hair for the lower 25% of the external part. The inner ear is mostly hairless except for a narrow fringe of dense hair along the lower 3/3 of the inside margin. The nose leaf is more complex than in other *Hipposideros* within the *diadema* group. The anterior leaf is large, almost covering the muzzle, with no median anterior emargination. The internarial septum is slightly bulbous, and the intermediate leaf is inflated and thickened, with the median projection greatly developed into a triangular lappet; The posterior leaf is very complex, with the upper part supported from behind by 2 narrow septa and divided in front into a narrow central lobe flanked on each side by a wider lateral lobe. There are 3 wide well developed supplementary leaflets. The skull represents *diadema* amongst the Thai forms but is rather smaller and has a broader interorbital region. There is no frontal depression, and the naso frontal region is rounded. The braincase is long and wide, with a moderate sagittal crest. The zygo mata strong and expanded. The upper incisors are irregularly bilobed, with the tips convergent; the lower incisors are bicuspid. The canines are not grooved and lack a posterior cusp. The upper first premolar is very small and variable in position, with the upper second premolar often in contact with the canine or nearly so.

Range Description:

This species is found in **Thailand,** Peninsular Malaysia, and the Philippines. In the Philippines, it is recorded only from one individual from Luzon (Isabela province) and one from Mindoro (Balete et al. 1995; Heaney et al. 1998). There are unverified and poorly documented records of this species from other islands in the Philippines. In **Thailand** it is usually found below 400 m asl.

Countries: Native:

Malaysia; Philippines**; Thailand**

Population:

The population status in Malaysia is unknown. In **Thailand,** two thirds of the known populations are outside protected areas, colonies comprise groups of up to three hundred individuals (S. Bumrungsri pers. comm. 2006). **A Thai population has undergone a decline from a few hundred individuals to 30 over a 20 year period** (S. Bumrungsri pers. comm.).
Population Trend: Decreasing

Habitat and Ecology:

All known sites are limestone cave areas; the species can survive in degraded forest and a population is known from a forest fragment within a sugar cane plantation in Thailand (S. Bumrungsri pers. comm.). The two known specimens from the Philippines were collected close to sea level in ultra basic forest, and in an agricultural area near a river (Balete et al. 1995).
Systems: Terrestrial

Major Threat(s):

The species is affected by tourism and religious disturbance to the caves, they are also hunted.

Conservation Actions:

Recommended actions include prevention of cave disturbance (S. Bumrungsri pers. comm. 2006

Shield-faced Roundleaf Bat *(Hipposideros lylei)*

Description:

HB:72-95T: 48-55;FA:78-84;E:30.

Brownish grey in colour, with buffy underparts, the nose leaf is brownish pink, the ears and wings are pale brown. The ears are broad and rectangular, with no antitragus. This is a species dimorphic species, with adult males having enormous wing like fleshy expansion terminated into 2 projections around the frontal sac; females and immature have a similar structure, but much smaller and very difficult to see behind the posterior leaf. The lateral margins of the anterior and posterior leaves are continuous. The skull has a low sagittal crest and moderate zygomata.

Range Description:

This species is found in Yunnan, China (Smith et al. 2008), extending to Myanmar, **Thailand,** Malaysia, and Vietnam. It can be expected in northern Lao PDR, but this has not been confirmed.

Countries: Native:

China; Malaysia; Myanmar; **Thailand;** Vietnam

Population:

This species tends to have small populations at each location. In **Thailand,** populations are typically comprised of less than a hundred individuals. The species is usually caught as individuals (S. Bumrungsri and P. Bates pers. comm.).

Population Trend: Decreasing

Habitat and Ecology:

It is associated with caves in limestone areas, the species can occur in disturbed and fragmented areas, including agricultural areas (S. Bumrungsri and P. Bates pers. comm.).

Systems: Terrestrial

Major Threat(s):

There are no major threats to this species, although cave disturbance affects the species throughout its range. In addition it is hunted at least in **Thailand** (S. Bumrungsri pers. comm.).

Conservation Actions:

It is known to occur in some protected areas.

Pomona Roundleaf Bat *(Hipposideros pomona)*

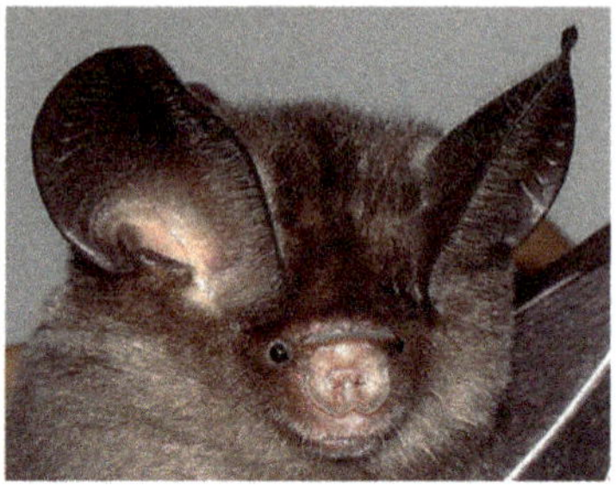

Description:

HB: 44-46; T: 30-38; FA:38-47; E: 16-18; HF:7-9.

The fur is long with upper parts smoky grey or brownish or reddish at the tip, with pale and white bases; the under parts are much paler, with a brown tinge. Some are even golden brown. The ears and wings are brown, the nose leaf brownish pink. The ears are rounded, with a distinct antitragus; they extend to the end of the muzzle when laid forwards and are haired for almost half their lenght. A band of skin connects the base of the ears across the crown of the head. The nose leaf is simple, the anterior leaf is almost square, without notches or indentations.. The posterior leaf as four poorly defined compartments with the upper margin semi circular. The skull is long and slender, with an enlarged braincase and narrow zygomata; the anterior half of the zygomata is massive. There is a low sagittal crest. The first upper premolar is very small and outside the too throw.

Range Description:

This very widespread species ranges from South Asia, through southern China, and much of mainland Southeast Asia. In South Asia it is known from Bangladesh (Sylhet division), India (Andhra Pradesh, Assam, Arunachal Pradesh, Karnataka, Kerala, Meghalaya, Nagaland, Nicobar Islands, Tamil Nadu, Sikkim and West Bengal) and Nepal (Western Nepal) (Molur et al. 2002). In China, it has been recorded from Yunnan, Guangxi, Guangdong, Fujian, Hunan, and Hainan island (Smith and Xie 2008). In Southeast Asia, the species ranges through Myanmar, **Thailand,** Lao PDR, Cambodia, Vietnam and northern Peninsular Malaysia.

Countries: Native:

Bangladesh; Cambodia; China; India; Lao People's Democratic Republic; Malaysia; Myanmar; Nepal; **Thailand;** Vietnam

Population:

The abundance, population size and trends for this species are not known in South Asia (Molur et al. 2002). In Southeast Asia the species is common throughout its range. In northern Myanmar, the population seems to be stable (K.M. Swe pers. comm., Molur et al. 2002).

Population Trend: Unknown

Habitat and Ecology:

Although *Hipposideros pomona* is widespread, little is known about the habitat or ecology of this species except that it roosts in small colonies of a few individuals in caves and crevices in subterranean habitats (Molur et al. 2002, Bates and Harrison 1997). In Southeast Asia it roosts predominantly though not exclusively in caves, and is considered to be very tolerant of modified habitats and can even occur in urban areas.

Systems: Terrestrial

Major Threat(s):

There are no major threats to this species as a whole. In South Asia, the habitat of this species is under threat from human activities in Nepal. In northern Myanmar some colonies are threatened by hunting for local consumption as food and medicine, and due to tourism related developmental activities (K.M. Swe pers. comm., Molur et al. 2002).

Conservation Actions:

The species is present in a number of protected areas in Southeast Asia. Research into the taxonomy, distribution, abundance, reproduction, ecology, population monitoring and habitat management are needed (Molur et al. 2002).

Pratt's Roundleaf Bat *(Hipposideros pratti)*

Description:

The Hipposideros pratti have a dorsal fur that is reddish brown in colour. However, its' ventral fur is lighter. The juveniles are lighter than the adults. The average forearm length of these bats are 83.4 - 90.8 mm and the average body mass is 53-68.9 grams. The H. pratti have a transverse bilobed shield behind the posterior nose leaf. In young males and females the shield may be small, but in old males it becomes an elaborate flashy structure.

Range Description:

The species is recorded from southern China (Smith et al. 2008) and Viet Nam (Robinson et al. 2003). According to Huynh et al. (1994), this species was reported in Vietnam only from Sa Pa (Lao Cai Province), however, there is evidence for the presence of this species considerably further South, in Phong Na (Quang Binh Province), where it possibly occurs together with related *H. lylei.* It is widespread, in southern China. Also seen in **Thailand but these reports are probably erroneous,**

Countries: Native:

China; Vietnam

Population:

The abundance and population size of this species are not known.

Population Trend: Unknown

Habitat and Ecology:

It is a cave-roosting species, and has been found roosting in the same caves as H. armiger.

Systems: Terrestrial

Major Threat(s):

The threats to this species are not known.

Conservation Actions:

It is not known if the species is present in any protected areas. Further studies are needed to determine the distribution, abundance, natural history and threats to this species.

Lesser Great Leaf-Nosed Bat *(Hipposideros turpis)*

Description:

HB: 66-79; T: 45-59; FA:67.2-80.

According to Hill (1936) *armiger* and *turpis* are essentially identical in their structural characteristics separated only by the much smaller size of *H. turpis* as compared to *H. armiger.* a situation not otherwise found in *Hipposideros.* The ears are large, broad, pointed and slightly thickened at the antitragal lobe,; they are haired for one third of their lenght. The nose leaf has tree lateral supplementary leaflets rather tan four found in *armiger.*

Range Description:

This species has a very disjunct distribution, with one subspecies *(H. t. turpis)* on the Yaeyama (Sakishima) Islands of Japan, including Iriomote, Ishigaki, Yonaguni, and Hateruma (Abe et al. 2005) and is not likely to occur more widely, and the other subspecies (*H. t. pendleburyi)* in **Southern Thailand** and Vietnam. The systematic status of the two subspecies remains unclear in view of its unusual distribution. Surveys have not found the species in areas between known locations.

Countries: Native:

Japan (Nansei-shoto); **Thailand;** Vietnam

Population:

In Japan, there are at least five colonies on Iriomote Island, with at least half of the total estimated population in a single colony that occurs on private land; on Yonaguni, there are no more than 1,000 individuals; on Ishigake, there are around 10 colonies, estimated at around 10,000 bats; and on Hateruma, there is no record of population size (this population is very distinct from other populations).

A study will be undertaken by Thong in Vietnam. The population is locally common in Cuc Phuong and Cat Ba island National Parks in Vietnam (N. Furey pers. comm.). Populations in **Thailand** are scattered (S. Bumrungri pers. comm.).

Population Trend: Decreasing

Habitat and Ecology:

During the day this species is found roosting in caves, often in large colonies of up to 1,000 individuals (Abe et al. 2005), and at night it forages in the surrounding woodland. The larger H. armiger feeds on large insects such as beetles (especially scarabeids), but smaller sympatric species take softer prey such as crickets, moths, Diptera, flying termites and Hymenoptera (Hutson et al. 2001). Four of the five known localities in central and northern Viet Nam are at karst sites and in protected areas (N. Furey pers. comm.), and it also occurs in karst areas in **Thailand** (S. Bumrungsri pers. comm.). The species uses disturbed forest, and it may persist in fragments of suitable habitat.

Systems: Terrestrial

Major Threat(s):

In general, human disturbance at roosting sites is the main threat. In Japan on the island of Iriomote the main breeding colonies are concentrated in two caves; most caves are subject to the risk of disturbance. The areas surrounding these caves are National Forest, but there has been pressure to develop the areas around the caves which would involve forest clearance. A nursing colony of 1,000 individuals in the western part of Iriomote Island has been disturbed by tourists since 2005. Development for agriculture and logging would remove key foraging areas and result in potential isolation of the cave roosts, causing declines of these principal populations. Similar declines and even extinctions have already been observed in other nearby islands. There has been similar pressure on Ishigaki to build an airport and holiday resort. Publicity about the plight of these bats resulted in increased public interest, including increased tourism to the breeding caves.

The species is affected by deforestation in **Thailand** (S. Bumrungsri pers. comm.), and in Vietnam there is ongoing decline in the quality of habitat through disturbance from tourism and habitat destruction (N. Furey pers. comm.). In Cat Ba in Vietnam, tourism disturbance is heavy (N. Furey pers. comm.). In **Thailand** areas around protected are-as are deforested and the areas themselves are degraded.

Conservation Actions:

In Japan the areas around some sites are national forests. There has been pressure to protect important roost sites and associated habitat. Fences were erected at the cave entrances to control tourists, and a small area originally designated for development was proposed as a reserve (Hutson et al. 2001). It is listed as Endangered (EN) in the Japanese Red List (2007).

In Vietnam three of the five known sites are protected and a new site is proposed (N. Furey pers. comm.).

The order Philodota comprises the eight species of pangolin. Pangolins are anteaters and have the powerful claws, elongated snout and long tongue seen in the other unrelated anteater species.

Family: Manidae:
Genus: Manis:

Sunda Pangolin *(Manis javanica)*

Range:

This species ranges over much of mainland Southeast Asia, from Southern Myanmar through Central and Southern Lao PDR, much of **Thailand,** Central and Southern Vietnam, Cambodia, to Peninsular Malaysia, to Sumatra, Java and adjacent islands (Indonesia) to Borneo (Malaysia, Indonesia, Brunei) (Schlitter 2005). The Northern and Western limits of its range are poorly known. It has been recorded from sea level up to 1,700 m asl.

This species is distributed in Southern Myanmar (Corbet and Hill 1992; Salter 1983), but is absent from lowland areas due to human agricultural expansion and hunting (Duckworth pers. comm.2006).
The species historically occurred throughout **Thailand** (Lekagul and McNeely 1977; Bain and Humphrey 1982; WCMC et al. 1999), but has since been lost from much of the lowland areas due to human agricultural expansion and hunting (J.W. Duckworth and R. Steinmitz pers. comm.2006).
In Vietnam, there are records from throughout the central and southern parts of the country. There are older records from Kontum Province, Tay Ninh Province and Quang Nam Province (Bourret 1942; Peenen et al. 1969). There are more recent records (summarised by (Newton 2007)) from: Ha Tinh Province (Timmins and Cuong 1999); Kein Giang and Ca Mau Provinces (in U Minh Thuong National Park) (CARE, 2004); Dong Nai, Bin Phuoc and Lam Dong Provinces (Cat Tien National Park) (Murphy and Phan 2001); Quang Binh (Le et al. 1997b); and Dak Lak (Le et al. 1997a; Dang et al. 1995).
The species is evidently widespread in Lao PDR, with recent records from a wide range of areas below around 600 m altitude, with the possibility that in Lao PDR the species is restricted to the Mekong plain and adjacent foothills to around 900 m, with a possible occurrence on the Bolaven Plateau, including Xe Pian National Biodiversity Conservation Area in the south at least as far north as Nam Kading (Deuve and Deuve 1963; Duckworth et al. 1999; J.W. Duckworth pers. comm. 2006).
The species is widespread in Peninsular Malaysia, primarily in forest, but also in gardens and plantations, including rubber (Medway 1977). It is also found on the island of Penang.
The species is still found in the wild in Singapore (CITES 2000; Lim and Ng 2007).
This species is reportedly widespread on Borneo, from sea level to 1,700 m on Mount Kinabalu in Sabah (Payne et al. 1985), although it appears to be absent from the extensive peat swamp forests of Sarawak (CITES 2000). In Sabah, the species is rarely seen, although is evidently widely distributed, being known by local people throughout Sabah (Davies and Payne 1982). The species is presumably present in Brunei (Medway 1977).

In Indonesia, the species is widespread on Sumatra, Java, Borneo, Kiau and Lingga archipelago, Bangka and Belitung,

In Indonesia, the species is widespread on Sumatra, Java, Borneo, Kiau and Lingga archipelago, Bangka and Belitung, Nias and Pagi islands, Bali, and adjacent islands (Corbet and Hill 1992).

In the northern part of the range, the species probably does not occur not above 600 m asl (J.W. Duckworth pers. comm. 2006). In Sabah it has been recorded up 1,700 m asl (Giman pers. comm. 2006). In Sumatra and Java it is found only up to about 400 m asl (Boeadi pers. comm. 2006), though there is a specimen in the Natural History Museum (London) at 1,500 m asl from Lombok (P. Newton pers. comm.). In the northern parts of its range, the s pecies overlaps with the range of Manis pentadactyla, which is generally said to occupy higher altitudinal habitats, though recent interviews with in Viet Nam suggest that they can be found in the same areas of forest, and that the differences between them are ecological, relating to diet and habitat use, rather than altitude (P. Newton pers. comm.).

Countries: Native:

Brunei Darussalam; Cambodia; Indonesia; Lao People's Democratic Republic; Malaysia; Myanmar; Singapore; **Thailand;** Vietnam

Population:

Virtually no information is available on population levels of any species of Asian pangolins. These species are rarely observed due to their secretive, solitary, and nocturnal habits, and there is not enough research on population densities or global population (WCMC et al. 1999; CITES 2000). There appear to be no comprehensive population estimates available, although records are reportedly rarer in many range states.

It is extremely rare in the northern part of its range (J.W. Duckworth pers. comm.), less so in the southern part (Boeadi pers. comm.). There have been massive declines in the northern part of its range (J.W. Duckworth pers. comm.). It is very common in parts of Singapore (J.W. Duckworth pers. comm.), where Lim and Ng (2007) estimated the range size of one individual, but made no estimate of total population size or density. In Sabah it is relatively common (Han and Giman pers. comm.).

In three areas of Vietnam where interviews were conducted (Khe Net Protected Area, Ke Go Nature Reserve and Song Thanh National Park), hunters reported that populations had massively declined in the last few decades, but particularly since about 1990 when the commercial trade in pangolins began to escalate (Newton 2007). In all three areas, the species was described as now being extremely rare. The intense biodiversity survey effort and extremely limited number of confirmed records of pangolins throughout Viet Nam?s protected areas adds weight to this observation (P. Newton pers. comm.).

In three separate areas within the range of Manis javanica in Lao PDR (Xe Pian, Dong Phou Veng and Khammouan Limestone NBCA), villagers have recently reported that pangolin populations have declined, in some areas to as little as one percent of the level 30 years ago due to hunting (Duckworth et al. 1999).

There is no recent data on the status of this species in Myanmar (WCMC et al. 1999).

M. javanica is considered threatened and becoming increasingly rare in **Thailand** (Bain and Humphrey 1982; Steinmitz pers. comm. 2006).

Population Trend: Decreasing

Description and Ecology:

The skin of its feet is granular, although there are pads on its front feet. Its tail has 30 scales. Head-body length : Up to 65 cm

Tail length : Up to 56 cm

Weight : up to 10 kg

In the past, this species has included the closely related Palawan pangolin *(Manis culionensis)* as both are in the subgenus Paramanis.It is closely related to the Chinese pangolin, although the Malayan species is larger, lighter in colour and has shorter fore claws. The Sunda pangolin's main predators are the tiger and the clouded leopard.

It has thick and powerful claws to dig into the soils in search of ant nests or to tear into termite mounds. Nocturnal in habits, the Pangolin will rest by day in burrows and tree holes. The species is also an adept climber, aided by its prehensile tail : it is also known to hide by day amongst the foliage of large epiphytes such as the Bird's Nest Fern. The nose is fleshy and possess a strong sense of smell. It has no teeth. Instead, its long, sticky tongue helps it collect ants and termites. Its body is covered by rows of scales and fibrous hair. Head-body length of pangolin is up to 65 cm, tail length is up to 56 cm and its weight is up to 10 kg. Males are larger than females. Its other common name is Scaly Anteater : it feeds wholly on ants and termites, which it locates by its strong sense of smell. It possesses thick, powerful claws which it uses to dig into the soil in search of ant nests or to tear into termite mounds. The insects are gathered with its long, sticky tongue and swallowed whole - the Pangolin has no need for teeth. It is estimated that on average a Pangolin might eat around 200,000 ants or termites per day.

Reproduction:

Pangolins give birth annually to one or two offspring. They breed in the autumn, and females give birth in the winter burrow. Parental care will be given for about three months. and the infants are carried astride the base of the mother's tail until such time as they are independent. Pangolins are sometimes found in pairs, but normally are solitary, nocturnal and behave timidly. They protect their soft underparts by rolling into balls when they feel threatened. They are strong diggers and will make burrows lined with vegetation for insulation near termite mounds and ant nests.

Threats:

Pangolins are hunted for their skins, scales and meat. Their parts are used for medicinal purposes. Scales are made into rings as charms against rheumatic fever, and meat is eaten by indigenous peoples. Skins are also used to make shoes. One of the main importers of pangolin skins from 1980–1985 was the United States of America.

Major Threat(s):

Threats to Asian pangolins include rapid loss and deterioration of available habitat and hunting for local use and for international trade in skins, scales, and meat. Evidence suggests that pangolins, in general, are able to adapt to modified habitats (e.g., secondary forests), provided their termite food source remains abundant and they are not unduly persecuted. However, whilst secondary habitats may be suitable, on the basis of hunters? reports in Vietnam and the evidence of Lim and Ng (2007) in Singapore, it seems that the availability of tree hollows, which is higher in undisturbed forest, is also extremely important for this species (P. Newton pers. comm.).

The species is intensively used, for its skin, meat and scales, and is evidently subject to heavy collection pressure in many parts of its range. The species may be harvested for local (i.e. national-level) use, as well as for international export either before or after processing. Observations in mainland Southeast Asia indicate that there is very heavy unofficial, or at least unrecorded, international trade in pangolins and pangolin products, although it is not possible at present to disentangle this trade from local use (WCMC et al. 1999; CITES 2000). The majority of utilization and trade data on pangolins in Asia do not distinguish reliably between the Asian species of pangolin *(Manis crassi-caudata, Manis culionensis, Manis javanica, Manis pentadactyla).* The two most commonly traded species (*Manis javanica, Manis pentadactyla)* have significant populations in some of the same countries (especially Lao PDR, Myanmar, and Vietnam), and because both species are imported into China, it is often not possible to determine which species is referred to in both local use and export (WCMCet al. 1999). The lack of accurate population and harvest data across this species? range, makes it difficult to assess the level and impact of harvest. The total from national use and international trade indicate that, at a minimum, several tens of thousands of animals were harvested and traded annually during the 1990s (WCMC et al. 1999). Figures, discussed in detail in Broad et al. (1988) and WCMC and IUCN SSC (1992), indicate that trade of this magnitude also took place at least up until the mid-1980s (e.g. over 185,000 skins reported in international trade by CITES in the period 1980-85 alone). An estimate in the late 1950s and early 1960s indicates that scales of some 10,000 pangolins *(Manis javanica)* per year were exported from Borneo (Harrison and Loh 1965).

animals are brought in from Indonesia and large numbers of live animals to be exported to China have been seized (GMA, Indonesia Workshop 2006). Indonesia has been illegally exporting great numbers of live animals, some of which come from east Kalimantan (Semiadi pers. comm. 2006).

which can smell it out, making hunting much more effective, such pangolin dogs are highly valued (up to USD 2000) (J.W. Duckworth pers. comm. 2006).

Every hunter interviewed in Vietnam (n = 84) reported that they now sell all pangolins that they catch (P. Newton pers. comm.). Prices are so high that local, subsistence use of pangolins for either meat or their scales has completely halted in favour or selling to the national/international trade (P. Newton pers. comm.). The only occasions on which a hunter might eat a pangolin is if it is already dead when they retrieve it from a trap, then they would use the meat and sell the scales (P. Newton pers. comm.). The price per kg of pangolin (in Vietnam, at least) has escalated rapidly (at a rate greater than that of annual inflation) since the commercial trade in wild pangolins began to expand in about 1990 (P. Newton pers. comm.). Prices paid to hunters now exceed US$95 per kg (Viet Nam, P. Newton pers. comm.); US$45 per kg (Cambodia, C. Phallika pers. comm. to P. Newton) and US$17 per kg (Indonesia, D. Martyr pers. comm. to P. Newton).

Conservation Actions:

This species is listed on CITES Appendix II; a zero annual export quota has been established for specimens removed from the wild and traded for primarily commercial purposes. It is protected by national legislation in Bangladesh, Indonesia, Malaysia, Myanmar, Philippines, Singapore and **Thailand.** It is found in protected areas in its range, but has been hunted out of some protected areas in its range, especially in **Thailand** (Anak pers. comm.). Much more effective enforcement of existing laws is critical for the conservation of this species (J.W. Duckworth pers. comm.). Some protected areas in Vietnam are heavily trapped for this and other species.

In Singapore, the species is protected under the Wild Animals and Birds Act (Domestic Law) 1904 and Endangered Species Act (Import/Export, CITES Law).

The species is legally protected in Vietnam for *Manis javanica.*

Manis javanica has been protected in Indonesia since 1931, under Wildlife Protection Ordinance No. 266 of 1931 (promulgated by the Dutch administration), as well as under Act. No. 5 of 1990, regarding Conservation of Natural Resources and Their Ecosystems; Decree of the Minister of Forestry No. 301/kpts-II/1991 and Decree of the Minister of Forestry No. 822/kpts-II/1992.

Manis javanica is completely protected in west Malaysia under the Protection of Wild Life Act, 1972; a protected species, banned from local trade, in Sarawak under the Wildlife Protection Ordinance 1998; and protected in Sabah under the Wildlife Conservation Bill, 1997.

In accordance with the Protection of Wildlife, Wild Plant and Conservation of Natural Areas Act 15(A), M. javanica is categorized as a Completely Protected Animal in Myanmar.

In Thailand, all Manis spp. are classified as Protected Wild Animals under the 1992 Wild Animals Reservation and Protection Act B.E. 2535.

The legal status of pangolins in Lao PDR is unclear, as a result of internal contradictions in Lao PDR laws applicable to wildlife and wildlife trading. However, Provincial and District Agricultural and Forestry Offices in Lao PDR have been confiscating large numbers of pangolins, so there is evidently a perceived legal basis for doing so (WCMC et al. 1999).

Chinese Pangolin *(Manis pentadactyla)*

Range Description:

This species occurs in the Himalayan foothills in Eastern Nepal, Bhutan and Northern India, Northeastern Bangladesh, across Myanmar to northern Lao PDR and Northern Vietnam, **Northern and Northeastern Thailand,** and through southern China (South of the Chiangjiang) to Hainan and Taiwan. The limits of its range are poorly known, and complicated by high levels of exploitation. The species exists at high altitudes, especially in the Southern and Western parts of its range, possibly occurring at much lower altitudes in northeast. Its latitudinal range is thought likely to overlap considerably with that of *Manis javanica*, with *Manis pentadactyla* tending to occur in hills and mountains and the latter more generally found at lower altitudes, though recent interviews with in Vietnam suggest that the two species can be found in the same areas of forest, and that the differences between them are ecological, relating to diet and habitat use, rather than altitude (P. Newton pers. comm.).

The species has been recorded in Northeastern India from Sikkim eastward (Tikader 1983). The species occurs in eastern Nepal and Bhutan at the foothills of the Himalayas, apparently confined to elevations below approximately 1,500 m in Nepal (Frick 1968; Mitchell 1975).

The species has been recorded in North and Central Lao PDR, however, there are too few locality records to determine the geographic and altitudinal range of the species in the country with any accuracy (Duckworth et al. 1999; Timmins and Evans 1996).

The species occurs throughout southeast China from the Southern border as far north as Changjiang (the Yangtze River), including on the island of Chusan at the mouth of the Changjiang (Allen and Coolidge 1940). This species is distributed widely in China in the provinces of Sichuan, Guizhou, Yunnan, Anhui, Jiangsu, Zhejiang, Jiangxi, Hunan, Guangdong, and Fujian, and in the Autonomous Regions of Hainan Island, Guangxi Zhuang, and Tibet (Zhang et al. 1997). It is recorded in several sites in central and northeast New Territories, as well as on Hong Kong (Lantau Island), although not on the smaller outlying islands (Reels 1996).

On Taiwan, the species occurs on the periphery of the Central Mountain Range, the Western Foothill Range, the Taoyuan Tableland, the Ouluanpi Tableland, the East Coast Mountain Range, the Tatun Volcano Group, Taipei Basin, Puli Basin, and the Pingtun Plain (Chao Jung-Tai 1989; Chao Jung-Tai et al. 2005). The upper limit of occurrence is around 2,000 m asl (Chao Jung-Tai 1989).

The species is probably widespread in northern Myanmar, although there are few records and the exact distribution is not well known (Salter 1983; Corbet and Hill 1992; J.W. Duckworth pers. comm. 2006).

The only records of the species in Thailand are from: Doi Inthanon (formerly Doi Angka) in Changwat, Chiang Mai (Northern Thailand), sometime in 1937 and 1939 (Allen and Coolidge 1940); Doi Sutep, Chiangmai (Northern Thailand) in 1901; and Mukdaharn in Northeastern Thailand.

All records of the species in Vietnam are from the Northern half of the country, as far south as Quang Tri Province, up to 1,000 m asl (Bourret 1942; Peenen et al. 1969; Do Tuoc pers. comm. 2006; P. Newton pers. comm.).

Countries: Native:

Bangladesh; Bhutan; China; Hong Kong; India; Lao People's Democratic Republic; Myanmar; Nepal; Taiwan, Province of China; **Thailand;** Vietnam

Population:

Virtually no information is available on population levels of any species of Asian pangolins. These species are rarely observed due to their secretive, solitary, and nocturnal habits, and there is not enough research on population densities or global population (WCMC et al. 1999; CITES 2000). There have been few documented records, and therefore there is very little information available on the population status anywhere in the species' range, but it appears to be decreasing over much of its range.

A 1993 survey conducted in the Royal Nagarjung Forest in Kathmandu, Nepal, determined that there was a healthy population there, however, the general trend elsewhere in Nepal was dramatically declining, due to increased access to hunting areas and loss of habitat (Gurung 1996).

This species was reported in the 1980s as common in the undisturbed hill forests of Arunachal Pradesh, however, little is known about the total population in India (Tikader 1983; Zoological Survey of India 1994).

Reports from the late 1980s and early 1990s suggest that the total population of this species in Taiwan was decreasing due to poaching and habitat destruction (Chao Jung-Tai 1989; Chao Jung-Tai et al. 2005).

The species is very rare in Vietnam (Do Tuoc pers. comm. 2006). There is a ?confirmed record? we found is for Ba Na National Park, which straddles the provinces of Quang Nam and Da Nang (Frontier Vietnam, 1994). Hunters in Viet Nam reported that they still find Manis pentadactyla in Cuc Phuong National Park (in Quang Binh province), in Khe Net Nature Reserve, and in Ke Go Nature Reserve (Ha Tinh province) (P. Newton pers. comm.). However, all hunters reported that the species is extremely rare, and that populations have declined dramatically in the last two decades (P. Newton pers. comm.). In 2007, P. Newton (pers. comm.) found recent (i.e., less than 1 month old) signs of pangolin activity (recently-dug burrows) in Cuc Phuong National Park ? these were almost certainly those of Manis pentadactyla, as the park is well outside of the range of Manis javanica. In Khe Net and Ke Go, hunters reported that numbers of Manis pentadactyla were lower than those of Manis javanica, probably because the former is easier to hunt. If this is the case, then in places where both species occur, populations of Manis pentadactyla are likely to be more heavily depleted.

The species has been so heavily hunted in Lao PDR that field sightings are exceptionally rare, and the only recent field sightings (during 1994-1995) was of an individual in Nam Theun Extension PNBCA (Proposed National Biodiversity Conservation Area) and one seen in a village in Nakai-Nam Theun NBCA during the same period (Duckworth et al. 1999). Manis pentadactyla is less often recorded in trade in Lao PDR than Manis javanica, perhaps due to its lower abundance in the wild (WCMC et al. 1999).
Population Trend: Decreasing

Habitat and Ecology:

This species is found in a wide range of habitats, including primary and secondary tropical forests, limestone forests, bamboo forests, grasslands and agricultural fields, and grasslands (Chao Jung-Tai 1989; Gurung 1996). The species digs its own burrows, or enlarges passages made by termites. Indications are that home range is relatively large, although concrete data are lacking (Heath and Vanderlip 1988). Hunters interviewed in Vietnam indicate that *Manis pentadactyla* digs its own, long burrows underground, which they use both to eat termites and in which to sleep (P. Newton pers. comm.). The main way in which they catch this species is to dig them out of these burrows, so this evidence is probably reliable (P. Newton pers. comm.).

This species is solitary, nocturnal (sometimes crepuscular), and largely terrestrial although it is fully capable of climbing trees and, like other pangolins, swims well (Heath and Vanderlip 1988; Chao Jung-Tai 1989). In addition, little is known of the species life history, although in China and Taiwan, young (normally one, occasionally two) are reportedly born in spring (Allen and Coolidge 1940; Chao Jung-Tai 1989). Hunters in Vietnam reported that they never find this species in trees, and so it seems likely that it is far more terrestrial than the more arboreal Manis javanica (P. Newton pers. comm.).

The diet consists of ants and termites (Heath and Vanderlip 1988). It has been noted that in China, there appears to be a close correlation between the its distribution and the distribution of two termite species *(Coptotermes formosanus* and *Termes (Cyclotermes) formosanus)* that are assumed to form a major component of its diet (Allen and Coolidge 1940).

Systems: Terrestrial The Chinese pangolin belongs to the burrowing family. It can dig up to 8 ft (2.5 m) deep in the ground with its strong and clawed forefeet, in four to five minutes. Once it enters the burrow, it blocks the opening. Some Chinese pangolins occupy burrows of other animals, as well. Its body mass is about 3.6 kg (8.0 lbs).

The Chinese pangolin appears like a scaly anteater. From head to body, it measures around 60 cm (24 in) and its tail measures about 18 cm (7 inches). A mature Chinese pangolin weighs about 2.4 kg (82.7 oz). A newborn pangolin weighs about 93 g (3.3 oz). It has 18 rows of overlapping scales accompanied by hair, a rare combination in mammals. It has a small, narrow mouth and a little, pointed head. Its nose is plump, with nostrils at its end. This is a bronze-colored animal with a round body equipped with extremely sharp claws.

The Chinese pangolin found in Nepal reproduces in April and May when the weather is a bit warm. The female gives birth to a single young at a time, and it weighs about 1 lb (450 g) and its length is about 45 cm (18 in). The young also has scales which remain very soft for two days. Although the young pangolin can walk on its very first day, the mother carries it on her back or tail. If the mother feels threatened, she immediately folds her baby onto her stomach with the help of her tail. Male pangolins have been noticed to show extraordinary parental instinct and allow the female and baby share the burrow.

Chinese pangolins are rather secretive, nocturnal creatures. They move very slowly and are known for their non aggressive behavior. Their hard scales work as a protective cover from predators, and when they feel endangered, they curl themselves into balls. Chinese pangolins are mainly terrestrial animals, but are observed in forests up to about 20 feet above the ground.

They mainly eat insects, such as termites and ants. Their sharp claws help them in digging up the ants and termite mounds and with the help of their sticky, long (25 cm) tongues, they can draw their prey into their mouths.

In Vietnam and Hong Kong, Chinese pangolins are considered a delicacy, and they are hunted on a large scale only for this purpose. Now, Chinese pangolins are being protected in the forests where they are generally found. Factors such as habitat destruction and hunting constantly challenge their survival. Since the forests they inhabit are difficult to patrol, hunters can hunt these animals without being caught.

Major Threat(s):

Threats to Asian pangolins include rapid loss and deterioration of available habitat and hunting for local use and for international trade in skins, scales, and meat. Evidence suggests that pangolins, in general, are able to adapt to modified habitats (e.g., secondary forests), provided their termite food source remains abundant and they are not unduly persecuted. The species is intensively used, for its skin, meat and scales, and is evidently subject to heavy collection pressure in many parts of its range. The species may be harvested for local (i.e. national-level) use, or for international export either before or after processing. Observations in mainland Southeast Asia indicate that there is very heavy unofficial, or at least unrecorded, international trade in pangolins and pangolin products. The species trade levels are significant across its range, although precise estimates are unknown (CITES 2000).

Of particular significance is that *Manis pentadactyla* is reported to be an easier species to locate and hunt in the wild (P. Newton pers. comm.). This is because it is more terrestrial, and is thus: easier to track their scent using specialised hunting dogs (the scent of *Manis javanica* is often lost at points at which the animal climbed a tree); and b) has conspicuous soil burrows that are more easily accessed than the tree hollows favoured by *Manis javanica* (P. Newton pers. comm.). For these reasons, the hunting threat to *Manis pentadactyla* is perhaps even greater than that to *Manis javanica* (P. Newton pers. comm.).

Every hunter interviewed in Vietnam (N = 84) reported that they now sell all pangolins that they catch (P. Newton pers. comm.). Prices are so high that local, subsistence use of pangolins for either meat or their scales has completely halted in favour or selling to the national/international trade (P. Newton pers. comm.). The only occasions on which a hunter might eat a pangolin is if it is already dead when they retrieve it from a trap ? then they would use the meat and sell the scales (P. Newton pers. comm.). The price per kg of pangolin (in Vietnam, at least) has escalated rapidly (at a rate greater than that of annual inflation) since the commercial trade in wild pangolins began to expand in about 1990 (P. Newton pers. comm.). Prices paid to hunters now exceed US$95 per kg (Vietnam, P. Newton pers. comm.).

Conservation Actions:

This species is listed on CITES Appendix II; a zero annual export quota has been established for specimens removed from the wild and traded for primarily commercial purposes. It is protected by national or sub national legislation in Bangladesh, China, India, Lao, Myanmar, Nepal, Taiwan, **Thailand,** and Vietnam. This wide ranging species is present in some protected areas, but protected area designation alone is not sufficient to protect this species. Greater enforcement and management to prevent poaching in protected areas is urgently needed. Large seizures of illegal animals do occur, but the trade continues largely unabated.

In Bangladesh, all pangolins are legally protected.

On Taiwan, all Manis species have been protected since August 1990 under the 1989 Wildlife Conservation Law.

This species is listed as a Class II protected species in China's Wild Animal Protection Law (1989), and also as a Class II protected species in China in the Regulations on the Conservation and Management of Wild Resources of Medicinal Plants and Animals (1987).

In Thailand, all Manis species are classified as Protected Wild Animals under the 1992 Wild Animals Reservation and Protection Act B.E. 2535.

In India, this species is completely protected, as it is included in Schedule I of the Wildlife Protection Act 1972.

Hunting of this species is prohibited in Nepal (Gaski and Hemley 1991).

The legal status of pangolins in Lao PDR is unclear, as a result of internal contradictions in Lao PDR laws applicable to wildlife and wildlife trading. However, Provincial and District Agricultural and Forestry Offices in Lao PDR have been confiscating large numbers of pangolins, so there is evidently a perceived legal basis for doing so (WCMC et al. 1999).

Minke Whale *(Balaenoptera acutorostrata)*

Description & Behavior:

There are two species of minke whales: minke whales, aka Common or Northern minke whales, *Balaenoptera acutorostrata* (Lacépède, 1804, acutorostrata means 'sharp snouted') and Antarctic or Southern minke whales, *Balaenoptera bonaerensis* (Burmeister, 1867).

Minke whales are the smallest of the baleen whales with 50-70 throat grooves. They are part of the largest group of baleen whales called the rorquals (Family Balaenopteridae, the family that includes: blue whales, Bryde's whales, fin whales, humpback whales, minke whales, and Sei whales). These long, slender whales are much more streamlined than other whales. They have a pointed snout that is distinctively triangular, narrow, and pointed (hence its nicknames "sharp-headed finner" and "little piked whale"), paired blowholes, and a broad, flat rostrum (upper part of the head). The throat grooves, in addition to streamlining the shape of the whale, allow the throat area (called the cavum ventrale) to expand tremendously during feeding.

Minkes are the most abundant baleen whale and have a characteristic white band on each flipper (this is absent on the southern minke whales though), contrasting with its very dark gray top colour, sometimes with pale trapezoidal stripes behind their flippers on the top. They have 2 blowholes, like all baleen whales. Minke whales grow to be about 7.8-9 m long, weighing about 5,400-6,800 kg. Females are about 0.6 m longer than males, as with all baleen whales. The largest minke whale measured was about 10.5 m long weighing 8,600 kg.

There are three subspecies of common or northern minke whales, *B. acutorostrata:*

1. North Atlantic minke whale, *B. acutorostrata acutorostrata (*Lacépède, 1804)
2. North Pacific or Scommon's minke whale, *B. acutorostrata scammoni (*Deméré, 1986)
3. dwarf-form or dwarf minke whale, *B. acutorostrata* subsp

Minke whales are stocky, having a layer of blubber several inches thick. They have 50-70 throat grooves, running from the chin to the mid-section. The minke whale has two long flippers (up to 1/8 of the body size), a small dorsal fin, and a series of small ridges along the its back near the flukes (tail). Minke whales either travel singly or congregated in small pods of about 2-3 whales. Minke whales can dive for up to 20-25 minutes, but usually make shorter dives, lasting about 10-12 minutes. Just before diving, minke whales arch their back to a great degree, but the flukes do not rise out of the water. At rest, minke whales spout (breathe) about 5-6 times per minute. The spout of the minke whale is a very low, almost inconspicuous stream that rises up to 2 m above the water. Minke whales begin exhaling before they reach the surface, which minimizes the blow. Minke whales makes very loud sounds, up to 152 decibels (as loud as a jet taking off). They make series (trains) of grunts, thuds, and raspy sounds, usually in the 100-200 Hertz range. These sounds may be used in communication with other minke whales and in echolocation .

World Range & Habitat:

Range Description: The Common Minke Whale is a cosmopolitan species found in all oceans and in virtually all latitudes, from 65°S to 80°N. In parts of its range it is very abundant, in other parts much less so. Its migration patterns are poorly known. It occurs in the North Atlantic, the North Pacific, and the Southern Hemisphere, but is not known from the northern Indian Ocean.

North Atlantic:

In summer, Minke Whales are common throughout the northern North Atlantic as far north as Baffin Bay, Greenland Sea, Svalbard (Norway), Franz Josef Land and Novaya Zemlya (Russian Federation), and as far south as 40°N (New Jersey) on the US east coast (Anon. 2005a), and as far south as the Hebrides (northwest British Isles) and the central North Sea in the east. In the mid-Atlantic summer concentrations of minke whales occur to at least as far south as 50°N (Sigurjónsson et al. 1991). It is likely that at least a part of the Minke Whale population over-winters in the summer range, but there has been very little observation effort in winter to confirm this.

Minke Whales also occur south of this range in the South-eastern North Atlantic, but with no obvious seasonality, and are not common, with the exception of the Canary Islands, where they appear to be frequent year-round (Van Waerebeek et al. 1999). There have been occasional sightings (Aguilar et al. 1983) and strandings (Van Waerebeek et al. 1999) off Spain and Portugal, Western Sahara, Mauritania and Senegal. Minke whales are rare in the Azores and not recorded from Madeira. The minke whale is considered a "visitor species" in the Mediterranean (average 1 record per year) with one vagrant recorded in the Black Sea (Reeves and Notabartolo di Sciara 2006).

There are very few winter records, but a summary by Mitchell (1991) indicates that they do occur in winter near Bermuda, the Bahamas and the Antilles, and the US coast south of 40°N. Ten strandings have been recorded in the Gulf of Mexico (Jefferson and Schiro 1997) but hardly any live sightings. A Norwegian winter expedition sent to the tropical Atlantic in 1989/90 to "find the breeding grounds of the minke whale" encountered just two minke whales, at 20°N and 10°N off West Africa in December (Folklow and Blix 1991).

North Pacific:

Minke whales occur in summer right across the North Pacific north of about 30°N in summer, with a tendency for the distribution to shift northward in high summer. They are particularly abundant in the Okhotsk Sea in August (Miyashita et al. 1995), and also occur in the Bering Sea, around the Aleutian Islands and in the Gulf of Alaska (Moore et al. 2002, Zerbini et al. 2006) and the Chukchi Sea (Ivashin and Votrogov 1981). In the eastern North Pacific, there appears to be a year-round population off California and Baja California and in the Gulf of California, and minke whales occur in summer off Oregon, Washington and British Columbia (Anon. 2003). They have been seen off Hawaii but are not common there (Anon. 2005b).

In the western North Pacific, there are at least two distinct subpopulations: the so-called "J stock," an autumn breeding population that occurs in the Yellow Sea, East China Sea and Sea of Japan, with some penetration into the Okhotsk Sea in summer; and the O-stock which, like most baleen whales, breeds in winter, and occurs in summer in the north-western Pacific, including the northeastern coasts of Japan, and in the Okhotsk Sea (Omura and Sakiura 1956, Kato 1992).

The winter distribution is poorly known. Japanese expeditions to look for wintering grounds in the southwestern North Pacific during 1993-95 failed to locate any minke whales (Miyashita et al. 1996). The timing of the arrival of minke whales in Korean and western Japanese waters is suggestive of migration from the south in spring and return in autumn (Ohsumi 1983). The wintering area in the eastern North Pacific has been identified acoustically to be primarily between 15 and 35 degrees N latitude (Rankin and Barlow 2005).

Southern hemisphere:

Much of the data on the occurrence of minke whales in the Southern Hemisphere is ambiguous with respect to identification as *B. acutorostrata* or *B. bonaerensis,* because the two species are partially sympatric. Japanese scouting vessel data indicated high abundance of minke whales in November between 10°-30°S in the central South Pacific and in much of the eastern and southern Indian Ocean down to 50°S (Miyashita et al. 1995), but their species identity is unclear. The limited information available from surveys in low and middle latitudes from the 1987/88 season onwards, when the two species were reliably distinguished, indicates that most of the minke whales are *B. bonaerensis* (Nishiwaki et al. 1991), probably on route to the Antarctic from (as yet unknown) low-latitude breeding grounds, but also that *B. acutorostrata* is present in these latitudes.

Dwarf minke whales occur at higher latitudes but are much less common than *B. bonaerensis*. Of more than 1,700 minke whales taken by Antarctic pelagic whaling from the 1987/88 to the 1992/93 season (when the two species have been reliably distinguished), only 16 were dwarf minke whales (Nishiwaki et al. 2005). One was taken at 65°S and the remainder at between 55-62°S, the northern limit of whaling operations.

In coastal waters, dwarf minke whales have been recorded off most of the South Atlantic coast of South America (Baldas and Castello 1986, Zerbini et al. 1996), in the Beagle Channel (Chile/Argentina) (Acevedo et al. 2005), off South Africa (Best 1985), Australia (Arnold et al. 1987, Bannister et al. 1996, Arnold 1997), New Zealand (Dawson and Slooten 1990), and New Caledonia (Garrigue and Greaves 2001). The most northerly confirmed Southern Hemisphere record is from 2°S, off the northern coast of Brazil (Magalhães et al. 2007). Three dwarf minke whales were caught in whaling operations off Costinha, Brazil in 1980 (along with 900 Antarctic minke whales) (da Rocha and Braga 1982).

Countries: Native:

Anguilla; Antarctica; Antigua and Barbuda; Argentina; Australia; Bahamas; Bangladesh; Belgium; Bermuda; Brazil; Canada; Cape Verde; Chile; China; Cuba; Denmark; Dominica; Dominican Republic; Ecuador; Faroe Islands; France; French Guiana; Gambia; Greece; Greenland; Guadeloupe; Iceland; Indonesia; Ireland; Israel; Italy; Japan; Korea, Democratic People's Republic of; Korea, Republic of; Mauritania; Mexico; Morocco; Mozambique; Netherlands; Netherlands Antilles; New Caledonia; New Zealand; Norway; Papua New Guinea; Portugal; Puerto Rico; Russian Federation; Saint Pierre and Miquelon; Senegal; South Africa; Spain; Svalbard and Jan Mayen; Sweden; Taiwan, Province of China; **Thailand**; Tunisia; Turks and Caicos Islands; United Kingdom; United States; Uruguay; Virgin Islands, British; Virgin Islands, U.S.; Western Sahara

Feeding Behavior:

Minke whales, Balaenoptera bonaerensis, (like all baleen whales) are seasonal feeders and carnivores. They sieve through the ocean water with their baleen . They filter out small plankton, krill, and small fish, even chasing schools of sardines, anchovies, cod, herring, and capelin. They have the same diet as blue whales. The baleen plates in their jaws have about 300 pairs of short, smooth plates. Their largest plates are less than 30 cm long and 13 cm wide. Their fine-textured baleen bristles are fringed and white.

Life History:

Minke whale breeding occurs mostly in the late winter to early spring while near the surface and in warm waters. Their gestation period is about 10 months and their calves are born near the surface. Their newborns instinctively swim to the surface within 10 seconds for their first breath; helped by their mother using her flippers. Within 30 minutes of their birth, baby minkes can swim. Newborn calves are about 2.8 m long and weigh about 454 kg. Baby minkes, like all mammals, feed on their mother's milk until weaning. Mother minkes and calves may stay together for a year or longer. Minke whales reach puberty at about 2 years of age. Minke whales live for over 50 years. It is estimated that there are about 800,000 minke whales worldwide. The Common Minke Whale occurs in both coastal and offshore waters and exploits a variety of prey species in different areas according to availability.

In the Northern Hemisphere some populations migrate to higher latitudes in summer, but minke whales are also found year-round in some areas. With the exception of the Sea of Japan – Yellow Sea – East China Sea population, conception and birth occur in winter. Most animals occur singly, and few in groups of more than two. In the North Atlantic, studies in the Barents and Norwegian Seas showed that minke whale diet varied greatly between areas and years, being dominated by krill in the northern areas, but by herring or capelin in other areas according to what was most abundant that year, with gadoids being taken when herring and capelin were scarce (Lindstrøm and Haug 2002). In the North Sea the diet consisted almost exclusively of sandeel (data from one year only). Minke whales taken off Iceland in 2003-04 contained mainly sandeel, with some capelin and gadoids (Víkingsson et al. 2006). Minke whales caught off Newfoundland during 1966-72 contained mainly capelin (Mallotus villosus) (Mitchell 1974). In the Northwest Pacific, Lindstrøm et al. (1998) found that krill Euphausia pacifica dominated the diet in coastal areas and in the Okhotsk Sea, while in the offshore Pacific, Pacific saury Cololabis saira dominated.

Systems: Marine

Conservation Status & Comments:

The story of this whale's name illustrates its blighted history. Minke was an 18th-century Norwegian whaler, Infamous for regularly breaking the rules concerning the sizes (and therefore species) of whales that he was permitted at that time to hunt. Soon all the small whales became known as "Minke's whales". Eventually, it was formally adopted as the name for this small species. In fact, because they are such a small whale, the whaling industry generally ignored the minkes until quite recently. As the larger whales became more scarce (and gained protected status), so minke whales became more economically attractive. In the North Atlantic, from the 1920s, whaling for this species has been conducted along the coast of Norway. When whale populations dwindled in the late 1940s, the International Whaling Commission (IWC) was established to manage whale stocks. The IWC announced how many whales could be taken in any year, however, populations continued to decline. In 1985-1986 a worldwide moratorium on whaling was put in place, banning all whaling. Norway, however, "took out a reservation", meaning that they could continue by IWC rules to legally hunt North Atlantic minke whales and, in 1993 and 1994, they killed several hundred. This has caused great dismay in other countries, including the U.K., who have officially objected to this hunt. In UK waters it is illegal to intentionally kill, injure, capture or harass any species of whale or dolphin. However, there is little in international law that can be done about hunting, except control of exports which is covered by CITES (the Convention for International Trade in Endangered Species). This is important because the main consumers of whale meat are still the Japanese and, if Norway could legally export whale meat to Japan, this might further promote whaling.

Sei Whale *(Balaenoptera borealis)*

Description:

The Sei whale is the third-largest Balaenopteridae, after the blue whale (up to 180 tonnes, 200 tons) and the fin whale (up to 70 tonnes, 77 tons)] Mature adults typically measure between 12–15 meters (39–49 ft) and weigh 20–30 tonnes (20–30 long tons; 22–33 short tons). The southern Sei whale is larger than the northern. Females are considerably larger than males. The largest known Sei whale measured 20 meters (66 ft), and weighed between 40–45 tonnes (39–44 long tons; 44–50 short tons). The largest specimens taken off Iceland were slightly longer than 16 meters (52 ft).[21] At birth, a calf typically measures 4–5 meters (13–16 ft) in length.

The Sei whale (/ˈseɪ/ or /ˈsaɪ/), Balaenoptera borealis, is a baleen whale, the third-largest rorqual after the blue whale and the fin whale. It inhabits most oceans and adjoining seas, and prefers deep offshore waters. It avoids polar and tropical waters and semi-enclosed bodies of water. The Sei whale migrates annually from cool and sub polar waters in summer to winter in temperate and subtropical waters.

Reaching 20 metres (66 ft) long and weighing as much as 28 tonnes (28 long tons; 31 short tons), the Sei whale daily consumes an average of 900 kilograms (2,000 lb) of food, primarily copepods, krill, and other zooplankton. It is among the fastest of all cetaceans, and can reach speeds of up to 50 kilometres per hour (31 mph) (27 knots) over short distances. The whale's name comes from the Norwegian word for pollock, a fish that appears off the coast of Norway at the same time of the year as the Sei whale.

Following large-scale commercial whaling during the late-nineteenth and twentieth centuries, when over 255,000 whales were taken, the Sei whale is now internationally protected, although limited hunting occurs under controversial research programmes conducted by Iceland and Japan. As of 2008, its worldwide population was about 80,000, nearly a third of its pre-whaling population.

Etymology

The species was first officially described by French naturalist René Primevère Lesson in 1828, but an earlier description was given by Karl Rudolphi in 1822 (although he assumed it was a minke whale, Balaenoptera acutorostrata), leading to occasional references to Sei whales as Rudolphi's rorqual. Additional names include Pollack whale, coalfish whale, sardine whale, or Japan finner. Additionally, it has been referred to as the lesser fin whale because it somewhat resembles the fin whale. The American naturalist Roy Chapman Andrews compared the Sei whale to the cheetah, because it can swim at great speeds "for a few hundred yards", but it "soon tires if the chase is long" and "does not have the strength and staying power of its larger relatives".

Sei is the Norwegian word for pollock, also referred to as coalfish, a close relative of codfish. Sei whales appeared off the coast of Norway at the same time as the pollock, both coming to feed on the abundant plankton. The specific name is the Latin word borealis, meaning northern. In the Pacific, the whale has been called the Japan finner; "finner" was a common term used to refer to rorquals. In Japanese, the whale was called iwashi kujira, or sardine whale, named for a fish that the whale has been observed to eat in the Pacific.

Anatomy:

The whale's body is typically a dark steel grey with irregular light grey to white markings on the ventral surface, or towards the front of the lower body. The whale has a series of 32–60 pleats or grooves along the bottom of the body that allow the throat area to expand greatly during feeding. The rostrum is pointed and the pectoral fins are relatively short, only 9%–10% of body length, and pointed at the tips. It has a single ridge extending from the tip of the rostrum to the paired blowholes that are a distinctive characteristic of baleen whales.

The whale's skin is often marked by pits or wounds, which after healing become white scars. These are believed to be caused by ectoparasitic copepods (Penella spp.), lampreys (family Petromyzontidae), or possibly "cookie-cutter" sharks (Isistius brasiliensis). It has a tall, sickle-shaped dorsal fin that ranges in height from 25–61 centimetres (9.8–24 in), about two-thirds of the way back from the tip of the rostrum. Dorsal fin shape, pigmentation pattern, and scarring have been used to a limited extent in photo-identification studies. The tail is thick and the fluke, or lobe, is relatively small in relation to the size of the whale's body.

A close-up view of baleen plates. The plates are used to strain food from the water

Adults have 300–380 ashy-black baleen plates on each side of the mouth, each about 48 centimetres (19 in) long. Each plate is made of fingernail-like keratin that frays out into whitish fine hairs on the ends inside the mouth near the tongue. The sei's very fine baleen bristles, about 0.1 millimetres (0.004 in) are the most reliable characteristic that distinguishes it from other baleen whales.

The Sei whale looks similar to other large baleen whales. The best way to distinguish between it and Bryde's whale, apart from differences in baleen plates, is by the presence of lateral ridges on the dorsal surface of the Bryde's whale's rostrum. Large individuals can be confused with fin whales, unless the fin whale's asymmetrical head coloration is clearly seen. The fin whale's lower jaw's right side is white, and the left side is grey. When viewed from the side, the upper edge of the sei's head has a small arch between the tip of the rostrum and eye, while the fin whale's profile is relatively flat.

Life history:

Sei whales usually travel alone or in groups of up to six individuals. Larger groups may assemble at particularly abundant feeding grounds. Very little is known about their social structure. Males and females may bond, but this is uncertain.

The Sei whale is among the fastest cetaceans. It can reach speeds of up to 50 kilometres per hour (27 kn) over short distances. However, it is not a remarkable diver, reaching relatively shallow depths for five to fifteen minutes. Between dives, the whale surfaces for a few minutes, remaining visible in clear, calm waters, with blows occurring at intervals of about 40–60 seconds. Unlike the fin whale, the Sei whale tends not to rise high out of the water as it dives. The blowholes and dorsal fin are often exposed above the water surface simultaneously. The whale almost never extends its flukes above the surface, and it rarely breaches.

Feeding:

Krill, shrimp-like marine invertebrate animals, are one of the Sei whale's primary foods.

This rorqual is a filter feeder, using its baleen plates to obtain its food by opening its mouth, engulfing large amounts of the water containing the food, then straining the water out through the baleen, trapping any food items inside its mouth.

The Sei whale feeds near the surface of the ocean, swimming on its side through swarms of prey to obtain its average of about 900 kilograms (2,000 lb) of food each day. For an animal of its size, for the most part, its preferred foods lie unusually relatively low in the food chain, including zooplankton and small fish. The whale's diet preferences has been determined from stomach analyses, direct observation of feeding behaviour., and analyzing fecal matter collected near them, which appears as a dilute brown cloud. The feces are collected in nets and DNA is separated, individually identified, and matched with known species. The whale competes for food against clupeid fish (herring and its relatives), basking sharks, and right whales.

In the North Atlantic, it feeds primarily on calanoid copepods, specifically *Calanus finmarchicus*, with a secondary preference for euphausiids, in particular *Meganyctiphanes norvegica* and *Thysanoessa inermis*. In the North Pacific, it feeds on similar zooplankton, including the copepod species *Neocalanus cristatus, Neocalanus plumchrus,* and *Calanus pacificus,* and euphausid species *Euphausia pacifica, Thysanoessa inermis, Thysanoessa longipes,* and *T. hysanoessa spinifera.* In addition, it eats larger organisms, such as the Japanese flying squid, *Todarodes pacificus pacificus*, and small fish, including members of the Engraulis (anchovies), *Cololabis (sauries), Sardinops (pilchards),* and *Trachurus* (jack mackerels) genera.[32][35] Some of these fish are commercially important. Off central California, the whale may feed on anchovies between June and August, and on krill *(Euphausia pacifica)* during September and October.[23] In the Southern Hemisphere, prey species include the copepods *Neocalanus tonsus, Calanus simillimus,* and *Drepanopus pectinatus,* as well as the *euphausids Euphausia* superba and *Euphausia vallentini.*

Reproduction:

Mating occurs in temperate, subtropical seas during the winter. Gestation is estimated to vary around 103⁄4 months, 111⁄4 months, or one year, depending which model of foetal growth is used. The different estimates result from scientists' inability to observe an entire pregnancy; most reproductive data for baleen whales were obtained from animals caught by commercial whalers, which offers only a single snapshot of fetal growth. Researchers attempt to extrapolate conception dates by comparing fetus size and characteristics with newborns.

A newborn is weaned from its mother at 6–9 months of age, when it is 11–12 meters (36–39 ft) long, so weaning takes place at the summer or autumn feeding grounds. Females reproduce every 2–3 years, with as many as six fetuses reported, but single births are far more common. The average age of sexual maturity of both sexes is 8–10 years,] at a length of around 12 meters (39 ft) for males and 13 meters (43 ft) for females. The whales can reach ages of up to 65 years.

Vocalizations

The Sei whale makes long, loud, low-frequency sounds. Relatively little is known about specific calls, but in 2003, observers noted Sei whale calls in addition to sounds that could be described as "growls" or "whooshes" off the coast of the Antarctic Peninsula. Many calls consisted of multiple parts at different frequencies. This combination distinguishes the their calls from those of other whales. Most calls lasted about a half second, and occurred in the 240–625 hertz range, well within the range of human hearing. The maximum volume of the vocal sequences is reported as 156 decibels relative to 1 micropascal (μPa) at a reference distance of one meter. An observer situated one meter from a vocalizing whale would perceive a volume roughly equivalent to the volume of a jackhammer operating two meters away.

Range and migration

Sei whales live in all oceans, although rarely in polar or tropical waters. The difficulty of distinguishing them at sea from their close relatives, Bryde's whales and in some cases from fin whales, creates confusion about their range and population, especially in warmer waters where Bryde's whales are most common.

In the North Atlantic, its range extends from southern Europe or North-western Africa to Norway, and from the southern United States to Greenland. The southernmost confirmed records are strandings along the northern Gulf of Mexico and in the Greater Antilles. Throughout its range, the whale tends to avoid semi-enclosed bodies of water, such as the Gulf of Mexico, the Gulf of Saint Lawrence, Hudson Bay, the North Sea, and the Mediterranean Sea.[5] It occurs predominantly in deep water, occurring most commonly over the continental slope, in basins situated between banks, or submarine canyon areas.

In the North Pacific, it ranges from 20°N to 23°N latitude in the winter, and from 35°N to 50°N latitude in the summer. Approximately 75% of the North Pacific population lives east of the International Date Line, but there is little information regarding the North Pacific distribution. Two whales tagged in deep waters off California were later recaptured off Washington and British Columbia, revealing a possible link between these areas,[46] but the lack of other tag recovery data makes these two cases inconclusive. In the Southern Hemisphere, summer distribution based upon historic catch data is between 40°S and 50°S latitude, while winter distribution is unknown.

Migration

In general, the Sei whale migrates annually from cool and sub polar waters in summer to temperate and subtropical waters for winter, where food is more abundant. In the northwest Atlantic, sightings and catch records suggest the whales move north along the shelf edge to arrive in the areas of Georges Bank, Northeast Channel, and Browns Bank by mid to late June. They are present off the south coast of Newfoundland in August and September, and a southbound migration begins moving west and south along the Nova Scotian shelf from mid-September to mid-November. Whales in the Labrador Sea as early as the first week of June may move farther northward to waters southwest of Greenland later in the summer. In the northeast Atlantic, the Sei whale winters as far south as West Africa, and follows the continental slope northward in spring. Large females lead the northward migration and reach the Denmark Strait earlier and more reliably than other sexes and classes, arriving in mid-July and remaining through mid-September. In some years, males and younger females remain at lower latitudes during the summer months.

Despite knowing some general migration patterns, exact routes are not known and scientists cannot readily predict exactly where groups will appear from one year to the next. F.O. Kapel noted a correlation between appearances west of Greenland and the incursion of relatively warm waters from the Irminger Current into that area. Some evidence from tagging data indicates individuals return off the coast of Iceland on an annual basis.

Whaling

The development of explosive harpoons and steam-powered whaling ships in the late nineteenth century brought previously unobtainable large whales within reach of commercial whalers. Initially their speed and elusiveness, and later the comparatively small yield of oil and meat partially protected them. Once stocks of more profitable right whales, blue whales, fin whales, and humpback whales became depleted, Sei whales were hunted in earnest, particularly from 1950 to 1980.

North Atlantic

In the North Atlantic between 1885 and 1984, 14,295 Sei whales were taken. They were hunted in large numbers off the coast of Norway and Scotland beginning in the late nineteenth and early twentieth centuries, and in 1885 alone, more than 700 were caught off Finnmark, Norway. Their meat was a popular Norwegian food. The meat's value made the hunting of this difficult-to-catch species profitable in the early twentieth century.

In Iceland, a total of 2,574 whales were taken from the Hvalfjörður whaling station between 1948 and 1985. Since the late 1960s to early 1970s, the Sei whale has been second only to the fin whale as the preferred target of Icelandic whalers, with meat in greater demand than whale oil, the prior target.

Small numbers were taken off the Iberian Peninsula, beginning in the 1920s by Spanish whalers, off the Nova Scotian shelf in the late 1960s and early 1970s by Canadian whalers, and off the coast of West Greenland from the 1920s to the 1950s by Norwegian and Danish whalers.

North Pacific

In the North Pacific, the total reported catch by commercial whalers was 72,215 between 1910 and 1975; the majority were taken after 1947. Shore stations in Japan and Korea, processed 300–600 each year between 1911 and 1955. In 1959, the Japanese catch peaked at 1,340. Heavy exploitation in the North Pacific began in the early 1960s, with catches averaging 3,643 per year from 1963 to 1974 (total 43,719; annual range 1,280–6,053). In 1971, after a decade of high catches, it became scarce in Japanese waters, ending commercial whaling in 1975.

Off the coast of North America, Sei whales were hunted off British Columbia from the late 1950s to the mid 1960s, when the number of whales captured dropped to around 14 per year. More than 2,000 were caught in British Columbia waters between 1962 and 1967.Between 1957 and 1971, California shore stations processed 386 whales. Commercial Sei whaling ended in the eastern North Pacific in 1971.

1960s. The catch peaked in 1964-65 at over 20,000 Sei whales, but by 1976, this number had dropped to below 2,000 and commercial whaling for the species ended in 1977.

Southern Hemisphere:

A total of 152,233 were taken in the Southern Hemisphere between 1910 and 1979. Whaling in southern oceans originally targeted humpback whales. By 1913, this species became rare, and the catch of fin and blue whales began to increase. As these species likewise became scarce, Sei whale catches increased rapidly in the late 1950s and early 1960s. The catch peaked in 1964-65 at over 20,000 Sei whales, but by 1976, this number had dropped to below 2,000 and commercial whaling for the species ended in 1977.

Post-protection whaling:

Since the moratorium on commercial whaling, some Sei whales have been taken by Icelandic and Japanese whalers under the IWC's scientific research programme. Iceland carried out four years of scientific whaling between 1986 and 1989, killing up to 40 Sei whales a year. Japanese scientists catch about 50 Sei whales each year for this purpose. The research is conducted by the Institute of Cetacean Research (ICR) in Tokyo, a privately-funded, non-profit institution. The main focus of the research is to examine what they eat and to assess the competition between whales and fisheries. Dr. Seiji Ohsumi, Director General of the ICR, said,

"It is estimated that whales consume 3 to 5 times the amount of marine resources as are caught for human consumption, so our whale research is providing valuable information required for improving the management of all our marine resources."

He later added,

"...Sei whales are the second most abundant species of whale in the western North Pacific, with an estimated population of over 28,000 animals. [It is] clearly not endangered."

Conservation groups, such as the World Wildlife Fund, dispute the value of this research, claiming that Sei whales feed primarily on squid and plankton which are not hunted by humans, and only rarely on fish. They say that the program is

"nothing more than a plan designed to keep the whaling fleet in business, and the need to use whales as the scapegoat for over fishing by humans."

At the 2001 meeting of the IWC Scientific Committee, 32 scientists submitted a document expressing their belief that the Japanese program lacked scientific rigour and would not meet minimum standards of academic review.

In 2010, a Los Angeles restaurant confirmed to be serving Sei whale meat was closed by its owners after prosecution by authorities for handling a protected species.

Conservation status:

The Sei whale did not have meaningful international protection until 1970, when the International Whaling Commission (IWC) first set catch quotas for the North Pacific for individual species. Before quotas, there were no legal limits. Complete protection from commercial whaling in the North Pacific came in 1976.

Quotas on Sei whales in the North Atlantic began in 1977. Southern Hemisphere stocks were protected in 1979. Facing mounting evidence that several whale species were threatened with extinction, the IWC established a complete moratorium on commercial whaling beginning in 1986.

In the late 1970s, some "pirate" whaling took place in the Eastern North Atlantic. There is no direct evidence of illegal whaling in the North Pacific, although the acknowledged misreporting of whaling data by the Soviet Union means that catch data are not entirely reliable.

The species remained listed on the IUCN Red List of Threatened Species in 2000, categorized as "endangered". Northern Hemisphere populations are listed as CITES Appendix II, indicating they are not immediately threatened with extinction, but may become so if they are not listed. Populations in the Southern Hemisphere are listed as CITES Appendix I, indicating they are threatened with extinction if trade is not halted.

The Sei whale is listed on both Appendix I and Appendix II of the Convention on the Conservation of Migratory Species of Wild Animals (CMS). It is listed on Appendix I as this species has been categorized as being in danger of extinction throughout all or a significant proportion of their range and CMS Parties strive towards strictly protecting these animals, conserving or restoring the places where they live, mitigating obstacles to migration and controlling other factors that might endanger them and also on Appendix IIas it has an unfavourable conservation status or would benefit significantly from international co-operation organised by tailored agreements.

Sei whale is covered by the Memorandum of Understanding for the Conservation of Cetaceans and Their Habitats in the Pacific Islands Region (Pacific Cetaceans MOU)and the Agreement on the Conservation of Small Cetaceans of the Baltic, North East Atlantic, Irish and North Seas (ACCOBAMS)

The species is listed as endangered by the U.S. government National Marine Fisheries Service under the U.S. Endangered Species Act.

Population estimates:

The current population is estimated at 80,000, nearly a third of the pre-whaling population. A 1991 study in the North Atlantic estimated only 4,000.[68][69] Sei whales were said to have been scarce in the 1960s and early 1970s off Northern Norway. One possible explanation for this disappearance is that the whales were overexploited. The drastic reduction in northeastern Atlantic copepod stocks during the late 1960s may be another culprit. Surveys in the Denmark Strait found 1,290 whales in 1987, and 1,590 whales in 1989. Nova Scotia's population estimates are between 1,393 and 2,248, with a minimum of 870.

A 1977 study estimated Pacific Ocean totals of 9,110, based upon catch and CPUE data.[56] Japanese interests claim this figure is outdated, and in 2002 claimed the western North Pacific population was over 28,000, a figure not accepted by the scientific community. In California waters, there was only one confirmed and five possible sightings by 1991 to 1993 aerial and ship surveys, and there were no confirmed sightings off Oregon and Washington. Prior to commercial whaling, the North Pacific hosted an estimated 42,000. By the end of whaling, the population was down to between 7,260 and 12,620.

In the Southern Hemisphere, population estimates range between 9,800 and 12,000, based upon catch history and CPUE. The IWC estimated 9,718 whales based upon survey data between 1978 and 1988. Prior to commercial whaling, there were an estimated 65,000.

Finless Porpoise *(Neophocaena phocaenoides)*

Range Description:

The type specimen was from the Yangtze River, Kiangsu [=Jiangsu] Province, China, ~129 km (=80 miles) northwest of Shanghai (Pilleri and Gihr 1972).

In general, the species occurs in a narrow strip of shallow (usually et al. 2010), but it probably represents an extralimital stray rather than an extension of the species' range.) The distribution also includes a few estuaries and rivers (e.g., the *N. a. asiaeorientalis* subspecies appears to be found exclusively in the Yangtze River system; Gao 1991; Gao and Zhou 1995). Human coastal development and other activities already may have substantially reduced and fragmented the distribution of finless porpoises (Reeves et al. 1997).

N. a. sunameri – Coastal waters from the Taiwan Strait, including the western coast of Taiwan (mainly the Matsu and Chinmen Islands), through the East China Sea north to the Bohai/Yellow Sea in China and the waters of Korea and Japan (Gao 1991, Gao and Zhou 1995). Korean and Japanese populations are geographically separate (Shirakihara et al. 1992,Yoshida et al. 2001, Yoshida 2002).Seen in **Thailand**

N. a. asiaeorientalis – Middle and lower reaches of the Yangtze River (= Chang Jiang), where it ranges (or did until recently) to 1,600 km upstream, i.e. to the gorges above Yichang (200 m above sea level). The range includes Poyang and Dongting lakes and their tributaries, the Gan Jiang and the Xiang Jiang (Gao 1991, Gao and Zhou 1995). The subspecies is thought to be restricted to fresh water.

Countries: Native:

China; Japan; Korea, Republic of; Taiwan, Province of China.

Narrow-ridged Finless Porpoises are found mainly in coastal waters, including shallow bays, possibly mangrove swamps, estuaries, and some large rivers. However, they can also occur in shallow waters
In Japanese waters, finless porpoises prefer shallow depths.

Systems: Freshwater; Marine

Description

The finless porpoise (Neophocaena phocaenoides) is one of six porpoise species. In the waters around Japan, at the northern end of its range, it is known as the sunameri (砂滑). A freshwater population found in the Yangtze River in China is known locally as the jiangzhu (江猪) or "river pig". There is a degree of taxonomic uncertainty surrounding the species, with the *N. p. phocaenoides* subspecies perhaps representing a different species from *N. p. sunameri* and *N. p. asiaeorientalis.*

Physical description:

The finless porpoise almost completely lacks a dorsal fin. Instead there is a low ridge covered in thick denticulated skin.

Adults are a uniform, light grey colour. Newborn calves are mostly black with grey around the dorsal ridge area, becoming fully grey after four to six months. Adults grow more than 1.55 m (5 ft) in length and up to 30–45 kg (65–100 lb) in weight. Males become sexually mature at around four to six years of age, and females at around six to 9 years of age.

Diet

The finless porpoise, subspecies *Neomeris kurrachiensis,* in the vicinity of Karachi, British India, now Pakistan, as drawn by R. A. Sterndale, in the Journal of the Bombay Natural History Society, volume 1, number 1, 1886.

Finless porpoises are reported to eat fish and shrimp in the Yangtze River, and fish, shrimp and squid in the Yellow Sea/Bohai area and off Pakistan. In Japanese waters, they are known to eat fish, shrimp, squid, cuttle fish and octopuses. They are opportunistic feeders using various kinds of available food items available in their habitat. Seasonal changes in their diets have not been studied. They also apparently ingest some plant material when living in estuaries, mangroves, and rivers, including leaves, rice, and eggs deposited on vegetation.

Schooling

Finless porpoises are generally found as singles, pairs, or in groups of up to 12, although aggregations of up to about 50 have been reported. Recent data suggest the basic unit of a finless porpoise pod is a mother/calf pair or two adults, and schools of three or more individuals are aggregations of these units or of solitary individuals. Social structure seems to be underdeveloped in the species, and the mother/calf pair is probably the only stable social unit

Behaviour and reproduction:

Like other porpoises, their behaviour tends to be not as energetic and showy as that of dolphins. They do not ride bow waves, and in some areas appear to be shy of boats. In the Yangtze River, finless porpoises are known to leap from the water and perform "tail stands". Breeding occurs in late spring and early summer, after a gestation period of 10–11 months. The calf clings to the denticulated area of skin on their mother's back and is carried by her as she swims. Calves are weaned at 6–15 months.

Swimming style:

Although they show no acrobatics in the water, finless porpoises are believed to be very active swimmers. They typically swim just beneath the surface of the water and roll to one side when surfacing to breathe. This rolling movement disturbs very little water on the surface, so they are often overlooked when rising to breathe. Surfacing generally lasts for one minute, as they take three to four quick successive breaths, then quickly submerge into the water. They often surface a great distance from the point where they dive beneath the water's surface.

Conservation:

On the IUCN "red list" database of endangered species, the finless porpoise is listed as "vulnerable" due to apparent declines in the best studied populations in the Sea of Japan, and the Yangtze river basin. The fresh water subspecies in the Yangtze river is characterized as endangered. Since this species remains in coastal waters, it has a high degree of interaction with humans, which often puts the finless porpoise at risk. Like other porpoises, large numbers of this species are killed by entanglement in gill nets.

Except for being briefly hunted after World War II due to the lack of seaworthy fishing boats, finless porpoises have never been widely hunted in Japan. It is a species protected since 1930 at the area around Awajima Island, Takehara and this coverage had since been extended to all Japanese coastal waters. The primary danger to the species is the environmental degradation. In addition, unlike other members of this family, finless porpoises have lived in captivity for over 15 years.

There are no well-established estimates of the animals' abundance. However, a comparison of two surveys, one from the late 1970s and the other from 1999–2000, shows a decline in population and distribution. Scientists believe this decline has been ongoing for decades, and the current population is just a fraction of its historical levels. A 2006 expedition estimated fewer than 400 animals survived in the Yangtze River.

At the end of 2006, an estimated 1400 porpoises were left living in China, with between 700 and 900 in the Yangtze and another 500 in Poyang and Dongting Lakes.

The 2007 population levels were less than half the 1997 levels, and the population is dropping at a rate of 7.3% per year. Current conservation efforts were undertaken alongside those for the recently functionally extinct baiji. In 1990, five individuals were relocated to the Tian-e-Zhou Oxbow Nature Reserve, and now a population of 28 currently inhabit the lake.

Sand dredging has become a mainstay of local economic development in the last few years, and it is an important source of revenue in the region that borders Poyang Lake. At the same time, though, high-density dredging projects have been the principal cause of the death of the local wildlife population.

Dredging makes the waters of the lake muddier, and the porpoises cannot see as far as they once could, and have to rely on their highly developed sonar systems to avoid obstacles and look for food. Large ships enter and leave the lake at the rate of two a minute, and such a high density of shipping means the porpoises have difficulty hearing their food, and also cannot swim freely from one bank to the other.

The Finless porpoise is listed on Appendix II[6] of the Convention on the Conservation of Migratory Species of Wild Animals (CMS). It is listed on Appendix II[6] as it has an unfavourable conservation status or would benefit significantly from international co-operation organised by tailored agreements.

Conservation Actions:

Neophocaena phocaenoides (meant at the time of listing to include *N. asiaeorientalis)* is listed in Appendix I of the Convention on International Trade in Endangered Species of Wild Fauna and Flora (CITES), and in Appendix II of the Convention on Migratory Species (CMS). Management measures are needed to reduce the threats, particularly incidental mortality in fisheries and vessel strikes, but also to protect (or if possible restore) the environmental conditions needed for these animals to survive and reproduce.

Sperm Whale *(Physeter macrocephalus)*

Description:

The sperm whale *(Physeter macrocephalus)* is a marine mammal species, order Cetacea, a toothed whale (odontocete) having the largest brain of any animal. The name comes from the milky-white waxy substance, spermaceti, found in the animal's head. The sperm whale is the only living member of genus Physeter. The now outdated synonym Physeter catodon refers to the same species. It is one of three extant species in the sperm whale superfamily, along with the pygmy sperm whale and dwarf sperm whale.

A mature male can grow to 20.5 metres (67 ft) long. It is the largest living toothed animal. For large males, the head can represent up to one-third of the animal's length. It has a cosmopolitan distribution across the oceans. The species feeds primarily on squid but to some extent on fish, diving as deep as 3 kilometres (9,800 ft), which makes it the deepest diving mammal. Its diet includes giant squid and colossal squid. The sperm whale's clicking vocalization is the loudest sound produced by any animal. The clicking is used for sonar and may also be used for other purposes. These whales live in groups called social units. Units of females and their young live separately from sexually mature males. The females cooperate to protect and nurse their young. Females give birth every three to six years, and care for the calves for more than a decade. The sperm whale has few natural predators, since few are strong enough to successfully attack a healthy adult; orcas attack units and are capable of killing the calves. The sperm whale can live for more than 70 years.

Historically, the sperm whale was also known as the common cachalot; "cachalot" is derived from an archaic French word for "tooth". Over most of the period from the early 18th century until the late 20th century, the sperm whale was hunted to obtain spermaceti and other products, such as sperm oil and ambergris. Spermaceti found many important uses, such as candles, soap, cosmetics and machine oil. Due to its size, the sperm whale could sometimes defend itself effectively against whalers. In the most famous example, a sperm whale attacked and sank the American whale ship Essex in 1820. As a result of whaling, the sperm whale is currently listed as vulnerable by the IUCN.

Appearance:

The sperm whale's unique body is unlikely to be confused with any other species. The sperm whale's distinctive shape comes from its very large, block-shaped head, which can be one-quarter to one-third of the animal's length. The S-shaped blowhole is located very close to the front of the head and shifted to the whale's left. This gives rise to a distinctive bushy, forward-angled spray.

The sperm whale's flukes are triangular and very thick. The whale lifts its flukes high out of the water as it begins a feeding dive. It has a series of ridges on the back's caudal third instead of a dorsal fin. The largest ridge was called the 'hump' by whalers, and can be mistaken for a dorsal fin because of its shape and size.

The sperm whale's flukes are triangular and very thick. The whale lifts its flukes high out of the water as it begins a feeding dive. It has a series of ridges on the back's caudal third instead of a dorsal fin. The largest ridge was called the 'hump' by whalers, and can be mistaken for a dorsal fin because of its shape and size.

In contrast to the smooth skin of most large whales, its back skin is usually wrinkly and has been likened to a prune by whale-watching enthusiasts. Skin is normally a uniform grey in colour, though it may appear brown in sunlight. Albinos have also been reported.

Jaws and teeth:

The sperm whales' lower jaw is very narrow and under slung. The sperm whale has 18 to 26 teeth on each side of its lower jaw which fit into sockets in the upper jaw. The teeth are cone-shaped and weigh up to 1 kilogram (2.2 lb) each. The teeth are functional, but do not appear to be necessary for capturing or eating squid, and well-fed animals have been found without teeth. One hypothesis is that the teeth are used in aggression between males. Mature males often show scars which seem to be caused by the teeth. Rudimentary teeth are also present in the upper jaw, but these rarely emerge into the mouth.

Respiration and diving:

Sperm whales, along with bottlenose whales and elephant seals, are the deepest-diving mammals. Sperm whales are believed to be able to reach 3 kilometres (1.9 mi) and remain submerged for 90 minutes. More typical dives are around 400 metres (1,300 ft) and 35 minutes in duration. At these great depths, sperm whales had sometimes become entangled in transoceanic telephone cables and drowned until improvements in laying and maintenance techniques were employed. The sperm whale has adapted to cope with drastic pressure changes when diving. The flexible ribcage allows lung collapse, reducing nitrogen intake, and metabolism can decrease to conserve oxygen. Myoglobin, which stores oxygen in muscle tissue, is much more abundant than in terrestrial animals. The blood has a high red blood cell density, which contain oxygen-carrying hemoglobin. The oxygenated blood can be directed towards the brain and other essential organs only when oxygen levels deplete. The spermaceti organ may also play a role by adjusting buoyancy (see below).

While sperm whales are well adapted to diving, repeated dives to great depths have long term effects. Bones show pitting that signals decompression sickness in humans. Older skeletons showed the most extensive pitting, whereas calves showed no damage. This damage may indicate that sperm whales are susceptible to decompression sickness, and sudden surfacing could be lethal to them.

Between dives, the sperm whale surfaces to breathe for about eight minutes before diving again. Odontoceti (toothed whales) breathe air at the surface through a single, S-shaped blowhole. Sperm whales spout (breathe) 3–5 times per minute at rest, increasing to 6–7 times per minute after a dive. The blow is a noisy, single stream that rises up to 2 metres (6.6 ft) or more above the surface and points forward and left at a 45° angle.[36] On average, females and juveniles blow every 12.5 seconds before dives, while large males blow every 17.5 seconds before dives.

A sperm whale killed 100 miles south of Durban, South Africa after a 1 hour, 50-minute dive and was found with two spiny dogfish sharks, usually found at the sea floor, in its belly.

Brain and senses:

The brain is the largest known of any modern or extinct animal, weighing on average about 8 kilograms (18 lb), though the sperm whale has a lower encephalization quotient than many other whale and dolphin species, lower than that of non-human anthropoid apes, and much lower than humans'.

Nasal complex and spermaceti functions:

Early on it was proposed that the nasal complex, which includes the spermaceti organ, the junk bodies, and other associated organs, was used as a battering ram (see below) or for buoyancy regulation (see below); however, researchers' current understanding suggest that the primary function of the spermaceti organ and the associated organs in the nose of the sperm whales are used as part of the world's most powerful natural sonar system.

Due to light absorption by water, most of the ocean is dark beyond a few hundred meters thus limiting visual range. As a result, sperm whales and the other toothed whales (suborder odontoceti) have evolved a system of echolocation as the main way to find food in the darkness of the ocean similar to that used by bats to find food in the darkness of the night sky. When echolocating, the sperm whale emits a directionally focused beam of broadband clicks. Clicks are generated by the forcing of air through a pair of phonic lips (also known as "monkey lips" or "museau de singe") at the front end of the nose, just below the blowhole. The sound then travels backwards along the length of the nose through the spermaceti organ. Most of the sound energy is then reflected off an air sac which sits against the skull and down into the Junk Bodies, where the sound is focused by the junk's lens-like structure. Some of the sound will reflect back into the spermaceti organ and back towards the front of the whale's nose where it will be reflected through the spermaceti organ a third time. This back and forth reflection which happens on the scale of a few milliseconds creates a multi-pulse click structure. This multi—pulse click structure actually allows researchers to measure the whale's spermaceti organ using only the sound of its clicks and given the size of the spermaceti organ relates to the size of the whale, biologists can measure the whales by recording their echolocation clicks. The lower jaw is the primary reception path for the echoes. A continuous fat-filled canal transmits received sounds to the inner ear.
The source of the air forced through the phonic lips is the right nasal passage. While the left nasal passage opens to the blow hole, the right nasal passage has evolved to supply air to the phonic lips. It is thought that the nostrils of the land-based ancestor of the sperm whale migrated through evolution to their current functions, the left nostril becoming the blowhole and the right nostril becoming the phonic lips.

The spermaceti organs may also help adjust the whale's buoyancy. It is hypothesized that before the whale dives, cold water enters the organ, and it is likely that the blood vessels constrict, reducing blood flow, and, hence, temperature. The wax therefore solidifies and reduces in volume. The increase in specific density generates a down force of about 392 Newtons (88 lbf) and allows the whale to dive with less effort. During the hunt, oxygen consumption, together with blood vessel dilation, produces heat and melts the spermaceti, increasing its buoyancy and enabling easy surfacing. However, more recent work have found many problems with this theory including the lack of anatomical structures for the actual heat exchange.

Herman Melville's Moby Dick suggests that the "case" containing the spermaceti had evolved as a kind of battering ram for use in fights between males. However, there are almost no modern accounts of fights between male sperm whales. Apart from a few famous exceptions of the well-documented sinking of the ships Essex and Ann Alexander by attackers estimated to weigh only one-fifth as much as the ships, this hypothesis is not well supported in current scientific literature.

Etymology:

The name sperm whale is an apocopation of spermaceti whale. Spermaceti, originally mistaken for the whales' "sperm", is the semi-liquid, waxy substance found in the spermaceti organ or case in front of and above the skull bone and also in the junk, the area below the spermaceti organ and just above the upper jaw. The case consists of a soft white, waxy substance saturated with spermaceti oil. The junk is composed of cavities filled with the same wax and spermaceti oil and intervening connective tissue. The sperm whale is also known as the "cachalot", which is thought to derive from the archaic French for "tooth" or "big teeth", as preserved for example in cachau in the Gascon dialect (a word of either Romance or Basque origin). The etymological dictionary of Corominas says the origin is uncertain, but it suggests that it comes from the Vulgar Latin cappula, plural of cappulum, sword hilt. According to Encarta Dictionary, the word cachalot came to English "via French from Spanish or Portuguese cachalote, perhaps from Portuguese cachola, 'big head'". The term is retained in the Russian word for the animal, кашалот (kashalot), as well as in many other languages.

Ecology:

Distribution.

The sperm whale is among the most cosmopolitan species. It prefers ice-free waters over 1,000 metres (3,300 ft) deep. Although both sexes range through temperate and tropical oceans and seas, only adult males populate higher latitudes.

It is relatively abundant from the poles to the equator and is found in all the oceans. It inhabits the Mediterranean Sea, but not the Black Sea, while its presence in the Red Sea is uncertain. The shallow entrances to both the Black Sea and the Red Sea may account for their absence. The Black Sea's lower layers are also anoxic and contain high concentrations of sulphur compounds such as hydrogen sulphide.

Populations are denser close to continental shelves and canyons. Sperm whales are usually found in deep off-shore waters, but may be seen closer to shore in areas where the continental shelf is small and drops quickly to depths of 310–920 metres (1,020–3,020 ft). Coastal areas with significant sperm whale populations include the Azores and the Caribbean island of Dominica.

Reproduction:

Sperm whales can live 70 years or more. They are a prime example of a species that has been K-selected, i.e., their reproductive strategy is associated with stable environmental conditions and comprises a low birth rate, significant parental aid to offspring, slow maturation, and high longevity.

How they choose mates has not been definitively determined. There is evidence that males have dominance hierarchies, and there is also evidence that female choice influences mating. Gestation requires 14 to 16 months, producing a single calf. Lactation proceeds for 19 to 42 months, but calves may suckle up to 13 years (although usually less). Calves can suckle from females other than their mothers. Females generally have birth intervals of three to six years.

Females reach sexual maturity between 7 and 13 years; males follow beginning at 18 years. Upon reaching sexual maturity, males move to higher latitudes, where the water is colder and feeding is more productive. Females remain at lower latitudes. Males reach their full size at about age 50.

Social behavior:

Females stay in groups of about a dozen individuals and their young. Mature males leave their "natal unit" somewhere between 4 and 21 years of age. Mature males sometimes form loose "bachelor groups" with other males of similar age and size. As males grow older, they typically live solitary lives. Mature males have beached themselves together, suggesting a degree of cooperation which is not yet fully understood.

The most common non-human attacker of sperm whales is the orca, but pilot whales and the false killer whale also sometimes harass them. Orcas prey on target groups of females with young, usually making an effort to extract and kill a calf. Female sperm whales repel these attacks by encircling their calves. The adults either face inwards to use their tail flukes against the orcas, or outwards, fighting with their teeth. This Marguerite formation, named after the flower, is also used by whales to support an injured unit member. Early whalers exploited this behavior, attracting a whole unit by injuring one of its members. If the orca pod is extremely large, its members may sometimes be able to kill adult female sperm whales. Large mature male sperm whales have no non-human predators, and are believed to be too large, powerful and aggressive to be threatened by orcas.

Feeding:

Sperm Whales usually dive between 300 to 800 metres (980 to 2,600 ft), and sometimes 1–2 kilometres (3,300–6,600 ft) to search for food.[69] Such dives can last more than an hour. They feed on several species, notably the giant squid, the colossal squid, octopuses, and diverse fish like demersal rays, but the main part of their diet consists of medium-sized squid. Some prey may be taken incidentally while eating other items. Most of what is known about deep sea squid has been learned from specimens in captured sperm whale stomachs, although more recent studies analysed fecal matter.

One study, carried out around the Galápagos, found that squid from the genera Histioteuthis (62%), Ancistrocheirus (16%), and Octopoteuthis (7%) weighing between 12 and 650 grams (0.026 and 1.4 lb) were the most commonly taken. Battles between sperm whales and colossal squid (which have been measured to weigh nearly 500 kilograms (1,100 lb)) have never been observed by humans; however white scars are believed to be caused by the large squid. One study published in 2010 collected evidence that suggests that female sperm whales may collaborate when hunting Humboldt squid.

An older study, examining whales captured by the New Zealand whaling fleet in the Cook Strait region, found a 1.69:1 ratio of squid to fish by weight. Sperm whales sometimes steal Sablefish and Toothfish from long lines. Long-line fishing operations in the Gulf of Alaska complain that sperm whales take advantage of their fishing operations to eat desirable species straight off the line, sparing the whales the need to hunt. However, the amount of fish taken is very little compared to what the sperm whale needs per day. Video footage has been captured of a large male sperm whale "bouncing" a long line, to gain the fish. Sperm whales are believed to prey on the megamouth shark, a rare and large deep-sea species discovered in the 1970s.In one case, three sperm whales were observed attacking or playing with a megamouth. The sharp beak of a consumed squid lodged in the whale's intestine may lead to the production of ambergris, analogous to the production of pearls. The irritation of the intestines caused by squid beaks stimulates the secretion of this lubricant-like substance. Sperm whales are prodigious feeders and eat around 3% of their body weight per day. The total annual consumption of prey by sperm whales worldwide is estimated to be about 100,000,000 short tons (91,000,000 t) — a figure greater than the total consumption of marine animals by humans each year.

It is not well understood why the sperm whale's head is so large in comparison to the lower jaw. One theory is that the sperm whale's ability to echolocate through its head aids in hunting. However, squid, its main prey, may have acoustic properties too similar to seawater to reflect sounds. The sperm whale's head contains a structure called the phonic lips, also known as the monkey lips, through which it blows air. This can create clicks that have a source level up to 176 decibels referenced to a distance of 1 metre (3.3 ft) – making it one of the loudest animals on Earth (compare to the pistol shrimp at 218decibels), and 10–14 dB louder than a powerful rifle sounds in air at 1 metre (3.3 ft) away. It has been hypothesised that clicks attempt to stun prey. Experimental studies attempting to duplicate this effect have been unable to replicate the supposed injuries, casting doubt on this idea.

Phylogeny:

The traditional view has been that Mysticeti (baleen whales) and Odontoceti (toothed whales) arose from more primitive whales early in the Oligocene period, and that the super-family Physeteroidea, which contains the sperm whale, dwarf sperm whale, and pygmy sperm whale, diverged from other toothed whales soon after that, over 23 million years ago.[86][88] In 1993–1996 molecular phylogenetics analyses by Milinkovitch and colleagues, based on comparing the genes of various modern whales, suggested that the sperm whales are more closely related to the baleen whales than they are to other toothed whales, which would have meant that Odontoceti were not monophyletic, in other words did not consist of a single ancestral toothed whale species and all its descendants. However more recent studies, based on various combinations of comparative anatomy and molecular phylogenetics, criticised Milinkovitch's analysis on technical grounds and reaffirmed that the Odontoceti are monophyletic.

These analyses also confirm that there was a rapid evolutionary radiation (diversification) of the Physeteroidea in the Miocene period. The Kogiidae (dwarf and pygmy sperm whales) diverged from the Physeteridae (true sperm whales) at least 8 million years ago.

Watching sperm whales:

Sperm whales are not the easiest of whales to watch, due to their long dive times and ability to travel long distances underwater. However, due to the distinctive look and large size of the whale, watching is increasingly popular. Sperm whale watchers often use hydrophones to listen to the clicks of the whales and locate them before they surface. Popular locations for sperm whale watching include the picturesque Kaikoura on New Zealand's South Island, Andenes and Tromsø in Arctic Norway; as well as the Azores, where the continental shelf is so narrow that whales can be observed from the shore. and Dominica where a long-term scientific research program, The Dominica Sperm Whale Project, has been in operation since 2005.

Current conservation status:

The number of sperm whales throughout the world is unknown, but is thought to be in the hundreds of thousands. The conservation outlook is brighter than for many other whales. Historically, Japan has taken ten sperm whales a year, and until 2006 tens of these whales were hunted off Indonesia. They are protected practically worldwide, and commercial whaling has ceased. Fishermen do not target the creatures that sperm whales eat. However, long-line fishing operations in the Gulf of Alaska have complained about sperm whales stealing fish from their lines.

Entanglement in fishing nets and collisions with ships represent the greatest threats to the sperm whale population currently. Other current threats include ingestion of marine debris, ocean noise, and chemical pollution. The IUCN regards the sperm whale as being "vulnerable". The species is listed as endangered on the United States Endangered Species Act.

The species is listed on Appendix I and Appendix II of the Convention on the Conservation of Migratory Species of Wild Animals (CMS). It is listed on Appendix I as this species has been categorized as being in danger of extinction throughout all or a significant proportion of their range and CMS Parties strive towards strictly protecting these animals, conserving or restoring the places where they live, mitigating obstacles to migration and controlling other factors that might endanger them. It is listed on Appendix II as it has an unfavourable conservation status or would benefit significantly from international co-operation organised by tailored agreements. It is also covered by the Agreement on the Conservation of Cetaceans in the Black Sea, Mediterranean Sea and Contiguous Atlantic Area (ACCOBAMS) and Memorandum of Understanding for the Conservation of Cetaceans and Their Habitats in the Pacific Islands Region (Pacific Cetaceans

Rough-toothed Dolphin *(Steno bredanensis)*

Description:

The Rough-toothed dolphin (Steno bredanensis) is species of dolphin that can be found in deep warm and tropical waters around the world.

The species was first described by Georges Cuvier in 1823. The genus name Steno, of which this species is the only member, comes from the Greek for 'narrow', referring to the animal's beak - which is a diagnostic characteristic of the species. The specific name honours van Breda, who studied Cuvier's writings. There are no recognised sub species.The rough-toothed dolphin is a relatively large species, with adults ranging from 209 to 283 centimetres (6.9 to 9.3 ft) in length, and weighing between 90 and 155 kilograms (200 and 340 lb); males are larger than females. Its most visible characteristic feature is its conical head and slender nose; other dolphins either have a shorter snout or a more visibly bulging melon on the forehead. As the common name for the species implies, the teeth are also distinctive, having a roughened surface formed by numerous narrow irregular ridges. They have been reported to have between nineteen and twenty-eight teeth in each quarter of the jaw.

The flippers are set back further along the body than in other similar dolphins, although, at sea this dolphin may be confused with spinner, spotted and bottlenose dolphins. The dorsal fin is pronounced, being from 18 to 28 centimetres (7.1 to 11 in) in height. The animal's flanks are a light gray, while the back and dorsal fin are a much darker gray. Older individuals often have distinctive pinkish, yellow, or white markings around the mouth and along the underside.

Population and distribution:

The Rough-toothed Dolphin is a tropical to subtropical species, which generally inhabits deep, oceanic waters of all three major oceans, rarely ranging north of 40°N or south of 35°S (Jefferson 2002). However, in some areas (such as off the coast of Brazil and West Africa), rough-toothed dolphins may occur in more shallow coastal waters. They are found in many semi-enclosed bodies of water (such as the **Gulf of Thailand,** Red Sea, Gulf of Mexico, Caribbean Sea, and Gulf of California), but they are regarded as visitors in the Mediterranean Sea (Watkins et al. 1987; Miyazaki and Perrin 1994; Reeves and Notarbartolo di Sciara 2006).

Countries: Native:

Albania; Algeria; American Samoa (American Samoa); Angola (Angola); Anguilla; Antigua and Barbuda; Aruba; Australia; Bahamas; Bahrain; Bangladesh; Barbados; Belize; Benin; Bosnia and Herzegovina; Brazil; British Indian Ocean Territory; Brunei Darussalam; Cambodia; Cameroon; Cape Verde; Cayman Islands; Chile; China; Christmas Island; Cocos (Keeling) Islands; Colombia; Comoros; Congo; Congo, The Democratic Republic of the; Cook Islands; Costa Rica; Côte d'Ivoire; Cuba; Cyprus; Djibouti; Dominica; Dominican Republic; Ecuador; El Salvador; Equatorial Guinea; Fiji; France; French Guiana; French Polynesia; Gabon; Gambia; Ghana; Gibraltar; Greece (Kriti); Grenada; Guadeloupe; Guam; Guatemala; Guinea;

Guinea-Bissau; Guyana; Haiti; Honduras; Hong Kong; India; Indonesia; Iran, Islamic Republic of; Iraq; Israel; Italy (Sardegna, Sicilia); Jamaica; Japan; Jordan; Kenya; Kiribati; Kuwait; Liberia; Libya; Macao; Madagascar; Malaysia; Maldives; Malta; Marshall Islands; Martinique; Mauritania; Mauritius; Mayotte; Mexico; Micronesia, Federated States of; Monaco; Montserrat; Morocco; Mozambique; Myanmar; Namibia; Nauru; Netherlands Antilles; New Caledonia; Nicaragua; Nigeria; Niue; Norfolk Island; Northern Mariana Islands; Oman; Pakistan; Palau; Panama; Papua New Guinea; Peru; Philippines; Pitcairn; Portugal; Puerto Rico; Qatar; Réunion; Saint Helena, Ascension and Tristan da Cunha; Saint Kitts and Nevis; Saint Lucia; Saint Vincent and the Grenadines; Samoa; Sao Tomé and Principe; Senegal; Seychelles; Sierra Leone; Singapore; Solomon Islands; Somalia; South Africa; Spain (Baleares, Canary Is.); Sri Lanka; Suriname; Taiwan, Province of China; Tanzania, United Republic of; **Thailand;** Timor -Leste; Togo; Tonga; Trinidad and Tobago; Turks and Caicos Islands; United Arab Emirates; United States; Venezuela, Bolivarian Republic of; Vietnam; Virgin Islands, British; Virgin Islands, U.S.; Wallis and Futuna; Western Sahara; Yemen

Habitat and Ecology:

Most often, Steno bredanensis is found in deep water far offshore, usually beyond the continental shelf (Maigret 1994), but may be seen close inshore in areas of steep bottom relief (Ritter 2002). In the eastern tropical Pacific, they tend to associate with other cetaceans (especially pilot whales and Fraser's dolphins) (Miyazaki and Perrin 1994). Rough-toothed Dolphins feed on cephalopods and fish, including large fish such as Coryphaena hippurus (Pitman and Stinchcomb 2002).

Systems: Marine

Population:

There are few estimates of abundance for this species. An estimated 145,900 (CV=32%) Rough-toothed Dolphins inhabit the eastern tropical Pacific (Wade and Gerrodette 1993), and about 2,746 (CV=36%) occur in the northern Gulf of Mexico (Waring et al. 2008), including an estimated 1,238 (CV=65%) on the continental shelf (Fulling et al. 2003). The US NMFS has estimated the regional population around Hawaii to be 19,904 (CV=52%), based on recent vessel surveys (Carretta et al. 2006).

Population Trend: Unknown

Behaviour and diet:

Rough-toothed dolphins are typically social animals, although solitary individuals are also sighted. An average group has between ten and twenty members, but they can vary from as few as two to as many as ninety.Such groups are thought to be temporary assemblages, composed of smaller, more permanent groups of two to eight closely related individuals that occasionally join together with others. They have also been reported to school together with other species of dolphin, and with pilot whales, false killer whales, and humpback whales.

Rough-toothed dolphins have been reported to bow-ride on a number of occasions, although apparently they do not do so as frequently as many other dolphin species. They do, however, commonly "skim", by swimming with their heads and chin above the surface of the water. They are known to be able to dive to at least 50 metres (160 ft) and be able to stay underwater for at least fifteen minutes. Their echolocation clicks are unusually brief, lasting no more than 0.2 seconds, and have a relatively low frequency, ranging from 2.7 to 256 kHz, with a maximum peak frequency of 25 kHz. They also make longer whistles with a frequency between 3 and 12 kHz.

Although details of their diet are sketchy, the stomach contents of stranded dolphins have included such fish such as silversides, sauries, hound fish, smelts, cutlass fish, and various squid and octopuses. Predators on rough-toothed dolphins are thought to include killer whales and sharks.

Reproduction:

Rough-toothed dolphins give birth to a single young, after an unknown period of gestation; it is also unknown whether or not they have a distinct breeding season. The young are about 100 centimetres (39 in) long at birth, and grow rapidly for the first five years of life. Females reach sexual maturity somewhere between six and ten years of age, and males between five and ten years.

Major Threat(s):

No fisheries are known to specifically target this species, but small numbers are taken in drive fisheries at Okinawa in the Ryukyus and other islands of Japan, the Solomon Islands and Papua New Guinea, by harpoon in Japan, St. Vincent in the Lesser Antilles, and in West Africa. They were possibly formerly were taken at St. Helena in the South Atlantic. However, only 23 Rough-toothed Dolphins were captured in Japan (Okinawa) during the period 1976-81 (Miyazaki and Perrin 1994). Recent information suggests catches in Taiwan (J. Wang pers. comm.).

A few Rough-toothed Dolphins are killed incidentally in tuna purse seines in the eastern tropical Pacific: 21 were estimated killed during the period 1971-75 and 36 died in a single net haul in 1982. Small numbers are also taken as by-catch in gillnet and driftnet fisheries in Sri Lanka, Brazil, the central North Pacific and probably elsewhere around the world in tropical and warm-temperate waters (Miyazaki and Perrin 1994). Monteiro-Neto et al. (2000) reported on fishery-related mortality along the coast of Ceara State, northeast Brazil, commenting on the possible conservation implications for the local subpopulations. Seasonally, incidental catches were more frequent during the austral spring (October-December). Rough-toothed Dolphins are also taken by gill nets, driftnets and pelagic long-lines in Taiwan (J. Wang pers. comm.).

Conservation Actions:

The species is listed in Appendix II of CITES.

The biology, life history, population size, and separation into subpopulations, as well as migratory behaviour are insufficiently known. Research on this species should be encouraged.

Indo-Pacific Hump-Backed Dolphin. *(Sousa chinensis)*

Common Name(s):

Indo-pacific Hump-backed Dolphin, Indo-Pacific Humpbacked Dolphin, Indo-pacific Humpback Dolphin, Chinese White Dolphin

Currently, all Indo-Pacific Humpback Dolphins are considered to be part of a single widespread and highly variable species, *Sousa chinensis*. Some biologists consider Humpback Dolphins in the Indo- Pacific to consist of two species: *S. plumbea* in the western Indian Ocean, from South Africa to at least the east coast of India, and *S. chinensis*, from the east coast of India to China and Australia. These two geographic forms differ markedly in their external morphology, and are moderately distinct in their skull morphometrics (see Jefferson and Van Waerebeek 2004 for a partial summary). However, the external morphological characters that have traditionally been used to differentiate these forms may be plesiomorphic (ancestral) or the result of convergent evolution, and as such, are not particularly informative phylogenetically. This was pointed out by Jefferson and Van Waerebeek (2004) when comparing the Atlantic *S. teuszii* and West Indian *plumbea* form, which are both uniformly gray in color and have dorsal humps, but may in fact not be that closely related. Similarly, a light gray to white adult coloration and lack of a prominent dorsal hump has traditionally been used to define the *chinensis form.*

Taxonomic Notes:

Recent mitochondrial (mt) DNA analyses indicate that Humpback Dolphins from Australia (*chinensis* form) are highly distinct from other Indo-Pacific populations (Frere *et al.* in press). Phylogenetic analyses comparing *chinensis* dolphins from Hong Kong and Australia, with *plumbea* dolphins from South Africa, as well as with Atlantic *S. teuszii*, strongly suggest that Hong Kong *chinensis*, South African *plumbea*, and Atlantic *S. teuszii* are more closely related to one another than to the '*chinensis*' dolphins in Australia. Previous mtDNA work by Cockcroft *et al.* (1997) indicated a similar pattern of relationships among these three Indo-Pacific regions.

It is clear that the taxonomy of the genus *Sousa* will not be resolved until comprehensive sampling and both genetic and morphological analyses have been conducted throughout the range. There is growing evidence that at least two putative species, and possibly more, are valid (see Rice 1998, Jefferson and Karczmarski 2001, Frere *et al.* in press). In the meantime, considering that there is a broad area of sympatry and no evidence of intermediate forms, the two morphological forms (*plumbea*-type and *chinensis*-type) are assessed separately in the account below. However, the reader should remain aware that the Australian *chinensis*-type is apparently not closely related to other *chinensis*-type dolphins.

***Chinensis*-type**

Indo-Pacific humpback dolphins of the *chinensis*-type are found in shallow, coastal waters from the east and west coasts of northern Australia and from southern China in the east, throughout the Indo-Malay Archipelago, and westward around the coastal rim of the Bay of Bengal to at least the Orissa coast of eastern India (Ross *et al.* 1994; Jefferson and Karczmarski 2001; Sutaria and Jefferson 2004). They regularly occur in some enclosed seas, such as the Gulf of Thailand. Their distribution appears to be limited to waters of the continental shelf, and the only places where they range far offshore are those where the water remains shallow (<100 m).

Range Description:

Native:

Australia; Bahrain; Bangladesh; Brunei Darussalam; Cambodia; China; Comoros; Djibouti; Egypt; Ethiopia; Hong Kong; India; Indonesia; Iran, Islamic Republic of; Iraq; Israel; Kenya; Kuwait; Macao; Madagascar; Malaysia; Mozambique; Myanmar; Oman; Pakistan; Papua New Guinea; Philippines; Qatar; Saudi Arabia; Singapore; Somalia; South Africa; Sri Lanka; Taiwan, Province of China; Tanzania, United Republic of; **Thailand;** Timor-Leste; United Arab Emirates; Viet Nam; Yemen Humpback dolphins occur in tropical to warm temperate coastal waters, including open coasts and bays, coastal lagoons, rocky and/or coral reefs, mangrove swamps and estuarine areas (Ross *et al.* 1994, Jefferson and Karczmarski 2001, Ross 2002). They are rarely encountered more than a few kilometres from shore. They sometimes enter rivers, but rarely move more than a few kilometres upstream and usually remain within the range of tidal influence.

Indo-Pacific humpback dolphins appear to be opportunistic feeders, consuming a wide variety of nearshore, estuarine, and reef fishes. They also eat cephalopods in some areas, but crustaceans are rare in their diet (Jefferson and Karczmarski 2001, Ross 2002).

Habitat and Ecology:

Chinensis-type dolphins often enter rivers, estuaries, and mangroves, preferring coasts with mangrove swamps, lagoons, and estuaries, as well as areas with reefs, sandbanks, and mud banks (Jefferson and Karczmarski 2001). In at least China and southern Asia, they are rarely found far from estuaries and mangrove habitats (Jefferson and Karczmarski 2001; Wang *et al.* 2007), and they show a strong preference for river mouths in northern Australia (Parra 2006; Parra *et al.* 2006b). Aerial surveys of the Great Barrier Reef region demonstrate that humpback dolphins occur mostly close to the coast but also in offshore waters that are relatively sheltered, and near reefs or islands (Corkeron *et al.* 1997). Fine- scaled resource partitioning between humpback and Australian snubfin dolphins (*Orcaella heinsohni*) has been documented off Queensland, where the two species favour river mouths and modified habitats but the humpback dolphins occur in slightly deeper (2-5 m deep) waters (Parra 2006).

Threats:

Chinensis-type dolphins are not known to be hunted directly in significant numbers anywhere in their range. However, they are often caught in fishing nets, such as gillnets and trawls, and in anti-shark nets set to protect bathing beaches from large sharks along the coasts of Queensland and New South Wales, Australia (Heinsohn 1979, Ross *et al.* 1994, Parra *et al.* 2004). Accurate catch data for humpback dolphins in the Australian nets are unavailable, but kills in anti-shark nets off Queensland are high relative to estimated abundance (Paterson 1990, Corkeron *et al.* 1997). The greatest direct sources of human-caused dolphin mortality in Hong Kong appear to be incidental catches in fishing gear (most likely pair trawls) and vessel collisions (Jefferson 2000, Parsons and Jefferson 2000). Between 1993 and 1998, at least 3 humpback dolphins were killed by boat strikes and another death was suspected of being caused by a boat strike. This represented 14% of all documented humpback dolphin strandings in Hong Kong during that period (Parsons and Jefferson 2000).

Concentrations of organochlorines in cetaceans from Hong Kong coastal waters are significantly higher than those found in cetaceans in other parts of the world (Parsons and Chan 1998, Minh *et al.* 1999) and it has been suggested that the reproductive success of Hong Kong's humpback dolphins (including neonatal survival) is being affected (Parsons 2004; Jefferson *et al.* 2006). In Hong Kong, high volumes of sewage discharge and the close proximity of contaminated mud pits means that there is considerable potential for trace metal contamination of local dolphins (Parsons 1997). Indeed, mercury concentrations in the tissues of Hong Kong humpback dolphins were found to be an order of magnitude higher than in prey items and in some cases, were high enough (max: 906 μg kg-1 dry weight) to be considered potentially health-threatening (Parsons 2004).

Hong Kong discharges over 2 billion litres of sewage into the surrounding waters daily. Parsons (1997) estimated that a humpback dolphin's minimum daily intake of sewage bacteria through ingestion of contaminated seawater could be up to 70,500 faecal coliforms. To put this in context, a one-off ingestion rate of 200-300 coliforms is considered unacceptable for humans (Parsons 2004).

The disposal of contaminated mud from Hong Kong's dredging and reclamation projects poses an indirect risk to humpback dolphins via their consumption of contaminated prey (Clarke *et al.* 2000). Humpback dolphins inhabit the waters of several coastal ports in Asia that host large volumes of ship traffic, such as Shanghai, Singapore and Hong Kong. Therefore, it is likely that they are highly contaminated with butyltin (BT) (see Tanabe *et al.* 1998, Tanabe 1999; Parsons 2004).

Underwater industrial activity, such as pile-driving during pier and bridge construction, are likely to cause acoustic disturbance (*e.g.* the development of Hong Kong's new airport). Boat traffic also might interfere with the dolphin's acoustic communication (Van Parijs *et al.* 2001).

Conservation Actions:

Conservation actions are currently either meagre or non-existent. Research is needed to help design effective conservation programmes.

THE BOTTLE-NOSED DOLPHIN (*Tursiops truncates)*

Recognition:

Short but distinct beak, tall dorsal fin, centrally placed.
Colour: darker, grey to brown, dorsally, with lighter flanks and even paler belly.
Maximum length: in the region of 4.0 m (males are larger than females).
Maximum weight: some 275 kg.

General Ecology

Bottle-nosed dolphins occur worldwide, including **Thailand,** except for polar regions. Two forms (or "ecotypes") are recognised: a generally larger offshore form and a smaller coastal form. Indeed, whether the widespread Bottle-nosed dolphin is really just one species is presently being debated. They are the most studied and familiar of all the dolphins but relatively little is known about them in the UK (most information is available from studies in the US). Historically they have been closely associated with estuaries, but present populations are believed to be reduced. The species is capable of a wide range of behaviour making it difficult to generalise about them. They are highly social animals, usually seen inshore in groups smaller than ten individuals, but far larger congregations are sometimes reported. Occasionally solitary bottle-nosed dolphins are reported, which seem to shun their own kind and, instead, favour contact with humankind.

The family groups are usually based around adult females. Mother-calf bonding can last many years and form the focus of family groups. Gestation lasts for twelve months and, once mature, females may have a single calf every second or third year. Females, other than the mother, may assist in caring for the calf. Adult males may occasionally join such families but, more frequently, may be seen in bachelor gatherings. Behaviour may also change seasonally but, as research into this is only at a very early stage.

This species feeds on a wide variety of fish and invertebrates, including squid, and they sometimes feed cooperatively herding their prey until it is concentrated near the water surface. Description[edit

Bottlenose dolphins are grey, varying from dark grey at the top near the dorsal fin to very light grey and almost white at the underside. This countershading makes them hard to see, both from above and below, when swimming. Adults range from 2 to 4 metres (6.6 to 13.1 ft), and 150 to 650 kilograms (330 to 1,430 lb).Males are, on average, slightly longer and considerably heavier than females. In most parts of the world, the adults are about 2.5 m (8.2 ft), and 200 to 300 kilograms (440 to 660 lb).Their size varies considerably with habitat. Except in the eastern Pacific, dolphins in warmer, shallower waters tend to be smaller than those in cooler, pelagic waters. Bottlenose dolphins can live for more than 40 years.

Senses.

The dolphin's search for food is aided by a form of sonar known as echolocation: it locates objects by producing sounds and listening for the echos. A broadband burst pulse of clicking sounds is emitted in a focused beam in front of the dolphin. To hear the returning echo, they have two small ear openings behind the eyes, but most sound waves are transmitted to the inner ear through the lower jaw. As the object of interest is approached, the echo becomes booming, and the dolphins adjust by decreasing the intensity of the emitted sounds.

(This contrasts with bats and sonar, which reduce sensitivity of the sound receptor.) The interclick interval also decreases as the animal nears the target. Evidently, the dolphin waits for each click's echo before clicking again. Echolocation details, such as signal strength, spectral qualities, and discrimination, are well-understood by researchers. Bottlenose dolphins are also able to extract shape information, suggesting they are able to form an "echoic image" or sound picture of their targets

Dolphins have sharp eyesight. The eyes are located at the sides of the head and have a tapetum lucidum, or reflecting membrane, at the back of the retina, which aids vision in dim light. Their horseshoe-shaped, double-slit pupils enable dolphins to have good vision both in air and underwater, despite the different indices of refraction of these media. When under water, the eyeball's lens serves to focus light, whereas in the in-air environment, the typically bright light serves to contract the specialized pupil, resulting in sharpness from a smaller aperture (similar to a pinhole camera).

By contrast, a bottlenose's sense of smell is poor, because its blowhole, the analogue to the nose, is closed when underwater and it opens only for breathing. It has no olfactory nerves or olfactory lobe in the brain. Bottlenose dolphins are able to detect salty, sweet, bitter (quinine sulphate), and sour (citric acid) tastes, but this has not been well-studied. Anecdotally, some individuals in captivity have been noted to have preferences for food fish types, although it is not clear if taste mediates this preference.

Communication:

Bottlenose dolphins communicate through burst pulsed sounds, whistles, and body language. Examples of body language include leaping out of the water, snapping jaws, slapping the tail on the surface and butting heads. Sounds and gestures help keep track of other dolphins in the group, and alert other dolphins to danger and nearby food. Lacking vocal cords, they produce sounds using six air sacs near their blow hole. Each animal has a uniquely identifying, frequency-modulated narrow-band signature vocalization (signature whistle). Researchers from the Bottlenose Dolphin Research Institute (BDRI), based in Sardinia (Italy) have now shown whistles and burst pulsed sounds are vital to the animals' social life and mirror their behaviors. The tonal whistle sounds (the most melodious ones) allow dolphins to stay in contact with each other (above all, mothers and offspring), and to coordinate hunting strategies. The burst-pulsed sounds (which are more complex and varied than the whistles) are used "to avoid physical aggression in situations of high excitement", such as when they are competing for the same piece of food, for example. The dolphins emit these strident sounds when in the presence of other individuals moving towards the same prey. The "least dominant" one soon moves away to avoid confrontation. Other communication uses about 30 distinguishable sounds, and although famously proposed by John Lilly in the 1950s, no "dolphin language" has been found. However, Herman, Richards, and Wolz demonstrated comprehension of an artificial language by two bottlenose dolphins (named Akeakamai and Phoenix) in the period of skepticism toward animal language following Herbert Terrace's critique.

Respiration and sleep:

The bottlenose dolphin has a single blowhole located on the dorsal surface of the head consisting of a hole and a muscular flap. The flap is closed during muscle relaxation and opens during contraction. Dolphins are voluntary breathers, who must deliberately surface and open their blowholes to get air. They can store almost twice as much oxygen in proportion to their body weight as a human can: the dolphin can store 36 milliliters (ml) of oxygen per kg of body weight, compared with 20 ml per kg for humans. This is an adaptation to diving. The bottlenose dolphin typically rises to the surface to breathe through its blowhole two to three times per minute, although it can remain submerged for up to 20 minutes. Dolphins can breathe while "half-asleep". During the sleeping cycle, one brain hemisphere remains active, while the other hemisphere shuts down. The active hemisphere handles surfacing and breathing behavior. The daily sleeping cycle lasts for about 8 hours, in increments of minutes to hours. During the sleeping cycle, they remain near the surface, swimming slowly or "logging", and occasionally closing one eye.

Reproduction:

Both sexes have genital slits on the underside of their bodies. The male can retract and conceal his penis through his slit. The female's slit houses her vagina and anus. Females have two mammary slits, each housing one nipple, one on each side of the genital slit. The ability to stow their reproductive organs (especially in males) allows for maximum hydrodynamics. The breeding season produces significant physiological changes in males. At that time, the testes enlarge, enabling them to hold more sperm. Large amounts of sperm allow a male to wash away the previous suitor's sperm, while leaving some of his own for fertilization. Also, sperm concentration markedly increases. Having less sperm for out-of-season social mating means it wastes less. This suggests sperm production is energetically expensive. Males have large testes in relation to their body size.

During the breeding season, males compete for access to females. Such competition can take the form of fighting other males or of herding females to prevent access by other males. In Shark Bay, male bottlenose dolphins have been observed working in pairs or larger groups to follow and/or restrict the movement of a female for weeks at a time, waiting for her to become sexually receptive.These coalitions, also known as male reproductive alliances, will fight with other coalitions for control of females.

Mating occurs belly to belly. Dolphins have been observed engaging in intercourse when the females are not in their estrous cycles and cannot produce young, suggesting they may mate for pleasure.The gestation period averages 12 months. Births can occur at any time of year, although peaks occur in warmer months. The young are born in shallow water, sometimes assisted by a (possibly male) "midwife", and usually only a single calf is born.Twins are possible, but rare. Newborn bottlenose dolphins are 0.8 to 1.4 m (2.6 to 4.6 ft) long and weigh 9 to 30 kg (20 to 66 lb), with Indo-Pacific bottlenose dolphin infants being generally smaller than common bottlenose dolphin infants.To accelerate nursing, the mother can eject milk from her mammary glands. The calf suckles for 18 months to up to 8 years, and continues to closely associate with its mother for several years after weaning. Females sexually mature at ages 5–13, males at ages 9–14. Females reproduce every two to six years. Georgetown University professor Janet Mann argues the strong personal behavior among male calves is about bond formation and benefits the species in an evolutionary context. She cites studies showing these dolphins as adults are inseparable, and that early bonds aid protection, as well as in locating females.

Feeding:

A dolphin's diet consists mainly of small fish, crustaceans, and squid.Although this varies by location, many populations share an appetite for fish from the mullet, the tuna and mackerel, and the drum and croaker families. Its cone-like teeth serve to grasp, but do not chew food. When they encounter a shoal of fish, they work as a team to herd them towards the shore to maximize the harvest. They also hunt alone, often targeting bottom-dwelling species. The bottlenose dolphin sometimes hits a fish with its fluke, sometimes knocking it out of the water, using a strategy called "fish whacking". "Strand feeding", is an inherited feeding technique used by bottlenose dolphins near and around coastal regions of Georgia and South Carolina. When a pod finds a school of fish, they will circle the school and trap the fish in a mini whirlpool. Then, the dolphins will charge at the school and push their bodies up onto a mud-flat, forcing the fish on the mud-flat, as well. The dolphins then crawl around on their sides, consuming the fish they washed up on shore. One type of feeding behavior seen in bottlenose dolphins is mud ring feeding. Bottlenose dolphins conflict with small-scale coastal commercial fisheries in some Mediterranean areas. Common bottlenose dolphins are probably attracted to fishing nets because they offer a concentrated food source.

Relations with other species:

Dolphins can exhibit altruistic behaviour toward other sea creatures. On Mahia Beach, New Zealand, on March 10, 2008, two pygmy sperm whales, a female and calf, stranded on the beach. Rescuers, including Department of Conservation officer Malcolm Smith, attempted to refloat them four times. Shortly, a playful bottlenose dolphin known to local residents as Moko arrived and, after apparently vocalizing at the whales, led them 200 m (660 ft) along a sandbar to the open sea, saving them from imminent euthanasia.

The bottlenose dolphin can behave aggressively. Males fight for rank and access to females. During mating season, males compete vigorously with each other through displays of toughness and size, with a series of acts, such as head-butting. They display aggression towards sharks and smaller dolphin species. At least one population, off Scotland, has practiced infanticide, and also has attacked and killed harbour porpoises. University of Aberdeen researchers say the dolphins do not eat their victims, but are simply competing for food. However, Dr. Read of Duke University, a porpoise expert researching similar cases of porpoise killings that had occurred in Virginia in 1996 and 1997, holds a different view. He states dolphins and porpoises feed on different types of fish, thus food competition is an unlikely cause of the killings. Similar behaviour has been observed in Ireland. In the first half of July, 2014, four attacks with three Porpoise fatalities were observed and caught on video by the Cardigan Bay Marine Wildlife Centre in the Cardigan Bay.The bottlenose dolphin sometimes forms mixed species groups with other species from the dolphin family, particularly larger species, such as the short-finned pilot whale, the false killer whale and Risso's dolphin. They also interact with smaller species, such as the Atlantic spotted dolphin and the rough-toothed dolphin. While interactions with smaller species are sometimes affiliative, they can also be hostile.

Predators:

Some large shark species, such as the tiger shark, the dusky shark, the great white shark and the bull shark, prey on the bottlenose dolphin, especially calves. The bottlenose dolphin is capable of defending itself by charging the predator; dolphin 'mobbing' behavior of sharks can occasionally prove fatal for the shark. Targeting a single adult dolphin can be dangerous for a shark of similar size. Killer whale populations in New Zealand and Peru have been observed preying on bottlenose dolphins, but this seems rare, and other orcas may swim with dolphins. Swimming in pods allows dolphins to better defend themselves against predators. Bottlenose dolphins either use complex evasive strategies to outswim their predators, or mobbing techniques to batter the predator to death or force it to flee.

Threats:

Bottlenose dolphins are still captured or killed in dolphin drive hunts for their meat, to eliminate competition for fish and for capture for marine parks. Bottlenose dolphins (and several other dolphin species) often travel with tuna, and can get caught in tuna nets, which can kill the dolphins.Boycotts of tuna products led to the concept of "dolphin-safe" labeling for fishing methods that avoid endangering dolphins.

The man-made chemical perfluorooctanesulfonic acid (PFOS) may be compromising the immune system of bottlenose dolphins.[131] PFOS affects the immune system of male mice at a concentration of 91.5 ppb, while PFOS has been reported in bottlenose dolphins in excess of 1 ppm. High levels of metal contaminants have been measured in tissues in many areas of the globe. A recent study found high levels of cadmium and mercury in bottlenose dolphins from South Australia, levels which were later found to be associated with kidney malformations, indicating possible health effects of high heavy metal concentrations in dolphins.

Conservation:

Bottlenose dolphins are not endangered. Their future is stable because of their abundance and adaptability. However, specific populations are threatened due to various environmental changes. The population in the Moray Firth in Scotland is estimated to consist of around 190 individuals, and are under threat from harassment, traumatic injury, water pollution and reduction in food availability. Likewise, an isolated population in Doubtful Sound, New Zealand, is in decline due to calf loss coincident to an increase in warm freshwater discharge into the fjord. Less local climate change, such as increasing water temperature may also play a role but has never been shown to be the case. One of the largest coastal populations of bottlenose dolphins in Shark Bay, Western Australia was forecast to be stable with little variation in mortality over time (Manlik *et al.* 2016).

In US waters, hunting and harassing of marine mammals is forbidden in almost all circumstances, from the passing of the Marine Mammal Protection Act of 1972

The Indo-Pacific bottlenose dolphin *(Tursiops aduncus)*

The Indo-Pacific bottlenose dolphin *(Tursiops aduncus)* is a species of bottlenose dolphin. This dolphin grows to 2.6 metres (8.5 ft) long, and weighs up to 230 kilograms (510 lb).It lives in the waters around India, northern Australia, South China, the Red Sea, and the eastern coast of Africa. Its back is dark grey and its belly is lighter grey or nearly white with grey spots.

Until 1998, all bottlenose dolphins were considered members of the single species *T. truncatus.* In that year, the Indo-Pacific bottlenose dolphin was recognized as a separate species. The Indo-Pacific bottlenose dolphin is generally smaller than the common bottlenose dolphin, has a proportionately longer rostrum, and has spots on its belly and lower sides. It also has more teeth than the common bottlenose dolphin — 23 to 29 teeth on each side of each jaw compared to 21 to 24 for the common bottlenose dolphin. There is evidence the Indo-Pacific bottlenose dolphin may actually be more closely related to certain dolphin species in the genera Stenella and Delphinus, especially the Atlantic spotted dolphin *(S. frontalis),* than it is to the common bottlenose dolphin.
Much of the old scientific data in the field combine data about the Indo-Pacific bottlenose dolphin and the common bottlenose dolphin into a single group, making it effectively useless in determining the structural differences between the two species. The IUCN lists both species as data deficient in their Red List of endangered species because of this issue.

Range Description:

The Indo-Pacific Bottlenose Dolphin has a discontinuous distribution in the warm temperate to tropical Indo-Pacific, from South Africa in the west, along the rim of the Indian Ocean (including the Red Sea, Persian Gulf and Indo-Malay Archipelago as far east as the Solomon Islands and possibly New Caledonia) to the southern half of Japan and southeast Australia in the east (Wells and Scott 2002; Möller and Beheregaray 2001). It is also found around oceanic islands distant from major land masses within this range.

Countries: Native:

Australia; Bahrain; Bangladesh; Brunei Darussalam; Cambodia; China; Comoros; Egypt; Eritrea; India; Indonesia; Iran, Islamic Republic of; Japan; Kenya; Korea, Republic of; Madagascar; Malaysia; Mayotte; Mozambique; Myanmar; New Caledonia; Oman; Pakistan; Papua New Guinea; Philippines; Saudi Arabia; Singapore; Solomon Islands; Somalia; South Africa; Sri Lanka; Taiwan, Province of China; Tanzania, **Thailand;** Timor-Leste; United Arab Emirates; Yemen

Possibly extinct:

Vietnam

Population:

Few estimates of abundance have been made. There are estimated to be 520-530 bottlenose dolphins off KwaZulu-Natal, South Africa, most of which are probably *T. aduncus (*the rest *T. truncatus* – Cockcroft et al. 1992). Between 136 and 179 (95% Cis = 124-212) are known off Zanzibar, Tanzania (Stensland et al. 2006); at least 1,200 in the Persian Gulf (Preen 2004); at least 400 photo-identified along the rim of the Swatch-of-No-Ground, Bangladesh (Rubaiyat Mansur Mowgli and Brian D. Smith pers. comm.); about 218 off western Kyushu, Japan (Shirakihara et al. 2002); 169 off Mikura Island, Japan (Kogi et al. 2004); low tens at the southern tip of Taiwan and about 50 from the Northeastern Philippines (J. Y. Wang pers. comm.); 1,099 off Queensland, eastern Australia (Chilvers and Corkeron 2003) and at least 2,000-3,000 in Shark Bay western Australia (Preen et al. 1997).
Population Trend: Unknown

Habitat and Ecology:

Indo-Pacific bottlenose dolphins generally occur over shallow coastal waters on the continental shelf or around oceanic islands. They sometimes occur in mixed groups with common bottlenose dolphins and other delphinid species. They feed on a wide variety of schooling, demersal and reef fishes, as well as cephalopods (Ross 1984; J.Y. Wang, unpubl. data).

Systems: Marine. Diet:

Indo-Pacific bottlenose dolphins feed on a wide variety of fish and cephalopods (particularly squid).
In a recent study conducted by Amir et al. (2005),[10] researchers looked at the feeding ecology of Indo-Pacific bottlenose dolphins by analyzing the stomach contents of ones that got caught in the gillnet fisheries off Zanzibar, Tanzania. The prey items found in the stomach contents included 50 species of bony fish and three species of squid. From their results, the researchers concluded the most important prey group was fish, which accounted for 87% of the total number of prey items consumed and occurred in 24 of 26 stomachs examined. Cephalopods comprised the other 13% of prey items and were found in 13 of the 26 stomachs.[10] The remains of some crustaceans were also found; they hypothesize, however, they were consumed secondarily, since a number were found intact in the fish prey stomachs and therefore were not included in the diet analysis.

Behaviour:

Indo-Pacific bottlenose dolphins live in groups that can number in the hundreds, but groups of five to 15 dolphins are most common. In some parts of their range, they associate with the common bottlenose dolphin[6] and other dolphin species, such as the humpback dolphin.
The peak mating and calving seasons are in the spring and summer, although mating and calving occur throughout the year in some regions. Gestation period is about 12 months. Calves are between 0.84 and 1.5 metres (2.8 and 4.9 ft) long, and weigh between 9 and 21 kilograms (20 and 46 lb). The calves are weaned between 1.5 and two years, but can remain with their mothers for up to five years. The interbirth interval for females is typically four to six years.
In some parts of its range, this dolphin is subject to predation by sharks; its life span is more than 40 years. Indo-Pacific bottlenose dolphins located in Shark Bay, Australia are thought to have a symbiotic relationship with sponges by doing what is called "sponging". A dolphin breaks a marine sponge off the sea floor and wears it over its rostrum, apparently to probe substrates for fish, possibly as a tool or simply for play.

Major Threat(s):

The species' near-shore distribution makes it vulnerable to environmental degradation, direct exploitation, and fishery conflicts (Curry and Smith 1997, Wells and Scott 1999; Reeves et al. 2003). Until hunting was outlawed in 1990, this species was hunted in a large-scale drive fishery in Taiwan's Penghu Islands. Some Indo-Pacific bottlenose dolphins are taken in the small cetacean fisheries of Sri Lanka.

Incidental catches occur in a number of fisheries throughout the range, including gillnets and purse seines. A Taiwanese shark gillnet fishery operated in northern Australian waters during the early 1980s and took up to 2,000 per year (Harwood and Hembree 1987). Incidental catch in Taiwan continues to be a serious problem. For example, multiple individuals have been seen observed in single catches there and throughout most of the species' range (J.Y. Wang pers. comm.). A large proportion of dolphins (~40%) off Bangladesh exhibit scars and mutilations consistent with rope and net entanglement in trawl and gill-net fisheries (Rubaiyat Mansur Mowgli and Brian D. Smith pers. comm.). In South Africa and Australia, Indo-Pacific Bottlenose dolphins also suffer considerable mortality in the large-mesh nets set to protect bathers from sharks (Peddemors 1999; Reeves et al. 2003).

Live-captures for oceanarium display have taken place in Taiwan, Indonesia and the Solomon Islands in recent years from unassessed populations; their preference as a captive display species makes them vulnerable to depletion from such catches (Wang et al. 1999, Reeves et al. 2003, Kahn).

Indo-Pacific bottlenose dolphins in coastal areas are exposed to a wide variety of threats in addition to direct and indirect takes. Threats that are cause for concern include: 1) the toxic effects of xenobiotic chemicals; 2) reduced prey availability caused by environmental degradation and overfishing (Jackson et al. 2001); 3) direct and indirect disturbance and harassment (e.g. boat traffic and commercial dolphin watching and interactive programs); 4)marine construction and demolition and 5) other forms of habitat destruction and degradation (including anthropogenic noise). Although these and other threats are technically challenging to quantify by comparison with takes, their cumulative impact is likely to result in longitudinal population declines. Lack of historical data in many cases hampers understanding of long-term trends, possibly resulting in shifting baselines.

Conservation Actions:

The species is listed in Appendix II of CITES.
More research is needed to establish the range and clarify the taxonomy of the genus Tursiops. More information is also needed on population size and the extent and magnitude of direct and indirect takes so that their impact on this species can be assessed.

Effects of whale watching:

Not much is known about the impact of whale watching on cetaceans, but research is being conducted at several locations.

Japan:

Morisaka et al. (2005)[15] conducted a study on three populations of Indo-Pacific bottlenose dolphins in Japan. The characteristics of acoustic signals are believed to be affected by the acoustic environments among habitats, and geographical variation in animal acoustic signals can result from differences in acoustic environments; therefore, the characteristics of the ambient noise in the dolphin's habitats and the whistles produced were compared. Ambient noise was recorded using a hydrophone located 10m below the surface and whistles were recorded by using an underwater video system.

The results showed dolphins produced whistles at varying frequencies with greater modulations when in habitats with less ambient noise, whereas habitats with greater ambient noise seem to cause dolphins to produce whistles of lower frequencies and fewer frequency modulations. Examination of the results suggest communication signals are adaptive and are selected to avoid the masking of signals and the decrease of higher-frequency signals. They concluded ambient noise has the potential to drive the variation in whistles of Indo-Pacific bottlenose dolphin populations.

Jervis Bay, Australia:

Small, motorized vessels have increased as a source of anthropogenic noise due to the rise in popularity of wildlife viewing such as whale watching. Lemon et al. (2006))[16] showed powerboat approaches within 100m altered the dolphin surface behaviour from traveling to milling, and changed their direction to travel away from the powerboat. When the powerboat left the area and its noise ceased, the dolphins returned to their preceding behaviour in the original direction.

Shark Bay, Australia:

Another study by Bejder et al. (2006). in Shark Bay, Western Australia on dolphin behavioural responses showed there were significant changes in the behaviour of targeted dolphins when compared with their behaviour before and after approaches by small watercraft. Dolphins in the low traffic site showed a stronger and longer-lasting response than dolphins in the high traffic site. These results are believed to show habituation of the dolphins to the vessels in a region of long-term vessel traffic. However, when compared to other studies in the same area, moderated responses, rather, were suggested to be because those individuals sensitive to vessel disturbance left the region before their study began.

Although these studies do show statistical significance for the effects of whale-watching boats, these results do not have biological significance and need to be researched further.

Pantropical Spotted Dolphin *(Stenella attenuata)*

The Pantropical spotted dolphin *(Stenella attenuata)* is a species of dolphin found in all the world's temperate and tropical oceans. The species was beginning to come under threat due to the killing of millions of individuals in tuna purse seines. The 1980s saw the rise of "dolphin-friendly" tuna capture methods in order to save millions of the species in the eastern Pacific Ocean and it is now one of the most abundant dolphin species in the world.

Taxonomy:

The species was first described by John Gray in 1846. Gray's initial analysis included the Atlantic spotted dolphin in this species. They are now regarded as separate. Both the genus and specific names come from Latin words meaning thin or thinning.

There are three subspecies recognised in Rice's 1998 survey of cetacean taxonomy. Two of these have not been formally named

S. a. subspecies A, the off-shore form found in the eastern Pacific

S. a. subspecies B, a form found around the Hawaiian islands.

S. a. graffmani, coastal form found from Mexico to Peru

Description:

The Pantropical Spotted Dolphin varies significantly in size and colouration throughout its range. The most significant division is between coastal and pelagic varieties. The coastal form is larger and more spotted. (These two forms have been divided into subspecies only in eastern Pacific populations. The most distinctive feature of the Pantropical spotted dolphin is, as its name suggests, the spots that speckle the body of adults. Newborn calves are unspotted, but by adulthood, a varying amount of light spots cover the upper surface, and dark spots cover the dolphin's underside. Underneath this spotting, the slender, stream-lined body is grey, with a darker grey cape extending back from the head and sweeping low underneath the dorsal fin. The dorsal fin is narrow and sickle-shaped.. The long, thin beak of the Pantropical spotted dolphin is separated from the melon by a distinct crease. In most adults, the tip of the beak is white . Male Pantropical spotted dolphins are slightly larger than females. A subspecies of the Pantropical spotted dolphin is recognised, Stenella attenuate graffmani, which inhabits more coastal areas and can be distinguished in appearance by its larger, stockier body, thicker beak and more extensive spotting.

Spots are key defining characteristics in adults, though immature individuals are generally uniformly coloured and susceptible to confusion with the bottlenose dolphin. Populations around the Gulf of Mexico may be relatively spot-free even in adulthood. In the Atlantic, confusion is possible with the Atlantic spotted dolphin.

Broadly speaking the dolphin has a long thin beak. The upper and lower jaws are darkly coloured but are separated by thin white "lips". The chin, throat and belly are white to pale grey with a limited amount of spots. The flanks are separated into three distinct bands of colour — the lightest at the bottom, followed by a thin grey strip in the middle of the flank and a dark grey back. The tall concave dorsal fin is similarly coloured. The thick tail stock matches the colour of the middle band.

The Pantropical spotted dolphin is very active and is prone to making large splashy leaps from the sea. It is a common breacher and will often clear the water for a second or more. Bow-riding and other play with boats is common.

In the eastern Pacific, the Dolphin is often found swimming with yellowfin tuna (hence the problem with dolphin deaths caused by tuna fishing — see the human interaction section). However they do not feed on that fish. In fact the two species have a similar diet of small epipelagic fish. In other areas the species may also feed on squid and crustaceans.

Population and distribution:

The Pantropical spotted dolphin, as its very name implies, is found across all tropical and sub-tropical waters around the world — roughly speaking all oceans and seas between 40° N and 40° S. The total world population is in excess of three million — the second most abundant cetacean after the bottlenose dolphin — of which two million are found in the eastern Pacific. However, this represents a decrease from at least 7 million since the 1950s. Centres of highest population density are the shallow warmest waters (water temperature in excess of 25 °C). There is also a tendency for groups to concentrate where there is a high temperature gradient.

Range Description:

S. a. attenuata is Pantropical, found in all oceans between about 40°N and 40°S, although it is much more abundant in the lower-latitude portions of its range. The range extends to some enclosed seas, such as the Red Sea and Persian Gulf, but does not include the Mediterranean Sea (Perrin 2001, 2002) al. 2005).

Countries: Native:

American Samoa (American Samoa); Argentina; Australia; Bahamas; Bangladesh; Belize; Brazil; Cambodia; Cape Verde; Cayman Islands; China; Colombia; Comoros; Cook Islands; Costa Rica; Côte d'Ivoire; Cuba; Djibouti; Dominica; Dominican Republic; Ecuador; Egypt; El Salvador; Equatorial Guinea; Fiji; French Polynesia; Gabon; Ghana; Grenada; Guadeloupe; Guam; Guatemala; Guinea; Haiti; Honduras; Hong Kong; India; Indonesia; Jamaica; Japan; Kenya; Kiribati; Madagascar; Malaysia; Maldives; Marshall Islands; Martinique; Mauritania; Mauritius; Mexico; Mozambique; Myanmar; Netherlands Antilles; New Caledonia; New Zealand; Nicaragua; Northern Mariana Islands; Oman; Pakistan; Panama; Papua New Guinea; Peru; Philippines; Puerto Rico; Réunion; Saint Helena, Ascension and Tristan da Cunha; Saint Lucia; Saint Vincent and the Grenadines; Saudi Arabia; Senegal; Seychelles; Singapore; Solomon Islands; Somalia; South Africa; Sri Lanka; Taiwan, Province of China; Tanzania, **Thailand;** Togo; Tonga; Trinidad and Tobago; Turks and Caicos Islands; Tuvalu; United Arab Emirates; United States; Uruguay; Vanuatu; Venezuela, Bolivarian Republic of; Vietnam

FAO Marine Fishing Areas: Native:

Atlantic – eastern central; Atlantic – northwest; Atlantic – southeast; Atlantic – southwest; Atlantic – western central; Indian Ocean – eastern; Indian Ocean – western; Pacific – eastern central; Pacific – northwest; Pacific – southeast; Pacific – southwest; Pacific – western central

Population:

In the eastern Pacific, there were an estimated 228,038 coastal spotted dolphins in 2000 (CV=34%; Gerrodette and Forcada 2002a). The north-eastern offshore spotted dolphin (the form most affected by the ETP tuna fishery) numbered about 737,000 in 2003 (CV=15%; Gerrodette et al. 2005), a reduction of 76% from original size in 1959 (Reilly et al. 2005). This population is not showing clear signs of recovery despite the dramatic decline in mortality in recent years (Gerrodette and Forcada 2005). The western/southern offshore stock (which is less affected by the fishery) numbered about 876,075 in 2000 (CV=31%; Gerrodette and Forcada 2002b). In Hawaiian waters, there are an estimated 8,978 (CV=48%) (Barlow 2006). About 438,000 inhabited Japanese waters in the early 1990s (Miyashita 1993). There are estimated to be 34,067 (CV=18%) in the northern Gulf of Mexico (Mullin 2006), and 4,439 (CV=49%) along the east coast of the United States (Waring et al. 2006). Dolar et al (2006) estimated about 14,930 (CV=41%) for the eastern Sulu Sea and 640 (CV=27%) for the Tañon Strait between the islands of Negros and Cebu.

Population Trend: Unknown

Biology:

The gregarious Pantropical spotted dolphin forms schools that can range in size from less than one hundred to thousands of individuals; although it has been observed that these impressively large herds are less common in the eastern tropical Pacific than they once were, as exploitation takes its toll. The Pantropical spotted dolphin is well known for its tendency to associate with schools of tuna in this region. While this may be due to an overlap in diet, other reasons for this association have also been suggested, such as increased protection from predators, as there is safety in numbers. In the eastern Pacific the Pantropical spotted dolphin is an inhabitant of the tropical, equatorial and southern subtropical water masses. The waters in which the animal occurs with greatest frequency are those underlain by a sharp thermocline at depths of less than 50 m and with surface temperatures over 25°C and salinities less than 34 parts per thousand. These conditions prevail year round in the region north of the Equator called the "Inner Tropical" waters of the eastern Pacific. Occurrence in this core habitat is correlated with apparent multi-species foraging and feeding behaviour. The species also occurs in similar waters south of the Equator that expand and contract greatly with season and year to year (Perrin and Hohn 1994). In the Atlantic, S. attenuata is primarily a dolphin of the high seas and oceanic islands, but in the eastern Pacific a large-bodied subspecies occurs along the coast from Mexico to Peru. Detailed analysis of oceanographic correlates of distribution will be necessary in order to understand fully the habitat requirements of these pelagic dolphins, often the most conspicuous elements of tropical cetacean communities around the world (Ballance and Pitman 1998).

Offshore spotted dolphins feed largely on small epi- and mesopelagic fishes, squids, and crustaceans that associate with the deep scattering layer (Robertson and Chivers 1997). In some areas, flying fish are also important prey. The diet of the coastal form is poorly known, but is thought to consist mainly of larger fishes, perhaps mainly bottom-living species (Perrin 2001, 2002).

Systems: Marine

his ocean mammal is a fast swimmer that often engages in a range of aerial acrobatics and will frequently ride the bow waves of boats, except for in tuna fishing grounds where it has learnt to avoid vessels. Juveniles in particular are known to make astoundingly high vertical leaps out of the water. The Pantropical spotted dolphin feeds mainly at night on small fish, squid and crustaceans that rise to near the surface at dusk, with flying fish forming a major part of the diet in some regions. In turn, this dolphin becomes prey for the killer whale *(Orcinus orca)* and a number of sharks.

While the breeding system of this species is not known, it is possible that it may be promiscuous, like that of the closely related spinner dolphin (Stenella longirostris) (2). Every two to three years, mature female Pantropical spinner dolphins give birth to a calf, after a gestation period of around 11 months. The calf is nursed for between one and two years. Females reach sexual maturity at 9 to 11 years, while males become sexually mature between the ages of 12 and 15 years. Birth length is 80-90 cm. Adults are about 2.5 m long and weigh 120 kg. Sexual maturity is reached at 10 years in females and 12 years in males. Lifespan is approximately 40 years.

Major Threat(s):

Offshore spotted dolphins bore the brunt of the massive dolphin kill by tuna seiners from the late 1950s to the 1980s in the eastern Pacific (although the coastal subspecies was also impacted). For example, in the period 1959 to 1972, nearly five million dolphins were killed, and of this number, about three million were from the north-eastern offshore population (Wade 1995). Since the Inter-American Tropical Tuna Commission (IATTC) implemented per-vessel mortality limits on the international fleet, the combined annual mortality for all spotted dolphins in the ETP has decreased greatly, e.g. to only 373 in 2005 (IATTC 2006). Although current mortality is greatly reduced, the north-eastern form appears to be recovering very slowly, if at all, and potential factors such as fishery-related stress, unobserved mortality due to calf separation and orphaning during fishing operations (Archer et al. 2001), possible mortality by small vessels that do not carry observers, under-reporting of mortality, and ecosystem change, have been suggested as possible reasons for the species' slow recovery (Gerrodette and Forcada 2005).

Spotted dolphins are also taken incidentally in local fisheries along the Central American coast (Palacios and Gerrodette 1996). Yang et al. (1999) also reported incidental mortality in Chinese fisheries, and Dolar 1994 found incidental spotted dolphin takes in the Philippines. An unknown but suspected large number of Pantropical spotted dolphins are taken by the large-mesh pelagic driftnet fishery off eastern Taiwan (J. Wang pers. comm.).

Japan takes large numbers of spotted dolphins for human consumption. The catch in 1982 was 3,799, and annual catches between 1994 and 1997 ranged from 23 to 449 (Perrin 2002). Between 1995 and 2004, the average annual catch was 129 animals (Kasuya 2007). The drive fishery for spotted dolphins began in 1959 and is thought to have caused a slight decline in the minimum age at attainment of sexual maturity in females (Kasuya 1985).

Pantropical spotted dolphins are also taken in hand-harpoon fisheries in the Philippines (Dolar et al. 1994); in Taiwan, where it is the locally preferred species of cetacean for human consumption (J. Wang pers. comm.); and regularly or opportunistically by gillnet and harpoon in India and Sri Lanka (Perrin and Hohn 1994). Drive hunts at Malaita in the Solomon Islands took several hundred or thousands of spotted dolphins annually in the 1960s; the hunts continue at present (Ross et al. 2003, Kahn 2006). Small numbers are taken in numerous small subsistence fisheries for dolphins and whales around the world, e.g. at St. Vincent in the Lesser Antilles (Perrin and Hohn 1994) and Lamalera in Indonesia (Kahn 2004). Most of these kills have not been adequately monitored and the effects on the subpopulations are usually not known.

Dolphins and small whales of several species, including S. attenuata, putatively interfere in hook-and-line fisheries for squid and yellowtail in the Iki Island region of Japan (Kishiro and Kasuya 1993). Bounties have been paid to fishermen for dolphins killed since 1957. During the period 1976-1982 a total of 538 spotted dolphins were killed. The effect of these takes on the regional population is not known (Perrin and Hohn 1994).

Conservation Actions:

The species is listed in Appendix II of CITES.

Spotted dolphins, as with other species impacted by the ETP tuna purse-seine fishery, are managed both nationally by the coastal countries and internationally by the IATTC. The IATTC has imposed annual stock mortality limits on each purse seine and promulgated regulations regarding the safe release of dolphins (Bayliff 2001).

As the species comprises several subspecies and regional populations, the conservation status of each of these should be assessed separately since the available estimates of abundance and removals suggest that some of them may fall into a Threatened category.

Spinner Dolphin *(Stenella longirostris)*

Description:

The spinner dolphin is a small cetaceans with a slim build. Adults are typically 129–235 cm and reach a body mass of 23–79 kg. This species has an elongated rostrum and a triangular or sub-triangular dorsal fin. Spinner dolphins generally have tripartite colour patterns. The dorsal area is dark gray, the sides light gray, and the underside pale gray or white. There is also a dark band that runs from the eye to the flipper, bordered above by a thin light line. However, the spinner dolphin has more geographic variation in form and coloration than other cetaceans. In the open waters of eastern Pacific, dolphins have relatively small skulls with short rostra. There is also a dwarf form of spinner dolphin that occurs around southeast Asia. In these same subspecies, a dark dorsal cape dims their tripartite colour patterns. Further offshore, subspecies tend to have a paler and less far-reaching cape. In certain subspecies, some males may have upright fins that slant forward. Some populations of spinner dolphin found in the eastern Pacific have bizarre backwards facing dorsal fins, and males with strange humps and upturned caudal flukes.

Ecology:

The spinner dolphin lives in nearly all tropical and subtropical waters between 40°N and 40°S. The species is primarily inhabit inshore waters, islands or banks. However, in the eastern tropical Pacific, dolphins live far from shore. Spinner dolphin may use different habitats depending on the season.

The spinner dolphin feed mainly on small mesopelagic fish, squids and sergestid shrimps and will dive 200-300m to feed on them. Spinner dolphins of Hawaii are nocturnal feeder and forage in deep scattering layers, which contain many species. The dwarf spinner dolphin may eat mostly on benthic fish in reefs and shallow water. Off Oahu, Hawaii, spinner dolphins foraging at night and cooperatively herd their prey in highly dense patches. They swim around circle of prey and a pair may swim through it to make a catch. Spinner dolphins are themselves preyed on by sharks. Other possible predators include then killer whale, the false killer whale, the pygmy killer whale and the short finned pilot whale. They are also susceptible to parasites, both external ones like barnacles and remoras, and internal ones like nematodes, trematodes, cestodes and acanthocephalans. Spinner dolphins move about the oceans in schools; groups that vary in size from just a few dolphins to over a thousand. They commonly school with other species such as Pantropical spotted dolphins, or small toothed whales. In such schools, spinner dolphins are known to undertake migrations, following prey or warm water currents. In Hawaii, spinner dolphins usually spend their days resting in shallow bays near deep water, and then move offshore at dusk and feed as they move substantial distances along the shore (8). Pelagic spinner dolphins feed primarily on small mesopelagic fish, squids and shrimps, and dive down to depths of 300 meters to catch their prey . The dwarf spinner dolphin feeds on reef fishes and other benthic organisms.

Mating in spinner dolphins appears to be promiscuous, and like many small dolphins, true courtship behaviour can be observed, such as mutual caressing between the male and female. The breeding system may vary geographically, with some populations showing a greater degree of polygyny than others . Calves are born every three years, after a gestation period of about ten months. The mother nurses the calf for up to two years, and they form a bond that lasts a lifetime. Females reach sexual maturity between four and seven years, whereas males do not reach maturity until between seven and ten years.

The purpose of the energetic spinning behaviour of the spinner dolphin is not known. It has been suggested that the large cloud of bubbles created by the powerful spin and splash landing may act as an echolocation target, to allow a widely dispersed school of dolphins to communicate. Another theory is that the spinning may dislodge hitch -hiking remoras, or the spinning may, at times, simply be play.

Behaviour and life history:

In certain regions, such as Hawaii and northern Brazil, dolphins spend the daytime resting in shallow bays near deep water. At dusk, they travel offshore to feed. They travel along the shore during foraging trips, and the individuals that occupy the same bay may change daily. Some individual dolphin do not anyways go to a Bay to rest. However, in Hawaii, dolphins do seem to return to the same site each trip.

Spinner dolphins live in an open and loose social organization. The spinner dolphins of Hawaii live in family groups but also have associations with others beyond their groups. Mothers and calves form strong social bonds. Spinner dolphins seem to have a promiscuous mating system, with individuals changing partners for up to some weeks. A dozen adult males may gather into coalitions. Vocalizations of spinner dolphins include whistles, which may be used to organization the structure of the school; burst-pulse signals, which may serve to be evocative and vocative and echolocation clicks. The spinner dolphin has a 10 months gestation period, and mothers nurse their young for 1–2 years. Females are sexuality mature at 4–7 years, with 3 year calving intervals, while males are sexually mature by 7–10 years. Breeding is seasonal, more so in certain regions than others.

Spinning behaviour:

Spinner dolphins are well known for their acrobatics and aerial behaviours. A spinner dolphin comes out of the water, front first, and twists its body as ascends in the air. After it reaches its maximum height the dolphin descends back into the water, landing on its side. A dolphin can make 2-5.5 spins in one leap. The swimming and rotational speed of the dolphin spinning underwater affects the number of spins it can do while airborne. These spins may serve multiple functions. Dolphins may also make nose-outs, tail slaps, flips, head slap, "salmon leaps" and side and back slaps.

Threats:

Spinner dolphins in the eastern tropical Pacific have been killed incidentally since the early 1960s by tuna purse seine fisheries. They were caught in such large numbers that the population of *S. l. orientalis* was reduced to less than one third of its original size. Following raised awareness of the number of dolphins killed in tuna purse seine fisheries, measures were implemented to reduce dolphin by-catch. Today spinner dolphins continue to be killed in this way, although in greatly reduced numbers. However, continued chase, capture and release of large numbers in the fishery may be preventing the population from recovering .

In Sri Lanka and the Philippines, large numbers of spinner dolphins have also been captured in gillnets and killed by harpoons for the past 20 years, and local harpoon fisheries exist in several more locations throughout the world. Incidentally captured dolphins are consumed by local people, or used as shark bait, and this has led to the development of markets and fisheries directed at dolphins. The takes in these fisheries may be unsustainable.

Conservation:

The eastern tropical Pacific and Southeast Asian populations of the spinner dolphin are listed on Appendix II of the Convention on Migratory Species (CMS). This means that the spinner dolphin is a migratory species that needs, or would significantly benefit from international co-operation, and the convention encourages the range states to conclude global or regional agreements. The spinner dolphin is also listed on Appendix II of the Convention on International Trade in Endangered Species (CITES), meaning that international trade in this species should be carefully regulated. The value of dolphins as a tourist attraction offers countries and communities an incentive to protect these beautiful animals. Fernando de Noronha National Marine Park; (an archipelago off equatorial Brazil) was established in 1988, to provide nominal protection to spinner dolphins and support dolphin watching tourism, and it has been reported that in Zanzibar, the value of spinner dolphins for tourism far exceeded that of using them as bait for sharks (8). It is hoped that with meaningful laws, the will and resources to enforce them, continued attention by non-governmental organizations and efforts to make the public aware of the intrinsic value of their endemic dolphins, this charismatic species will continue spinning in our oceans forever.

Taxonomy:

The spinner dolphin is sometimes referred to as the **long-snouted dolphin**, particularly in older texts, to distinguish it from the similar Clymene dolphin, which is often called the short-snouted spinner dolphin. The species was described by John Gray in 1828. The four named subspecies are:

- **Eastern spinner dolphin** (*S. l. orientalis*), from the tropical eastern Pacific.
- **Central American** or **Costa Rican spinner dolphin** (*S. l. centroamericana*), also found in the tropical eastern Pacific.
- **Gray's** or **Hawaiian spinner dolphin** (*S. l. longirostris*), from the central Pacific Ocean around Hawaii but represents a mixture of broadly similar subtypes found worldwide.
- **Dwarf spinner dolphin** (*S. l. roseiventris*), first found in the **Gulf of Thailand.**

The species, though, displays greater variety than these subspecies might indicate. A hybrid form characterized by its white belly inhabits the eastern Pacific. Other less distinct groupings inhabit other oceans.

Fraser's Dolphin (Lagenodelphis hosei)

Taxonomy:

In 1895, Charles E. Hose found a skull on a beach in Sarawak, Borneo. He donated it to the British Museum. The skull remained unstudied until 1956 when Francis Fraser examined it and concluded that it was similar to species in both the Lagenorhynchus and Delphinus genera but not the same as either. A new genus was created by simply merging these two names together. The specific name is given in Hose's honour. It wasn't until 1971 that the whole body of a Fraser's Dolphin, as it was by then becoming known, was discovered. At that time washed-up specimens were found on Cocos Island in the eastern Pacific, in South Australia and in South Africa.

Description:

Fraser's dolphins are about 1 metre (3 ft 3 in) long and 20 kg weight at birth, growing to 2.75 metres (9 ft 0 in) and 200 kg at adulthood. They have a stocky build, a small fin in relation to the size of the body, conspicuously small flippers. The dorsal fin and beak are also insubstantial. The upper side is a gray-blue to gray-brown. A dirty cream colored line runs along the flanks from the beak, above the eye, to the anus. There is a dark stripe under this line. The belly and throat are usually white, sometimes tinged pink. The lack of a prominent beak is a distinguishing characteristic of the dolphin. From a distance however it may be confused with the Striped Dolphin which has a similar coloration and is found in the same regions. Fraser's dolphins swim quickly in large tightly packed groups of about 100 to 1000 in number. Often porpoising, the group chop up the water tremendously. The sight of seeing a large group fleeing from a fishing vessels has been reported as "very dramatic". It is also marked by having the smallest genitalia of any open sea dolphin. The species feeds on pelagic fish, squid and shrimp found some distance below the surface of the water (200 metres (660 ft) to 500 metres (1,600 ft)). Virtually no sunlight penetrates this depth, so feeding is carried out using echolocation alone.

Biology and Ecology:

The preference of Fraser's dolphin for deep waters is due to the prey on which it feeds; fish, squid and crustacean species that inhabit the deeper waters of the oceans. Feeding on such food requires Fraser's dolphin to dive down to depths of at least 250 to 500 metres to hunt. It is thought that Fraser's dolphin itself may be occasional prey for killer whales, false killer whales and large sharks, and circular wounds caused by the peculiar cookie-cutter shark (Isistius brasiliensis) have been found on this species.

Fraser's dolphins are highly sociable mammals that swim around in tightly-bonded schools of 100 to 1,000 individuals, often together with schools of melon-headed whales *(Peponocephala electra)*, other dolphin species, or in some areas, such as the Sulu Sea, with short-finned pilot whales *(Globicephala macrorhynchus)*. A school of Fraser's dolphins moves quickly, on very rare occasions riding the bow waves of boats, and with members of the school frequently porpoising; the term used to describe a dolphin leaping clear of the water when surfacing to breathe. It is an oceanic species that prefers deep offshore waters, but it can be seen near shore in some areas where deep water approaches the coast (such as the Philippines, Taiwan, and some islands of the Caribbean and the Indo-Malay archipelago) (Perrin et al. 1994). In the eastern tropical Pacific, it occurs more often in Equatorial - southern subtropical surface water and other waters typified by upwelling and generally more variable conditions (Au and Perryman 1985). Off South Africa, records are associated with the warm Agulhas Current that moves south in the summer (Perrin et al. 1994). Fraser's dolphins feed on midwater fish (especially myctophids), squid, and crustaceans (Dolar et al. 2003). Physiological studies indicate that Fraser's are capable of quite deep diving (and it is thought that they do most of their feeding deep in the water column – in waters up to 600 m deep), but they have been observed to feed near the surface as well (Watkins et al. 1994).
Systems: Marine Mating in Fraser's dolphin is believed to be promiscuous, and mature females give birth approximately every two years to a metre-long calf, after a gestation period of 12.5 months. Males reach sexual maturity at an age of seven to ten years, while females are able to reproduce at five to eight years of age.

Population and distribution:

Though only accounted for relatively recently, the number of reported sightings has become substantial—indicating that the species may not be as rare as thought as recently as the 1980s. However the species is still not nearly as well understood as its more coastal cousins. No global population estimates exist. The Dolphin is normally sighted in deep tropical waters; between 30° S and 20° N. The Eastern Pacific is the most reliable site for viewings. Groups of stranded dolphins have been found as far a field as France and Uruguay. However these are regarded as anomalous and possibly due to unusual oceanographic conditions, such as El Niño. The species is also relatively common in the Gulf of Mexico but less so in the rest of the Atlantic Ocean. Inhabits deep, oceanic waters except in places where deep water approaches the coast, such as in the Philippines and Indonesia. The exact distribution of this species is poorly known. Fraser's dolphin has a Pantropical distribution, largely between 30°N and 30°S in all three major oceans (Jefferson and Leatherwood 1994; Dolar 2002). Strandings in temperate areas (Victoria in Australia, Brittany and Uruguay) may represent extralimital forays connected with temporary oceanographic anomalies such as the world-wide El Niño phenomenon in 1983-84, during which a mass stranding occurred in France (Perrin et al. 1994). Bones et al. (1998) reported on a stranding on the coast of Scotland.

Countries: Native:

Angola (Angola); Argentina; Aruba; Australia; Brazil; Brunei Darussalam; Cape Verde; China; Colombia; Comoros; Cook Islands; Costa Rica; Dominica; Ecuador; French Polynesia; Ghana; Indonesia; Japan; Kenya; Kiribati; Madagascar; Malaysia; Maldives; Mayotte; Micronesia, Federated States of; Nauru; Oman; Palau; Panama; Papua New Guinea; Philippines; Puerto Rico; Saint Vincent and the Grenadines; Samoa; Senegal; Solomon Islands; South Africa; Sri Lanka; **Thailand;** United States; Uruguay; Venezuela, Bolivarian Republic of; Vietnam

Vagrant:

France; United Kingdom

Population:

There are estimated to be about 289,300 (CV=34%) Fraser's dolphins in the eastern tropical Pacific (Wade and Gerrodette 1993; Perrin et al. 1994), and 16,836 (CV=11%) in Hawaiian waters (Carretta et al. 2006). In the eastern Sulu Sea, Dolar et al. (2006) estimated a total abundance of 13,518 (CV=27%) Fraser's dolphins. About 726 (CV=70%) were estimated present in the northern Gulf of Mexico (Waring et al. 2006).

Population Trend: Unknown

Major Threat(s):

Small numbers of Fraser's dolphins are taken regularly or opportunistically by harpoon in the Lesser Antilles, Sri Lanka, Indonesia (Kahn 2004), the Philippines, Taiwan and probably elsewhere in the Indo-Pacific (Jefferson and Leatherwood 1994). A few have been taken in drive fisheries in Taiwan and Japan (Perrin et al. 1994). Dolar et al. (1994) investigated directed fisheries for marine mammals in central and southern Visayas, northern Mindanao and Palawan, Philippines. Some of the hunters take only dolphins, for bait or human consumption and the species taken include Fraser's dolphins. Around 800 cetaceans are taken annually by hunters at the seven sites, mostly during the inter-monsoonal period of February-May. Some Fraser's dolphins are killed incidentally in the tuna purse-seine fishery in the eastern tropical Pacific (Gerrodette and Wade 1991): 26 were estimated taken during the period 1971 - 75. A few are also taken in gill nets in Sri Lanka, the Philippines, and likely in other tropical gillnet fisheries as well. Some are killed by anti-shark nets in South Africa (Perrin et al. 1994; Cockcroft 1990). Other incidental catches in purse seines (Philippines), gillnets, driftnets (Taiwan), and trap nets (Japan) are also known (Jefferson and Leatherwood 1994).

Conservation Actions:

The species is listed in Appendix II of CITES
The Southeast Asian subpopulations are listed in Appendix II of CMS. Subpopulation structure and the impact of direct and incidental takes require further investigation.

Risso's Dolphin *(Grampus griseus)*

Taxonomy:

Risso's dolphin is named after Antoine Risso, whose description formed the basis of the first public description of the animal, by Georges Cuvier, in 1812. Another common name for the Risso's dolphin is grampus (also the species' genus), although this common name was more often used for the orca. The etymology of the word grampus is unclear. It may be an agglomeration of the Latin grandis piscis or French grand poisson both meaning big fish. The specific epithet griseus refers to the mottled (almost scarred) grey colour of its body.

Description:

Risso's have a relatively large anterior body and dorsal fin, while the posterior tapers to a relatively narrow tail. The bulbous head has a vertical crease in front.

Infants are dorsally gray to brown and ventrally cream-colored, with a white anchor-shaped area between the pectorals and around the mouth. In older calves, the non-white areas darken to nearly black, and then lighten (except for the always dark dorsal fin.) Linear scars mostly from social interaction eventually cover the bulk of the body. Older individuals appear mostly white. Most individuals have 2–7 pairs of teeth, all in the lower jaw. Length is typically 10 feet (3.0 m) although specimens may reach 14.1 feet (4.3 m). Like most dolphins, males are typically slightly larger than females. This species weighs 300–500 kilograms (660–1,100 lb) making it the largest species called "dolphin".

Range and habitat:

They are found worldwide in temperate and tropical waters, usually in deep waters rather than close to land. As well as the tropical parts of the Indian, Pacific and Atlantic Oceans, Rissos are also found in the Persian Gulf, Mediterranean and Red Seas, but not the Black Sea. They range as far north as the Gulf of Alaska and southern Greenland and as far south as Tierra del Fuego.Range Description: This is a widely-distributed species, inhabiting primarily deep waters of the continental slope and outer shelf (especially with steep bottom topography), from the tropics through the temperate regions in both hemispheres (Kruse et al. 1999). It also occurs in some oceanic areas, beyond the continental slope, such as in the eastern tropical Pacific. It is found from Newfoundland, Norway, the Kamchatka Peninsula, and Gulf of Alaska in the north to the tips of South America and South Africa, southern Australia, and southern New Zealand in the south. Its range includes many semi-enclosed bodies of water, such as the Gulf of Mexico, Gulf of California, Red Sea, Persian Gulf, Sea of Japan, and Mediterranean Sea.

Countries: Native:

Algeria; American Samoa (American Samoa); Anguilla; Antigua and Barbuda; Argentina; Aruba; Australia; Bahamas; Bangladesh; Barbados; Belgium; Belize; Benin; Bermuda; Brazil; British Indian Ocean Territory; Brunei Darussalam; Cambodia; Cameroon; Canada; Cape Verde; Cayman Islands; Chile; China; Cocos (Keeling) Islands; Colombia; Comoros; Congo; Congo, The Democratic Republic of the; Cook Islands; Costa Rica; Côte d'Ivoire; Croatia; Cuba; Denmark; Djibouti; Dominica; Dominican Republic; Ecuador; Egypt; El Salvador; Equatorial Guinea; Estonia; Fiji; France; French Guiana; French Polynesia; Gabon; Gambia; Germany; Ghana; Greece; Greenland; Grenada; Guadeloupe; Guam; Guatemala; Guernsey; Guinea; Guinea-Bissau; Guyana; Haiti; Honduras; Hong Kong; India; Indonesia; Iran, Islamic Republic of; Iraq; Ireland; Isle of Man; Israel; Italy; Jamaica; Japan; Jersey; Jordan; Kenya; Kiribati; Kuwait; Lebanon; Liberia; Madagascar; Malaysia; Maldives; Malta; Marshall Islands; Martinique; Mauritania; Mayotte; Mexico; Monaco; Morocco; Mozambique; Myanmar; Namibia; Netherlands; Netherlands Antilles; New Caledonia; New Zealand; Nicaragua; Nigeria; Niue; Northern Mariana Islands; Norway; Oman; Pakistan; Palau; Panama; Papua New Guinea; Peru; Philippines; Pitcairn; Portugal; Puerto Rico; Qatar; Russian Federation; Saint Kitts and Nevis; Saint Lucia; Saint Pierre and Miquelon; Saint Vincent and the Grenadines; Samoa; Sao Tomé and Principe; Saudi Arabia; Senegal; Serbia (Serbia); Sierra Leone; Singapore; Slovenia; Solomon Islands; Somalia; South Africa; Spain; Sri Lanka; Sudan; Suriname; Sweden; Syrian Arab Republic; Tanzania,; **Thailand;** Timor-Leste; Togo; Tonga; Trinidad and Tobago; Turkey; Turks and Caicos Islands; United Arab Emirates; United Kingdom; United States; United States Minor Outlying Islands; Uruguay; Vanuatu; Venezuela, Bolivarian Republic of; Viet Nam; Virgin Islands, British; Virgin Islands, U.S.; Wallis and Futuna; Western Sahara; Yemen

Vagrant:

Finland; Lithuania; Poland

Population:

There are no estimates of global abundance, but there are some estimates for specific areas. Forney and Barlow (1998) observed that the estimated abundance of Risso's dolphins off California was almost an order of magnitude higher in winter (N= 32,376) than in summer (N= 3,980). However, the California, Oregon, Washington subpopulation is now estimated at only 16,066 (CV=28%) whales (Barlow 2003). Hawaiian waters are estimated to contain 2,351 (CV=65%) Risso's dolphins (Barlow 2006). Abundance estimates off Sri Lanka ranged from 5,500 to 13,000 animals (Kruse et al. 1999). In the eastern Sulu Sea, Dolar et al. (2006) estimated the abundance at 1514 (CV=55%) individuals. There are an estimated 20,479 (CV=59%) Risso's dolphins off the eastern United States (Waring et al. 2006), 2,169 (CV=32%) in the northern Gulf of Mexico (Mullin and Fulling 2004), 83,300 (CV=17%) in three areas of concentrated occurrence off Japan (Miyashita 1993), and 175,000 in the eastern tropical Pacific (Wade and Gerrodette 1993). In relative terms, there are several examples of long term changes in abundance and distribution, e.g. in the Southern California Bight (Kruse et al. 1999). In the late 1950s, Risso's dolphins were rarely encountered in this area, and between 1975 and 1978 they were still considered to be a minor constituent of the cetacean fauna of the Bight, representing only 3% of the cetaceans observed. After the El Niño of 1982/83, however, numbers of Risso's dolphins increased, especially around Santa Catalina Island where they came to be considered common (Shane 1995). There is no information on global trends in the abundance.

Population Trend: Unknown

Their preferred environment is just off the continental shelf on steep banks with water depths varying from 400–1,000 meters (1,300–3,300 ft) and water temperature at least 10 °C (50 °F) and preferably 15–20 °C (59–68 °F). The population around the continental shelf of the United States is estimated in excess of 60,000. In the Pacific a census recorded 175,000 individuals in eastern tropical waters and 85,000 in the west. No global estimate exists.

Ecology:

They feed almost exclusively on neritic and oceanic squid, mostly nocturnally. Predation does not appear significant. Mass strandings are infrequent. These dolphins typically travel in groups of 10–51, but that may reach 400. Smaller, stable subgroups exist within larger groups. They also travel with other cetaceans. They harass and surf the bow waves of gray whales as well as ocean swells.Risso's dolphins inhabit deep oceanic and continental slope waters, generally 400-1,000 m deep (Baird 2002; Jefferson et al. 1993), mostly occurring seaward of the continental slope. They frequent subsurface seamounts and escarpments, where they are thought to feed on vertically migrant and mesopelagic cephalopods. In Monterey Bay, California, Risso's dolphins are concentrated over areas with steep bottom topography (Kruse 1989). Currents and upwelling causing local increases in marine productivity may enhance feeding opportunities, resulting in the patchy distribution and local abundance of this species worldwide (Kruse et al. 1999). Davis et al. (1998) and Baumgartner (1997) reported that in the Gulf of Mexico, Risso's dolphins were mostly found over deeper bottom depths, concentrating along the upper continental slope, which may reflect squid distribution. Most records of *Grampus griseus* in Britain and Ireland are within 11 km of the coast. In certain areas, such as in the southwest English Channel, Risso's dolphins are known to occur seasonally in shallow coastal waters to feed on cuttlefishes *Sepia officinalis* (Kiszka et al., 2004). Long-term changes in the occurrence of Risso's dolphins in some areas (e.g., off Catalina Island and in central California) have been linked to oceanographic conditions and movements of spawning squid (Kruse et al. 1999). Risso's dolphins feed on crustaceans and cephalopods, but seem to prefer squid. Squid bites may be the cause of at least some of the scars found on the bodies of these animals. In the few areas where feeding habits have been studied, they appear to feed mainly at night.

Systems: Marine

Reproduction:

Gestation requires an estimated 13–14 months, at intervals of 2.4 years. Calving reaches seasonal peaks in the winter in the eastern Pacific and in the summer and fall in the western Pacific. Females mature sexually at ages 8–10, and males at age 10–12. The oldest specimen reached 34.5 years.

Major Threat(s):

Occasional direct killing of Risso's dolphins has occurred. This is generally as a result of the dolphins removing fish from loglines, or in multi-species small cetacean fisheries, such as those that occur in Sri Lanka, the Caribbean, and Indonesia. One regular hunt occurs in Japan, where about 250-500 are taken per year in a drive fishery. Some Risso's dolphins have been captured for live display in oceanaria, although there are not many of them in oceanaria. In Sri Lanka, Risso's dolphins are apparently the second most commonly taken cetacean in fisheries, providing fish and meat for human consumption and fish bait; subpopulations there may be adversely affected (see Jefferson et al. 1993; Kruse et al. 1991). An estimated 1,300 Risso's dolphins may be landed annually as a result of this fishery, and abundance estimates in these waters range only from 5,500 to 13,000 animals (Kruse et al. 1999). In Japan, Risso's dolphins are taken periodically for food and fertilizer in set nets and as a limited catch in the small-type whaling industry (Kruse et al. 1999), with reported catches in recent years ranging from about 250–500. They are also a major target of artisanal hunting, and are taken often in gillnets and other fishing gear in the Philippines (Dolar 1994, Dolar et al. 1994). Off eastern Taiwan, Risso's dolphins are also taken by harpoon opportunistically and oceanic large-mesh driftnets for large pelagic fish appear to take considerable numbers incidentally (Wang pers. comm.). There are reports of bycatches from the North Atlantic, the Mediterranean Sea, the southern Caribbean, the Azores, Peru, and the Solomon Islands. They are also a rare bycatch in the US tuna purse seine industry, and are taken occasionally in coastal gill net and squid seining industries off the US coast, or shot by aggravated fishermen (Kruse et al. 1999). This species, like beaked whales that are also deep-divers that feed on squid, is likely to be vulnerable to loud anthropogenic sounds, such as those generated by navy sonar and seismic exploration (Cox et al. 2006). Predicted impacts of global climate change on the marine environment may affect this species of whale, although the nature of impacts is unclear (Learmonth et al. 2006).

Conservation Actions:

The species is listed in Appendix II of CITES. The North and Baltic Sea subpopulations are included in Appendix II of CMS. This is a circumglobal species, which migrates between summering and wintering grounds. Off California, where these movements are best known, they may cross between US and Mexican waters. Data on abundance, bycatch, and behavior needed in order to develop conservation measures that will enable protection of the natural habitat of the species

Pygmy Killer Whale *(Feresa attenuata)*

Description & Ecology:

Pygmy killer whales, *Feresa attenuata* (Gray, 1874), are very similar in appearance to melon-headed whales and to juvenile false killer whales. From a distance it is very difficult to tell these three apart. The shape of the head, the dorsal fin and the flippers of pygmy killer whales are different from the other two. The head of *Feresa attenuata* is rounded and lacks a beak. They have an under slung jaw and white lips, and usually a white patch on the tip of the lower jaw. The skull is asymmetrical and the right jaw is smaller and usually has one less tooth than the left jaw. The teeth are large and conical. There are usually 8-11 pairs of teeth in the upper jaw and 11-13 pairs of teeth in the lower jaw. Pygmy killer whales have a sub-triangular, long based, high dorsal fin with a tip that points backward. The dorsal fin is located near the centre of the body and lacks rigidity, often inclining to the side. The flippers of pygmy killer whales are moderate in length and have rounded tips. The body of *Feresa attenuata* is slender, though the midsection forward is slightly more robust than the midsection back. Pygmy killer whales are not whale-sized at all; rather, they are average-sized dolphins and part of the Family Delphinidae. An adult ranges in length from 2.1-2.6 m and weighs between 110-170 kg. The colouring is dark gray to black with some paler markings on the underside, as well as some white on the belly. Feresa attenuata has a groove on the skin of its belly that extends from up front to their umbilicus (navel) to their anus. In both males and females this groove contains the umbilicus, the anus and the genitals. Longevity in this species is poorly understood. Pygmy killer whales are described as aggressive animals that have been seen snapping their jaws, beating their flippers and flukes on the surface of the water, and growling. In captivity they elicit fear reactions from other cetaceans. They will charge, bite, and snap their jaws at other cetaceans as well as their trainers. They will often kill all other cetaceans that are in the tank with them. Though aggressive toward other animals pygmy killer whales are shy of vessels. At the water's surface, pygmy killer whales are quite acrobatic. They are often seen leaping, spy hopping, tail slapping, and occasionally bow riding. Social groups of this species are usually made up of approximately 25 individuals. In areas where they are more frequently sighted, such at the Hawaiian Islands, herds contain up to 50 individuals and on rare occasions a few hundred. The pygmy killer whale occurs in deep, warm waters, generally beyond the edge of the continental shelf, and rarely close to shore (except near some oceanic island groups where the water is deep and clear). This species is mainly tropical, but occasionally strays into warm temperate regions. Little is known of the diet of this species, although it is known to eat fish and squid. It has occasionally been recorded attacking dolphins, at least those involved in tuna fishery interactions in the eastern tropical Pacific (Perryman and Foster 1980). Reproductive biology is poorly known in this species. Pygmy killer whales usually occur in groups of 50 or less. Both sexes may remain in their birth groups. They are generally less active than other oceanic dolphins; frequently they are seen "logging"--resting in groups at the surface with all animals oriented the same way.

Systems: Marine

Range Description:

This is a tropical/subtropical species that inhabits oceanic waters around the globe generally between 40°N and 35° S. It does not generally approach close to shore, except in some areas where deep, clear waters are very close to the coast (such as around oceanic archipelagos like Hawaii). Reports of its occurrence in the Mediterranean Sea, while common in the literature, are not supported by authenticated records. It is also doubtful whether it occurs regularly in the Red Sea or Persian Gulf (Leatherwood et al. 1991). A few high-latitude strandings and sightings are thought to be extralimital records, and are generally associated with incursions of warm water (Ross and Leatherwood 1994; Donohue and Perryman 2002; Williams et al. 2002).

Countries: Native:

Algeria; Anguilla; Antigua and Barbuda; Aruba; Australia; Bahamas; Bangladesh; Barbados; Belize; Benin; Brazil; Brunei Darussalam; Cambodia; Cameroon; Cayman Islands; China; Cocos (Keeling) Islands; Colombia; Comoros; Congo; Congo, The Democratic Republic of the; Cook Islands; Costa Rica; Côte d'Ivoire; Cuba; Djibouti; Dominica; Dominican Republic; Ecuador; El Salvador; Equatorial Guinea; Fiji; France; French Guiana; French Polynesia; Gabon; Gambia; Ghana; Grenada; Guam; Guatemala; Guinea; Guinea-Bissau; Guyana; Haiti; Honduras; India; Indonesia; Iran, Islamic Republic of; Italy; Jamaica; Japan; Kenya; Kiribati; Korea, Republic of; Liberia; Madagascar; Malaysia (Peninsular Malaysia, Sabah, Sarawak); Maldives; Marshall Islands; Mauritania; Mayotte; Mexico; Micronesia, Federated States of; Morocco; Mozambique; Myanmar; Nauru; Netherlands Antilles (Bonaire, Curaçao, Netherlands Leeward Is.); New Caledonia; Nicaragua; Nigeria; Niue; Northern Mariana Islands; Oman; Pakistan; Palau; Panama; Papua New Guinea; Peru; Philippines; Pitcairn; Portugal; Puerto Rico; Saint Kitts and Nevis; Saint Vincent and the Grenadines; Samoa; Senegal; Sierra Leone; Singapore; Solomon Islands; Somalia; South Africa; Spain; Sri Lanka; Suriname; Taiwan, Province of China; Tanzania, **Thailand;** Timor-Leste; Togo; Tokelau; Tonga; Trinidad and Tobago; United States; Venezuela, Bolivarian Republic of; Vietnam; Virgin Islands, British; Virgin Islands, U.S.; Wallis and Futuna; Western Sahara; Yemen

Population.

Although there is little information on the population biology of this species, the pygmy killer whale appears to be naturally uncommon. Wade and Gerrodette (1993) estimated that there were about 38,900 (CV=31%) of these whales in the eastern tropical Pacific. There are estimated to be 817 (CV=112%) in the Hawaiian portion of the US EEZ, and 408 (CV=60%) in the northern Gulf of Mexico (Barlow 2006; Mullin and Fulling 2004).

Population Trend: Unknown

Major Threat(s):

Because of their relatively low abundance, even small takes in localized areas could be significant. Although there is considerable controversy regarding the absolute level of declines, there is good evidence of large-scale reductions in many predatory fish populations (e.g., Baum et al. 2003, 2005; Sibert et al. 2006; Polacheck 2006) and over-fishing and collapse of several important "prey" fish stocks world-wide (e.g., Jackson et al. 2001). The effects of such fish population reductions and subsequent ecosystem changes on world-wide populations of pygmy killer whales are unknown but could result in population declines. Predicted impacts of global climate change on the marine environment may affect pygmy killer whales, although the nature of impacts is unclear (Learmonth et al. 2006). This species, like beaked whales, is likely to be vulnerable to loud anthropogenic sounds, such as those generated by navy sonar and seismic exploration (Cox et al. 2006) and have been a part of multi-species unusual stranding events in Taiwan (Wang and Yang, 2006). Pygmy killer whales have been killed directly in both harpoon and driftnet fisheries (Caribbean islands, Sri Lanka, Taiwan and Indonesia) and incidentally in various types of fishing gear (most areas of the species' range). A few individuals are known to be taken in drives and in driftnets in various regions, most notably Japan and Sri Lanka (Ross and Leatherwood 1994). Reports on the small-cetacean fisheries of St Vincent and Lamalera suggest that pygmy killer whales form a very small proportion of the catch and that catches probably have little impact on the subpopulations in those areas. In Sri Lanka, there is mortality of this species due to harpooning of dolphins for use as bait on long-lines for sharks, billfish, and other oceanic fishes (Ross and Leatherwood 1994). Although they comprise less than 2% of all cetaceans in monitored by-catches in gillnet fisheries in Trincomalee, Sri Lanka and in villages on the south-west coast of Sri Lanka, this may amount to 300 - 900 pygmy killer whales each year (Ross and Leatherwood 1994). The numbers of animals killed incidentally in net fisheries, such as those in Sri Lanka, may be much higher than is so far documented, because monitoring of these widespread activities is incomplete. In the long term, such takes may have a significant impact on pygmy killer whales where their distribution overlaps with extensive gillnetting operations (Ross and Leatherwood 1994). Small incidental catches are known in fisheries in other areas including the Philippines and Taiwan (Ross and Leatherwood 1994, Dolar 1994, J. Wang pers. comm. 2007). Evidence from stranded individuals of several similar species indicates that they have swallowed discarded plastic items, which may eventually lead to death (e.g. Scott et al. 2001); this species may also be at risk. This species does not appear to be particularly abundant anywhere that it has been sighted. In Hawaii, subpopulations appear to be small, and this, along with their limited movements, suggests the species may be particularly vulnerable to human impacts regionally.

Conservation Actions:

The species is listed on Appendix II of CITES. Research is needed to determine the impact of potential threats on this species.

Irrawaddy Dolphin *(Orcaella brevirostris)*

Etymology and taxonomic history:

The Irrawaddy dolphin was first described by Sir Richard Owen in 1866 based on a specimen found in 1852, in the harbour of Visakhapatnam on the east coast of India.[3] It is one of two species in its genus. It has sometimes been listed variously in a family containing just itself and in Monodontidae and in Delphinapteridae. There is now widespread agreement to list it in the Delphinidae family.

Genetically, the Irrawaddy dolphin is closely related to the killer whale (orca). The species name brevirostris comes from the Latin meaning short-beaked. In 2005, genetic analysis showed the Australian snub fin dolphin found at the coast of northern Australia forms a second species in the Orcaella genus. Overall, the dolphins' colour is grey to dark slate blue, paler underneath, with no distinctive pattern. The dorsal fin is small and rounded behind the middle of the back. The forehead is high and rounded; the beak is lacking. The flippers are broad and rounded. The species found in Borneo, the finless porpoise, Neophocaena phocaenoides, is similar and has no back fin; the humpback dolphin, *Sausa chinensis,* is larger, has a longer beak and a larger dorsal fin.

Vernacular names for *O. brevirostris* include the following.

Burmese: labai

Chilika dialect: baslnyya magar or bhuasuni magar (lit. oil-yielding dolphin)

Filipino: lampasut

Indonesia: pesut

Khmer: □□□□□ ph'sout

Lao: pha'ka

Oriya: khem or khera

Thai: pla loma hua bat ("alms-bowl dolphin", due to the shape of their head)

Irrawaddy dolphins are similar to the beluga in appearance, though most closely related to the orca. They have a large melon and a blunt, rounded head, and the beak is indistinct. The dorsal fin, located about two-thirds posterior along the back, is short, blunt and triangular. The flippers are long and broad. It is lightly coloured all over, but slightly more white on the underside than the back. Adult weight exceeds 130 kg (290 lb) and length is 2.3 m (7.5 ft) m at full maturity. Maximum recorded length is 2.75 m (9.0 ft) of a male from **Thailand.**

Ecology:

Irrawaddy dolphins communicate with clicks, creaks and buzzes at a dominant frequency of about 60 kilohertz, which is thought to be used for echolocation. Bony fish and fish eggs, cephalopods, and crustaceans are taken as food. Observations of captive animals indicate food may be taken into the mouth by suction. Irrawaddy dolphins sometimes spit streams of water, sometimes while spy hopping, during feeding, apparently to expel water ingested during fish capture or possibly to herd fish. Some Irrawaddy dolphins kept in captivity have been trained to do spy hopping on command.

The Irrawaddy dolphin is a slow swimmer, but swimming speeds of 20–25 km/hour were reported when dolphins were being chased in a boat. It surfaces in a rolling fashion and lifts its tail fluke clear of the water only for a deep dive. Deep dive times range from 70–150 seconds to 12 minutes. When 277 group dives were timed (time of disappearance of last dolphin in group to emergence of first dolphin in the group) in Laos, mean duration was 115.3 seconds with a range of 19 seconds to 7.18 minutes. They make only occasional low leaps and never bow -ride. Groups of fewer than six individuals are most common, but sometimes up to 15 dolphins are seen together. Intraspecific competition has been observed when Orcaella was forced inshore and excluded by more specialised dolphins. When captive humpback dolphins *(Sonsa chinensis)* and Irrawaddy dolphins were held together, reportedly the Irrawaddy dolphins were frequently chased and confined to a small portion of the tank by the dominant humpbacks. In Chilika Lake, local fishers say when Irrawaddy dolphins and bottlenose dolphins meet in the outer channel, the former get frightened and are forced to return toward the lake. These dolphins are thought to reach sexual maturity at seven to nine years. In the Northern Hemisphere, mating is reported from December to June. Its gestation period is 14 months; cows give birth to a single calf every two to three years. Length is about 1 m (3.3 ft) at birth. Birth weight is about 10 kg (22 lb). Weaning is after two years. Lifespan is about 30 years. Irrawaddy dolphins prefer coastal areas associated with the muddy, brackish waters at river mouths, ranging offshore as far as the extent of the freshwater plume – often only a few km but more than 60 km at the Meghna River mouth in Bangladesh (Smith et al. 2005). In rivers and mangrove channels, the species is most often observed at channel confluences and divergences and downstream of sharp meanders. They have been seen in the same area as finless porpoises in coastal waters of Bangladesh and Myanmar (Smith et al. 2005), and Ganges River dolphins in the waterways of the Sundarbans mangrove forest (Smith et al. 2006).

Systems: Freshwater; Marine

Range:

Irrawaddy dolphins have a discontinuous distribution in the tropical and subtropical Indo-Pacific, almost exclusively in estuarine and fresh waters (Stacey and Arnold 1999; Arnold 2002). They occur from Borneo and the central islands of the Indonesian Archipelago north to Palawan, Philippines, and west to the Bay of Bengal, including the Gulf of Thailand. There are freshwater subpopulations in three large rivers: Ayeyarwady (up to 1,400 km upstream) in Myanmar, Mahakam (up to 560 km upstream) in Indonesia, and Mekong (up to 690 km upstream) in Viet Nam, Cambodia and Lao PDR, and two marine-appended brackish water bodies or lakes: Chilika in India and **Songkhla in Thailand.** The fine-scale range of the species is poorly documented throughout much of its range in estuarine waters.

Countries: Native:

Bangladesh; Brunei Darussalam; Cambodia; India; Indonesia; Lao People's Democratic Republic; Malaysia; Myanmar; Philippines; Singapore; **Thailand;** Vietnam

Although sometimes called the Irrawaddy river dolphin, it is not a true river dolphin, but an oceanic dolphin that lives in brackish water near coasts, river mouths and in estuaries. It has established subpopulations in freshwater rivers, including the Ganges and the Mekong, as well as the Irrawaddy River from which it takes its name. Its range extends from the Bay of Bengal to New Guinea and the Philippines.

It is often seen in estuaries and bays in Borneo Island, with sightings from Sandakan in Sabah, Malaysia, to most parts of Brunei and Sarawak, Malaysia. A specimen was collected at Mahakam River in East Kalimantan.
No range-wide survey has been conducted for this vulnerable species; however, the worldwide population appears to be over 7,000, with over 90% occurring in Bangladesh. Populations outside Bangladesh and India are classified as critically endangered. Known subpopulations of Irrawaddy dolphins are found in eight places, listed here in order of population, including conservation status.

Chilka Lake, Orissa, India, habitat of Irrawaddy dolphins

Bangladesh; 5,832 (VU) in coastal waters of the Bay of Bengal[8] and 451 (VU) in the brackish Sundarbans mangrove forest

India; 138 (VU) in the brackish water Chilka Lake

Laos and Cambodia; 78-91 (CR) in a 190 km (118.1 mi) freshwater stretch of the Mekong River

Philippines; about 42 (CR) in the brackish inner Malampaya Sound. Researchers are studying the recent discovery of 30-40 dolphins sighted in the waters of Bago City and Pulupandan town in the province of Negros Occidental, in Western Visayas

Burma; about 58-72 (CR) in a 370 km (229.9 mi) freshwater stretch of the Ayeyarwady River

Thailand: less than 50 (CR) in the brackish Songkhla Lake.

Population:

No range-wide survey has been conducted for this species; nor is there a synoptic estimate of total numbers from local or regional surveys. Statistically rigorous abundance estimates are available for only a few portions of the range: 77 (CV = 27.4%) in Malampaya Sound, Philippines (Smith et al. 2004); at least 125 (95% CI = 114-152) in the Mekong River (Beasley et al. 2007); 70 (CV = 10%; 95% CL = 58-79) in the Mahakam River, Indonesia (Kreb et al. 2007); 58-72 in the Ayeyarwady River, Myanmar (Smith et al. 2007-a); 5,383 (CV=40%) in coastal waters of Bangladesh (Smith et al. 2005); and 451 (CV=9.6%) in the Sundarbans mangrove forest of Bangladesh (Smith et al,/i>. 2006).

Major Threat(s):

The estuarine and freshwater occurrence of this species makes it particularly vulnerable to threats from the human activities that occur in these environments. Threats include direct mortality from fisheries interactions (particularly gillnet entanglement), vessel strikes, and habitat loss and degradation (e.g. declining or altered freshwater flows due to dam and embankment construction, environmental contamination). Live captures for aquarium display also have been a conservation issue in some local areas. Irrawaddy dolphins have been hunted directly in the past, at least in the Mekong and Mahakam Rivers, but are revered by local people in many areas of Asia.

Irrawaddy dolphins are caught accidentally in fishing nets in almost all areas where they have been studied (Smith et al. 2007-b). The most detailed information on bycatch comes from the Mekong River where, of 15 confirmed human-caused deaths 2001-2005, 13 (87%) were due to gillnet entanglement (Beasley et al. 2007). Based on reports from local fishermen and the retrieval of eight carcasses along the Mahakam River between 1995 and 2005, Kreb et al. (2007) documented 48 deaths, 66% of them from entanglement in large-mesh (10 –17.5 cm) gillnets. Mortality also has been recorded in drift gillnets targeting elasmobranches in coastal waters of Bangladesh (Smith et al. 2005) and bottom-set gillnets targeting crabs in Malampaya Sound (Smith et al. 2004). Fishermen in some areas report the dolphins are released if found still alive (Smith and Hobbs 2002, Kreb et al. 2007), but in the case of drowned animals, the oil may be used for medicinal purposes or the flesh eaten (Smith et al. 2004).

There have been no systematic observer schemes in freshwater or coastal regions, but evidence of bycatch and increased use of gillnets is cause for concern (IWC 2000). Fishing with electricity is considered a dire threat to the Ayeyarwady subpopulation (Smith et al. 2007-a).

Many dams have been proposed that are likely to degrade the channels inhabited by Irrawaddy dolphins in the Mekong River Basin. Of greatest concern are the large run-of-the-river dams (dams without a reservoir that generally preserve a relatively natural flow regime) proposed for the Mekong main stem near Stung Treng and Sambor (Perrin et al. 1996; Mekong Secretariat 1995). Dam projects in Lao PDR, Cambodia and Viet Nam threaten not only dolphins but also fisheries and therefore human livelihoods (Smith et al. 2007-b). A recent report of a high dam planned for the headwaters of the Ayeyarwady River, Myanmar, in Myitsone just below the confluence of the Mali Hka and N'Mai Hka tributaries, provides reason for concern about its effects on the population of Irrawaddy dolphins downstream (Anon. 2007).

Deforestation and gold, sand and gravel mining are causing major changes to the geomorphologic and hydraulic features of rivers and marine-appended lakes where Irrawaddy dolphins occur (Smith et al. 2007-b). Increased sedimentation resulting from deforestation in surrounding watersheds has resulted in declining water depths in Songkhla, Chilika and Semayang Lakes. The last of these water bodies is appended to the Mahakam River and previously supported dolphins throughout most of its breadth. Now it contains suitable habitat only in a small area near the channel connecting it with the main stem (Kreb et al. 2007). Between 1992 and 1997 the maximum depth of Chilika Lake declined from 3.4 to 1.4 meters and the accumulation of sediments led to shrinkage of the opening channel and a dramatic decline in salinity. A new channel dredged in the northern portion of the lake in 2000 apparently has mitigated at least some of the problems caused by sedimentation (Pattnaik et al. 2007).

Habitat loss and population fragmentation in several areas have resulted from the proliferation of fixed fishing gears. In the middle and southern portions of Songkhla Lake about 27,000 Sai nong or sitting traps and 13,000 Sang sai or barrier traps create more than 8000 linear km of barrier in multiple rows. These fishing structures are left in place year-round and restrict dolphin movements such that their habitat is substantially reduced and the potential for demographic interaction with individuals in the Gulf of Thailand is eliminated (Smith et al. 2004).

Conservation Actions:

The species is listed in Appendix I of CITES.

The Action Plan for the Conservation of Freshwater Populations of Irrawaddy Dolphins (Smith et al. 2007-c) notes that multiple-use protected areas will play a key role for conserving freshwater populations of Irrawaddy dolphins. Protected areas could be a particularly effective conservation tool due to the fidelity of the species in freshwater systems to relatively circumscribed areas, as this can facilitate management. The Action Plan also provided details on strategies for mitigating bycatch that included establishing core conservation areas where gillnetting is banned or severely restricted; promoting net attendance rules and providing training on the safe release of entangled dolphins; initiating a program to compensate fishermen for damage caused to their nets by entangled dolphins that are safely released; providing alternative or diversified employment options for gillnet fishermen; encouraging the use of fishing gears that do not harm dolphins by altering or establishing fee structures for fishing permits to make gillnetting more expensive while decreasing the fees for non-destructive gears; and experimenting with acoustical deterrents and reflective nets.

Order: Carnivora (carnivores)

The carnivores include over 260 species, the majority of which eat meat as their primary dietary item. Carnivores have a characteristic skull shape and dentition.

Asian Golden Cat *(Catopuma temminckii)*

The Asian golden cat *(Pardofelis temminckii, syn. Catopuma temminckii),* also called the Asiatic golden cat and Temminck's cat, is a medium-sized wild cat of South Eastern Asia. In 2008, the IUCN classified Asian golden cats as Near Threatened, stating that the species comes close to qualifying as Vulnerable due to hunting pressure and habitat loss, since Southeast Asian forests are undergoing the world's fastest regional deforestation rates. The Asian golden cat was named in honour of the Dutch zoologist Coenraad Jacob Temminck, who first described the African golden cat in 1827.

Description:

The Asian golden cat is heavily built, with a typical cat-like appearance. It has a head-body length of 66 to 105 cm (26 to 41 in), with a tail 40 to 57 cm (16 to 22 in) long, and is 56 cm (22 in) at the shoulder. The weight ranges from 9 to 16 kg (20 to 35 lb), which is about two or three times the size of a domesticated cat. The pelage is uniform in colour, but highly variable ranging from red to golden brown, dark brown to pale cinnamon, gray to black. Transitional forms among the different colorations also exist. It may be marked with spots and stripes. White and black lines run across the cheeks and up to the top of the head, while the ears are black with a central grey area. Golden cats with leopard-like spots have been found in China, resembling a large leopard cat. This spotted fur is a recessive characteristic.

Distribution and Habitat:

Asian golden cats live throughout Southeast Asia, ranging from Tibet, Nepal, Bhutan, India and Bangladesh to Myanmar, **Thailand,** Cambodia, Laos, Vietnam, Southern China to Malaysia and Sumatra. They prefer forest habitats interspersed with rocky areas, and are found in dry deciduous, subtropical evergreen and tropical rainforests. Sometimes, they are found in more open terrain such as the grasslands of Assam's Manas National Park. In altitude, they range from the lowlands to over 3,000 m (9,800 ft) in the Himalayas.

I

n Laos, they also occur in bamboo regrowth, scrub and degraded forest from the Mekong plains to at least 1,100 m (3,600 ft).Surveys in Sumatra and in the Nam Et-Phou Louey National Protected Area in northern Laos indicated that they are more common than sympatric small cats, suggesting that they are more numerous than previously believed. **Surveys in Thailand,** Northern Myanmar and India's western Arunachal Pradesh revealed fewer numbers. In Bhutan's Jigme Singye Wangchuck National Park, they were recorded by camera traps at an altitude of 3,738 m (12,264 ft). In Sikkim's Khangchendzonga Biosphere Reserve, they were photo-trapped at elevations up to 3,960 m (12,990 ft).

Since Hodgson's description of 1831 of a male individual in Nepal under the binomial *Felis moormensis,* the country is believed to be the westernmost part of the felid's range. But no specimen has been recorded in the country, until in May 2009 a camera trap survey yielded the first photographic record of a melanistic Asian golden cat in Makalu Barun National Park at an altitude of 2,517 m (8,258 ft).

Three subspecies have been recognized.

Pardofelis temminckii temminckii found in the Himalayas, Southeast Asian mainland and Sumatra
Pardofelis temminckii dominicanorum found in southeast China
Pardofelis temminckii tristis found in southwest China

These trinomials do not yet reflect the taxonomic re-classification accepted since 2006.

Native:

Bangladesh; Bhutan; Cambodia; China; India; Indonesia (Sumatera); Lao People's Democratic Republic; Malaysia; Myanmar; Nepal**; Thailand;** Vietnam.

Population:

Some surveys have found more records for the Asiatic golden cat than for some other sympatric small felids (including the marbled cat, flat-headed cat, and fishing cat), suggesting that it may be more common than previously thought (Duckworth et al. 1999, Holden 2001, Duckworth et al. 2005, J. Sanderson pers. comm. 2007). Other survey efforts, however, have turned up fewer records (Rao et al. 2005, Lynam et al. 2006, Mishra et al. 2006). Given that its distribution (Nowell and Jackson 1996), home range (Grassman et al. 2005) and camera trap encounter rates (Holden 2001, Lynam et al. 2006) are similar to those of clouded leopards (*Neofelis nebulosa and Neofelis diardi diardi),* it may be of roughly equivalent abundance.
Population Trend: Decreasing

Ecology and behavior:

Asian golden cats are territorial and solitary. Previous observations suggested that they are primarily nocturnal, but a field study on two radio-collared specimens revealed arrhythmic activity patterns dominated by crepuscular and diurnal activity peaks, with much less activity late at night. The male's territory was 47.7 km2 (18.4 sq mi) in size and increased more than 15% during the rainy season. The female's territory was 32.6 square kilometres (12.6 sq mi) in size. Both cats travelled between only 55 metres (180 ft) to more than 9 kilometres (5.6 mi) in a day and were more active in July than in March.
Asian golden cats can climb trees when necessary. They hunt birds, large rodents and reptiles, small ungulates such as Muntjac and young Sambar deer. They are capable of bringing down prey much larger than themselves, such as domestic water buffalo calves. In the mountains of Sikkim, they reportedly prey on ghoral. Their vocalizations include hissing, spitting, meowing, purring, growling, and gurgling. Other methods of communication observed in captive Asian Golden cats include scent marking, urine spraying, raking trees and logs with claws, and rubbing of the head against various objects, much like a domestic cat.

Not much is known about the reproductive behavior of this rather elusive cat in the wild. Most of what is known, has been learned in captivity. Female Asian golden cats are sexually mature between 18 and 24 months, while males mature at 24 months. Females come into estrus every 39 days, at which times they leave markings and seek contact with the male by adopting receptive postures. During intercourse, the male will seize the skin of the neck of the female with its teeth. After a gestation period of 78 to 80 days, the female gives birth to a litter of one to three kittens in a sheltered place. The kittens weigh 220 to 250 grams (7.8 to 8.8 oz) at birth, but triple in size over the first eight weeks of life. They are born already possessing the adult coat pattern, and open their eyes at six to twelve days. In captivity, they live for up to twenty years. The Asian golden cat bears a great resemblance to the African golden cat, but it is unlikely that they are closely related because the forests of Africa and Asia have not been connected in over 20 million years. Their similarity is more an example of convergent evolution.
The Asian golden cat is similar to the bay cat of Borneo in both appearance and behavior. Genetic studies revealed that the two species are very closely related. The Asian golden cat is found in Sumatra and Malaysia, which only separated from Borneo about 10,000 to 15,000 years ago. These observations led to the assumption that the Borneo bay cat is an insular subspecies of the Asian golden cat. Genetic analysis has shown that the Asian golden cat, along with the bay cat and the marbled cat, diverged from the other felids about 9.4 million years ago, and that the Asian golden cat and bay cat differed as long as four million years ago, suggesting that the bay cat was a different species long before the isolation of Borneo. Because of the evident close relationship with the marbled cat, it has recently been suggested that all three species should be grouped in the genus Pardofelis. In some regions of **Thailand,** the Asian golden cat is called Seua fai ("fire tiger"). According to a regional legend, the burning of an Asian Golden Cat's fur drives tigers away. Eating the flesh is believed to have the same effect. The Karen people believe that simply carrying a single hair of the cat will be sufficient. Many indigenous people believe this cat to be fierce, but in captivity it has been known to be very docile and tranquil.T he Asiatic golden cat is primarily found in forest habitats, ranging from tropical and subtropical evergreen to mixed and dry deciduous forest (Nowell and Jackson 1996). Less frequently it is found in more open habitats such as shrub and grasslands (Choudhury 2007). Grassman et al. (2005) found golden cats used closed forest and more open habitats in proportion to their occurrence, showing no significant preference. Some studies have suggested it may be less common in montane forest: in Sumatra's Kerinci Seblat National Park, all records for this species were from lowland forest with none from montane forest, unlike the clouded leopard and marbled cat (Holden 2001). Mishra (2006) also found clouded leopard and marbled cat, but no Asiatic golden cat, in the hill forests of India's western Arunachal Pradesh province. However, Wang (2007) obtained camera trap photos of the Asiatic golden cat at an elevation of 3,738 m in Bhutan's Jigme Sigye Wangchuk National Park in an area of dwarf rhododendron and grassland, an elevation record for the species. Activity readings from two radio-collared golden cats in **Thailand's** Phu Khieu National Park showed daytime and crepuscular activity peaks (Grassman et al. 2005). Forty-seven per cent of 15 camera trap records in Sumatra's Kerinci Seblat National Park were in daytime (Holden 2001). This suggests that the species is not primarily nocturnal, as thought previously. An adult female Asiatic golden cat in **Thailand's Phu Khieu National Park** had a home range of 32.6 km^2, overlapped 78% by a male whose home range was 47.7 km^2. Golden cat home ranges were larger than clouded leopard home ranges, although they were similar in activity and mean daily distance moved (Grassman et al. 2005).One confirmed scat contained the remains of Indochinese ground squirrel (Grassman et al. 2005). Scats from Sumatra contained rat and Muntjac remains, and the stomach contents of a carcass in **Thailand's Kaeng Krachan National Par**k included the remains of a small snake (Grassman 1998). While the reddish-gold pelage the cat is named for is the most common form, there are also are spotted (Wang 2007) and melanistic morphs (Holden 2001, Grassman et al. 2005).
Systems: Terrestrial

Major Threat(s):

The Asiatic golden cat is threatened primarily by habitat loss to deforestation, but it threatened by indiscriminate snaring (Holden 2001), and there is illegal trade in its pelt and bones (Nowell and Jackson 1996, Duckworth et al. 1999, Lynam et al. 2006). It has been reported killed for depredating livestock, including poultry but also larger animals such as sheep, goats and buffalo calves (Sunquist and Sunquist 2002). The Asiatic golden cat is capable of taking larger prey, and is threatened by declining ungulate abundance in many parts of Southeast Asia.

Conservation Actions:

P. temminckii is listed under Appendix I of CITES (as *Catopuma temminckii).* It is fully protected over most of its range by National legislation. Hunting is prohibited in Bangladesh, Cambodia, China, India, Indonesia, Peninsular Malaysia, Myanmar, Nepal, **Thailand** and Vietnam and is regulated in Lao PDR. There is no legal protection outside protected areas in Bhutan (Nowell and Jackson 1996). Its range includes many protected areas.

Jungle Cat *(Felis chaus)*

Description:

The jungle cat (Felis chaus) is a medium-sized cat native to Asia from southern China in the east through Southeast and Central Asia to the Nile Valley in the west. It is listed as Least Concern in the IUCN Red List of Threatened Species as it is widespread and common particularly in India. Population declines and range contraction are of concern, particularly in Egypt, in the Caucasus, and in southwestern, Central and Southeast Asia. Geographic variation in the jungle cat is quite considerable. Due to the small tuft on the ears it is also called the jungle lynx but is not a member of the Lynx genus. Sometimes more than twice as heavy as domestic cat, this species rivals the Chinese mountain cat as the largest member of the Felis genus. True to Bergmann's Rule, the species becomes smaller-bodied closer to the tropics and largest at the northern limits of the range. Jungle cats can range from 50 to 94 cm (20 to 37 in) in length, plus a relatively short 20 to 31 cm (7.9 to 12 in) tail, and stand about 36 cm (14 in) tall. Weight varies across the range from 3 to 16 kg (6.6 to 35 lb), with a median weight of around 8 kg (18 lb). Females are slightly smaller than males. Males are slightly larger than females. The face is relatively slender. Because of its long legs and short tail, and (in winter a tuft of black hair on its ears, this cat resembles a small lynx. The colour of the fur varies with subspecies, yellowish-grey to reddish-brown or tawny-grey, and is ticked with black. Vertical bars are visible on the fur of kittens, which disappear in adult cats, although a few dark markings may be retained on the limbs or tail. The muzzle is white, and the underside is paler in colour than the rest of the body. The skull is fairly broad in the region of the zygomatic arch, which leads to its appearance of having a rounder head than some other cats. The ears are quite long, and relatively broad at the base, pointed towards the end, and set quite high. Small tuft of long hairs occurs on ear tips in winter. These hairs form an indistinct tassel ranging from 7 to 20 mm (0.28 to 0.79 in) in length. The fur grows to about 4000 hairs/cm² on the back, and 1700 hairs/cm² on the abdomen, and generally becomes a shade of greyish-ochre in winter. The paw prints measure about 5×6 cm, and a typical pace is 29 to 32 cm (11 to 13 in). The most distinctive feature of a jungle cat is the presence of equal-sized claws on both fore and hind legs (unlike those of common domestic cats, for example, where hind claws are normally longer and stronger than fore). These allow it to climb down trees as easily as up, with its head facing downward.

Distribution and habitat:

Jungle cats are largely oriental in distribution and found in Egypt, West and Central Asia, but also in South Asia, Sri Lanka and Southeast Asia. In India they are the most common small cats among the felidae found there. They inhabit savannas, tropical dry forests and reed beds along rivers and lakes in the lowlands, but, despite the name, are not found in rainforests. Although they are adaptable animals, being found even in dry steppe, they prefer wetland environments with tall grasses or reeds in which to hide. They do not survive well in cold climates, and are not found in areas where winter snowfall is common.

They have been observed from sea levels to altitudes of 8,000 ft (2,400 m) or perhaps higher in the Himalayas. They frequent jungles or open country, and are often seen in the neighbourhood of villages. Jungle cats were known to be absent from south of the Isthmus of Kra in the Malayan peninsula, the possibility of their occurrence was reported from a highly fragmented forest in the Malaysian state of Selangor in 2010. Although never truly domesticated, a small number of jungle cats have been found among the cat mummies of Ancient Egypt (the vast majority of which are domestic cats), suggesting that they may have been used to help control rodent populations.

Distribution of subspecies:

When Johann Anton Güldenstädt travelled in the Russian empire's southern frontier during 1768–1775 at the behest of Catherine II of Russia, he was the first naturalist to catch sight of a Kirmyschak in the Caucasus. In his Latin description of 15 pages, published in 1776, he names the animal Chaus – a name retained for the cat by all subsequent zoologists.

Today, the trinomial *Felis chaus chaus* still refers to the jungle cat subspecies living in the Caucasus, Turkestan, Iran, Baluchistan and Yarkand, Chinese Turkestan. The other recognized subspecies are listed by year of first description:
Felis chaus affinis (Gray, 1830) – inhabits the Himalayan region ranging from Kashmir and Nepal to Sikkim and Yunnan.
Felis chaus kutas (Pearson, 1832) – ranges from Bengal westwards to Kutch.
Felis chaus nilotica (de Winton, 1898) – inhabits Egypt.
Felis chaus furax (de Winton, 1898) – inhabits Palestine, southern Syria, and Iraq.
Felis chaus fulvidina (Thomas, 1929) – inhabits Southeast Asia ranging from Myanmar and **Thailand** to Laos, Cambodia and Vietnam.
Felis chaus prateri (Pocock, 1939) – inhabits western India and Sindh.
Felis chaus kelaarti (Pocock, 1939) – lives in Sri Lanka and southern India south of the Kistna River.
Felis chaus oxiana (Heptner, 1969),lives along the right tributaries of the Amu Darya River, in the lower courses of the Vakhsh River ranging eastwards to the Gissar Valley and slightly beyond Dushanbe.

The jungle cat has a broad but patchy distribution. In Africa, it is found only in Egypt, along the Nile River Valley south to Aswan, and in El Faiyum, Farafara, Dakhla and Kharga oases (Glas in press). Into southwest Asia, it occurs through Israel, southern Lebanon, north-western Jordan, western Syria, and into Turkey and western Iraq (Abu-Baker et al., 2003). In this region its occurrence is highly localized around riparian vegetation and permanent water sources. In central Asia, it is found in the Caucasus mountains (up to 1,000 m), around the Caspian and Aral Seas and associated river valleys, and through Iran west to Pakistan. It occurs widely in tropical and sub-tropical Asia, including almost all of India as well as Sri Lanka, ranging up to 2,400 m in the Himalayan foothills, and through Southeast Asia to southern China, but absent from Malayan peninsula south of the Isthmus of Kra (Nowell and Jackson, 1996). Its range in Indochina is poorly known, especially in Myanmar. Duckworth et al. (2005) review the few historical distribution records from Indochina and add several recent ones from Lao PDR, Cambodia and Vietnam.

Countries: Native:

Afghanistan; Armenia (Armenia); Azerbaijan; Bangladesh; Bhutan; Cambodia; China; Egypt; Georgia; India; Iran, Islamic Republic of; Iraq; Israel; Jordan; Kazakhstan; Kyrgyzstan; Lao People's Democratic Republic; Lebanon; Myanmar; Nepal; Pakistan; Russian Federation; Sri Lanka; Syrian Arab Republic; Tajikistan; **Thailand;** Turkey; Turkmenistan; Uzbekistan; Vietnam

Population:

The jungle cat is considered common in some parts of its range, primarily in India (Mukherjee 1988), but also Pakistan and Bangladesh (Duckworth et al., 2005).

In southern China and Southeast Asia (with the exception of northeastern Cambodia), however, it appears quite rare in comparison to sympatric small cats (Duckworth et al. 2005). This rarity appears to be a relatively recent phenomenon associated with unselective trapping and snaring, especially in Lao PDR and **Thailand** as well, where it was described as common by Lekagul and McNeely (1977) but has since suffered drastic declines and is rarely encountered (Duckworth et al. 2005; Lynam et al. 2006, W. Duckworth and R. Steinmetz, Southeast Asia mammal assessment 2003).

In Europe, it is of marginal occurrence, with small populations in Cis-Caspian region and the Caucasus along the Caspian Sea. The European population has been rapidly declining since the 1960s. There was no records of this species in Astrakhan State Reserve (Russian Federation) since the 1980s. In Northern Ossetia (Russian Federation) only about 150 animals were recorded (Kuryatnikov and Varziev 1983). Marked population fluctuations are characteristic of this species in this region, probably because of absence of adaptations to cold winters. Despite these fluctuations the long-term trend in Europe is of decline in both population and area of occupancy. Data from Russia suggest that there are about 500 animals left in the wild (Prisazhnyuk and Belousova 2007). A very small population persists in Georgia (I. Macharashvili pers. comm. to K. Tsysulina 2007). This species is considered threatened in a number of range states in Europe and the Caucasus, and is included in the Red Books of the Russian Federation, Armenia, Azerbaijan and Georgia (IUCN 2007).

In Southwest Asia the species is also considered rare and threatened (Abu-Baker et al. 2003, Habibi 2004).

Population Trend: Decreasing

Ecology:

Jungle cats are solitary in nature. They rest in other animals' abandoned burrows, tree holes, and humid coves under swamp rocks, or in areas of dense vegetation. Although often active at night, they are less nocturnal than many other cats, and in cold weather may sun themselves during the day. They have been estimated to travel between 3 and 6 kilometres (1.9 and 3.7 mi) per night, although this likely varies depending on the availability of prey. Territories are maintained by urine spraying and scent marking. Jungle cats can climb trees. Like most cats, they utilize not only sight and hearing while hunting, but also their sense of smell. While running, they tend to sway from side to side. They mostly hunt for rodents, frogs, birds, hares, squirrels, juvenile wild pigs, as well as various reptiles, including turtles and snakes. Near human settlements, they feed on domestic chickens and ducks. They catch fish while diving, but mostly swim in order to disguise their scent trails, or to escape threats, such as dogs or humans. They are generally hard to tame, even if taken into captivity at a young age. Like most other cats, they hunt by stalking and ambushing their prey, and they use reeds or tall grass as cover. They are adept at leaping, and sometimes attempt to catch birds in flight. Although they can run at up to 32 kilometres per hour (20 mph), they rarely pursue prey that escapes their initial pounce. The jungle cat's main competitors are the jackal and forest cat. Their most common predators include crocodiles, bears, wolves, and other larger felines such as tigers. When countered by a threat, the jungle cat will vocalize before engaging in attack, producing small roars, a behavior uncommon for domestic cats. The meow of the jungle cat is also somewhat lower than that of a typical domestic cat. In some cases, they jump on their attacker, but will usually retreat upon encountering larger threats. There have been known cases of jungle cats attacking curious humans near their habitat, but their attack seems to pose no medical significance besides wound infection from clawing.

they have been observed to be capable of swimming as much as 1.5 km at a time. Females are sexually mature at the age of 11 months; estrus appears to last from January through to mid-April. In males, spermatogenesis occurs mainly in February and March. In southern Turkmenistan, mating occurs in January to early February. Females give birth to litters of three to five kittens, usually only three. They sometimes raise two litters in a year. Gestation lasts 63–66 days and is remarkably short for an animal of this size. Birth generally takes place between December and June, depending on the local climate, although females can sometimes give birth to two litters in a year. Before birth, the mother prepares a den in an abandoned animal burrow, hollow tree, or reed bed. Kittens weigh 43 to 160 grams (1.5 to 5.6 oz) at birth, tending to be much smaller in the wild than in captivity. Initially blind and helpless, they open their eyes at ten to thirteen days of age, and are fully weaned by around three months. Males usually do not participate in the raising of kittens, but in captivity have been observed to be very protective of their offspring, more than the females, or males of other cat species. Kittens begin to catch their own prey at around six months, and leave the mother after eight or nine months. The jungle cat's median life expectancy in captivity is ten to twelve years. In the wild, however, some jungle cats have been known to live for as long as twenty years. The jungle cat, despite its name, is not strongly associated with the classic rainforest "jungle" habitat, but rather with wetlands - habitats with water and dense vegetative cover, especially reed swamps, marsh, and littoral and riparian environments. Hence its other common and more applicable name, the swamp cat. Water and dense ground cover can be found in a variety of habitats, ranging from desert (where it is found near oases or along riverbeds) to grassland, shrubby woodland and dry deciduous forest, as well as cleared areas in moist forest (Nowell and Jackson 1996). Areas with extensive deciduous dipterocarp forest and at least scattered surface water are the species predominant known habitat in Indochina. However, areas such as the Nakai Plateau which support other forms of savannah-like vegetation may support the species. It is probably absent from all closed canopy forests, including rainforest. The species may make use of agricultural areas with a low intensity of human use and which retain patches of scrub (Duckworth et al., 2005). Jungle cats have adapted well to irrigated cultivation, having been observed in many different types of agricultural and forest plantations throughout their range, such as sugar cane plantations in India. In Israel they can be found around pisciculture ponds and irrigation ditches. Vereschagin (1959) noted that the jungle cat's use of the semi-arid plains of Azerbaijan increased with the development of a local irrigation system and decreased with its abandonment. However, mowing the seasonally flooded riverine tugai vegetation (trees and shrubs with dense stands of tall reeds and grasses) of this region for livestock fodder, as well as ploughing it under for agriculture, is known to be associated with the decline of jungle cat populations in the European-central Asian parts of its range (Nowell and Jackson, 1996). Jungle cats feed mainly on prey that weighs less than one kilogram. Small mammals, principally rodents, are the prey most frequently found in feces and stomach contents. A study in India's Sariska reserve estimated that jungle cats catch and eat three to five rodent per day (Mukherjee et al. 2004). Birds rank second in importance, but in southern Russia waterfowl are the mainstay of jungle cat diet in the winter. With over wintering populations of waterfowl congregating in large numbers on unfrozen rivers and marshes, the jungle cat hunts among reed beds and along edges of wetlands, searching for injured or weakened birds. Other prey species are taken more opportunistically, including hares, nutria, lizards, snakes, frogs, insects, and fish. In India, they have been seen to scavenge kills of large predators such as the Asiatic lion. In a study in southern Uzbekistan, the fruits of the Russian olive made up 17% of their diet in winter. While jungle cats specialize on small prey, they are large and powerful enough to kill young swine, subadult gazelles, and chital fawns (Sunquist and Sunquist, 2002). Density estimates from natural tugai habitat in central Asia range from 4-15 individuals per 10 km^2, but where this vegetation type has declined due to development density does not exceed two cats per 10 km^2 (Nowell and Jackson, 1996).

Systems: Terrestrial

Threats:

Some populations of jungle cat subspecies are declining in several countries and areas: Since the 1960s, populations of the Caucasian jungle cat living in the Cis-Caspian region, along the Caspian Sea and in the Caucasus range states have been rapidly declining. Only some small populations persist today. There has been no record in the Astrakhan Nature Reserve in the Volga Delta since the 1980s. This subspecies is considered threatened and included in the Red Books of the Russian Federation, Armenia, Azerbaijan and Georgia. In the 1970s, Southeast Asian jungle cats still used to be the most common wild cats near villages in certain parts of **Northern Thailand** and occurred in many protected areas of the country. But since the early 1990s, jungle cats are rarely encountered and have suffered drastic declines due to hunting and habitat destruction. Today, their official **Thai** status is critically endangered. In Cambodia, Laos and Vietnam, jungle cats probably once occurred widely using secondary habitats, which is easily accessible to hunters and where hunting pressure is now very heavy. Due to unselective trapping and snaring, jungle cats appear quite rare nowadays in comparison to sympatric small cats. Skins are occasionally recorded in border markets, and live individuals, possibly taken from Myanmar or Cambodia, occasionally turn up in the **Khao Khieo and Chiang Mai zoos of Thailand.** Jungle cats are rare in the Middle East. In Jordan, they are highly affected by the expansion of agricultural areas around the river beds of Yarmouk and Jordan rivers, where they are hunted and poisoned by farmers for attacking poultry. In Afghanistan they are also considered rare and threatened. Jungle cats can do well in cultivated landscapes (especially those that lead to increased numbers of rodents) and artificial wetlands. However, reclamation and destruction of natural wetlands, ongoing throughout its range but particularly in the arid areas, still pose a threat to the species, as density in natural wetlands is generally higher (Nowell and Jackson 1996). Unselective trapping, snaring and poisoning around agricultural and settled areas have caused population declines in many areas throughout its range (Abu-Baker et al. 2003, Duckworth et al. 2005). India formerly exported large numbers of jungle cat skins before the species came under legal protection (over 300,000 were declared as being held by traders there when export was banned in 1979), and some illegal trade continues there (Sunquist and Sunquist 2002), as well as in Egypt (Glas in press) and Afghanistan (Habibi 2004).

Conservation Actions:

The jungle cat is listed on CITES Appendix II. It is protected from hunting in some range states (India), but in many receives no legal protection outside protected areas (Nowell and Jackson 1996). The species now receives legal protection from all hunting and trading within Afghanistan after being placed on the country's 2009 Protected Species List. Furthermore, given the amount of habitat loss occurring in riparian and wetland areas in Afghanistan, this species should be considered a research priority. The ecology and status of the jungle cat is poorly known (Nowell and Jackson 1996, Sunquist and Sunquist 2002). In Southwest and Southeast Asia, where it is considered rare and declining, more research needs to be undertaken to gain knowledge of current distribution, both in and outside of protected areas (Abu-Baker et al. 2003, Duckworth et al. 2005). Some farmers consider the jungle cat a pest which takes poultry (Abu-Baker et al. 2003), and conservation measures should include better protection for domestic fowl and halting of indiscriminate poisoning and trapping. The jungle cat would also benefit from improved protection of natural wetlands and reed beds, particularly in the more arid parts of its range, and improved legislation prohibiting fur trade.

Marbled Cat *(Pardofelis marmorata)*

Description:

The marbled cat *(Pardofelis marmorata) is* a small wild cat of South and Southeast Asia. Since 2002 it has been listed as vulnerable by IUCN as it occurs at low densities, and its total effective population size is suspected to be fewer than 10,000 mature individuals, with no single population numbering more than 1,000. The species was once considered to belong to the pantherine lineage of "big cats". Genetic analysis has shown that it is closely related with the Asian golden cat and the bay cat, all of which diverged from the other felids about 9.4 million years ago The marbled cat is similar in size to a domestic cat, with a more thickly furred tail (which may be longer than the body), showing adaptation to its arboreal life-style, where the tail is used as a counterbalance. Marbled cats range from 45 to 62 centimetres (18 to 24 in) in head-body length, with a 35 to 55 centimetres (14 to 22 in) tail. Recorded weights vary between 2 and 5 kilograms (4.4 and 11 lb). The coat is thick and soft, and varies in background colour from dark grey-brown through yellowish grey to red-brown. Spots on the forehead and crown merge into narrow longitudinal stripes on the neck, and irregular stripes on the back. The back and flanks are marked with dark, irregular dark-edged blotches. The legs and underparts are patterned with black dots, and the tail is marked with black spots proximally and rings distally. In addition to its long tail, the marbled cat can also be distinguished by its large feet. It also possesses unusually large canine teeth, resembling those of the big cats, although these appear to be the result of parallel evolution. When standing or resting, marbled cats assume a characteristic position with their backs arched. The marbled cat *(Pardofelis marmorata)* possesses an unusual mixture of small and big cat characteristics. Although just three kilograms and about the size of a domestic cat, this species superficially resembles the much larger clouded leopard in its broad feet, enlarged canines and strikingly similar, blotched coat pattern . The thick, soft, brownish-yellow fur is covered on the back and sides in large, mottled, irregular-shaped blotches margined with black. However, these markings are less well-defined in the marbled cat than those of its larger cousin, tending to be more broken and marbled (hence the name), while the black spots on the limbs are more numerous . The bushy tail is extremely long, reflecting the cat's arboreal lifestyle, and similarly marked with black spots and rings. Prominent black lines occur on the head, neck and back, starting as dark, interrupted bands running from the corner of each eye up and over the forehead. Distinctive dark stripes also mark the cheeks, while the chin, upper lip, cheeks and patches around the eyes are contrastingly white or buff in colour. The eyes are amber or golden, and the ears are short, rounded and black, with a conspicuous white to buff spot on the back .

Range Description:

The marbled cat is found in tropical Indo Malaya westward along the Himalayan foothills westward into Nepal and eastward into Southwest China, and on the islands of Sumatra and Borneo. There are few locality records of this species (Nowell and Jackson, 1996, Sunquist and Sunquist, 2002).
Bhutan; Brunei Darussalam; Cambodia; China; India; Indonesia (Kalimantan, Sumatera); Lao People's Democratic Republic; Malaysia (Peninsular Malaysia, Sabah, Sarawak); Myanmar; Nepal; **Thailand;** Vietnam

Population:

The marbled cat appears relatively rare compared to sympatric felids, based on the paucity of historical as well as recent records (Nowell and Jackson 1996, Duckworth et al. 1999, Holden 2001, Sunquist and Sunquist 2002, Grassman et al. 2005, Azlan and Sharma 2006, Lynam et al. 2006, Mishra et al. 2006, Yasuda et al. 2007), although Cambodia stands out for having a relatively high encounter rate (13 camera trap records, compared to 12 for the Asiatic golden cat and 4 for the clouded leopard: Duckworth et al. 2005).
Population Trend: Decreasing

Ecology:

In May 2000, a female marbled cat was trapped along an animal trail in a hill-evergreen / bamboo mixed forest in **Thailand's Phu Khieu Wildlife Sanctuary.** This first-ever radio-tracked marbled cat had an overall home range of 5.8 km2 (2.2 sq mi) at an elevation of 1,000 to 1,200 m (3,300 to 3,900 ft) and was active primarily during nocturnal and crepuscular time periods. It is probable that forest canopies provide the marbled cat with much of its prey: birds, squirrels, other rodents and reptiles. A few marbled cats have been bred in captivity, with gestation estimated at between 66 and 82 days. In the few recorded instances, two kittens were born in each litter, and weighed from 61 to 85 g (2.2 to 3.0 oz). The eyes open at around twelve days, and the kittens begin to take solid food at two months, around the time that they begin actively climbing. Marbled cats reach sexual maturity at 21 or 22 months of age, and have lived for up to twelve years in captivity. Very little is known about the biology, behaviour and diet of the marbled cat, except what has been observed in captivity. The species is believed to be primarily nocturnal and more arboreal than most other cats, which would help explain its relative obscurity , although recent studies have shown activity during both day and night. With its long, slender body, extremely long tail, short legs and broad feet, the marbled cat is well-adapted for tree-climbing and has been observed in trees in the wild, once stalking a bird, and is an adept climber in captivity. Birds are thought to constitute a major part of the diet, and there have also been records of squirrels and rats being eaten, while lizards and frogs may also be taken . Little is known about how far this secretive cat ranges, although one female, radio tracked in **Thailand** for a month, was found to have a home range of 5.3 square kilometres . What is known of this cat's reproductive behaviour comes from observations of just a few captive individuals. Two litters of two kittens have been recorded from January and February, and one litter of unknown size was born in September. Gestation is estimated to last somewhere between 66 and 82 days. Young attain sexual maturity at 21 to 22 months and individuals in captivity have lived up to 12 years and three months. The marbled cat is primarily associated with moist and mixed deciduous-evergreen tropical forest (Nowell and Jackson 1996), and may prefer hill forest (Duckworth et al. 1999, Holden 2001, Grassman et al. 2005). A few sightings have been made in secondary forest or cleared areas near forest, but it is likely forest-dependent (Nowell and Jackson 1996). Grassman and Tewes (2002) reported the observation of a pair of adult marbled cats in a salt lick in **Thailand's Phu Khieu National Park.** It has never been studied, although Grassman et al. (2005) made a preliminary home range estimate of 5.3 km^2 for an adult female who was radio-collared and tracked for one month in **Thailand's Phu Khieu National Park.** The marbled cat probably preys primarily on rodents, including squirrels (Nowell and Jackson, 1996), and birds.
Systems: Terrestrial

Major Threat(s):

The marbled cat appears to be forest-dependent, and its habitat in Southeast Asia is undergoing the world's fastest deforestation rate (1.2-1.3% a year since 1990: FAO 2007), due to logging, oil palm and other plantations, and human settlement and agriculture. Although infrequently observed in the illegal Asian wildlife trade (Nowell and Jackson 1996), it is valued for its skin, meat and bones, and indiscriminate snaring is prevalent throughout much of its range and is likely to pose a major threat (IUCN Cats Red List workshop, 2007). They have been reported as poultry pests (Nowell and Jackson 1996, Mishra et al. 2006).

Conservation Actions:

Included on CITES Appendix I. Hunting of this species is prohibited in Bangladesh, Cambodia, China (Yunnan only), India, Indonesia, Malaysia, Myanmar, Nepal and **Thailand.** Hunting regulations are in place in Lao PDR, Singapore (Nowell and Jackson 1996). It occurs in a number of protected areas. Further research is needed on its ecology, distribution and status (IUCN Cats Red List workshop, 2007)

Genus: Prionailurus:

Leopard Cat (*Prionailurus bengalensis*)

Description:

The leopard cat *(Prionailurus bengalensis)* is a small wild cat of South and East Asia. Since 2002 it has been listed as Least Concern by IUCN as it is widely distributed but threatened by habitat loss and hunting in parts of its range. There are twelve leopard cat subspecies, which differ widely in appearance. The leopard cat's name is derived from the leopard-like spots prevalent in all subspecies, but its relation to the leopard is distant. Leopard cats are about the size of a domestic cat, but more slender with longer legs and well-defined webs between the toes. Their small head is marked with two prominent dark stripes, their short and narrow muzzle white. There are two dark stripes running from the eyes to the ears, and smaller white streaks running from the eyes to the nose. The backs of their moderately long and rounded ears are black with a central white spot. Body and limbs are marked with black spots of varying size and colour, and along the back are two to four rows of elongated spots. The tail is about half the size of their head-body-length and spotted with a few indistinct rings near the black tip. The background colour of their spotted fur is tawny with a white chest and belly. But in their huge range, they vary so much in coloration and size of spots as well as in body size and weight that initially they were thought to be several different species. The fur colour is yellowish brown in the southern populations, but pale silver-grey in the northern ones. The black markings may be spotted, rosetted, or even forming dotted streaks, depending on the subspecies. In the tropics, leopard cats weigh 0.55 to 3.8 kg (1.2 to 8.4 lb), have a head-body-length of 38.8 to 66 cm (15.3 to 26 in) with a 17.2 to 31 cm (6.8 to 12 in) long tail. In northern China and Siberia, they weigh up to 7.1 kg (16 lb), and have a head-body -length of up to 75 cm (30 in); generally, they put on weight before winter and become thinner until spring. Shoulder height is about 41 cm (16 in). In 1792, Robert Kerr first described a leopard cat under the binominal *Felis bengalensis* in his translation of Carl von Linné's Systema Naturae as being native to southern Bengal. Between 1829 and 1922, different authors of 20 more descriptions classified the cat either as Felis or Leopardus.Owing to the individual variation in fur colour, leopard cats from British India were described as *Felis nipalensis* from Nepal, Leopardus ellioti from the area of Bombay, *Felis wagati* and *Felis tenasserimensis* from Tenasserim. In 1939, Reginald Innes Pocock subordinated them to the genus Prionailurus. The collection of the Natural History Museum in London comprised several skulls and large amounts of skins of leopard cats from various regions. Based on this broad variety of skins, he proposed to differentiate between a southern subspecies *Prionailurus bengalensis bengalensis* from warmer latitudes to the west and east of the Bay of Bengal, and a northern *Prionailurus bengalensis horsfieldi* from the Himalayas, having a fuller winter coat than the southern. His description of leopard cats from the areas of Gilgit and Karachi under the trinomen *Prionailurus bengalensis trevelyani* is based on seven skins that had longer, paler and more greyish fur than those from the Himalayas. He assumed that trevelyani inhabits more rocky, less forested habitats than bengalensis and horsfieldi. Between 1837 and 1930, skins and skulls from China were described as *Felis chinensis, Leopardus reevesii, Felis scripta, Felis microtis, decolorata, ricketti, ingrami, anastasiae and sinensis,* and later grouped under

Oldfield Thomas classified these as *Felis microtis,* which had been first described by Henri Milne-Edwards in 1872. Two skins from Siberia motivated Daniel Giraud Elliot to write a detailed description of *Felis euptilura* in 1871. One was depicted in Gustav Radde's illustration cum description of a wild cat; the other was part of a collection at the Regent's Park Zoo. The ground colour of both was light brownish-yellow, strongly mixed with grey and covered with reddish-brown spots, head grey with a dark-red stripe across the cheek. In 1922, Tamezo Mori described a similar but lighter grey spotted skin of a wild cat from the vicinity of Mukden in Manchuria that he named *Felis manchurica.* Later both were grouped under the trinomen *Felis bengalensis euptilura.* In the 1970s and 1980s, the Russian zoologists Geptner, Gromov and Baranova disagreed with this classification. They emphasized the differences of skins and skulls at their disposal and the ones originating in Southeast Asia, and coined the term Amur forest cat, which they regarded as a distinct species. In 1987, Chinese zoologists pointed out the affinity of leopard cats from northern China, Amur cats and leopard cats from southern latitudes. In view of the morphological similarities they did not support classifying the Amur cat as a species.

The initial binomial euptilura given by Elliott[22] was eventually changed to euptilurus referring to the ICZN Principle of Gender Agreement; at present, both terms are used.

Distribution and Habitat:

Leopard cats are the most widely distributed Asian small cats. Their range extends from the Amur region in the Russian Far East over the Korean Peninsula, China, Indochina, the Indian Subcontinent, to the West in northern Pakistan, and to the south in the Philippines and the Sunda islands of Indonesia. They are found in agriculturally used areas but prefer forested habitats. They live in tropical evergreen rainforests and plantations at sea level, in subtropical deciduous and coniferous forests in the foothills of the Himalayas at altitudes above 1,000 m (3,300 ft). [3] In 2009, a leopard cat was camera trapped in Nepal's Makalu-Barun National Park at an altitude of 3,254 m (10,676 ft). At least six individuals inhabit the survey area, which is dominated by associations of rhododendron, oak and maple.[4] In the northeast of their range they live close to rivers, valleys and in ravine forests, but avoid areas with more than 10 cm (3.9 in) of snowfall. They are rare in Pakistan's arid treeless areas.

In Sabah's Tabin Wildlife Reserve leopard cats had average home ranges of 3.5 km2 (1.4 sq mi).

In Thailand's Phu Khieu Wildlife Reserve 20 leopard cats were radio-collared between 1999 and 2003. Home ranges of males ranged from 2.2 km2 (0.85 sq mi) to 28.9 km2 (11.2 sq mi), and of the six females from 4.4 km2 (1.7 sq mi) to 37.1 km2 (14.3 sq mi) The leopard cat is a widespread species in Asia. It is found throughout most of India west into Pakistan and Afghanistan (Habibi 2004), through the Himalayan foothills, across most of China, and north to the Korean peninsula and into the Russian Far East (Nowell and Jackson 1996). It is found throughout Southeast Asia, and on the islands of Sumatra, Java, Borneo and Taiwan. It is found on numerous small offshore islands of mainland Asia (Nowell and Jackson 1996, Sunquist and Sunquist 2002). The leopard cat is the only wild felid found in the Japan, where it occurs on the small islands Tsushima and Iriomote, and the Philippines, where it occurs on the islands of Palawan, Panay, Negros and Cebu. In the Philippines, there are recent (2007) unconfirmed reports from the island of Masbate. It should be present in Guimaras due to proximity to Negros and Panay, but no presence was reported, and is therefore presumed to be extinct (R. Lorica and W. Oliver, unpub.).

Countries: Native:

Afghanistan; Bangladesh; Bhutan; Brunei Darussalam; Cambodia; China; Hong Kong; India; Indonesia (Jawa, Kalimantan, Sumatera); Japan (Nansei-shoto); Korea, Democratic People's Republic of; Korea, Republic of; Lao People's Democratic Republic; Malaysia; Myanmar; Nepal; Pakistan; Philippines; Russian Federation; Singapore; Taiwan, Province of China; **Thailand;** Vietnam

Distribution of subspecies:

As of 2009, the following subspecies are recognized:

Prionailurus bengalensis bengalensis (Kerr, 1792) — India, Bangladesh, Myanmar, **Thailand,** the Malay Peninsula, Indochina to Yunnan in China.

Prionailurus bengalensis javanensis (Desmarest, 1816) — Java and Bali.

Prionailurus bengalensis sumatranus (Horsfield 1821) — Sumatra and Tebingtinggi.

Prionailurus bengalensis chinensis (Gray 1837) — Taiwan and China except Yunnan.
Prionailurus bengalensis horsfieldi (Gray 1842) — Kashmir, Punjab, Kumaon, Nepal, Sikkim and Bhutan.
Prionailurus bengalensis euptilurus/euptilura, the Amur cat, (Elliott 1871) — eastern Siberia, in Manchuria, in Korea and on the Tsushima Island in the Korea Strait.
Prionailurus bengalensis borneoensis (Brongersma 1936) — Borneo.
Prionailurus bengalensis trevelyani (Pocock 1939) — northern Kashmir and Punjab, and in Southern Baluchistan.
Prionailurus bengalensis alleni (Sody, 1949) — Hainan Island.
Prionailurus bengalensis iriomotensis (Imaizumi, 1967) — found exclusively on the tiny island of Iriomote, one of the Ryukyu Islands in the Japanese Archipelago.
Prionailurus bengalensis heaneyi (Groves 1997) — the Philippine island of Palawan.
Prionailurus bengalensis rabori (Groves 1997) — the Philippine islands of Negros, Cebu, and Panay.
The Iriomote cat *(P.b. iriomotensis)* has been proposed as a species since 1967, but following mtDNA analysis in the 1990s is now considered a subspecies of the leopard cat.
The Tsushima leopard cat lives exclusively on Tsushima Island. Initially regarded as belonging to the Chinese leopard cat subspecies, it is now considered an isolated population of the **Amur cat *P. b. euptilurus/euptilura***

Population:

The leopard cat is the most frequently recorded small cat across most of its wide range, in comparison with sympatric species (Nowell and Jackson 1996, Duckworth et al. 1999, Holden 2001, Duckworth et al. 2005, Lynam et al. 2006, Yasuda et al. 2007), and with its broad distribution has an abundant population. However, it is probably declining due to habitat loss and hunting. Large numbers of leopard cat furs were exported from China (averaging 200,000 skins per year in the late 1980s) (Nowell and Jackson 1996). Higher survival rates (92%) were recorded in a protected area with little human influence, compared with lower rates in areas with greater human activity (53-82%) (Haines et al. 2004). While the leopard cat is more tolerant of disturbed areas than other small Asian felids, it likely undergoes higher mortality in such areas. Island populations are most at risk of extinction, with the Iriomote cat *P.b. iriomotensis l*isted as Critically Endangered, and the Visayan leopard cat *P.b. rabori* of the Philippine islands of Panay, Negros, Cebu and possibly Masbate listed a Vulnerable. The small population (approximately 100) on Japan's 710 km^2 Tsushima Island, considered the same subspecies as occurs in northeastern mainland Asia, has decreased over the last 30-40 years (Izawa et al. 2007).
Population Trend: Stable

Ecology and behavior:

Leopard cats are solitary, except during breeding season. Some are active during the day, but most hunt at night, preferring to stalk murids, tree shrews and hares. They are agile climbers and quite arboreal in their habits. They rest in trees, but also hide in dense thorny undergrowth on the ground. In the oil palm plantations of Sabah, they have been observed up to 4 m (13 ft) above ground hunting rodents and beetles. In this habitat, males had larger home ranges than females, averaging 3.5 km2 (1.4 sq mi) and 2.1 km2 (0.81 sq mi) respectively. Each male's range overlapped one or more female ranges. Leopard cats can swim, but seldom do so. They produce a similar range of vocalisations to the domestic cat. Both sexes scent mark their territory by spraying urine, leaving feces in exposed locations, head rubbing, and scratching. he species can range up to 3,000 m in parts of its range, which extends into the Himalayas along river valleys. It occurs in a broad spectrum of habitats, from tropical rainforest to temperate broadleaf and, marginally, coniferous forest, as well as shrub forest and successional grasslands. The northern boundaries of its range are limited by snow cover; the leopard cat avoids areas where snow is more than 10 cm deep. It is not found in the cold steppe grasslands, and generally does not occur in arid zones, although there are a few records from relatively dry and treeless areas in Pakistan. Leopard cats occur commonly in dense secondary growth, including logged areas, and have been found in agricultural and forest (rubber tree, oil palm, sugarcane) plantations. The species can live close to rural settlements. Leopard cats are excellent swimmers, and have successfully colonized offshore islands throughout their range (Nowell and Jackson 1996; Sunquist and Sunquist 2002). In the most comprehensive study, based on a large sample size of 20 radio-collared cats in **Thailand's Phu Khieu Wildlife Sanctuary,** mean home range size (95% MCP) was 12.7 km^2, larger than in other are-as of **Thailand** (4.5 km^2) (Grassman et al. 2005), on Borneo (3.5 km^2: Rajaratnam 2000), or on Japan's Iriomote island (Schmidt et al. 2003). There was no significant difference between male and female home range size. Open and closed forest habitats were used in proportion to their occurrence, and activity patterns showed crepuscular and nocturnal peaks. On Borneo, Rajaratnam et al. (2007) found that leopard cats hunted rodents in oil palm plantations, and used forest fragments for resting and breeding. Murids dominate the diet (85-90%: Grassman et al. 2005b, Rajaratnam et al. 2007). Other small mammals, eels and fish have also been reported, as well as occasional scavenging of carrion (Nowell and Jackson 1996).

Systems: Terrestrial

Diet:

Leopard cats are carnivorous, feeding on a variety of small prey including mammals, lizards, amphibians, birds and insects. In most parts of their range, small rodents such as rats and mice form the major part of their diet, which is often supplemented with grass, eggs, poultry, and aquatic prey. They are active hunters, dispatching their prey with a rapid pounce and bite. Unlike many other small cats, they do not "play" with their food, maintaining a tight grip with their claws until the animal is dead. This may be related to the relatively high proportion of birds in their diet, which are more likely to escape when released than are rodents.

Reproduction and development:

There is no fixed breeding period in the southern part of its range; in the colder northern range they tend to breed around March or April, when the weather is mild enough to support newborn kittens. The estrus period lasts for 5–9 days. If the kittens do not survive, the female may come into heat again and have another litter that year. After a gestation period of 60–70 days, two to four kittens are born in a den, where they remain until they are a month old. The kittens weigh about 75 to 130 grams (2.6 to 4.6 oz) at birth and usually double their weight by age of two weeks; at five weeks, they are four times their birth weight. The eyes open at ten days, and the kittens start to eat solid food at 23 days. At the age of four weeks, the permanent canines appear, and the kittens begin to eat solid food. Leopard cats have lived for up to thirteen years in captivity. Leopard cats usually pair for life and raise their kittens together for about 7 to 10 months. Full maturity is reached at 18 months, but in captivity, the male can become ready to breed at 7 months, and the female at 10 months.

Threats:

In China, leopard cats are hunted mainly for their fur. Between 1984 and 1989, about 200,000 skins were exported yearly. A survey carried out in 1989 among major fur traders revealed more than 800,000 skins on stock. Since the European Union imposed an import ban in 1988, Japan has become the main buyer, and imported 50,000 skins in 1989. Although commercial trade is much reduced, the species continues to be hunted throughout most of its range for fur, for food, and as pets. They are also widely viewed as poultry pests and killed in retribution. In Myanmar, 483 body parts of at least 443 individuals were observed in four markets surveyed between 1991 and 2006. Numbers were significantly larger than non-threatened species. Three of the surveyed markets are situated on international borders with China and **Thailand,** and cater to international buyers, although the leopard cat is completely protected under Myanmar's national legislation. Effective implementation and enforcement of CITES is considered inadequate.

Major Threat(s)

China, the centre of its range, commercial exploitation has been heavy: hundreds of thousands of Leopard Cat skins per year were exported in the 1980s. Although commercial trade is much reduced, the species continues to be hunted throughout most of its range for fur, for food, and as pets. They are also widely viewed as poultry pests and killed in retribution. Island populations are small and seriously threatened in the Philippines and Japan. Leopard cats can hybridize with domestic cats, as is shown by the popular domestic breed, the "safari cat". Hybridization in the wild has been reported, but is not considered a significant threat. Although the species is less dependent on forest cover than others, habitat loss and fragmentation is still a major threat across most of its Asian range (Nowell and Jackson 1996).

Conservation Actions:

Included on CITES Appendix II; populations in Bangladesh, India and Thailand are included on Appendix I (as Prionailurus bengalensis bengalensis). The species is protected at the national level over part of its range, with hunting prohibited in Bangladesh, Cambodia, Hong Kong, India, Indonesia, Japan, Malaysia, Myanmar, Nepal, Pakistan, Philippines, Russia, Thailand and Taiwan, and hunting and trade regulations in place in South Korea, Lao PDR and Singapore (Nowell and Jackson 1996, A. Wilting pers. comm. 2008). The species is on Afghanistan's 2009 Protected Species List, banning all hunting and trading of this species within the country. It is found in numerous protected areas.

Flat-headed Cat *(Prionailurus planiceps)*

The flat-headed cat *(Prionailurus planiceps)* is a small wild cat patchily distributed in the **Thai-Malay Peninsula,** Borneo and Sumatra. Since 2008, it has been listed as Endangered by the IUCN due to destruction of wetlands in their habitat. It is suspected that the effective population size could be fewer than 2,500 mature individuals, with no subpopulation having an effective population size larger than 250 adult individuals. Like some other small cats, it was originally placed in the genus Felis, but is now considered one of the five species in Prionailurus. Flat-headed cats are very rare in captivity, with less than 10 individuals, all kept in Malaysian and **Thai** zoos as recorded by ISIS.

Description:

The flat-headed cat is distinguished at once by the extreme depression of the skull, which extends along the nose to the extremity of the muzzle, the sides of which are laterally distended. The general habit of body is slender, and the extremities are delicate and lengthened. The head itself is more lengthened and cylindrical than in the domestic cat. The distance between the eyes and the ears is comparatively great. The cylindrical form and lateral contraction of the head is contrasted by an unusual length of the teeth. The canine teeth are nearly as long as in an individual of double its size. The thick fur is reddish-brown on top of the head, dark roan brown on the body, and mottled white on the underbelly. The face is lighter in colour than the body, and the muzzle and chin are white. Two prominent buff whitish streaks run on either side of the nose between the eyes. The ears are rounded. The eyes are unusually far forward and close together, compared with other cats, giving the felid improved stereoscopic vision. The teeth are adapted for gripping onto slippery prey, and the jaws are relatively powerful. These features help the flat-headed cat to catch and retain aquatic prey, to which it is at least as well adapted as the fishing cat. Legs are fairly short. Claws are retractable, but the covering sheaths are so reduced in size that about two-thirds of the claws are left protruding. The anterior upper premolars are larger and sharper relative to other cats. The inter-digital webs on its paws help the cat gain better traction in muddy environments and water, and are even more pronounced on this cat than those on the paws of the fishing cat. It has a head-and-body length of 41 to 50 cm (16 to 20 in) and a short tail of 13 to 15 cm (5.1 to 5.9 in).[4] It weighs 1.5 to 2.5 kg (3.3 to 5.5 lb).

Ecology:

Flat-headed cats are presumably solitary, and probably maintain their home ranges by scent marking. In captivity, both females and males spray urine by walking forward in a crouching position, leaving a trail on the ground. Anecdotal historical accounts report that flat-headed cats are nocturnal, but an adult captive female was crepuscular and most active between 8:00 and 11:30 and between 18:00 and 22:00 hours. Very little is known about this species, with only a handful of observations and camera trap records. Available information suggest that, like its close relative the fishing cat, the flat-headed cat is strongly associated with wetlands and preys primarily on fish. Stomach contents of two dead animals contained mostly fish, and also shrimp shells. They may also take birds and small rodents, and have been reported to prey on domestic poultry (Nowell and Jackson 1996). Most records for the Flat-headed Cat are from swampy areas, lakes and streams, and riverine forest (Nowell and Jackson 1996, Yasuda et al. 2007).

They also occur in peat-swamp forest (Bezuijen 2000), and have been observed in secondary forest (Bezuijen 2000, Bezuijen 2003, Meijaard et al. 2005, Mohamed et al. 2009). All published observations of live animals have taken place at night or early morning, near water (Nowell and Jackson 1996, Bezuijen 2000, Bezuijen 2003, Meijaard et al. 2005, Yasuda et al. 2007). Over 80% of the records gathered by Wilting et al. (2010) were from elevations below 100 m asl, and over 70 % were recorded within a distance of 3 km to larger water sources. The Flat-headed Cat takes its name from its unusually long, sloping snout and flattened skull roof, with small ears set well down the sides of its head. It has large, close-set eyes, and relatively longer and sharper teeth than its close relatives. Its claws do not fully retract into their shortened sheaths, and its toes are more completely webbed than the fishing cat's, with long narrow foot pads. Muul and Lim (1970), commenting on the cat's feet and other features, termed it the ecological counterpart of a semi-aquatic mustelid.

Systems: Terrestrial; Freshwater

The stomach contents of an adult shot on a Malaysian riverbank consisted only of fish. They have been observed to wash objects, raccoon-style. Live fish are readily taken, with full submergence of the head, and the fish were usually carried at least two meters away, suggesting a feeding strategy to avoid letting aquatic prey escape back into water. Captive specimens show much greater interest in potential prey in the water than on dry land, suggesting a strong preference for riverine hunting in their natural habitat. Their morphological specialization suggest that their diet is mostly composed of fish, but they are reported to hunt for frogs, and are thought to catch crustaceans. They also catch rats and chicken. Vocalizations of a flat-headed cat kitten resembled those of a domestic cat. The vocal repertoire of adults has not been analyzed completely, but they purr and give other short-ranged vocalizations. Their gestation period lasts about 56 days. Of three litters recorded in captivity one consisted of two kittens, the other two were singletons. Two captive individuals have lived for fourteen years.

Range Description:

The Flat-headed Cat has a restricted distribution, found only on Sumatra, Borneo and the Malayan peninsula (Malaysia and extreme southern Thailand). It is a lowland species strongly associated with wetlands (Nowell and Jackson 1996, Sunquist and Sunquist 2002). The map is adapted from Wilting et al. (2010); this online publication includes other more detailed maps as well.

Countries: Native:

Brunei Darussalam; Indonesia (Kalimantan, Sumatera); Malaysia (Peninsular Malaysia, Sabah, Sarawak); **Thailand**

Population:

The Flat-headed Cat is closely associated with wetlands and lowland forests, habitats which are increasingly being occupied and modified by people (Wilting et al. 2010). It has never been studied, there are few records of the species, and it is generally considered rare, with a highly localized distribution around bodies of water (Nowell and Jackson 1996, Anon 1999, Bezuijen 2000, Sunquist and Sunquist 2002, Meijaard et al. 2005, Yasuda et al. 2007, Barita and Boeadi pers. comm. 2006, Mohamed et al. 2009). Although fishermen along the Merang river in south Sumatra (which has relatively intact peat forests) described it as common (Bezuijen 2000), they tend to use a single generic term for both flat-headed and leopard cats Prionailurus bengalensis, a more abundant species (Bezuijen 2003). Most of the recent records come from Sabah in north-eastern Borneo, where it can be frequently be observed along the Kinabatangan River (Wilting et al. 2010), and where it has been several times photographed by camera-traps in Deramakot Forest Reserve (Mohamed et al. 2009).

Population Trend: Decreasing

Major Threats:

Wetland and lowland forest destruction and degradation is the primary threat faced by the species (Nowell and Jackson 1996, Wilting et al. 2010). Causes of this destruction include human settlement, forest transformation to plantations, draining for agriculture, pollution, and excessive hunting, wood-cutting and fishing. In addition, clearance of coastal mangroves over the past decade has been rapid in Tropical Asia. The depletion of fish stocks from over-fishing is prevalent in many Asian wetland environments and is likely to be a significant threat. Expansion of oil palm plantations is currently viewed as the most urgent threat (IUCN Cats Red List workshop assessment 2007). Trapping, snaring and poisoning are also threats: E. Bennett (in Sunquist and Sunquist 2002) reported that skins were frequently seen in longhouses in the interior of Sarawak, and Flat-headed Cats have been captured in traps set out to protect domestic fowl (Nowell and Jackson 1996).

Conservation Actions:

Included on CITES Appendix I. The species is fully protected by national legislation over its range, with hunting and trade prohibited in Indonesia, Malaysia, and Thailand (Nowell and Jackson 1996). It is known to occur in Berbak (Bezuijen 2000), Way Kambas (Anon 1996), Batang Gadis (Barita pers. comm.2006) and Kerinci Seblat National Parks (M. Linkie pers. comm. 2008) in Sumatra; the Danum Valley (Hearn et al. 2007), Tabin Wildlife Reserve (Yasuda et al. 2007), Kinabatangan Wildlife Sanctuary in Sabah (Lackman-Ancrenaz and Ancrenaz 1997), Deramakot Forest Reserve (Mohamed et al. 2009), Tangkulap Forest Reserve and Maliau Basi (Wilting et al. 2010); Bukit Sarang Conservation Area (Giman pers. comm. 2006) and Loagan Bunut National Park (Wilting et al. 2010) in Sarawak; Kutai and Kayan Mentarang National Parks (Wulfraat and Samso 2000), Bukit Suharto Protection Forest and Sabangau Peat Swamp Forest (Cheyne et al. 2009) in Kalimantan; Selangor and Pahang Peat Swamp Forest in Peninsula Malaysia (Wilting et al. 2010) and the **Phru Tao Dang Peat Swamp Forest protected area in Southern-most Thailand (**Nowell and Jackson 1996). In Brunei it has been recorded from the Task Merim-bun Heritage Park (Yasuda et al. 2007). Conservation of this species depends on adequate habitat protection, and better understanding of its ecology and status in the remaining lowland and wetland forests. Therefore species specific field surveys focusing on these wetland and lowland areas are needed.

Fishing Cat *(Prionailurus viverrinus)*

The fishing cat *(Prionailurus viverrinus)* is a medium-sized wild cat of South and Southeast Asia. In 2008, the IUCN classified the fishing cat as endangered since they are concentrated primarily in wetland habitats, which are increasingly being settled, degraded and converted. Over the last decade, the fishing cat population throughout much of its Asian range declined severely. Like its closest relative, the leopard cat, the fishing cat lives along rivers, streams and mangrove swamps. It is well adapted to this habitat, being an eager and skilled swimmer.

Range Description:

The Fishing Cat has a broad but discontinuous distribution in Asia, with large gaps - some the result of its association primarily with wetlands, some the result of recent extirpation, and some supposed due to a lack of confirmed records. In Pakistan, the only known population was in the Indus river valley (Roberts 1977), but there are no recent records to confirm it still occurs. The fishing cat has been extirpated in recent years from parts of India, including the Bharatpur region of Western India (Shomita Mukherjee, Jamal Khan pers. comms. 2007), home to Keoladeo National Park, one of the few areas in India were fishing cats were studied (Mukerjee 1989, Haque and Vijayan 1993). It has possibly disappeared also from the southern Western Ghats (Nowell and Jackson 1996; Shomita Mukherjee and Jamal Khan pers. comms. 2007). However, there is also a new record from Umred, near Nagpur in central India, an area well outside of the fishing cat's known range, when a Fishing Cat that had been killed by a vehicle was found (Anon 2005). It is primarily found in the terai region of the Himalayan foothills, and eastern India into Bangladesh, where it is widely distributed and locally common in some areas (Khan 2004), although in eastern India few prime habitats remain (Kolipaka 2006). On the island of Sri Lanka, it occurs apparently all over the island, and has been found on waterways near the capital city of Colombo in degraded habitats (S. Mukherjee pers. comm. 2007). In Southeast Asia, its distribution appears very patchy, with few recent records (Anak, W. Duckworth and R. Steinmetz, Southeast Asia mammal assessment, 2003). There are no confirmed records of the fishing cat from Peninsular Malaysia, but a 1999 camera trap image from Taman Negara National Park, an incomplete image showing only the animal's hindquarters, suggests the species occurrence here (Kawanishi and Sunquist 2003). However, the fishing cat never occurred on Taiwan, where it was mistakenly reported in the past (Nowell and Jackson, 1996; Sunquist and Sunquist, 2002). Its possible occurrence (based on an old, unsubstantiated record) in southwestern China is unknown (J. Sanderson pers. comm. 2007). On the island of Java, it has become scarce and apparently restricted to a few coastal wetlands (Melisch et al. 1996). Although commonly considered to occur on the island of Sumatra, there are no definite historic records, recent records have been shown to be erroneous, and its presence there remains to be confirmed (Duckworth et al. 2009).

Countries: Native:

Bangladesh; Bhutan; Cambodia; India; Indonesia (Jawa); Lao People's Democratic Republic; Myanmar; Nepal; Sri Lanka; **Thailand;** Vietnam

Possibly extinct:

Pakistan

Population:

There is concern about the species status in Southeast Asia where it is very infrequently encountered and believed to be declining (Southeast Asia regional mammal assessment, 2003). There are very few records from camera trapping in Lao (Duckworth pers. comm. 2003) or Cambodia, although there are a sizeable number of confiscated live captive animals there (Duckworth et al. 2005). There have been declines in **Thailand** (Anak pers. comm. 2003) where it is very rarely encountered (Steinmetz pers. comm. 2003) and was more common in the past (Anak pers. comm. 2003). The fishing cat could not be confirmed in any reserves in Vietnam during a survey of wildlife officers (Johnsingh and Nguyen 1995). In 2004, the Fishing Cat SSP and the Cincinnati Zoo and Botanical Garden funded a field survey by Thai biologists Namfon Boontua and Budsabong Kanchanasaka to locate fishing cats in prime wetland areas in **Southern Thailand.** Four months of camera trapping failed to find any sign of fishing cats despite confirmed presence of numerous other wildlife species. There have also been big declines in Lao PDR (W. Duckworth pers. comm), as well as on the island of Java, where the population, possibly a valid subspecies *Prionailurus viverrinus rizophoreus* (Sody 1936), may qualify as Critically Endangered (Boeadi pers. comm.; Melisch et al. 1996). In India it has apparently been extirpated from large parts of its range in recent years (S. Mukherjee and J.A. Khan pers. comm. 2007), and it may no longer occur in Pakistan.
Population Trend: Decreasing

Description:

This is easily the largest of the cats in the genus Prionailurus, often weighing twice as much as the roughly house cat -sized leopard cat. Adult fishing cats may weigh from 6 to 16 kg (13 to 35 lb). The head-and-body length ranges from 66 to 86 cm (26 to 34 in), plus a relatively short tail length of 25 to 33 cm (9.8 to 13 in). Fishing cats are powerfully built, with short, muscular limbs and a robust body. This species has a distinctly long, narrow face distinctive from all other cats. The coarse fur is brownish-gray overlaid with somewhat variable dark markings. Black spots run longitudinally across the body while 6 to 8 dark stripes run from behind the eyes to the nape. The underside fur is paler and longer, often being overlaid with spots. The paws are webbed. The face is spotted and the ears are short and rounded, being black on the backside and having a white spot in the middle of the front.

Ecology and behavior:

The solitary living fishing cats are thought to be primarily nocturnal. They are very much at home in the water and can swim long distances, even under water. Females have been reported to range over areas of 4 to 6 km2 (1.5 to 2.3 sq mi), while males range over 16 to 22 km2 (6.2 to 8.5 sq mi). Adults have been observed to make a "chuckling" sound and likely have other calls similar to those of domestic cats.
As the name implies, fish is their main prey. A one-year study of scats in India's Keoladeo National Park found that fish comprised approximately three-quarters of the diet, with the remainder consisting of birds, insects, and small rodents. Molluscs, reptiles, including snakes, amphibians and carrion, up to the size of full-grown domestic cattle, may supplement the diet. They hunt along the edges of watercourses, grabbing prey from the water, and sometimes diving in to catch prey further from the banks. They mark their territory using cheek-rubbing, head rubbing, chin rubbing, neck rubbing and urine-spraying to leave scent marks. They also sharpen their claws and display flehmen. Fishing cats may mate at any time of the year, although most commonly between January and February. The female constructs a den in a secluded area such as a dense thicket of reeds, and gives birth to two to three kittens after a gestation period of 63–70 days. The kittens weigh around 170 g (6.0 oz) at birth, and are able to actively move around by the age of one month. They begin to play in water and to take solid food at about two months, but are not fully weaned for six months. They reach full adult size at around eight and a half months, acquire their adult canine teeth at eleven months, and are sexually mature at fifteen months. They live for up to ten years in captivity Fishing Cats are strongly associated with wetland. They are typically found in swamps and marshy areas, oxbow lakes, reed beds, tidal creeks and mangrove areas and are more scarce around smaller, fast-moving watercourses.

Along watercourses they have been recorded at elevations up to 1,525 m, but most records are from lowland areas. Although fishing cats are widely distributed through a variety of habitat types (including both evergreen and tropical dry forest: Rabinowitz and Walker 1991), their occurrence tends to be highly localized (Nowell and Jackson 1996). Fishing cats are good swimmers, and unlike most other small cats may prey primarily on fish rather than small mammals. A one-year study of scats in India's Keoladeo National Park found that fish comprised 76% of the diet, followed by birds (27%), insects (13%) and small rodents last (9%) (Haque and Vijayan 1993). Molluscs, reptiles and amphibians are also taken (Haque and Vijayan 1993, Mukherjee 1989). However, they are capable of taking large mammal prey, including small chital fawns (Nowell and Jackson 1996, Sunquist and Sunquist 2002), and have been seen scavenging livestock carcasses and tiger kills (Nowell and Jackson 1996). Predation on small domestic livestock and dogs has also been reported (Nowell and Jackson 1996). The only radio-telemetry study took place in Nepal's Chitwan National Park in the early 1990s. Cats were active only at night and spent most of their time in dense tall and short grasslands, sometimes well away from water. Home ranges of three females were 4?6 km^2; that of a single male was larger at 16?22 km^2 (JLD Smith pers comm. in Sunquist and Sunquist 2002).

Fishing cats have been observed in degraded habitats, such as near aquaculture ponds with little vegetation outside the Indian city of Calcutta (P. Sanyal in Anon. 1989).

Systems: Terrestrial; Freshwater

Major Threat(s):

Wetland destruction and degradation is the primary threat faced by the species (Nowell and Jackson 1996). Over 45% of protected wetlands and 94% of globally significant wetlands in Southeast Asia are considered threatened (Dugan 1993). Threats to wetlands include human settlement, draining for agriculture, pollution, and excessive hunting, wood-cutting and fishing. In addition, clearance of coastal mangroves over the past decade has been rapid in Tropical Asia. The depletion of fish stocks from over-fishing is prevalent in many Asian wetland environments and is likely to be a significant threat. While fishing cats appear relatively tolerant of modified habitats, they are also vulnerable to accidental snaring, while generally not being a commercially valued species (Nowell and Jackson 1996). Widespread indiscriminate snaring, trapping and poisoning are believed to underlie recent declines in Southeast Asia, where fishing cats have not been found even in seemingly intact wetland habitats (Southeast Asia regional mammal assessment, 2003). Kolipaka (2006) reported that fishermen have killed and eaten fishing cats which they say had taken fish from their nets. Wetlands are under-represented in the matrix of Asian protected areas (W. Duckworth pers. comm.). Fishing cat skins have been found in illegal trade in India for many years (Sunquist and Sunquist 2002, Anon 2005).

Conservation Actions:

Included on CITES Appendix II. Protected by national legislation over most of its range. Hunting prohibited: Bangladesh, Cambodia, China, India, Indonesia, Myanmar, Nepal, Pakistan, Sri Lanka, **Thailand.** Hunting regulations apply in Lao PDR. No protection outside protected areas: Bhutan, Vietnam (Nowell and Jackson 1996). The Fishing Cat is confirmed to occur in protected areas including the Sundarbans (Bangladesh and India), Chitwan (Nepal), Corbett, Dudwha, and Kaziranga (India) (IUCN Cats Red List workshop 2007),

Khao Sam Roi Yot and Thale Noi Thailand: (Cutter and Cutter 2009), Botum-Sakor (Cambodia: Royan 2009) and Ujung Kulon and Pulau Dua (Java, Indonesia: A. Compost in Duckworth et al. 2009).

Conservation of the species depends on adequate protection of remaining wild wetlands in Asia, and prevention of indiscriminate trapping, snaring and poisoning.

Clouded Leopard *(Neofelis nebulosa)*

Etymology:

The scientific name of the genus Neofelis is a composite of the Greek word νεο- meaning "new", and the Latin word feles meaning "cat", so it literally means "new cat.

The clouded leopard *(Neofelis nebulosa)* is a felid found from the Himalayan foothills through mainland Southeast Asia into China, and has been classified as vulnerable in 2008 by IUCN. Its total population size is suspected to be fewer than 10,000 mature individuals, with a decreasing population trend and no single population numbering more than 1,000 adults. The Sunda clouded leopard *(Neofelis diardi)* found on Sumatra and Borneo is genetically distinct and considered a separate species since 2006.

Description:

The fur of clouded leopards is of a dark grey or ochreous ground-colour, often largely obliterated by black and dark dusky-grey blotched pattern. There are black spots on the head, and the ears are black. Partly fused or broken up stripes run from the corner of the eyes over the cheek, from the corner of the mouth to the neck, and along the nape to the shoulders. Elongated blotches continue down the spine and form a single median stripe on the loins. Two large blotches of dark dusky-grey hairs on the side of the shoulders are each emphasized posteriorly by a dark stripe, which passes on to the fore leg and breaks up into irregular spots. The flanks are marked by dark dusky-grey irregular blotches bordered behind by long, oblique irregularly curved or looped stripes. These blotches yielding the clouded pattern suggest the English name of the cat. The underparts and legs are spotted, and the tail is marked by large irregular paired spots. Melanistic clouded leopards are not common. Clouded leopards weigh between 11.5 and 23 kg (25 and 51 lb). Females vary in head-to-body length from 68.6 to 94 cm (27.0 to 37 in), with a 61 to 82 cm (24 to 32 in) long tail. Males are larger at 81 to 108 cm (32 to 43 in) with a 74 to 91 cm (29 to 36 in) long tail. Their shoulder height varies from 25 to 40 cm (9.8 to 16 in). Their legs are short and stout, with broad paws. They have exceptionally long, piercing canine teeth, the upper being about three times as long as the basal width of the socket. The upper pair of canines may measure 4 cm (1.6 in) or longer.

Range Description:

The clouded leopard is found from the Himalayan foothills in Nepal through mainland Southeast Asia into China (Nowell and Jackson 1996). The clouded leopard historically had a wide distribution in China, south of the Yangtze, but recent records are few, habitat is fast disappearing, illegal hunting of this species has been prolific and its current distribution in China is poorly known (Wozencraft et al. 2008). The clouded leopard is extinct on the island of Taiwan (Anon. 1996). It still occurs marginally in Bangladesh: Khan (2004) reported that local people still see clouded leopards in the mixed-evergreen forests of the northeastern and South Eastern parts of the country.

The clouded leopards of Sumatra and Borneo were recently diagnosed as a separate species Neofelis diardi (Buckley-Beason et al. 2006, Kitchener et al. 2006, Eizirik et al. submitted), the Sundaland clouded leopard. Sundaland refers to the Malay peninsula and the islands of Sumatra, Borneo and Java. Clouded leopards do not occur on Java. Because of limited samples from Peninsular Malaysia, it is unclear which species of clouded leopard occur here - on the basis of a single skin, Kitchener et al. (2006) ascribed Peninsular Malayasia to the mainland clouded leopard, but indicated that more samples were needed for confirmation.

Countries: Native:
Bangladesh; Bhutan; Cambodia; China; India; Lao People's Democratic Republic; Malaysia (Peninsular Malaysia); Myanmar; Nepal; **Thailand;** Vietnam
Regionally extinct:
Taiwan, Province of China

Population:

The clouded leopard is most strongly associated with primary tropical forest which is rapidly disappearing across its range (Nowell and Jackson 1996), and clouded leopard skins have been observed in large numbers in illegal wildlife trade in Southeast Asia (Nowell 2007). Increasing use of camera traps has helped to better document its distribution and recent research efforts should help improve understanding of its population status (Grassman et al. 2005, Austin et al. 2007).
Population Trend: Decreasing

Ecology:

Clouded leopards are the most talented climbers among the cats. In captivity, they have been observed to climb down vertical tree trunks head first, and hang on to branches with their hind paws bent around branchings of tree limbs. They are capable of supination and can even hang down from branches only by bending their hind paws and their tail around them. When jumping down, they keep hanging on to a branch this way until the very last moment. They can climb on horizontal branches with their back to the ground, and in this position make short jumps forward. When balancing on thin branches, they use their long tail to steer. They can easily jump up to 1.2 m (3.9 ft) high.
Clouded leopards have been observed to scent mark in captivity by urine spraying and head-rubbing on prominent objects. Presumably such habits are used to mark their territory in the wild, although the size of their home ranges is unknown. Like other big cats, they do not appear able to purr, but they otherwise have a wide range of vocalisations, including mewing, hissing, growling, moaning, and snorting. Apart from information stemming from observations of captive clouded leopards, little is known of their natural history and behavior in the wild. Early accounts depict them as rare, secretive, arboreal and nocturnal denizens of dense primary forest. More recent observations suggest that they may not be as arboreal and nocturnal as previously thought. They may use trees as day time rest sites but also spend a significant proportion of time on the ground. Some daytime movement has been observed suggesting that they are not strictly nocturnal but crepuscular. However, the time of day when they are active depends on their prey and the level of human disturbance. Their partly nocturnal and far-ranging behaviour, their low densities, and the fact they inhabit densely vegetated habitats and remote areas makes the censuring and monitoring of clouded leopards extremely difficult. Consequently, little is known about their behaviour and status. Available information on their ecology is anecdotal, based on local interviews and a few sighting reports.

Home ranges have only been estimated in Thailand:
Four individuals were radio-collared in Phu Khieo Wildlife Sanctuary from April 2000 to February 2003. Home ranges of two females were 25.7 km2 (9.9 sq mi) and 22.9 km2 (8.8 sq mi), and of two males 29.7 km2 (11.5 sq mi) and 49.1 km2 (19.0 sq mi).
Two individuals were radio-collared during a study from 1997 to 1999 in the Khao Yai National Park. The home range of one female was 39.4 km2 (15.2 sq mi), of the one male 42 km2 (16 sq mi). Both individuals had a core area of 2.9 km2 (1.1 sq mi).

Little is known of the feeding ecology of clouded leopards. Their prey likely includes both arboreal and terrestrial vertebrates.Pocock presumed that they are adapted for preying upon herbivorous mammals of considerable bulk because of their powerful build, the deep penetration of their bite, attested by their long canines. Confirmed prey species include hog deer, slow loris, brush-tailed porcupine, Malayan pangolin and Indochinese ground squirrel. Known prey species in China include barking deer and pheasants. They most likely also hunt civets, birds, and domestic livestock. Captive clouded leopards also eat eggs and some vegetation. Both males and females average 26 months at first reproduction. Estrus last 6 days in average, estrus cycle averages 30 days. After a gestation period of 93±6 days females give birth to a litter of one to five, most often three cubs. Initially, the young are blind and helpless, much like the young of many other cats, and weigh from 140 to 280 grams (4.9 to 9.9 oz). Unlike adults, the kittens' spots are "solid" — completely dark rather than dark rings. The young can see within about 10 days of birth, are active within five weeks, and are fully weaned at around three months of age. They attain the adult coat pattern at around six months, and probably become independent after around ten months. Females are able to bear one litter each year. In captivity they have an average lifespan of 11 years. One individual has lived to be almost 17 years old. The clouded leopard is intermediate in size between large and small cats, with wild females from **Thailand** weighing 11.5 (Austin and Tewes 1999) to 13.5 kg (Grassman et al. 2005), and males 16 (Grassman et al. 2005) to 18 kg (Austin and Tewes 1999). Its coat is patterned with distinctive large cloud shaped markings, its canines are exceptionally elongated, as is its tail - for a large cat, the clouded leopard is highly arboreal (Sunquist and Sunquist 2002). They are strongly associated with forest habitat, particularly primary evergreen tropical rainforest, but there are also records from dry and deciduous forest, as well as secondary and logged forests. They have been recorded in the Himalayas up to 2,500 m and possibly as high as 3,000 m. Less frequently, they have been found in grassland and scrub, dry tropical forests and mangrove swamps (Nowell and Jackson 1996). Radio-tracking studies in Thailand have showed a preference for forest over more open habitats (Austin et al. 2007). A study in **Thailand's Phu Khieu National Park** found that clouded leopards preyed upon a variety of arboreal and terrestrial prey, including hog deer, slow loris, bush-tailed porcupine, Malayan pangolin and Indochinese ground squirrel (Grassman et al. 2005). Other observations include mainly primate prey, but also Muntjac and Argus pheasant (Nowell and Jackson 1996). Clouded leopards are primarily nocturnal, with crepuscular activity peaks (Grassman et al. 2005, Austin et al. 2007). Two radio-telemetry studies in different parks in **Thailand** have found that adult male and female clouded leopards had similar home range sizes between 30-40 km^2 in size (95% fixed kernel estimators), with smaller intensively used core areas of 3-5 km^2 (Grassman et al. 2005, Austin et al. 2007). While both studies found substantial home range overlap between males and females, as is typical of most felids, Grassman et al. (2005) also found that the ranges of their two radio-collared males overlapped by a significant amount (39%). Although both studies found similar home ranges, clouded leopards in Phu Khieu National Park travelled approximately twice the average daily distance (average 2 km) than clouded leopards in Khao Yai National Park (Grassman et al. 2005, Austin et al. 2007). Clouded leopards may occur at higher densities where densities of the larger cats, tigers and leopards, are lower (Lynam et al. 2001, Grassman et al. 2005, Rao et al. 2005).

Systems: Terrestrial

Major Threat(s):

Clouded leopards prefer closed forest (Grassman et al. 2005, Austin et al. 2007), and their habitat in Southeast Asia is undergoing the world's fastest deforestation rate (1.2-1.3% a year since 1990: FAO 2007). The clouded leopard is hunted for the illegal wildlife trade - large numbers of skins have been seen in market surveys, and there is also trade in bones for medicines, meat for exotic dishes and live animals for the pet trade. Wild animals are likely the primary source, but there is also some illegal trade from captive animals (Nowell 2007).

Conservation Actions:

Included on CITES Appendix I and protected by national legislation over most of its range (Nowell and Jackson 1996). Hunting is banned in Bangladesh, Brunei, Cambodia, China, India, Indonesia, Malaysia, Myanmar, Nepal, Taiwan, **Thailand,** and Vietnam, and hunting regulations apply in Lao PDR (Nowell and Jackson 1996). It occurs in many protected areas.

Leopard *(Panthera pardus)*

The Leopard (/ˈlɛpərd/), Panthera pardus, is a member of the Felidae family and the smallest of the four "big cats" in the genus Panthera, the other three being the tiger, lion, and jaguar. The leopard was once distributed across Eastern and southern Asia and Africa, from Siberia to South Africa, but its range of distribution has decreased radically because of hunting and loss of habitat. It is now chiefly found in sub-Saharan Africa; there are also fragmented populations in the Indian subcontinent, Sri Lanka, Indochina, Malaysia, Indonesia, and China. Because of its declining range and population, it is listed as a "Near Threatened" species on the IUCN Red List. Compared to other members of the Felidae family, the leopard has relatively short legs and a long body with a large skull. It is similar in appearance to the jaguar, but is smaller and more slightly built. Its fur is marked with rosettes similar to those of the jaguar, but the leopard's rosettes are smaller and more densely packed, and do not usually have central spots as the jaguars do. Both leopards and jaguars that are melanistic are known as black panthers. The species' success in the wild is in part due to its opportunistic hunting behavior, its adaptability to habitats, its ability to run at speeds approaching 58 kilometres per hour (36 mph), its unequalled ability to climb trees even when carrying a heavy carcass, and its notorious ability for stealth. The leopard consumes virtually any animal that it can hunt down and catch. Its habitat ranges from rainforest to desert terrains.

Etymology:

In antiquity, a leopard was believed to be a hybrid of a lion and a panther, as is reflected in its name, which is a Greek compound of λέων leōn (lion) and πάρδος pardos (male panther). The Greek word is related to Sanskrit पृदाकु pṛdāku (snake, tiger, panther), and probably derives from a Mediterranean language, such as Egyptian. A panther can be any of several species of large felids: the term can refer to cougars and jaguars in the American continents but (given the European origins of the word) it is largely thought to define the leopard at its source. Black panther refers to leopards with melanistic genes, which are not uncommon in rainforest habitats. The generic component of its modern scientific designation, Panthera pardus, derives from Latin via Greek πάνθηρ (pánthēr). Folk etymology saw the name as a compound of παν (pan, all) and θηρ (beast). However, it is believed instead to be derived from an Indo-Iranian word meaning "white-yellow, pale"; in Sanskrit, this word's reflex was पाण्डर pāṇḍara, which was derived from पुण्डरीक puṇḍárīka (tiger, among other things), then borrowed into Greek.

Description:

Leopards are agile and stealthy predators. Although smaller than other members of the Panthera genus, they are able to take large prey due to their massive skulls that facilitate powerful jaw muscles. Head and body length is usually between 90 and 165 cm (35 and 65 in). The tail reaches 60 to 110 cm (24 to 43 in) long, around the same length as the tiger's tail and relatively the longest tail in the Panthera genus (though snow leopards and the much smaller marbled cats are relatively longer tailed). Shoulder height is from 45 to 80 cm (18 to 31 in). The muscles attached to the scapula are exceptionally strong, which enhance their ability to climb trees. They are very diverse in size. Males are about 30% larger than females, weighing 30 to 91 kg (66 to 200 lb) compared to 23 to 60 kg (51 to 130 lb) for females. Large males of up to 91 kg (200 lb) have been documented in Kruger National Park in South Africa; however, males in South Africa's coastal mountains average 31 kg (68 lb) and the females from the desert-edge in Somalia average 23 to 27 kg (51 to 60 lb).

This wide variation in size is thought to result from the quality and availability of prey found in each habitat. The most diminutive leopard subspecies overall is the Arabian leopard *(P. p. nimr)*, from deserts of the Middle East, with adult females of this race weighing as little as 17 kg (37 lb). Other large subspecies, in which males weigh up to 91 kg (200 lb), are the Sri Lankan leopard *(P. p. kotiya)* and the Anatolian leopard *(P. p. tulliana)*. Such larger leopards tend to be found in areas which lack tigers and lions, thus putting the leopard at the top of the food chain with no competitive restriction from large prey items. The largest verified leopards weighed 96.5 kg (213 lb) and can reach 190 cm (75 in) in head-and-body length. Larger sizes have been reported but are generally considered unreliable. The leopard's body is comparatively long, and its legs are short. Leopards show a great diversity in coat colour and rosette patterns. Their rosettes are circular in East Africa but tend to be squarer in southern Africa and larger in Asian populations. Their yellow coat tends to be more pale and cream colored in desert populations, more gray in colder climates, and of a darker golden hue in rainforest habitats. Overall, the fur under the belly tends to be lighter coloured and of a softer, downy type. Solid black spots in place of open rosettes are generally seen along the face, limbs and underbelly. Leopards may sometimes be confused with two other large spotted cats, the cheetah, with which it may co-exist in Africa, and the jaguar, a neotropical species that it does not naturally co-exist with. However, the patterns of spots in each are different: the cheetah has simple black spots, evenly spread; the jaguar has small spots inside the polygonal rosettes; while the leopard normally has rounder, smaller rosettes than those of the jaguar. The cheetah has longer legs and a thinner build that makes it look more streamlined and taller but less powerfully built than the leopard. The jaguar is more similar in build to the leopard but is generally larger in size and has a more muscular, bulky appearance.

Melanistic leopards are commonly called black panthers, a term that also applies to melanistic jaguars. *Pseudomelanism (abundism)* also occurs in leopards. Melanism in leopards is inherited as a Mendelian, monogenic recessive trait relative to the spotted form. Pairings of black animals inter se have a significantly smaller litter size than other possible pairings. The black colour is caused by recessive gene loci. The black panther is common in the equatorial rainforest of Malaya and the tropical rainforest on the slopes of some African mountains such as Mount Kenya. Between January 1996 and March 2009, Indochinese leopards were photographed at 16 sites in the Malay Peninsula in a sampling effort of more than 1000 trap nights. Of 445 photographs of melanistic leopards taken, 410 came from study sites south of the Isthmus of Kra, where the non-melanistic morph was never photographed. These data suggest the near fixation of the dark allele in the region. The expected time to fixation of this recessive allele due to genetic drift alone ranged from about 1,100 years to about 100,000 years. Melanism in leopards has been hypothesized to be causally associated with a selective advantage for ambush.

Distribution and habitat:

Leopards have the largest distribution of any wild cat, occurring widely in eastern and central Africa, although populations have shown a declining trend and are fragmented outside of sub-Saharan Africa. Within sub-Saharan Africa, the species is still numerous and even thriving in marginal habitats where other large cats have disappeared. But populations in North Africa may be extinct. Data on their distribution in Asia are not consistent. Populations in southwest and Central Asia are small and fragmented; in the northeast, they are critically endangered. In the Indian subcontinent, Southeast Asia, and China, leopards are still relatively abundant. Of the species as a whole, its numbers are greater than those of other Panthera species, all of which face more acute conservation concerns. Leopards live mainly in grasslands, woodlands, and riverine forests. They are usually associated with savannah and rainforest, but leopards are exceptionally adaptable: in the Russian Far East, they inhabit temperate forests where winter temperatures reach a low of −25 °C (−13 °F). Leopards in west and central Asia avoid deserts, areas with long -duration snow cover and areas that are near urban development. The leopard occurs across most of sub-Saharan Africa, as remnant populations in North Africa, and then in the Arabian peninsula and Sinai/Judean Desert (Egypt/Israel/Jordan), south-western and eastern Turkey, and through Southwest Asia and the Caucasus into the Himalayan foothills, India, China and the Russian Far East, as well as on the islands of Java and Sri Lanka (Nowell and Jackson 1996; Sunquist and Sunquist 2002; Hunter et al. in press).

In sub-Saharan Africa, leopards remain widely, albeit now patchily, distributed within historical limits (see Hunter et al. in press, and references therein). Ray et al. (2005) estimated that leopards have disappeared from at least 36.7% of their historical range in Africa. The most marked range loss has been in the Sahel belt, as well as in Nigeria and South Africa. They have been locally extirpated from areas densely populated with people or where habitat conversion is extreme (Hunter et al. in press). They are likely extinct on Zanzibar, where there have been no confirmed records since 1996 (Hunter et al. in press).

In North Africa, a tiny relict population persists in the Atlas Mountains of Morocco (Cuzin 2003), and there was a probable observation on the Morocco-Algerian border in Figuig in 2007 (F. Cuzin pers. comm.), while a population was recently found in the Ahaggar of south-eastern Algeria, a region from which they had not previously been recorded (Busby et al. 2006). Leopard are likely extinct in Egypt, although they may occur in the Eastern Desert (Hunter et al. in press).

A 2006 Arabian Fauna Conservation Workshop estimated there were fewer than 200 leopards remaining on the Arabian peninsula, in three confirmed separate subpopulations: the Negev desert, the Wada'a mountains of Yemen, and the Dhofar mountains of Oman. Presence in Saudi Arabia is uncertain (Breitenmoser 2006, Spalton and Al Hikmani 2006). I. Khorozyan (pers. comm. 2008) compiled detailed country information on the distribution of the Endangered Persian leopard as follows:

Russian North Caucasus: mountain ridges in the headwaters of the Avarskoe Koisu and Andiiskoe Koisu rivers (Republic of Dagestan). Possibly exists in the Chegem River canyon (Kabardino-Balkarian Republic); Erzi Reserve, Assa River valley (Republic of Ingushetia); Armkhi River basin (Republic of North Osetia-Alania), headwaters of the Sharoargun and Argun rivers (Chechen Republic) (Akkiev and Mokaev, 2006; Khorozyan and Abramov, 2007; Lukarevsky et al., 2007).

Georgia: Vashlovani Reserve in the south-east; Arkhoti River canyon in the upper part of the Assa River basin and the headwaters of the Andiiskoe Koisu River in the north-east (Lukarevsky et al., 2007). Some anecdotal records from south-western Georgia are either unreliable or can be attributed to individuals coming from north-eastern Turkey (Arabuli, 2006; Khorozyan and Abramov, 2007).

Armenia: South-Western and southern parts of the country from Khosrov Reserve to the Armenian-Iranian state border throughout the Geghama, Zangezur, Aiotsdzor, Bargushat and Meghri ridges. The range boundaries are the Azat River in the north-west; Vardenis Ridge in the north; semi-desert of the Ararat Valley in the west; state border with Azerbaijan and the alpine meadow/nival belt transition zone in the south-west and east; Arax River basin along the Armenian-Iranian border in the south. Until the early 1970s it lived also in north-eastern parts of Armenia (Khorozyan et al., 2005; Khorozyan and Abramov, 2007).

Azerbaijan: Talysh Mts. in the extreme south-east, Akhar-Bakhar Ridge of the Iori-Mingechaur Highland in the north -west and the Zangezur Ridge in the Nakhichevan Republic along the state border with Armenia in the west (Lukarevsky et al., 2007). Nagorno-Karabakh Republic: distribution in Shushi, Mardakert and Hadrut districts and in the adjoining Kelbajar district was recorded in 1941-1967 (Alekperov, 1966; Sludsky, 1973). Up-to-date information on leopard status is impossible to obtain for the political tension between Armenia and Azerbaijan over this republic, even though it plays a vital role as a corridor between the southern (Armenia, Azerbaijan?s Nakhichevan Republic), central (Iori-Mingechaur Highland in Azerbaijan and Vashlovani Reserve in Georgia) and northern (Russian North Caucasus) parts of the Caucasus (Khorozyan and Abramov, 2007).

Iran: virtually all country, except the vast deserts of Desht-e-Kevir and Desht-e-Lut in central and eastern parts. Particularly common in the Alborz Mts. along the southern fringe of the Caspian Sea. Quite common in protected areas (e.g., Tandooreh, Sarigol, Bafgh, Golestan, Kolah?ghazy, Touran, Kavir, Khojir, Khabr and Bamu national parks; Kiamaki and Naybandan wildlife refuges; Jahan Nama, Central Alborz, Varjin, Arasbaran, Dena and Bahram/gur protected areas) and some unprotected lands (Chapur-Ghoymeh, Safee Abad-Dozain or Minoo Dasht, Ramsar, Khaeez and Darestan-Rudbar) (Joslin and Shoemaker 1988; Kiabi et al., 2002; Farhadinia et al. 2007; Abdoli et al. 2008; M Farhadinia and A. Ghoddousi pers. comms., 2008).

Turkey: north-east (around the Artvin city), east (vicinities of Mt. Ararat or Agri) and south-east (Bitlis Ridge). Possibly exists in the mountains of the Black Sea coast and south-westwards to the Taurus Mts. (H. Diker pers. comm., 2008). It is unclear whether leopards still survive in western Turkey. Turkmenistan: western Kopetdag Ridge, central Kopetdag Ridge, eastern Kopetdag Ridge, Badkhyz Reserve and Giaz-Gyadyk Ridge (Lukarevsky, 2001).

Afghanistan: (Hindu Kush, Kohe Baba, Kohe Paghman and Safed Koh ranges of the central highlands), north-eastern (Wakhan corridor) and northern (Darkad peninsula of Badakshan) parts of the country (Habibi, 2004).

In the Central Asian republics, leopard distribution is poorly known. Historically, leopards had a wider distribution in Turkmenistan, and were found in parts of Uzbekistan and Tajikistan. An old male leopard was killed by a local hunter in January 2000 in Kazakhstan, the first record of the species in this country, in a location over 600 km from possible occurrences in Uzbekistan and Tajikistan, and over 1,200 km from known occurrences in Turkmenistan. It is aplso possible that the leopard travelled along the foothills of the Pamirs, then proceeded via the Ugam and Pskem ranges into the Talas river valley. Habitat appears to be suitable, but the existence of any leopard subpopulation in any of these three countries is uncertain (Shakula 2004).

In Pakistan, the leopard is thinly distributed in montane areas, and there have only been a handful of confirmed records in recent years (Ahmed 2001).

Leopards occur widely in the forests of the Indian sub-continent, through Southeast Asia and into China, although they are becoming increasingly rare outside protected areas. They are not found on the islands of Borneo or Sumatra (Nowell and Jackson 1996).

Countries: Native:

Afghanistan; Algeria; Angola (Angola); Armenia (Armenia); Azerbaijan; Bangladesh; Benin; Bhutan; Botswana; Burkina Faso; Burundi; Cambodia; Cameroon; Central African Republic; Chad; China; Congo; Congo, The Democratic Republic of the; Côte d'Ivoire; Djibouti; Egypt; Equatorial Guinea; Eritrea; Ethiopia; Gabon; Gambia; Georgia; Ghana; Guinea; Guinea-Bissau; India; Indonesia (Jawa); Iran, Islamic Republic of; Israel; Jordan; Kenya; Korea, Democratic People's Republic of; Lao People's Democratic Republic; Liberia; Malawi; Malaysia; Mali; Morocco; Mozambique; Myanmar; Namibia; Nepal; Niger; Nigeria; Oman; Pakistan; Russian Federation; Rwanda; Saudi Arabia; Senegal; Sierra Leone; Somalia; South Africa; Sri Lanka; Sudan; Swaziland; Tajikistan; Tanzania, **Thailand;** Togo; Turkey; Turkmenistan; Uganda; United Arab Emirates; Uzbekistan; Viet Nam; Yemen; Zambia; Zimbabwe

Regionally extinct:

Hong Kong; Kuwait; Libya; Singapore; Syrian Arab Republic; Tunisia

Population:

The leopard is an adaptable, widespread species that nonetheless has many threatened subpopulations. While still numerous and even thriving in some marginal habitats from which other big cats have disappeared in many parts of sub-Saharan Africa, in North Africa leopards are on the verge of extinction.

There are no reliable continent-wide estimates of population size in Africa, and the most commonly cited estimate of over 700,000 leopards in Africa (Martin and de Meulenaar 1988) is flawed. In India, based on pugmark censuses (a methodology which has been criticized as inaccurate), 9,844 leopards were estimated in 2001. Many populations are believed to be increasing (Singh 2005), and there are high levels of human-leopard conflict (Singh et al. 2008).

Several Asian subspecies are included on the Red List, with population information as follows:

Amur leopard *P.p. orientalis* CR C2a(ii),D: 14-20 (Anon. 2007)

Arabian leopard P.p. nimr CR C2a(I): Javan leopard *P.p. melas* CR C2a(i): 323-525, with Sri Lankan leopard *P.p. kotiya* EN C2a(i): 700-950 (Kittle and Watson 2007)

Persian leopard *P.p. saxicolor* EN C2a(i): 871-1290 (Khorozyan et al., 2005; Lukarevsky et al. 2007)

Distribution of subspecies

Since Carl Linnaeus published his description of leopards in the Systema Naturae in 1758, as many as 27 leopard subspecies were subsequently described by naturalists from 1794 to 1956. In 1996, according to DNA analysis carried out in the 1990s, only eight subspecies are considered valid. Later analysis revealed a ninth valid subspecies, the Arabian leopard *(P. p. nimr)*. Because of limited sampling of African leopards, this number might be an underestimation.

The nine subspecies recognised by IUCN are:

African leopard *(P. p. pardus)* (Linnaeus, 1758) inhabits sub-Saharan Africa

Indian leopard *(P. p. fusca)* (Meyer, 1794) inhabits the Indian Subcontinent

Javan leopard *(P. p. melas)* (Cuvier, 1809) inhabits Java, Indonesia

Arabian leopard *(P. p. nimr)* (Hemprich and Ehrenberg, 1833) inhabits the Arabian Peninsula
Amur leopard *(P. p. orientalis)* (Schlegel, 1857) inhabits the Russian Far East, Korean Peninsula and Northeast China
North Chinese leopard *(P. p. japonensis)* (Gray, 1862) inhabits Northern China
Persian leopard *(P. p. saxicolor)* (Pocock, 1927), initially described as Caucasian leopard *(P. p. ciscaucasica) (Satunin,* 1914), inhabits the Caucasus, Turkmenistan and northern Iran[33]
Indochinese leopard *(P. p. delacouri)* (Pocock, 1930) inhabits mainland Southeast Asia
Sri Lankan leopard *(P. p. kotiya)* (Deraniyagala, 1956) inhabits Sri Lanka

A morphological analysis of characters of leopard skulls implies the validity of two more subspecies:
Anatolian leopard *(P. p. tulliana)* (Valenciennes, 1856) inhabits Western Turkey
Balochistan leopard *(P. p. sindica)* (Pocock, 1930) inhabits Pakistan, and possibly also parts of Afghanistan and Iran .

Ecology and behavior:

Leopards are elusive, solitary and largely nocturnal. They have primarily been studied in open savanna habitats, which may have biased common descriptions. Activity level varies depending on the habitat and the type of prey that they hunt. Radio-tracking and scat analysis in West Africa showed that rainforest leopards are more likely to be diurnal and crepuscular. Forest leopards are also more specialized in prey selection and exhibit seasonal differences in activity patterns.

Leopards are known for their ability in climbing, and have been observed resting on tree branches during the day, dragging their kills up trees and hanging them there, and descending from trees headfirst.[37] They are powerful swimmers, although are not as disposed to swimming as some other big cats, such as the tiger. They are very agile, and can run at over 58 kilometres per hour (36 mph), leap over 6 metres (20 ft) horizontally, and jump up to 3 metres (9.8 ft) vertically.[38] They produce a number of vocalizations, including grunts, roars, growls, meows, and "sawing" sounds.[39]

Social structure and home range:

Home ranges of male leopards vary between 30 km2 (12 sq mi) and 78 km2 (30 sq mi), and of females between 15 to 16 km2 (5.8 to 6.2 sq mi).[5] Virtually all sources suggest that males do have larger home ranges. There seems to be little or no overlap in territory among males, although overlap exists between the sexes; one radio-collar analysis in the Ivory Coast found a female home range completely enclosed within a male's. Research in a conservation area in Kenya showed similar territory sizes and sex differential: 32.8 km2 (12.7 sq mi) average ranges for males, and 14 km2 (5.4 sq mi) for females. In Nepal, somewhat larger male ranges have been found at about 48 km2 (19 sq mi), while female ranges at 17 km2 (6.6 sq mi); female home ranges decreased to 5 to 7 km2 (1.9 to 2.7 sq mi) when young cubs were present, while the sexual difference in range size seemed to be in positive proportion to overall increase. Studies of leopard home range size have tended to focus on protected areas, which may have led to skewed data; as of the mid-1980s, only 13% of the leopard range actually fell within a protected area. However, significant variations in the size of home ranges have been suggested across the leopard's range. Research in Namibia that focused on spatial ecology in farmlands outside of protected areas revealed ranges that were consistently above 100 km2 (39 sq mi) with some more than 300 km2 (120 sq mi). Admitting that their data were at odds with others, the researchers found little or no sexual variation in the size of territories. Aggressive encounters have been observed. Two of five males studied over a period of a year at a game reserve in South Africa died, both violently. One was initially wounded in a male–male territorial battle over a carcass; taken in by researchers, it was released after a successful convalescence only to be killed by a different male a few months later. A second was killed by another predator, possibly a spotted hyena. A third of the five was badly wounded in intraspecific fighting, but recovered. Leopards are versatile, opportunistic hunters, and have a very broad diet. They feed on a greater diversity of prey than other members of the Panthera species, and will eat anything from dung beetles to 900 kg (2,000 lb) male giant elands, though prey usually weighs considerably less than 200 kg (440 lb) . Their diet consists mostly of ungulates, followed by primates, primarily monkeys of various species. However, they will also opportunistically eat rodents, reptiles, amphibians, insects, birds (especially ground-based types like the Vulturine Guinea fowl), fish and sometimes smaller predators (such as foxes, jackals, martens and smaller felid species). In at least one instance, a leopard has predated a sub-adult Nile crocodile that was crossing over land. Leopards are the only natural predators of adult chimpanzees and gorillas, although the cat may sometimes choose to avoid these as they are potentially hazardous prey, especially large male silverback gorillas. They stalk their prey silently, pounce on it at the last minute, and strangle its throat with a quick bite. In Africa, mid-sized antelopes provide a majority of their prey, especially impala and Thomson's gazelles.

In the open savanna of Tsavo National Park, they kill most of their prey while hunting between sunset and sunrise. In Kruger National Park, males and females with cubs are more active at night. At least 92 prey species have been documented in their diet. They focus their hunting activity on locally abundant medium-sized ungulate species in the 20 to 80 kg (44 to 180 lb) range, while opportunistically taking other prey. Analysis of leopard scats found that 67% contained ungulate remains, of which 60% were impala, the most abundant antelope, with adult weights of 40 to 60 kg (88 to 130 lb). Small mammal remains were found most often in scats of sub-adult leopards, especially females. Average daily consumption rates was estimated at 3.5 kg (7.7 lb) for adult males and 2.8 kg (6.2 lb) for females. In Asia, the leopard primarily preys on deer such as chitals and muntjacs, as well as various Asian antelopes and ibex. Prey preference estimates in southern India showed that the most favoured prey of the leopard were chitals. A study at the Wolong Reserve in China revealed how adaptable their hunting behaviour is. Over the course of seven years, the vegetative cover receded, and the animals opportunistically shifted from primarily consuming tufted deer to pursuing bamboo rats and other smaller prey. They select their prey focusing on small herds, dense habitat, and low risk of injury, preferring prey weights of 10 to 40 kg (22 to 88 lb) such as impala, chital, bushbuck and common duiker with an average body weight of 25 kg (55 lb).

In search of safety, leopards often stash their young or recent kills high up in a tree, which can be a great feat of strength considering that they may be carrying prey heavier than themselves in their the mouth while they climb vertically. One leopard was seen to haul a young giraffe, estimated to weigh up to 125 kg (280 lb), more than twice the weight of the cat, up 5.7 m (19 ft) into a tree. Leopards must compete for food and shelter with other large predators such as lions, tigers, spotted hyenas, striped hyena, up to 5 species of bear and both African and Asiatic wild dogs. These animals may steal the leopard's kill, devour its young or even kill adult leopards, although lions are most likely to kill and not eat young leopards if they are discovered. Conflictions over carcasses often end without bloodshed, because the leopard will often begrudgingly cede their kills if they encounter a larger-bodied or roup-living predator that can potentially kill them. In some areas of Africa, troops of large baboon species (potential leopard prey themselves) will kill and sometimes eat leopard young if they discover them. Occasionally, Nile crocodiles may predate on leopards of any age. Leopards co-exist alongside these other predators by hunting for different types of prey and by avoiding areas frequented by them. Lions are occasionally successful in climbing trees and fetching leopard kills. In the Kalahari desert, leopards frequently lose kills to the brown hyena, if the leopard is unable to move the kill into a tree. Single brown hyenas have been observed charging at and displacing male leopards from kills. Burmese Pythons have been known to prey on leopards, with an adult cat having been recovered from the stomach of an 18-foot specimen. Resource partitioning occurs where leopards share their range with lions or tigers. Leopards tend to take smaller prey, usually less than 75 kg (170 lb), where the larger cats are present. In the Chitwan National Park in Nepal, leopards killed prey ranging from less than 25 kg (55 lb) to 100 kg (220 lb) in weight with most kills in the 25–50 kg (55–110 lb) range; tigers killed more prey in the 50–100 kg (110–220 lb) range. In the tropical forests of India's Nagarhole National Park, tigers selected prey weighing more than 176 kg (390 lb), whereas leopards selected prey in the 30–175 kg (66–390 lb) range. The average weights of leopard prey was 37.6 kg (83 lb), and of tiger prey was 91.5 kg (202 lb) with a bias towards adult males of chital, Sambar and wild pig, and young gaur. In tropical forest they do not always avoid the larger cats by hunting at different times. With relatively abundant prey, tigers and leopards seem to successfully coexist without competitive exclusion or inter-species dominance hierarchies that may be more common to the leopard's co-existence with the lion in savanna habitats. In areas with high tiger populations, such as in the central parts of India's Kanha National Park, leopards are not permanent residents, but transients. They were common near villages at the periphery of the park and outside the park. Depending on the region, leopards may mate all year round. In Manchuria and Siberia, they mate during January and February. The oestrous cycle lasts about 46 days and the female usually is in heat for 6–7 days. Gestation lasts for 90 to 105 days. Cubs are usually born in a litter of 2–4 cubs.[63] But mortality of cubs is estimated at 41–50% during the first year. Females give birth in a cave, crevice among boulders, hollow tree, or thicket to make a den. Cubs are born with closed eyes, which open four to nine days after birth. The fur of the young tends to be longer and thicker than that of adults. Their pelage is also more gray in colour with less defined spots. Around three months of age, the young begin to follow the mother on hunts. At one year of age, leopard young can probably fend for themselves, but remain with the mother for 18–24 months. Leopards have been reported to reach 21 years of age in captivity. he leopard has the widest habitat tolerance of any Old World felid, ranging from rainforest to desert. In Africa, they are most successful in woodland, grassland savanna and forest but also occur widely in mountain habitats, coastal scrub, swampy areas, scrubland, semi-desert and desert. They range from sea level to as much as 4,600 m on Mt Kenya (Hunter et al. in press). In Southwest and Central Asia, leopards formerly occupied a range of habitats, but now are confined chiefly to the more remote montane and rugged foothill areas.

through India and Southeast Asia, Leopard are found in all forest types, from tropical rainforest to the temperate deciduous and alpine coniferous (up to 5,200 m in the Himalaya), and also occur in dry scrub and grasslands (Nowell and Jackson 1996). Leopards have extremely catholic diets including more than 90 species in sub-Saharan Africa, ranging from arthropods to large antelope up to the size of adult male *Eland Tragelaphus oryx* (Hunter et al. in press). Densities vary with habitat, prey availability, and degree of threat, from fewer than one per 100 km^2 to over 30 per 100 km^2, with highest densities obtained in protected East and southern African mesic woodland savannas (Hunter et al. in press). A study in **Thailand** found a home range of 8.8 km^2 for a radio-collared female, and 17.3-18 km^2 for two adult males (Grassman 1998). Important prey species were hog badger *Arctonyx collaris* (45.9%), Muntjac *Muntiacus muntjak* (20.9%) and wild pig Sus scrofa (6.3%).
Systems: Terrestrial

Threats:

Although the leopard is an amazingly adaptable species that has an extensive range, the future of this big cat is far from certain, and several subspecies are on the verge of extinction . The major threats to the leopard are habitat conversion for agriculture and development, and persecution by humans . Across Asia, this species is also threatened by habitat destruction and fragmentation due to logging. The leopard is subjected to intense persecution in many parts of its range, largely in response to loss of livestock. Poisoning of carcasses by livestock owners to target large carnivores is a rapidly increasing threat to this species. Elimination of the leopard's natural prey is creating more conflict with humans, putting this big cat at even greater risk. The leopard is one of the 'Big Five', meaning that along with the lion *(Panthera leo)*, the African elephant *(Loxodonata africana),* the African buffalo *(Syncerus caffer)* and the rhinoceros, it is one of the most highly prized targets of sport hunters. Trophy-hunting therefore poses a severe threat to some leopard populations. Furthermore, trade in leopard skins and teeth for use in traditional rituals and ceremonial dress is common in some Central and West African countries, and illegal trade in Europe and Asia also poses a major problem.

Conservation:

The leopard is included on Appendix I of the Convention for International Trade in Endangered Species (CITES), meaning that a quota system limits the legal exports of skins and hunting trophies. The leopard is also protected under national legislation across much of its range (1). In Afghanistan, the leopard has been placed on the country's Protected Species List, which prohibits all hunting and trading of this species within Afghanistan. The leopard occurs in numerous protected areas across Africa and Asia, and growing wildlife tourism throughout this species' range may make a significant contribution to its survival. However, despite this, many populations occur outside of protected areas, and therefore improved mitigation methods to resolve conflicts between leopards and landowners are desperately required. In West Asia, the leopard is almost entirely restricted to protected areas. However, while these areas may receive protection, most of them are too small to support large enough populations of this species to ensure its survival in the region, and therefore need to be expanded and connected to other areas using buffer zones and wildlife 'corridors'. Several conservation organisations are closely involved in the conservation of the leopard. Over the past few years, scientists at Panthera have led the Munyawana Leopard Project, which is the longest and most comprehensive leopard study to date. The findings from this project have resulted in an effective conservation plan for this beautiful and charismatic wild cat.

Tiger *(Panthera tigris)*

Range Description:

Taxonomy and etymology:

In 1758, Linnaeus first described the species in his work Systema Naturae under the scientific name Felis tigris. In 1929, the British taxonomist Reginald Innes Pocock subordinated the species under the genus Panthera using the scientific name Panthera tigris.

The word Panthera is probably of Oriental origin and retraceable to the Ancient Greek word panther, the Latin word panthera, the Old French word pantere, most likely meaning "the yellowish animal", or from pandarah meaning whitish-yellow. The derivation from Greek pan- ("all") and ther ("beast") may be folk etymology that led to many curious fables.

The word "tiger" is retraceable to the Latin word tigris meaning a spotted tiger hound of Actaeon. The Greek word tigris is possibly derived from a Persian source.

General details:

The Tiger once ranged widely across Asia, from Turkey in the west to the eastern coast of Russia (Nowell and Jackson 1996). Over the past 100 years Tigers have disappeared from Southwest and Central Asia, from two Indonesian islands (Java and Bali) and from large areas of Southeast and Eastern Asia. Tigers have lost over 93% of their historic range (Sanderson et al. 2006, Walston et al. 2010b). Tigers are currently found in thirteen Asian range states: Bangladesh, Bhutan, Cambodia, China, India, Indonesia, Lao PDR, Malaysia, Myanmar, Nepal, Russia, **Thailand** and Vietnam. They may still persist in North Korea, although there has been no recent confirmed evidence.

In 1994, the first comprehensive assessment to delineate Tiger range was carried out (Dinerstein et al. 1997). Priority areas for Tiger conservation were estimated to total 1.64 million km^2 in 159 Tiger Conservation Units (TCUs), roughly equivalent to discrete meta-populations, not including Russia (later estimated at 270,0000 km^2: Sanderson et al. 2006) and China. While this was generally considered representative of current distribution, Tiger presence was confirmed in just 47% the TCUs, and 89% were scored as undergoing medium to high levels of poaching of Tigers and their prey.

This exercise was revised and updated ten years later, and in delineating Tiger Conservation Landscapes (TCLs), greater emphasis was placed on actual records of Tiger presence and breeding (Sanderson et al. 2006). TCLs were defined as areas where there is sufficient habitat to conserve at least five Tigers, and Tigers have been confirmed to occur in the past decade. Tiger range was estimated at 1.1 million km^2 in 76 TCLs (again, roughly equivalent to discrete meta-populations). This represented a 41% decline from the range described a decade earlier (in South and Southeast Asia, a drop from 1.55 million km^2 to 914,000 km^2: Sanderson et al. 2006: 63), attributed primarily to poaching pressure (Dinerstein et al. 2007). Habitat loss due to deforestation was also to blame, notable particularly in Sumatra and Myanmar (Wikramanayake et al. 2010). In India, landscapes with Tigers found to be much smaller and more fragmented than in the original assessment (Sanderson et al. 2006: 63)

Records of Tigers were collected over a ten-year period (1995–2004), a period which may have been too liberal for places like Cambodia which underwent a sharp rise in poaching pressure in the 1990s (Sanderson et al. 2006: Appendix 6). While 53% of the TCU survey respondents reported evidence of Tiger breeding in the time period 1995 –2004, out of over 2,500 point records collected in 2005, just 8% had confirmed evidence of breeding Tigers (Sanderson et al. 2006: 11-17). Large areas of habitat were defined as Tiger landscapes based on suitability, but given data paucity on Tiger presence there were often few records of breeding and actual Tiger occupancy to substantiate these (Sanderson et al. 2006:).

A review of land management within Tiger Conservation Landscapes described the TCLs as "potential habitat for Tigers" and found only 21% of their area to be legally protected. Management effectiveness was generally poor in the protected areas, with regulatory, budgetary and enforcement constraints, and hunting cited as the main threat. Significant portions of the TCLs are designated concessions for resource extraction (timber, oil and gas, minerals, etc.) (Forrest et al. 2011).

Tiger range was revisited again in 2009, by which time the extent of the Tiger's range collapse had become evident. "Vast areas of Southeast Asia [were] recently found to be void of Tigers and depleted of prey by hunters" (Walston et al. 2010a: 5). The exercise used a different methodology to prioritize areas for Tiger conservation. Source Sites were defined as areas with confirmed current presence of Tigers and evidence of breeding, population estimates of >25 breeding females, legal protection, and embedded in a larger habitat landscape with the potential to hold >50 breeding females. An extensive review of scientific literature as well as correspondence with Tiger scientists and protected area managers resulted in the identification of just 42 source sites totalling approximately 90,000 km^2. Many Southeast Asian countries, previously considered to have large areas with Tigers, are now considered, on the basis of extensive survey effort over the past decade or more, to have essentially no healthy breeding populations: Myanmar, Cambodia, Vietnam and Bhutan have no confirmed source sites (although some sites with potential), and Laos just one (Walston et al. 2010a,b).

On the range map accompanying this Red List assessment, the Source Sites are delineated as Extant range, and the Tiger Conservation Landscapes as Probably Extant, as they are based primarily on "realistic inferences (e.g., based on distribution of suitable habitat at appropriate altitudes and proximity to areas where the taxon is known or thought very likely to remain extant," as defined by the IUCN Red List of Threatened Species).

Countries: Native:

Bangladesh; Bhutan; Cambodia; China (Anhui - Regionally Extinct, Beijing - Regionally Extinct, Chongqing - Regionally Extinct, Fujian - Possibly Extinct, Guangdong - Possibly Extinct, Guangxi - Regionally Extinct, Guizhou - Regionally Extinct, Hebei - Regionally Extinct, Heilongjiang, Henan - Regionally Extinct, Hubei - Regionally Extinct, Hunan - Possibly Extinct, Jiangsu - Regionally Extinct, Jiangxi - Possibly Extinct, Jilin, Liaoning - Regionally Extinct, Shaanxi - Possibly Extinct, Shandong - Regionally Extinct, Shanghai - Regionally Extinct, Shanxi - Regionally Extinct, Sichuan - Regionally Extinct, Tianjin - Regionally Extinct, Tibet [or Xizang], Xinjiang - Regionally Extinct, Yunnan, Zhejiang - Possibly Extinct); India; Indonesia (Bali - Regionally Extinct, Jawa - Regionally Extinct, Sumatera); Lao People's Democratic Republic; Malaysia (Peninsular Malaysia); Myanmar; Nepal; Russian Federation; **Thailand;** Vietnam

Possibly extinct:

Korea, Democratic People's Republic of

Regionally extinct:

Afghanistan; Iran, Islamic Republic of; Kazakhstan; Kyrgyzstan; Pakistan; Singapore; Tajikistan; Turkey; Turkmenistan; Uzbekistan

Description:

Tigers have muscular bodies with particularly powerful forelimbs and large heads. The pelage coloration varies between shades of orange or brown with white ventral areas and distinctive black stripes. The face has long whiskers, which are especially long in males. The pupils are circular with yellow irises. The small, rounded ears have black markings on the back, surrounding a white spot. These spots, called ocelli, play an important role in intraspecific communication.

The pattern of stripes is unique to each animal, these unique markings can be used by researchers to identify individuals (both in the wild and captivity), much in the same way that fingerprints are used to identify humans. It seems likely that the function of stripes is camouflage, serving to help tigers conceal themselves amongst the dappled shadows and long grass of their environment as they stalk their prey. The stripe pattern is also found on the skin of the tiger. If a tiger were to be shaved, its distinctive camouflage pattern would be preserved.

The tiger are the most variable in size of all big cats, even more so than the leopard and much more so than lions. The Bengal, Caspian and Siberian tiger subspecies represent the largest living felids, and rank among the biggest felids that ever existed. An average adult male tiger from Northern India or Siberia outweighs an average adult male lion by around 45.5 kg (100 lb). Females vary in length from 200 to 275 cm (79 to 108 in), weigh 65 to 167 kg (140 to 370 lb) with a greatest length of skull ranging from 268 to 318 mm (10.6 to 12.5 in). Males vary in size from 250 to 390 cm (98 to 150 in), weigh 90 to 306 kg (200 to 670 lb) with a greatest length of skull ranging from 316 to 383 mm (12.4 to 15.1 in). Body size of different populations seems to be correlated with climate—Bergmann's Rule—and can be explained from the point of view of thermoregulation. Large male Siberian tigers can reach a total length of more than 3.5 m (11.5 ft) "over curves", 3.3 m (10.8 ft) "between pegs" and a weight of 306 kg (670 lb). This is considerably larger than the size reached by the smallest living tiger subspecies, the Sumatran tiger, which reach a body weight of 75 to 140 kg (170 to 310 lb). Of the total length of a tiger, the tail comprises 0.6 to 1.1 m (2.0 to 3.6 ft).At the shoulder, tigers may variously stand 0.7 to 1.22 m (2.3 to 4.0 ft) tall. The accepted record weight, per the Guinness Book of World Records, for a wild tiger was 389 kg (860 lb) for a Bengal tiger shot in 1967, though its weight may have been boosted by the fact that it had eaten a water buffalo the previous night. Tigresses are smaller than the males in each subspecies, although the size difference between male and female tigers tends to be more pronounced in the larger tiger subspecies, with males weighing up to 1.7 times more than the females. In addition, male tigers have wider forepaw pads than females. Biologists use this difference to determine gender based on tiger tracks. The skull of the tiger is very similar to that of the lion, though the frontal region is usually not as depressed or flattened, with a slightly longer postorbital region. The skull of a lion has broader nasal openings. However, due to the amount of skull variation in the two species, usually, only the structure of the lower jaw can be used as a reliable indicator of species.

Subspecies:

There are nine subspecies of tiger, three of which are extinct. Their historical range in Bangladesh, Siberia, Iran, Afghanistan, India, China, and southeast Asia, including three Indonesian islands is severely diminished today. The surviving subspecies, in descending order of wild population, are:

The Bengal tiger *(Panthera tigris tigris)*, also called the Indian tiger, lives in India, Nepal, Bhutan and Bangladesh, and is the most common subspecies with populations estimated at less than 2,500 adult individuals. In 2011 the total population of adult tigers is estimated at 1,520–1,909 in India, 440 in Bangladesh, 155 in Nepal and 75 in Bhutan. It lives in alluvial grasslands, subtropical and tropical rainforests, scrub forests, wet and dry deciduous forests, and mangroves. Male Bengal tigers had a total length, including the tail, of 270 to 310 cm (110 to 120 in), while females range from 240 to 265 cm (94 to 104 in). The weight of males range from 175 to 260 kg (390 to 570 lb), while that of the females range from 100 to 181 kg (220 to 400 lb).In northern India and Nepal, tigers tend to be of larger size.

Males often average 235 kilograms (520 lb), while females average 141 kilograms (310 lb). In 1972, Project Tiger was founded in India aiming at ensuring a viable population of tigers in the country and preserving areas of biological importance as a natural heritage for the people. But the illicit demand for bones and body parts from wild tigers for use in Traditional Chinese medicine is the reason for the unrelenting poaching pressure on tigers on the Indian subcontinent. Between 1994 and 2009, the Wildlife Protection Society of India has documented 893 cases of tigers killed in India, which is just a fraction of the actual poaching and illegal trade in tiger parts during those years. An area of special conservation interest lies in the Terai Arc Landscape in the Himalayan foothills of northern India and southern Nepal, where 11 protected areas comprising dry forest foothills and tall grass savannas harbour tigers in a 49,000 square kilometres (19,000 sq mi) landscape. The goals are to manage tigers as a single metapopulation, the dispersal of which between core refuges can help maintain genetic, demographic, and ecological integrity, and to ensure that species and habitat conservation becomes mainstreamed into the rural development agenda. In Nepal, a community-based tourism model has been developed with a strong emphasis on sharing benefits with local people and on the regeneration of degraded forests. The approach has been successful in reducing poaching, restoring habitats, and creating a local constituency for conservation.

The Indochinese tiger *(Panthera tigris corbetti)*, also called Corbett's tiger, is found in Cambodia, China, Laos, Burma, **Thailand**, and Vietnam. These tigers are smaller and darker than Bengal tigers: Males weigh from 150–190 kg (330–420 lb) while females are smaller at 110–140 kg (240–310 lb). Their preferred habitat is forests in mountainous or hilly regions. According to government estimates of national tiger populations, the subspecies numbers around a total of 350 individuals. All existing populations are at extreme risk from poaching, prey depletion as a result of poaching of primary prey species such as deer and wild pigs, habitat fragmentation and inbreeding. In Vietnam, almost three-quarters of the tigers killed provide stock for Chinese pharmacies.

The Malayan tiger *(Panthera tigris jacksoni)*, exclusively found in the southern part of the Malay Peninsula, was not considered a subspecies in its own right until 2004. The new classification came about after a study by Luo et al. from the Laboratory of Genomic Diversity Study, part of the National Cancer Institute of the United States. According to official government figures, the population in the wild may number around 500 individuals, but is under considerable poaching pressure. The Malayan tiger is the smallest of the mainland tiger subspecies, and the second smallest living subspecies, with males averaging about 120 kg (260 lb) and females about 100 kg (220 lb) in weight. The Malayan tiger is a national icon in Malaysia, appearing on its coat of arms and in logos of Malaysian institutions, such as Maybank.

The Sumatran tiger *(Panthera tigris sumatrae)* is found only on the Indonesian island of Sumatra, and is critically endangered. It is the smallest of all living tiger subspecies, with adult males weighing between 100–140 kg (220–310 lb) and females 75–110 kg (170–240 lb). Their small size is an adaptation to the thick, dense forests of the island of Sumatra where they reside, as well as the smaller-sized prey. The wild population is estimated at between 400 and 500, seen chiefly in the island's national parks. Recent genetic testing has revealed the presence of unique genetic markers, indicating that it may develop into a separate species, if it does not go extinct. This has led to suggestions that Sumatran tigers should have greater priority for conservation than any other subspecies. While habitat destruction is the main threat to existing tiger population (logging continues even in the supposedly protected national parks), 66 tigers were recorded as being shot and killed between 1998 and 2000, or nearly 20% of the total population.

The Siberian tiger *(Panthera tigris altaica)*, also known as the Amur tiger, inhabits the Amur-Ussuri region of Primorsky Krai and Khabarovsk Krai in far eastern Siberia. It ranks among the biggest felids that have ever existed with a head and body length of 160–180 cm (63–71 in) for females and 190–230 cm (75–91 in) for males, plus a tail of about 60–110 cm (24–43 in) and an average weight of around 227 kg (500 lb) for males. Siberian tigers have thick coats and a paler golden hue and fewer stripes.[4] The heaviest wild Siberian tiger weighed 384 kg (850 lb) but according to Mazák this record is not reliable. In 2005, there were 331–393 adult-subadult Siberian tigers in the region, with a breeding adult population of about 250 individuals. The population has been stable for more than a decade, but partial surveys conducted after 2005 indicate that the Russian tiger population is declining. At the turn of the century, the phylogenetic relationships of tiger subspecies was re-assessed, and a remarkable similarity between the Siberian and Caspian tiger observed indicating that the Siberian tiger population is the genetically closest living relative of the extinct Caspian tiger, and strongly implying a very recent common ancestry for the two groups.

The South China tiger *(Panthera tigris amoyensis)*, also known as the Amoy or Xiamen tiger, is the most critically endangered subspecies of tiger and is listed as one of the 10 most endangered animals in the world. One of the smaller tiger subspecies, the length of the South China tiger ranges from 2.2–2.6 m (87–100 in) for both males and females. Males weigh between 127 and 177 kg (280 and 390 lb) while females weigh between 100 and 118 kg (220 and 260 lb). From 1983 to 2007, no South China tigers were sighted.[42] In 2007 a farmer spotted a tiger and handed in photographs to the authorities as proof. The photographs in question, however, were later exposed as fake, copied from a Chinese calendar and digitally altered, and the "sighting" turned into a massive scandal. In 1977, the Chinese government passed a law banning the killing of wild tigers, but this may have been too late to save the subspecies, since it is possibly already extinct in the wild. There are currently 59 known captive South China tigers, all within China, but these are known to be descended from only six animals. Thus, the genetic diversity required to maintain the subspecies may no longer exist. Currently, there are breeding efforts to reintroduce these tigers to the wild.

Hybridisation among the big cats, including the tiger, was first conceptualised in the 19th century, when zoos were particularly interested in the pursuit of finding oddities to display for financial gain. Lions have been known to breed with tigers (most often the Amur and Bengal subspecies) to create hybrids called ligers and tigons. Such hybrids were once commonly bred in zoos, but this is now discouraged due to the emphasis on conserving species and subspecies. Hybrids are still bred in private menageries and in zoos in China.

The liger is a cross between a male lion and a tigress. Because the lion sire passes on a growth-promoting gene, but the corresponding growth-inhibiting gene from the female tiger is absent, ligers grow far larger than either parent. They share physical and behavioural qualities of both parent species (spots and stripes on a sandy background). Male ligers are sterile, but female ligers are often fertile. Males have about a 50% chance of having a mane, but, even if they do, their manes will be only around half the size of that of a pure lion. Ligers are typically between 10 to 12 feet in length, and can be between 800 and 1,000 pounds or more.

The less common tigons is a cross between the lioness and the male tiger.

Colour variations

White tigers

There is a well-known allele that produces the white tiger, technically known as chinchilla Calvinistic, an animal which is rare in the wild, but widely bred in zoos due to its popularity. Breeding of white tigers will often lead to inbreeding (as the trait is recessive). Many initiatives have taken place in white and orange tiger mating in an attempt to remedy the issue, often mixing subspecies in the process. Such inbreeding has led to white tigers having a greater likelihood of being born with physical defects, such as cleft palates and scoliosis (curvature of the spine). Furthermore, white tigers are prone to having crossed eyes (a condition known as strabismus). Even apparently healthy white tigers generally do not live as long as their orange counterparts. Recordings of white tigers were first made in the early 19th century. They can only occur when both parents carry the rare gene found in white tigers; this gene has been calculated to occur in only one in every 10,000 births. The white tiger is not a separate sub-species, but only a colour variation; since the only white tigers that have been observed in the wild have been Bengal tigers (and all white tigers in captivity are at least part Bengal), it is commonly thought that the recessive gene that causes the white colouring is probably carried only by Bengal tigers, although the reasons for this are not known. Nor are they in any way more endangered than tigers are generally, this being a common misconception. Another misconception is that white tigers are albinos, despite the fact that pigment is evident in the white tiger's stripes. They are distinct not only because of their white hue; they also have blue eyes.

Golden tigers

In addition, another recessive gene may create a very unusual "golden" or "golden tabby" colour variation, sometimes known as "strawberry." Golden tigers have light gold fur, pale legs and faint orange stripes. Their fur tends to be much thicker than normal.[61] There are extremely few golden tigers in captivity, around 30 in all. Like white tigers, golden tigers are invariably at least part Bengal. Some golden tigers carry the white tiger gene, and when two such tigers are mated, can produce some stripe less white offspring. Both white and golden tigers tend to be larger than average Bengal tigers.

Other colour variations

There is no authenticated case of a black tiger, with the possible exception of one dead specimen examined in Chittagong in 1846.There are unconfirmed reports of a "blue" or slate-coloured tiger, the Maltese tiger. Largely or totally black tigers are assumed, if real, to be intermittent mutations rather than distinct species.

Population:

The thirteen Tiger Range Countries have come together in an unprecedented pledge to double the world's Tiger population by 2022, the next Year of the Tiger on the Asian lunar calendar, with a goal of achieving at least 6,000 Tigers. This figure was based on a baseline global population of 3,200, agreed upon at a preparatory workshop held in Kathmandu, Nepal in October 2009; 3,200 Tigers was the IUCN Red List population estimate at that time. Since then, Tiger Range Countries have adjusted their baseline national Tiger estimates, finalized in the Global Tiger Recovery Program adopted at the International Tiger Forum in St Petersburg, Russia in November 2010 (GTRP 2010). These estimates now total approximately 4,000 adult Tigers (see Table, which updates the GTRP's figure for India based on its most recent national census results [Jhala et al. 2011]).

However, estimates of the Tiger populations in 42 protected source sites where there is evidence of breeding total 2,154 Tigers (Walston et al. 2010a, modified from their 2,126 to include the most recent published estimates from Nepal of Tiger populations in Chitwan, Bardia and Shuklaphanta). Although this is not a complete estimate of the global Tiger population (for example, most Amur Tigers in Russia are found in unprotected areas), generally Tiger status outside the source sites is poor and poorly known. IUCN Guidelines (IUCN Standards and Petitions Subcommittee 2010) define population as the number of mature individuals, defined as "individuals known, estimated or inferred to be capable of reproduction." While in general this refers to all reproductive-age adults in the population, the Guidelines also "stress that the intention of the definition of mature individuals is to allow the estimate of the number of mature individuals to take account of all the factors that may make a taxon more vulnerable than otherwise might be expected." Tigers require large populations to persist, and the survival rate of breeding adult females is a key parameter, with models suggesting population declines when mortality of breeding females rises over 15% (Chapron et al. 2008). Population declines in recent years have been most pronounced outside protected areas (Walston et al. 2010b). The IUCN Guidelines advise that "mature individuals that will never produce new recruits should not be counted." For the purposes of Red List assessment, the estimated population in the Source Sites is used as a proxy for the breeding population of adult Tigers.

In 1998, the global Tiger population was estimated, less rigorously, at 5,000 to 7,000 Tigers (Seidensticker et al. 1999). Although to some extent the new numbers represent improved knowledge, it is clear that there have been substantial population declines, with Tigers all but eliminated from much of their recent forest range, particularly in Southeast Asia.

Habitat and Ecology:

Tigers are found mainly in the forests of tropical Asia, although they historically occurred more widely in drier and colder climes. One subspecies, the Amur Tiger *P.t. altaica,* persists in the Russian Far East. Photos of Tigers up to 4,500 m have been obtained in Bhutan (Wang 2008). Availability of a sufficient prey base of large ungulates is the Tiger's major habitat requirement: "wild pigs and deer of various species are the two prey types that make up the bulk of the Tiger's diet, and in general Tigers require a good population of these species in order to survive and reproduce" (Sunquist and Sunquist 2002). Based on studies, Karanth et al. (2004) estimate that Tigers need to kill 50 large prey animals per year. Tigers are opportunistic predators, however, and their diet includes birds, fish, rodents, insects, amphibians, reptiles in addition to other mammals such as primates and porcupines.Tigers can also take ungulate prey much larger than themselves, including large bovids (water buffalo, gaur, banteng), elephants and rhinos (Nowell and Jackson 1996). Tigers are generally solitary, with adults maintaining exclusive territories, or home ranges. Adult female home ranges seldom overlap, whereas male ranges typically overlap from 1–3 females, a typical felid pattern of social organization. Tiger home ranges are small where prey is abundant - e.g., female home ranges in Chitwan averaged 20 km^2, while in the Russian Far East they are much larger at 450 km^2 (Sunquist and Sunquist 2002). Similarly, reported Tiger densities range from 11.65 adult Tigers per 100 km^2 where prey is abundant (India's Nagarhole National Park) to as low as 0.13–0.45 per 100 km^2 where prey is more thinly distributed, as in Russia's Sikhote Alin Mountains (Nowell and Jackson 1996). In the past, tigers were found throughout Asia, from the Caucasus and the Caspian Sea to Siberia and the Indonesian islands of Java, Bali and Sumatra. During the 20th century, tigers have been extirpated in western Asia and became restricted to isolated pockets in the remaining parts of their range. Today, their fragmented and partly degraded range extends from India in the west to China and Southeast Asia. The northern limit of their range is close to the Amur River in south eastern Siberia. The only large island inhabited by tigers today is Sumatra.

tigers were extirpated on the island of Bali in the 1940s, around the Caspian Sea in the 1970s, and on Java in the 1980s. Loss of habitat and the persistent killing of tigers and tiger prey precipitated these extirpations, a process that continues to leave forests devoid of tigers and other large mammals across South and Southeast Asia. Since the beginning of the 21st century, their historical range has shrunk by 93%. In the decade from 1997 to 2007, the estimated area known to be occupied by tigers has declined by 41%.

Fossil remains indicate that tigers were present in Borneo and Palawan in the Philippines during the late Pleistocene and Holocene.

Tigers can occupy a wide range of habitat types but will usually require sufficient cover, proximity to water, and an abundance of prey. Bengal tigers live in many types of forests, including wet, evergreen, the semi-evergreen of Assam and eastern Bengal; the swampy mangrove forest of the Ganges Delta; the deciduous forest of Nepal, and the thorn forests of the Western Ghats. In various parts of their range they inhabit or have inhabited additionally partially open grassland and savanna as well as taiga forests and rocky habitats. Compared to the lion, the tiger prefers denser vegetation, for which its camouflage colouring is ideally suited, and where a single predator is not at a disadvantage compared with the multiple felines in a pride. A further habitat requirement is the placement of suitably secluded den locations, which may consist of caves, large hollow trees or dense vegetation.

Territorial behaviour:

Adult tigers lead solitary lives and congregate only on an ad hoc and transitory basis when special conditions permit, such as plentiful supply of food. They establish and maintain home ranges. Resident adults of either sex tend to confine their movements to a definite area of habitat, within which they satisfy their needs, and in the case of tigresses, those of their growing cubs. Those sharing the same ground are well aware of each other's movements and activities. The size of a tiger's home range mainly depends on prey abundance, and, in the case of male tigers, on access to females. A tigress may have a territory of 20 km2 (7.7 sq mi), while the territories of males are much larger, covering 60 to 100 km2 (23 to 39 sq mi). The range of a male tends to overlap those of several females. Tigers are strong swimmers, and are often found bathing in ponds, lakes, and rivers. Among fellow big cats, only the jaguar shares with the tiger a similar fondness for and capability in the water. They may also cross rivers up to 6 to 7 km (3.7 to 4.3 mi) across and can swim a distance of up to 29 km (18 mi) in a day. During the extreme heat of the day, they often cool off in pools. They are able to carry prey through or capturing it in the water. The relationships between individuals can be quite complex, and it appears that there is no set "rule" that tigers follow with regards to territorial rights and infringing territories. For instance, although for the most part tigers avoid each other, both male and female tigers have been documented sharing kills, usually with others of the opposite sex or cub. George Schaller observed a male tiger share a kill with two females and four cubs. Females are often reluctant to let males near their cubs, but Schaller saw that these females made no effort to protect or keep their cubs from the male, suggesting that the male might have been the father of the cubs. In contrast to male lions, male tigers will allow the females and cubs to feed on the kill first. Furthermore, tigers seem to behave relatively amicably when sharing kills, in contrast to lions, which tend to squabble and fight. Unrelated tigers have also been observed feeding on prey together. The following quotation is from Stephen Mills' book Tiger, as he describes an event witnessed by Valmiki Thapar and Fateh Singh Rathore in Ranthambhore National Park: A dominant tigress they called Padmini killed a 250 kg (550 lb) male nilgai – a very large antelope. They found her at the kill just after dawn with her three 14-month-old cubs and they watched uninterrupted for the next ten hours. During this period the family was joined by two adult females and one adult male – all offspring from Padmini's previous litters and by two unrelated tigers, one female the other unidentified. By three o'clock there were no fewer than nine tigers round the kill.

When young female tigers first establish a territory, they tend to do so fairly close to their mother's area. The overlap between the female and her mother's territory tends to wane with increasing time. Males, however, wander further than their female counterparts, and set out at a younger age to mark out their own area. A young male will acquire territory either by seeking out a range devoid of other male tigers, or by living as a transient in another male's territory until he is old and strong enough to challenge the resident male. The highest mortality rate (30–35% per year) amongst adult tigers occurs for young male tigers who have just left their natal area, seeking out territories of their own.

Male tigers are generally more intolerant of other males within their territory than females are of other females. For the most part, however, territorial disputes are usually solved by displays of intimidation, rather than outright aggression. Several such incidents have been observed, in which the subordinate tiger yielded defeat by rolling onto its back, showing its belly in a submissive posture. Once dominance has been established, a male may actually tolerate a subordinate within his range, as long as they do not live in too close quarters. The most violent disputes tend to occur between two males when a female is in oestrus, and may result in the death of one of the males, although this is a rare occurrence. To identify his territory, the male marks trees by spraying of urine and anal gland secretions, as well as marking trails with scat. Males show a grimacing face, called the Flehmen response, when identifying a female's reproductive condition by sniffing their urine markings. Like the other Panthera cats, tigers can roar. Tigers will roar for both aggressive and non-aggressive reasons. Other tiger vocal communications include moans, hisses, growls and chuffs.

Tigers have been studied in the wild using a variety of techniques. The populations of tigers were estimated in the past using plaster casts of their pugmarks. This method was criticized as being inaccurate. Attempts were made to use camera trapping instead. Newer techniques based on DNA from their scat are also being evaluated. Radio collaring has also been a popular approach to tracking them for study in the wild.

Hunting and diet:

In the wild, tigers mostly feed on larger and medium sized animals, with most studies indicating a preference for native ungulates averaging 90 kg (200 lb) at a minimum. Sambar, chital, barasingha, wild boar, gaur, nilgai and both water buffalo and domestic buffalo, in descending order of preference, are the tiger's favoured prey in India. Sometimes, they also prey on other predators, including other large species such as leopards, pythons, sloth bears and crocodiles. In Siberia the main prey species are manchurian wapiti and wild boar (the two species comprising nearly 80% of the prey selected) followed by sika deer, moose, roe deer, and musk deer. In Sumatra, sambar, Muntjac, wild boar, and Malayan tapir are predominantly preyed upon. In the former Caspian tiger's range, prey included saga antelope, camels, Caucasian wisent, yak, and wild horses. Like many predators, they are opportunistic and will eat much smaller prey, such as monkeys, peafowl's, other largish, ground-based birds, hares, porcupines and fish.

Adult elephants are too large to serve as common prey, but conflicts between tigers and elephants, with the huge elephant typically dominating the predator, do sometimes take place. A case where a tiger killed an adult Indian Rhinoceros has been observed, although adult rhinoceros are often ignored as potential prey due to a combination of very large size, a short temper and very thick skin, which render them a laborious and very difficult kill. Young elephant and rhino calves are occasionally taken. Tigers also sometimes prey on domestic animals such as dogs, cattle, horses, and donkeys. These individuals are termed cattle-lifters or cattle-killers in contrast to typical game-killers.

Old tigers, or those wounded and rendered incapable of catching their natural prey, have turned into man-eaters; this pattern has recurred frequently across India. An exceptional case is that of the Sundarbans, where healthy tigers prey upon fishermen and villagers in search of forest produce, humans thereby forming a minor part of the tiger's diet. Tigers will occasionally eat vegetation for dietary fiber, the fruit of the Slow Match Tree being favoured. Tigers are thought to be nocturnal predators, hunting at night. However, in areas where humans are typically absent, they have been observed via remote controlled, hidden cameras hunting during the daylight hours. They generally hunt alone and ambush their prey as most other cats do, overpowering them from any angle, using their body size and strength to knock the prey off balance. Successful hunts usually require the tiger to almost simultaneously leap onto their quarry, knock it over and grab the throat or nape with its teeth. Even with their great masses, tigers can reach speeds of about 49–65 kilometres per hour (35–40 miles per hour), although they can only do so in short bursts, since they have relatively little stamina; consequently, tigers must be relatively close to their prey before they break their cover. If the prey catches wind of the tiger's presence before the moments of the pounce, the tiger will usually abandon the hunt rather than chase prey or battle it head-on. Tigers have great leaping ability; horizontal leaps of up to 10 m (33 ft) have been reported, although leaps of around half this amount are more typical. However, only one in twenty (or 5–10%) hunts, including any instances of stalking in proximity to potential prey, ends in a successful kill. An adult tiger can go up to two weeks without eating but then can gorge on up to 34 kg (75 lb) of flesh at one sitting. In captivity, adult tigers are fed 3 to 6 kg (6.6 to 13 lb) of meat a day. Due to their low hunting success rate, ability to go prolonged periods without food and naturally low population densities, tigers typically have little to no deleterious effect on the populations of the species they prey on. Several other large carnivores, such as gray wolves, spotted hyenas and lions, live in groups and need to capture relatively greater quantities of prey in order to feed and maintain stability in their respective packs, clans or prides.

When hunting large prey, tigers prefer to bite the throat and use their extremely powerful forelimbs to hold onto the prey, often simultaneously wrestling it to the ground. The tiger remains latched onto the neck until its prey dies of strangulation. By this method, gaurs and water buffalos weighing over a ton have been killed by tigers weighing about a sixth as much. Although they can kill healthy adults of large bovids weighing at least 1,000 kg (2,200 lb), tigers often select the calves or infirm of very large species. Large prey can be quite dangerous to tackle, with the great bulk and massive horns of large bovids, the strong legs and antlers of mature deer and the long, powerful tusks of boars all being potentially fatal to the tiger. No other extant land predator routinely takes on prey this large on their own. Whilst hunting sambars, which comprise up to 60% of their prey in India, tigers have reportedly called out a passable impersonation of the male sambar's rutting call in order to attract them.With small prey such as monkeys and hares, the tiger bites the nape, often breaking the spinal cord, piercing the windpipe, or severing the jugular vein or common carotid artery. Though rarely observed, some tigers have been recorded to kill prey by swiping with their paws, which are powerful enough to smash the skulls of domestic cattle, and break the backs of sloth bears. After killing their prey, tigers sometimes drag their prey to conceal it in vegetative coverage, usually pulling it by grasping with their mouths at the site of the killing bite (on the throat in large prey, on the nape in smaller prey). This too can require great physical strength. In one case, after it had killed an adult gaur, a tiger was observed to drag the massive carcass over a distance of 12 m (39 ft). When 13 men simultaneously tried to drag the same carcass later, they were unable to move it. During the 1980s, a tiger named "Genghis" in Ranthambhore National Park was observed frequently hunting prey through deep lake water, a pattern of behaviour that had not been previously witnessed in over 200 years of observations. Moreover, he appeared to be extraordinarily successful for a tiger, with as many as 20% of hunts ending in a kill. Tigers usually prefer to eat prey they've caught themselves but are not above eating carrion in times of scarcity and may even pirate prey from other large carnivores. Although predators typically avoid one another, if a prey item is under dispute or a serious competitor is encountered, displays of aggression are a regular occurrence. If these are not sufficient, the conflicts may come turn violent and tigers may kill such formidable competitors as leopards, striped hyenas, pythons and even crocodiles on occasion, In some cases, rather than being strictly competitive, the attacks by tigers on other large carnivores seem to be predatory in nature. Situations where smaller predators, such as badgers, lynxes and foxes are attacked, are almost certainly predatory. Interestingly, this species' closest living relative, the lion, deals with competing predators very differently, undoubtedly due to the fact that it lives in large prides. Lions do not treat other predators as prey, as does the tiger, but invests a good deal of time proactively track down other predators and killing them, then leaving their bodies uneaten. Lions kill competitors from honey badgers to spotted hyenas and, in protected areas of Africa, are the leading cause of mortality for African wild dogs and cheetahs. The tiger does not spend as much time tracking down other predators.

Occasionally, a large crocodile may attempt to predate a tiger. When seized by a crocodile, a tiger will strike at the reptile's eyes with its paws. When killing crocodiles, after stunning them about the face, tigers will flip the reptile's body over and disembowel it through the softer belly rather try to penetrate the thick, well-armoured upper hide. Eighteenth century Physician Oliver Goldsmith described the frequent conflicts between mugger crocodiles and tigers that occurred during that time. Thirsty tigers would frequently descend to the rivers to drink and on occasion were seized and killed by the muggers, though more often the tiger escaped and the reptile was disabled. Mature mugger crocodiles may target much the same prey as the tiger, including sambar and water buffalo. Occasionally, a mugger and a tiger will try to claim a carcass killed by either one, resulting in a "tug of war" at the water's edge until one of them comes away with it. A potentially more formidable foe is the larger, more aggressive Saltwater Crocodile, which the tiger rarely encounters outside of estuarial regions of eastern India. The first confirmed case of a saltwater crocodile predating an adult tiger occurred in that region in 2011. There is a second-hand account of a tiger killing a "small" saltwater crocodile. he considerably smaller leopard dodges competition from tigers by hunting in different times of the day and hunting different prey. In India's Nagarhole National Park, most prey selected by leopards was from 30 to 175 kg (66 to 390 lb) against a preference for prey weighing over 176 kg (390 lb) in the tigers.

The average prey weight in the two respective big cats in India was 37.6 kg (83 lb) against 91.5 kg (202 lb). With relatively abundant prey, tigers and leopards were seen to successfully coexist without competitive exclusion or inter-species dominance hierarchies that may be more common to the savanna (where the leopard may co-exist with the lion). Tigers have been known to suppress wolf populations in areas where the two species coexist, mainly via competitive exclusion. There four proven records of Siberian tigers killing wolves and not eating them. Dhole packs have been observed to challenge the big cats in disputes over food and have even killed tigers in rare cases. However, tigers have also been observed killing multiple dholes at once and the dhole will typically only attack a tiger directly if the pack is quite large. Lone golden jackals expelled from their pack have been known to form commensal relationships with tigers. These solitary jackals, known as kol-bahl, will attach themselves to a particular tiger, trailing it at a safe distance in order to feed on the big cat's kills. A kol-bahl will even alert a tiger to a kill with a loud pheal. Tigers have been known to tolerate these jackals: one report describes how a jackal confidently walked in and out between three tigers walking together a few feet away from each other. When in the presence of a tiger, a golden jackal pack will emit a howl very different from its normal vocalization that is thought to function as a warning to other jackals.

Other than the rare large crocodile or large dhole pack, the only serious competitors to tigers are bears. Some bears, especially the brown bear of the North, will try to steal tigers' kills, although the tiger will sometimes defend its kill. However, in some cases, bears (especially cubs) are preyed upon by tigers. Although it hunts all its prey via ambush, tigers are especially cautious when handling bears, as many bears are capable of killing a tiger whilst defending themselves. Predation seems especially prevalent in India, where tigers may attack sloth bears. The sloth bears can be quite aggressive and will sometimes displace young tigers away from their kills or successfully defend themselves via counterattack. Despite this, sloth bears are killed with some regularity and react fearfully to the presence of tigers or even stimuli related to them (i.e. the call of the sambar deer due to the tiger's impersonation of it). Bears (Asiatic black bears and brown bears) make up 5–8% of the tiger's diet in the Russian Far East. Some accounts claim that the black bears more successfully avoid predation by tigers because they are skilled tree-climbers, although dietary research has contrarily indicated that the smaller, less aggressive black bear (comprising 4–6.5% of the tiger's local diet) is the more common prey species than the brown bear (at 1–1.5% of the diet).Siberian tigers and brown bears usually avoid confrontation but can sometimes be competitors, with dominance seemingly determined by the age, sex and size of the rivals rather than species. Older and larger males of both species tend to dominate in this interspecies conflict. Some brown bears, upon emerging from hibernation, follow tigers habitually in order to steal their kills. Tigers will kill brown bear cubs and even adults on some occasions, especially if they find the bears in their dens during the hibernation cycle or in periods of low prey density in the fall. There are also records of brown bears killing tigers up to the size of adult males, either in self-defence or in disputes over kills. Tigers may additionally predate the other bear species it encounters (or had encountered historically), which includes Giant pandas and sun bears, but information is very limited on such interactions.
Systems: Terrestrial

Reproduction:

Mating can occur all year round, but is generally more common between November and April. A female is only receptive for three to six days and mating is frequent during that time period. A pair will copulate frequently and noisily, like other cats. The gestation period can range from 93 to 112 days, although the average is 104–106 days. The litter size usually consists of anywhere from 1 to 6 cubs, though 2 to 3 is usually the norm. Cubs can weigh from 680 to 1,400 g (1.5 to 3.1 lb) each at birth and are born blind and helpless. The females rear them alone, with the birth site and maternal den being sheltered locations such as thickets, caves and rocky crevices. The father of the cubs generally takes no part in rearing them. Unrelated wandering male tigers may even kill cubs to make the female receptive, since the tigress may give birth to another litter within 5 months if the cubs of the previous litter are lost. The mortality rate of tiger cubs is fairly high – approximately half do not survive to be more than two years old. Few other predators attack tiger cubs due to the diligence and ferocity of the mother tiger. Beyond humans and other tigers, common causes of cub mortality are starvation, freezing and accidents.
There is generally a dominant cub in each litter, which tends to be male but may be of either sex. This cub generally dominates its siblings during play and tends to be more active, leaving its mother earlier than usual. The cubs open their eyes at 6 to 14 days old. At 8 weeks, the cubs may make short ventures out of the den with their mother, although they do not travel with her as she roams her territory until they are older. The cubs are nursed in total for a period of 3 to 6 months. Around the time they are weaned, they start regularly engaging in territorial walks with their mother.

During this stage, the tigress's young are also taught how to hunt. The cubs are often capable (and nearly adult size) hunters by the time they are 11 months old. The cubs become independent around 18 months of age, but it is not until they are around 2–2½ years old that they fully separate from their mother. Females reach sexual maturity at 3–4 years, whereas males reach sexual maturity at 4–5 years.

Over the course of her life, a female tiger will typically give birth to an approximately equal number of male and female cubs. Tigers breed well in captivity, and the captive population in the United States may rival the wild population of the world.[88] The known limit for lifespan in captivity is 26 years, and while captive animals usually outlive wild ones, it is thought that a wild adult tiger, with no natural predators as long as doesn't run afoul of humans, can likely live to a comparable age.

Intraspecific predatory relationships

Man-eating tigers:

Normally wild tigers, especially if they have no prior contact with humans, will actively avoid interactions with humans. However, according to some sources, tigers are thought to be responsible for more human deaths through direct attack than any other wild mammal. Attacks are occasionally provoked, as tigers will lash out after being injured while they themselves are hunted. Occasionally, attacks are provoked accidentally, as when a human surprises a tiger or inadvertently comes between a mother and her young. Occasionally human behavior will inadvertently provoke tiger attacks by triggering their natural instincts. In one case, a postman who delivered mail on foot in a rural region of India where interactions with tigers are commonplace, was not bothered by them for several years despite many interactions. Soon after the postman started to use a bicycle, the man was attacked by a tiger, theatrically having been instinctively provoked by the chase. Although humans are not regular prey for tigers, occasionally tigers will come to view people as prey. Such attacks tend to be particularly prevalent in areas where population growth, logging, and farming have put pressure on tiger habitats and reduced wild prey for them. Most man-eating tigers are old and missing teeth, acquiring a taste for humans because of their inability to capture their preferred prey. This was the case in the Champawat Tiger, a tigress found in Nepal and then India, that was found to have had two broken canines. She was responsible for an estimated 430 human deaths, the most attacks known to be perpetrated by a single wild animal per the Guinness Book, by the time she was shot in 1907 by Jim Corbett. Unlike man-eating leopards, even established man-eating tigers will seldom enter human settlements, usually remaining at village outskirts. Nevertheless, attacks in human villages do occur. Tigers treat humans as they do other potential prey, engaging in a length stalking phase before pouncing from close range. Since they are nocturnal, many attacks occur at night while people are usually sleeping. Attacks are also common when people are working outdoors and are physically engaged in distracting tasks. Thanks to their natural predatory instincts, such as their use of stealth and surprise and their tendency to attack partially isolated people, early writings tend to profile man-eating tigers and other similarly-disposed big cats as "cowardly". Due to the size and power of the tiger, few humans survive when a predatory attack is carried out.Reportedly, in the Singapore area (where tigers are now extirpated) in the 1840s, an estimated 1,000 fatalities occurred from tiger attacks. Man-eaters have been a particular problem in recent decades in India and Bangladesh, especially in Kumaon, Garhwal and the Sundarbans mangrove swamps of Bengal, where some healthy tigers have been known to hunt humans. Because of rapid habitat loss attributed to climate change, tiger attacks have increased in the Sundarbans.The Sundarbans area reportedly had 129 human deaths from tigers from 1969 to 1971. In the 10 years prior to that period, according to Chakrabarti (1984), humans were predated at an estimated rate of 100 per year in the Sudarban region, with a possible high of around 430 in some years of the 1960s. Unusually, in some years in the Sundarbans, more humans are killed by tigers than vica versa. In the year of 1972, India's production of honey and beeswax dropped by 50% when at least 29 people who gathered these materials were devoured.Almost all tigers that are identified as man-eaters are quickly captured, shot, or poisoned. Current Indian wildlife protection laws state that animals must be saved unless the tiger is a repeat offender and no hope exists for rehabilitation. However, man-eating attacks may still lead to revenge killing of several tigers, including those not involved in the attack.

On occasion, man-eating tigers are relocated to large nature preserves, with mixed success. In 1986 in the Sundarbans, since tiger almost always attack from the rear, the idea was implemented that masks with human faces on them be worn to the back of the head, on the theory that tigers will usually not carry through attacks if seen by their prey. This temporarily decreased the number of attack, though the tigers appeared to become habituated to the masks and attacks again increased in the following years.

Tigers kept in captivity retain wild instinct and, especially those in privately-owned collections where improper handling is more common, may attack humans. An estimated 1.75 fatal attacks occur per year in captivity, with at least 27 people killed or seriously injured in the United States by tigers from 1998 to 2001.In large, well-kept public zoos, tiger attacks on humans are very rare and tigers who associate with their zookeepers from birth may be docile and even affectionate towards their handlers once fully grown. However, most zoos are rightfully cautious and, when the tigers must be handled closely (such as medical procedures), it is a necessity to assure that tigers are fully unconscious from anesthesia.Tatiana, a female tiger, escaped from her enclosure in the San Francisco Zoo, killing one person and seriously injuring two more before being shot and killed by the police. The enclosure had walls that were lower than they were legally required to be, allowing the tiger to climb the wall and escape

Major Threat(s):

Illegal trade in high-value Tiger products including skins, bones, meat and tonics is a primary threat to Tigers, which has led to their recent disappearance from broad areas of otherwise suitable habitat, and continues at unsustainable rates.

Asia is a densely populated and rapidly developing region, bringing huge pressures to bear on the large wild areas required for viable Tiger populations. Conversion of forest land to agriculture and silviculture, commercial logging, and human settlement are the main drivers of Tiger habitat loss. With their substantial dietary requirements, Tigers require a healthy large ungulate prey base, but these species are also under heavy human subsistence hunting pressure and competition from domestic livestock.

Tiger attacks on livestock and people can lead to intolerance of Tigers by neighbouring communities and presents an ongoing challenge to managers to build local support for Tiger conservation. In some areas there have been many human deaths - for example, 41 people were killed by Tigers in the Sundarbans mangrove forest of Bangladesh during an 18-month period in 2001–2003 (Khan 2004).

Commercial hunting and traditional medicine:

Historically, tigers have been hunted at a large scale so their famous striped skins could be collected. The trade in tiger skins peaked in 1960s, just before international conservation efforts took effect. By 1977, a tiger skin in an English market was considered to be worth $4,250 American dollars.

Many people in China and other parts of Asia have a belief that various tiger parts have medicinal properties, including as pain killers and aphrodisiacs. There is no scientific evidence to support these beliefs. The use of tiger parts in pharmaceutical drugs in China is already banned, and the government has made some offences in connection with tiger poaching punishable by death. Furthermore, all trade in tiger parts is illegal under the Convention on International Trade in Endangered Species of Wild Fauna and Flora and a domestic trade ban has been in place in China since 1993. However, the trading of tiger parts in Asia has become a major black market industry and governmental and conservation attempts to stop it have been ineffective to date. Almost all black marketers engaged in the trade are based in China and have either been shipped and sold within in their own country or into Taiwan, South Korea or Japan. The Chinese subspecies was almost completely decimated by killing for commerce due to both the parts and skin trades in the 1950s through the 1970s. Contributing to the illegal trade, there are a number of tiger farms in the country specialising in breeding the cats for profit.

It is estimated that between 5,000 and 10,000 captive-bred, semi-tame animals live in these farms today. However, many tigers for traditional medicine black market are wild ones shot or snared by poachers and may be caught anywhere in the tiger's remaining range (from Siberia to India to the Malay Peninsula to Sumatra). In the Asian black market, a tiger penis can be worth the equivalent of around $300 U.S. dollars. In the years of 1990 through 1992, 27 million products with tiger derivatives were found

Conservation Actions:

In 2010, the Year of the Tiger on the Asian lunar calendar, Tigers were the focus of substantial conservation effort and investment. At a "Tiger Summit" held in St. Petersburg, Russia in November 2010, the 13 Tiger Range Countries adopted a Global Tiger Recovery Program (GTRP 2010). The goal is to effectively double the number of wild Tigers by 2022 through actions to: i) effectively preserve, manage, enhance and protect Tiger habitats; ii) eradicate poaching, smuggling and illegal trade of Tigers, their parts and derivatives; iii) cooperate in transboundary landscape management and in combating illegal trade; iv) engage with indigenous and local communities; v) increase the effectiveness of Tiger and habitat management; and vi) restore Tigers to their former range. The Tiger Summit was attended by Heads of State including Russia, China, Laos, Nepal and Bangladesh, and represents a policy commitment to Tiger conservation of unprecedented significance.

The future of Tiger range depends upon the Asian governments creating effective Tiger landscapes by conserving large areas of suitable habitat. Within these landscapes, the most urgent need is to first secure the source sites—protected areas with viable Tiger populations—where most of the global Tiger population is now clustered, and many of which are currently too threatened to deliver their potential as the wellspring of species recovery (Walston et al. 2010b

Binturong *(Arctictis binturong)*

Description:

HB: 610-965; T: 500-840; HF: 130-180; E: 49-53; W: 9-20kg

The Binturong is the largest of the *Viverridae,* and the long shaggy hair makes it look much larger that it already is. The general colour is black, by this may vary in each individual., this being dependant on the number of hairs with white tips.. In some individuals the whole head is greyish to almost white in in others the grey may extend to behind the shoulders. The ears are round and edges conspicuously with white, the hair behind each ear are long and protrude beyond the ear tip as a long tuft. (a characteristic not found in any other *Viverridae*). The tail is long and prehensile. The feet are plantigrade, with short, slightly curved, semi retractile claws; the soles are naked to the heels. There are two pairs of mammae.The skull closely resemble that of *Paguma,* but is somewhat larger. and has a palate considerably more curved from front to back(resembling *Arctogalidia i*n this character). The post orbital area is not as constricted.. The dentition contains the full Viverridae compliment of 40 teeth, but the upper second premolar is sometimes missing, and the first lower premolar in minute and occasionally lacking. The upper incisors are rather more separated than in other civets.

Range Description:

The binturong is widespread in South and Southeast Asia occurring in Bangladesh, Bhutan, Myanmar, China (Yunnan), India (including Sikkim), Indonesia (Kalimantan, Java, Sumatra), Lao PDR, Malaysia, Nepal, Philippines (Palawan), **Thailand,** and Vietnam (Heaney et al. 1998; Wozencraft 2005). Records from outside this range include a 1928 record from Guangxi, China (Zhang 1997) and record from Calauit Island, Philippines (Corbet and Hill 1992) and several from Cambodia (Walston 2001).

Countries: Native:

Bangladesh; Bhutan; Brunei Darussalam; Cambodia; China; India; Indonesia (Jawa, Kalimantan, Sumatera); Lao People's Democratic Republic; Malaysia (Peninsular Malaysia, Sabah, Sarawak); Myanmar; Nepal; Philippines; **Thailand**; Vietnam.

Population:

Historically, the binturong was often thought to be abundant within its distribution range, but is now uncommon or rare over much of the range. Lekagul and McNeely (1977) reported this species as rare in **Thailand.** In Lao PDR, there were only three sightings in extensive wildlife surveys into some of the remotest parts of the country between 1992 and 1999, two from Nam Kading National Biodiversity Conservation Area (Duckworth et al. 1999), and one from Hin Namno National Biodiversity Conservation Area in early 1998 (Walston and Vinton 1999, Duckworth et al. 1999). While Deuve (1972) reported this species to be common in Lao PDR, it is thought that this might be due to its striking appearance, rather than natural abundance (Duckworth et al. 1999).

Work in **Thailand** by Nettelbeck (1997) suggested that binturongs can be seen frequently when the threat of hunting is removed; however, as hunting is common in most areas, it is unclear whether the observed is only applicable to the site where this observation was noted (**Near Khao Yai National Park headquarters in Thailand). Grassman et al. (2005) recorded 31 individuals in Phu Khieo Wildlife Sanctuary in Thailand between 1998 and 2002.** In the Philippines, the populations are restricted and uncommon (Heaney et al. 1998). Populations are thought to be decreasing as a result of collection for the pet trade. In Assam, India, the binturong has been noted as not uncommon in forested areas, and is most common in regions with good tree cover (Choudhury 1997).
Population Trend: Decreasing

Habitat and Ecology:

Binturongs are primarily arboreal, but do descend to the ground and some are captured (Duckworth et al. 1999); in fact the number of pictures coming up on camera-traps across its mainland range reveals an unexpectedly high level of ground activity. This is no doubt because it is heavy and ponderous, and where other animals would leap between trees, it must descend to the ground to go from one tree to another, i.e. quite often when commuting (Than Zaw et al. in press). The ecology of this species is poorly clarified and may vary between areas, as publications about activity are conflicting. Grassman et al. (2005) noted the binturong to be crepuscular and nocturnal and Nowak (1991) reported them to be predominantly nocturnal, whereas Nettelbeck (1997) reports them to often be active during the day, and there are many other day-time field sightings made incidentally during forest research (e.g. Lambert 1990; Datta 1999). Activity patterns have also been described as cathemeral or arrhythmic (Than Zaw et al. in press). **In Thailand, Grassman et al. (2005) found that this species has a mean annual range size of 6.2 km² with a mean overlap of 35% in a study on this species conducted in Phu Kieo Wildlife Sanctuary. Within this range, the binturong is confined to tall forest, where it feeds on fruits and small animals like insects, birds, and rodents, as well as fish (Lekagul and McNeely 1977).** In Lao PDR, recent records are from extensive evergreen forest, while in other countries a variety of tall forests are used (Duckworth et al. 1999). In the Philippines, the species is found in primary and secondary lowland forest, including grassland/forest mosaic from sea level to 400 m (Rabor 1986; Esselstyn et al. 2004). It was recorded in secondary forest, that was logged in the 1970s, and which surrounds a palm estate, in Malaysia in 2000-01 by Azlan (2003).The litter size is about one to three, with a gestation of 92 days, reaching adult size in one year, and they may live as long as 18 years (Lekagul and McNeely 1977).
Systems: Terrestrial.

Major Threat(s):

Habitat loss and degradation are a major threat to the binturong (Schreiber et al. 1989). Throughout this species' range, there has been loss and degradation of forests through logging and conversion of forests to non-forest land-uses. Forest conversion has been extremely high in the lower altitude parts of its Sundaic range in the last 20 years (e.g. Birdlife International, 2001; Holmes, 2000; Jepson et al., 2001; Cornrow and Talip, 2001; Lambert and Collar, 2002; Curran et al. 2004; Fuller, 2004; Eames et al. 2005, Aratrakorn et al. 2006; Kinnaird et al. 2003). Choudhury (1997) notes that large-scale deforestation in Indian portions of the species range could be contributing to its increased rarity, since many records come from areas where forests are now being degraded. In China, rampant deforestation and opportunistic logging practices have fragmented suitable habitat or eliminated sites altogether (Pu et al. 2007). In Borneo, the overall density of civets (including the binturong) in logged forests was found to be significantly lower than in primary forests (Heydon and Bulloh, 1996).

In the Philippines this species is harvested for the pet trade, and in the south of its range it is also taken for human consumption (GMA Philippines 2006). In Lao PDR, this species is one of most frequently displayed caged live carnivores and skins are traded frequently in at least Vientiane (R. Tizard pers. comm.). Since many of the animals being traded are young, there is the possibility that trees are deliberately felled to allow individuals to be caught (I. Johnson pers. comm.. 1996). Considered a delicacy in parts of Lao PDR, the binturong is taken for food and is also traded as a food item to Vietnam (I. Johnson pers. comm. 1999). Given recent camara trapping evidence in mainland Asia, it has become clear that the binturong descends to the ground more frequently than previously thought (Duckworth pers. comm. 2007); therefore the threat of snaring when this species descends to the ground may be more serious than previously considered (Duckworth et al 1999). Non-specific hunting of large mammals is very high across most of the species' mainland range. Duckworth (1997) speculated that hunting was unlikely to be the cause of the few recent sightings of binturong in Lao PDR, citing the many Black Giant Squirrels and gibbons in several areas lacking Binturong records.

However, given the possibility of intraspecific differences in population dynamics, these species may likewise have differing resilience to hunting pressure (Duckworth et al. 1999). Given that the binturong is relatively unafraid of humans and is sometimes active during the day, the species is often conspicuous both to surveyors (suggesting that the few encounters reflect a low population) and to hunters (thus exposing it to elevated risk) (Duckworth et al 1999).

Conservation Actions:

India included the binturong in CITES Appendix III in 1989 (UNAPT-WCMC Species Database: CITES-Listed Species 2006). In the Philippines, the Environmental Legal Assistance Centre has been involved in controlling and enforcing wildlife laws (applicable for all Palawan species). The species is protected in Malaysia (Azlan 2003), and is listed as Critically Endangered on the China Red List (Wang and Xie 2004). The species would benefit from effective controls on the trade. The binturong occurs in protected areas across its current range, however, the effectiveness of these reserves at protecting the species is variable. Stricter enforcement of legislation against hunting, poaching, encroachment, habitat degradation, and deforestation is required to achieve the necessary protective status for this species.

This species has been recorded from several studies in protected areas, such as the following examples. A study of the range, habitat use, and activity patterns of this species was conducted by Grassman et al (2005) in **Phu Kieo Wildlife Sanctuary in Thailand (16° 05? to 16° 35? N, 101° 20? to 101° 55 E). It has also been studied in Khao Yai National Park in Thailand by Nettelbeck (1997) and Austin (2002).** This species has been recorded recently in Nam Kading National Biodiversity Conservation Area in Lao PDR (Duckworth et al, 1999. This species was recorded from Jerangau Forest Reserve in Peninsular Malaysia in 2000-01 (04° 55.5? N, 103° 05.7? E; Azlan 2003), and Krau Wildlife Reserve (Laidlaw 2001).

Small-toothed Palm Civet *(Arctogalidia trivirgata)*

Description:

HB: 432-532; T: 510-660; HF: 75-95; E: 42-50; W: 2-2.5kg.

This small species of civet is also called the Three-striped Palm Civet. It occurs in primary rainforest and adjacent secondary forest, is nocturnal and almost exclusively arboreal. Its fur colour is predominantly greyish (sometimes reddish-brown), becoming paler and somewhat yellowish around the neck and on the belly. Its facial fur is black, but in some specimens or populations there is a narrow pale stripe on the forehead. Three dark stripes, or rows of broken spots, extend from the neck to the base of the tail : these may be hard to see when the animal is in profile. Its tail is dark, longer than its head-body length, and somewhat prehensile. Only the female as scent glands (both sexes have scent glands in other Paradoxurinae. The female as two pair of mammae. The skull as a very long postorbital process, although they do not form a complete bony orbits as in the mongooses; the zygomatic arch rises sharply to form part of the lower rim of the orbit. The sagittal crest is very low and incomplete in some individuals indicating relatively little development of the masseter muscles controlling the light mandible, but the occipital crests are well developed and form a right angle with each other when viewed from the rear. The muzzle is narrow, but widens slightly directly above the canines.

Range Description:

The small-toothed palm civet range covers parts of the Sundaic region and northern Southeast Asia. It is currently known to occur in Cambodia (Walston and Duckworth 2003), Lao PDR (Duckworth 1997b; Johnson pers. comm.), Borneo (Malaysia) (Duckworth 1997a, Wells et al. 2005, Belden et al. 2007), **Thailand** (Rabinowitz 1991, Conforti 1996, Duckworth and Nettelbeck in press), Viet Nam (Borissenko et al. 2004, Long and Minh Hoang 2006), Indonesia (Java: Suyanto 2003; Sumatra: Holden 2006; Kalimantan), and Peninsular Malaysia (Ratnam et al. 1995).This species was recorded from Kadan Kyun Island off the coast of Myanmar in the early part of the twentieth century (Meiri 2005), no surveys using methodology suitable for this species have been conducted in other areas in Myanmar, where it was historically recorded quite widely (Than Zaw et al. in press); a record of a released animal in the outskirts of Yangon (Su 2005) suggests populations are still extant in the country. Given the extent of remaining suitable habitat, it is likely that Myanmar supports large populations. It was first recorded in Cambodia from Keo Seima Biodiversity Conservation Area (150 masl) in 2003, confirming previously inferred extension of the species range into this country. Duckworth (1997b) reports that this species is widespread in south and central Lao PDR; there has been no suitable survey carried out in the country north of approximately 19°N. In **Thailand** there are many records in protected areas (Duckworth and Nettelbeck in press). The current status of this species in China is unclear, as no proper spotlighting surveys have been conducted and nothing can be inferred on its status (W. Duckworth in litt. 2006).

There is a record of a skin collected from Darjeeling, India by Dr. Steward in 1856, although there seem to have been no recent confirmed records from this region. Pocock (1939) gave several records for India. The type locality of A. t. millsi is Mokokchung in the Naga hills (26°20'N, 94°32'E). This species has been reported by Suyanto (2003) from Gunung Halimun, West Java, and there are records from the 1930s from Gunung Gede National Park and Ujung Kulon National Park in Java (see Schreiber et al. 1989). However, it is unknown whether these populations are still extant (Meijaard pers. comm.). It is also presumed to occur still in Indonesian Borneo and Brunei. Current status in Bangladesh is unclear but suitable habitat remains in the northern and eastern hill tracts. Reliance on recent records to define its current distribution will produce an underestimate because the species is readily overlooked by conventional survey techniques (Duckworth and Nettelbeck in press).

Countries: Native:

Bangladesh; Brunei Darussalam; Cambodia; China; India; Indonesia (Jawa, Kalimantan, Sumatera); Lao People's Democratic Republic; Malaysia (Peninsular Malaysia, Sabah, Sarawak); Myanmar; Singapore; **Thailand;** Vietnam

Population:

Very few surveys using appropriate methodology have been carried out through much of its range, thus the small-toothed palm civet population trend is unclear. Past authors have noted this as a rare species, however, as pointed out by several authors, the paucity of record reflects the arboreal, nocturnal habits of this species (Pocock 1939; van Bummed 1952; Duckworth 1997a; Duckworth and Nettelbeck in press). Therefore, the species is not found by any conventional survey method other than spot-lighting, and the latter is rarely carried out in most places. In one exception with extensive spotlighting in Lao PDR, the species was found, often commonly, in most semi-evergreen and evergreen forests surveyed in the southern two–thirds of the country in the 1990s (Duckworth 1997b). The great paucity of camera-trap photographs reflects the rarity of the species coming to ground, as has been shown by its Neotropical ecological analogue, the Kinkajou (Schipper 2007). Hence, several well camera-trapped areas have failed to record the species (Seima Biodviersity Conservation Area, Cambodia, Walston and Duckworth 2003; Cat Tien, Vietnam, Borissenko et al. 2004; **Khao Yai NP, Thailand,** Duckworth and Nettelbeck in press; Sarawak Planted Forests Concession, Borneo, Belden et al. 2007); yet when even quite modest levels of spot-lighting have taken place in these areas, the species has been found, sometimes even exceptionally commonly (e.g. ten animals in the first three hours at Sarawak Planted Forests Concession; Belden et al. 2007). It is a reasonable inference, therefore, that the species is not just locally common, but widely numerous.

Population Trend: Decreasing

Habitat and Ecology:

The small-toothed palm civet is nocturnal and strongly arboreal (Pocock 1939; van Bemmel 1952; Duckworth 1997a; Duckworth and Nettelbeck in press). Although consumption of a wide range of animals has been recorded, this species is primarily frugivorous (Harrison 1962; Payne 1995; Duckworth and Nettelbeck in press). It likely breeds throughout the year, and there may be two litters per year, with two to three young in each litter (Lekagul and McNeely 1977). The lifespan of this species is 10 to 12 years (Lekagul and McNeely 1977). In Cambodia, Lao PDR and **Thailand**, the small-toothed palm civet is found in evergreen and semi-evergreen forest, including degraded areas provided some contiguity of canopy remains; elevation range extends up to at least 1200 m (Duckworth et al. 1999; Duckworth and Nettelbeck in press). In the Sundaic part of its range, it inhabits evergreen rainforest and possibly others forest types (Payne et al. 1985; Duckworth 1995; Payne 1995; Wells et al. 2005; Holden 2006; Belden et al. in 2007). It was recorded from lowland, mature semi-evergreen forest surrounded by disturbed secondary growth in Vietnam in 2003 (Borissenko et al. 2004). All Bornean civets (except *Diplogale hosei)* have been recorded in disturbed forest areas, though abundance declines in this habitat (Heydon and Bulloh 1996).

Systems: Terrestrial.

Major Threat(s):

Habitat loss and degradation are a major threat to the small-toothed palm civet (Schreiber et al. 1989). Deforestation has led to the decline of this species in **Thailand** (Lekagul and McNeely 1977), and in Borneo, the overall density of civets (including the small-tooth palm civet) in logged forests was found to be significantly lower than in primary forests (Heydon and Bulloh 1996).

Nevertheless, forest loss in Northern Southeast Asia is not enough to drive declines even at rates sufficient for NT status, and its wide altitudinal range in the Sundaic region (to 1500 m; Payne et al. 1985) means that while lowland populations have been and are continuing to decline, those occupying the higher altitudinal range are not declining as rapidly. This species is hunted (Walston and Duckworth 2003) and there has been an increased demand for civet meat in Chinese and Vietnamese markets (Bell et al. 2004; Lynam et al. 2005). Given that this species is arboreal and nocturnal in nature, it is one of the least affected by current high levels of hunting in Southeast Asia. In Cambodia, Lao PDR and Vietnam, government gun confiscation programs enacted during the 1990s were thought to result in a decrease in hunting by locals; however, the removal of guns has resulted in an upsurge in snaring and other trapping practices. As a result, there has been increased trapping of small carnivores overall, particularly nocturnal species and those that reside on the ground. As the small-toothed palm civet is arboreal, snares and other ground-level traps do not pose a major threat in this region (Walston and Duckworth 2003). Active night hunting appears to be the only effective means of hunting this civet and, as this is labour-intensive, is rarely undertaken for arboreal mammals in most of this species' range (Walston and Duckworth 2003). In Lao PDR, Vietnam and Myanmar, spotlight hunting does occur, but again the impact on this species is minimal as hunters focus on taking higher market value ungulates and concentrate for these almost exclusively at ground level (Walston and Duckworth 2003).

Conservation Actions:

This small-toothed palm civet is Totally Protected in Peninsular Malaysia, but only Protected in Sabah and Sarawak under the Wildlife Protection Act of 1972 (WPA 1972). In Thailand this species is protected, while it is not in Vietnam or Cambodia (GMA Small Carnivore Workshop 2006). The subspecies *A. t. trilineata* from Java has been classified as 'Endangered' by the IUCN (IUCN Red List, 1994) and as 'Threatened' in the IUCN Action Plan for the Conservation of Mustelids and Viverrids (Schreiber et al., 1989). Given the patchiness of suitable surveys for this species, it is important to note that many large protected areas have been established in northern Southeast Asian evergreen and semi-evergreen forest and presumably support the species. Some protected areas where the species occurs are listed as follows. Extensive conservation measures with reasonably effective law enforcement in Cat Tien National Park in Vietnam have ensured that there has been no logging for the past two decades and little or no hunting pressure (Borissenko et al. 2004). The small-toothed palm civet has been confirmed in a total of six protected areas in Vietnam (Roberton in prep), and many protected areas in Lao PDR (Duckworth 1997b). This species was recorded from Tawau Hills National Park in Borneo in 2003-04 (Wells et al. 2005), from Temengor Forest Reserve in Malaysia by Ratnam et al. (1995), and from Keo Seima Biodiversity Conservation Area in Cambodia in 2003 (Walston and Duckworth, 2003). Further studies would allow a better understanding of habitat requirements and ecology.

Masked Palm Civet *(Paguma larvata)*

Description:

In morphology the masked palm civet resembles other civets. Unlike most civets though, its orange-brown to gray fur completely lacks spots, stripes, and other patterns besides a mask. That mask consists of a prominent white stripe stretching from nose to forehead (sometimes extends farther but has greatly reduced thickness) that halves a black mask that extends laterally to the far edges of the cheeks and caudally up the forehead, past the ears, and down the back of the neck before stopping just under the shoulder blades. The eyes are surrounded by white fur that can vary from faint, incomplete outlines to well-defined blotches. The lips, chin, and throat are white. In some, white stripes of fur, comparable to sideburns on humans due to shape and location, curve up from the throat. These curves vary in thickness and have ends that terminate either in small blotches at the ear base or large blotches that surround the base of both darkly furred ears. No matter which coat it sports, masked palm civet's feet are always dark, often black, and the melanism usually extends partway up the legs in varying distances and intensities depending on the individual. The end of a masked palm civet's tail is sometimes darker than the majority of its coat. This difference in pigmentation can vary from a few shades darker than its coat to solid black and can cover a fourth to half of the tail. The main body varies from 51 to 76 cm (20 to 28 in) in length, to which is added a tail of 51 to 63 cm (20 to 25 in). It weighs between 3.6 and 6 kg (8 and 13.2 lb).

Range Description:

Recent records include China (Hubei province, Zhang pers. comm.), Lao PDR (Duckworth, 1997), Peninsular Malaysia (Azlan, 2003; Kawanishi and Sunquist, 2004; Laidlaw pers. comm.), Borneo (Simons, 1987; Blundell, 1996; Gonner, 1997; van Strien, 2001), Sumatra (Berghaeir, 1995), **Thailand** (Rabinowitz, 1991; Grassman, 1998) and Vietnam (Long and Minh Hoang 2006, Veron, pers. obs.), Myanmar (Than Zaw et al. in press), Cambodia (J. L. Walmart pers. comm.), India (Choudhury 1997). In India it extends west to Kashmir (Pocock 1939); recent records for the western portion of its range (including Nepal) have not been checked for. Brooks and Dutson (1994) saw the species in the Gede/Pangrango National Park in Java, however, the species had not otherwise been recorded for the island and the authors suggested that it was probably introduced. Payne et al. (1985) underestimated the range in southern Borneo. A specimen from Barito River (van Strien, 2001) suggest that this is part of the range. Also records from S. Wain and Belayan River (G. Frederiksson in litt to E. Meijaard), and Mahakam Lages (Gonner 1997); probably also in Danau Sentarum NP (Jeanes and Meijaard, 2000); and reported from G. Niut (Simons, 1987) and G. Palung (Blundell, 1996).

This species is found in China from Szechuan to Fukien and the southeast coast of Yunnan, west along the Himalayas to Kashmir, in Indochina, **Thailand,** Malaya, Sumatra, Borneo, and the Andaman Islands, it was also reported from Japan in the 1980s (Lekagul and McNeely, 1988). It has also been recorded from Koh yao (9 00 N, 98 00 E) off **Thailand** and Rutland (11 25 N, 92 40 E) of India (Meiri, 2005).

Countries: Native:

Bhutan; Brunei Darussalam; Cambodia; China; India (Andaman Is.); Indonesia (Kalimantan, Sumatera); Lao People's Democratic Republic; Malaysia (Peninsular Malaysia, Sabah, Sarawak); Myanmar; Nepal; **Thailand;** Vietnam

Population:

The species seems to be common across its range, except in southern China where it has been hunted/trapped for the commercial food trade (M. Lau pers. comm. 2006). In Indochina it is associated with evergreen forests and in Lao it is common in hill evergreen forests (Duckworth 1997), and in India with deciduous forests and in open areas near towns (Muddapa pers. comm.). There are few records from Cambodia (Olsson pers. comm.). In **Thailand,** it is common in evergreen forests (Kanchanasaka pers. comm.). Populations are difficult to assess by camera-trapping, as shown in Myanmar where comparing results of this method at several sites where there were also serious searches for hunted remains in villages reveals that even high levels of camera-trapping may fail to detect the species. This presumably reflects its semi-arboreal lifestyle (Than Zaw et al. in press.). The rather few camera-trapping records for Cambodia (J. L. Walstone pers. comm.) and Myanmar should not therefore be seen as evidence that the species lives at low population densities.

Population Trend: Decreasing

Habitat and Ecology:

Masked palm civets have been recorded in both evergreen and deciduous forest, and in disturbed habitat (Rabinowitz, 1991, Ratnam et al. 1995; Heydon and Bullon, 1996; Grassman, 1998; Duckworth 1997; Azlan, 2003; Roberton pers. comm.). There seems to be some difference in habitat usage across its range. It has been found up to 2500 m (Thinley et al. 1994). There are two published radio-telemetry studies on this species (Rabinowitz, 1991; Grassman, 1998). In **Thailand,** the home-range was 5.9 km^2 for an adult male (Grassman, 1998) and 3.7 km^2 for an adult female (Rabinowitz, 1991). Both studies showed that masked palm civets are nocturnal with occasional diurnal activity; this finding based on only two individuals is probably general sable to the species as a whole, based on field observations of many individuals (e.g. Duckworth 1997) and camera-trapping (e.g. Than Zaw et al. in press). The species is often very common in appropriate habitats. In Lao PDR, "recent records of this species come only from large blocks of evergreen forest above 500 m, and the three historical sites listed in Osgood (1932) are all also above this altitude (Duckworth et al, 1999)" - however, the species can be found at lower elevations in Cambodia and Viet Nam (sea level) (Roberton S., pers comm.). This species is nocturnal, partly arboreal, and omnivorous (Lekagul and McNeely 1977). It is often found near secondary growth, where it may hunt for rats, and chickens, as well as forage through rice field dumps, and it eats figs, mangoes, bananas, and leaves (Lekagul and McNeely 1977). Many records are known from evergreen forests and deciduous forest and it is common in degraded forest. In China it is a generalist. It is known from peat swamp forests in Malaysia, and a wide variety of natural and human modified habitats, including rural agricultural areas (e.g. Wang and Fuller 2003). In Vietnam, Lao PDR, Cambodia, and **Thailand,** all records are from evergreen forests.Up to four young can be born per litter, with two breeding seasons per year (Lekagul and McNeely 1977). Lifespan in captivity is up to 15 years (Lekagul and McNeely 1977) All Bornean civets (except Diplogale hosei) have been recorded in disturbed forest areas, though abundance declines in this habitat (Heydon and Bulloh, 1996; Colon, 2002; pers. comm.).

It was recorded in disturbed habitat in Malaysia by Ratnam et al. (1995). It was recorded in secondary forest, that was logged in the 1970s, and which surrounds a palm estate, in Malaysia in 2000-01 by Azlan (2003). Wang and Fuller (2001) conducted a study on the ecology of this species near Taohong Village in northern Jiangxi Province, South Eastern China, from April 1993 to November 1994. Wang and Fuller (2003) conducted a study on the food habits of this species in and around a rural agricultural area of south-eastern China (Taohong Village, Jiangxi Province) by analyzing its scats, the study was conducted between June 1992 and November 1994, and found that this species ate some mammals and insects, but mostly fruit.

Systems: Terrestrial

Major Threat(s):

The impact of habitat loss and degradation on masked palm civet populations is unknown. Heydon and Bulloh (1996) found that the overall density of civet species (including the masked palm civet) in logged forest was significantly lower than in primary forest. It is likely that the species cannot survive total deforestation, but that it persists in degraded and fragmented areas. In a study on the dynamics of trade in live wildlife across the Guangxi border between China and Vietnam during 1993 to 1996, Li and Li (1998) "noted that the volume of trade in Paguma lavata was small, but the percentage of [it] in the restaurants was very large. This suggested that the supply of [this] species in the restaurants was partly from the domestic wildlife trade in China." It is farmed in China and elsewhere, and used for food throughout its range. SARS is a major issue for this species, causing the trade and sale of it to be banned in order to control the disease, as well as many breeding farms being closed. It is hunted in Vietnam for local consumption. It is also hunted in Lao PDR, for local consumption and in trade to Vietnam (W. Duckworth in litt. 2006), but there is no evidence from either of these countries that these activities are threatening populations. Because it is quite arboreal, it spends that proportion of its time outside the range of snares and most traps, and so is less exposed to them than are the more ground-dwelling small carnivores.

Conservation Actions:

The species occurs in protected areas throughout its range. This species was recorded from Jerangau Forest Reserve in Peninsular Malaysia in 2000-01 (Azlan, 2003). It was also recorded from Temengor Forest Reserve in Malaysia by Ratnam et al. (1995). This species is protected in Malaysia (Azlan pers. comm.). It is protected by law in Gansu, China (Li et al, 2000). In China, it is listed as Near Threatened (A2cd). It is protected in Peninsular Malaysia and **Thailand.** The population of India is listed on CITES Appendix III.

Asian Palm Civet *(Paradoxurus hermaphroditus)*

Description:

The Asian palm civet is a small, mottled gray and black viverrid weighing 2 to 5 kg (4.4 to 11 lb). It has a body length of about 53 cm (21 in) with a 48 cm (19 in) long tail. Its long, stocky body is covered with coarse, shaggy hair that is usually greyish in color. There is a white mask across the forehead, a small white patch under each eye, a white spot on each side of the nostrils, and a narrow dark line between the eyes. The muzzle, ears, lower legs, and distal half of the tail are black, with three rows of black markings on the body. The tail is without rings, unlike in similar civet species. Anal scent glands emit a nauseating secretion as a chemical defence when threatened or upset. Despite its species name hermaphroditus, the civets (like all other mammals) have two distinct sexes and are not hermaphrodites.

Range Description:

The species has a widespread distribution in Central, South and South Eastern Asia occurring in Borneo (Wells et al. 2005), India (Krishnakumar and Balakrishnan 2003); Lao (Duckworth 1997), Peninsular Malaysia (Azlan 2003; Kawanishi and Sunquist 2004; Laidlaw pers. comm.), Myanmar (Su Su 2005), Siberut Island (Mentawai, Indonesia; Abegg 2003), Philippines (Heaney et al. 1991), **Thailand** (Austin and Tewes 1999), Bhutan, Cambodia (J.L. Walstone pers. comm.), southern China (including Hainan), Nepal, Singapore (B.P.Y.H. Lee pers. comm.), Sri Lanka , Viet Nam (Roberton 2007); and with scattered records in Sulawesi, Moluccas, and the Aru Islands (New Guinea), probably resulting from introductions (Wozencraft 2005). It was also introduced to Japan in the late 1800s, and still persists there today (S. Roy in litt. 2006). It has also been recorded from the islands of Bawean (Indonesia), Con Son (Vietnam), **Koh Samui (Thailand), Koh Yao (Thailand),** Samar (Philippines), and Tele-bon **(Thailand)** (Meiri, 2005), in addition to many others (Pocock 1939). *Paradoxurus lignicolor* (included in Paradoxurus hermaphroditus by Wilson and Reeder 2005) was recorded by Abegg (2003) on Siberut of the Mentawai Islands of Indonesia. In addition it has been found on the Philippine islands of Biliran, Maripipi (Rickart et al. 1993) Mindoro, Catanduanes (Heaney et al. 1991), Cebu, Masbate, Polillo, Ilin, Samar, Dumaran and Panay (Timm and Birney 1980; Lastimosa pers. comm.).

Countries: Native:

Bangladesh; Bhutan; Brunei Darussalam; Cambodia; China; India; Indonesia (Irian Jaya - Introduced, Jawa, Kalimantan, Lesser Sunda Is. - Introduced, Maluku - Introduced, Sulawesi - Introduced, Sumatera); Lao People's Democratic Republic; Malaysia (Peninsular Malaysia, Sabah, Sarawak); Myanmar; Nepal; Philippines; Singapore; Sri Lanka; **Thailand;** Vietnam

Introduced:

Japan

Population:

The population status is poorly known. However, there is sufficient evidence to indicate that across its wide range this is generally one of the most common species of civets, except in southern China were it is extensively hunted/trapped (M. Lau pers. comm. 2006) . It is probably the most common mammalian carnivore on Palawan Island, the Philippines (Esselstyn et al. 2004) and Su Su (2005) found that it was the most common species of small carnivore recorded in Hlawga Wildlife Park in Myanmar, a high degraded and heavily hunted small fragment of forest from which most wild carnivores have been eradicated. In mainland (non-Sundaic) Southeast Asia it occurs commonly at almost any site that is surveyed using suitable methodology, including the most degraded, isolated among human environments, and hunted small sites such as Houay Nhang in Lao PDR and Hlawga in Myanmar (Duckworth 2007, Su Su 2005, Than Zaw et al. in prep., Roberton et al. in prep., R. J. Timmins pers. comm., J. L.Walston pers. comm.).

Population Trend: Stable

Ecology and Behaviour:

The Asian palm civet is an omnivore utilizing fruits such as berries and pulpy fruits as a major food source, and thus help to maintain tropical forest ecosystems via seed dispersal. They eat chiku, mango, rambutan and coffee, but also small mammals and insects. Ecologically, they fill a similar niche in Asia as Common Raccoons in North America. It also feeds on palm flower sap, which when fermented becomes toddy, a sweet liquor. Because of this habit it is called the toddy cat. It plays a role in the germination of the *Pinanga kuhlii* and *P. zavana* palm trees. The Asian palm civet is believed to lead a solitary lifestyle, except for brief periods during mating. It is both terrestrial and arboreal, showing nocturnal activity patterns with peaks between late evening until after midnight. It is usually active between 6:00 pm and 4:00 am, being less active during nights when the moon is brightest. It performs scent marking using anal glands, urine, and feces. The most common marking behaviour is dragging the anal glands on a surface to leave a scent. It is able to identify animal species, sex, and whether the animal that left the scent is known or unknown by smelling an anal scent secretion. This species has been found in a wide range of habitats including evergreen and deciduous forest (primary and secondary), plantations and near humans, in habitats up to 2,400 m (Ratnam et al., 1995; Heydon and Bulloh, 1996; Duckworth 1997; Azlan, 2003; Heaney et al. 2004; Su Su, 2005; Wells et al. 2005; Than Zaw et al. in press). Radio-tracking studies have revealed home-ranges of up to 17 km^2 for males and 1.6 km^2 for females (Dhungle and Edge, 1985; Rabinowitz, 1991; Joshi et al. 1995; Grassman, 1998). In the Philippines the species has been recorded in agricultural (including coffee plantations) and forested areas from sea level up to at least 2,400 m asl (Balete and Heaney in press, Heaney et al. 1991 in press, Hoogstraal 1951, Rabor 1986, Thomas 1898) and in montane and mossy forest from 925-2150 m asl in Balbalasang, Kalinga Province (Heaney et al.2004). In Lao PDR, this species has been found in all habitats surveyed, from Mekong lowlands to montane areas, evergreen to deciduous forest to scrub (Duckworth et al. 1999). This species is adapted for forest living, yet it tolerates living in areas near humans; sleeping in barns, drains, or roofs during the day, and coming out at night to catch rats or forage for mango, coffee, pineapples, melons, and bananas, it also eats insects and molluscs (Lekagul and McNeely 1977). In Myanmar, it was recorded in mixed deciduous forest and a wide range of evergreen forest-dominated sites (Su Su, 2005, than Zaw et al. in press). This species was recorded in primary lowland rainforest in Tawau Hills National Park in Borneo by Wells et al. (2005). All Bornean civets (except Diplogale hosei) have been recorded in disturbed forest areas, though abundance declines in this habitat (Heydon and Bulloh, 1996; Colon, 2002; pers. comm.). It was recorded in disturbed habitat in Malaysia by Ratnam et al. (1995). It was recorded in secondary forest, that was logged in the 1970s, and which surrounds a palm estate, in Malaysia in 2000-01 by Azlan (2003). This species is largely arboreal (Payne et al. 1985), crepuscular (Azlan, 2005) and nocturnal (e.g. Duckworth 1997). There is interesting variation across its mainland range in habitat use. In Lao PDR it occurs commonly deep within old-growth evergreen and semi-evergreen forest (Duckworth 1997) but it seems to avoid such habitat in the Western Ghats (Mudappa in press).

Systems: Terrestrial

Major Threat(s):

In some parts of its range this species is hunted for bush meat and the pet trade. In South China it is extensively hunted and trapped (Lau pers. comm.). It is also persecuted as a pest (Gupta, 2004, Su Su 2005, GMA Philippines, 2006) though it seems able to tolerate very high levels of persecution (e.g., Duckworth 1997). Dead individuals of this species were found with local tribes during a visit to Coimbatore, Tamil Nadu and Agra, Uttar Pradesh in India between 1998 and 2003, where it is killed for its meat (Gupta, 2004). While these pressures are certainly having localized effects on populations in highly fragmented and 'humanized' areas, e.g. Houay Nhang (Lao PDR), there is no evidence in mainland Southeast Asia of them significantly reducing the population levels in large tracts of natural and semi-natural habitat, even in the heavily hunted countries of Lao PDR and Vietnam; while in India it is a common urban commensal (e.g., Gupta 2004).

Conservation Actions:

It is found within protected areas throughout its range (e.g. Lao PDR - Duckworth 1997, Vietnam – Roberton et al. in prep.; Cambodia–J. L. Walmart pers. comm.; Myanmar – Than Zaw et al. in press). It was recorded from Hlawga Wildlife Park in Myanmar between 2000 and 2003 (Su Su, 2005). This species was recorded from Tawau Hills National Park in Borneo in 2003-04 (Wells et al, 2005). This species was recorded from Jerangau Forest Reserve in Peninsular Malaysia in 2000-01 (Azlan, 2003). It was also recorded from Temengor Forest Reserve in Malaysia by Ratnam et al. (1995). It has been found in Bawangling, Wuzhishan and Yinggeling Nature Reserves in Hainan in the last few years (Lau pers. comm. 2006). This species is protected in Malaysia (Azlan pers. comm.). It is protected by law in Sichuan, China (Li et al, 2000), and it is listed as Vulnerable on the China Red List (Wang and Xie, 2004). It is listed on CITES Appendix III (India). Field surveys, ecological studies, habitat protection and monitoring of threats are needed, especially in areas where it may be reduced due to human depredation (ie China).

Otter Civet *(Cynogale bennettii)*

Description:

HB: 600-680; T: 120-200; HF: 102-110; W: 3-5 kg.

Adapted for an aquatic life, with a cylindrical, otter like shape, partially webbed feet, and nostrils and ears that can be closed when the head is submerged. The general colour is dark brown, but the under cat is light duff; the tips of the hairs are grey, giving a grizzled appearance. The lips are white, with the upper lip rounded, fleshy and heavy., the vibrissae arising from this muscular muzzle are remarkably long. The ears are small, round and far apart on the head. The nostrils open upward, and the civet lies in ambush at watering spots with only its eyes and nostrils above the surface of the water very similar to a crocodile. Scent glands are present but consist of only three pores in the skin. The female as four mammae. The skull is long and low, with only a slight sagittal crest, a low ascending ramus, with a flat zygomatic arch; there is only a hint of a post orbital process. The muzzle is long, with the plate over half the condylobasal lenght. The infrorbital foramina are quite large in order to supply the blood vessels nerves to the very sensitive facial region. The dentition is also specialised for an aquatic diet, with the canines jus a little longer than the premolars. The premolars are set close together and are triangular, long and sharp, rather like the teeth of fish catching seals. The molars have retained a more generalised form, with broad surfaces and rather low cusps,

Range Description:

The otter civet has a Sundaic distribution and is found in Malaysia, Indonesia (Sumatra, Borneo), **Thailand** and apparently, far from the Sundaic region, northern Vietnam (Tonkin, which is the type locality of Cynogale lowei) (Veron et al. 2006). There is one very old record from Singapore (Meiri 2005) but there is no evidence that this involved an animal taken there, rather than one with a trade locality (B.Y.P.H. Lee in litt. to J. W. Duckworth 2007).

Lekagul and McNeeley (1977) listed the species in southern Thailand, but there were no confirmed records until 1992 (Kanchanasaka pers. comm.). Another record in Kaeng Krachan National Park in 1998 confirms its presence in southern Thailand (Anon 1998). There is also an unconfirmed sighting from northern Thailand from Phu Kradung National Park in 1986 (Schreiber 1989; Veron et al. 2006), which should be viewed with scepticism until corroborated, as it would take the species range out of the purely Sundaic pattern is shows by all other records and it is based on an unverified local report.

Recent records of this species from Borneo include sightings in Sarawak (Sebastian 2005; Belden et al. 2007s) and in Sabah (Veron et al. 2006). The otter civet has been recorded as high as 1,370 m above Bario in Sarawak, but the majority of records are from lowland forest (Veron et al, 2006).There are some camera trap pictures of this species in Sumatra (Veron et al. 2006). There is also a possible record from a fisherman's house near Yilong Lake in southern Yunnan, China in 1973, which is of unknown provenance (Schreiber et al. 1989). If the species does actually occur in both northern Thailand and Yunnan (China), then it is likely to also occur in Lao PDR (Duckworth et al. 1999).

Countries: Native:

Brunei Darussalam; Indonesia; Malaysia; **Thailand**

Population:

There is very little known about the otter civet's population trends and local abundance, and further studies are required (GMA Small Carnivore Workshop 2006). Sebastian (2005) reports that this species' apparent rarity is puzzling, including a low detection frequency in heavily surveyed areas. There is a need to determine whether the species' rarity is due to sampling or methodology bias (possibly due to its habitat specificity) or due to naturally low densities (GMA Small Carnivore Workshop 2006). It is known from 75 total museum individuals: 40 from Borneo, 12 from Sumatra, and eight from Peninsular Malaysia (Schreiber et al. 1989; Veron et al. 2004). There have been 19 other sightings of this species in Indonesia, Malaysia, and Thailand (Veron et al. 2006). There was an intensive survey that recorded this species 59 times during 3,920 camera trap days in Way Kamabas National Park in Sumatra between January 1996 and December 1998 (Veron et al. 2006).

Population Trend: Unknown

Habitat and Ecology:

Little is known of the habitat and ecology of the otter civet and further studies are required. This species is thought to be largely confined to peat swamp forests, though there are recent records from lowland dry forest (Sebastian 2005). Preferred habitat appears to be lowland primary forest, but it has been recorded in secondary forest, bamboo, and logged forest (Heydon and Bulloh 1996; Franklin 2001; Veron et al. 2006), however, the long-term persistence of this species in these habitats is unknown (Veron et al. 2006). The otter civet has also been recorded from freshwater swamp forest and limestone forest, surrounded by acacia plantation in Bukit Sarang Conservation Area, Sarawak (Belden et al. 2007).

As the otter civet is semi aquatic (Veron et al, 2006) and known to forage in the water (Medway, 1978; Lekagul and McNeely 1977; Yasuma, 1994), it is assumed that this species feeds on fish, crabs, molluscs, small mammals, and birds (Lekagul and McNeely 1977). It is also thought to be nocturnal (Sebastian, 2005), though there is data indicating it is also occasionally active during the day (see data from N. Franklin; G. Veron pers. comm. 2006; Sebastian 2005).

Systems: Terrestrial

Major Threat(s):

Reduction in primary forest habitat has proceeded very fast throughout the lowland Sundaic region in the last 20 years (e.g. Birdlife International, 2001; Holmes, 2000; Jepson et al., 2001; McMorrow and Talip, 2001; Lambert and Collar, 2002; Curran et al. 2004; Fuller, 2004; Eames et al. 2005, Aratrakorn et al. 2006; Kinnaird et al. 2003). This has probably reduced populations of otter civets, and threatens its persistence (Veron et al. 2006). Clear-cut logging is one of the major factors contributing to decline in suitable habitat, and even selective logging may sufficiently alter habitat such that it is the species can no longer occupy it; combined, this loss of primary forest may be responsible for the current rarity of the otter civet (Veron et al. 2006). Heydon and Bulloh (1996) found that the abundance of civets (including palm civets, banded palm-civet, otter civet, terrestrial civets and linsangs) in northern Borneo was significantly lower in logged forest than in primary forest, with the most specialized civets, including the otter civet, being less tolerant of logged forests than generalist civets. Conversion of peat swamp forests to oil palm plantations is a major threat particularly to this species. There is no evidence that the species is specifically hunted, but as a ground-dwelling species it will be exposed to the snares and other ground-level taps set for other species. The numbers of animals caught and effects on populations, if any, are unknown.

Conservation Actions:

The otter civet is listed in Appendix II of CITES and as 'Threatened' in the IUCN Action Plan for the Conservation of Mustelids and Viverrids (Schreiber et al. 1989). Conservation of this species requires protection of forest and riverine habitat, and policing against illegal harvesting of timber and hunting (Veron et al. 2006). There is a need to survey for this species to determine its tolerance to secondary habitats, including riverine areas in plantations and other areas that maintain some natural vegetation, as well as to further assess its distribution and monitor its populations. Live trapping and camera trapping are being done in Krau Wildlife Reserve in Peninsular Malaysia in order to study different species of small carnivores, and hopefully to detect the presence of the otter civet (Malaysia Carnivore Project 2005). The absence of recent records in Peninsular Malaysia is of great concern, and it is very important to know if some protected areas still harbour this species (G. Veron pers. comm., 2006).

This species is found in many protected areas throughout its range. Including Samunsam Wildlife Sanctuary in western Sarawak in 2000 (Sebastian, 2005), **Kaeng Krachan National Park (12 57 N, 99 23 E) in Thailand** in 1998 (Anon, 1998), Bukit Sarang Conservation Area in Sarawak (Belden et al. 2007), Way Kambas National Park in Sumatra (Veron et al, 2006), Danau Sentarum National Park (Jeanes and Meijaard 2000), and Leuser National Park in Sumatra (van Strien, 1996

Banded Palm Civet *(Hemigalus derbyanus)*

Description:

The ground colour is pale brown, with 7 or 8 wide black bands running across the back,. These bands resemble long triangles having their bases on the back and the vertex towards the abdomen. There is a black band running along the midline of the nose and the forehead, disappearing on the neck. Two other black lines running from the long thin muzzle through each eye to merge with a long black patch on the neck. The under parts and legs are lighter than the upper parts and lack any markings. The hairs on the neck are reversed, pointing forward. The long, tapered tail is black except at the root, which is brown with a few faint black rings. The ears are well developed and long, and the eyes are larger. The tongue is said to be rougher than in other civets. The claws are fully retractable. The skull is long a narrow with a constricted postorbital area and a very low crest. The zygomatic arch is relatively flat and the postorbital process is short and blunt. The auditory bullae is low and flat. The dentition is rather small and widely spaced upper incisors, with the lower incisors closer together and forming a continuous row with the canines. The first premolars are relatively well developed, with small accessory cusps; the remaining teeth in the molar tooth row are complex and multi cusped. The forth premolar and first molar in both jaws resemble each other rather more closely than in other civets.

Range Description:

The banded palm civet occurs in the Sundaic region and is found in Peninsular Myanmar, Indonesia (Sipora Island, South Pagi Island, Kalimantan, Sumatra; Holden 2006), Borneo (Azlan 2004; Wells et al. 2005), Peninsular Malaysia (Ratnam et al. 1995; Kawanishi and Sunquist 2004; Laidlaw pers. comm.), and peninsular Thailand (Wozencraft 2005, A.J. Lynam pers. comm). The distribution implies that Brunei may be included in this range but a specific record has not been traced.

The type locality of one race has been reported in Bankachon, Myanmar, but there are no known current records in Myanmar (Than Zaw et al. in press.). According to Payne et al. (1985) this species has been recorded in many localities in Borneo, and there are many subsequent records from the island including Mount Kinabulu National Park in Borneo, near Poring Hot Spring (600 m asl) by Wells et al. (2005) and Similajau National Park (Duckworth 1997). It is found at elevations up to 1,200 m (Payne et al. 1985). In Sumatra, Holden (2006) had only a few records, all from lowland primary forest (sea level to a few hundred meters, with a maximum of 800 m) in the region of Kerinci Salbat. Other records may exist from the island and need to be collated. However, these data suggest the species may perhaps be confined to lowlands in Sumatra, and does not occur in hills or mountains.

Countries: Native:

Indonesia (Kalimantan, Sumatera); Malaysia (Peninsular Malaysia, Sabah, Sarawak); Myanmar; **Thailand**

Population:

The population status of the banded palm civet is poorly known. Holden (in press) has speculated that this species may be rare where it is found. However, Payne et al (1985) states that it was the second most common viverrid in the forests of Sabah, and it occurs in tall and secondary forests. In the 25 years since the data were gathered to make the assessment of Payne et al. was made, habitat landscape in Sabah has changed greatly, and the banded palm civet's current status in Sabah might therefore differ greatly from prior estimations. In peninsular Malaysia, this species has not been found in secondary forests, but was found in Taman Negara National Park (Kawanishi and Sunquist 2004), indicating populations may also be reduced across the species range in mainland South-east Asia, where forest conversion has been extensive.
Population Trend: Decreasing

Habitat and Ecology:

Little is known on the ecology of the banded palm civet and further studies are required. This species has been recorded from primary lowland rainforest, but also in disturbed habitat, peat swamp forest and acacia plantations (Ratnam et al. 1995; Azlan, 2004; Wells et al., 2005; Kanchanasaka pers. comm.; B. Giman pers. comm.). In Borneo, it was found at elevations up to 1,200 m (Payne et al., 1985).
It is nocturnal (Lekagul and McNeely 1977). Medway (1969) suggests that it is confined to the ground under tall forest. Davis (1962) found in Borneo that over 90% of its diet was insects, and no stomachs contained fruit or vegetables. All Bornean civets (except Diplogale hosei) have been recorded in disturbed forest areas, though abundance declines in this habitat (Heydon and Bulloh 1996; Colon pers. comm. 2002).
Systems: Terrestrial

Major Threat(s):

Habitat loss and degradation have been assumed to be major threats to the banded palm civet (Schreiber et al. 1989). Reduction in primary forest habitat has proceeded very fast throughout the lowland Sundaic region in the last 20 years, particularly in the lower altitudes which evidently support the bulk of this species' population (e.g. Birdlife International, 2001; Holmes, 2000; Jepson et al., 2001; McMorrow and Talip, 2001; Lambert and Collar, 2002; Curran et al. 2004; Fuller, 2004; Eames et al. 2005, Aratrakorn et al. 2006; Kinnaird et al. 2003). This has surely lead to steep population declines. In Borneo, the overall density of civets (including the banded palm civet) in logged forests was found to be significantly lower than in primary forests (Heydon and Bulloh 1996). From observation in Thailand there is clear no evidence that the banded palm civet can survive in plantations or other areas outside of evergreen forests (Kanchanasaka pers. comm.). Additionally the Mentawi populations are thought to be impacted by economic development as human settlements expand into civet habitat, resulting in conflicts since this species will prey on domestic livestock such as chickens (Schreiber et al. 1989). Hunting and trade are also threats for this species. Because Banded Civet spends a lot of time on the ground, it is more exposed to snares and other traps than are the partly and largely arboreal palm civets. It is hunted in Sarawak for food. In **Thailand,** this civet is hunted, and in the last five years, there have been less than five live individuals brought to a zoo (Kanchanasaka pers. comm.).

Conservation Actions:

The banded palm civet is listed on CITES Appendix II. The Mentawai subspecies was listed as Threatened? in the IUCN Action Plan for the Conservation of Mustelids and Viverrids (Schreiber et al. 1989). This species is protected in Malaysia (Azlan pers. comm.), as well as in **Thailand,** Brunei, Indonesia and in Myanmar. This species was recorded from Mount Kinabalu National Park in Borneo in 2003-04 (Wells et al. 2005), Temengor Forest Reserve in Malaysia by Ratnam et al. (1995), Similajau National Park in Sarawak (Duckworth 1997), and many other protected areas throughout its range (W. Duckworth in litt. 2006).

Banded Linsang *(Prionodon linsang)*

Description:

HB: 350-450; T: 310-420; HF: 50-70; E: 25-30; W: 6-7kg.

Has a very long, slim body and tail and very short legs so that when it moves on a branch, it looks more like snake than a mammal. It differs from the Spotted Linsang *P. particolor* in having the spots on the back merge together to form five irregular bands across the back; along the sides of the neck and body the spots merge together to form 2 irregular lines. The tail has seven complete black and white rings. There is no mane or erectile crest, and the scent glands are lacking or underdeveloped. The claws are completely sheathed and retractile as the cats, indicating strongly carnivorous habits. As in other *Viverrinae,* the skill is long and low, but the skull is rather delicate than in *Viverra* and *Viverricula* with the sagittal crest low and incomplete. The post orbital area is slightly constricted. As in other Feloidea the auditory bulla comprises of two bones; in P. linsang the tympanic bone is relatively small, the entotympanic rather large9tympanic bone larger is *P. pardicolor.*

Range Description:

The species is found in Peninsular Malaysia (Ratnam et al., 1995; Azlan, 2003; Kawanishi and Sunquist, 2004; Laidlaw pers. comm.), southern Myanmar (Than Zaw et al, in press), Borneo (Wells et al., 2005) and Sumatra (Holden, 2006), **Southern Thailand,** peninsular Malaysia, Sumatra, Java, and Borneo, as well as on Bangka and Belitung Islands (Van Rompaey 1993). On Java, it has been suggested to be probably restricted to the mountains in the west (e.g. Mt. Halimun, see Suyanto, 2003), but there are two historical specimens from the Ijang plateau in East Java (held in Museum für Naturkunde, Humboldt Universität zu Berlin), and, given very low levels in recent decades of small carnivore survey in Java (see, e.g., Riffel 1991), the species could still occur on other mountains in Java. On Borneo and Sumatra, it is probably widespread, although it is unclear whether the species occurs in Central Kalimantan swamps (not reported from Sebangau or Tanjung Puting). It was found up to 1100 m on Java by Bartels (1941) . This species was recorded up to 1800 m in Mount Kinabulu National Park in Borneo Payne et al. (1998). It was recorded by Azlan (2003) in Jerangau Forest Reserve (04 55.5 N, 103 05.7 E) in Peninsular Malaysia in 2000-01. In Sumatra, it has been found up to 2,400 m (Holden 2006) in Kerinci Seblat NP.

The most northerly record is at 15° 20'N (Steinmetz and Simcharoen 2006). The record at 17° 54'N (Humphrey and Bain 1990) in **Northern-Central Thailand is probably erroneous** (Steinmetz and Simcharoen 2006). On the basis of portrayal in picture-guides alone, this species can be readily confused with spotted linsang.

Countries: Native:

Brunei Darussalam; Indonesia (Jawa, Kalimantan, Sumatera); Malaysia (Peninsular Malaysia, Sabah, Sarawak); Myanmar; **Thailand**

Population:

The population status is poorly known, but various authors have considered this species uncommon or rare (Chasen, 1940; Ellerman and Morrison-Scott, 1961; Davis, 1962; Lim, 1973; Lekagul and McNeely, 1977; Van Rompaey 1993).

In Myanmar, it is known from only two individuals recorded in 1878 by Blanford (Van Rompaey pers. comm.), also from 5 recent records in southern Myanmar (Than Zaw et al. in press); the records included presence at all surveyed sites within its historical range. It is considered rare and localized in **Thailand** (Chasen, 1940; Ellerman and Morrison-Scott, 1961; Davis, 1962; Lim, 1973; Lekagul and McNeely 1977, Steinmetz and Simcharoen 2006). In Peninsular Malaysia it is widespread at all elevations but is nowhere common (Medway, 1969), two years of camera trapping revealed only a single photo in Taman Negara National Park (Kawanishi and Sunquist 2004), partly because it is small and semi-arboreal, making it difficult to capture on camera. Morphs may be confused with other species (Azlan 2006). On Java, this species is considered rare (or difficult to observe) by Bartels (1941). On Borneo, it is also infrequently recorded (Medway, 1965; Payne et al, 1985). In Sumatra (Holden 2006), it was photo trapped infrequently, but this is not indicative of rarity. In summary, throughout its range, it is widespread, but due to its small size and semi arboreal nature, it has been infrequently recorded, and thus its population status is difficult to assess.

Population Trend: Decreasing

Habitat and Ecology:

Little is known about the ecology of this species. It has been recorded in primary and secondary forest and occasionally in human inhabited areas (Van Rompaey, 1993; Ratnam et al., 1995; Azlan, 2003; Kawanishi and Sunquist, 2004; Wells et al., 2005). It has been found up to 2,400 m (Bartels, 1941; Payne et al., 1998;. Holden, 2006).

It is nocturnal and generally arboreal, though it does come to the ground in search of food; its diet consists of birds, tree rats, snakes, and any other small animal it can catch (Lekagul and McNeely 1977; Van Rompaey pers. comm.). An analysis of stomach contents by Lim (1973) indicates that this species hunts mostly in disturbed forests and forest edge habitats (Lekagul and McNeely 1977). According to Lim (1973) it may respond favourably to secondary growth and ecotonal habitats. It is found in primary and secondary forest, plus fringe and occasionally human inhabited areas (Van Rompaey pers. comm.). It breeds in February and August, with litters of two young (Nowak, 1999). It has lived up to ten years and eight months in captivity (Jones, 1982). This species was recorded from primary lowland rainforest in Mount Kinabulu National Park by Wells et al (2005). It was recorded in disturbed habitat in Malaysia by Ratnam et al (1995).

In northeastern Peninsular Malaysia and in Sarawak, it is found in secondary forest, including forest adjacent to palm plantations (Azlan pers. comm.).

Systems: Terrestrial

Major Threat(s):

Habitat loss and degradation were assessed as a threat to this species (Schreiber et al., 1989), and deforestation has been said to be a threat to this species in Thailand (Van Rompaey 1993), however, little is known on its tolerance to disturbance. In Borneo, the overall density of civets (including the banded linsang) in logged forests was found to be significantly lower than in primary forests (Heydon and Bulloh, 1996). Various observers have found the species in edge and degraded habitats, both historically and recently, with Lim (1973) even speculating that it benefits from secondary growth and ecotonal areas.

This makes it difficult to infer a population trend based around the recent and ongoing rapid forest degradation over much of its range. Old-growth forest has been cleared very rapidly throughout the lowland Sundaic region in the last 20 years (e.g. Birdlife International, 2001; Holmes, 2000; Jepson et al. 2001; McMorrow and Talip, 2001; Lambert and Collar, 2002; Curran et al. 2004; Fuller, 2004; Eames et al. 2005, Aratrakorn et al. 2006; Kinnaird et al. 2003) but this has primarily been in the lower altitudes; this species' wide altitudinal range means that significant proportion of the population is outside these areas of very rapid clearance. Also in Thailand, live animals have been traded in the past, as seven were reported by CITES in 1980 (from Thailand to USA) (Van Rompaey 1993), in the last decade, however, there has been no commercial trade reported through CITES.

Conservation Actions:

It is listed on CITES Appendix II. In **Thailand,** the Wild Animals Preservation Protection Act (WAPPA, 1992) protects this species from hunting and regulates trade (Van Rompaey, 1993). In Peninsular Malaysia and in Sabah it receives full protection, but in Sarawak it is only partially protected (Aslan pers. comm.). It is a totally protected species in Myanmar (WPA 1994). This species has been recorded in some protected areas (Ratnam et al., 1995; Azlan, 2003; Wells et al., 2005; Azlan and Engkamat, 2006). Field surveys, ecological studies and monitoring of threats are needed. This species was recorded from Mount Kinabulu National Park in Borneo in 2003-04 (Wells et al. 2005). It was also recorded from Temengor Forest Reserve in Malaysia by Ratnam et al. (1995). This species has also been recorded in Lambir Hill National Park in Sarawak (Azlan and Engkaamat, 2006). It was recorded from Jerangau Forest Reserve in Peninsular Malaysia in 2000-01 (Azlan, 2003).

Spotted Linsang *(Prionodon pardicolor)*

Range Description:

This species is found throughout much of southeast Asia, including eastern Nepal, India (Sikkim, Assam, Bengal and other states of north-east India), Bhutan (although no records have been traced), northern and central Myanmar, Northern and Central Thailand, Lao PDR, Cambodia, Northern, central and probably Southern Vietnam, and Central to southern China (Western Sichuan, Yunnan, Guizhou, Xizang, Hunan, Jiangxi, Guangdong, and Guangxi) (Van Rompaey, 1995; Walston 2001, Zhang, 1997, Roberton et al. in prep., Duckworth et al. in prep.). Two stuffed individuals were recorded by Choudhury (2002), one in Dzulake (25°3'8 N, 93°55'E) and one in Chizami (25°36'N, 94°24'E). **It was recorded from Doi Inthanon National Park in Thailand in 1995 (Tizard, 2002).** It was also sighted for sale on the side of a road between Mengla and Shangyong in Xishuanbanna Autonomous Region, Yunnan China in 1997 (Tizard, 2002). It was photographed by camera-traps in Nonggang, Jiuwanshan and Cenwanglaoshan Nature Reserves in Guangxi from 2002 to 2004, and also at Chebaling and North Guangdong Tiger Nature Reserve in Guangdong in 2003 (Lau et al., in prep.). Another recent record from China was reported by Yangsheng (1998) for Jiangxi province. F. Debruyne and A. Schoofs (in litt. 2001 to Tizard 2002) report it from Oudomxay Market, in Lao PDR. Walston (2001) reports it from the hills of southwestern Cambodia. Van Rompaey (1995) gave an elevational range from 150 m to 2700 m, there are records from up to nearly 4000 m (Pocock 1939), although it is not confirmed that these animals were truly caught at such altitudes (Duckworth et al. in prep.). It has been sighted in Nakai-Nam Theun National Biodiversity Conservation Area, Lao PDR (Evans et al. 1994, Duckworth 1997a). Dead animals have been seen in Xekong and Xaignabouli Provinces (Bergmans 1995).

Countries: Native:

Cambodia; China; India; Lao People's Democratic Republic; Myanmar; Nepal; **Thailand;** Vietnam.

Population:

This species is rather rarely recorded; which is likely an artefact of sampling techniques. For example, one of the most intensive direct observation surveys of carnivores in its range, that of Duckworth (1998) did not record it during the main survey, although it was found incidentally during bat survey at the same site. This survey gave no evidence of high numbers/high encounter rates in pristine habitat far from human habitation in the heart of its range, implying that this is the natural state of affairs for this species; at this site, various key quarry species (e.g. diurnal primates) showed little fear of people and even heavily decreased species such as Tiger *Panthera tigris* and Gaur *Bos gaurus* were observed directly, making it inconceivable that the linsang populations had been artificially reduced there.

With the advent of camera-trapping, records have increased in many areas. The relatively low numbers of records of the species in most of its range may be a result of its lurking and rather arboreal nature (Duckworth et al. in prep.). It was formerly thought to be uncommon to rare throughout its range (Van Rompae, 1995), but even historically there were indications that the rarity of specimens belied the species' real status. Hodgson (1841) states that it is "sufficiently common in the mountains of Nepal", and in 1847 he found "the species very numerous in the eastern half of Nepal and Sikkim." In the 1970s four observations were made in Chitwan National Park (Sunquist 1982). In the 1880s it was not considered rare in Sikkim (Blanford 1888-1891). The species is near its southern distributional limit in Thailand, and has been considered to be very rare and localized there (Lekagul and McNeely 1977) but there is no evidence it is not simply overlooked. Pham-chong-Ahn (1980) found the species to be uncommon in northern Viet Nam, but, judging from sale in markets, it seems to have been common in 1988 (Schreiber et al. 1989), and a more comprehensive collation of small carnivore records from the country confirms this impression (Roberton et al. in prep.). Bergmans (1995) speculated that it would be found to be common in Lao PDR and subsequent camera-trapping in the country, while rather limited, has found the species in most survey blocks of both main areas so far surveyed (Johson et al. 2006, Johnson and Johnson 2007). This species is the relatively rarely recorded in northeastern India (Choudhury, 1999) but there has been little camera-trapping or spotlighting. There are very few records in Cambodia (Olsson pers. comm.). It is relatively common in Yunnan (China). In sum recent records show it is not as rare as was often previously thought, although it probably lives at low density (Duckworth et al. in prep.).

Population Trend: Unknown

Habitat and Ecology:

Little is known on the ecology of this species. It has been recorded in lowland, hill and mountain forest, bamboo forest, secondary growth, dense grassland and along rivers (Pham-chong-Ahn, 1980; Sunquist, 1982; Van Rompaey, 1995; Tizard, 2002; Wozencraft, 2005). It is partly arboreal and preys mostly on small vertebrates (Lekagul and McNeely, 1977; Pham-chong-Ahn, 1980; Van Rompaey, 1995). This species is most often recorded in both primary and secondary forests, mainly in mountain and hill areas, up to at least 2,700 m, probably to 4,000 m (Van Rompaey 1995, Duckworth et al. in prep). It is found in primary and secondary forests at elevations between 150 and at least 2,700 m (Van Rompaey, 1995). In Chitawan National Park in Nepal, it was observed in a mosaic of lowland riverine forest and tall, dense grassland, as well as in an area of sal (Shorea robusta) forest interspersed with dense grasses (Sunquist, 1982). In Viet Nam it is found in moist, mixed, and bamboo forest, and along mountain rivers (Pham-chong-Ahn, 1980). It is quite arboreal, though it is equally at home in trees and on the ground (Lekagul and McNeely 1977). It has been suggested that it mainly inhabits the lower shrub layer (Kuznetzov and Baranauskas, 1993). It preys mostly on small birds (Hodgson, 1847), remains of rodents, frogs, and snakes have also been found in the stomachs of some specimens (Pham-chong-Ahn, 1980). It has also been spotted feeding on the carcass of a tiger, indicating that it is an opportunistic scavenger (Sunquist, 1982). Each litter consists of two young (Lekagul and McNeely 1977), twice a year, breeding in February and August (Hodgson, 1847). They are seldom maintained in captivity, and no lifespan has been recorded (Lekagul and McNeely 1977). Two stuffed individuals were recorded by Choudhury (2002), one in Dzulake (25 38 N, 93 55 E, 1750 m) in excellent primary subtropical broadleaf forest and one in Chizami (25 36 N, 94 24 E, 1400 to 1700 m), with degraded secondary forest in the lower areas and some primary forest in the upper areas (all subtropical broadleaf). It was recorded in moist secondary growth along water courses through a mosaic of cultivation, scrub, and pine plantations at about 1400 m in **Thailand** (Tizard, 2002). It is found in limestone forests in China and preys on rats (Lau pers. comm.).

Systems: Terrestrial

Major Threat(s):

The species was previously considered to be threatened by habitat loss and degradation, hunting and trade across much of its range (Schreiber et al., 1989). Although the impacts of these threats on populations are unknown they may be causing declines (Choudhury, 2002; Tizard, 2002) and further research is needed. Across its range, there has been loss and degradation of forests through logging and conversion of forests to agriculture. There has been an increased demand for civet meat in Chinese and Vietnamese markets (Bell et al. 2004), but this species does not seem to be a specifically sought one in the trade: extensive examination of trade seizures through Vietnam (the main route to China hardly ever found the species (U. Streicher in Duckworth et al. in prep.), nor is it sought in trade within China (M. W. N. Lau pers. comm.). Nor is it preferentially sought at the local, subsistence level, although throughout its range when it is caught it is eaten. Generalised mammal hunting is extremely heavy in much of its range. As a semi-arboreal species which is almost exclusively nocturnal, it is less exposed to risks of incidental killing and bycatch in snares and traps. The concern expressed in the IUCN Action Plan for Conservation of Mustelids and Viverrids (Schreiber et al. 1989), was made at a time when most of its geographic range was out of access to western biologists, through political factors (Viet Nam, Cambodia, Lao PDR, Myanmar, the relevant parts of China, most of north-east India, Bhutan), and capacity-building programmes for national conservation personnel were only just beginning in some areas. Hence, there was a very weak information base upon which to make this listing: indirect information from most range states and direct information from Nepal and Thailand, both at the margins of the species' range. In the intervening two decades, most of this area has opened up, and there have been many records of this species (Duckworth et al. in prep.), indicating that this 1989 assessment was unduly pessimistic.

Conservation Actions:

This species is listed on CITES Appendix I. It has been sighted in Nakai-Nam Theun National Biodiversity Conservation Area (Evans et al. 1994, Duckworth 1997a). In recent years, it has been recorded from Mouling National Park (Singh et al. 1996), Mehao Wildlife Sanctuary (Chakraborty and Sen 1991) and the border of Pakhui Wildlife Sanctuary and Nameri National Park in 1999 (P. Saikia pers. comm.). It was also recorded from Khonoma Nature Conservation and Tragopan Sanctuary (a local protected area not covered by the national government) by Choudhury (2002). It was recorded from Doi Inthanon National Park in Thailand in 1995 (Tizard 2002). Tizard (2002) reports a record of this species from Xishuanbanna Biosphere Reserve and recently it has also been camera trapped from several nature reserves in Guangdong and Guangxi (Lau, pers comm) in China. Many other recent records from South-east Asia are listed in Duckworth et al. (in prep.) This species is listed on Category II of the China Wildlife Protection Law (1988) (Li et al. 2000). The species is totally protected in Myanmar, on the list of protected species in Nepal, and protected **in Thailand,** Vietnam and India. **In Thailand is protected by the Wild Animals Reservation and Protection Act (WARPA) of 1980 (Van Rompaey, 1995).** This species is listed on Category II of the China Wildlife Protection Law (1988) (Li et al. 2000), is totally protected in Myanmar, **Thailand,** Vietnam and India, and is on the list of protected species in Nepal (Van Rompaey, 1995). It is recorded from many protected areas across its range (Chakraborty and Sen, 1991; Evans et al., 1994, Singh et al., 1996, Duckworth, 1997; Walston 2001, Tizard, 2002, Johnson et al. 2006, Roberton et al. in prep., Than Zaw et al. in press, Duckworth et al. in prep.).

Large-spotted Civet *(Viverra megaspila)*

Description:

HB: 720-850; T: 300-369; HF: 130-138; E: 45-48; W: 8-9kg.

This large civet is distinguished by the large black spots on the flanks, thighs and hind legs on a ground colour which varies from grey to buff. As in *V. zibetha* there is an erectile crest beginning at the neck,, but in megaspila the crest continues down the tail as a strip of black hairs, so that the white rings cannot completely encircle the tail. Also the head is more massive than in *zibetha,* and the muzzle is longer and more swollen. The feet are solid brown. The skull resembles that of *zibetha* but as larger bullae, a more post orbital process constriction and smaller post orbital process, located behind the mid point of the total lenght of the skull. The suborbital foramina has the long axis horizontal. The dentition is similar to *zibetha* but with longer maxillary tooth row and smaller canines and incisors.

Range Description:

This species is found in Southern China (last record in 1998; Ying-xiang pers. comm.), Cambodia (Walston 2001; Olson pers. obs.), Lao PDR (Duckworth 1997; Khounboline 2005), Peninsular Malaysia (last record in 1985; Asakawa et al. 1986), Myanmar (Lynam et al. 2005), **Thailand** (Lynam et al 2005) and Vietnam (Roberton et al. in prep.). Only one record from Cambodia was reported by Walston (2001), but the advent of camera-trapping led to many more recent records (CI, WCS, WWF, unpublished per J. Walston pers. comm., Olson pers. comm. 2006). There is some confusion as to whether this species has been found on Singapore, and some authors are explicit that it does not occur there (Harrison 1966), and the only specimen checked has not turned out to represent this species (Lyman et al. 2005). There are historical records from Peninsular Malaysia (including one from Penang Island) (Veron 2004). However, the only recent record from Peninsular Malaysia appears to be a road-kill from Sungai Petani in 1985 (Asakawa et al. 1986). There are also several records from southern China (Southern Yunnan and south western Guangxi; Wang Ying-xiang 1987, 2003; Zhang Yong-zu 1997; Wang Sung 1998; Sheng Helin et al. 1999). It was recorded by Lynam et al. (2005) in Htuang Pru Reserve Forest and Hukaung Valley Wildlife Sanctuary in Myanmar, and **Taphyra National Park in Thailand**. In **Thailand,** the species has been found in several protected areas, and there is a southern record from Bam Nang Nom (Ra Nong Province).This species is a lowland species, with almost all field records from below 300 m (Lyman et al. 2005). In fact, Lyman et al. (2005) report that they "are not aware of any site in non-Sundaic Southeast Asia lying predominantly under 300 m, supporting 500+ sq. km of (semi-) evergreen forest, and having received heavy camera trapping or spotlighting effort, that has not recorded the species."

A freshly-killed (by hunters) specimen was reported from Ban Thalang (Thalang village; formerly Ban Namtheun; 17º51'N, 105º03'E, ca. 520 m) on the Nakai Plateau of central Lao PDR significantly higher in altitude than other recent records of this species, and this level area also hold population of plains birds at anomalously high altitude (Khounboline 2005). The species is potentially more widespread in Myanmar, as there have been few surveys below 300 m using appropriate techniques (Than Zaw et al. in press). There are only five confirmed records from Vietnam, with the furthest north being Phong Nha NP (Roberton et al. In prep). If the species was once present further north in Viet Nam, it is doubtful that any significant populations could still survive (Timmins and Roberton pers. comm.2006). The lack of records in Viet Nam seems to reflect a genuine scarcity in the species and not a lack of appropriate surveys. There are no recent records from China, with the last record from 1998 (Wang Ying-xiang pers. comm.2006). The species is potentially more widespread in Lao PDR, as there have been few surveys below 300 m using appropriate techniques (Duckworth pers comm. 2006). The records from Xe Pian National Protected area suggest that its reasonably common in level lowland forest (Austin 1999). It is probably rare on the Nakai Plateau of central Lao PDR (Khounboline 2005). In Cambodia, there are three records from camera traps in the southwest. The species has been commonly photo-trapped in several sites in northern and eastern Cambodia (CI, WCS, WWF, unpublished per J. Walstone pers. comm.). Albeit considered widely distributed geographically in Peninsular Malaysia, it was considered rare (Medway 1977). There are no recent records from the area. The species is likely to be very localized on Peninsular Malaysia, as there has been appropriate surveys at low altitudes without results (Azlan pers comm. 2006).

Countries: Native:

Cambodia; China; Lao People's Democratic Republic; Malaysia (Peninsular Malaysia); Myanmar; **Thailand**; Vietnam

Population:

The population status is poorly known. What is know are from relatively few recent records anywhere in its range (Lynam et al. 2005). In Lao PDR, from 1997-1999 there were four singles camera trapped in Xe Pian National Biodiversity Conservation Area (main block and Dong Kalo) in early 1997 (Astain 1999). Recent records of this species in Lao come only from Phou Xang He and Xe Pian National Biodiversity Conservation Areas (Duckworth et al. 1999) and the Nakai Plateau in 2002 (Khounboline 2005). There are no Lao records from hill and montane forest, but the species has been found in forest below 300 m which suggests that it is genuinely patchy in occurrence (Lynam et al. 2005). In Thailand, it used to be found all over the country and was rather common (Lekagul and McNeely 1977). Little over a decade later, a global review of all species of Viverridae found that there was very little known about this species, and traced no records from protected areas, and urged for surveys to assess its current status (Schreiber et al 1989). This disparity is probably due to heavy logging of lowland forest in the interim, and indicates an actual decline and fragmentation in population (Lynam et al, 2005). There are few recent records from Viet Nam (R. J. Timmins in litt, 2004). It has been found at several sites in Cambodia (J. L. Walston in litt, 2004).

Population Trend: Decreasing

Habitat and Ecology:

This species has been recorded in evergreen and deciduous forest, and dry dipterocarp forest, all below 300 m of altitude (Duckworth, 1994, Austin, 1999; Duckworth et al., 1999; Lynam et al. 2005; Kanchanasaka pers. comm.). In China, the species is found below 800 m, although the exact habitat is unknown. In Myanmar, there are records from evergreen forest, including forest-grassland edges at 300 m, and there is no information on whether the species is found in some of the dry lowland forests in Myanmar which are the more prevalent habitat within the species' known elevational range (Than Zaw et al. in press). In Lao PDR, the species occurs in lowland evergreen/semi-evergreen forest (including degraded areas) with one in open dry dipterocarp forest, all below 300 m altitude (Duckworth et al, 1999), with one outlier at 520 m (Khounboline 2005). In Thailand, it is found in deciduous forest and dry evergreen forest (Kanchanasaka pers. comm.2006). Within northern and eastern Cambodia, the species is found in mosaic deciduous forest, along with semi-evergreen patches and riverine gallery forest (WCS, WWF, CI pers. comm.2006). Lyman et al (2005) report three records of this species from evergreen forest, the predominant habitat of other recent records, though some have come from deciduous dipterocarp forest (Duckworth, 1994; Austin, 1999). The species is not dependent on primary forest, and can probably persist in degraded forest that has forest structure (Duckworth, 1994).

New information from Cambodia and elsewhere suggests that it can live in fragmented areas, but that it might only persist in large forest blocks (Timmins and Duckworth pers. comm.), as was previously suggested by Lynam et al (2005). This species can potentially be misidentified as either Viverra zibetha or Viverra tangalunga and it is important that wherever possible records are verified through photographs (J.W. Duckworth pers. comm.2006).
Systems: Terrestrial

Major Threat(s):

Throughout much of this species range, terrestrial small carnivores are exposed to heavy hunting, particularly with snares and dogs. This is occurring in much of its range, including Lao PDR (Duckworth et al. 1999) and **Thailand,** with snaring found even in some protected areas such as Ta Pra Ya National Park. There has been an increased demand for food of civets in Chinese and Vietnamese markets (Bell et al., 2004; Lyman et al. 2005). From the 1970s and the 1990s, large areas of lowland forest were logged across parts the species' range, particularly in China, **Thailand** and Vietnam, including conversion to non-forest land-uses (Lynam et al. 2005; Wang Ying-xiang pers. comm. 2006). This increased fragmentation of habitat increases the threat of hunting as well as the direct loss in area able to support the species (Lynam et al. 2005). In Cambodia, at present there is a massive trend in deforestation of lowlands, particularly for local agriculture (Timmins pers. comm. 2006). In Myanmar, there is a major trend in conversion of forest to plant oil agriculture, particularly in lowlands (Duckworth and Than Zaw pers. comm. 2006). As a ground-dwelling species, and by analogy with V. zibetha, this species should be readily snared. Despite the massive levels of civet hunting, and the often-taken opportunities to check large numbers of civets in trade, this species is not recorded in the widespread trade in China (Lau et al.,1997; Kadoorie Farm and Botanic Garden, 2004) and Vietnam, which suggests that the populations are already reduced to overall extremely low levels (Scott Roberton and Wang Ying-xiang, Nguyen Xuan Dang, Michael Lau pers. comm.).

Conservation Actions:

The species is found in some protected areas in its range, including two in Viet Nam (Roberton et al, In prep), as well as Phou Xang He and Xe Pian National Biodiversity Conservation Areas in Lao PDR (Duckworth et al, 1999). It was recorded by Lyman et al (2005) in Htuang Pru Reserve Forest and Hukaung Valley Wildlife Sanctuary in Myanmar, and Taphyra National Park in **Thailand.** In Vietnam, the trade in the species is nominally regulated, and it is protected in Group 2b (Roberton et al, in prep). This species is protected in Myanmar under the Wildlife Act of 1994. The species is protected in Peninsular Malaysia and **Thailand,** but there is no protection in China. Though there have presumably been recent advances in reduction of gun usage by civilians for hunting, there have been increased snaring efforts (in compensation), and there is now a need to reduce, and preferably eradicate, this form of hunting (Lyman et al, 2005).

Malayan Civet *(Viverra tangalunga)*

Description:

V. tangalunga measures 585 to 950 mm in head and body length; tail length is 300 to 482 mm. Coloration is composed of black spots on a background of tawny or greyish body colour. There are usually three black and two white collars on the sides of the neck and throat. The fur is long and loose, and usually elongated along the spinal area forming a low crest or mane. This mane is marked by a black stripe running from the shoulders to the tail. The tail is also banded with black and white. The feet of the Oriental civet are all black.Viverra have five toes on each foot. On the third and fourth digit of the forefeet are lobes of skin which sheath and protect their retractile claws. The dental formula is I 3/3 C 1/1 PM 3-4/3-4 M 1-2/1-2. (Nowak 1983)

Range Description:

The species is known to occur in Peninsular Malaysia (Azlan, 2003; Kawanishi and Sunquist, 2004; Malaysia Carnivore Project, 2006; Laidlaw pers. comm.), Indonesia, Philippine islands (Heaney and Tabaranza 1991; Heaney et al. 1991) and Sulawesi (Buton island) (Jennings et al. 2006). In Indonesia, it is found in Borneo (Colon 2002), Sumatra, Rhio-Lingga Archipelago, Bawal Island, Bangka Island, Karimata Island, Sulawesi, Telok Pai, Amboina and the Moluccas (Meiri, 2005; Wozencraft, 2005). Two specimens have been recorded from Java (Meiri, 2005) but there is no evidence of a native population. In the Philippines: Bohol, Busuanga, Culion, Leyte, Luzon, Mindanao, Mindoro, Negros, Palawan, Samar and Sibuyan. It is also reported from Camiguin (Heaney and Tabaranza. 1991), Catanduanes (Heaney et al., 1991), Panay (Timm and Birney, 1980, Lastimosa pers. comm.) and Siguijor (Timm and Birney, 1980). In Malaysia, it is found in Borneo, Banggi Island, Langkawi Island, Penang Island and in Peninsular Malaysia (Corbet and Hill, 1992; Nor, 1996; Meiri, 2005). It was introduced to several islands in Southeast Asia (Jennings et al., 2006). The historical range of the species includes Indonesia, Malaysia, Brunei, the Philippines and Singapore (Corbet and Hill, 1992; Nowak, 1999; Wozencraft, 2005). **Although it is also listed from Cambodia, China and Thailand in Wozencraft (1993), there is no evidence it occurs in these countries.**

Countries: Native:

Indonesia (Kalimantan, Maluku - Introduced, Sulawesi, Sumatera); Malaysia (Peninsular Malaysia, Sabah, Sarawak); Philippines; Singapore

Population:

Although the Malay civet is a widespread species, little is still known about its population levels in countries where it is native or has been introduced. Colon (2002) found lower population densities in logged forest than in unlogged forest, and suggested that this may be because of lower fruit availability in logged forest. The species is widespread in Asia and is moderately common in forest and rare in other habitats.

Population Trend: Stable

Habitat and Ecology:

The Malay civet occurs in a variety of habitats including primary and secondary forests, cultivated land and the outskirts of villages (Nowak, 1999; Colon, 2002; Jennings et al., 2006). It is found from sea level to at least 1,200 m (Rabor, 1955; Payne et al., 1985; Rickart et al., 1993; Heaney et al., in press). Malay civets are solitary, omnivorous and primarily terrestrial (Kanchanasakha et al., 1998). A wide range of home-ranges for Malay civets has now been documented on Sulawesi (24 – 189 ha) and Borneo (27 – 283 ha) (MacDonald and Wise, 1979; Nozaki et al., 1994; Colon, 2002; Jennings et al., 2006). Mean home-range size for adults of both sexes was 110 ha in Sabah and 70 ha on Sulawesi (Colon, 2002; Jennings et al., 2006). Colon (2002) considered that the Malay civet was not territorial in Sabah but Jennings et al. (2006) found low intra-sexual overlap on Buton Island. Malay civets are mainly nocturnal (Colon, 2002; Jennings et al., 2006). Day rest sites are situated at ground level and associated with some form of cover (Colon, 2002; Jennings et al., 2006). Malay civets are solitary, omnivorous, and primarily terrestrial (Kanchanasakha et al. 1998). A wide range of home-ranges for Malay civets have now been documented on Buton Island, Sulawesi (24– 89 ha) and Borneo (27–283 ha) (MacDonald and Wise 1979; Nozaki et al. 1994; Colon 2002; Jennings et al. 2006). Mean home-range size for adults of both sexes was 110 ha in Sabah, East Malaysia and 70 ha on Buton Island, Sulawesi (Colon 2002; Jennings et al. 2006). Colon (2002) considered that the Malay civet was not territorial in East Malaysia but Jennings et al.(2006) found low intra-sexual overlap on Buton Island. Malay civets are most active at night from 18h00 to 07h00, although Malay civets were more active during the day on Buton Island, Sulawesi than in Sabah, East Malaysia (Colon 2002; Jennings et al. 2006). Day rest sites are situated at ground level and associated with some form of cover such as logs, dense brush pile, or thick herbaceous vegetation (Colon 2002; Jennings et al. 2006). The species' habitat is primary and secondary lowland, montane, and mossy forest from sea level to at least 1,200 m asl (Rabor 1955, Rickart et al. 1993, Heaney et al. in press). It is also found also in agricultural areas and near human settlements in the proximity of forest (Wemmer and Watling, 1986; Nowak, 1999). In a study on home range behaviour of this species on Buton Island, Jennings et al. (2006) found a home range size of 70 ha, with smaller home ranges for females as compared to those found in logged forest on Borneo. It is an adaptable species that seems to thrive in a variety of environmental conditions, including disturbed areas (Jennings et al. 2006). This species was recorded in primary lowland rainforest in Tawau Hills National Park in Borneo by Wells et al. (2005). All Bornean civets (except Diplogale hosei) have been recorded in disturbed forest areas, though abundance declines in this habitat (Heydon and Bulloh, 1996; Colon, 2002; pers. comm.). It was recorded in secondary forest, that was logged in the 1970s, and which surrounds a palm estate, in Malaysia in 2000-01 by Azlan (2003). This species is ground-living (Medway, 1978) and predominantly crepuscular (Azlan and Gulan Azad, 2005).

Systems: Terrestrial.

Major Threat(s):

As a ground-living species it is exposed to snaring and other forms of ground-level trapping, and hunting with dogs, however, the limited survey in areas heavily used by people suggests it is rather well able to persist at general levels of threat. The species is occasionally hunted for food and treated as a pest as it raids poultry.

Conservation Actions:

Malay civets are found in a number of protected areas throughout its range. This species is protected in Malaysia under the Wildlife Protection Act of 1972 (WPA, 1972). Field surveys, ecological studies, habitat protection and monitoring of threats are needed. The species is found in a number of protected areas throughout its range. This species was recorded from Tawau Hills National Park in Borneo in 2003-04 (Wells et al. 2005). This species was recorded from Jerangau Forest Reserve in Peninsular Malaysia in 2000-01 (Azlan, 2003). This species is partially protected in Malaysia under the Wildlife Protection Act of 1972 (WPA 1972), meaning that anyone found killing this species will be liable to a fine not exceeding three thousand ringgit (Approx. USD 790) or a term of imprisonment not exceeding three years, or both (Azlan, 2003). However, Section 55 of this Act allows farmer to shoot any wild animal that causes damage to their property, as long as reasonable efforts have been made to frighten the animal away, and many civets are considered a pest in Peninsular Malaysia, as the prey on small livestock and raid fruit orchards (Azlan, 2003).

Large Indian Civet *(Viverra zibetha)*

Description:

HB: 750-850; t: 380-462; HF: 109-140; E: 47- 52; W: 8-9kg.

A sturdily built animal with conspicuous black and white patches on the throat and sides of the rather long neck; a black crest of erectile hairs runs down over the neck and along the back, ending at the root of the tail.. The tail is ringed alternately with white and black. The flanks, thighs and hind legs are covered with confluent stripes, not large spots as in *V. megaspila.*The new born are mostly black. The skull is long and low, with moderate crests and a low, flat zygomatic arch; there is a marked depression between the nasals. The post orbital processes are small, but more prominent than in *megaspila*, they are located in the front of the skull(behind the mid point in *megaspila.* The suborbital foramina are relatively small, with the long axis vertical, (horizontal in *megaspila)* The mandibular ramus has the posterior edge vertical inclined backwards in *megaspila* (inclined forward in *Viverricula malaccensis).* The bullae are small, shorter than the width across the occipital condyles. In comparison with *V. megaspila t*he dentition is more robust, with longer and stronger canines. The first upper premolar, however is smaller than in *megaspila.*

Range Description:

species is found in Lao PDR (Duckworth, 1997), Peninsular Malaysia (Azlan, 2003; Kawanishi and Sunquist, 2004), **Thailand** (Rabinowitz, 1991; Austin and Tewes, 1999), Vietnam (Boonratana , 2004; Long, and Minh Hoang 2006), Cambodia (J.L. Walston pers. comm.), China (Anhui, Shaanxi, Ganus, Hubei, Hunan, Jiangxi, Sichuan, Guizhou, Yunnan, Xisang, Guangxi, Gunagdong, Hainan, Fujian Zhejiang and Jiangsu), northeast India, Myanmar (Than Zaw et al. in press), Nepal, Bhutan, Singapore, (Pocock 1939, Corbet and Hill, 1992; Wozencraft, 2005). Introduced to the Andaman Islands (Lever, 1985).

Countries: Native:

Bhutan; Cambodia; China; India; Lao People's Democratic Republic; Malaysia; Myanmar; Nepal; Singapore; **Thailand;** Vietnam.

Population:

This species lives at a naturally fairly high density for a carnivorous animal and was almost universally considered common by historical collectors (Pocock 1939). It remains common in much of its range: it is possible to see several in a single night of spotlighting on foot even in heavily-hunted Lao PDR (Duckworth 1997), it is among the most common mammals camera-trapped across Cambodia (J. L. Walmart pers. comm.) and Myanmar (Than Zaw et al. in press), and is one of the most commonly recorded civets in Vietnam (Roberton et al. in prep.). However, in some areas (such as Southeastern China) it has become effectively extinct over large areas (M. W. N. Lau pers. comm.). Few other parts of the range are as severely impacted by habitat fragmentation and degradation coupled with hunting.

However, in some areas (such as Southeastern China) it has become effectively extinct over large areas (M. W. N. Lau pers. comm.). Few other parts of the range are as severely impacted by habitat fragmentation and degradation coupled with hunting. It is likely that populations are widely reduced in the most heavily hunted parts of its range where habitat has been heavily fragmented, e.g. much of northern Vietnam and lowland Lao PDR. Recent camera-trapping in the Nakai–Nam Theun national protected areas, Central Lao PDR, found rather few animals (Johnson and Johnston 2007), suggesting the possibility for very heavy ground-level trapping as occurs in much of this area) to reduce populations greatly even in large tracts of little-encroached forest.
Population Trend: Decreasing

Habitat and Ecology:

This species has been recorded in primary forest (both evergreen and deciduous), secondary forest and plantations (Duckworth et al. 1997; Azlan, 2003) and is often said to have even wider habitat use (e.g. Lekagul and McNeely 1977). It has been recorded up to 1,600 m (Than Zaw et al. in press). It is solitary, nocturnal although there are occasional day-time records of active animals (e.g. Than Zaw et al. in press) and it is usually active on the ground (Lekagul and McNeely, 1977, Duckworth 1997). An adult male was radio-tracked in **Thailand** and had a home-range of 12 km^2 (Rabinowitz, 1991). Occupancy of suitable habitats varies within Indochina. Its diet consists of a wide range of animals, including fish, birds, lizards, frogs, insects, scorpions (and other arthropods) and crabs, as well as poultry and garbage (Lekagul and McNeely 1977). In Lao PDR, this species is found in tall forest, both evergreen and deciduous, and adjacent degraded areas, over at least 200 to 1000 m, with few recent records from below 400 m (Duckworth et al. 1999); however, there are many records from other countries, e.g. Myanmar, below this altitude (Than Zaw et al. in press). They are believed to breed throughout the year, with two litters per year, and two to four young per litter (Lekagul and McNeely 1977). Breeding and resting dens are usually holes in the ground which were originally dug by other species. It was recorded in secondary forest, that was logged in the 1970s, and which surrounds a palm estate, in Malaysia in 2000-01 by Azlan (2003). Like Viverricula and Civettictis, but to a generally much lesser extent, this civet has been used as a source of civetone, an oil-like substance secreted by the perineal gland used by the animal for territorial marking.
Systems: Terrestrial

Major Threat(s):

Habitat loss and degradation are a threat to this species (Schreiber et al., 1989). Across its range, there has been loss and degradation of forests through logging and conversion of forests to non-forest land-uses. It is hunted for food, probably throughout its range, and certainly in Vietnam, China, Lao PDR, Myanmar, NE India and **Thailand,** and for scent glands in Vietnam and China. Ground-living small carnivores are exposed to high levels of non-specific hunting, particularly with snares, throughout most of South-east Asia. Dogs are widely likely to be a problem for this ground-dwelling species, even though it is largely within burrows by day. Snaring and other forms of ground-level trapping occur in much of its range, including Lao PDR (Duckworth et al. 1997), Vietnam, and **Thailand,** with trapping found both inside and outside protected areas (Kanchanasaka, pers. comm.). There has been an increased demand for civet meat in Chinese and Vietnamese markets (Bell et al. 2004; Lynam et al. 2005).

Conservation Actions:

This species is totally protected in Malaysia under the Wildlife Protection Act of 1972 (WPA 1972) (Azlan, 2003). This species is listed on Category II of the China Wildlife Protection Law (1988) (Li et al, 2000). China listed it as 'Endangered' under criteria A2acd, and it is a class II protected State species (due to trapping for food and scent glands). It is protected in **Thailand,** Vietnam and Myanmar (GMA Small Carnivore Workshop, 2006). It is found in several protected areas throughout its range (Duckworth, 1997; Azlan, 2003; Kawanishi and Sunquist, 2004). The population of India is listed on CITES Appendix III.

Small Indian Civet *(Viverricula indica)*

Range Description:

This species is currently known to occur in South and Central China (Wang and Fuller, 2001, 2003), Hong Kong (Suen, 2002), most of India (Mudappa, 2002), Lao PDR (Duckworth, 1997), Myanmar (Su Su, 2005), **Thailand** (Rabinowitz, 1991; Austin and Tewes, 1999), Viet Nam (Roberton 2007), Cambodia (J. L. Walston pers. comm.) and Sri Lanka (Ratnayeke pers. comm.). No search has been made for recent records from Nepal, Bhutan, Bangladesh, Peninsular Malaysia, Java or Bali (Indonesia), areas where it was historically recorded, mostly commonly. It was formerly known from Singapore but its current status there is unclear (B. P. Y.H. Lee pers. comm.). Corbet and Hill (1992) include all of Sumatra for the species' distribution range, but only four individuals are known, all from one locality in the far north. Sody (1931) described these as a distinct subspecies, *Viverricula malaccensis atchinensis.* Other records from Sumatra have not been seen, and it is suggested to restrict its range to the north (Meijaard pers. comm.), and the lack of recent records from Sumatra is puzzling (W. Duckworth in litt. 2006). It has been introduced to Madagascar (Lekagul and McNeely 1977; Zhang 1997), Socotra (Yemen; Pocock 1939) and Zanzibar (Tanzania; W. Duckworth in litt. 2006). The current status of these introduced populations is poorly known.

Countries: Native:

Bangladesh; Bhutan; Cambodia; China; India; Indonesia (Jawa, Sumatera); Lao People's Democratic Republic; Malaysia (Peninsular Malaysia); Myanmar; Nepal; Sri Lanka; **Thailand;** Vietnam

Introduced:

Madagascar

Population:

The population status of this species is less well known that that of some other Southeast Asian small carnivores, because recent survey efforts have mostly not been appropriate for assessing the species. This is best shown by records from Myanmar. Su Su (2005) found that it was the second most common species of small carnivore recorded in Hlawga Wildlife Park in Myanmar, a secondary small isolate of semi-natural habitat in the outskirts of Yangon, a former capital city, subject to barely controlled snaring and other forms of encroachment, where only one other species of small carnivore persists in significant numbers. Recent camera-trapping showed that it is also abundant (by far the most frequently photographed nocturnal small carnivore) in Alungdaw Kathapa national park (Thaint Thaint Myo pers. comm. 2007), although intensive camera-trapping for Tigers a few years earlier at this site had recorded the species only twice.

This reflected the focus of the Tiger survey on little-disturbed evergreen forest, a habitat little-used by the species in South-east Asia, rather than the edge, disturbed and secondary areas the focus of the latter survey (Than Zaw et al. in press). The habitat degradation and hunting patterns of these two sites in Myanmar where the species is common are representative of much of non-Sundaic Southeast Asia, and incidental records from various other sites (e.g. Duckworth and Robichaud 2005) imply large populations in total, although it is likely that in Vietnam and Lao at least populations will have been somewhat depleted by snaring. This factor (the tendency for camera-trap surveys to go to the least encroached habitat blocks) has certainly much reduced the number of recent records of the species from South-east Asia. In Cambodia, where more camera-trapping has taken place in deciduous forest areas, it is commonly camera trapped (J. L. Walston pers. comm.). The same factor was suspected to be responsible for the relatively few recent records from Lao PDR (Duckworth et al. 1999), and may also have generated earlier remarks such as in **Thailand,** this species is rather rare (Lekagul and McNeely 1977); in fact, it is common in the degraded parts of Khao Yai national park but difficult to fin within the forest itself (J. W. Duckworth and A. Nettelbeck pers. comm.). It is also abundant over large areas of India. It is common in deciduous forests of Dak Lak, Vietnam (Le Xuan Canh et al. 1997) and more widely in the country (Roberton 2007). No information has been sought for the Sundaic populations except for Sumatra, where the species status has always been unclear; no recent records were traced.

Population Trend: Stable

Habitat and Ecology:

This species has been recorded in semi-evergreen and deciduous forest, mixed deciduous forest, bamboo forest, scrubby areas, grasslands and riverine habitat (Duckworth 1997, Le Xuan Canh et al. 1997; Mudappa, 2002; Su Su, 2005). This species is nocturnal and mostly terrestrial (Mudappa, 2002). An adult male was radio-tracked in **Thailand** and had a home range of 3.1 km^2 (Rabinowitz, 1991). In Lao PDR, this species is found in Semi-evergreen (rarely) and deciduous forest, including adjacent degraded areas (Duckworth et al. 1999). In other countries it is tolerant of habitat degradation and lives in proximity to human communities (Lekagul and McNeely 1988) and an active avoidance of closed evergreen forest was shown in Myanmar (Than Zaw et al. in press; see 'population'). In **Thailand,** it is mostly found in long grass or scrub, particularly in areas near villages, where it may live in drains, out-houses, and roofs, eating domestic poultry (Su Su, 2005) as well as rats, mice, birds, snakes, fruit, and roots, as well as carrion (Lekagul and McNeely 1988). It has litters of three to five, and the life span is eight to nine years (Lekagul and McNeely 1988). In Myanmar this species was recorded from both mixed deciduous forest and bamboo forest (Su Su, 2005). This species was rarely seen in the undisturbed rainforests of Kalakad-Mundanthurai Tiger Reserve (KMTR) in India, and was mostly seen near garbage dumps (Mudappa, 2002). In KMTR they were not camera-trapped frequently in rainforest, but were the most camera-trapped species in grasslands and in a riverine habitat (Mudappa, 2002). This species is nocturnal, and mostly terrestrial and insectivorous (Mudappa, 2002). Wang and Fuller (2001) conducted a study on the ecology of this species near Taohong Village in northern Jiangxi Province, Southeastern China, from April 1993 to November 1994. Wang and Fuller (2003) conducted a study on the food habits of this species in a rural agricultural area of Southeastern China (Taohong Village, Jiangxi Province) by analyzing its scats, the study was conducted between June 1992 and November 1994, and found that this species ate mostly mammals, with moderate insects and plants.

Systems: Terrestrial

Major Threat(s):

The extent to which extensive habitat loss and degradation are a threat to this species remains unclear as in most or all of its range areas it is more common in altered landscapes than in closed-canopy old-growth evergreen forest. This species is hunted for its meat and scent (Gupta 2004) in some portions of its range which potentially might reduce populations. Ground-dwelling small carnivores are exposed to hunting, particularly with snares. This is occurring in much of its range, including Lao PDR (Duckworth et al. 1997) and **Thailand,** with snaring found even in some protected areas (Kanchanasaka pers. comm.). There has been an increased demand for civet meat in Chinese and Vietnamese markets (Bell et al. 2004, Lynam et al. 2005). In India animals are caught for captivity for collection of 'civet', a fixative used internationally in the perfume industry and domestically for various purposes; even in areas of heavy collection, the animals remain common in the degraded forest, scrub and agricultural landscapes covering most of peninsular India. There is no evidence (at least from the non-Sundaic parts of its range) that it is not well able to survive high combined pressures of forest conversion and harvesting. The converse seems to be true: it remains more common than most other species of small carnivore in heavily encroached areas of southern China (M.W.N. Lau pers. comm.), as it does in heavily encroached Myanmar (Su Su 2005).

Conservation Actions:

In Myanmar, this species is totally protected under the Wildlife Act of 1994 (Su, 2005). This species is listed on Category II of the China Wildlife Protection Law (1988) (Li et al, 2000) and as 'Vulnerable' in the China Red List (Wang and Xie, 2004). It is listed on CITES Appendix III (India). It has been recorded in many protected areas (Duckworth, 1997; Mudappa, 2002; Su Su, 2005).

Small Asian Mongoose *(Herpestes javanicus)*

Description:

This is the only mongoose found in **Thailand.** The general colour is grizzled dark brown, with the individual hairs having alternate white and dark brown rings; when excited , the mongoose erects its body hairs and the rings on the hairs form a alternate light and dark stripes, possibly for camouflage against strikes from snakes and other dangerous prey. The head is reddish, with a darker nose; the legs are the same colour as the body or sometimes a little darker. The tail is 2 thirds the lenght of the head an body. The female as 3 sets of mammae. The skull is similar to that of *H. urva,* but is smaller , narrower and less robust. The crest are relatively longer than in *H.urva.* The orbit is usually complete in adults, The auditory bullae are relatively small and flat. The dentition is typical for mongooses, with the premolars and molars complex and trenchant, well adapted for a carnivorous diet.

Range Description:

The small Asian mongoose occurs across a wide range from Iran through northern India and into Indochina (Lekagul and McNeely 1977). In Asia, this species ranges from sea level to 2,100 m (Simberloff et al. 2000). Outside of its natural range, this species has many well established populations. Introduced mongoose has been implicated in the devastation of the native fauna, especially on islands (Baldwin et al. 1952, Seaman and Randall 1962, Nellis and Everard 1983, Coblentz and Coblentz 1985). The IUCN lists the Small Asian Mongoose as one of the world's 100 worst invasive alien species (Lowe et al. 2000). This species was introduced to the West Indies, the Hawaiian Islands, Mauritius, the Fijian Islands, and Okinawa (Simberloff et al. 2000), as well as the Comores and Amami-Oshima Island, Japan (Abe 2005). The reasoning behind these introductions was primarily control of rat and snake populations (S. Abe pers. comm. 2006). The Small Asian Mongoose is also often taken aboard ships, indirectly introducing them to new areas (J.W. Duckworth pers. comm.). The species recently reached Hong Kong (M. Lau pers. comm. 2006), and has also been recorded from the island of Madura, Indonesia (Meiri 2005), but it is not known whether this was due to human introduction or natural dispersal.There are several individuals from northern Sumatra (see van Strien 2001), which were described by Sody (1949) as *H. javanicus tjerapai.* Within its introduced range, the Small Asian Mongoose has been recorded from sea level to maximum elevations of 3,000 m on the Hawaiian Islands (Baldwin et al. 1952).

Countries: Native:

Afghanistan; Bangladesh; Bhutan; Cambodia; China; India; Indonesia; Lao People's Democratic Republic; Malaysia; Myanmar; Nepal; Pakistan; **Thailand;** Vietnam

Introduced:
Bosnia and Herzegovina; Croatia; Cuba; Dominican Republic; Fiji; Jamaica; Japan; Mauritius; Puerto Rico; Saint Kitts and Nevis; Saint Lucia; Suriname; Trinidad and Tobago; United States (Hawaiian Is.); Virgin Islands, British; Virgin Islands, U.S.

Population:

The Small Asian Mongoose has a relatively large population across its range, and tolerates a wide degree of habitat conversion perhaps preferring degraded habitats in some areas. In mainland Southeast Asia, this species reaches high densities in well-watered naturally open deciduous forests, scrublands and grassy areas as in southern Lao PDR, Viet Nam and much of Cambodia. Given adequate habitat the species is locally common (Le Xuan Canh et al. 1997), and widespread (Robertson et al. in prep.).

Snaring and other trapping pressures keep population numbers low in degraded habitats of large parts of Lao PDR - where the species might, if unmolested, be quite common (J.W. Duckworth pers. comm.). In India, it occurs at low densities and is not as common as *Hespestes edwardsii* (Shekhar 2003). In Myanmar, where village harvest of small carnivores and other similar-sized mammals seems to be much lower than in Lao PDR and Vietnam, the species is so common in places that it is trapped by conservation agencies as a pest (Su 2005).
Population Trend: Unknown

Habitat and Ecology:

The ecology of the Small Asian Mongoose has been studied in introduced parts of its range (Nellis 1989), but its ecology in its native range is known only to the level of coarse habitat correlations. The species is known to occur in a variety of habitats but appears to prefer well-watered naturally open deciduous forests, shrub lands and grasslands (Shekar 2003; J.W. Duckworth pers. comm.). While it tends to avoid closed evergreen forests, it will utilize secondary forest, degraded sites and areas of former evergreen forest opened by logging or similar practices (Lekagul and McNeely 1977; J.W. Duckworth pers. comm.). Where it has been introduced in the West Indies (Pimentel 1955, Nellis and Everard 1983) and the Hawaiian Islands (Baldwin et al. 1952) the species is found in grasslands, crops, and forest of various kinds, coastal areas, and even settled suburbs (Simberloff et al. 2000). It tends to prefer edge habitat in most areas.
This species is terrestrial, seldom climbing trees and feeds, during both the day and the night, on a wide diet, which includes rats, birds, reptiles, frogs, crabs, insects, and even scorpions (Lekagul and McNeely 1977). It produces litters of two to four at short intervals, with a gestation period of about 7 weeks (Lekagul and McNeely 1977).
Systems: Terrestrial

Major Threat(s):

The Small Asian Mongoose faces heavy exploitation in localized parts of its range, such as the Mekong Delta, but on the whole appears to be quite common and adaptable. This species is often captured and sold as pets (Shekhar 2003) and there is some commercial trade in China, India and Nepal. (A. Choudury pers. comm.). In Northern Vietnam it is hunted and sold in wild meat markets in both Vietnam and China (S. Robertson pers. comm.). None of these threats seem to be contributing to the decline of the species globally.

Conservation Actions:

The Small Asian Mongoose is listed under CITES Appendix III in India (as *Herpestes javanicus auropunctatus)* (Wozencraft 2005). It is totally protected in Peninsular Malaysia and **Thailand,** and was listed as Vulnerable on the Chinese Red List (A1cd). In central India people consider the mongoose to be sacred, and thus it is not killed there (Shekhar 2003).

Crab-eating Mongoose *(Herpestes urva)*

Description:

HB:440-480; T: 265-310; HF: 90-109; E: 29-35; W: 3-4kg.

The crab eating mongoose is usually dark brown to grey in colour, with the individual hairs ringed black and white. The chin is white, the front of the throat is blackish, the chest is dark brown and the belly is a lighter brown. The legs are brownish black. The tail is ochraceous towards the tip. There is a long white patch running from the corner of the mouth up to the shoulder. This is a large mongoose, with the tail less than two thirds the lenght of the head and body. It as anal glands about the size of a cherry, and as the power to squirt out a foetid fluid from then backwards at great force. The skull is large and heavy, but with relatively small crests. the bullae is less flat than in H. javanicus and extends to well below the occipital condyle. Mature individuals have a complete bony orbit, and immatures have an incomplete one.The dentition is typical of mongooses, but the lower first premolar may be lacking immatures. The last lower molar is small but complex.

Range Description:

This species is found in Bangladesh (Ashan, 1989), SE China (Wang and Fuller, 2001, 2003), Lao PDR (Duckworth, 1997), Peninsular Malaysia (Lim, 2001, Wells and Francis, 1988, Laidlaw pers. comm.), Myanmar (Than Zaw et al. in press.), Lao PDR, Nepal, North eastern India, Taiwan, **Thailand,** Vietnam and Cambodia (Wilson and Reeder, 2005, Van Rompaey, 2001). It is rarely found on high mountains (Van Rompaey, 2001), but it has been collected at 1650 m (Kurseong, Bengal, India; Pocock, 1941). From 16 records at 10 sites in Lao PDR, 14 are from over 450 m, suggesting that this species occurs, at least in south and central Lao PDR, mostly in hills and mountains (Van Rompaey, 2001). In Cambodia and Vietnam, its range extends down to the plains (150 m) and to sea level in Hong Kong.

Countries: Native:

Bangladesh; Cambodia; China; India; Lao People's Democratic Republic; Malaysia; Myanmar; Nepal; Taiwan, Province of China; **Thailand;** Vietnam

Population:

In watered edge habitats amid largely natural vegetation, this is a common species in Vietnam, Lao, Cambodia, Myanmar. However, outside of this preferred habitat and in some portions of its range the species is uncommon or rare (Gyldenstolpe, 1916; Inglis et al., 1919; Delacour, 1940; Khan, 1982; Duckworth, 1997; Choudhury, 1997a; 1997b; 1999). However, it is uncommon in the Jalpaiguiri District, Bengal, India (Inglis et al. 1919), while it is common in Assam, Arunachal Pradesh, and North Bengal, India (Choudhury, 1997a; 1997b; 1999). It's rare in Bangladesh (Khan, 1982) and relatively common in Vietnam (Delacour, 1940). Today in Lao it is common in forested areas in north and central (Duckworth 1997, Duckworth and Robichaud 2005). It is widespread in northern Vietnam north of Thanh Hoa, but in small numbers (Pham Trong Anh, 1980, 1982).

It is common in the forests between **Thailand** and Tenasserim (Peninsular Myanmar) (Gyldenstolpe, 1916). Very common in the Hukaung valley of North Myanmar, widespread and at least frequent in the Northern two-thirds of the country (Than Zaw et al. in press.).
Population Trend: Stable

Habitat and Ecology:

This species has been recorded mainly near water, in evergreen and deciduous forest, scrubby areas, in plantations and near human settlements (Pham Trong Anh, 1980; Duckworth et al, 1997; Van Rompaey, 2001, Than Zaw et al. in press). It has been recorded up to 1,650 m (Pocock, 1941; Duckworth, 1997; Van Rompaey, 2001). It is diurnal in at least Lao PDR, Vietnam, Cambodia and Myanmar, despite earlier statements it was nocturnal, which were based on very little real evidence (reviewed in Than Zaw et al. (in press) (Pham Trong Anh, 1992; Duckworth, 1997, Long and Minh Hoang 2006, Roberton et al. in prep., Than Zaw et al. in press., J. Walstone pers. comm.). Wang and Fuller (2001) conducted a study on the food habits of this species in Southeastern China and found that it ate mammals, reptiles, insects and crustaceans. In Lao PDR, this species is found in evergreen forest (including degraded areas), mainly near water; the most recent records are from hill and mountainous areas (Duckworth et al, 1999). In **Thailand,** Cambodia and southern Viet Nam this species is found also in deciduous forest. In India this species is found in lowland wet-evergreen forests, secondary forest and areas around industrial areas (i.e. oil refineries). There are records from rice fields and other agricultural areas, and even near human settlements (Pham Trong Anh, 1980). However, in Assam, India, it has not been observed near human habitations (Choudhury, 1997). Little is known about its breeding, though the gestation period is thought to be about nine weeks; probably meaning that this species reproduces more slowly than Herpestes javanicus (Lekagul and McNeely 1977). It feeds on fish, frogs, crabs, molluscs, insects and crayfish (Van Rompaey, 2001).

It is readily approached by humans due to its apparent nearsightedness (Van Rompaey, 2001), and fearlessness (Pocock, 1941). It has lived up to 13 years and 4 months in captivity (Jones, 1982). Wang and Fuller (2001) conducted a study on the ecology of this species near Taohong Village in northern Jiangxi Province, South eastern China, from April 1993 to November 1994. Wang and Fuller (2003) conducted a study on the food habits of this species in a rural agricultural area of South eastern China (Taohong Village, Jiangxi Province) by analyzing its scats, the study was conducted between June 1992 and November 1994, and found that this species ate mammals, reptiles, insects, and crustaceans.
Systems: Terrestrial

Major Threat(s):

The impact of habitat loss and degradation on this species is not fully understood, is can be found in secondary forest and selectively logged areas, however, there is no evidence that it can live in totally converted areas. Hunting and retaliatory killing for raiding poultry on farms may also be threats in areas near human settlement, but the impacts of this killing at population scales are unknown.

On the China Red List it is considered near threatened, nearly meeting vulnerable due to past population declines (A1c). There is no demand for its meat in restaurants in Viet Nam (Roberton, S. pers. comm.). Incidental capture in snares is also a threat. Retaliatory killing for raiding poultry on farms in India might be a threat, but this is not considered a major threat to the population. Hunting is probably the main threat in Lao PDR, but the species persists widely (and, despite being a ground-dwelling animal, and thus potentially suffering from incidental trapping, there is equivocal evidence for only localized population reduction (Duckworth, W. pers. comm.).

Conservation Actions:

This species is protected in China ('Near Threatened'), **Thailand,** Myanmar and Peninsular Malaysia. It is listed in Schedule IV of the Indian Wildlife (protection) Act, 1972, and in Appendix III.

Golden Jackal (*Canis aureus)*

Description:

HB: 600-750; T:200-250; HF: 130-156; E: 75-85; W: 8-9 kg.

This animal, in **Thailand,** may be confused with the Asian Wild Dog, *Cuon alpinus,* but differs in being generally greyish brown, not reddish brown, and also having the shoulder hairs tipped black, forming a saddle like pattern. Also, the muzzle is not blackish as in *alpinus,* the tail is short and blackish only on the distal third, also the female has 10 mammae. The skull is typical of *Canis,* with a long muzzle, rather strong, wide zygomatic arches, a low sagittal crest, and large simple bullae. The skull differs from *Cuon* in having a narrower muzzle, a less inflated frontal region, and the orbits relatively higher on the head. The dentition has the full 42 teeth of typical dogs; the second upper and lower molars are normal, not reduced and in *Cuon*. The lower carnassial has two cusps on its heel (one cusp in *Cuon*)

Range Description:

The Golden Jackal is widespread in North and north-east Africa, occurring from Senegal on the west coast of Africa to Egypt in the east, in a range that includes Morocco, Algeria, and Libya in the north to Nigeria, Chad and Tanzania in the south. They also occur in the Arabian Peninsula and have expanded their range into Europe, where they have a patchy distribution, being resident in the Balkans and, since recent times, in Hungary and south-western Ukraine. It is regularly found as a vagrant in Austria, Slovakia, Slovenia and north-eastern Italy (Kryštufek 1999). Eastwards they range into Turkey, Syria, Iraq, Iran, Central Asia, the entire Indian subcontinent, then east and south to Sri Lanka, Myanmar, Thailand and parts of Indo-China.

Countries: Native:

Afghanistan; Albania; Algeria; Bahrain; Bhutan; Bosnia and Herzegovina; Bulgaria; Central African Republic; Croatia; Djibouti; Egypt; Eritrea; Ethiopia; Greece; India; Iran, Islamic Republic of; Iraq; Israel; Jordan; Kenya; Kuwait; Lebanon; Libya; Mali; Mauritania; Morocco; Myanmar; Nepal; Niger; Nigeria; Oman; Pakistan; Qatar; Saudi Arabia; Senegal; Somalia; Sri Lanka; Sudan; Syrian Arab Republic; Tanzania, United Republic of; **Thailand;** Tunisia; Turkey; Turkmenistan; United Arab Emirates; Vietnam; Western Sahara; Yemen

Vagrant:

Austria; Italy; Slovakia; Slovenia

Population:

The Golden Jackal is fairly common throughout its range. High densities are observed in areas with abundant food and cover. In several parts of India, high densities of low-quality cattle are maintained. Due to religious beliefs, most people do not consume beef, and cattle carcasses are freely available for scavenging. In India, jackal populations achieve high densities in pastoral areas such as Kutch, Maharashtra, Rajasthan, and Haryana. Based on intensive observations on breeding pack units and radio-collared individuals, jackal densities in the semi-arid Velavadar National Park were estimated between one and two jackals per km² (Y. Jhala et al., unpubl.); see Sharma (1998) for densities quoted for the Thar Desert in India. On the African continent, in the Serengeti National Park, densities can range as high as four adults per km² (Moehlman 1983, 1986, 1989). Based on known density estimates for parts of India and considering that about 19% (i.e., about 637,000 km²) of the geographical area of India has forest cover with jackal populations (and that jackals are also found outside forested habitats), a minimum population estimate of over 80,000 Golden Jackals would not be unreasonable for the Indian sub-continent. Population estimates for Africa are not available.

Population Trend: Increasing

Habitat and Ecology:

Due to their tolerance of dry habitats and their omnivorous diet, the Golden Jackal can live in a wide variety of habitats. These range from the Sahel Desert to the evergreen forests of Myanmar and **Thailand.** They occupy semi-desert, short to medium grasslands and savannas in Africa; and forested, mangrove, agricultural, rural and semi-urban habitats in India and Bangladesh (Clutton-Brock et al. 1976; Poche et al. 1987; Y. Jhala, pers. obs.). Golden Jackals are opportunistic and will venture into human habitation at night to feed on garbage. Jackals have been recorded at elevations of 3,800 m in the Bale Mountains of Ethiopia (Sillero-Zubiri 1996) and are well established around hill stations at 2,000 m in India (Prater 1980).

Systems: Terrestrial

Major Threat(s):

Over its entire range, except in protected areas like National Parks and Sanctuaries, the jackal population is steadily declining. Traditional land use practices, like livestock rearing and dry farming that were conducive to the survival of jackals and other wildlife, are being steadily replaced by industrialization and intensive agriculture; wilderness areas and rural landscapes are being rapidly urbanized. Jackal populations adapt to some extent to this change and may persist for a while, but eventually disappear from such areas like other wildlife. There are no other known threats, except for local policies of extirpation and poisoning (for example, Israel and Morocco). Jackals may occasionally be hunted as a game species and eaten, as has been recorded in Morocco (F. Cuzin pers. comm. 2007). There is no significant trade in jackal products, although skins and tails are occasionally sold.

Conservation Actions:

Golden jackals are present in all protected areas of India except for those in the high elevation regions of the Himalayas. In East Africa, they occur in the Serengeti-Masai Mara-Ngorongoro complex, as well as numerous other conservation units. Thus they have a wide coverage in terms of protected populations.

The species is included in CITES Appendix III (in India). Jackals feature on Schedule III of the Wildlife Protection Act (1972) of India and are afforded the least legal protection (mainly to control trade of pelts and tails). However, no hunting of any wildlife is permitted under the current legal system in India. The golden jackal could be considered as a "species requiring no immediate protection" with caution and knowledge that populations throughout its range are likely declining.

Besides being represented in a wide array of protected areas covering several landscapes, no special species targeted conservation efforts have been undertaken. Almost all zoos in India have golden jackals.

Current or planned research projects include ongoing, long-term studies in the Serengeti, Tanzania; ongoing studies on wolves, jackals, and striped hyenas in Bhal and Kutch areas of Gujarat, India; and investigation into crop damage, densities and ranging patterns of golden jackals in Bangladesh.

Gaps in knowledge

Little quantitative information is available on jackal densities, habitat use, and ranging patterns in relation to food availability. Information on dispersal, survival and mortality factors of adults, pups and dispersing individuals is needed. Jackal ecology needs to be studied in forested ecosystems of Southeast Asia where a different set of factors are likely to operate affecting food availability, ranging patterns and survival. Aspects of canid diseases in relation to population dynamics of jackals and transmission need to be better understood.

Dhole *(Cuon alpinus)*

Description:

HB:800-900; T: 305-345; HF: 78-87; E: 95-105; W: 10-20kg

In **Thailand** this animal may be confused with the Asian Wild Dog, *Cuon alpinus,* but differs with in being generally greyish brown, not reddish, and having the shoulder hairs tipped black, forming a saddle like pattern. Also, the muzzles not blackish as it is in *alpinus,* the tail is short and blackish only on the distal third, the female as 10 mammae. The skull is typical of a *Canis* skull, with a long muzzle, rather wide, strong zygomatic arches, a low sagittal crest, and large , simple bullae. The skull differs from *Cuon in* having a narrower muzzle, a less inflated frontal region, and orbits relatively higher on the head. The dentition has the full 42 teeth of the typical dog; the second upper and lower molars are normal, but not reduced as in *Cuon.* The lower carnassial has two cusps on its heel. (one cusp in *Cuon.*)

Range Description:

In Central and Eastern Asia, there have been no confirmed, recent reports of dholes from Russia, Mongolia, Kazakhstan, Kyrgyzstan (where they were found formerly in the Tian-Shan area) or Tajikistan (where they were found formerly in the eastern Pamir area) (A. Poyarkov and N. Ovsyanikov in litt. D. Miquelle pers. comm.). There is a recent report of a Dhole that was captured in Jiangxi district, south China (C. Bellamy pers. comm.). Dholes were once present in parts of western China in the Tian-Shan Range, but the species' current status in this area is unclear; they do at least still persist, perhaps in low numbers, in parts of the Qilian Shan in north-western Gansu Province (Harris 2006). The species is still found in Tibet today, particularly in areas bordering the Leach region of India (R. Wangchuk pers. comm.), and the Tibet Forestry Bureau has reported that dholes are still "common" in parts of southeast Tibet (S. Chan, in litt.). Dholes occurred in northern Korea (Won Chang Man and Smith 1999) and a few small populations may still exist. There have been no records from Pakistan, but the species occurred on the alpine steppes of Ladakh, Kashmir, and India (Johnsingh 1985) that extend into the region termed Pakistan-occupied Kashmir by India.

Dholes are still found throughout much of India south of the river Ganges, and especially in the Central Indian Highlands and the Western and Eastern Ghats of the southern states. They are also found throughout north-east India, in the states of Arunachal Pradesh, Assam, Meghalaya, and West Bengal (A. Venkataraman, A.J.T. Johnsingh and L. Durbin pers. comm.). In the Himalaya and north-western India, the status of dholes seems more precarious with a much more fragmented distribution. Dholes reportedly still occur in the Ladakh area of Kashmir, which is contiguous with the Tibetan highlands in China (R. Wangchuk pers. comm.).

The species formerly was recorded in the Terai region of the Indo-gangetic plain, including the Royal Chitawan National Park in Nepal, but there have been few recent reports. There is an unconfirmed report of dholes in Dhorpatan Hunting Reserve in the late 1990s (R.C. Kandel pers. comm.).

In Bhutan, there have been recent press reports that dholes have recovered from a government-initiated mass poisoning campaign in the 1970s and there have apparently been numerous recent incidents of dholes killing livestock in the lower Kheng region. Two recent, independent, eye-witness reports identify dholes in six protected areas in Bhutan (S. Wangchuk pers. comm., T. Wangchuk pers. comm.). In some regions, Dhole predation on wild boar (Sus scrofa) may be viewed in a positive light by local people (T. Wangchuk pers. comm.).

In Bangladesh, dholes were thought to occur in the forested tracts of the Chittagong and Sylhet Districts (Johnsingh 1985). It is not certain whether any remain in Bangladesh.

In Myanmar, dholes were recorded by camera trapping at 11 of 15 survey areas scattered across the country, only four of which were protected. Dholes and/or leopards have apparently replaced tigers as the top predator in these areas (Myanmar Forest Department 2003).

In Indochina, dholes probably ranged over all or almost all of Lao PDR, Cambodia, Vietnam and **Thailand,** although reliable site-specific information is scarce. Present distribution is highly fragmented and large parts, particularly of Vietnam and **Thailand,** are without any regular occurrence of dholes, although they persist in a number of protected areas (Duckworth et al. 1999, Waltson 2001, M. Baltzer and R. Shore in litt., A. Lynam pers. comm.).

The species' historical range probably included all or most of the Malaysian peninsula and the Indonesian islands of Sumatra and Java, but reliable information is scarce. Current distribution is poorly known but is thought to be highly fragmented. On the Malaysian peninsula, dholes are known to occur in four sites in northern and central areas of the peninsula (from recent camera-trap surveys; J.B. Abdul pers. comm.). On Java, dholes appear to be most common in the protected areas at the eastern and western ends of the island. On Sumatra, very little is known, but dholes are known to occur in major protected areas in the southern, central, and northern parts of the island (e.g., from camera trapping; D. Martyr pers. comm.).

There is no reliable evidence of the presence of Dhole in Turkey (Kryštufek and Vohralík 2001; Can 2004).

Countries: Native:

Bangladesh; Bhutan; Cambodia; China; India; Indonesia; Kazakhstan; Kyrgyzstan; Lao People's Democratic Republic; Malaysia; Mongolia; Myanmar; Nepal; Russian Federation; Tajikistan; **Thailand;** Vietnam

Population Trend: Decreasing

The Dhole is found in a wide variety of vegetation types, including: primary, secondary and degraded forms of tropical dry and moist deciduous forest; evergreen and semi-evergreen forests; dry thorn forests; grassland–scrub–forest mosaics; and alpine steppe (above 3,000 m). They are not recorded from desert regions. In India, tropical dry and moist deciduous forest may represent optimal habitats, based on the regions thought to hold the largest Dhole populations. Ungulate biomass, particularly that of cervid species, is highest in these vegetation types when compared to others in the same region (A. Venkataraman and V. Narendra Babu, unpubl.). In India, tropical dry and moist deciduous forests are subject to seasonal monsoon climates. Important factors that may influence habitat selection include the availability of medium to large ungulate prey species, water, the presence of other large carnivore species, human population levels and suitability of breeding sites (proximity to water, presence of suitable boulder structures and sufficient prey).

Systems: Terrestrial

Major Threat(s):

Depletion of prey base: Across almost all of Cambodia, Lao PDR, and Vietnam, as well as within protected areas, ungulates occur at levels well below natural. All species of ungulate except Muntjacs (Muntiacus spp.), pigs (Sus spp.) and in some areas southern Serow *(Naemorhedus sumatraensis)* are ecologically or fully extinct across extensive parts of the region. Only a few of the largest wildernesses support nearly intact species assemblages and even in these, the larger species (Bos spp., Cervus spp., hog deer Axis porcinus) are very rare. This situation will likely hinder any possibility of recovery by the region's Dhole populations, even if the other issues could be addressed. While not as depressed as in Indochina, prey levels in Indonesia also exist at levels much below carrying capacity (because of illegal hunting and habitat degradation). In protected areas in southern and central India, where Dhole numbers are stable, prey densities are high. In north-east India, prey densities are very low in protected areas with dholes.

Habitat loss and transformation: Currently, extensive areas of natural or semi-natural vegetation remain in Lao PDR and Cambodia, some areas encompassing many hundreds of square kilometres of potential Dhole habitat. However, habitat conversion and fragmentation are proceeding apace. In Vietnam, very few natural areas of over 50 km^2 remain. Habitat loss and fragmentation is a major threat to protected areas in Indonesia, particularly those on Sumatra. Habitat loss and degradation are also serious threats to dholes in South Asia and the disappearance of dholes from many of the forested tracts in India has been attributed in large part to loss of habitat. Persecution: This certainly occurs in Indochina, although it is unclear how often. In Indonesia, too, it is a threat but again its significance is unknown. In India, such persecution can play a serious role in limiting local populations. Dholes living outside or on the edge of core protected areas are particularly vulnerable to human kleptoparasitism, snaring (non-selective) and direct persecution. For example, during a radio-tracking study in 2000, in the buffer zone of Kanha Tiger Reserve, central India, at least 16 out of 24 dholes in one pack died from a sudden strychnine poisoning (L. Durbin pers. obs.). In southern India, such persecution is moderate to low and often occurs indirectly when cattle graziers and others inadvertently go close to Dhole dens and disturb adults and pups, disrupting breeding and rearing (A. Venkataraman pers. obs.). "By-catch" in snares and other traps is probably a significant threat to dholes across Indochina at least. Competition with other species: Apparently, free-living dogs have been seen and/or camera trapped in many parts of Indochina, but there is no evidence for existence of large populations. Undoubtedly, the main competitor for prey species in Indochina is people. There is no evidence that feral dogs are significant competitors with dholes in Indonesia. In many parts of their range, dholes are sympatric with tigers and leopards and so the potential for significant intraspecific competition for prey exists, especially if the prey populations are reduced as a result of hunting by people.

Disease and pathogens: Particularly those transmitted by feral and/or domestic dogs (e.g., mange, canine distemper, parvovirus and rabies). The significance of disease is unclear in Indochina, but diseases are a significant threat in South Asia and probably in parts of Indonesia. There is no widespread exploitation for fur or other purposes, though medicinal use should be investigated in China.

Conservation Actions: Included in CITES – Appendix II (2003).

In Cambodia, the current wildlife decrees give the Dhole protection from all hunting. A new forestry law is under preparation, and a proposal to list the species as a fully protected species is under discussion. In India, the Dhole is protected under Schedule 2 of the Wildlife Act of 1972 (permission is required to kill any individual unless in self defence or if an individual is a man killer). The creation of Project Tiger Reserves in India has provided some protection for populations of the dukhunensis subspecies (A.J.T. Johnsingh pers. comm., L. Durbin, pers. obs.). In the Russian Federation, dholes received the status of "protected animal" in 1974 (A. Poyarkov and N. Ovsyanikov in litt.); however, the poisoning of grey wolves may inadvertently affect any remnant Dhole populations (V. Puzanskii pers. comm.). In Vietnam, the Dhole is protected by Decree 18/HDBT (17/01/1992) and the amendment Decree 48/2002/ND-DP (22/04/2002) under category IIB, which limits extraction and utilization. However, the levels of extraction or utilization are not quantified (B. Long in litt. 2003). Dholes are listed as a category II protected species under the Chinese wildlife protection act of 1988. The species occurs in protected areas throughout its range. No conservation measures specifically focused on dholes have been reported for most range states. In India, Project Tiger could potentially maintain Dhole prey bases in areas where tigers and dholes coexist. There do not appear to be any specific measures for Dhole conservation in Indochina, although the declaration of relatively large protected area networks in Cambodia, Lao PDR, and Viet Nam will, when these areas become functional realities on the ground, form a suitable conservation system for the species in at least Cambodia and Lao PDR. There are at least 110 dholes in captivity, and the sex ratio is approximately even. Except for some captive populations in India heterozygosity appears to be good, but there is little chance of breeding the putative subspecies as animals from diverse geographical origins have been widely interbred (M. Boeer pers. comm.).

Asiatic Black Bear *(Ursus thibetanus)*

Description:

HB: 1200-1500; T: 65-100; HF: 175-195; E: 120-180; W: 100-160kg.

Asian black bears are similar in general appearance to brown bears, but are more lightly built and are more slender limbed. The skulls of Asian black bears are relatively small, but massive, particularly in the lower jaw. Adult males have skulls measuring 311.7–328 mm (12.3–13 in) long and 199.5–228 mm (7.9–9 in) wide, while females have skulls measuring 291.6–315 mm (11.5–12.4 in) long and 163–173 mm (6.4–6.8 in) wide. Compared to other bears of the genus *Ursus*, the projections of the skull are weakly developed; the sagittal crest is low and short, even in old specimens, and does not exceed more than 19–20% of the total length of the skull, unlike in brown bears, which have sagittal crests comprising up to 41% of the skull's length. Although mostly herbivorous, the jaw structure of Asian black bears is not as specialised for plant eating as that of pandas: Asian black bears have much narrower zygomatic arches, and the weight ratio of the two pterygoid muscles is also much smaller in Asian black bears. However, the lateral slips of the temporal muscles are thicker and stronger in black bears. In contrast to polar bears, Asian black bears have powerful upper bodies for climbing trees, and relatively weak hind legs, which are shorter than those in brown bears and American black bears. A black bear with broken hind legs can still climb effectively. They are the most bipedal of all bears, and have been known to walk upright for over a quarter mile. The heel pads on the forefeet are larger than those of most other bear species. Their claws, which are primarily used for climbing and digging, are slightly longer on the fore foot (30–45 mm) than the back (18–36 mm), and are larger and more hooked than those of the American black bear. The ears, which are bell shaped, are proportionately longer than those of other bears, and stick out sideways from the head. The lips and nose are larger and more mobile than those of brown bears. On average, adult Asian black bears are slightly smaller than American black bears, though large males can exceed the size of several other bear species. They measure 70–100 cm (28–40 in) at the shoulder, and 120–195 cm (47–77 in) in length. The tail is 11 cm (4.4 inches) long.Mature males typically weigh between 100–200 kg (220-440 lbs), with an average weight of about 135 kg (about 300 lbs). Females weigh about 65–90 kg (143–198 lbs), with large ones up to 140 kg (308 lbs). The famed British sportsman known as the "Old Shekarry" wrote of how a black bear he shot in India probably weighed no less than 363 kg (800 lbs) based on how many people it took to lift its body, though Gary Brown, author of *The Great Bear Almanac* writes that the largest Asian black bear on record weighed 200 kg (440 lbs). Zoo-kept specimens can weigh up to 225 kg (500 lbs). Although their senses are more acute than those of brown bears, their eyesight is poor, and their powers of hearing moderate, the upper limit being 30 kHz. Easily distinguished from the sun bear by its very long pelage. It is also twice as heavy as the sun bear and as relatively longer ears. The white area of the muzzle seldom extends as far as the orbits. As in the sun bear, there is a white mark on the chest though it is more V shaped than U shaped as in the Sun Bear. The skull is flatter and narrower than in Helarctos, with the mastoid width seldom exceeding the length of the palate. The bullae is flat, not swollen as in *H. malayanus*. The muzzle is longer and narrower, and not as much extended over the canines.

The incisors and canines are smaller, but the cheek teeth are further apart and the molars are longer, more premolars are generally retained. Asian black bears are diurnal, though they become nocturnal near human habitations. They may live in family groups consisting of two adults and two successive litters of young. They will walk in a procession of largest to smallest. They are good climbers of rocks and trees, and will climb to feed, rest, sun, elude enemies and hibernate. Some older bears may become too heavy to climb. Half of their life is spent in trees and they are one of the largest arboreal mammals. In the Ussuri territory, black bears can spend up to 15% of their time in trees. Asian black bears break branches and twigs to place under themselves when feeding on trees, thus causing many trees in their home ranges to have nest-like structures on their tops. Asian black bears will rest for short periods in nests on trees standing fifteen feet or higher. Asian black bears do not hibernate over most of their range. They may hibernate in their colder, northern ranges, though some bears will simply move to lower elevations. Nearly all pregnant sows hibernate. Black bears prepare their dens for hibernation in mid October, and will sleep from November until March. Their dens can either be dug out hollow trees (sixty feet above ground), caves or holes in the ground, hollow logs, or steep, mountainous and sunny slopes. They may also den in abandoned brown bear dens. Asiatic black bears tend to den at lower elevations and on less steeper slopes than brown bears. Female black bears emerge from dens later than do males, and female black bears with cubs emerge later than barren females. Asian black bears tend to be less mobile than brown bears. With sufficient food, Asian black bears can remain in an area of roughly 1–2 sq km, and sometimes even as little as 0.5–1 sq km.Asian black bears have a wide range of vocalisations, including grunts, whines, roars, slurping sounds (sometimes made when feeding) and "an appalling row" when wounded, alarmed or angry. They emit loud hisses when issuing warnings or threats, and scream when fighting. When approaching other bears, they produce "tut tut" sounds, thought to be produced by bears snapping their tongue against the roof of their mouth. When courting, they emit clucking sounds

Range Description:

Fossil remains of the Asiatic black bear have been found as far west as Germany and France, but in historic times the species has been limited to Asia. This species occupies a narrow band from Southeastern Iran (Gutleb and Ziaie 1999) eastward through Afghanistan and Pakistan, across the foothills of the Himalayas, to Myanmar. It occupies all countries in mainland Southeast Asia except Malaysia. It has a patchy distribution in southern China, and is absent in much of east-central China. Another population cluster exists in northeastern China, the southern Russian Far East, and into North Korea. A small remnant population exists in South Korea. They also live on the southern islands of Japan (Honshu and Shikoku) and on Taiwan and Hainan. The species now occurs very patchily through much of its former range, especially in Iran, Afghanistan, Pakistan, mainland southeast Asia and China. Its distribution in parts of China and Myanmar remains very poorly known. The distribution of the Asiatic black bear roughly coincides with forest distribution in southern and eastern Asia (FAO 2006), except that in central and southern India this species is replaced by the sloth bear *(Melursus ursinus),* in **Southern Thailand** and into Malaysia it is replaced by the sun bear (Helarctos malayanus) and north and west of the Russian Far East it is replaced by the brown bear *(Ursus arctos)*. However, the Asiatic black bear overlaps the ranges of each of these species, especially the sun bear in a large portion of Southeast Asia.

Countries: Native:

Afghanistan; Bangladesh; Bhutan; Cambodia; China; India; Iran, Islamic Republic of; Japan; Korea, Democratic People's Republic of; Korea, Republic of; Lao People's Democratic Republic; Myanmar; Nepal; Pakistan; Russian Federation; Taiwan, Province of China; **Thailand;** Vietnam

Range Map: Click here to open the map viewer and explore range.

Population:

No rigorous population estimates exist for this species. Japan formerly posed estimates of 8–14,000 bears on Honshu Island, but these are no longer considered valid. Russian biologists have presented a number of density estimates, yielding a range wide estimate of about 5–6,000, but the reliability of these is unclear (Aramilev 2006).

Likewise, rough density estimates have been made for some portions of India and Pakistan, which have been extrapolated country-wide (7-9000 in India: Sathyakumar 2006, 1000 in Pakistan: Sheikh 2006), but without corroborating methodology or data. A host of recent countrywide estimates have been posed for Asiatic black bears in China, ranging from 15–46,000 (summarized by Garshelis 2002, Gong and Harris 2006), with an official government estimate (in 2003) of about 28,000; none of these estimates have been substantiated.
Population Trend: Decreasing

Habitat and Ecology:

Asiatic black bears occupy a variety of forested habitats, both broad-leaved and coniferous, from near sea level to an elevation of 4,300 m (in northeastern India, A. Choudhury, Rhino Foundation for Nature pers. comm.). They also infrequently use open alpine meadows. Individual bears move to different habitats and elevations seasonally (Izumiyama and Shiraishi 2004), tracking changes in food abundance. Foods include succulent vegetation (shoots, forbs and leaves) in spring, turning to insects and a variety of tree and shrub-borne fruits in summer, and finally nuts in autumn (Bromlei 1965, Reid et al. 1991, Huygens et al. 2003). In some places the diet contains a sizeable portion of meat from mammalian ungulates (which they either kill or scavenge, Hwang et al. 2002) In temperate forests, Asiatic black bears rely heavily on hard mast in autumn, in part to put on sufficient fat reserves for winter denning (hibernation). Therefore, these bears tend to focus their activities in habitats with high abundance of oak acorns, beechnuts, walnuts, chestnuts, hazelnuts, or stone pine seeds (Schaller et al. 1989, Hashimoto et al. 2003). When Asiatic black bears feed in hard mast trees they often break branches and pile them up in the canopy, forming what appears to be a platform or "nest". Males may socially exclude females from rich stands of hard mast (Huygens and Hayashi 2001, Hwang 2003). In northern latitudes, where food becomes unavailable in winter, both sexes hibernate. In the most northerly parts of their range, bears enter dens as early as October and exit as late as the end of May (Seryodkin et al. 2003). They den in rock crevices, in hollow trees or stumps, under upturned trees, in dug-out earthen dens, or in ground nests. In Russia, Asiatic black bears have been reported to select flat river bottoms for denning (Seryodkin et al. 2003), whereas in central China they move to high elevation rocky outcrops on steep slopes (Reid et al. 1991). Hunters often have knowledge of the sorts of places and types of dens that the bears tend to use. Denning and active black bears are also subject to predation by other Asiatic black bears, brown bears, and tigers (Seryodkin et al. 2005).

In the tropics, Asiatic black bears generally do not hibernate, except females giving birth during winter (Hwang and Garshelis 2007). They still make use of hard mast, but additionally consume numerous species of soft fruits. In **Thailand,** for example, Asiatic black bears were found to feed on 160 species of tree-borne fruits. Sympatric sun bears also eat most of these same fruits. Both species most often climb (apparently for feeding) trees in the cinnamon (Lauraceae) and teak (Labiatae) families. Both species live together in lowland habitats. Asiatic black bears also use regenerating forests, which may have a high production of berries or young bamboo shoots. They also feed in plantations, where they may damage trees by stripping the bark and eating cambium, and in cultivated areas, especially corn and oat fields and fruit orchards (Carr et al. 2002, Yamazaki 2003, Mizukami et al. 2005, Gong and Harris 2006, Vinitpornsawan et al. 2006). Asiatic black bears generally breed during June–July and give birth during November–March; however, timing of reproduction is not known for all portions of the range. Age of first reproduction is 4–5 years, and they normally produce litters of 1 or 2 cubs every other year (at most). Maximum lifespan is over 30 years, but average lifespan is less in the wild.

Systems: Terrestrial

Major Threat(s):

Habitat loss due to logging, expansion of human settlements, roadway networks, and hydro-power stations, combined with hunting for skins, paws and especially gall bladders are the main threats to this species. Habitat loss and degradation is most severe in the southern portion of the range. In India, et al. 2006). Forest area has recently been increasing in Viet Nam, but much of the present remaining forest is highly degraded from both legal and illegal lumbering (Nguyen Xuan Dang 2006). Forest area is increasing rapidly in China, which is now first in the world in terms of area gained per year. This increasing forest area stems from mandated government programs aimed mainly toward reducing flooding and erosion; the replanted trees may or may not be particularly suitable for bears. However, good forest habitat does persist in northeastern China, Taiwan, Korea, Russia, and Japan. In Japan, black bear range has expanded with increasing forest area and diminishing rural human populations (Oi and Yamazaki 2006). Meanwhile, the number of people killed or injured by Japanese black bears has been on the rise (presumably reflective of the increasing bear population), and the same may be true in some parts of China (J. Gong, Sichuan Forestry Dept., Chengdu pers. comm.). The major threat to bears in China and Southeast Asia is the commercial trade in live bears and bear parts, especially gall bladders (bile). China initiated commercial bear farming in 1984, ostensibly to satisfy the demand for bile by practitioners of Traditional Chinese Medicine (TCM; and also Traditional Korean Medicine, TKM). The bile is periodically drained, so the captive bears do not have to be killed; it was claimed that this practice would thereby reduce the taking of wild bears. However, these farms were initially stocked with wild bears, and although the Chinese farms are purportedly now mainly self-propagating (with some continuing exceptions), there is no evidence that their existence has reduced the killing (poaching) of wild bears. In Viet Nam, many small-scale bile farms have been started, which were stocked by several thousand bears removed from the wild (from Viet Nam as well as from neighbouring countries). The condition in which these bears are kept precludes successful breeding and cub rearing; in fact, most of these farms do not attempt to breed their bears. Moreover, although this practice has been illegal since 1992, with regulations strengthened in 2002, the number of wild-caught farmed bears in Vietnam is estimated to have increased by an order of magnitude in less than a decade (J. Robinson and G. Cochrane, Animals Asia Foundation pers. comm.). A surplus of bile is produced by the 8000–10,000 bears currently kept on Chinese bear farms, spurring efforts to find markets in non-traditional uses of bile (e.g., lotions, shampoos, cosmetics); meanwhile, many practitioners of TCM/TKM believe that bile from wild bears is more effective at healing various ailments, and are thus willing to pay higher prices for this product and may be disinclined to use substitutes (Chang et al. 1995, Kang and Phipps 2003). The market for bear paws also appears to be increasing commensurate with an increasing number of wealthy people who find it within their means to indulge in this very expensive delicacy. The demand for these bear products has fuelled a growing network of international trade throughout Southeast Asia, and has turned many subsistence hunters into commercial hunters. Most commercial trade routes eventually terminate in China (Saw Htun 2006; C. Shepherd, TRAFFIC SE Asia pers. comm.). However, it is difficult to assess the true extent of this trade because only a small fraction of the parts are confiscated. Moreover, with no reliable population estimates or monitoring system it is not possible to evaluate the actual impacts on populations. Nevertheless, it seems highly probable that this commercially-driven trade in parts is unsustainable and therefore causing populations to decline. The capture of live bears presents yet another threat to this species. In several Southeast Asian countries Asiatic black bears are routinely confiscated from people attempting to raise them as pets. In Pakistan, several thousand bears were taken from the wild for exhibitions (referred to as bear baiting) in which individual bears (with canines and claws removed) fight with dogs. This practice was made illegal in 2001, but continues to some extent.

Conservation Actions:

The most beneficial conservation measure for Asiatic black bears would be to substantially lessen the demand for bear products, and thus reduce hunting and trade. The species is protected under both international and national laws, but often these laws are not enforced. It has been included on CITES Appendix I since 1979. The so-called Baluchistan bear, a subspecies *(U. thibetanus gedrosianus)* living in the arid thorn forest in the Baluchistan region of southern Pakistan and Iran, was listed as Critically Endangered (B1+2abc, C2a) in the 1996 IUCN Red List, and is nationally listed as critically endangered in Pakistan. Authorities have proposed a protected area to assist in the recovery of this very small, isolated population (Sheikh 2006). In most range countries Asiatic black bears are listed as a protected species. For example, they are protected under Class 2 of China's Wildlife Protection Law (a limited number of permits are issued to kill nuisance animals), and under Schedule I of the Indian Wildlife (Protection) Act. In South Korea they are designated as a national monument (No. 329) within the Cultural Properties Protection Law and also as an Endangered Wild Animal. In Japan, this species is listed under the Law for Conservation of Endangered Species of Wild Fauna and Flora, which for trade requires certification of legal take; however, gall bladders and paws are exempted. Throughout Southeast Asia this species is totally protected in every range country, with the exception of Myanmar, where this species is classified as "normally protected", meaning that it may be killed with a special license (although such licenses are rarely issued; Saw Htun, Wildlife Conservation Society, Myanmar pers. comm.). In Afghanistan, U. thibetanus is listed as a protected species, imposing a Government ban on all hunting and trading of this species within the country. Sport hunting of Asiatic black bears is legal only in Japan and Russia. Russia reports a legal harvest of 75-100 bears/year and an estimated illegal take of about 500 bears/year. Sport harvests of black bears in Japan average about 500/year and have been slowly declining since the late 1980s due to diminishing interest in hunting (Oi and Yamazaki 2006). However, a high number (generally 1,000–2,000, but as many as 4,000) of nuisance black bears are killed annually (using guns, traps, and snares) in towns or agricultural areas of Japan. Farming bears for bile presents another conservation difficulty that needs to be resolved. In Vietnam, bears are still being removed from the wild to supply farms. In China, whereas the farms themselves may not require restocking from the wild, the excessive bile produced may fuel the market, and thus may actually increase demand for bile from wild bears. In South Korea, where wild Asiatic black bears have been nearly extirpated, 2000 bears are kept and propagated in captivity and it is believed that bile and other parts from this captive population supply an illicit market. Efforts are underway in South Korea to restore the wild bear population through restocking, initially with captive-born bears, but more recently with orphaned wild bears from Russia. Some Southeast Asian countries, like Cambodia and **In Thailand are also considering reintroducing bears from captivity.** Throughout much of the southern portion of the range of this species, efforts to reduce habitat degradation outside PAs and to increase the number and/or area of PAs would be highly beneficial. An increasing number of PAs are being established in China, India, and a few other countries within the range of Asiatic black bears (Chape et al. 2003), mainly to protect other species, but serving as well to increase protection for bears. Additionally, the recently amended (2003) Indian Wild Life (Protection) Act provides options for new categories of PAs that could be established to form travel corridors between existing PAs.

Sun Bear *(Helarctos malayanus)*

Description:

HB: 1000-1400; T: 30-70; HF: 180-210; E: 40-60; W: 25-65 kg.. In the wild anything above 50 kg is considered heavy.

This bear is the smallest bear in the world. The hair is sort and black all over the body except for the muzzle and the U shaped. marking on the upper breast. The white muzzle is more conspicuous than that of the black bear and in most cases the white area extends far above the eyes. The ears are small and round. The forehead and eras behind the ears have whorls of hair which is lacking in *Selenarctos*. The skull is round and wide, with the mastoid width greater than the lenght of the palate; the skull becomes even more wider are the animal get older. The jaws are shorter than in *S. thibetanus* and the bullae are more swollen; the muzzle is more expanded over the canines. Along with the more protractible upper lip of the Sun Bear, the incisors and canines are better developed, indicating that the front of the mouth is of greater importance than it is in the Asiatic Black bear. However the molars are relatively smaller and the cheek teeth are more tightly packed, often lacking some of the smaller premolars.

Range Description:

Sun bears occur in mainland Southeast Asia as far west as Bangladesh and northeastern India (Chauhan 2006), as far north as southern Yunnan Province in China, and south and east to Sumatra and Borneo, respectively. It now occurs very patchily through much of its former range, and has been extirpated from many areas, especially in mainland southeast Asia. Its current distribution in eastern Myanmar and most of Yunnan is unknown. Reports of sun bears formerly occupying Nepal appear to be erroneous. Sun bear fossils from the Pleistocene have been found much further north into China and on the island of Java (Erdbrink 1953), but sun bears did not occur there in historical times.

Sun bears are uncommon at the northern and western edges of their range (southern Yunnan province, Southeastern Tibet, northeast India, and Bangladesh; (Chauhan 2006, Gong and Harris 2006); this lower abundance was apparent in historical times (e.g., in India; Higgins 1932) so is probably a natural gradient unrelated to human exploitation.

Countries: Native:

Bangladesh; Brunei Darussalam; Cambodia; China; India; Indonesia; Lao People's Democratic Republic; Malaysia; Myanmar; **Thailand;** Vietnam

Regionally extinct:

Singapore

Population:

Reliable estimates of sun bear populations are lacking. However, rapid loss of forests throughout their range and an active trade in wild bears and their parts is strong evidence of a declining trend. Attempts to extrapolate population size (e.g., Meijaard 2001) from anecdotal information on bear density (derived from occasional bear sightings and sign surveys, e.g. Davies and Payne 1981) have led to unreliable estimates (Garshelis 2002). Augeri (2005) used occupancy modelling (based on camera trapping) to estimate proportions of Indonesian Protected Areas that were occupied by sun bears. The proportion of an area occupied (occupancy) is likely correlated with population size, so finer-scale population trends might be gleaned from changes in occupancy.

Population Trend: Decreasing

Habitat and Ecology:

Sun bears rely on tropical forest habitat. Two ecologically distinct categories of tropical forest occur within its range, distinguished by differences in climate, phenology, and floristic composition. Tropical evergreen rainforest is the sun bear's main habitat in Borneo, Sumatra, and Peninsular Malaysia. This seasonal habitat receives high annual rainfall that is relatively evenly distributed throughout the year. Tropical evergreen rainforest, includes a wide diversity of forest types used by sun bears, including lowland dipterocarp, peat swamp, freshwater swamp, limestone/karst hills, hill dipterocarp, and lower montane forest. In contrast, sun bears in mainland Southeast Asia inhabit seasonal ecosystems with a long dry season (3–7 months), during which rainfall is 1,000 m). The range of sun bears overlaps that of Asiatic black bears (Ursus thibetanus) in this seasonal forest mosaic. Sun bears also have been reported in mangrove forest, although their occurrence in this forest type probably depends on proximity to other, more favoured habitats. Sun bears use selectively logged areas (Wong et al. 2004, Meijaard et al. 2005), and oil palm plantations near forest edges (Nomura et al. 2004). However, there is no evidence that sun bears can survive in deforested or agricultural areas in the absence of nearby forest (Augeri 2005). Sun bears occur from near sea level to over 2,100 m elevation, but appear to be most common in lower elevation forests. In Indonesia and western **Thailand**, for example, sun bears occur primarily below 1,200 m (Augeri 2005, Vinitpornsawan et al. 2006). Sun bears have been observed up to 2,100 m in Myanmar (Saw Htun 2006), 1,600 m in Lao PDR (Steinmetz et al. 1999), and 2,143 m in Sumatra (Augeri 2005). Sun bears are omnivores, feeding primarily on termites, ants, beetle larvae, bee larvae and honey, and a large variety of fruit species, especially figs (Ficus spp.), when available (McConkey and Galetti 1999, Wong et al. 2002, Augeri 2005, Fredriksson et al. 2006). Occasionally, growth shoots of certain palms and some species of flowers are consumed (Fredriksson et al. 2006), but otherwise vegetative matter rarely occurs in the diet. In Bornean forests, fruits of the families Moraceae, Burseraceae and Myrtaceae make up more than 50% of the fruit diet (Fredriksson et al. 2006), whereas in Western Thailand fruits of Lauraceae and Fagaceae are the most commonly consumed (Vinitpornsawan et al. 2006). In Thailand sun bears and Asiatic black bears use many of the same habitats and have extensive overlap in diet. However, in montane forests >1,200 m elevation (where ground cover is sparse) Asiatic black bears are more abundant than sun bears (Vinitpornsawan et al. 2006). Little is known about social structure or reproduction in sun bears. Except for females with their offspring, sun bears are usually solitary. They may congregate to feed from large fruiting trees, but this behavior appears to be rare. Sun bears do not seem to have a defined breeding season anywhere in their range and usually give birth to only one cub (less commonly two; Schwarzenberger et al. 2004). Female bears use cavities of either standing or fallen large hollow trees as birthing sites. As sun bears occur in tropical regions with year-round available foods, they do not hibernate.

Systems: Terrestrial

Major Threat(s):

The two major threats to sun bears are habitat loss and commercial hunting. These threats are not evenly distributed throughout the range of the species. In areas where deforestation is actively occurring, sun bears are mainly threatened by the loss of forest habitat and forest degradation arising from: clear-cutting for plantation development, unsustainable logging practices (Augeri 2005, Meijaard et al. 2005, Tumbelaka and Fredriksson 2006, Wong 2006), illegal logging both within and outside protected areas (Fuller et al. 2004), and forest fires (Fredriksson et al. 2007). These threats are prevalent in Indonesia and Malaysia on the islands of Sumatra and Borneo (Sundaland), where large-scale conversion of forest to oil palm *(Elaeis guineenis)* or other cash crops is proceeding at the rate of 1,000s of km² per year (Holmes 2002). Human-caused fires in parts of Sundaland are also diminishing habitat quality for sun bears. These fires are more extensive during El Niño-related droughts. On Borneo, periods of prolonged drought have disrupted fruiting patterns (e.g., Harrison 2000), which in combination with reduced habitat availability due to logging and fires, resulted in starvation among sun bears, even in primary forest areas (Wong et al. 2005, Fredriksson et al. 2006b). Commercial poaching of bears for the wildlife trade is a considerable threat in most countries (Meijaard 1999, Nea and Nong 2006, Nguyen Xuan Dang 2006, Saw Htun 2006, Tumbelaka and Fredriksson 2006, Wong 2006), and is the main threat where deforestation is currently negligible (for example in **Thailand** where nearly all remaining forest is within protected areas; (Vinitporsawan et al. 2006). Killing bears is illegal in all range countries but is largely uncontrolled. In **Thailand,** local hunters in one area estimated that commercial poaching reduced the abundance of sun bears by 50% in 20 years (Steinmetz et al. 2006). In Myanmar, **Thailand**, Lao PDR, Cambodia and Vietnam, sun bears are commonly poached for their gall bladders (i.e., bile) and bear-paws; the former is used as a Traditional Chinese Medicine, and the latter as an expensive delicacy. In China and Vietnam, bile is milked from commercially-farmed bears; however, as there are few sun bears in China, farms there contain mainly Asiatic black bears. Conversely, both sun bears and Asiatic black bears are farmed in Vietnam, in small private enterprises. Bears are routinely removed from the wild to stock or restock these small farms (Nguyen Xuan Dang 2006, B. Long, MOSAIC and WWF-Vietnam pers. comm.). Other motivations for killing bears include: preventing damage to crops (Fredriksson 2005), subsistence use, fear of bears near villages, and capture of cubs for pets (the mother being killed in the process). Although few sun bears exist in India, villagers there still kill sizeable numbers (Chauhan and Singh 2006). Despite significant poaching within extant forest areas, sun bear populations appear to persist longer than some other heavily-exploited large carnivores. For example, tiger *(Panthera tigris)* populations have been severely reduced or extirpated in 12 of 15 protected areas surveyed in Myanmar, whereas sun bears were still encountered relatively frequently in 13 of these areas (Lynam 2003, Saw Htun 2006). Similarly, in Thailand tigers are close to extirpation in the Khao Yai forest complex, but sun bears and their signs are still consistently encountered there (Lynam et al. 2006, Vinitpornsawan et al. 2006).

Conservation Actions:

Killing of sun bears is strictly prohibited under national wildlife protection laws throughout their range. However, little enforcement of these laws occurs. The sun bear has been listed on CITES Appendix I since 1979. Conservation measures and priorities vary by country. None of the range countries have established specific conservation measures for sun bears, and some taking is permitted (Servheen 1999). General measures to reduce forest loss and poaching would help conserve the species. The most beneficial conservation measure in Indonesia and Malaysia would be protection of remaining forests from conversion to other land-uses, eliminating unsustainable logging practices, and prevention of forest fires. Establishment of new and effectively managed protected areas in Indonesia and Malaysia should be promoted in order to pre-empt land conversion (Augeri 2005, Tumbelaka and Fredriksson 2006, Wong 2006). Reducing the trade in bear parts would be highly beneficial for the survival of the species in mainland Southeast Asia. However, given available resources, the patrolling and monitoring of entire protected areas is currently an overwhelming task. To make this problem more manageable, a network of small bear recovery zones (100–200 km²) could be established within key protected areas. The patrolling efforts of rangers could then be focused on these zones. Recovery zones should be locations with plentiful bear foods such as trees from the families Lauraceae, Moraceae, Burseraceae, Myrtaceae and Fagaceae. Such zones would provide a biologically meaningful, geographically focused, and logistically realistic way for the efforts of protected area staff to be translated into population recovery for bears (and other wildlife species). Recently, the Bear Specialist Group mapped the current, range-wide distribution of sun bears. Important habitat blocks for long-term survival of sun bears were identified (Bear Conservation Units-BCUs). Anti-poaching efforts within these BCUs should be a high priority. Trends in bear occurrence and relative abundance within BCUs could be monitored using standardized sign surveys and camera trapping. Results of such monitoring could indicate which management or ecological conditions promote successful bear conservation, and which do not, and provide a means to assess the results of conservation efforts (e.g., future range expansion and/or increased bear density being indicative of effective conservation efforts). Additional field studies also would be helpful in this regard; few intensive studies have been conducted on this species.

Yellow-bellied Weasel (*Mustela kathiah)*

Description:

The yellow-bellied weasel is named for its yellow-coloured underbelly. The top of its body and the tail are dark brown. Yellow-Bellied Weasels have a body length of 9.8-10.6 inches (25–27 cm.) and a tail length of 4.9-5.9 inches (12.5–15 cm.). The tail is about half the length of the body. Yellow-Bellied Weasels weigh approximately 3.3 pounds (1.5 kg.).

Range Description:

This species is found in "Myanmar, Southern and Eastern China, the Indochinese peninsula, and Nepal" (Wozencraft, 2005) and in addition Northern **Thailand,** India and Bhutan (Pocock 1941, Duckworth and Robichaud 2005, P. Yonzon pers. comm.). Corbet and Hill (1992) report a lower altitudinal limit of 1,000 m. In the Himalayas, this species is found between the altitude of 1,800 and 4,000 m (http://www.wii.gov.in/envis/envisdec99/yellowweasel.htm). In Bhutan to 3,800 m (Yonzon pers. comm.). In India found as low as 1,000 m (Choudhury pers. comm.). In Hong Kong, it is found from close to sea-level to over 200 m (Lau pers. comm. 2006). In Western Himalaya from 3,000-5,200 m in the cold deserts (Muddapa pers. comm.). This species is little known in Lao PDR, where one recent sighting was in Fokienia forest at about 1,500 m in Nam Xam National Biodiversity Conservation Area (within 20°02' to 14' N and 104° 18' to 53' E) in January 1998 (Showler et al. 1998), though historical records came from Xiangkhouang town (19° 20' N and 103° 22' E) in 1926, Phongsaly town in 1929 (Osgood 1932, Delacour 1940), and the Bolaven Plateau (within 14° 42' to 15° 30' N and 106° 15' to 50' E) at about 1,200 to 1,400 m in 1932 (Tizard 2002, Legendre 1932). In 2004 and 2005 it was recorded in Phongsaly, the northernmost province in Lao, on Phou Fa (=Fa mountain, 21° 41' N and 102° 06' E, 1,550 m) and close to Ban Bakong (=Bakong village, 21° 37' N 102° 05' E, 1,200 m) (Duckworth and Robichaud 2005). Historically, most of the records of this species in Southeast Asia come from the northern highlands of Vietnam, though there are two recent records to the south, in the Annamite mountains of central Viet Nam (Duckworth and Robichaud, 2005), one was recorded in primary evergreen forest at 1,000 m in A Vuong proposed Nature Reserve, Tay Giang district, Quang Nam Province (16° 00' N and 107° 30' E) in 2003 (Long et al, 2004) and another in the Ngoc Linh highlands at about 15 degree N latitude in the late 1990s (J. C. Eames in litt. 1999). It may yet be found to extend even farther south in Indochina (Duckworth and Robichaud, 2005). In Thailand, there are no historical records, though it apparently occurs in the northern highlands (Duckworth and Robichaud, 2005), as Vikorn (2001) reported a road kill specimen of *Mustela nivalis* from Doi Pha Luang station, which Duckworth later determined was *Mustela katiah.* (and not *M. nivalis* as it was previously identified). There are multiple skins of this species from Myanmar, all collected in the north (Pocock, 1941: 359, Than Zaw et al. in press.).

Countries: Native:

Bhutan; China; India; Lao People's Democratic Republic; Myanmar; Nepal; **Thailand;** Vietnam

Population:

It is possible that the low number of records from Lao PDR may reflect generally low encounter rates of tropical weasels (Duckworth and Robichaud, 2005). There is no reason to assume that this species is rare, and the low number of species in collection may be a relict of collection and trapping techniques (Duckworth per comm.). Recent camera trap records suggest that this species is fairly common in forested areas in southern China (Lau pers. comm. 2006), where few other small carnivores survive in meaningful numbers.
Population Trend: Unknown

Habitat and Ecology:

It is generally associated with large extents of high elevation (1,000 m+) terrain. It may occur well below 1,000 m in such areas, but unlikely that the species occurs at lower elevations in areas away from high altitude terrain (except in parts of China and Vietnam). In Lao PDR, this little known species was sighted in Fokienia forest at about 1,500 m (Showler et al. 1998). In Lao PDR, this species was recorded from "relict and degraded montane evergreen forest, linked through extensive scrub and grassland to various other small and degraded forest patches (Duckworth and Robichaud, 2005)." It seems likely that this species is tolerant of quite severe habitat degradation, and it appears to persist in the face of heavy hunting (Duckworth and Robichaud, 2005), supporting findings in heavily disturbed areas of southern China (M. W. N. Lau pers. comm. 2006). It was recorded in primary evergreen forest in Vietnam in 2003 (Long et al. 2004). It probably occurs at lower elevations in the northeast the country (Timmins pers. comm.) and there is one record from the Ke Go lowlands (Roberton et al. in prep). In India, it is found from 1,800 to 2,200 m in temperate forests and cold deserts (Choudhury pers com).
Systems: Terrestrial

Major Threat(s):

As with most other weasels, there are no major threats known or suspected to this species. In particular, the effects of the current high levels of forest degradation within much of its range cannot be assumed to be strongly negative, given the number of records from degraded and fragmented areas.

Conservation Actions:

This species was sighted in Nam Xam National Biodiversity Conservation Area in 1998 (Showler et al. 1998). It is listed in Schedule II part II of the Indian Wildlife (Protection) Act, 1972 and Appendix III of CITES (http://www.wii.gov.in/envis/envisdec99/yellowweasel.htm). More information is needed on this species in Lao PDR before a conservation status can be certainly determined (Duckworth and Robichaud, 2005). In Vietnam, this species is protected in group 2b, because it is an enemy of rats (GMA Small Carnivore Workshop 2006). This species is listed as Near Threatened on the China Red List (Wang and Xie, 2004).

Malayan Weasel *(Mustela nudipes)*

Description:

HB: 300-360;

T:240-260; HF: 49-55; E: 23-24; W: 1-3 kg.

M. nudipes is a typical weasel, with variable body colouration from pale greyish white to reddish brown and a head which is much lighter than the body. The tails of the young have white tips. The female has 4 mammae. The soles of the feet are naked around the pads. As in *M. sibirica* the post orbital area is long and parallel sided, but the occipital crests form a curved line. The sagittal crest is small. As in *strigidorsa* the baculum is bifid at the tip, but the tip is hammer shaped rather than sharply curved

Range Description:

This species is found in **Southern Thailand,** Malaysia, Indonesia (Sumatra, Borneo) and Brunei (Duckworth et al. 2006). Confusion about the existence of this species on Java is due to an error in the original description, which is occasionally still repeated today, e.g. by Wilson and Reeder (2005), where the heliotype was said to come from Java (Duckworth et al. 2006). An individual from southern **Thailand** was collected at Khao Chong, Trang Province (Lekagul and McNeely 1977), also recently recorded records (about 5) extend the known range farther north (Specimen in Thailand Institute of Science and Technical Research), up to 10 degrees North Latitude (Duckworth et al. 2006). This species was recorded from Kerinci Seblat National Park in Sumatra in 1994, at altitudes of 400 m and 800 m (Franklin and Wells, 2005). Attitudinally wide-ranging, with records up to 1,700 m (Payne et al. 1985, Duckworth et al. 2006), with many records up to 1,400 m (Duckworth et al. 2006). There are wide-spread records of this species on Borneo, across the island both historically and recently, as well as on Sumatra (Duckworth et al. 2006).

Countries: Native:

Brunei Darussalam; Indonesia (Kalimantan, Sumatera); Malaysia (Peninsular Malaysia, Sabah, Sarawak); **Thailand.**

Population:

There are no studies on this species, however, it appears to be widespread but difficult to see, at low density, and/or patchily distributed. This species has not been camera trapped regularly, as there is currently only one known record of camera trapping within the range, and the aggregate camera-trapping at known sites suggests that as currently used camera-traps are not an efficient way to find the species (Duckworth et al. 2006).

Population Trend: Decreasing

Habitat and Ecology:

Little is known about the habitat and ecology of this species, though it is probably similar to other weasels in that it is ground-dwelling and so potentially exposed to generalised snaring and other forms of trapping. However, the distribution of recent records in deforested areas, even urban sites, indicate a high tolerance to human activities (Duckworth et al. 2006). Little is known about the habitat and ecology of this species, though it is probably similar to other weasels in that it is ground-dwelling and so potentially exposed to generalised snaring and other forms of trapping. However, the distribution of recent records in deforested areas, even urban sites, indicate a high tolerance to human activities (Duckworth et al. 2006).
Systems: Terrestrial

Major Threat(s):

Currently no major threats to this species have been traced. It is eaten in parts of Sarawak and there is some evidence of medicinal use, but no evidence that these activities are major threats (Duckworth et al. 2006)..

Conservation Actions:

This species is protected in **Thailand,** Peninsular Malaysia and Sabah, but not in Sarawak nor in Indonesia (based ARCBC database). This species has been reported from many protected areas within its range (Duckworth et al. (2006).

Siberian Weasel *(Mustela sibirica)*

Description:

The pelage colour is widely variable, but darker in the summer; the underparts are only a little lighter than the upperparts. There is usually a distinctive dark mask, often confined to just the muzzle. The throat may be reddish, whitish or white with red spots. There are 4 pairs of mammae, 2 more than *M. strigidorsa*.This species has marked sexual dimorphism with the female smaller. The soles of the feet are hairy around the pads. The skull is a typical Mustela skull, long and narrow, with a well filled out braincase. The post orbital area is long and parallel sided. The forehead may be swollen due the presence of filariae in the frontal sinuses, an infection found in all weasels. The occipital crests form a straight line (curved in *M. nudipes*. The baculum is hooked with the tip not noticeably bifid. The dental formula as one less lower premolar than in *M. stigidorsa*.

Range Description:

This species is found in Northern Myanmar, Lao PDR, China, Japan, DPR Korea, Republic of Korea, Pakistan, Nepal, India (Himalayas), Bhutan, Russia (from Kirov Province, Tataria and western Ural Mountains throughout Siberia to Far East), Taiwan, and **Northern Thailand** (Pocock 1941, Duckworth 1997, Wozencraft, 2005). In Japan, it has been introduced to Honshu, Shikoku, and Kyushu Islands (Abe, 2005). It is native on the islands of Sakhalin, Kamishima, and Jeju (Abe, 2005). The distribution in southeast Asia is poorly known (A. Abramov pers. comm. 2006). The species was recorded recently in two locations in national parks in **Thailand** (Kanchanasaka pers. comm) and in one location in Lao PDR, Nakai-Nam Theun National Biodiversity Conservation Area in 1996 (Duckworth, 1997; sight record only). Gao and Sun (2005) conducted a study of the effects of this species on Liadong oak (*Quercus wutaishanica)* in Beijing Forest Ecosystem Research Station (BFERS, 40 00 N, 115 26 E), one of the research stations affiliated to the Chinese Ecosystem Research Network (CERN) of the Chinese Academy of Sciences (CAS). This species seems to prefer mountains from 1500 to 5000 m in the southern part of its range (Lekagul and McNeely, 1988) but there is one provisional sighting in Khammouan Limestone National Biodiversity Conservation Area in valley semi-evergreen forest amid karst at about 500 m in 1998 (Robinson and Webber, 1998).

Countries: Native:

Bhutan; China; India; Japan; Korea, Democratic People's Republic of; Korea, Republic of; Lao People's Democratic Republic; Mongolia; Myanmar; Nepal; Pakistan; Russian Federation; Taiwan, Province of China; **Thailand;** Vietnam.

Population:

The species is widespread and abundant in Siberia and China (Abramov pers. comm). It is also common in northern central Korea as well, where few other mammals other than rats and squirrels are currently easily seen (W. Duckworth in litt. 2006).

Population Trend: Stable

Habitat and Ecology:

This species is found in "a wide variety of habitats, including dense forest, dry areas, and human villages and towns, where it dens in any convenient shelter, including burrows of other animals (Lekagul and McNeely, 1988)." It occurs in primary and secondary deciduous, coniferous and mixed forests, as well as open areas with small patches of forest enclaves and forest steppe. It is also found along river valleys. It is found from to over 3000 m in Nepal. In Bhutan, it is found as low as 1500 m and up to 4800 m (Thinley, 2004). In China, its altitude can reach up to 5000 m. It feeds on small mammals, such as voles, squirrels, mice and pikas, amphibians, fish, and carrion. During the summer time, it feeds on pine nuts. In Lao PDR this little known species has been observed in primary evergreen forest at 1000 m (Duckworth 1997a), and possibly in valley semi-evergreen forest amid karst at about 500 m (Robinson and Webber 1998a; M. F. Robinson, 1998). It occurs commonly down to sea-level in the north of its range, e.g. Korea (J. W. Duckworth pers. comm.).

Systems: Terrestrial

Major Threat(s):

There are no major threats known to this species. It is legally hunted in Russia for its fur (A. Abramov pers. comm. 2006)."Unfortunately, the small forest carnivores are not well protected by either the local forest managers or the residents living in the forest areas in China, though most of them are listed as protected animals by the national or local governments. Weasels and badgers are largely hunted for their hides and meat. (Gao and Sun, 2005)". Unsustainable hunting for skins, for international trade. However, hunting levels are low at present, reflecting the low commercial value of skins. Competition for resources with sable *(Martes zibellina)* and natural wildfires also constitute minor threats.

Back-striped Weasel *(Mustela strigidorsa)*

Description:

HB: 275-325; T: 145-205; HF: 47-54; E: 20-23; W: 1-2 kg.

Generally brown on the back, tail and legs, with the upper lip, cheeks, chin and throat pale yellowish. The tails is rather bushy. The feet have well developed footpads, and the area around them is entirely naked, as in *M nudipes*, but different from *M. sibirica*. The female as 2 pairs of mammae. The skull is similar *M.sibirica* but as no sagittal crest and has the post orbital area constricted rather than parallel sided. The occipital crest form a straight line as in *M. sibirica*. The baculum has the tip divided into two processes, one on each side of a groove; the tip is sharply curved. The dention has 4 lower premolars, one more than *M..sibirica* and *M. nudipes*.

Range Description:

This species is found in "Myanmar, China (Yunnan, Guizhou, Guangxi), India, Lao, **Thailand**, and Vietnam" (Wilson and Reeder 2005; Zhang 1997). There are also two old specimens catalogued as from 'Nepal', but they probably did not come from within the boundaries of the modern country (Hinton and Fry 1923) and the species has not been found in Bhutan (Yonzon pers. comm.), which lies between Nepal and the species' known range. The occurrence of *M. strigidorsa* has been confirmed from scattered localities in and around northeastern India, northern and central Myanmar, southern China, northern Thailand, northern and central Lao and Vietnam (Abramov et al. 2008). This species has a wide altitudinal range of almost sea-level to 2,500 m (Abramov et al. 2008). In Lao PDR, this species was historically found only at Phongsali (Delacour 1940), however, there are recent records from all but a couple of the remaining blocks of hill semi- and evergreen forest which have had biodiversity surveys exceeding a few weeks (Duckworth et al. 1999, Tizard 2003, Abramov et al. 2008). In India, it has been recorded from Dampa in 1994, and in Namdapha Reserve (Datta 1999). **It has been recorded from Phu Khieo Wildlife Sanctuary (PKWS) and various other sites recently in Thailand** (Grassman et al. 2002, Abramov et al. 2008).

Countries: Native:

China; India; Lao People's Democratic Republic; Myanmar; **Thailand;** Vietnam

Population:

In view of limited survey effort, the number of recent records indicates that the species is less rare than previously believed (Grassman et al. 2002, Abramov et al. 2008). Globally, there are several dozen historical specimens, their number significantly under-estimated in various 1960s-1990s texts thereby giving a misleading impression of rarity.

Population Trend: Unknown

Habitat and Ecology:

The stripe-backed weasel probably lives mainly in evergreen forests in hills and mountains, but has been recorded from other biotopes including dense scrub, secondary forest, grassland and farmland (Abramov et al., 2008). Most records come from in or near larger extents of high elevation (1,000 m+) terrain. It sometimes occurs well below 1,000 m in such areas (down almost to sea-level in Vietnam), but there are no records, except in Vietnam, from at lower elevations in areas away from high altitude terrain (Abramov et al. 2008).
Systems: Terrestrial

Major Threat(s):

Although usually perceived as a species under threat, there is no real evidence that it is. However, the habitat requirements of tropical Mustela populations remain effectively unknown, and it may be rash to extrapolate from primarily Holarctic information. About 3,000 to 4,000 pelts were harvested annually in China in the 1970s, with 50 skins were purchased in Nanning, Guangxi, in 1973 (Sheng Helin, 1998). The skins/dried corpses of M. strigidorsa were seen "2-5 times" (the middle class of frequency) in a survey of wildlife trade along the Yunnan-Vietnam border in June-August 1997; the source was said to be Yunnan (Li and Wang, 1999). Outside China, the species is sold at least occasionally in Lao PDR (Hansel and Tizard, 2006) and Vietnam. Even though this weasel is not known to have high economic value, hunting or harvesting for trade could still drive declines because many harvest methods (notably snares) are non-selective. As remaining natural habitats are yet further encroached, the proportion of the species' occupied area in which it faces such threats will increase. However, the number of records from areas with already high hunting pressure indicate a resilience to such activities (Abramov et al. 2008).

Conservation Actions:

This species is listed as Endangered on the China Red List (Wang and Xie, 2004). It is protected in **Thailand.** It is not a targeted species, as it is small and not sought for use for food or medicine, though it is utilized if unintentionally captured (Abramov et al. 2008). It has been recorded in scattered protected areas across its range and likely occurs in many more unsampled parks and reserves (Abramov et al. 2008). More field research would help determine the status, distribution, and conservation of this species (Grassman et al, 2002).

Yellow-throated Marten *(Martes flavigula)*

Description:

HB: 450-600; T: 380-430; HF: 90-110; E: 28-40; W: 2-3 kg.

With a long weasel like body and the longest tail of the Martens, which is an indicator of its arboreal habits. The upper half of the head and neck, the legs and the tail are black; the rest of the body is light brown except for the bright yellow patch from the chin to the throat and chest. The ears are rounded and whitish at the rims. The female has 4 mammae. There is a clinical variation in body size of both males and females, with the largest ones found in the northernmost part of the rage and the smallest found south of the equator. The skull is large and broad with a short facial region. The braincase is expanded, with only a slight post orbital constriction. The zygomatic arch is slender and flat. The sagittal crest varies with age and sex, being most strongly developed in old males. There is a sexual dimorphism in the dentition, with males having considerably larger canines. The upper molar is distinctive, with the upper lobe not expanding nor the waste constricted, and the outer lobe emarginate rather than convex. There is one more premolar in each half of each jaw than in *Mustela*

Range Description:

The yellow-throated marten has an Asian and Sundaic distribution, and countries where this species is found include China, India, Indonesia (Islands of Sumatra, Java, and Borneo), DPR Korea, Republic of Korea, Pakistan, Russia, Taiwan, Vietnam (Wozencraft 2005; Le Xuan Canh et al. 1997; Roberton et al. in prep), Lao PDR (Duckworth 1997), **Thailand** (Grassman et al. 2005), Myanmar (Than Zaw et al. in press), Malaysia (Azlan 2003), Cambodia (J. L. Walston pers. comm.), and possibly Singapore (Meiri 2005). The species elevational range extends from sea-level to 3,000 m (Lekagul and McNeely 1977, Duckworth 1995, Than Zaw et al. in press).

Countries: Native:

Bangladesh; Bhutan; Brunei Darussalam; Cambodia; China; India; Indonesia (Jawa, Kalimantan, Sumatera); Korea, Democratic People's Republic of; Korea, Republic of; Lao People's Democratic Republic; Malaysia (Peninsular Malaysia, Sabah, Sarawak); Myanmar; Nepal; Pakistan; Russian Federation; Taiwan, Province of China; **Thailand;** Vietnam

Population [top] Population: Few population assessments of the yellow-throated marten exist. Grassman et al. (2005) recorded 40 individuals in Phu Khieo Wildlife Sanctuary in Thailand between 1998 and 2002. In Sikhote-Alinsky Nature Reserve (Russian Far East) the population density was estimated to be 1-5 per 100 square kilometres (Matyushkin 1993). The total amount in Russia is estimated as 2500-3500 specimens (Alexei Abramov pers. comm. 2006). It is evidently common across Lao PDR (Duckworth 1997) and Myanmar (Than Zaw et al. in press) and probably widely in at least South-east Asia (Parr and Duckworth 2007).

Population Trend: Stable

Habitat and Ecology:

In the Russian Far East the yellow-throated marten prefers mixed (spruce and broad-leaved) forests of the Manchurian type, while it occurs rarely in the dark coniferous taiga of the upper mountain zone and in the oak forests zone (Matyushkin, 1993). In Lao PDR, Myanmar and **Thailand** this species is found in forests and various other adjacent habitats across a wide altitudinal range (Duckworth et al.1999, Lekagul and McNeely 1977, Than Zaw et al. in press), but it clearly favours forests. It was recorded in secondary forest, that was logged in the 1970s, and which surrounds a palm estate, in Malaysia in 2000-01 by Azlan (2003) and there are many records from other areas of secondary forest, even areas well isolated from old-growth stands. Although sometimes said to be largely or entirely nocturnal, the species is primarily diurnal, but also hunts at night increasing nocturnal activity during lunar nights (plus or minus 7 days from full moon) (Duckworth 1997, Grassman et al. 2005, Than Zaw et al. in press, Parr and Duckworth 2007, J. L. Walston pers. comm. (for Cambodia)). Common food items include squirrels, birds, snakes, and lizards, though insects, eggs, frogs, fruit, nectar, and berries are also taken, as well as honey and bees (Lekagul and McNeely 1977) and in fact it probably has a very wide diet (Parr and Duckworth 2007). In nature, groups of two to three or more rarely, five to seven individuals can be seen; in the Russian Far East the species hunts in groups for musk deer (Matyushkin 1993). It is also usually found in small groups, rather than as single Individuals, at least in tropical parts of its range (Parr and Duckworth, 2007). Grassman et al. (2005) found that this species has a mean annual range size of 7.2 km^2 with a mean overlap of 34% in a study on this species conducted in Phu Kieo Wildlife Sanctuary in **Thailand.** The litter size is up to five, and the gestation period is 220-290 days, and it has life span of up to 14 years (Lekagul and McNeely 1977).

Systems: Terrestrial

Major Threat(s):

As the yellow-throated marten is tied to forest areas, at least in the Southeast Asian parts of its range, forest conversion there over the last few decades will have resulted in some overall population reduction. Nevertheless, the species is surviving well within remaining forests (including secondary stands), perhaps because it is less preferred as food by most residents and its scansorial nature reduces its exposure to snares and other traps, as well as allows easy escape from dogs. Therefore, no significant threats at the population level are known to the species in Southeast Asia, although it is no doubt below carrying capacity in heavily hunted areas such as Lao PDR. It is occasionally hunted in Siberia (Russia) and DPR Korea for its fur (A. Abramov pers. comm. 2006, J. W. Duckworth pers. comm. 2006) but this does not constitute a global threat, rather it affects local populations at most. It can habituate to close approach of many people and take food from human waste (Parr and Duckworth 2007).

Conservation Actions:

The yellow-throated marten is protected in many parts of its range. In Myanmar, this species is protected all year under the Wildlife Act of 1994 (Su Su 2005) and in Peninsular Malaysia it is protected under the Wildlife Protection Act of 1972 (WPA 1972; Azlan, 2003). This species is listed on CITES Appendix III (India) and Category II of the China Wildlife Protection Law (1988) (Li et al. 2000). This species is listed as Near Threatened on the China Red List (Wang and Xie, 2004). This species is known from many protected areas across its range. Grassman et al. (2005) studied it in Phu Khieo Wildlife Sanctuary in north central **Thailand.** It was recorded by Azlan (2003) in Jerangau Forest Reserve in Peninsular Malaysia in 2000-01.

Hog Badger *(Arctonyx collaris)*

Description:

The world's largest extant badger.The hog badger (*Arctonyx collaris*) is aptly named for its unusual pig-like snout. The fur on the body of the hog badger is generally dark brown in colour but the exact shade varies between individuals and may sometimes have a yellow or grey tint. In contrast, the head is yellow-white, as is the thick tail, and two distinctive dark stripes run from the snout, past the eyes, to the back of the head. The throat, ears and tail are white, the belly is black, and the short, black legs bear long, white claws.

The distinctive black and white stripes on the face of the hog badger may be interpreted as aposematic, meaning they may act to warn potential predators of the hog badger's ability to release noxious odours from its anal scent glands, or its ferocity when threatened . The hog badger is predated by the dhole (*Cuon alpinus*), tiger (*Panthera tigris*) and leopard (*Panthera pardus*). It is likely that the secretions from the anal scent glands of hog badgers are also used to mark territory .

Range Description:

The hog badger occurs in Central to Southeast Asia. It is found in Mongolia, India (Sikkim, Terai, Assam, Arunacha Pradesh), throughout southern China, Indochina (Vietnam, Lao PDR and Cambodia), Myanmar, in Indonesia (Sumatra), throughout **Thailand** and possibly in Perak, Malaysia (Lekagul and McNeely 1977; Duckworth 1997; Pocock 1941; Holden 2006; Roberton et al. in prep.; Than Zaw et al. in press). There is one isolated record in eastern Mongolia (Aimak Dornod) (Stubbe et al. 1998). According to Holden (2006) in Sumatra the hog badger appears to occur primarily above 2,000 m with one record at 700 m, and historical records also indicate a montane range (Miller 1942). Corbet and Hill's (1992) map suggest that on Sumatra the species is restricted to the southern part of the island, whereas, in fact, individuals have been found in many mountainous locations in the north as well (van Strien 2001). In Lao PDR, most recent records are from the central part of the country, with some from the north, although historic records come also from the south (Duckworth 1997, Duckworth et al. 1999). There are recent indirect reports (unsubstantiated villager reports) from many survey areas in Lao PDR, but few documented records (R.J. Timmins and J.W. Duckworth pers. comm. 2006); in aggregate, these suggest that this species was present in the recent past, but has more or less been hunted out from quite wide areas (J.W. Duckworth in litt. 2006). Deuve (1972) considered that the species occurred in the southern region of Lao PDR, listing several lowland sites; however, Deuve's Lao range information is often faulty (e.g. Timmins and Duckworth 1999), so this cannot be taken as complete confirmation the animal was formerly widespread in Lao lowlands. All seven records in 1992-1996 were from in and around the Nam Theun catchment at sites above 500 m (Duckworth 1997), while both historical sites listed by Delacour (1940) are in mountainous areas: Phongsali and the Bolaven Plateau. The post-1996 records are also from hills and mountains (Duckworth et al. 1999).

Countries: Native:

Bhutan; Cambodia; China; India (Assam); Indonesia (Sumatera); Lao People's Democratic Republic; Mongolia; Myanmar**; Thailand;** Vietnam

Population:

Population trends for the hog badger may vary across its range. In Lao PDR, this species can be locally common, as indicated by its presence during most surveys in and around the Nam Theun catchment (Duckworth et al. 1999). The lack of sightings elsewhere indicates that this species is either naturally patchy in abundance or under widespread decline (Duckworth et al. 1999). Occurrence in Myanmar is also patchy without obvious natural explanation (Than Zaw et al. in press). In Thailand, it is fairly common and found in both the north and the south (B. Kanchanaska pers. comm.). It is also very common in the high montane zone of Sumatra (Holden 2006), and in southwestern and eastern Cambodia (J. L. Walston and R. J. Timmins pers. comm.). In India, this species is fairly common in Terai. The hog badger is historically widespread in Viet Nam, but sightings seem to be declining (Roberton et al. in prep.).

Population Trend: Decreasing

Habitat and Ecology:

The hog badger is active by day, terrestrial, and not very wary of humans (Duckworth et al. 1999). This species if often referred to as nocturnal, however, analysis of numerous camera-trap pictures from Myanmar show no peak at either day or night; it can be active at any time (Than Zaw et al. in press). It is usually found in forested areas as high as 3,500 m, and it feeds on "tubers, roots, earthworms, insects, and other small living creatures" (Lekagul and McNeely 1977). Wang and Fuller (2003) conducted a study on the food habits of this species in a rural agricultural area of Southeastern China (Taohong Village, Jiangxi Province), and found that this species ate more mammals and gastropods than other species studied. Little is known about its breeding habits, though litter size seems to be two to three young, and individuals have lived up to seven years in captivity (Lekagul and McNeely 1977). In Lao PDR, the hog badger is found in forested areas, and mainly now on hills and mountains (Duckworth et al. 1999), however, this altitudinal restriction may be a secondary effect of over hunting. In contrast to Lao PDR, this species in Cambodia occurs in level lowlands, in mosaics of deciduous and semi-evergreen forests, which is a further line of supposition that its current Lao distribution reflects anthropogenic restriction. In India, this species is fairly common within grassland habitats of Terai, as well as in dense, tropical evergreen and semi-evergreen forests, and tall grassland -woodland mosaic. In **Thailand** it is also found in rubber plantations adjacent to forests (B. Kanchanaska pers. comm.). In Myanmar, the hog badger has been recorded in forest including bamboo stands under tree cover (Than Zaw et al. in press), and it is also found in limestone forests in Viet Nam (Roberton et al. in prep.). It mainly occurs in upper montane forest in Sumatra (Holden 2006).

Systems: Terrestrial

Major Threat(s):

Major threats to the hog badger are hunting by dogs as well as snaring, primarily for human consumption and as bycatch. In Lao, the palatability of hog badger varies among ethnic groups, with some groups disliking the taste, whereas groups in parts of the Nam Theun basin (and perhaps widely elsewhere) seek the species specifically for food (J. Baker pers. comm. and J. Chamberlain per. comm. in Duckworth et al. 1999). This species is also eaten by some groups in India, and is hunted as well as farmed for food in China (M.W.N. Lao pers. comm.). Field surveys in China generated very few records of wild animals in Southeastern China (M.W.N. Lao pers. comm.), and the species is also hunted at the local level in Vietnam (Roberton et al. in prep.). In all of Indochina, this species is threatened by the use of hunting dogs (J. Baker pers. comm. 1999). The snaring intensity in Cambodia is considerably lower than that in Lao PDR and Vietnam, and the relatively larger number of recent records from Cambodia than from Vietnam and Lao PDR is strong indirect evidence that trapping levels are driving reductions in these latter countries. In Vietnam and presumably elsewhere, gun-hunting poses another threat to the species (Timmins et al. 1999). While threats similar to those in Lao PDR and Vietnam are known to exist in **Thailand**, it is generally thought that the hunting is operating at much lower intensities and are therefore not as serious. In Sumatra as well, the threats are minimal, because the zone of occurrence is above where the majority of hunting takes place (Holden 2006)

Conservation Actions:

Throughout its range, this species is found in a number of protected areas. In **Thailand** this species is protected by law, and in India this species is protected under the highest level of protection. It is not protected in Vietnam or Cambodia and is the largest-bodied unprotected mammal, except for Eurasian Wild Hog Sus scrofa, in Myanmar (Than Zaw et al. in press). The China Red List has listed the hog badger as Vulnerable under C1 and A2c.

Burmese Ferret-badger *(Melogale personata)*

Description:

The Burmese ferret-badger has a head and body length of 35–40 centimetres (14–16 in), a tail length of 15–21 centimetres (5.9–8.3 in) and a body weight of 1.5–3 kilograms (3.3–6.6 lb). The fur ranges from fawn brown to dark brown, with a white dorsal stripe. The face is marked with black and white patches, which are unique to each individual. The rear part of the tail is whitish.It is the smallest of the badger family, only reaching a weight of 2.5 lbs (1.1 kg) to 7 lbs. (3.1 kg). Their bodies reach a length of 13 inches (33 cm) to 17 inches (43 cm). They are also called tree badgers, or pahmi. Aside from the Burmese and Chinese ferret badger there is Everett's ferret badger *Melogale everetti,* which is only found on Mount Kinabalu at the northern tip of the island of Borneo .Cubs can be born year round, but the average season is in late spring, usually in May or June until late fall, around September and October.. The average litter size is two to three babies. Ferret badgers are born in burrows and fed by the mother for two to three months.

Range Description:

This species is found in northeast India, Myanmar**, Thailand,** Indochina, and southern Yunnan (China) (Lekagul and McNeely 1977). There is an old record from western Guangdong (China). In India it is found from 30 to 1,950 m (Choudhury pers. comm.) and Datta (1999) confirmed current occurrence at about 27°N in India. Records from Nepal are historic and may differ from what is today recognized as "Nepal" (Hodgson 1836- heliotype for subspecies) - and there have been no subsequent records since (Pralad Yonzon, verbally; see also Hinton and Fry 1923). Records from northern Vietnam need to be checked in Museums and further investigated (Roberton pers. comm.), although there is a record confirmed from Yanbai (Thomas 1922); other records may not have ruled out *M. moschata* from the identification. Distributions of species in Cambodia and Lao PDR are highly speculative and based on few verifiable specimens (Duckworth et al. 1999 pers. comm.): A skull of this species was found Khammouan Limestone National Biodiversity Conservation Area in early 1998 (Robinson and Webber 1998a), and "the species was previously common on the Bolaven Plateau (Osgood 1932, Delacour 1940)." [from Duckworth et al. 1999]. The distribution in Myanmar north to 22N is well supported by validated individuals, but there seem to be no post-1950 records identified to species (Pocock 1941, Than Zaw et al. in press). This is the only species of the genus confirmed to occur in Thailand, and hence records of the genus are routinely assumed to relate to this species. In fact, *M. moschata* might also occur too (and was even mapped, apparently predicatively, by Storz and Wozencraft 1999). *Melogale moschata* and *M. personata* are very similar in external morphology, and field records, including skin specimens lacking an associated skull, need to be considered as identifiable only as Melogale sp. throughout the parts its range where *M. moschata* is known to occur or might plausibly do so (effectively, this is all the known range). Only reference to skull characteristics should be used for species-level identification. Thus, in Myanmar, **Thailand,** Lao PDR, Vietnam, Cambodia and India there is continued difficulty of identifying recent records of ferret badgers to species, most of which lack skulls or did not have the skull characters checked and were not preserved. Hence it is impossible to say anything about this species' current distribution, status or ecology in these regions.

Countries: Native:

China; India; Lao People's Democratic Republic; Myanmar; **Thailand;** Vietnam

Population:

The species seems to be patchy in occurrence and generally uncommon but in some localised parts of South-east Asia, ferret badgers (species not known) seem to be more common. The genus is uncommon in **Thailand** (Kanchanasaka pers. comm.). The species was previously common on the Bolaven Plateau (Osgood 1932, Delacour 1940), however, there are very few recent records from Lao PDR – it is unknown if this is due to elusiveness or rarity (W. Duckworth in litt. 2006).

Habitat and Ecology:

It is nocturnal and feeds primarily on small animals such as insects, earthworms, snails, frogs, and sometimes carcasses of small birds and mammals, eggs, and fruit (Chian and Sheng, 1976; Long and Killingley, 1983; Ewer, 1985; Neal, 1986; Chuang, 1994). This species sleeps during the day in its burrow, and comes out at night to feed on cockroaches, grasshoppers, and earthworms (Lekagul and McNeely 1977). It is found in forest, grassland, and even rice fields (Lekagul and McNeely 1977), but it is unclear in which habitats populations can persist; records from other habitats may involve sink populations or dispersing individuals. Not much is known about the breeding of this species, though it does have an average litter size of about three (Lekagul and McNeely 1977). This species is fossorial and lives in pre-existing holes, rather than digging new ones (Taylor, 1989).In Lao PDR, little is known about the habitat use of this species (Duckworth et al. 1999). In **Thailand,** records were found in hill evergreen forests, pine forests and grasslands (Kanchanasaka pers. comm.).

Systems: Terrestrial

Major Threat(s):

In Lao PDR, parts of all badgers are used in traditional medicine (Baird 1995b), however, there is no evidence that there is a big enough demand to cause declines of ferret badgers (W. Duckworth in litt. 2006). There are few records recently from Lao PDR of ferret badgers, but this may well simply be due to inadequate search effort. In northeastern India the genus is hunted for food (A. Choudhury pers. comm.). Because this genus does not prey on poultry or livestock, nor cause to damage to property or farm facilities, it is not threatened by humans, despite its close proximity to them (Wang and Fuller, 2003). In addition, the value of an individual pelt is not high, and the meat is eaten in some areas (Wang and Fuller, 2003).

Conservation Actions:

A skull of this species was found Khammouan Limestone National Biodiversity Conservation Area in early 1998 (Robinson and Webber 1998a). This species may be found to occur in many protected areas across its range (W. Duckworth in litt. 2006); until there are surveys known to be using appropriate methodology, it is difficult to speculate on current presence in protected areas. In India, it is protected in Schedule 2, Part 1. **In Thailand it (the Genus) was found in Doi Chieng Doi Wildlife Sanctuary and Huai Kha Khaeng Wildlife Sanctuary (Kanchanasaka pers. comm.). It has also been recorded from Phu Khieo Wildlife Sanctuary (PKWS) in Thailand (L. Grassman pers. comm.).**

Eurasian otter (*Lutra lutra)*

Description:

Brown fur, often pale on underside; long slender body; small ears; long thick tail; webbed feet
Head/body length: about 60-120cm; tail about 40-45cm
Weight: average 10.1 kg for males; 7.0 kg for females. The Eurasian otter is a typical species of the otter subfamily. Brown above and cream below, these long, slender creatures are well-equipped for their aquatic habits. Their bones show osteosclerosis, increasing their density to reduce buoyancy. This otter differs from the North American river otter by its shorter neck, broader visage, the greater space between the ears and its longer tail. However, the Eurasian otter is the only otter in much of its range, so it is rarely confused for any other animal. Normally, this species is 57 to 95 cm (22.5 to 37.5 in) long, not counting a tail of 35–45 cm (14–17.5 in). The female is shorter than the male. The otter's average body weight is 7 to 12 kg (15 to 26 lb), although occasionally a large old male may reach up to 17 kg (37 lb). The record-sized specimen, reported by a reliable source but not verified, weighed over 24 kg (53 lb).

Range and habitat.

The Eurasian otter is the most widely distributed otter species, its range including parts of Asia and Africa, including **Thailand** as well as being spread across Europe. Though currently believed to be extinct in Liechtenstein and Switzerland, they are now very common in Latvia, along the coast of Norway, and across Great Britain, especially Shetland, where 12% of the UK breeding population exists.[8] Ireland has the highest density of Eurasian otters in Europe. In Italy, they can be found in southern parts of the peninsula. The South Korean population is endangered. In general, their varied and adaptable diets mean they may inhabit any unpolluted body of fresh water, including lakes, streams, rivers, and ponds, as long as the food supply is adequate. Eurasian otters may also live along the coast, in salt water, but require regular access to fresh water to clean their fur. When living in the sea, individuals of this species are sometimes referred to as "sea otters", but they should not be confused with the true sea otter, a North American species much more strongly adapted to a marine existence.

Ecology:

The Eurasian otter's diet mainly consists of fish. Fish is their most preferred choice of food in Mediterranean and temperate freshwater habitats. During the winter and in colder environments, though, fish consumption is significantly lower, and the otters use other sources of food, including amphibians, crustaceans, insects,birds and sometimes small mammals, including young beavers. Otters can travel over large areas. Some are known to use 20 kilometres or more of river habitat. Otters deposit faeces (known as spraints, with a characteristic sweet musky odour) in prominent places around their ranges. These probably serve to mark an otter's range and help neighbouring animals keep in social contact with one another. Eurasian otters are strongly territorial, living alone for the most part. An individual's territory may vary between about one and 40 km long (about one-half to 25 mi), with about 18 km (about 11 mi) being usual. The length of the territory depends on the density of food available and the width of the water suitable for hunting (it is shorter on coasts, where the available width is much wider, and longer on narrower rivers).

The Eurasian otter uses its feces, spraints, to mark its territory and prioritize the use of resources to other group members. The territories are only held against members of the same sex, so those of males and females may overlap. Mating takes place in water. Eurasian otters are nonseasonal breeders (males and females will breed at any time of the year) and it has been found that their mating season is most likely determined simply by the otters' reproductive maturity and physiological state. Female otters are sexually mature between 18 and 24 months old and the average age of first breeding is found to be 2.5 years old. Gestation for *L. lutra* is 60–64 days, the litter weighing about 10% of the female body mass. After the gestation period, one to four pups are born, which remain dependent on the mother for about 13 months. The male plays no direct role in parental care, although the territory of a female with her pups is usually entirely within that of the male. Hunting mainly takes place at night, while the day is usually spent in the Eurasian otter's holt (den) – usually a burrow or hollow tree on the riverbank which can sometimes only be entered from under water. Though long thought to hunt using sight and touch only, evidence is emerging that they may also be able to smell underwater – possibly in a similar manner to the star-nosed mole.

Conservation:

The Eurasian otter declined across its range in the second half of the 20th century primarily due to pollution from pesticides such asorganochlorine and polychlorinated biphenyls. Other threats includedhabitat loss and hunting, both legal and illegal. Eurasian otter populations are now recovering in many parts of Europe. In the United Kingdom, the number of sites with an otter presence increased by 55% between 1994 and 2002. In August, 2011, the Environment Agency announced that otters had returned to every county in England since vanishing from every county except the West Country and parts of Northern England. Recovery is partly due to a ban on the most harmful pesticides that has been in place across Europe since 1979, partly to improvements in water quality leading to increases in prey populations, and partly to direct legal protection under the European Union Habitats Directive [23] and national legislation in several European countries. In Hong Kong, it is a protected species under Wild Animals Protection Ordinance Cap 170. It is listed as Near Threatened by the IUCN Red List. In Pakistan-administered Kash-mir, it is listed as an endangered species.

Most species that are victims of population decline or a loss of habitat tend to eventually lose their genetic difference due to inbreeding from small populations. A study conducted in 2001, examined whether or not the populations of Eurasian otters suffered from a lack of genetic variability. In the study, they examined teeth of otter skulls that had been collected by the Zoological Museum, Copenhagen and the Natural History Museum, Aarhus. The samples were collected between 1883 and 1963 on the islands of Funen and Sealand, and the Jutland Peninsula. The study examined the tissue on the teeth of the skulls and determined the genetic variability based on DNA analysis. In conclusion, the study discovered that despite the population declines, the Eurasian otter was not a victim of declining genetic variability.

Hairy-nosed Otter (*Lutra sumatrana)*

Description:

The **Hairy-Nosed Otter** (*Lutra sumatrana*) is a semiaquatic mammal native to Southeast Asia. It is one of the rarest otter species. Until 1998, it was thought to have been extinct, but small populations have been discovered since then.

- Weight: 11 to 13 lb (5.0 to 5.9 kg)
- Head-body length: 20 to 32 in (51 to 81 cm)
- Tail length: 14 to 20 in (36 to 51 cm)

The hairy-nosed otter is the least known of the Asian otters, and is also the most difficult to identify in the field. It gets its name from the hairs on the end of its rhinarium (moist part of its nose); in most other respects, it is similar to European otter, *Lutra lutra*. Hairy-nosed otters are entirely brown, except for lips, chin, and upper throat, which are whitish. Their fur is rather rough but short. Their tails are flattened and oval in cross section, and their feet are fully webbed between the digits, with prominent claws. The penis of the adult male is not visible externally. The contact call between otters is a single-syllabic chirp; adult females call to pups with a staccato chatter. Large otters are very similar and can be positively distinguished only by close inspection of the nose and fur, or the skull. In this species, the skull is flatter than that of smooth otter, *Lutrogale perspicillatal* and has smaller teeth.

Range Description:

The hairy-nosed otter *(Lutra sumatrana,)* Gray 1865 is the rarest and least known among the five species of otters occurring in Asia. It is endemic to South Asia. The type specimen came from Sumatra. Once believed to be extinct, the hairy nosed otter has been rediscovered from many parts of Southeast Asia such as **Thailand,** Vietnam and Cambodia. Historically, it occurred from southern Indochina to Malaysia (Sebastain 1994, 1995) and **Thailand** (Lekagul and Mcneely 1988, Kanchanasaka 2000). **In Thailand it has been reported from Phru Toa Daeng Peat Swamp Forest (Kanchanasaka et al. 1998).** From Vietnam it has also been reported from U Minh Thuong Nature Reserve in Mekong Delta (Nguyen et al. 2001, Nguyen 2005, Nguyen et al. 2007) and from Cambodia it has been reported from Tonle Sap wetlands (Long 2000, Poole 2003, Olsson 2007). It has been reported from Malaysia at Terengganu in 2003 and from Maur in 1995.

It has been recently reported in Sumatra, Indonesia (Lubis 2005). Historical records of its occurrence comes from the coast off of Penang Island (Harrison 1984), and from accidental road kill in Brunei in 1997 (Sasaki 2006). Historically it has also been reported from Myanmar as evident from the skin present in the British museum. From this, the possible range of hairy-nosed otter can be worked out which is eastward from Northeast India (Indo-china), Myanmar, **Thailand,** Cambodia, Vietnam, Lao PDR, Brunei, Malaysia and Indonesia.

Countries: Native:

Cambodia; Malaysia; **Thailand;** Vietnam

Population:

In its distribution range the species is considered as rare. It has been studied in Phru Toa Daeng Peat Swamp Forest in **Thailand** (Kanchanasaka et al. 1998), U Minh Thuong Nature Reserve in Vietnam (Nguyen et al. 2001, Sage et al. 2004) and in Tonle Sap from Cambodia (Poole 2003, Heng 2007). However, their abundance or group sizes are relatively not known. In U Minha National Park though it was estimated that there were around 50-230 individuals. The species is believed to be extremely rare in Peninsular Malaysia (Sebastian 1995) and reported from scattered localities in the Borneo (Payne et al. 1985).

Population Trend: Decreasing

Habitat and Ecology:

In **Thailand** it lives in Phru Toa Daeng Peat Swamp Forests (Kanchanasaka et al. 1998). In Vietnam it has also been reported from the low lying peat swamp forests of U Minh Thuong Nature Reserve in Mekong Delta (Nguyen et al. 2001, Sage et al. 2004) and from Cambodia it has been reported from Tonle Sap wetlands (Poole 2003, Heng 2007). It inhabits freshwater and coastal areas, especially mangroves in Indonesia. Wayre (1974, 1978) considered that the Hairy-nosed otter mainly inhabited mountain streams above 300 m, Medway (1969) recorded it in the sea off Penang. The hairy nosed otter principally predate on fish (85.5%) followed by water snake and they also supplement their diet with frog, lizard, turtle, crab, mammal and insect, although these may not be that important in its diet (Kanchanasaka 2007). Fish belonging to the families Channidae, Belontiidae, Anabantidae, Notopteridae, Synbranchidae, Clariidae, Nandidae, were identified in the fecal samples from **Thailand.** The main prey selected were three-spot gourami *(Trichogaster trichopterus),* common climbing perch *(Anabas testudineus),* and snakeheads (Channa sp) (Kanchanasaka 2007). Not much is known about its breeding behaviour but there is indication that it breeds in November-December in the Mekong delta. Kanchanasaka et al. (2003) found that gestation was around 2 months as with other otters, and cubs were seen in December to February, and one family observed consisted of both parents and a cub.

Systems: Terrestrial; Freshwater; Marine.

Major Threat(s):

The hairy-nosed otter has limited distribution range. **In Thailand, the otters are found in the Toa Daeng peat swamp forests, and also from near the mouth of the Bang Nara River which is low lying and tidal.** In Vietnam, the species have been reported from low lying peat swamp forests dominated by *Melaleuca cajuputi* in lower Mekong. In Cambodia, the species mainly live around the Tonle Sap Lake where the otters live mainly in the flood-ed forest and scrub surrounding the lake. Like many predators the hairy-nosed otter occurs in low density and the number and frequency of sightings are very few. In recent years, the tropical peat swamp forests are under severe threat due to fire and other anthropogenic activities such as plantation for oil palm, food crops such as rice, corn and soybean, and fish farming. In Vietnam the entire Mekong Delta has been converted into rice fields reducing the habitat of otter and other wildlife species into few pocks.

In Malaysia, fire reduced 70% of the Binsulok Forest Reserve and 10% of the Klias Forest Reserve. This has affected the surrounding environment and the biodiversity. In Indonesia over the last 20 years, the ecosystem has been reduced from almost 30 million hectares to only about 15 million hectares, and most of what remains has already been logged selectively. Such levels of habitat modification have profound effect on the native biodiversity. In its entire range the hairy-nosed otter is under increasing pressure due to intensive poaching (Yoxn 2007). In Cambodia, around the Tonle Sap Lake, poaching of otters and other wildlife are common practice (Somanak 2007). In Vietnam otters are hunted for illegal wildlife trade, and also for meat and medical use. Similar is the case in other range countries. The principal threat to the fauna of Southeast Asian region is the burgeoning human population, and resultant biomass demand which puts pressure on natural resources. Unavailability of adequate prey species and suitable undisturbed habitat are putting additional pressure on all wildlife species including hairy-nosed otter.

Conservation Actions:

Lutra sumatrana is listed in Appendix II of the CITES. It is legally protected in all the range countries. In **Thailand** all the otter species have been protected since 1961 under Wild Animals Preservation and Protection Act and are listed as endangered species in Thailand Red Data Book (Nabhitabhata and Chan-ard, 2005). In Viet Nam, otters are protected and their hunting and use is strictly banned. In Cambodia, the hairy-nosed otter is listed as Rare and is fully protected. In Sarawak all otter species are protected by the First Schedule [Section 2(1)] PART II on Protected Animals from the Wild Life Protection Ordinance, 1998.

Smooth-coated Otter *(Lutrogale perspicillata)*

Description:

Smooth-coated otters are relatively large for otters, from 7 to 11 kg (15 to 24 lb) in weight and 59 to 64 cm (23 to 25 in) in head-body length, with a tail 37 to 43 cm (15 to 17 in) long. They may be distinguished from other species of otters by a more rounded head and a hairless nose in the shape of a distorted diamond. The tail is flattened, in contrast to the more rounded tails of other species. The legs are short and strong, with large webbed feet bearing strong claws. As their name suggests, they have unusually short and sleek fur; this is dark to reddish brown along the back, while the underside is light brown to almost grey in color. Females have two pairs of teats.

Range Description:

The smooth-coated otter is distributed throughout South Asia and Southeast Asia. Its distribution is continuous from Indonesia, through Southeast Asia, and westwards from southern China to India and Pakistan, with an isolated population in Iraq (Pocock 1941, Medway 1969, Hussain 1993). Though its current status in the Middle East is not known, its presence has been confirmed from Pakistan, India, Nepal, Bhutan, Bangladesh, Southwest China, Myanmar, **Thailand,** Vietnam, Malaysia, Sumatra, Java and Borneo (Mason and Macdonald 1986, Corbet and Hill 1992, Wozencraft 1993).

Countries: Native:

Bangladesh; Bhutan; Brunei Darussalam; Cambodia; China; India; Indonesia; Iraq; Lao People's Democratic Republic; Malaysia; Myanmar; Nepal; Pakistan; **Thailand;** Vietnam.

Population:

Because of secretive and nocturnal behaviour of smooth-coated otter, reliable estimates of its population are not available. In Southeast Asia it is found in large groups (Foster-Turly 1992, Hussain 1993, Hussain, 1996) where the basic family group consisted of adult female and her offspring, father and older siblings often join the group (Lekagul and McNeely 1988, Hussain 1995). The group size of otters varies considerably between months and seasons, the group being large during the monsoon (Hussain 1996). Along the Chambal river in Central India the group size ranges from 1-9 individuals (mean=4.62). During a study conducted in the Corbett Tiger Reserve, in North India an estimate of 41 individuals (35 adults and 6 juveniles) with a mean of 5.1±1.55 was recorded from the 85 km of river stretch within the Reserve (Nawab and Hussain 2007).

Population Trend: Decreasing

Habitat and Ecology:

The smooth-coated otter is essentially a plain's otter. In the Indian subcontinent they are adapted to live even in the semiarid region of north-western India and Deccan plateau (Prater 1971). Generally, it uses large rivers and lakes, peat swamp forests, mangroves and estuaries, and it even uses the rice fields for foraging (Foster-Turly 1992). In south-east Asia rice fields appear to be one of the most suitable habitats in supporting its viable populations (Melisch et al. 1996). However, they were more abundant in the mangroves of Kuala Gula, Malaysia as compared to the rain forest rivers (Shariff 1984). Along the large rivers in India, the smooth-coated otters prefer rocky stretches since these stretches provide sites for den and resting. River stretches with bank side vegetation and marshes are used in proportion to their availability especially in summer as they provide ample cover while travelling or foraging. Open clayey and sandy banks are largely avoided as they lack escape covers (Hussain 1993, Hussain and Choudhury 1995, 1997). In the Tarai areas of the upper Gangetic plains the smooth-coated otters use seasonally flooded swamps during monsoon (July-September) and in early winter (October-February). Winter being the breeding season, the swamps are extensively used as natal den sites and nursery. By February-March the swamps begin to dry and the fish biomass appears to be depleted, consequently the otters move to the perennial rivers (Hussain 1997). In west Java, the smooth-coated otters prefer mangroves and tidal stretches of the rivers and rice fields. Freshwater swamps and coastal stretches lacking vegetation are avoided (Melisch et al. 1996). In rice fields and pond areas they prefer sites having moderate diversity of vegetation. Rivers with moderate to slow or stagnant water and water bodies having a width of 10-40 m are preferred. In Huai Kha Khaeng, Thailand, the smooth-coated otters extensively use the slowly meandering river near the dam and the dam itself (Kruuk et al. 1994). The smooth-coated otter is predominantly a fish eater, but supplements its diet with shrimp/crayfish, crab and insects, and other vertebrates such as frog, mudskippers, birds and rats (Prater 1971, Foster-Turly 1992, Hussain and Choudhury 1998). In general, its diet is similar to that of Eurasian otter. The range of fish in the diet varies from 75% to 100% (Tiler et al. 1989, Foster-Turly 1992, Hussain 1993, Melisch et al. 1996, Hussain and Choudhury 1998). The smooth-coated otters exhibit the typical opportunistic feeding behaviour similar to the Eurasian otter. Along the major rivers in India they eat more fish, often making up to 94% of the total diet (Hussain and Choudhury 1998), while along the coast in mangrove habitats and in rice fields it ranges between 75-100% (Foster-Turly 1992, Melisch et al. 1996). In west Java otters inhabiting mangrove areas eat more of crustaceans, around 22% of the scats contain remains of crab and shrimp where as in the paddy fields in Malaysia, along with fish they tend to eat more rice field rats Rattus argentiventer, represented in 23% of the sprains' (Foster-Turly 1992). Hussain (1993) identified 12 fish species from the spraint of smooth-coated otter from Chambal river, of which seven species were eaten throughout the year. *Rhinomugil corsula* (Mugilidae) and *Rita rita* (Bagaridae) were the preferred species. At least eight species of fish were identified from the spraints from Malaysia (Foster-Turly 1992). Some of these include Gourami (Trichogater spp.), climbing perch (*Anabis testudineus)*, catfish (Clarius spp.), snakehead (Channa spp.) and mudskipper (Gobioidei). Along the rice field they ate large amount of most common fish (Trichogater and Anabis spp.). The size of the fish consumed varied from 5-46 cm, often ranging between 15-30 cm (Wayre 1978). No significant relationship was found between the calorific values of the fish and the quantities consumed. Thus, the prey selection by smooth otter is mainly influenced by its availability (Hussain 1993). The smooth-coated otter attains sexual maturity at twenty-two months in captivity (Desai 1974). Yadav (1967) observed first litter at four years of age. In captivity they mate during August to October in water followed by prolonged playful bouts between partners (Desai, 1974, Naidu and Malhotra 1989). Copulation takes place several times daily. Males are polygamous mating with up to four females (Desai 1974). In the wild in northern India mating occurs in August-September and littering in November-December (Hussain 1993). Desai (1974) observed litter size of 2-5 (mean = 3.25). Similarly, Naidu and Malhotra (1989) observed litter size 1-5. The breeding schedule of smooth otters in the wild in India corresponds with the captive records (Hussain 1993, 1996). Evidences of littering have been recorded in October and the cubs were seen out of the den in February. The mean litter size along the Chambal river was three (Hussain 1996). Based on the percentage of active telemetry recordings, Hussain (1993) observed that the overall activity pattern of smooth-coated otters along the Chambal River in India was in the form of a bimodal curve in which two period of high activity were separated by a period of relative inactivity. This pattern of activity varied considerably among seasons. In summer, the relative period of inactivity was greater than in winter or monsoons.

A significant difference was also found between day and night time activities in different seasons. Otters were more diurnal during winter than in summer or monsoon. Hussain (1993) observed daily movements of four radio-implanted smooth otters of different age and sex within their home range. Two types of movements were identified; small-scale movements associated with foraging in a restricted area close to dens, and more extensive travels between dens and foraging sites. Most of the movement was restricted between 250 to 1500 m. A typical group of smooth-coated otter consisting of male, female, and up to four young ones require 7 to 12 km of river for their territory and an even longer stretch of shoreline if living along the coast (Wayre 1974). During a radio-tracking study along the Chambal River, India, the home range of all the otters tracked overlapped intensively. Among the radio-implanted otters, the maximum home range was observed in sub-adult male and the minimum in juvenile female and male. Among the non-tagged otters, the home range of female with cubs was estimated as 5.5 km. In case of the adult male it was estimated to be approximately 17 km (Hussain 1993). Along the Chambal River, the home range length and area of smooth-coated otter was less than that of Eurasian otter, in Perthshire, Scotland and Northern river otter *Lutra candensis* in Idaho (Melquist and Hornocker, 1983), but larger than that of Eurasian otter (Erlinge 1967) in Sweden.

Systems: Terrestrial; Freshwater; Marine

Major Threat(s):

Major threats to Asian otter population are loss of wetland habitats due to construction of large-scale hydroelectric projects, reclamation of wetlands for settlements and agriculture, reduction in prey biomass, poaching and contamination of waterways by pesticides. In most Asian countries increased human population during the last century, inadequate and ineffective rural development programmes have not been able to address the problems of poverty, forcing people to be more and more dependent on natural resources (Badola, 1997). Consequently, most of the wetlands and waterways do not have adequate prey base for sustaining otter populations. Wetlands and waterways are polluted by eutrophication and accumulation of persistent pesticides such as chlorinated hydrocarbons and organophosphates through agricultural runoffs. Increased pesticide use is not only regarded as a major obstacle to the development of rice-fish culture, but also poses a danger to all predators feeding on aquatic prey in the area (Melisch 1996). In the entire south and southeast Asia there is severe conflict between otters and humans, because of poverty and recent increase in aquaculture activities leading to indiscriminate killing of otters. Many important habitats of smooth otter have been lost to development activities. In south east Asian countries, there does not seem to be any intentional otter trapping (Melisch et al. 1996) though it is prevalent in India, Nepal and Bangladesh.

Conservation Actions:

: Since 1977, the smooth-coated otter is listed on Appendix II CITES. However, most range countries are not able to control the clandestine trade leading to extensive poaching. Nevertheless, it is a protected species in almost all the range countries which prohibits its killing. The smooth-coated otter, once common in the wetlands and low lying areas of South Asia, is now restricted to a few protected areas. Creation of networks of Protected Areas, identification of sites as wetland of national and International importance under Ramsar Convention has to some extent halts the degradation of its habitat.

Oriental Small-clawed Otter *(Aonyx cinereus)*

Description:

The **oriental small-clawed otter** (*Amblonyx cinerea*), also known as the **Asian small-clawed otter**, is the smallest otter species in the world, weighing less than 5 kg. It lives in mangrove swamps and freshwater wetlands of Bangladesh, Burma, India, southern China, Laos, Malaysia, Indonesia, Singapore, the Philippines, **Thailand,** and Vietnam. Its paws are a distinctive feature, its claws not extending beyond the fleshy end pads of its partially webbed fingers and toes. This gives it a high degree of manual dexterity so that it can use its paws to feed on molluscs, crabs and other small aquatic animals. The oriental small-clawed otter lives in extended family groups with only the alpha pair breeding; offspring from previous years help to raise the young.Oriental small-clawed otters form monogamous pairs for life. The estrous cycle in the female is 28 days with a three-day period of estrus. The mated pairs can have two litters of one to six young per year and the gestation period is about 60 days. The newborn pups are relatively undeveloped; when they are born, they weigh around 50 g, are toothless, practically immobile and their eyes are still closed. They remain in their birthing dens and spend their first few weeks nursing and sleeping. The pups nurse every three to four hours for 10 to 15 minutes at a time. They open their eyes after 40 days and are fully weaned at 14 weeks. In the next 40 days, the young start to eat solid food and can swim three months later. Young otters will stay with their mother until the next litter is born. The male otter assists the female building the nest before birth and in food procurement after parturition. The life span of this species is around 11 to 16 years.This charismatic mammal, equally at home in the water and on land, is the smallest of the world's otters . As well as its size, the Asian short-clawed otter can be distinguished from other otters by its small claws , after which it is named, and the incomplete webbing between digits. These tiny claws, which do not protrude beyond the ends of the fingers, enhance the manual dexterity of this otter as it handles prey.. The Asian short-clawed otter typically has brown fur with a paler underside, although cream-coloured individuals have also been known . The edges of the upper lip, chin, throat, sides of neck and face are generally greyish-white.. Like other otters, the Asian short-clawed otter has a small head, short legs and flattened tail, creating a streamlined silhouette that can move easily through the water.

Range Description:

The Asian small-clawed otter has a large distribution range, extending from India in South Asia east ward through entire Southeast Asia up to Palawan (Philippines) and Taiwan and southern China in the north (Mason and Macdonald 1986, Wozencraft 1993, Hussain 2000). In recent years, in England it has established itself in the wild after escaping from captivity (Jefferies 1990 and 1991). In India it occurs in the Himalayan foothills of Himachal Pradesh (Kulu), West Bengal, Assam and Arunachal Pradesh as well as in southern Indian hill ranges of Coorg (Karnataka), Ashambu, Nilgiri and Palni hills (Tamil Nadu) and some places in Kerala (Pocock 1941, Prater 1971, Hussain 2000).

Countries: Native:

Bangladesh; Bhutan; Brunei Darussalam; Cambodia; China; India (Arunachal Pradesh, Assam, Himachal Pradesh, Karnataka, Kerala, Tamil Nadu, West Bengal); Indonesia; Lao People's Democratic Republic; Malaysia; Myanmar; Nepal; Philippines; Singapore; Taiwan, Province of China; **Thailand;** Vietnam

Population:

In most of their range the Asian small-clawed otter is sympatric with smooth-coated and Eurasian otters. In India, all the three species occur in Western Ghats and in the Northeast India where the species occur in small group of two to four individuals. In Western Ghats they are mostly found along the hill streams. They were once common in the mangroves of east Calcutta and Sunderbans (Sanyal 1991). The Asian small-clawed otters occur in freshwater and peat swamp forests, rice fields, lakes, streams, reservoirs, canals, mangrove and along the coast (Sivasothi and Nor 1994). In Malaysia and Indonesia they occur in coastal wetlands, and along the banks of paddy fields. Comparable data from Java, Myanmar, and India revealed that the Asian small-clawed otters have a high climatic and trophic adaptability in south and Southeast Asian tropics, occurring from coastal wetlands up to mountain streams (Melisch et al., 1996). A reliable population estimate of the Asian small-clawed otter is lacking. As many as 15 individuals were seen in a group in Malaysia (Wayre 1978), four to eight in coastal Sabha (Mason and Macdonald 1986) and two to four in India.

Population Trend: Decreasing

Habitat and Ecology:

The typical habitats of the Asian small-clawed otter in west Java are wetland systems having pools and stagnant water, including shallow stretches, with depth less than 1 m. These habitats are represented by freshwater swamps, meandering rivers, mangroves and tidal pools. Muller (1839) observed small-clawed otter from slow-flowing lowland streams to submontane streams dominated by rocks and boulders in forested areas. Irrigated rice fields with many crab species (Brachyura) are extensively used by small-clawed otter if proper shelter for them is available. These can act as suitable man made habitats (Melisch et al., 1996). In **Thailand,** the rapid-flowing upper areas of the Huay Kha Khaeng are dominated by *L. lutra,* the slowly meandering river near the dam and the dam itself were used by *L. perspicillata* while the Asian small-clawed otter occurred mostly in the middle sections, but also at the upper reaches. When different otter species occurred in the same site there was evidence of difference in use of the habitats. Signs of the of small-clawed otter were found wandering further away from the river than the two other species, between patches of reeds and river debris where crabs were more likely to be found (Kruuk et al. 1994). In west Java, its presence is positively correlated with slow flowing and stagnant broad rivers and smaller streams, depicting a distinct decline in preference from slow to deep-water bodies. On the other hand, they also use shallow fast-flowing mountain creeks narrower than 5 m, particularly when the course of the streams includes natural pools. In rice fields, they chose slow-flowing irrigation channels narrower than 2 m and with a varied, moderate or low vegetation structure. Like smooth-coated otter the Asian small-clawed otter dislike bare and open areas that do not offer any shelter (Melisch et al. 1996). It prefers pond areas and rice fields than the rivers, whereas it uses mangroves and lakes in proportion to their availability (Melisch et al. 1996). In riverine systems it prefers moderate and low vegetation structure, though their presence was also observed from banks with poor vegetation cover. Neither in ponds nor in rice field areas did they show preference for any of the vegetation structure categories, though poor nor bare structural conditions were the least favoured both in riverine and pond areas and along the rice fields. The small-clawed otter is adapted to feed on invertebrates as evident from the last two upper teeth (pm4 and m3) which are larger in size for crushing the exoskeleton of crabs and other hard shelled prey. The small clawed otter feeds mainly on crabs, snails and other molluscs, insects and small fish such as gourami and catfish (Pocock 1941, Wayre 1978).They supplement their diet with rodents, snakes and amphibians too. During a study in Malaysia, Foster-Turly (1992) examined 328 scats and found that around 80.8% of the scats consisted of crabs, 77.8% fish, 12.5% insects and 4.0% snails. This is the first study in which quantitative information on the diet of wild small-clawed otter was made. This study revealed that though the small-clawed otter is adapted for an invertebrate diet it substantiates its diet with large quantity of fish. Apart from crabs, the major prey item for small-clawed otter was the mudskipper (Gobioidei). This was recorded in the 48% of the scats. The other important prey was Trichogaster spp. and Anabantidae fish, which were represented in 27.4% scats. As evident from the scats the major fish prey were Trichogaster spp 20.7%, *Anabis testudineus* 5.2%, Clarius spp 2.4% and *Channa striatus* 1.5%. Apart from these the small-clawed otter in Malaysia also fed on snakes, frogs and insects.

Foster-Turly (1992) also examined the diet composition at four different times of the year coinciding with different water levels in the rice fields and concluded that the diet of the small-clawed otter was significantly different at different times of the year. Only the relatively rare dietary components of rodents, snails and snakehead fish (Clarius spp.) showed no significant difference among seasons. Crabs were always the most prevalent food items, but the frequency of occurrence in scats varied from 70.4% to 93.2%. Similarly, though the mudskippers were the second most important food items, they were consumed in significantly different amounts in different seasons from a low of 27.3% to 63.6%. The amount of Trichogaster, Anabis and the Anbantidae also varied considerably. This difference in the use of these prey are most likely due to difference in the life cycle and availability of these prey at different times of the year. Preliminary analysis of the small-clawed otter spraints from west Java showed their preference for crabs in both natural and man-made habitats (Melisch et al. 1996). In 87% of all collected spraints, crabs formed the dominant prey. Remaining part of the spraints consisted of fish bones and scales, ribs and vertebrae, unidentified mammalian hair, shrimps, insects and snake scales. In the Huay Kha Khaeng, **Thailand** almost 90% of the spraints of small-clawed otter contained remains of crabs *Potamon smithianus,* whereas 5% scats contained each of Fish and Amphibians. Apart from this, in few scats evidences of rodents and other arthropods were also found. Kruuk et al. (1994) estimated the preference for various size classes of crabs eaten by small-clawed otter. Of the 92 scats, 14 scats had crabs size 10-14 cm, 42 scats had 15-19 cm, 26 had 20-24 cm, 12 had 25-29 cm, 4 had 30-34 and 1 had 40-44 cm. The size distribution of crabs taken by small-clawed otter was similar to what was available, and there was not much evidence for selection of specific size. In west Java a preliminary estimate of preferred size confirmed an average of 3-4 cm carapace width (Melisch et al. 1996). The sexual behaviour of small clawed otter has been observed in as young as 18 months old. In captivity, successful breeding has been reported for 2.1 year females and 2.8 year males. The youngest animal to reproduce was a female of 13 months captive born at Bronx Zoo, and the oldest was a 15 years male at the National Zoo, USA (Foster-Turley and Engfer 1988). In the females oestrous cycle has duration of anywhere from 28 to 30 days, with breeding occurring the year round (Lancaster 1975). Some facilities report this cycle extending to ?every few months? with older animals. Oestrus lasts from one to thirteen days. Behavioural signs of the onset of oestrus may include increased rubbing and marking. In captivity mating usually takes place in the water, but has also been observed on land on a few occasions. In most cases the exact gestation period could not be ascertained but it is believed to be around 60-86 days (Lancaster 1975, Sobel 1996). The litter size ranges from 2-7. Life span in captivity is around 11 years (Crandall 1964).

Systems: Terrestrial; Freshwater; Marine

Major Threat(s):

The threat to Small-clawed otter is similar to that of smooth-coated and Eurasian otters. Throughout Asia the potential threat to its continued survival is destruction of its habitats due to changing land use pattern in the form of developmental activities. In many parts of Asia, the habitats have been reduced due to reclamation of peat swamp forests and mangroves, aquaculture activities along the intertidal wetlands and loss of hill streams. In India the primary threats are loss of habitats due to tea and coffee plantations along the hills, in the coastal areas loss of mangroves due to aquaculture and increased human settlements and siltation of smaller hill streams due to deforestation. Increased influx of pesticides into the streams from the plantations reduces the quality of the habitats. The next important threats to Asian small-clawed otter are reduction in prey biomass due to over-exploitation, which make its remaining habitats unsustainable. Pollution is probably the single most factors causing decline in the population of many fish species (Dehadrai and Ponniah 1997). Reduction in prey biomass affects otter population, and organochloric and heavy metal contamination interferes with their normal physiology leading to the decline in population. The threats to small-clawed otter is prominent in its western range so much so that since last 60 years its range has been shrunk considerable moving west to east from Himachal Pradesh to Assam (Hussain 2007). Once common in the mangroves of east Calcutta and Sunderbans (Sanyal 1991) now it is believed to be locally extinct. It is likely that the present range boundary at the western limit is Assam and in the Western Ghats of Southern India.

Conservation Actions:

Since 1977, the Asian small clawed otter is listed on Appendix II CITES which indicates that the species is not necessarily threatened with extinction, but the trade on its pelt must be controlled in order to avoid utilization incompatible with their survival. However, most range countries are not able to control the clandestine trade leading to extensive poaching. Nevertheless, it is a protected species in almost all the range countries which prohibits its killing. The Asian small-clawed otter is once common in the streams and wetlands of south and south east Asia is now restricted to a few Protected Areas. Creation of networks of Protected Areas, identification of sites as wetland of national and International importance under Ramsar Convention has to some extent halts the degradation of its habitat. Over the years the Otter Specialist Group has developed a cadre of biologist across Asia to conduct field surveys and has popularize otter conservation by promoting otter as ambassador of the wetlands. However, concerted effort to conserve this species is needed. For the long term survival of the species, policy based action, research on factors affecting its survival, habitat based action on creation and where required expansion of protected areas and communication and awareness building actions are needed.

Order: Perissodactyla (odd-toed ungulates)

The odd-toed ungulates are browsing and grazing mammals. They are usually large to very large, and have relatively simple stomachs and a large middle toe.

Family: Tapiridae (tapirs)
Genus: Tapirus:
Malayan Tapir *(Tapirus indicus)*

General appearance and characteristics:

The animal is easily identified by its markings, most notably the light-colored "patch" which extends from its shoulders to its rear. The rest of its hair is black, except for the tips of its ears, which, as with other tapirs, are rimmed with white. This pattern is for camouflage; the disrupted coloration makes it more difficult to recognize it as a tapir, and other animals may mistake it for a large rock rather than prey when it is lying down to sleep. Malayan tapirs grow to between 1.8 and 2.4 m (5 ft 10 in and 7 ft 10 in) in length, not counting a stubby tail of only 5 to 10 cm (2.0 to 3.9 in) in length, and stand 90 to 107 cm (2 ft 10 in to 3 ft 6 in) tall. They typically weigh between 250 and 320 kg (550 and 710 lb), although some adults can weigh up to 540 kg (1,200 lb). The females are usually larger than the males. Like the other types of tapirs, they have small, stubby tails and long, flexible proboscises. They have four toes on each front foot and three toes on each back foot. The Malayan tapir has rather poor eyesight, but excellent hearing and sense of smell. They have a large sagittal crest, a bone running along the middle of the skull that is necessary for muscle attachment. They also have unusually positioned orbits, an unusually shaped cranium with the frontal bones elevated, and a retracted nasal incision. All of these modifications to the normal mammal skull are, of course, to make room for the proboscis. This proboscis caused retraction of bones and cartilage in the face during the evolution of the tapir, and even caused the loss of some cartilages, facial muscles, and the bony wall of the nasal chamber. Malayan tapirs have very poor eyesight, making them rely greatly on their excellent sense of smell and hearing to go about in their everyday lives. They have small, beady eyes with brown irises on either side of their faces. Their eyes are often covered in a blue haze, which is corneal cloudiness thought to be caused by repetitive exposure to light. Corneal cloudiness is when the cornea starts to lose its transparency. The cornea is necessary for the transmitting and focusing of outside light as it enters the eye, and cloudiness can cause vision loss. This causes the Malayan tapir to have very inadequate vision, both on land and in water, where they spend the majority of their time. Also, as these tapirs are most active at night and since they have poor eyesight, it is harder for them to search for food and avoid predators in the dark.

Brevetianus variation:

A small number of melanistic (all-black) Malayan tapirs have been observed. In 1924, an all-black tapir was sent to Rotterdam Zoo and was classified as a subspecies called *Tapirus indicus brevetianus* after its discoverer, Captain K. Brevet.[8] In 2000, two melanistic tapirs were observed during a study of tigers in the Jerangau Forest Reserve in Malaysia. The cause of this variation may be a genetic abnormality similar to that of black panthers that appear in populations of spotted jaguars. However, unless and until more T. i. brevetianus individuals can be studied, the precise explanation for the trait will remain unknown.

Range Description:

Tapirus indicus occurs in southern and central parts of Sumatra (Indonesia), and on the Asian mainland in Peninsular Malaysia, **Thailand** (along the western border and on the Peninsula south to the Malaysian border, and in Huai Kha Khaeng Wildlife Sanctuary in the north), and Myanmar (south of latitude 18°N). Its populations are now highly fragmented within its former range. It was listed as occurring in southern Cambodia and possibly southern Viet Nam by Brooks et al. (1997). It was reported from Hongquan district, eastern Cochin China, Vietnam, in 1944 (Harper, 1945), and there was an authentic-sounding record from Lao PDR in 1902 (Duckworth et al., 1999). It is presumed to be extinct in all three countries. However, further investigation of these historical records and of other indications from Lao, Vietnam, Cambodia, **Northern Thailand** and even southern China have found none that has any compelling evidence in its support. In some cases (e.g. the 1902 Lao PDR record; Cheminaud 1939), review of the statement in the context of the same author's wider work means that records sounding, on the face of it, strong need to be dismissed (Duckworth and Hedges 1998, Duckworth et al. 1999, in prep., J. W. Duckworth pers. comm. 2008, G. J. Galbreath pers. comm. 2008). In sum, there is no credible historical-era record from north of the Thai/Malay peninsula, although fossil remains do come from Vietnam and China and indicate a much wider range under different climatic scenarios. The species habitat distribution at the northern edge of its Thai range, where the climate develops a more marked dry season, and the tapir occupancy changes from attitudinally wide-ranging to being restricted to the most humid altitudes, strongly supports a climatic limitation (Steinmetz et al. in press), thereby casting further empirical doubt on the 20th century reports from Lao PDR, Vietnam and Cambodia: these reports showed no association with where, climatically, they ought to have been (Annamite wet areas supporting other Sundaic species like Annamite striped rabbit *Nesolagus timminsi* and crested Argus *Rheinardia ocellata,* and in fact some came from the driest parts of Indochina, least plausible to support tapirs.

Countries: Native:

Indonesia (Sumatera); Malaysia; Myanmar; **Thailand**

Population:

Tapirus indicus occurs in two disjunctly and isolated populations - one occurring on mainland Southeast Asia and the other occupying on the island of Sumatra. The species is more widespread and common on the mainland, and it is declining rapidly in Sumatra due to extensive loss of habitat, accidental and deliberate trapping for meat and removal of animals for zoos in Indonesia.

In Malaysia there are approximately 1,500-2,000 individuals (C. Traeholt pers. comm.). Further research efforts are needed to determine the total population size. In **Thailand** it is one of the least-affected large mammals by recent heavy levels of hunting that have caused severe cross-country declines in many large mammal species (Steinmetz et al. in press). The situation in **Thailand** is similar, although with much less habitat available, and the Thai populations are is likely to be quite fragile since it is severely fragmented, and most subpopulations are unlikely to reach more than 50-100 individuals at the most. In many places there are only 10-15 individuals left with no chance of linking up to other protected areas and suitable habitats.

Population Trend: Decreasing.

Ecology:

The gestation period of the Malayan tapir is about 390–395 days, after which a single offspring, weighing around 15 pounds (6.8 kg), is born. Malayan tapirs are the largest of the four tapir species at birth and grow more quickly than their congeners. Young tapirs of all species have brown hair with white stripes and spots, a pattern which enables them to hide effectively in the dappled light of the forest. This baby coat fades into adult coloration between four and seven months after birth. Weaning occurs between six and eight months of age, at which time the babies are nearly full-grown, and the animals reach sexual maturity around age three. Breeding typically occurs in April, May or June, and females generally produce one calf every two years.

Malayan tapirs can live up to 30 years, both in the wild and in captivity. Malayan tapirs are primarily solitary creatures, marking out large tracts of land as their territory, though these areas usually overlap with those of other individuals. Tapirs mark out their territories by spraying urine on plants, and they often follow distinct paths which they have bulldozed through the undergrowth. Exclusively herbivorous, the animal forages for the tender shoots and leaves of more than 115 species of plants (around 30 are particularly preferred), moving slowly through the forest and pausing often to eat and note the scents left behind by other tapirs in the area. However, when threatened or frightened, the tapir can run quickly, despite its considerable bulk, and can also defend itself with its strong jaws and sharp teeth. Malayan tapirs communicate with high-pitched squeaks and whistles. They usually prefer to live near water and often bathe and swim, and they are also able to climb steep slopes. Tapirs are mainly active at night, though they are not exclusively nocturnal. They tend to eat soon after sunset or before sunrise, and they will often nap in the middle of the night. This behaviour characterizes them as crepuscular animals. *T. indicus* is restricted to tropical moist forest areas and occurs in both primary and secondary forest. The more seasonal climate in northern Myanmar, northern (= most of non-peninsula) **Thailand,** Lao PDR, Vietnam and Cambodia and the harsher dry season of the forest (even in evergreen areas, excepting the eastern flanks and adjacent Vietnamese lowlands of the Annamite chain) there is likely to be the main reason this species is not found there. The Malayan tapir is also predominantly found in the lowlands and the lower montane zone in some parts of the range, although it remains common to the highest peaks in its Thai range (Steinmetz et al. in press). Because the lowland forests are disappearing at a faster rate than the montane forests, an accelerated reduction in range and population is suspected (N. van Strien pers. comm.)

Systems: Terrestrial; Freshwater

Major Threat(s):

Tapirus indicus is threatened throughout most of its range. The primary threats to the species are large scale deforestation and increasingly, hunting. Tapir population have declined by well over 50% in **Thailand** and Malaysia, whereas it is suspected to be slightly less than 50% in Sumatra. The main reason for declines in the past is habitat conversion, with large tracts land being converted into palm oil plantations. However, increasingly as other large 'prey" species decline in the area hunters are beginning to look towards tapir as a food source.

Destruction of habitat is the main threat to the species: in central Sumatra much of the remaining habitat is outside of any protected area and uncontrolled illegal logging continues; in Thailand, almost all remaining intact forest now lies within protected areas, with mostly degraded lands outside; in contrast, Myanmar's protected areas make up 3.2% of land area (data provided by Myanmar Forest Department) and most tapir habitat lies outside these protected areas. In Malaysia forest loss is extremely severe, especially for expanding oil palm plantations.

Tapirus indicus are shy animals and appear to be highly sensitive to forest fragmentation. In Halabala Wildlife Sanctuary on the Thai-Malaysia border, Kaewsirisuk (2001) found that the species does not venture within a few hundred meters of forest-plantation edges. At Khao Sok National Park, tapirs are interior forest species that avoid forest edges (Lynam 1996). Kawanishi (2002), however, found in Taman Negara, the largest national park in Malaysia, that although the human traffic level was heavier in area closer to the park boundary, tapirs showed no edge effects. While forest loss continues in **Thailand**, forests in protected areas remain relatively stable in size and composition to other countries because of a ban on commercial logging that has been in place since 1989. For this reason, while tapirs may indeed be threatened in general by forest loss, populations in **Thailand** and Malaysia are probably more stable. Large-scale habitat destruction has continued in Sumatra, historically the species main stronghold, and most remaining habitat in central Sumatra is outside protected areas. In Sumatra, populations have declined by slightly less than 50% simply because the onslaught of habitat only started to be serious in the late 1980s. However, the rate of decline is continuing to escalate in this region. In fact Sumatra has only 60% of the forest cover that it had 15 years ago, so things are developing fast there and future declines of the species are likely well over 50% in the next 30 years. Given the uncontrolled illegal logging situation in Sumatra, they are becoming increasingly threatened island-wide. Localized hunting also occurs and is suspected elsewhere in its distribution range. Unless serious efforts to stem illegal logging and forest encroachment are made, all Sumatran forests outside conservation areas will be lost over the next few decades.

In Malaysia the current forestry trend seems to be stabilized at approx. 43% remaining forest cover (57% lost), of which at least half can be considered tapir habitat. In **Thailand,** 40% of the remaining forest is outside protected areas and only 5% of Myanmar's land area is protected forest (Lynam pers. comm.).

The species has uncertain status and future in Myanmar due to security issues and forest clearance for rubber and oil palm plantations. However, two new protected areas have been designated in the Tenasserims: Taninthayi National Park and Lenya River Wildlife Sanctuary. If these areas can be protected, they will preserve valuable tapir habitat in the future.

In the past, several Indonesian zoos, especially Pekanbaru, traded in live tapirs for sale to other Indonesian zoos or private collections, or for sale as meat in local markets. Fifty tapirs are reported passing through the Pekanbaru Zoo since 1993. Some of these animals are suspected of having originated from protected areas. Elsewhere, extraction may not be very high but it is uncertain how many individuals are actually hunted every year. Hunting is specifically known to be comparatively (by comparison with other mammals of similar size) very low in **Thailand** and at least parts of Sumatra (Holden et al. 2003, Stienmetz et al. in press)

There are indications that live tapirs have been traded through several Indonesian zoos, with some destined for private collections or for sale as meat in local markets to the non-Muslim community. Some of these animals are suspected to have originated from protected areas.

Hunting has been a minor threat to *Tapirus indicus i*n the past, but is has been increasingly a cause of concern as more and more hunting of the species is discovered. Some localized hunting has been reported in Sumatra, however, and historically tapirs are not hunted for subsistence or commercial trade in **Thailand** or Myanmar, since their flesh is considered distasteful. Some hill tribes believe that killing a tapir brings bad luck, so they are not hunted.

Conservation Actions:

The species is legally protected in all range states and the habitat of large parts of the range is protected, including several National Parks in **Thailand,** Myanmar, Peninsula Malaysia and Sumatra. The impact of habitat reduction/destruction on the tapir is not fully understood and needs further investigation. It is listed on CITES Appendix I.

Thailand supports one of the most comprehensive systems of protected areas in Southeast Asia. Over 200 National Parks, Marine National Parks, Wildlife Sanctuaries and Non-hunting areas cover 17% of land area (Prayurasiddhi et al. 1999). Since most existing *Tapirus indicus* habitat is already protected, the future for conservation of the species in **Thailand** is quite positive. In contrast, Myanmar's protected areas make up 5% of land area (Lynam 2003) and most tapir habitat lies outside these protected areas. In Myanmar, Malayan tapirs are entirely restricted to rainforests in the Tenasserim Ranges along the Thai-Myanmar border (Yin, 1993). The tenure of these lands on the Myanmar side of the border is disputed and due to civil unrest, has been inaccessible for wildlife survey until now. A team of Myanmar Forest Department staff working the Tenasserim border during 2001 detected tapirs from camera-traps, track and scat at two sites: the Minmoletka Taung area and the Hitaung Pru Reserve Forest (Lynam 2003).In **Thailand**, *Tapirus indicus* is recorded from forest areas in the west and south of the country (Lekagul and McNeely 1988), including Tran boundary forest areas in border areas, and large isolated forest remnants. The Tran boundary forests represent the most extensive, contiguous habitats for large mammals left in the country (Prayurasiddhi et al. 1999). They include the Western Forest Complex **(Thai-Myanmar border),** which includes 12 protected areas, and covers over 18,730 sq km including both dry and wet forests, and the Kaeng Krachan/Chumpol complex which covers 4,373 sq km, mostly wet evergreen forest on the **Thai-Myanmar border**. The Balahala Forest is an expanse of 1,850 sq km of tropical rainforest on the Thai-Malaysia border. All areas are contiguous with larger forest areas on opposite sides of the country border. Recent survey efforts (Lynam 1999; Lynam 2000; WCS 2001; Kaewsirisuk 2001; A. Pattanavibool pers. comm.) suggest that tapirs are present though uncommon in each of these Tran boundary forest areas.

Javan Rhinoceros *(Rhinoceros sondaicus)*

Description:

The Javan rhinoceros (more precisely *Sunda rhinoceros*) or lesser one-horned rhinoceros (*Rhinoceros sondaicus)* is a member of the family Rhinocerotidae and one of five extant rhinoceroses. It belongs to the same genus as the Indian rhinoceros, and has similar mosaic ked skin which resembles armour, but at 3.1–3.2 m (10–10.5 feet) in length and 1.4–1.7 m (4.6–5.8 ft) in height, it is smaller (in fact, it is closer in size to the black rhinoceros of the genus Diceros). Its horn is usually less than 25 cm (10 inches), smaller than those of the other rhino species. Once the most widespread of Asian rhinoceroses, the Javan rhinoceros ranged from the islands of Java and Sumatra, throughout Southeast Asia, and into India and China. The species is critically endangered, with only one known population in the wild, and no individuals in captivity. It is possibly the rarest large mammal on earth, with a population of as few as 40 in Ujung Kulon National Park at the western tip of Java in Indonesia. A second population in Cat Tien National Park in Vietnam was confirmed as extinct in 2011. The decline of the Javan rhinoceros is attributed to poaching, primarily for their horns, which are highly valued in traditional Chinese medicine, fetching as much as US$30,000 per kilogramme on the black market. Loss of habitat, especially as the result of wars, such as the Vietnam War, in Southeast Asia, has also contributed to the species' decline and hindered recovery. The remaining range is within one nationally protected area, but the rhinos are still at risk from poachers, disease and loss of genetic diversity leading to inbreeding depression. The Javan rhino is smaller than the Indian rhinoceros, and is close in size to the black rhinoceros. The body length of the Javan rhino (including its head) can be up to 2 to 4 m (6.6 to 13 ft), and it can reach a height of 1.4–1.7 m (4.6–5.8 ft). Adults are variously reported to weigh between 900 and 2,300 kg (2,000 and 5,100 lb), although a study to collect accurate measurements of the animals has never been conducted and is not a priority because they are endangered. There is not a substantial size difference between genders, but females may be slightly bigger. The rhinos in Vietnam appeared to be significantly smaller than those in Java, based on studies of photographic evidence and measurements of their footprints. Like the Indian rhino, the Javan rhinoceros has a single horn (the other extant species have two horns). Its horn is the smallest of all extant rhinos, usually less than 20 cm (7.9 inches) with the longest recorded only 27 cm (10.5 in). The Javan rhinoceros does not appear to often use its horn for fighting, but instead uses it to scrape mud away in wallows, to pull down plants for eating, and to open paths through thick vegetation. Similar to the other browsing species of rhino (the black, Sumatran and Indian), the Javan rhino has long, pointed, upper lips which help in grabbing food. Its lower incisors are long and sharp; when the Javan rhino fights, it uses these teeth. Behind the incisors, two rows of six low-crowned molars are used for chewing coarse plants. Like all rhinos, the Javan rhino smells and hears well, but has very poor vision. They are estimated to live for 30 to 45 years. Its hairless, splotchy gray or gray-brown skin falls in folds to the shoulder, back and rump. The skin has a natural mosaic pattern which lends the rhino an armoured appearance. The neck folds of the Javan rhinoceros are smaller than those of the Indian rhinoceros, but still form a saddle shape over the shoulder. Because of the risks of interfering with such an Endangered species, however, the Javan rhinoceros is primarily studied through fecal sampling and camera traps. They are rarely encountered, observed or measured directly

Taxonomy and Naming:

The first studies of the Javan rhinoceros by naturalists from outside of its region took place in 1787, when two animals were shot in Java. The skulls were sent to the renowned Dutch naturalist Petrus Camper, who died in 1789 before he was able to publish his discovery that the rhinos of Java were a distinct species. Another Javan rhinoceros was shot on the island of Sumatra by Alfred Duvaucel, who sent the specimen to his stepfather Georges Cuvier, the famous French scientist. Cuvier recognized the animal as a distinct species in 1822, and in the same year it was identified by Anselme Gaëtan Desmarest as Rhinoceros sondaicus. It was the last species of rhinoceros to be identified. Desmarest initially identified the rhino as being from Sumatra, but later amended this to say his specimen was from Java.

The genus name Rhinoceros, which also includes the Indian rhinoceros, is derived from the ancient Greek words ῥίς (rhis), which means "nose", and κέρας (ceras), which means "horn"; sondaicus is derived from sunda, the biogeographically region that comprises the islands of Sumatra, Java, Borneo, and surrounding smaller islands. The Javan rhino is also known as the lesser one-horned rhinoceros (in contrast with the greater one-horned rhinoceros, another name for the Indian rhino)

There are three distinct subspecies, of which only one is still extant:

Rhinoceros sondaicus sondaicus, the type subspecies, known as the Indonesian Javan rhinoceros, once lived on Java and Sumatra. The population is now confined to as few as 40 animals in the wild, Ujung Kulon National Park on the western tip of the island of Java. One researcher has suggested the Javan rhino on Sumatra belonged to a distinct subspecies, *R.s. floweri,* but this is not widely accepted.

Rhinoceros sondaicus annamiticus, known as the Vietnamese Javan rhinoceros or Vietnamese rhinoceros, once lived across Vietnam, Cambodia, Laos and into **Thailand** and Malaysia. The sub specific annamiticus is derived from the Annamite Mountain Range in Southeast Asia, part of this subspecies' range. In 2006, a single population, estimated at fewer than 12 remaining rhinos, lived in an area of lowland forest in the Cat Tien National Park in Vietnam. Genetic analysis suggested this subspecies and the Indonesian Javan rhinoceros last shared a common ancestor between 300,000 and 2 million years ago. The last individual of this population was shot by a poacher in 2010.

Rhinoceros sondaicus inermis, known as the Indian Javan rhinoceros, once ranged from Bengal to Burma, but is presumed to have gone extinct before 1925. The term inermis means "unarmed", as the most distinctive characteristic of this subspecies is the small horns in males, and evident lack of horns in females. The original specimen of this species was a hornless female. The political situation in Burma has prevented assessment of the species in that country, but its survival is considered unlikely.

Ecology:

The Javan rhino can live approximately 30–45 years in the wild. It historically inhabited lowland rain forest, wet grasslands and large floodplains. The Javan rhino is mostly solitary, except for courtship and offspring-rearing, though groups may occasionally congregate near wallows and salt licks. Aside from humans, adults have no predators in their range. The Javan rhino usually avoids humans, but will attack when it feels threatened. Scientists and conservationists rarely study the animals directly due to their extreme rarity and the danger of interfering with such an endangered species. Researchers rely on camera traps and fecal samples to gauge health and behavior. Consequently, the Javan rhino is theTwo adult rhinos with their calves were filmed in a motion-triggered video released on February 28, 2011 by WWF and Indonesia's National Park Authority, which proved it is still breeding in the wild. In April 2012, the National Parks Authority released video showing 35 individual Javan rhinos, including offspring. least studied of all rhino species. The Javan rhinoceros is a solitary animal with the exception of breeding pairs and mothers with calves. They will sometimes congregate in small groups at salt licks and mud wallows. Wallowing in mud is a common behavior for all rhinos; the activity allows them to maintain cool body temperatures and helps prevent disease and parasite infestation.

The Javan rhinoceros does not generally dig its own mud wallows, preferring to use other animals' wallows or naturally occurring pits, which it will use its horn to enlarge. Salt licks are also very important because of the essential nutrients the rhino receives from the salt. Male home ranges are larger at 12–20 km^2 (5–8 $miles^2$) compared to the female, which are around 3–14 km^2 (1–5 mi^2). Male territories overlap each other less than those of the female. It is not known if there are territorial fights. Males mark their territories with dung piles and by urine spraying. Scrapes made by the feet in the ground and twisted saplings also seem to be used for communication. Members of other rhino species have a peculiar habit of defecating in massive rhino dung piles and then scraping their back feet in the dung. The Sumatran and Javan rhinoceros, while defecating in piles, do not engage in the scraping. This adaptation in behavior is thought to be ecological; in the wet forests of Java and Sumatra, the method may not be useful for spreading odours. The Javan rhino is much less vocal than the Sumatran; very few Javan rhino vocalizations have ever been recorded. Adult Javan rhinos have no known predators other than humans. The species, particularly in Vietnam, is skittish and retreats into dense forests whenever humans are near. Though a valuable trait from a survival standpoint, it has made the rhinos difficult to study. Nevertheless, when humans approach too closely, the Javan rhino becomes aggressive and will attack, stabbing with the incisors of its lower jaw while thrusting upward with its head. Its comparatively antisocial behavior may be a recent adaptation to population stresses; historical evidence suggests they, like other rhinos, were once more gregarious. The Javan rhinoceros is herbivorous and eats diverse plant species, especially their shoots, twigs, young foliage and fallen fruit. Most of the plants favoured by the species grow in sunny areas in forest clearings, scrubland and other vegetation types with no large trees. The rhino knocks down saplings to reach its food and grabs it with its prehensile upper lip. It is the most adaptable feeder of all the rhino species. Currently, it is a pure browser, but probably once both browsed and grazed in its historical range. The rhino eats an estimated 50 kg (110 lb) of food daily. Like the Sumatran rhino, it needs salt in its diet. The salt licks common in its historical range do not exist in Ujung Kulon, but the rhinos there have been observed drinking seawater, likely for the same nutritional need. An estimated 40-60 animals live in the area on the western tip of Java in Ujung Kulon National Park. Another smaller population occurs in and around the Cat Loc part (Dong Nai province) of the Cat Tien National Park in of Vietnam, with maybe as few as six individuals remaining (R. Steinmetz, M. Khan bin Momin Khan pers. comm.). These populations have not significantly declined over the last few decades, and the current trend is not known (Sectionov and Waladi pers. comm.), but no breeding has been observed in the Cat Loc population for many years (M. Khan bin Momin Khan pers. comm.). There are no animals currently in captivity, and a total of only 22 individuals have ever been known to exist in captivity (Rookmaaker et al., 1998).

Population Trend: Unknown The Javan Rhinoceros currently occurs in lowland tropical rainforest areas, especially in the vicinity of water (Schenkel and Schenkel, 1969). The species formerly occurred in more open mixed forest and grassland and on high mountains. Because of its rarity, little is known about its preferred habitat, but it is certainly not naturally restricted to dense tropical forest water (Schenkel and Schenkel, 1969). Little is known about the species' biology and the habitats in which the two remaining populations are found may not be optimal. The home range size of females is probably no more than 500 ha, while males wonder over larger areas, with likely limited dispersal distance. The species is generally solitary, except for mating pairs and mothers with young (Nowak, 1999). Its life history characteristics are not well known, with longevity estimated at about 30-40 years, gestation length of approximately 16 months (as with other rhino species), and age at sexual maturity estimated at 5-7 years for females and 10 years for males (Nowak, 1999; Systems: Terrestrial

Major Threat(s):

The cause of population decline is mainly attributable to the excessive demand for rhino horn and other products for Chinese and allied medicine systems (Foose and van Strien 1997). The bulk of the remaining population occurs as a single population within a national park and the population size in Ujung Kulon National Park is probably limited to the effective carrying capacity of the area (around 50 animals).One possible threat to this population is disease. In addition, such a small population faces a constant threat from poachers, although there is evidence that current poaching levels are under control (Sectionov and Waladi pers. comm.). The Cat Loc population may be too small to be viable, and no breeding has been observed for many years, and it is possible that the animals are too old to breed. The population is so small that all the animals could be of the same sex.

Conservation Actions:

It is legally protected in all range states. The species has been on CITES Appendix I since 1975.

A Rhino Protection Unit (RPU) has been established for the protection of this species on Java (Sectionov and Waladi pers. comm.). It occurs in two protected areas: Ujung Kulon National Park on Java and the Cat Loc part (Dong Nai province) of the Cat Tien National Park in Viet Nam.

There is an urgent need to review the feasibility of a reintroduction/translocation program, since the only known viable population occurs in a geographically restricted area of Java. There is also a need to survey parts of its historical range for the very remote possibility that small remnant populations exist, especially in parts of Lao PDR or Cambodia. The population in Cat Loc is probably no longer viable, and requires intensive management measures in order to survive (perhaps including captive breeding and re-introductions).

Sumatran Rhinoceros *(Dicerorhinus sumatrensis)*

This species is now extinct in Thailand:

Description:

The Sumatran rhinoceros *(Dicerorhinus sumatrensis)* is a member of the family Rhinocerotidae and one of five extant rhinoceroses. It is the only extant species of the genus Dicerorhinus. It is the smallest rhinoceros, although is still a large mammal. This rhino stands 112–145 cm (3.67–4.76 ft) high at the shoulder, with a head-and-body length of 2.36–3.18 m (7.7–10.4 ft) and a tail of 35–70 cm (14–28 in). The weight is reported to range from 500 to 1,000 kg (1,100 to 2,200 lb), averaging 700–800 kg (1,500–1,800 lb), although there is a single record of a 2,000 kg (4,400 lb) specimen. Like the African species, it has two horns; the larger is the nasal horn, typically 15–25 centimetres (5.9–9.8 in), while the other horn is typically a stub. A coat of reddish-brown hair covers most of the Sumatran rhino's body.

Members of the species once inhabited rainforests, swamps and cloud forests in India, Bhutan, Bangladesh, Myanmar, Laos, **Thailand,** Malaysia, Indonesia, and China. In historical times they lived in southwest China, particularly in Sichuan. They are now critically endangered, with only six substantial populations in the wild: four on Sumatra, one on Borneo, and one in the Malay Peninsula. Their numbers are difficult to determine because they are solitary animals that are widely scattered across their range, but they are estimated to number fewer than 275 Survival of the Peninsular Malaysia population is in doubt, and one of the Sumatran populations may already be extinct. Total numbers today may be as low as 200. The decline in the number of Sumatran rhinoceros is attributed primarily to poaching for their horns, which are highly valued in traditional Chinese medicine, fetching as much as US$30,000 per kilogram on the black market. The Sumatran rhino is a mostly solitary animal except for courtship and offspring-rearing. It is the most vocal rhino species and also communicates through marking soil with its feet, twisting saplings into patterns, and leaving excrement. The species is much better studied than the similarly reclusive Javan rhinoceros, in part because of a program that brought 40 Sumatran rhinos into captivity with the goal of preserving the species. The program was considered a disaster even by its initiator; most of the rhinos died and no offspring were produced for nearly 20 years, representing an even worse population decline than in the wild.

Taxonomy and naming:

The first documented Sumatran rhinoceros was shot 16 kilometres (9.9 mi) outside Fort Marlborough, near the west coast of Sumatra, in 1793. Drawings of the animal, and a written description, were sent to the naturalist Joseph Banks, then president of the Royal Society of London, who published a paper on the specimen that year. It was not until 1814, however, that the species was given a scientific name, by Johann Fischer von Waldheim, a German scientist and curator of the State Darwin Museum in Moscow, Russia.

The scientific name *Dicerorhinus sumatrensis* comes from the Greek terms di (δι, meaning "two"), cero (κέρας, meaning "horn"), and rhinos (ρινος, meaning "nose"). Sumatrensis signifies "of Sumatra", the Indonesian island where the rhinos were first discovered. Carolus Linnaeus originally classified all rhinos in the genus Rhinoceros; therefore the species was originally identified as Rhinoceros sumatrensis. Joshua Brookes considered the Sumatran rhinoceros, with its two horns, a distinct genus from the one-horned Rhinoceros, and gave it the name Didermocerus in 1828. Constantin Wilhelm Lambert Gloger proposed the name Dicerorhinus in 1841. In 1868, John Edward Gray proposed the name Ceratorhinus. Normally the oldest name would be used, but a 1977 ruling by the International Commission on Zoological Nomenclature established the proper genus name as Dicerorhinus.

The three subspecies are:

D. s. sumatrensis, known as the western Sumatran rhinoceros, has only 170 to 230 rhinos remaining, mostly in the national parks of Bukit Barisan Selatan and Gunung Leuser in Sumatra. Around 75 may also live in Peninsular Malaysia. The main threats against this subspecies are habitat loss and illegal poaching. There is a slight genetic difference between the western and eastern Sumatran rhinos. The rhinos in Peninsular Malaysia were once known as *D. s. niger,* but were later recognized to be similar to the rhinos on western Sumatra.

D. s. harrissoni, known as the eastern Sumatran rhinoceros or Bornean rhinoceros, was once common throughout Borneo; now only about 50 individuals are estimated to survive. The known population on Borneo lives in Sabah. There are unconfirmed reports of animals surviving in Sarawak and Kalimantan. This subspecies is named after Tom Harrisson, who worked extensively with Bornean zoology and anthropology in the 1960s. The Bornean subspecies is markedly smaller than the other two.

D. s. lasiotis, known as the northern Sumatran rhinoceros, once roamed in India and Bangladesh, but has been declared extinct in these countries. Unconfirmed reports suggest there may be a small population still surviving in Burma, but the political situation in the country has prevented verification. The name lasiotis is derived from the Greek for "hairy-ears". Later studies showed their ear-hair was not longer than other Sumatran rhinos, but *D. s. lasiotis* remained a subspecies because it was significantly larger than the other subspecies.

Ecology:

Ancestral rhinoceroses first diverged from other perissodactyls in the Early Eocene. Mitochondrial DNA comparison suggests the ancestors of modern rhinos split from the ancestors of Equidae around 50 million years ago. The extant family, the Rhinocerotidae, first appeared in the Late Eocene in Eurasia, and the ancestors of the extant rhino species dispersed from Asia beginning in the Miocene

The Sumatran rhinoceros is considered the least derived of the extant species, as it shares more traits with its Miocene ancestors. Paleontological evidence in the fossil record dates the genus Dicerorhinus to the Early Miocene, 23–16 million years ago. Many fossils have been classified as members of Dicerorhinus, but no other recent species are in the genus.Molecular dating suggests a split of Dicerorhinus from the four other extant species as far back as 25.9 ± 1.9 million years. Three hypotheses have been proposed for the relationship between the Sumatran rhinoceros and the other living species. One hypothesis suggests the Sumatran rhinoceros is closely related to the black and white rhinos in Africa, evidenced by the species having two horns, instead of one. Other taxonomists regard the Sumatran rhinoceros as a sister taxon of the Indian and Javan rhinoceros because their ranges overlap so closely. A third hypothesis, based on more recent analyses, however, suggests the two African rhinos, the two Asian rhinos and The Sumatran rhinoceros represent three essentially separate lineages that split around 25.9 million years ago, and it may therefore be unclear which group diverged first. Because of morphological similarities, the Sumatran rhinoceros is believed to be closely related to the extinct woolly rhinoceros *(Coelodonta antiquitatis)*. The woolly rhinoceros, so named for the coat of hair it shares with the Sumatran rhinoceros, first appeared in China, and by the Upper Pleistocene, ranged across the Eurasian continent from Korea to Spain. The woolly rhinoceros survived the last Ice Age, but like the woolly mammoth, most or all became extinct around 10,000 years ago. Although some morphological studies questioned the relationship, recent molecular analysis has supported the two species as sister taxa.A mature Sumatran rhino stands about 120–145 centimetres (3.9–4.76 ft) high at the shoulder, has a body length of around 250 centimetres (8.2 ft) and weighs 500–800 kilograms (1,100–1,800 lb), though the largest individuals in zoos have been known to weigh as much as 1,000 kilograms (2,200 lb).

Like the African species, it has two horns. The larger is the nasal horn, typically only 15–25 centimetres (5.9–9.8 in), though the longest recorded specimen was much longer at 81 centimetres (32 in). The posterior horn is much smaller, usually less than 10 centimetres (3.9 in) long, and often little more than a knob. The larger nasal horn is also known as the anterior horn; the smaller posterior horn as the frontal horn. The horns are dark gray or black in colour. The males have larger horns than the females, though the species is not otherwise sexually dimorphic. The Sumatran rhino lives an estimated 30–45 years in the wild, while the record time in captivity is a female *D. lasiotis* which lived for 32 years and 8 months before dying in the London Zoo in 1900. Two thick folds of skin encircle the body behind the front legs and before the hind legs. The rhino has a smaller fold of skin around its neck. The skin itself is thin, 10–16 millimetres (0.39–0.63 in), and in the wild the rhino appears to have no subcutaneous fat. Hair can range from dense (the most dense hair in young calves) to scarce and is usually a reddish-brown. In the wild, this hair is hard to observe because the rhinos are often covered in mud. In captivity, however, the hair grows out and becomes much shaggier, likely because of less abrasion from walking through vegetation. The rhino has a patch of long hair around the ears and a thick clump of hair at the end of the tail. Like all rhinos, they have very poor vision. The Sumatran rhinoceros is fast and agile; it climbs mountains easily and comfortably traverses steep slopes and riverbanks. Genetic analysis of Sumatran rhino populations has identified three distinct genetic lineages. The channel between Sumatra and Malaysia was not as significant a barrier for the rhinos as the Barisan Mountains along the length of Sumatra, for rhinos in eastern Sumatra and Peninsular Malaysia are more closely related than the rhinos on the other side of the mountains in western Sumatra. In fact, the eastern Sumatra and Malaysia rhinos show so little genetic variance that the populations were likely not separate during the Pleistocene, when sea levels were much lower and Sumatra formed part of the mainland. Both populations of Sumatra and Malaysia, however, are close enough genetically that interbreeding would not be problematic. The rhinos of Borneo are sufficiently distinct that conservation geneticists have advised against crossing their lineages with the other populations. Conservation geneticists have recently begun to study the diversity of the gene pool within these populations by identifying micro satellite loci. The results of initial testing found levels of variability within Sumatran rhino populations comparable to those in the population of the less endangered African rhinos, but the genetic diversity of Sumatran rhinos is an area of continuing study. Sumatran rhinoceroses are solitary creatures except for pairing before mating and during offspring rearing. Individuals have home ranges; bulls have territories as large as 50 km2 (19 sq mi), whereas females' ranges are 10–15 km2 (3.9–5.8 sq mi). The ranges of females appear to be spaced apart; males' ranges often overlap. There is no evidence Sumatran rhinos defend their territory through fighting. Marking their territory is done by scraping soil with their feet, bending saplings into distinctive patterns, and leaving excrement. The Sumatran rhino is usually most active when eating, at dawn, and just after dusk. During the day, the rhino wallows in mud baths to cool down and rest. In the rainy season, they move to higher elevations; in the cooler months, they return to lower areas in their range. The rhino spends a large part of its day in wallows. When mud holes are unavailable, the rhino will deepen puddles with its feet and horns. The wallowing behaviour helps the rhino maintain its body temperature and protect its skin from ectoparasites and other insects. Captive specimens, deprived of adequate wallowing, have quickly developed broken and inflamed skins, suppurations, eye problems, inflamed nails, hair loss and have eventually died. One 20-month study of wallowing behavior found they will visit no more than three wallows at any given time. After two to 12 weeks using a particular wallow, the rhino will abandon it. Typically, the rhino will wallow around midday for two to three hours at a time before venturing out for food. Although in zoos the Sumatran rhino has been observed wallowing less than 45 minutes a day, the study of wild animals found 80–300 minutes (an average of 166 minutes) per day spent in wallows. There has been little opportunity to study epidemiology in the Sumatran rhinoceros. Ticks and gyrostigma were reported to cause deaths in captive animals in the 19th century. The rhino is also known to be vulnerable to the blood disease surra, which can be spread by horse-flies carrying parasitic trypanosomes; in 2004, all five rhinos at the Sumatran Rhinoceros Conservation Centre died over an 18-day period after becoming infected by the disease. The Sumatran rhino has no known predators other than humans. Tigers and wild dogs may be capable of killing a calf, but calves stay close to their mothers, and the frequency of such killings is unknown.

Although the rhino's range overlaps with elephants and tapirs, the species do not appear to compete for food or habitat. Elephants (Elephas maximus) and Sumatran rhinos are even known to share trails, and many smaller species such as deer, boar and wild dogs will use the trails the rhinos and elephants create. The Sumatran rhino maintains trails across its range. The trails fall into two types. Main trails will be used by generations of rhinos to travel between important areas in the rhino's range, such as between salt licks, or in corridors through inhospitable terrain that separates ranges. In feeding areas, the rhinos will make smaller trails, still covered by vegetation, to areas containing food the rhino eats. Sumatran rhino trails have been found that cross rivers deeper than 1.5 meters (5 ft) and about 50 meters (165 ft) across. The currents of these rivers are known to be strong, but the rhino is a strong swimmer. A relative absence of wallows near rivers in the range of the Sumatran rhinoceros indicates they may occasionally bathe in rivers in lieu of wallowing. The Sumatran rhino eats a wide range of plants, such as: *Mallotus, mangosteens, Ardisia, and Eugenia.* Most feeding occurs just before nightfall and in the morning. The Sumatran rhino is a browser, with a diet of young saplings, leaves, fruits, twigs and shoots. The rhinos usually consume up to 50 kg (110 lb) of food a day. Primarily by measuring dung samples, researchers have identified more than 100 food species consumed by the Sumatran rhinoceros. The largest portion of the diet is tree saplings with a trunk diameter of 1–6 cm (0.4–2.4 inches). The rhinoceros typically pushes these saplings over with its body, walking over the sapling without stepping on it, to eat the leaves. Many of the plant species the rhino consumes exist in only small portions, which indicates the rhino is frequently changing its diet and feeding in different locations. Among the most common plants the rhino eats are many species from the Euphorbiaceae, Rubiaceae and Melastomataceae families. The most common species the rhino consumes is Eugenia. The vegetal diet of the Sumatran rhinoceros is high in fiber and only moderate in protein. Salt licks are very important to the nutrition of the rhino. These licks can be small hot springs, seepages of salty water or mud-volcanoes. The salt licks also serve an important social purpose for the rhinos—males visit the licks to pick up the scent of females in oestrus. Some Sumatran rhinos, however, live in areas where salt licks are not readily available, or the rhinos have not been observed using the licks. These rhinos may get their necessary mineral requirements by consuming plants rich in minerals.

The Sumatran rhinoceros is the most vocal of the rhinoceros species. Observations of the species in zoos show the animal almost constantly vocalizing, and it is known to do so in the wild, as well The rhino makes three distinct noises: eeps, whales, and whistle-blows. The eep, a short, one-second-long yelp, is the most common sound. The whale, named for its similarity to vocalizations of the humpback whale is the most song-like vocalization and the second-most common. The whale varies in pitch and lasts from four to seven seconds. The whistle-blow is named because it consists of a two-second-long whistling noise and a burst of air in immediate succession. The whistle-blow is the loudest of the vocalizations, loud enough to make the iron bars in the zoo enclosure where the rhinos were studied vibrate. The purpose of the vocalizations is unknown, though they are theorized to convey danger, sexual readiness, and location, as do other ungulate vocalizations. The whistle-blow could be heard at a great distance, even in the dense brush in which the Sumatran rhino lives. A vocalization of similar volume from elephants has been shown to carry 9.8 km (6.1 mi) and thus the whistle-blow may carry as far. The Sumatran rhinoceros will sometimes twist the saplings they do not eat. This twisting behavior is believed to be used as a form of communication, frequently indicating a junction in a trail.

Females become sexually mature at the age of six to seven years, while males become sexually mature at about 10 years old. The gestation period is around 15–16 months. The calf, which typically weighs 40–60 kg (88–132 lb), is weaned after about 15 months and stays with the mother for the first two to three years of its life. In the wild, the birth interval for this species is estimated to be four to five years; its natural offspring-rearing behavior is unstudied. The reproductive habits of the Sumatran rhinoceros have been studied in captivity. Sexual relationships begin with a courtship period characterized by increased vocalization, tail raising, urination and increased physical contact, with both male and female using their snouts to bump the other in the head and genitals. The pattern of courtship is most similar to that of the black rhinoceros.Young Sumatran rhino males are often too aggressive with females, sometimes injuring and even killing them during the courtship. In the wild, the female could run away from an over-ly aggressive male, but in their smaller captive enclosures, they cannot; this inability to escape aggressive males may partly contribute to the low success rate of captive breeding programs. The period of oestrus itself, when the female is receptive to the male, lasts about 24 hours, and observations have placed its recurrence between 21–25 days. Rhinos in the Cincinnati Zoo have been observed copulating for 30–50 minutes, similar in length to other rhinos; observations at the Sumatran rhinoceros Conservation Centre in Malaysia have shown a briefer copulation

As the Cincinnati Zoo has had successful pregnancies, and other rhinos also have lengthy copulatory periods, a lengthy rut may be the natural behavior. Though researchers observed successful conceptions, all these pregnancies ended in failure for a variety of reasons until the first successful captive birth in 2001; studies of these failures at the Cincinnati Zoo discovered the Sumatran rhino's ovulation is induced by mating and it had unpredictable progesterone levels. Breeding success was finally achieved in 2001, 2004 and 2007 by providing a pregnant rhino with supplementary progestin. Recently, a calf was born in captivity of an endangered female in western Indonesia, only fifth such birth in one and a quarter century. Aside from those few individuals kept in zoos and pictured in books, the Sumatran rhinoceros has remained little known, overshadowed by the more common Indian, black and white rhinos. Recently, however, video footage of the Sumatran rhinoceros in its native habitat and in breeding centres has been featured in several nature documentaries. Extensive footage can be found in an Asia Geographic documentary The Littlest Rhino. Natural History New Zealand showed footage of a Sumatran rhino, shot by freelance Indonesian-based cameraman Alain Compost, in the 2001 documentary The Forgotten Rhino, which featured mainly Javan and Indian rhinos. Though documented by droppings and tracks, pictures of the Bornean rhinoceros were first taken and widely distributed by modern conservationists in April 2006, when camera traps photographed a healthy adult in the jungles of Sabah in Malaysian Borneo.[54] On April 24, 2007 it was announced that cameras had captured the first-ever video footage of a wild Bornean rhino. The night-time footage showed the rhino eating, peering through jungle foliage, and sniffing the film equipment. The World Wildlife Fund, which took the video, has used it in efforts to convince local governments to turn the area into a rhino conservation zone. Monitoring has continued; 50 new cameras have been set up, and in February 2010, what appeared to be a pregnant rhino was filmed. A number of folk tales about the Sumatran rhino were collected by colonial naturalists and hunters from the mid-19th century to early 20th century. In Burma, the belief was once widespread that the Sumatran rhino ate fire. Tales described the fire-eating rhino following smoke to its source, especially campfires, and then attacking the camp. There was also a Burmese belief that the best time to hunt was every July, when the Sumatran rhinos would congregate beneath the full moon. In Malaya, it was said that the rhino's horn was hollow and could be used as a sort of hose for breathing air and squirting water. In Malaya and Sumatra, it was once believed that the rhino shed its horn every year and buried it under the ground. In Borneo, the rhino was said to have a strange carnivorous practice: after defecating in a stream, it would turn around and eat fish that had been stupefied by the excrement. The species inhabits tropical rainforest and montane moss forest, and occasionally occurs at forest margins and in secondary forest (Nowak, 1999). The species occurs mainly in hilly areas nearby water sources, and exhibits seasonal movements, moving uphill in times of lowland flooding (van Strien, 1975). This shy species is dependent on salt licks, and occurs mostly in primary forest in protected areas, but wandering into secondary forests outside protected areas, especially in the dry season in search of water (Van Strien, 1975; Boeadi pers. comm.). Males are primarily solitary, but can have overlapping territories with females, which are commonly found with offspring (Nowak, 1999). The home range size of females is probably no more than 500 ha, while males wonder over larger areas, with likely limited dispersal distance. The species is generally solitary, except for mating pairs and mothers with young (Nowak, 1999). Its life history characteristics are not well known, with longevity estimated at about 35-40 years, gestation length of approximately 15-16 months, and age at sexual maturity estimated at 6-7 years for females and 10 years for males (Nowak, 1999; IRF website (www.rhinos-irf.org), 2007). Home range: Males up to 5,000 ha, females 1,000 -1,500 ha. Daily movements between feeding sites and wallows are probably only a few kilometres per day. Longer treks are made when males and females go to saltlicks (5-10 km) and by males exploring their large ranges. Dispersal appears to be mainly by sub-adult animals (4-7 years) old. In this period they may be found rather far from the home grounds. Adults are very traditional in the use of their ranges and will not move away unless severely disturbed. Water is never very far away in the habitats occupied by the Sumatran rhino.

Systems: Terrestrial

The Taman Negara National Park contains the only known concentrated population of Sumatran rhinoceros on mainland Asia.

The Sumatran rhinoceros lives in both lowland and highland secondary rainforest, swamps and cloud forests. It inhabits hilly areas close to water, particularly steep upper valleys with a lot of undergrowth. The Sumatran rhinoceros once inhabited a continuous range as far north as Burma, eastern India and Bangladesh. Unconfirmed reports also placed it in Cambodia, Laos and Vietnam. All known living animals occur in Peninsular Malaysia, the island of Sumatra and Sabah, Borneo. Some conservationists hope Sumatran rhinos may still survive in Burma, though it is considered unlikely. Political turmoil in Burma has prevented any assessment or study of possible survivors. The last reports of stray animals from Indian limits were in 1990s. The Sumatran rhino is widely scattered across its range, much more so than the other Asian rhinos, which has made it difficult for conservationists to protect members of the species effectively. Only six areas are known to contain communities of more than a handful of Sumatran rhinoceros: Bukit Barisan Selatan National Park, Gunung Leuser National Park, and Way Kambas National Park on Sumatra; Taman Negara National Park in Peninsular Malaysia; and the Tabin Wildlife Reserve in Sabah, Malaysia on the island of Borneo. The Kerinci Seblat National Park, Sumatra's largest, was estimated to contain a population of around 500 rhinos in the 1980s, but due to poaching, this population is now considered extinct. The survival of any animals in Peninsula Malaysia is also in doubt. Population: The total population is estimated at fewer than 275 individuals, though probably more than 220. Until the early 1990's the numbers continued to decline at a rapid rate with estimated losses of 50% or more of the population per decade (Foose and van Strien 1997). Over the last decade the decrease has been halted or slowed in most of the larger populations because of better protection, but animals are still being lost in the small remnant populations. The subspecies *Dicerorhinus sumatrensis sumatrensis* now occurs mainly on Sumatra, where there are 170 to 230 individuals. It has its largest populations remaining in Bukit Barisan Selata, Way Kambas, and Gunung Leuser National Park (Foose et al., 1997). There are about 60 to 80 animals in Gunung Leuser, about 60 to 80 Bukit Barisan Selatan, and 15-25 in Way Kambas, with some local reports of rhinos occurring outside of protected areas in Aceh Province (Sectionov and Waladi pers. comm.). There are also a few small, non-viable populations, including no more than a few individuals in Kerinci-Seblat National Park. Some populations are decreasing due to poaching, with very steep decreases in some areas (Sectionov and Waladi pers. comm.). Poaching has ceased in Bukit Barisan Selata and Way Kambas National Parks recently (Sectionov and Waladi pers. comm.). Populations in Peninsular Malaysia are now very small, but the species possibly survives in Taman Negara National Park and in Tamon Besor/Belum area. It probably no longer survives in Endau Rompin National Park (Malaysia).

The majority of the few remaining individuals of the subspecies *Dicerorhinus sumatrensis harrissoni* occur in Tabin National Park in Sabah (Malaysia), with some also in the Danum Valley (also in Sabah). The total population in Sabah is likely to be about 50 individuals (Han pers. comm.). A two year survey from 2000-2002 indicated 6 known individuals, 10 probable individuals, and an additional 35 possible (Van Strien, 2005). The population status of the subspecies *Dicerorhinus sumatrensis lasiotis* is unknown, with the very slight possibility that a small number of individuals survive in the Lassai Tract in Myanmar. There are over 20 animals in captivity, mostly in Indonesia and Malaysia, with a few in the United States.

Population Trend: Decreasing

Major Threats:

Sumatran rhinoceroses were once quite numerous throughout Southeast Asia. It is now estimated that fewer than 275 individuals remain. The species is classed as critically endangered, primarily due to illegal poaching. Until the early 1990s, the population decline was estimated at more than 50% per decade, and the small, scattered populations now face high risks of inbreeding depression. Most remaining habitat is in relatively inaccessible mountainous areas of Indonesia. Poaching of Sumatran rhinoceros is a cause for concern, as the price of its horn has been estimated as high as $30,000 per kilogram. This species has been over hunted for many centuries, leading to the current greatly reduced – and still declining – population.

The rhinos are difficult to observe and hunt directly (one field researcher spent seven weeks in a tree hide near a salt lick without ever observing a rhino directly), so poachers make use of spear traps and pit traps. In the 1970s, uses of the rhinoceros's body parts among the local people of Sumatra were documented, such as the use of rhino horns in amulets and a folk-belief that the horns offer some protection against poison. Dried rhinoceros meat was used as medicine for diarrhoea, leprosy and tuberculosis. "Rhino-oil", a concoction made from leaving a rhino's skull in coconut oil for several weeks, may be used to treat skin diseases. The extent of use and belief in these practices is not known. It was once believed that rhinoceros horn was widely used as an aphrodisiac; in fact traditional Chinese medicine never used it for this purpose. Nevertheless, hunting in this species has primarily been driven by a demand for rhino horns with supposedly medicinal properties

The rainforests of Indonesia and Malaysia, which the Sumatran rhino inhabits, are also targets for legal and illegal logging because of the desirability of their hardwoods. Rare woods like merbau, meranti and semaram are valuable on the international markets, fetching as much as $1,800 per m3 ($1,375 per cu yd). Enforcement of illegal-logging laws is difficult because humans live within or near many of the same forests as the rhino. The 2004 Indian Ocean earthquake has been used to justify new logging. Although the hardwoods in the rainforests of the Sumatran rhino are destined for international markets and not widely used in domestic construction, the number of logging permits for these woods has increased dramatically because of the tsunami. However, while it has been suggested that this species is highly sensible to habitat disturbance, it appears this is of little importance compared to hunting, as it can withstand more or less any forest condition.

Conservation Actions:

The species has been included on CITES Appendix I since 1975, and legally protected in all range states. An extensive international co-operative programme for the conservation of this species is being implemented with in situ activities being conducted in Indonesia and Malaysia. The primary objectives are to develop and deploy effective anti-poaching teams and to provide the co-ordination capacity to manage and sustain the programme. Rhino Protection Units (RPU) have been a force majeur in stopping poaching in Sumatra. Many organizations are involved with these units, including the Government of Indonesia (Sectionov and Waladi pers. comm.). The expansion and reinforcement of anti-poaching programmes is the top priority if this species is to survive. There are also ongoing efforts to develop managed breeding centres for the species in Indonesia and Malaysia. There have been recent advances in captive breeding techniques for this species, including a successful births at the Cincinnati Zoo in 2001 and 2004 (Khan et al., 2004). One of these offspring was transferred back to a breeding centre in Sumatra. There is a need for further surveys in northern Myanmar to determine the status of any remaining populations.

Order: Artiodactyla (even-toed ungulates)

The even-toed ungulates are ungulates whose weight is borne about equally by the third and fourth toes, rather than mostly or entirely by the third as in perissodactyls. There are about 220 artiodactyl species, including many that are of great economic importance to humans.

Family: Suidae (pigs)
Subfamily: Suinae
Genus: *Sus:*
Boar (*Sus scrofa*)

Range Description:

The Eurasian wild pig has one of the widest geographic distributions of all terrestrial mammals, and this range has been greatly expanded by human agency. The species now occurs in pure wild or barely modified feral form on all continents excepting Antarctica, and on many oceanic islands. It is the ancestor of most (but not all) ancient and modern domestic pig breeds, and there is evidence to suggest that it was independently domesticated in several different parts of its range, including Southeast Asia, the Far East and Asia Minor. As a wild form, it has constituted a primary resource of subsistence hunters since the earliest times, and it is one of the most important targets for recreational hunting wherever it remains sufficiently abundant. Over-hunting and changes in land use have resulted in the fragmentation of its range and its extermination throughout the British Isles, Scandinavia, parts of North Africa, and relatively extensive parts of its range in the former Soviet Union. and northern Japan. Nevertheless, the species remains widely distributed and is often locally abundant. As a result of its depredations on crops it is regarded as a pest in many countries, where it remains unprotected outside designated wildlife reserves or is managed as a game animal.

S. scrofa has by far the largest range of all pigs. It occurs throughout the steppe and broadleaved forest regions of the Palaearctic, from western Europe to the Russian Far East, extending southwards as far North Africa, the Mediterranean Basin and the Middle East, through India, Indo-China, Japan (including the Ryukyu Chain), Taiwan and the Greater Sunda Islands of South-east Asia. Populations east of Bali are probably all introduced. It has been extinct in the British Isles since sometime in the 17th century, despite attempted introductions of new stock from Europe (Harting, 1880) (though see below for more recent information). It is also extinct in southern Scandinavia (but see below), over extensive portions of its recent range in west-central and eastern parts of the former Soviet Union (Heptner et al., 1961), and in northern Japan (Chiba, 1964, 1975). The species was last reported in Libya in the 1880s, and it became extinct in Egypt in about 1902 (Hufnagl, 1972).

Groves and Grubb (1993) distinguished four 'subspecies groupings', based on both geographic and morphological criteria, as follows:

1. 'Western races' of Europe (scrofa and meridionalis), North Africa (algira) and the Middle East (lybicus), extending at least as far east as Soviet Central Asia (attila and nigripes);

2. 'Indian races' of the sub-Himalayan region from Iran in the west (davidi) to north India and adjacent countries as far east as Myanmar and west **Thailand** (cristatus), and south India and Sri Lanka (affinis and subsp. nov.);

3. 'Eastern races' of Mongolia and the Soviet Far East (*sibiricus and ussuricus*), Japan (*leucomystax and riukiuanus*), Taiwan (*taivanus),* to south-east China and Vietnam (*moupinensis);* and

4. 'Indonesian race' (or banded pig) from the Malay Peninsular, Sumatra, Java, Bali and certain offshore islands (*vittatus).*

In Europe, it is widespread in most continental areas, with the exception of northern Fennoscandia and European Russia. As mentioined above, it disappeared from the British Isles and Scandinavia in the 17th century, although it has now been reintroduced to Sweden and escaped animals have established themselves in the wild in Britain (Spitz 1999). Animals have escaped from captivity in the UK and have established themselves in the wild. There are at least three small wild populations in England, on the Kent/East Sussex border, in Dorset, and in Hereford (Battersby 2005). It is native to Corsica and Sardinia, but the population in Sicily was introduced (Spitz 1999).

Countries: Native:

Afghanistan; Albania; Algeria; Andorra; Armenia (Armenia); Austria; Azerbaijan; Bangladesh; Belarus; Belgium; Bhutan; Bosnia and Herzegovina; Bulgaria; Cambodia; China; Croatia; Cyprus; Czech Republic; Estonia; Finland; France; Georgia; Germany; Greece; Hong Kong; Hungary; India; Indonesia (Bali, Irian Jaya, Jawa, Lesser Sunda Is., Sumatera); Iran, Islamic Republic of; Iraq; Israel; Italy (Sardegna, Sicilia - Introduced); Japan; Jordan; Kazakhstan; Korea, Democratic People's Republic of; Korea, Republic of; Kyrgyzstan; Lao People's Democratic Republic; Latvia; Lebanon; Liechtenstein; Lithuania; Luxembourg; Macedonia, the former Yugoslav Republic of; Malaysia; Moldova; Monaco; Mongolia; Montenegro; Morocco; Myanmar; Nepal; Netherlands; Pakistan; Palestinian Territory, Occupied; Poland; Portugal; Romania; Russian Federation; San Marino; Serbia (Serbia); Slovakia; Slovenia; Spain; Sri Lanka; Switzerland; Syrian Arab Republic; Taiwan, Province of China; Tajikistan; **Thailand;** Tunisia; Turkey; Turkmenistan; Ukraine; Uzbekistan; Vietnam

Regionally extinct:

Denmark; Egypt; Ireland; Libya; Norway

Reintroduced:

Sweden; United Kingdom

Introduced:

Antigua and Barbuda; Argentina; Australia; Brazil; Colombia; Cuba; Dominican Republic; Ecuador (Galápagos); Fiji; Haiti; Jamaica; New Zealand; Papua New Guinea; South Africa; Sudan; United States (Arizona, Florida, Georgia, Georgia - Native, Hawaiian Is., Kentucky, Mississippi, New Mexico, North Carolina, Tennessee, Texas, Virginia, West Virginia); Virgin Islands, U.S.

Population:

This species is abundant in many parts of its range, though populations can be depressed in places where hunting intensity is high (for example in eastern and Southeastern Asia). Eurasian wild pig populations in Europe increased markedly during the latter part of the 20th century (Spitz 1999), but are now thought to be stable in most areas. Populations in England, southern Sweden and Finland may still be increasing (Battersby 2005). Although there is no global population estimate, numbers can be high in many places. For example, in 1982 in Mongolia, there was a population density of 0.9 per 1,000 ha in the Khovsgol area. By 1989, it was estimated that 34,000 individuals were living in the Khentii and Khangai Mountain regions of Mongolia, with an average density of 1 - 2 per 1,000 ha.

Population Trend: Unknown

Habitat and Ecology:

The Eurasian wild pig occupies a wide variety of temperate and tropical habitats, from semi-desert to tropical rain forests, temperate woodlands, grasslands and reed jungles; often venturing onto agricultural land to forage. It is found in a variety of habitats. In Europe, it prefers broadleaved forests and especially evergreen oak forests, but may also be found in more open habitats such as steppe, Mediterranean scrubland, and farmland, so long as there is water and tree cover nearby (Spitz 1999). In Europe it is found from sea level to 2,400 in the Pyrenees (Palomo and Gisbert 2002), but it can be found at higher elevations in Asia.

The species is omnivorous, though stomach and fecal contents analyses indicate that vegetable matter, principally fruits, seeds, roots and tubers, constitutes about 90% of the diet (Spitz, 1986). A field study of the Indonesian wild pig, S. s. vittatus, in Ujung Kulon National Park in Java, indicated that these animals are predominately frugivorous, feeding on about 50 species of fruits, especially those of strangling figs (Ficus spp.), and that they are important seed dispersal agents (Pauwels, 1980). By comparison, analyses of the stomach contents of wild pigs (also *S. s. vittatus*) in agricultural areas of West Malaysia by Diong (1973), revealed that sugar cane, tapioca and rice were the commonest food items, but that usually more than one type of food had been eaten, even where a single cultivated crop was abundant. Other items commonly consumed by these pigs included soil, earthworms, roots and other vegetable matter and, in mangrove areas, molluscs, crabs and other arthropods and even fishes. The consumption of invertebrate and small vertebrate prey may be a necessary component of the diet, since a study of free-ranging domestic pigs in Papua New Guinea revealed that animals fed ad libitum lost weight when denied earthworms (Rose and Williams, 1983). In common with its feral derivatives (Oliver and Brisbin, 1993), *S. scrofa* has also occasionally been reported to predate larger vertebrates, such as deer fawns and (tethered) goats (Hoogerwerf, 1970), though it is possible that such incidents involve only a few individuals in the population; an aspect also noted by Pauwels (1980) when referring to the predation of sea turtle nests by wild pigs in Ujung Kulon. Similarly, a large boar (*S. s. cristatus*) in Royal Chitawan National Park, Nepal, which was seen to displace an adult leopard from its kill, a domestic buffalo calf, which it then partly consumed (W. Oliver, pers. obs.), was reported by Park staff to regularly commandeer such kills, but that no other individual pigs had been seen to do this.

Wild pigs are normally most active in the early morning and late afternoon, though they become nocturnal in disturbed areas, where activity usually commences shortly before sunset and continues throughout the night. A total of 4 to 8 hours are spent foraging or travelling to feeding areas. Feeding is generally a social activity (even solitary males may join feeding groups) which also provides an opportunity for display and other agonistic behaviours (Beuerle, 1975). Radio telemetry studies in southern France indicate that they generally travel between 2 and 15 km per night, though this is often within an area of only 20 to 150 ha. However, the home range estimates for adult females and adult males over a 2-3 month period varied from 500-1,000 ha and 1,000-2,000 ha, respectively. During this same period, subadult covered an area of 500- 5,000 ha, and after 6 to 12 months they may have covered more than 10,000 ha; the larger home ranges of these animals being related to their expulsion from their natal groups and then undergoing a wandering phase. Movements over long distances (50 to 250 km) have also been recorded in Europe, but the extent and purpose of these movements has yet to be studied (Spitz, 1986). Experiments in which tagged animals are released and subsequently recovered provide evidence that they disperse freely over even larger areas (500 to 750 km^2), which may also indicate the area occupied by large population units. The density of free-ranging S. s. scrofa in Europe rarely exceeds 5 individuals/km^2 (Spitz, 1986), though much higher concentrations have been reported elsewhere, e.g. 27-32/km^2 on Peucang Island in Ujung Kulon National Park, Java (Pauwels, 1980) and 32.2-72.1/km^2 in sugarcane areas in the Punjab, Pakistan (Shafi and Khokhar, 1985).

Wild pigs are gregarious, forming herds or 'sounders' of varying size depending on locality and season, but usually of between 6-20 individuals, though aggregations of over 100 have been reported (Prater, 1971; Legakul and McNeely, 1977; Briedermann, 1990)). The basic social unit is a nucleus of one or more females and their last litters. Animals peripheral to this comprise sub adults from previous litters, and adult males during the mating season. However, the latter tend to stay in relatively close contact with 1 or 2 female groups at other times of the year, and subadult males or mixed sex groups of sub adults may also form longer-term associations (Spitz and Janeau, 1990). The dynamics of the basic group include the isolation of the preparturient female, her re-entry with young, entry of nulliparous females, the arrival of adult males with the simultaneous departure of subadult animals (Spitz, 1986). In contrast to its domestic derivatives, reproductive activity in S. scrofa tends to be seasonal and positively correlated with the relative availability of principal foodstuffs or related climatic factors. In tropical countries, such as Sri Lanka, peak estrus activity has been recorded during the wettest months of November and December (Santiapillai and Chambers, 1980). However, social organization may also play a role in modulating the timing of reproductive events, since farrowing is often synchronized amongst females in the same social groups, which suggests a mechanism for synchronizing the onset of estrus (Spitz, 1986).

Wide fluctuations in the numbers of animals killed by hunters, particularly in the (former) U.S.S.R. and in France, suggest cyclic changes in the numbers of wild pigs available for hunting. Annual recruitment into the total population depends on reproductive rate (i.e. the number and prolificacy of females) and juvenile mortality, both of which factors may be influenced by the availability of foodstuffs and other external factors (Spitz, 1986). In western Europe, litter size is usually between 4 and 7 piglets (Briedermann, 199), though Harrison and Bates (1991) cite reports of 5 and 7-10 piglets per litter as being usual in Iraq and Armenia, respectively. Pauwels (1980) recorded an average litter size of 6-10 piglets at the beginning of the breeding season in Java, but this number dropped to only 2 -4 piglets per litter towards the end of the breeding season. In the Ryukyu Islands, S. Japan, there is evidence that the wild pigs (*S. s. riukiuanus*) have two breeding seasons per year, though it remains uncertain whether individual sows normally produce litters twice a year (Yasuma, 1984). Juvenile mortality averages 15% in the first three months in western Europe, though between 50% and 75% mortality have been reported by the end of the first year of life (Jezierski, 1977; Briedermann, 1990). These mortality rates are thought to be highly dependant on such external factors as predation and climatic hazards, at least in wilderness areas (Spitz, 1986). Similarly, Pauwel (1982) suggested that the principal causes of juvenile mortality in wild pigs in Ujung Kulon were predation (particularly as a consequence of the accidental separation of infants from their mothers), along with differences in the relative rate of development of litter-mates and various parasite-related causes. These factors resulted in only about 15% of all progeny surviving to independence.

Systems: Terrestrial .

Major Threat(s):

At a global level, there are no major threats to the species. However, there are many threats at a more local level, principally habitat destruction and hunting pressure, either for food, sport or in reprisal for crop damage, particularly in areas near human habitation. In Afghanistan, Hassinger (1973) cited reports of the decrease in the numbers of wild pigs in the Pul-i-Khumri District in the 1950s as a result of the draining of marshlands for agricultural purposes, and hunting by Europeans; but noted that they were still numerous in other districts where: "they invade the fields and cause serious damage during the harvest". In parts of Perak and Johore in West Malaysia, Diong (1973) reported that the numbers of wild pigs (*S. s. vittatus*) had diminished drastically as a result of increased hunting pressure, particularly for commercial purposes, and that most hunting methods, whether with guns, dogs or snares, was entirely non-selective. Nonetheless, wild pigs are still included in Malaysia's 'Agricultural Pest Ordinance, 1977', on account of their damage to a variety of crops, including sugar cane, tapioca, rice and even coconuts. Lay (1967) also remarked upon the damage to crops by wild pigs in Iran which: "...brings great wrath upon them, usually ineffectual, from the local farmers". In Pakistan, the expansion of the sugar cane industry in the 1960s and early 1970s brought about local increases in the numbers of wild pigs (*S. s. davidi*), whose depredations in the cane fields (estimated at an annual loss rate of Rs.5 million in 1978) led to the development of control measures, including the use of poison baits (Shafi and Khokhar, 1985).

Although actually referring to wild pigs in India, it is clear that Prater's (1971) opinion that: "No animal is more destructive to crops and, in cultivated areas, it is impossible to make a plea for its protection", would find popular support in many other countries. In the Ryukyu Islands, S. Japan, the endemic *S. s. riukiuanus* is actively regarded as an environmental pest, and bounties are paid to farmers for killing them by the Japanese Government, despite the fact that this taxon is recognized internationally as being seriously threatened throughout it extremely restricted range (see later text). In various places in Indonesia, most notably in Java, deliberate attempts have been made in the past to eradicate wild pigs altogether by means of organized shooting parties and poisoning campaigns. However, despite many thousands of wild pigs being destroyed in this way, it is clear that this has had little lasting effect on these animals which, according to Hoogerwerf (1970), have: "remained further removed from extinction than any other species of game in Java". This situation evidently obtains in a number of other countries, including Vietnam (R. Ratajszsak pers. comm.) and Taiwan (K. Nowell pers. comm.), where wild pigs are reported to survive in a number of nominally protected areas where all or most of the other principal 'game' species have been eradicated by sustained hunting pressure.

In these and many other (non-Islamic) countries or local communities, wild pigs often constitute the single most important game animal to subsistence and/or recreational hunters. Sport hunting accounts for about 30 to 50% of animals heavier than 20 kg in southern France, though this figure may be as high as 50 to 75% in some heavily populated countries, where hunting remains largely or wholly uncontrolled (Spitz, 1986). For example, the wild pig population inhabiting the 25,000 ha of broadleaved woodlands around Monticiano in Italy, has apparently been able to sustain its numbers despite an annual hunter-kill rate of about 50% (c. 500 animals) of the population (Devitt, 1984).

According to figures produced by the Japanese Environment Agency a total of 1,279,453 (an average of 63,973 per annum) wild pigs were harvested in Japan between 1970 and 1989 (incl.), of which 1,083,858 (85%) were taken by hunters, and the remaining 195,596 were destroyed as crop pests during the closed season. The great majority of these animals are S. s. leucomystax from the main islands of Honshu, Shikoko and Kyushu, and this subspecies is evidently able to withstand high levels of harvesting to judge from contemporary reports of marginal expansions in the overall range of this subspecies (Hanai, 1982; Takahashi, 1980). Unfortunately, however, these totals do not differentiate between the numbers of S. s. leucomystax and those of S. s. riukiuanus from the three small islands of Amami Oshima, Tokuno Shima and Kakeroma Jima, which are also included in these figures. As previously indicated, the latter subspecies is declining rapidly in numbers on these islands, and on the neighbouring island of Okinawa, largely as a result of over-hunting (Barber et al., 1984; H. Obara pers. comm.). This conclusion is also borne out by official figures on the annual harvest of S. s. riukiuanus from two other islands, Iriomote and Ishigaki, in the Ryukyu chain between 1965 and 1981, which indicate a downward trend in the numbers of these animals (Yasuma, 1984) and a marked increase in the amount of effort required by local hunters to obtain smaller catches (Hanai, 1982). Nonetheless, according to these figures an average of 530 and 190 wild pigs were still being taken annually on Iriomote and Ishigaki, respectively; of which slightly more than 50% were killed as 'agricultural pests', i.e. mostly out of season, often with a Government bounty being paid. Moreover, it is certain that these totals are a serious underestimate - perhaps by as much as 50-100% - of the actual numbers killed, since many hunters operate without licenses and do not declare their catches (Barber et al., 1984; Yasuma, 1984). Formerly, virtually all of these pigs were consumed locally but, following the restoration of electricity (and, hence, refrigeration) in 1972, commercial traders began operating on the islands, and the majority of carcasses are now exported to gourmet markets in Osaka, where they realize relatively high prices and ensure a continued demand for wild-caught meat (Yasuma, 1984).

Wild pigs are also susceptible to a variety of highly contagious diseases which can decimate their populations. Santiapillai et al. (1991) referred to a catastrophic crash in the wild pig population of Ruhuna National Park in Sri Lanka in 1989 owing to an outbreak of swine fever, but that population numbers had apparently recovered by June 1991.

Many wild pigs are also reported to have died during an outbreak of disease in the Kanto Region on Honshu Island, Japan in 1877 (Yanagida, 1913). It is possible that this was hog cholera, contracted from domestic pigs which were imported to the region at that time (Chiba, 1975), and this may have been one of the factors contributing to the extirpation of *S. s. leucomystax* throughout the north of their range on this island. An outbreak of severe skin disease (tentatively identified as sarcoptic mange) on Iriomote, which was first noted in 1976, had spread throughout the Island by 1980, and infected up to 83% of the *S. s. riukiuanus* population according to data based on hunter-killed samples (Yasuma, 1984).

The Ryukyu pigs are also threatened by genetic contamination through contact with free-ranging domesticates. On Ishigaki Island, interbreeding with domestic pigs has been so prevalent that it is doubtful if the remaining wild stock can be considered genetically pure (Obara, 1982). Moreover, it is unlikely that this problem will be confined to Ishigaki, given that the meat from wild x domestic hybrids or 'ino-buta' is popular and local farmers will continue to produce it. Indeed, ino-buta have been farmed commercially on Iriomote for many years, often in poorly controlled conditions. As a result, feral ino-buta have become established on the adjacent, smaller islands of Sotobanare and Uchibanare, and it is claimed that a number of these animals have crossed onto Iriomote at low tide; thereby posing a further, serious threat to the remaining wild stock there (Yasuma, 1984; Takahashi and Tisdell, 1992).

Similarly, Genov et al. (1991) reported that the traditional practice of rearing 'domestic' pigs in semi-wild conditions in Bulgaria has resulted in their hybridizing with the wild boar populations in the eastern and north-eastern parts of that country, and that genetically pure wild boars now occur only in the Rila-Pirin-Rhodopes Mountains in the south of this country, where domestic pigs are not reared in the wild. The principal threat to the survival of the Monticiano population in Italy, lies not so much with the high annual hunter-kill rate but with the fact that the genetic integrity of the native population has been seriously compromised by the introduction of Central European boars after World War II, which have interbred with these animals. Indeed, the introduction of non-native wild pigs, by hunting interest groups in the belief that these were more desirable because of their larger size and reputed greater fecundity, has been so widespread in Italy that the only surviving pure-bred native stock is thought to be in the presidential reserve in Castel Porziano, near Rome (Devitt, 1984).

Conservation Actions:

Generally, few, if any, practical measures have been taken in any country for the specific purpose of conserving wild populations of any subspecies of S. scrofa, except in the sense of maintaining stock levels for hunting, particularly for sport hunting. Thus Genov et al. (1991) reported a gradual increase of the wild boar population in Bulgaria in the late 1950s following legal measures to regulate hunting, and because of the reintroduction of animals from breeding farms in the north-east of the country, in the Balkan Range and Sredna Gora. Even in this instance, however, other factors may also be operating, since similar increases in the local wild pig populations in Spain, France, Switzerland, Czechoslovakia and eastern Russia have been described by Sáez-Royuela and Tellería (1986), who simply attributed this trend to a progression to a more temperate climate and, hence, progressively milder winters of benefit to these animals.Local wild pig populations are also reported to be increasing in numbers in several countries as a consequence of the spread of Islam. For example, the north-west African race, *S. s. algira*, although extinct in Libya, is thought to be expanding in neighboring Algeria where it is designated a game animal but, as the meat is now seldom eaten in this country, hunting pressure has been greatly reduced since colonial times (Kowalski and Rzebik-Kowalski, 1991). These authors also cite earlier reports of wild boars hybridizing with domestic pigs, but the latter are no longer kept in Algeria.

S. s. riukiuanus has been included in the IUCN Red List since 1982, where its 'Vulnerable' status reflected a widespread concern about the environmentally destructive development of the Ryukyu Islands and the growing number of threats to the ecosystems and endemic species of Iriomote and other islands in this region. In 1978, these concerns led to the passing of a resolution at the 14th IUCN General Assembly in Ashkhabad, which called upon the Government of Japan to take immediate steps to ensure the conservation of Iriomote Island and its endemic taxa, and for the subsequent designation of the whole of the Nansei Shoto Region (often referred to as the 'Galápagos of the East' on account of the high percent endemicity of its fauna and flora) as a bio-geographical priority region in the World Conservation Strategy. In 1982, the 'Nansei Shoto Conservation Project' was launched by WWF-Japan, with initial emphasis being placed on the conduct of wildlife and socio-economic surveys on Iriomote and Amami Oshima (Obara, 1984a,b). In the interim, however, the desired increased protection of the Nansei Shoto Region was clearly at odds with the Japanese Government's regional policies, and the development of the islands continued - a circumstance which led to the passage of a another resolution, highlighting Japan's treatment of the region, at the 16th General Assembly of IUCN in Costa Rica in 1986. This also appears to have had little affect, as evinced by further major development plans, including the controversial Ishigaki Airport extension, and recent plans to construct a new dam and road on Iriomote Island. Although one third of Iriomote has been designated as a national park, the proposed road will cut through the park and will also increase access by poachers who have already reduced the wild pig to less than half of its numbers before the park was created (Brazil, 1988).

Of all wild pigs species, *S. scrofa* is by far the most widely maintained and bred in captivity, though general perceptions about the relative abundance of this species has resulted in diminished interest in its propagation. Many zoos, particularly in western Europe, have therefore disposed of their stocks of these animals, mostly to 'wild boar breeding farms', which have escalated in number to meet growing demands in the gourmet meat markets and for commercial diversification. In addition, increasingly restrictive quarantine and other veterinary regulations appertaining to the international movements of all live suids, has made it extremely difficult to establish new breeding programs.

Contrarily, since these animals are most easily acquired in their countries of origin, local stocks are most likely to be pure-bred and, hence, most valuable for research and conservation purposes. In this situation, priority should be given to the establishment of national or regional breeding programmes for each of the rarest subspecies, especially S. s. riukiuanus.

As a general rule, *S. scrofa* is both highly adaptable and highly resistant to a variety of degradative processes, and may thrive under conditions of habitat modification and hunting pressure which have devastated other forms of wildlife. In addition, most subspecies are well represented in protected areas in their relatively extensive ranges. Nonetheless, there are reasons for concern about a number of distinct populations of S. scrofa, of which the Ryukyu pigs are undoubtedly the most seriously threatened. Indeed, riukiuanus is the only subspecies of S. scrofa to be included in the IUCN Red List of Threatened Animals (IUCN, 1990), where it has been accorded the status of 'Vulnerable' since 1982, but is already thought to be 'Endangered' on at least four of the six islands in the Ryukyu Chain which constitute its entire, and therefore extremely restricted, range. This particular form is also the most distinct, and (with an average adult male body weight of only about 45 kg; Barber et al., 1984) by far the smallest, subspecies. It is therefore of some interest as a potential genetic resource, as well as being one of the principal 'flagship' animals for increased conservation activity in the crucially important Nansei Shoto Region. For all of these reasons, the following recommendations (taken from Oliver et al 1993) are mostly confined to the immediate and longer-term management requirements of this taxon.

Objectives

1. To promote local interest in, and the enhanced future protection and management of, the Ryukyu Islands' dwarf wild pig, S. s. riukiuanus, and any other threatened native subspecies may be identified in the future;

2. To promote further applied research on the taxonomy, distribution, conservation status and biology of the species, particularly the least known forms; 3. To discourage any future releases of any specimens of any subspecies of this species, or its domestic and feral derivatives, outside their known former ranges, and encourage the eradication or control of such introduced populations as may exist at present; and

4. To investigate the cultural and economic significance of these animals to local people, particularly in southeast Asia, and their potential genetic importance in terms of: a) possibilities for their further domestication and/or b) an enhanced understanding of the origins and relationships of present day domestic and wild populations.

Priority Projects:

1. Provide all appropriate support to any existing or proposed future conservation initiatives in the Nansei Shoto Region, relevant to the enhanced future management and protection of the Ryukyu Island's dwarf wild pig, *S. s. riukiuanus*.

2. Conduct (preliminary or repeat) status surveys on all (six) islands in the Ryukyu Group known to support remnant populations of *S. s. riukiuanus* (i.e. Iriomote, Ishigaki, Okinawa, Tokunoshima, Amamioshima and Kakerome), and implement emergent recommendations.

3. Promote development of a coordinated breeding programmed for *S. s. riukiuanus,* both nationally and internationally.

4. Collect, evaluate and disseminate data on the taxonomic relationships, distribution and conservation status of other, less-known forms, with a view to the resolution of certain outstanding taxonomic problems and the development of recommendations for their enhanced future protection, if so required. Additional data is needed on the distribution and status of some, poorly-known subspecies and populations, including a tentatively assigned form (subsp. nov.) from the central highlands of Sri Lanka and the insular taivanus (Taiwan). Efforts should also be made to resolve outstanding questions relating to the genetic distinctness and range limitations of many of the continental races, which are poorly understood at present.

5. Promote further research on the biology and ecology of all less-studied populations, particularly those in extreme habitats, or in areas of sympatry with other species of Sus. Most of the information on this species has originated from studies conducted in Europe or on the feral populations in the U.S.A., Australia and New Zealand. Priority should therefore be given to studies of other populations elsewhere, particularly those in semi-desert regions or other extreme habitats, and/or studies focused on the ecological and genetic relationships between this and other species of Sus in those areas where sympatry occurs (e.g. with S. barbatus in the Malaysian Peninsular and Sumatra, and with *S. verrucosus* on Java).

Lesser Mouse Deer *(Tragulus javanicus)*

Range Description:

Tragulus javanicus as here defined is endemic to the island of Java, Indonesia, according to Meijaard and Groves (2004). The latter authors did not mention the island of Bali, but a sighting was reported from Bali Barat National Park, Bali, in a bird watching trip report (Bird quest 2006). The genus was not listed for Bali in the exhaustive reviews of Chasen (1940; of the genus) or Meijaard (2003; mammals of Indonesian islands), nor by Grubb (2005). Given the live-animal trade in the genus on Java, further records are needed to confirm whether or not there is a native population of the genus on Bali. If there is, bio geographic considerations suggest it would be most closely related to the Javan population and quite probably conspecific.

Hoogerwerf (1970) wrote that "in Java *T. javanicus* was encountered in all provinces, fairly intensively distributed from sea-level to high in the mountains". Present-day occurrence in East Java is questionable: S. Hedges, M. Tyson and E. Meijaard (pers. comm. 2008) know of no certain records (as distinct from listings in collation documents which do not cite information to primary source) from areas like Baluran or Alas Purwo national parks, despite high survey effort during 1991–2000.

Meijaard and Groves (2004) gave only one specific locality for the species: Cheribon (= Cirebon) on the north coast of West Java Province, which is the type locality of *T. j. pelandoc* Dobroruka, 1967. Dobroruka (1967) also mentioned the western part of Java, to the southern coast, for what he called *T. j. focalinus (*which is *T. javanicus s .s.).* Meijaard and Groves (2004) did not list the localities for the many specimens they examined. Dobroruka (1967) and Van Dort (1986) both discuss variation within Javan chevrotains; neither has yet been checked for specific localities. More recent localities, although not of specimen records, include: Gunung Halimun (reportedly camera-trapped some time before 2003 [Suyanto 2003], but the photograph is not reproduced in the appendix and a painting is used for the species instead); Ujung Kulon (1991–1993; van Schaik and Griffiths 1996; C.P. Groves pers. comm. 2008), and the Dieng Plateau (during 1999–2000; V. Nijman pers. comm. 2008). There has been no collation of records from the various surveys over the last 20 years or so, and some observers no doubt are aware of other localities. Dubious sighting have been recorded in southern **Thailand but these could be erroneous**

Countries: Native:

Indonesia (Jawa, Kalimantan, Sumatera); Malaysia (Peninsular Malaysia, Sabah, Sarawak)

Population:

There appears to have been no field study specifically of chevrotains on Java but current wildlife surveyors agree that they are rarely seen compared with results from similar styles of observation in Kalimantan

Even Hoogerwerf (1970), who clearly found the species with some regularity (explicitly not daily, but enough for him to learn its calls and thereby establish dates of a probable mating season), complained that "it is particularly difficult to obtain any insight into the situation of this species in Java", but concluded that "it is improbable that the species is in danger of extinction". The basis for an unattributed statement in Whitten et al. (1996), that "it is still numerous and can be seen easily in many areas such as the tourist park part of Pangarandaran Nature Reserve" is unclear. The genus was recorded at five of ten sites on the Dieng Plateau surveyed in 1999–2000, but mostly only through single observations (V. Nijman pers. comm. 2008). S. van Balen (pers. comm. 2008) points out that in Java the genus seems very shy (compared with animals in Malaysia and Kalimantan), so most records come as footprints. This contrasts with the assignment by Hoogerwerf (1970) that the genus was common and widespread. It is therefore quite plausible that a major decline has taken place, although other explanations remain to be explored which could have bolstered Hoogerwerf's sighting rates, such as his having a dog with him which flushed the chevrotains, or his spending a lot of time in the species' favoured microhabitats. Specifically, neither M. Tyson nor S. Hedges (pers. comm. 2008) saw chevrotains during a rhinoceros survey in Ujung Kulon in 1992, whereas this was Hoogerwerf's (1970) main site in assessing the species as relatively common and readily found. It was camera-trapped there five times during 1991–1993 (van Schaik and Griffiths 1996), a rate comparable with that of many other species in the study and certainly not suggesting out-and-out rarity. Very recently, numbers in trade in Java have dropped sharply in most cities, and it is plausible that this reflects increasing difficulty in procuring the animal (G. Semiadi pers. comm. 2008).

Population Trend: Unknown

Habitat and Ecology:

Hoogerwerf (1970) wrote of chevrotains on Java occurring "from sea-level to high in the mountains". In the Dieng plateau area, V. Nijman (pers. comm. 2008) found them only a few times in the lowlands (400–700 m asl), where most survey took place, and had no records from above about 1,500 m asl. They have been found on Gunung Gede –Pangangro up to about 1,600 m asl (V. Nijman pers. comm. 2008). Hoogerwerf's (1970) description of favoured habitats on Java suggests that chevrotains there might be an 'edge' species, certainly seeming to prefer areas with thick understory vegetation, such as that along riverbanks. This would not be unusual within the genus (see other Tragulus accounts).

Systems: Terrestrial

Major Threat(s):

Java has highly fragmented natural habitats and has done for centuries, reflecting longstanding high human population densities. Many protected areas were established during the Dutch colonial period but from independence up until the 1970s they were largely under-funded and neglected. After hosting the World Parks Conference in 1982, the Indonesian government gazetted a swathe of national parks and more structured conservation planning began, funded by the World Bank and other donors. The focus was largely on the 'multi-function' national parks and much money was spent on infrastructure, some staff training and increased personnel. The 'lesser' protected areas such as "game reserves" and "nature reserves" still had few staff and resources, and that has continued to the present. During the 1980s to the mid 1990s, guns were tightly controlled and the military and police were feared and respected. However, the strong culture of caged bird keeping meant that hunting, including that within protected areas, was primarily for birds and some small game, through various forms of trapping, including snaring; this latter could well have included chevrotains. There was some habitat loss from protected areas through illegal logging, agricultural encroachment and other offtake, but the national parks of Java remained remarkably intact for much of this period. Socio-political changes from 1997 led to a reduction in the respect for the police and military and the rise of a viewpoint that protected areas were the peoples' resources and would therefore benefit from decentralised management (S. Hedges and M. Tyson pers. comm. 2008). This policy change, which risked a 'tragedy of the commons', has indeed led to increased destruction and poaching in the past decade. There is some evidence that species readily uses edge and secondary vegetation, meaning that effects of forest encroachment will be less serious than for old-growth forest obligates. Moreover, chevrotains already seemed rare for surveyors operating in the early–mid 1990s, when habitats had been relatively stable since 1970s or early 1980s. For any decline which may have occurred in the 1980s and early–mid 1990s, therefore, habitat factors are an unlikely driver.

Chevrotains occur regularly in markets in towns such as Jakarta, Surabaya, Yogyakarta, and Malang, but often they are cramped in small cages, and can therefore be overlooked, and may even be more common than observations suggest; numbers in trade are surprisingly high given the small numbers to be seen in the field nowadays (V. Nijman pers. comm. 2008). They have been traded at high levels for many decades: Hoogerwerf (1970) wrote of "numerous reports of mouse deer being regularly trapped and offered for sale alive" in Java. They are hunted and traded both for pets and as wild meat (S. Hedges pers. comm. 2008). Numbers passing through markets in Jakarta, Bogor, and Sukabumi have recently declined sharply, perhaps because of tightened control by the forest police in those markets; but there is the possibility that falling trade might indicate, at least in part, increasing difficulty to catch the animal and thus a decline in populations. In the Malang area, it is still "relatively easy" to procure one, although it is "getting time consuming" (G. Semiadi pers. comm. 2008). Hunting is probably largely with snares; dogs are also likely to be a serious threat (M. Tyson pers. comm. 2008). They are vulnerable to active hunting at night through a propensity to freeze when spot lit. However, the effects of these comparably high hunting levels on the genus in Java have not been studied empirically. The continued presence of many animals in markets suggests that significant populations remain somewhere on Java (assuming that the animals are not now imports from elsewhere in Indonesia), and thus that a major decline may not have occurred, despite current indications. A comparable situation was found with in ornate squirrel *Callosciurus inornatus* in Lao PDR, when extensive 1990s surveys found few animals in the field, in contrast with historic statements of abundance and ongoing substantial numbers being traded in fresh meat markets. This led to conservation concern for the species (e.g. Duckworth et al. 1999); but later field survey of degraded and edge areas found out that the species was indeed common and evidently a species associated with degraded areas, and had hence been severely under-recorded by the 1990s surveys (Timmins and Duckworth in press).

Conservation Actions:

Chevrotains on Java occur in some protected areas, e.g. Ujung Kulon (Hoogerwerf 1970; van Schaik and Griffiths 1996) and were earlier said to "occur in all game sanctuaries in Java and in most of the nature reserves established on that island...." (Hoogerwerf 1970). The species has been officially protected since 1931, yet it is still hunted. The taxonomic revision of Meijaard and Groves (2004) means that *T. javanicus* is endemic to Java, and, if it occurs, Bali. There are fair indications of a decline, perhaps a major one, and thus this newly-revealed endemic species should be swiftly removed from the anonymity in which it has lain for decades. An urgent first step is a collation of existing information, because it is possible that many more records exist than were traced during the preparation of this assessment. This should include trawling likely observers and examination of collections not covered by Meijaard and Groves (2004). If insufficient number of specimens are found to clarify the number of taxa on Java, more should be obtained. Some may come from markets but, because of the pre-eminence of locality in determining systematics among very similar taxa (e.g. Groves in press: discussion under Wapiti group), specimens of known locality origin must form the basis of analysis. Whatever the number of species on Java, the difference in sighting rate between Hoogerwerf (1970) and observers from the 1990s onwards is suggestive of a major decline, although other explanations are possible (see Population and Threats). Current status needs to be clarified through specific surveys for the genus (camera-trapping, but undertaken in a way more suitable for smaller species than is usual, spotlighting, and hunting surveys may all play a role). Surveys must take care to investigate secondary and edge areas which are often eschewed by general wildlife surveys in favour of the less encroached areas.

Napu *(Tragulus napu)*

Description:

Although very small for an ungulate, the greater mouse-deer is one of the largest members of its genus. It is rivalled in size by Williamson's mouse-deer. It weighs 5 to 8 kg (11 to 18 lb). Its head-and-body length is 70 to 75 cm (2.30 to 2.46 ft) and its tail length is 8 to 10 cm (3.1 to 3.9 in). Its shoulder height is about 30 to 35 cm (12 to 14 in) It has a small, triangular head with a small, pointed, black nose and large eyes. Its long legs are as thin as a pencil. The hind legs are visibly longer than the front legs. The body is rounded. The fur on the upper part of its body is grey-buff to orange-buff. On the sides, the fur is quite pale, but darker along the midline. It is white underneath, especially on the neck, stomach, chest and chin. The male has neither horns nor antlers, but has small "tusks" – elongated canines in the upper jaws. The greater mouse-deer is solitary and nocturnal. It uses small trails through thick brush in the forest. When the male is ready to mate, he rubs a large gland on his lower jaw against the female to determine whether she is ready to mate. If she is not ready, she responds by walking away. The male is very territorial, marking his territory with feces, urine and secretions from the intermandibular gland under the chin. When angry, the male will beat the ground with his hooves at a rate of four times per second. They are rather trusting but delicate animals. They feed on fallen fruits, aquatic plants, buds, leaves, shrubs and grasses. Greater mouse-deer breed throughout the year; The female spends most of her adult life pregnant. They usually produce one young per birth, after a gestation of 152–155 days. Newborn animals are well-developed and immediately able to stand; they are fully active after 30 minutes. The young stand on three legs while nursing. Both male and female become mature at age 4½ months. Their life span is up to 14 years.

Range Description:

Tragulus napu, as constituted here (that is, excluding *T. versicolor* of Indochina and *T. nigricans* of the Philippines) occurs in the Sundaic sub region, extending some way up the Thai–Malay peninsula, in the following countries: Brunei, Indonesia (Kalimantan, Sumatra, and many small islands), Malaysia (West Malaysia, Sabah, Sarawak, and many small islands), Myanmar (far south only), Singapore (Pulau Ubin only), and **Thailand (south only)** (Meijaard and Groves 2004, Chua et al. 2009). Grubb (2005) also included Cambodia, Lao PDR, and 'Indochina' in the range of *T. napu;* but this was founded on the assumption that because *T. versicolor,* formerly considered a subspecies of T. napu, occurred in Vietnam, *T. napu* must presumably occur in intervening Lao PDR and Cambodia. There is, for example, a map of such a conterminous range for *T. napu* in Corbet and Hill (1992). There is no evidence of such a distribution and, given the distribution patterns of other species, absolutely no reason to expect that it would have such a range, even if *T. napu* and *T. versicolor* are closely allied. Recent surveys in both countries have not suspected the species (Duckworth et al. 1999, R.J. Timmins pers. comm. 2008).

The northern limit on the Thai–Malay peninsula is not well clarified. Despite fairly intensive camera-trapping in Kuiburi National Park, **Thailand** (12°N and thus within the generally-assumed range for T. napu), Greater Chevrotain has not been photographed there. It must be scarce in this park if it occurs at all (Steinmetz et al. 2007, R. Steinmetz pers. comm. 2008).

Caution is needed in interpreting modern records of *T. napu* from anywhere outside the specimen- and photograph -validated range (as presented in Meijaard and Groves 2004); some camera-trap studies seem to have taken an essentially arbitrary attitude on naming their photographs of chevrotains, resulting in published listings of *T. napu* from, for example, northern **Thailand**. Because such listings never highlight the significance of such records, let alone discuss them, all are assumed to be in error, and are not discussed further, here.

Countries: Native:

Brunei Darussalam; Indonesia (Kalimantan, Sumatera); Malaysia (Peninsular Malaysia, Sabah, Sarawak); Myanmar; Singapore; **Thailand**

Population:

Greater Oriental Chevrotain populations are not well studied, in part reflecting a general lack of research interest in the genus and in part its almost complete sympatry with the Lesser Oriental Chevrotain, from which separation requires care under field conditions (Duckworth 1997; Matsubayashi and Sukor 2005). It seems to be relatively common on at least Borneo (Payne et al. 1985; Belden Giman pers. comm. 2008).

The genus (presumably including this species) was ubiquitous in a study of virgin jungle reserves (VJRs) of West Malaysia which paired each of seven VJRs with a nearby unprotected area with seriously encroached habitat (Laidlaw 2000). The ability of *T. napu* to use plantations is probably low, because all (the admittedly relatively few) sources agree that it does not persist well in secondary or logged forest. Moreover, Belden Giman (pers. comm. 2008), in extensive observations at Sarawak Planted Forests, Bintulu, Sarawak, has never recorded *T. napu* in any of the blocks of monoculture plantations there, despite many records of *T. kanchil*. Given the pace of conversion of forests to plantations across the rane of *T. napu,* clarification of its degree of plantation use is urgent. Competition with *T. kanchil* is not pronounced. In Tabin Wildlife Reserve, Sabah, *T. napu* was largely confined to mature forest with *T. kanchil* only in immature forest (Matsubayashi and Sukor 2005). In unlogged forest at Danum Valley, Sabah, *T. napu* was significantly more common than *T. kanchil,* but following logging, the numerical dominance of *T. napu* declined, and in 2–5-year-old regrowth T. kanchil was the commoner species. By 12 years post-logging, T. napu had reasserted its numerical predominance (Heydon 1994). Tragulus spp. are generally thought to be nocturnal, and at two sites in Borneo, they comprised over half the contacts of spot-lighting surveys (Duckworth 1997), but Matsubayashi and Sukor (2005) found substantial amount of both day- and night-time activity in this species. Duckworth (1997: 186–187) pointed out that most chevrotains spotlight in Similajau National Park, Sarawak, (identified as *T. napu*) were sitting around resting. The related *T. kanchil* was considered highly solitary by Matsubayashi et al. 2006) and this may also be somewhat true for T. napu; Duckworth (1995: 177) observed chevrotains identified as T. napu 14 times, of which 12 records involved singles and two involved duos in apparent association. Collation of information from general sources suggests the following, but the contribution of information from *T. nigricans* to these conclusions is unknown, so their reliability for *T. napu s.s.* should be considered preliminary. Females gestate for 152–155 days and give birth to one or, rarely, two young. Young wean at 2–3 months, and reach sexual maturity at 4.5 months. Animals live up to 14 years. The female can be pregnant throughout her adult life, often having just an 85–155 minute 'breather' between giving birth and becoming pregnant again. The young are born fully developed, and can stand within 30 minutes of birth. Mothers nurse their young standing on three legs. They are generally solitary.

Systems: Terrestrial.

Major Threat(s):

Hunting and habitat loss occur at high levels almost throughout Greater Chevrotain's range, but the species is said to be relatively resilient. However, few data to support this assessment have been traced, particularly because most key studies have not within the Sundaic region identified chevrotains to species. Thus, while clear that the genus remains widespread and common in the face of these threats, there seems to be little publicly available evidence that T. napu specifically does so, and reasonable evidence that it is not very adaptable to forest degradation (see Habitat and ecology). However, on the assumption that most chevrotains above 100 m asl on Borneo are this species rather than *T. kanchil,* Borneo at least retains large populations. Greater Oriental Chevrotain is extinct on Singapore (K. Lim pers. comm. 2008), although the relative contributions of hunting versus habitat perturbation are not known. Poaching pressure in Kerinci-Seblat National Park, Sumatra, among snared ungulates was greatest for muntjacs and lowest for chevrotains, as assessed by diameter of the cable (Linkie et al. 2003). Chevrotains were scarcer in areas of higher human use (and thus hunting) of Bukit Barisan Selatan National Park, Sumatra (O'Brien et al. 2003; see Habitat and ecology). Bennett et al. (2000) profiled the effects of hunting in both Bornean states of Malaysia (Sarawak and Sabah), each with eight study areas, upon chevrotains (not distinguished to species) through a lengthy hunting study in February 1993 to June 1995. They constituted 8–9% of animals killed. No direct or index measure of chevrotain densities was possible at these sites to determine what effects, if any, such heavy hunting was having on chevrotains. Chevrotains are among the most favoured wild meat by inhabitants of the Sarawak Planted Forests area, Bintulu, Sarawak, after Sambar Rusa unicolor and wild pigs Sus (Belden Giman pers. comm. 2008).

Conservation Actions:

Hunting and habitat loss occur at high levels almost throughout this species' range. Adequate legal basis exists in all range states to control both threats (through protected areas and regulations governing wildlife trade and hunting). The chief need is for effective translation of these laws into action. There are no obvious species-specific needs additional to general consolidation of protected areas and enforcement of hunting and wildlife trading laws. The paucity of Sundaic studies determining chevrotains to species shows admirable caution but hinders assessments of this species's status. Better data would be useful to confirm the opinion-based conclusion here that this is a resilient species, and to investigate further habitat use, in areas where it overlaps with *T. kanchil* and in those (probably, primarily at higher altitude) where it does not.

Eld's Deer *(Cervus eldii)*

Description:

The following measurements have been reported for the Eld's deer:

- Head–body length: 150–180 cm (59–71 in)
- Shoulder height: 110–125 cm (43–49 in)
- Tail length: 20–30 cm (8–12 in)
- Weight: 125–175 kg (276–386 lb)
- Antler length: 99 cm (39 in)

The deer are generally of medium size and are similar to the size and shape of the barasingha. The species has a very regal and graceful *Cervus* physique. Its legs are thin and long, and has a long body with a large head on a thin neck. The throat of a male has a thick mane of long hair. Males (stags) are taller and heavier than the females (hinds or does). Their coats, rough and coarse, change colour with the season; in summer the colour is reddish-brown, while in winter, it turns dark brown, with males tending to be darker than the females. The tail is short in length and the rump has no distinct patch. Despite these features, they are actually related to the Père David's deer. The antlers, bow- or lyre-shaped, do not grow upwards, but tend to grow outwards and then inwards; a smaller branch grows towards the front of the head. The brow tines are especially long and noticeable. The brow-antlered deer is so named because they have long brow tines. They shed their antlers every year, with the largest size attained during the breeding season.

Range Description:

This species was formerly widely distributed across suitable habitats of South and Southeast Asia, from the Manipur region of northeastern India through much of Myanmar, **Thailand,** Lao PDR, Cambodia, and Vietnam to the island of Hainan (China) in the east (Salter and Sayer 1986; Grubb 2005). The historical range was broken into four major components, consisting of the Manipur region of India inhabited by *R. e. eldii; R. e. thamin* on the central plains of Myanmar*; R. e. siamensis* populations in the lowlands of **Thailand,** Cambodia, Lao PDR and Vietnam (separated by the mountains along the Thai/ Myanmar border from *R. e. thamin);* and the population on Hainan and former populations in mainland southern China, which appear to have been disjunct outliers of *R. e. siamensis*, separated from its main range by mountainous terrain in Lao PDR and Vietnam.

The global Eld's deer population is currently very localised to small areas within the species' former range. *R. e. eldii* is now confined to a single small population at the southern end of Loktak Lake in Manipur, India (Singh 2004). *R. e. thamin* still occurs in several localised areas of central Myanmar, as well as there being introduced populations in **Thailand** (McShea et al. 2000; Aung 2004; Naris Bhumpakphan et al. 2004). *R. e. siamensis* occurs in one or two small localised populations in Lao PDR (Johnson et al. 2004), and as scattered small subpopulations mainly in the northern and eastern lowlands of Cambodia (Tordoff et al. 2005), and occurs in a relatively wild state in one protected area on Hainan, with additionally several other managed herds on that island (Pang et al. 2003).

Countries: Native:

Cambodia; China; India; Lao People's Democratic Republic; Myanmar

Possibly extinct:

Thailand; Vietnam

Population:

Based on both habitat extent and size of remaining patches, Myanmar and Cambodia are the pre-eminent countries for Eld's deer (McShea et al. 2005). However, the actual status of remaining populations is more related to hunting levels. In Cambodia Eld's deer still occurs over a relatively wide area, although localised within this, surviving as small remnant groups in the lowland forests of the north and east. The total population could well be several thousand animals in a forest area that probably exceeds 20,000 km^2, but densities are extremely low (Tordoff et al. 2005; R.J. Timmins pers. comm. 2008). Survey work in appropriate habitat in the last decade has detected populations in at least 15 separate areas, with anecdotal evidence from several more areas (McShea et al. 2005; Tordoff et al. 2005; Bird et al. 2006; Timmins 2006; Bezuijen et al. in press). A few may remain in the southwest, although there is little evidence other than for one small population in the Phnom Aural area which persisted until at least the late 1990s (Henshaw et al. 2002; Tordoff et al. 2005). Declines since the late 1950s, when large herds were still readily seen (Wharton 1957), have been dramatic. Even in the last decade or so, the species has declined by 90% or more, given comparisons of aerial counts in 1994 with ground survey results from 1998 to the present (Olivier and Woodford 1994; Timmins and Ou 2001; Timmins et al. 2003; R.J Timmins pers. comm. 2008). Declines are likely to continue into the future, especially as most remnant groups and animals live either outside protected areas, or in protected areas with little active management. At least a 50% decline in the next 10 to15 years seems likely. Populations in the Srepok Wilderness Area, the Siema Biodiversity Conservation Area, the Preah Vihear Protection Forest, Ang Trapeang Thmor Conservation Area and Kulen Wildlife Sanctuary are likely to remain stable and may potentially increase due to ongoing externally funded conservation area management projects (WWF unpublished data; WCS unpublished data; R. J. Timmins pers. comm. 2008). *R. e. thamin* was still relatively widespread and abundant during the mid-1980s on the plains of central and northern Myanmar, where its range centred on the Irrawaddy Plain, including the Bago or Sittang Plain to the east.It was said to be present to the southeast, along the **Thai border (**Salter and Sayer 1986), and along the western border with Bangladesh (Lekagul and McNeely 1977), but these records are doubtful and a questionnaire in 1992 and surveys in 1997 could only find evidence for the species in the northeast (McShea et al. 2000). In 1992 a countrywide questionnaire was distributed by the then Wildlife Department of Myanmar, this resulted in reports of Eld's deer from 28 Myanmar townships, and 2,200 Eld's deer were estimated to remain within Myanmar, with the largest population (over 1,200 deer) in Chatthin Wildlife Sanctuary (Myint Aung 1994; McShea et al. 2000). Regular transect surveys in Chatthin Wildlife Sanctuary between 1983 and 1996 indicated a population decline of upwards of 40% (McShea et al. 2000), with a population estimated at about 500 deer in the latter years (Myint Aung pers. comm. 1996). A nationwide survey of Myanmar in 1997 found evidence of Eld's deer within 23 of 24 townships surveyed, out of the 28 that were reported to contain Eld's deer in 1992. The four unsurveyed townships were considered very unlikely to still harbour deer because of the little remaining habitat within them (McShea et al. 2000).

By 2003 a repeat of the 1992 surveys concluded that deer had disappeared from four townships and had declined in at least another four, but there were signs that in some townships populations were stable or even increasing (Myint Aung 2004). Increases and decreases were in part put down to reforestation efforts and habitat loss respectively (Myint Aung 2004). Conditions in Myanmar for Eld's deer have deteriorated in the most recent years. At Shwesettaw Wildlife Sanctuary, favourable habitat has been taken over by agriculture. Chatthin Wildlife Sanctuary is now the only viable population with nominal protection, and habitat currently appears more secure, but protection efforts appear to be failing, at least partly reflecting changes in management. Since funding from the Smithsonian Institution ceased in 2003, the amount of patrolling and the Eld's deer population have both decreased every year (W. McShea pers. comm. 2008); a causal link seems inescapable. A population detected on the outskirts of Alaungdaw Kathapa National Park was never protected by inclusion in the national park, and elsewhere in the country other populations and large blocks of suitable habitat have not been gazetted into the protected area system despite hopes that they would be (McShea et al. 2000; W. McShea pers. comm. 2008). No large populations have been found outside of Chatthin WS, and indeed all new populations detected in recent years have been based on sporadic records in planted teak forests, with numbers too low to attempt any empirical estimate of population (W. McShea pers. comm. 2008). The rate of net removal of forest is high within Eld's deer range, even taking into consideration efforts to plant teak, giving further cause for concern (W. McShea pers. comm. 2008). In Lao PDR there are one or two known populations. One (of probably under 100 animals) is in Savannakhet province (central Lao PDR) and has been the focus of a small-scale species-specific conservation project, with regular subjective monitoring of the population (Johnson et al. 2004; A. Johnson pers. comm. 2007). Activities lapsed in 2007, but options are being assessed for restarting support. In the late 1990s another small population, of probably the low dozens of animals, was found from Champasak Province, far south-western Lao PDR, in areas adjacent to the Dong Khanthung Provincial Protected Area, but there has been no recent re-evaluation of the site and it is quite plausible that the species has been hunted out there (Round 1998; Duckworth et al. 1999; A. Johnson pers. comm. 2007; R. J. Timmins pers. comm. 2008). It is unlikely, given the distribution of suitable habitat, the distribution of the human population, and the pervasive hunting culture in Lao PDR that any more significant populations of the species remain to be found (R.J. Timmins pers. obs.); lowland Savannakhet province was the only unprotected part of Lao PDR where in Duckworth et al. (1999: 41) recommended additional surveys of large mammals as likely to reveal new conservation priorities. One population was indeed found but since that was written, options for others have greatly declined due to extensive road-building, conversion of remaining dry forest habitat and assignment of much of the rest to plantation concessions (particularly on the western plains of the province) and continued extremely high levels of trade-driven large mammal hunting make the chances of any others remaining now very low (J.W. Duckworth in litt. 2008 from field assessment in late 2007). Survival in Vietnam is now very doubtful. Eld's deer is still reported by local inhabitants (e.g. as reported in Nguyen Xuan Dang and Nguyen Thi Thuy (2004), but field surveys of such areas have failed to find the species (Do Tuoc pers. comm. 2006). Perhaps a few animals remain along the border with Cambodia (Do Tuoc pers. comm. 2006), but repeated recent surveys of one of the more promising areas, Yok Don National Park, has not produced good evidence (Le Xuan Canh et al. 1997; Eames et al. 2004). The reported distributions given by Dang Huy Huynh (1990) and Ratajszczak (1991) were either overly optimistic and based on unverified evidence, or indicate a recent swift decline in the species in Vietnam; probably both factors are responsible. Continued presence of ancestral wild populations in **Thailand is also now very unlikely**. Very occasional reports, which are largely unverifiable, continue to be made, primarily in the **Thailand**/Lao PDR/Cambodia triborder area, and the Huai Kha Khaeng WS (e.g. Naris Bhumpakphan et al. 2004). At best a few remnant individuals may survive or immigrate. The Thai Royal Forestry Department has a captive breeding programme for both R. e. thamin and R. e. eldii and there have been attempts at introduction of the former in two areas (where it was never a native subspecies); numbers living in a ?wild state? are in the low tens (Naris Bhumpakphan et al. 2004). Although once distributed in the tropical zone of southwest Yunnan and further east, the South China Eld's deer has long been extirpated (Smith et al. in press; Zeng et al. 2005). By the 1970s the only remaining Chinese populations were on Hainan and had declined to some 40 animals in Dong Fang and Bai Sha Counties, primarily due to poaching, from a reported 500 individuals in 1950s (Zeng et al. 2005).

By the 1970s the only remaining Chinese populations were on Hainan and had declined to some 40 animals in Dong Fang and Bai Sha Counties, primarily due to poaching, from a reported 500 individuals in 1950s (Zeng et al. 2005). In 1976 two protected areas, the Datian Nature Reserve, and the Bangxi NR, West Hainan, were founded around the two last remaining isolated herds of deer, with 26 and 20 animals respectively (Song and Zeng 2004; Pang et al. 2003; Zeng et al. 2005). By 1981 the Bangxi NR population had been exterminated by poaching (Zeng et al. 2005). The animals in Datian NR were kept in fenced enclosures while the population grew, and at the current time the whole c. 13 km^2 of the reserve is enclosed (Song 1996; Pang et al. 2003; Zeng et al. 2005). Initial problems with conservation efforts at Datian included encroachment of almost half its habitat by agriculture since its establishment, deer deaths through drought and overcrowding in small enclosures, and some poaching (Pang et al. 2003; Zeng et al. 2005). By 1991 the Chinese population had rebounded to 346 animals, of which 261 were within the Datian Nature Reserve (Yuan et al. 1993). By 2003 there were about 1,000 animals in Datian NR, and beginning in 1986 captive and semi-wild populations were established at other facilities, the total population in which was approximately 260 animals in 2003 (Song and Zeng 2004; Zeng et al. 2005). A semi-wild herd was founded once again in Bangxi NR, and by 2002 the population there was reportedly 115 animals (Zeng et al. 2005). In 2003 a third ?wild? herd was founded in Mihouling Reserve (Song and Zeng 2004; Zeng et al. 2005). In India, R. e. eldii was thought to be extinct in the early 1950s but was subsequently rediscovered. By 1975, the only remaining wild population had declined to about 14 animals in the swamps of Loktak Lake, Manipur (Ranjitsinh 1978). On this basis, a floating marsh on the southern end of Loktak Lake was gazetted in 1977 as the Keibul Lamjao National Park (Ranjitsinh 1978). The population was reported to have increased to about 137 by 1994 (Singsit 1994; Singh 2004), and by 2003 censuses indicated a total of about 180 animals (Singh 2004). There are additionally similar numbers in captive conditions in India (Salter and Sayer 1986; Decoux 1993).
Population Trend: Decreasing

Habitat and Ecology:

Populations in Lao PDR, Vietnam and Cambodia seems to have occurred in a variety of primarily open, grass dominated habitats. Most evidence of presence comes from Deciduous Dipterocarp Forests which primarily occur in the highly monsoonal areas of the Mekong plains (Wharton 1957; Tordoff et al. 2005; R.J. Timmins pers. comm. 2008), but historical populations known from, for instance, the Nakai and Bolaven plateaux of Lao PDR appear to have used a mosaic of small grasslands and wetlands amongst pine, semi-evergreen and other forest types, and those known from the Dalat and Kirirom plateaux would seemingly have occurred in more extensive pine savanna habitats (R.J. Timmins pers. comm. 2008). The species may also have occurred in northern Lao PDR, where suitable habitat is localised to a few plains amid the generally unsuitable rugged densely forested terrain but no conclusive records have been traced (Duckworth et al. 1999), and would have presumably inhabited Mixed Deciduous Forests and grasslands of anthropogenic origin, as Deciduous Dipterocarp Forest does not occur in such northern areas (R.J. Timmins pers. comm. 2008). Populations in Deciduous Dipterocarp Forest appear to favour open-canopy formations, with grass dominated understory, especially areas with extensive grassland patches (mainly hydrological in origin), and avoid closed-canopy formations where small deciduous bamboos predominate in the understory (Timmins and Ou 2001; Timmins et al. 2003; Tordoff et al. 2005; R.J. Timmins pers. comm. 2008). Given the highly seasonal nature of such forest areas, permanent wetlands, usually in the form of forest pools or stream bed pools, are likely to be very significant to the species and the densities of animals that a given area can support (R.J. Timmins pers. comm. 2008).Despite suggestions to the contrary in some sources, there is no evidence that Eld's deer in these countries are particularly wetland associated, nor is there evidence that the species reached high densities in floodplain tall grasslands (contrary to the case with hog deer) (R.J. Timmins pers. comm. 2008). The hydrological grasslands that the species appears often to associate with are favoured over forest for conversion to rice paddies, and the paddies likewise appear attractive to the deer. Males appear to be fond of wallowing. Anecdotal evidence suggests that deer do not make substantial movements and tend to remain year round within relatively small areas (R.J. Timmins pers. comm. 2008). Lekagul and McNeely (1988) suggested that Eld's deer had been driven into drier areas by hunting and habitat destruction.

However, this is clearly not the case in Cambodia where historically Eld's deer were considered one of the most abundant species in the open seasonally very dry forests of the north and east (Wharton 1957). The Loktak lake population in Manipur, India, inhabits an area of floating marsh called locally ?humid? (Singh 1983; 2004). This population has adaptations of the feet which are thought to help the animals move easily in their marshland habitat (Pocock 1943), but Pocock (1943) speculated that elsewhere in north-east India Eld's deer would also have occupied drier plains. In Myanmar most Eld's deer occur in indaing forest, which is usually dominated by the tree *Dipterocarpus tuberculatus,* and is structurally and ecologically fairly equivalent to the Deciduous Dipterocarp Forest of Indochina and Thailand. There are two other types of deciduous forests used by the deer in Myanmar, dry (thandahat), and mixed (teak). All three forest types receive 100?200 cm of rainfall a year (Prescott 1987; Bronson 1989; McShea et al. 1999, 2001; Myint Aung et al. 2001). Pristine habitat is now absent within the Myanmar range of Eld's deer and all populations use habitats at various stages of secondary succession (McShea et al. 2005). The ecology of animals in Myanmar is likely to be similar to those in Indochina and **Thailand** reflecting the similarities in habitat, with dry-season movement closely correlated with the locations of water sources (Prescott, 1987; McShea et al. 1999, 2001). At the Hlawga wildlife park just outside Yangon, Myanmar, many species of Myanmar's native wildlife have been introduced. In contrast to hog deer, sambar and northern red Muntjac, which have all increased to good populations, Eld's deer has failed to establish itself within this small fenced area. This is attributed by staff, including a past veterinarian, to the area being too humid (it is said to be outside the native range of Eld's deer) although this has not been confirmed. The typical habitat of Eld's deer in Hainan Island is scrubland and dry grassland together with sparse trees in hills below 200 m asl in altitude (Zeng et al. 2005). Eld's deer feeds on grass and some browse and also take fallen fruits and flowers, and reportedly can live without water for several days. Eld's deer regularly visits salt licks. Stags are generally solitary except during the rut, while hinds congregate throughout most of the year (Gee 1961; Myint Aung pers. comm. 1996). Wharton (1957) recorded large herds in Cambodia in the 1950s, and Lekagul and McNeely (1988) stated that, prior to hunting reductions, herds of over 50 formed. In Keibul Lamjao National Park, R. e. eldii is thought to move from the ?humid? to island hillocks during periods of flooding (Green 1990). In China, the rut is during February/ June, with a single fawn born from September to January. In India, calving occurs from mid-October to the end of December. Gestation period is between 237 and 240 days, and sexual maturity is reached at approximately 1.5?2.0 years (Wemmer and Grodinsky 1988).

Systems: Terrestrial; Freshwater

Major Threat(s):

In Cambodia, Lao PDR and Vietnam the most obvious threat to Eld's deer is hunting, which in addition to local consumption of meat is driven by a thriving and probably increasing trade in bush meat, a national, regional and East Asian market for traditional medicinal products derived from the species, and a regional international market for trophy antlers (Duckworth et al. 1999; Timmins and Ou 2001; Tordoff et al. 2005; Zeng et al. 2005). The influence of traditional medicinal trade is uncertain, but may be very significant as products derived from the species are reportedly more valuable than are those from other deer (Zeng et al. 2005). Some hunting in Cambodia is probably also driven by a demand for captive animals especially from zoos and menageries in **Thailand** and Cambodia (Salter and Sayer 1986; R.J. Timmins pers. comm. 2008). The species is apparently easy to hunt compared with other sympatric deer, and fawns are likely to be very vulnerable to dogs which almost always accompany human parties during forest visits, even when the visits are not for hunting (Tordoff et al. 2005; R.J. Timmins pers. comm. 2008). The effects of hunting have been exacerbated by forest loss and fragmentation which is rapidly accelerating with human population in-migration, infrastructural developments (especially roads), commercial agricultural expansion, economic land speculation and mineral extraction. Selective logging was initially one of the main driving forces, but it is now of relatively minor concern (Tordoff et al. 2005; R. J. Timmins pers. comm. 2008).Human in-migration and subsequent agricultural encroachment into forest areas tends to concentrate on those habitats most used by Eld's deer, and thus the species has suffered more than most from hunting (R.J. Timmins pers. comm. 2008; Tordoff et al. 2005). As an inhabitant of relatively open habitat through which travel and transport are easy, it is also likely to have been more heavily affected by a given level of hunting effort compared with species of rugged, dense hill evergreen forest; a similar scenario has been postulated for Jungle Cat Felis chaus (Duckworth et al. 2005; R. J. Timmins pers. comm. 2008). Habitat loss per se is generally a lesser threat (especially in recent decades), and although huge tracts of suitable habitat have been lost in the last century, at least in Lao PDR, Cambodia and Vietnam this has probably largely occurred after Eld's deer have been hunted out (Duckworth et al. 1999). Probably the most significant challenge to conserving the species in Lao PDR, Cambodia and Vietnam is the uncertainty involved with long-term protected-area based conservation management. Protected areas, even those that at present have relatively active and effective management, face an uncertain future with the possibility of excision of conservation status for parts or all of some, the lack of long-term security in external funding adequate to maintain high standards of management, fluctuations in political support necessary to uphold high protection standards and the consequential difficulties of maintaining a motivated and well-trained staff.

Currently most protected areas that could or do support Eld's deer in Lao PDR, Cambodia and Viet Nam offer the species little if, any, protection (exceptions are listed under Conservation Measures). Protected areas, primarily those along the triborder area of **Thailand** where it adjoins both Lao PDR and Cambodia, and the Huai Kha Khaeng WS, were unable to prevent the decline and likely extirpation of Eld's deer from the wild in **Thailand** (see Bhumpakphan et al. 2004). This decline was likely to have been driven by much the same reasons as in Indochina. If any Eld's deer do remain in the wild there, it seems unlikely that they will ever recover unassisted. There are few data on the introduced populations of R. e. thamin to non-native areas (see Bhumpakphan et al. 2004). Threats to the species in Myanmar appear to be similar to those in Indochina, with perhaps more emphasis on exploitation for meat (Salter and Sayer 1986; Myint Aung pers. comm. 1997). Habitat loss in some range areas is still ongoing (Myint Aung 2004) and Koy et al. (2005) found that ?patterns of percentage tree canopy-cover were negatively correlated with human population density, suggesting further threats to Eld's deer populations as the human population continues to grow?. A support programme for the main site, Chatthin Wildlife Sanctuary, had an encouraging start but following cessation of external funding the population is now in decline (see Population), while other protected areas in the country either failed to include appropriate habitat for the species (e.g. Alaungdaw Kathapa National Park) or failed to protect the habitat even within the protected area (e.g. Shwesettaw Wildlife Sanctuary). Overall, the challenges of generating functional protected areas which will sustain Eld's deer are similar to, and under current conditions even more daunting than, those outlined for Cambodia, Lao PDR and Vietnam (above). In general, protected areas in Myanmar are themselves highly threatened by major shortfalls in funding and political will (Rao et al. 2002). The population on Hainan faces an uncertain future as a wild animal without hands-on management because protected area management has not been effective in preventing poaching (see Pang et al. 2003; Zeng et al. 2005). Loss of suitable habitat has been extensive and is continuing, but large enough areas probably remain for further population expansion, provided that herds in such situations could be adequately protected from poaching (see Zeng et al. 2005). Genetic analyses of the population suggest that there is low diversity resulting from a population bottleneck, and this could in the future threaten the population's viability (Pang et al. 2003; Balakrishnan et al. 2003). Indian populations of Eld's deer declined due to some combination of hunting and habitat loss. Even at the Keibul Lamjao National Park, the habitats of the deer have been encroached for grazing, cultivation, and fish farming (Green 1990), but this seems to be mainly now under control (Singh 2004). This population is now most threatened by a hydro-electric project in Logtak Lake which maintains unnaturally high water levels during the dry season and has changed water flow patterns and water quality in the lake (Singh 2004). These in turn appear to be affecting the ecology of the Keibul Lamjao National Park floating marsh (Singh 2004; Sanjit et al. 2005). Floods could also seriously affect the population, which has no high ground to which it can easily escape (Singh 2004). There may also be the threat of low genetic diversity resulting from the effect of a very small founder population (Balakrishnan et al., 2004) and as a small isolated population it must also be considered at risk from disease transfer from domestic livestock, and potentially other factors. Eld's deer readily grazes on rice and some other agricultural products (McShea and Myint Aung 2001). This provides some potential for human/ wildlife conflict, but under current conditions populations seem to be too low for this to be a serious problem (e.g. at the Chonbuly district population, Savannakhet province, Lao PDR; J.W. Duckworth in litt. 2008). Retributory killing may become more of a threat if populations are successfully rebuilt in areas close to agriculture.

Conservation Actions:

This species is listed on CITES Appendix I.

Most Cambodian Eld's deer are within protected areas, although a significant (but decreasing) number occur outside (Timmins et al. 2003; McShea et al. 2005; Tordoff et al. 2005; Timmins 2006). In several protected areas mammals effectively receive no protection other than a decreased likelihood of landscape-scale conversion of habitat (R.J. Timmins pers. comm. 2008). Conservation management of the Siema Biodiversity Conservation Area, Kulen Wildlife Sanctuary, Preah Vihear Protection Forest, and, particularly, Ang Trapeang Thmor Conservation Area and Srepok Wilderness Area may be stabilising and perhaps even increasing Eld's deer populations, through externally funded conservation management projects supported by the Wildlife Conservation Society and The World Wide Fund for Nature (WWF unpublished data; WCS unpublished data; R.J. Timmins pers. comm. 2008). These five areas cover well over 3,000 km² of relatively suitable Eld's deer habitat, and thus the most pressing conservation need is the continued support and consolidation of these protected areas. Only once this is in place would it be sensible to divert resources to protect Eld's deer in further areas of Cambodia currently with little to no active protection. The central Lao population in Chonbuly District, Savannakhet Province, is under nominal protection in a roughly 200 km² provincial protected area specifically designated for the species (Johnson et al. 2004). The area had an ongoing conservation management project supported by the Wildlife Conservation Society (Johnson et al. 2004) but despite a very promising start, this stalled in 2007.

Options are being investigated for resumption of collaboration, because this population is highly threatened by ongoing agricultural expansion and unfettered dogs, and probably by poaching (A. Johnson pers. comm. to J.W. Duckworth 2008). The single southern Lao population found during extensive surveys in the 1990s was small and inhabited a mosaic of farmland amidst forest, adjacent to a subsequently declared provincial protected area (Round 1998). Attempts to upgrade the provincial protected area to national status have failed (A. Johnson pers. comm. 2007). The most important conservation measures for the species in Lao PDR are resumption and consolidation of the Chonbuly District Eld's deer project, and perhaps assisting protected area management in southern Lao PDR, especially of Xe Pian National Protected Area, in order to maintain significant areas of suitable habitat, which might one day be recolonised. In **Thailand** the protected area system failed to prevent the species' effective national extinction some years ago. Therefore, recent conservation efforts there have focused on the captive breeding of large numbers of *R. e. thamin* Bhumpakphan et al. 2004), which is not, however, native to most of **Thailand.** Attempts to introduce *R. e. thamin* to non-native areas have had limited success, although there is a residual herd at Phu Khieo Wildlife Sanctuary and Propagation Station (Bhumpakphan et al. 2004). There is a current plan to introduce 40 of them into Hauai Kha Khaeng WS, in 2008, with 100 more in 2009 (W.J. McShea pers. comm. 2008). R. e. siamensis was the subject of an abortive Smithsonian Institution reintroduction program initiated in 1985. By 1989, progress was limited to construction of holding facilities at Phu Khieo Wildlife Sanctuary and Propagation Station. As with other range states the most pressing conservation measures seem to be strengthening protected area management in those areas that cover the species' former range, so that reintroduction might be possible in the future, jointly with measures to align all stakeholders to ensure that introduction programmes do not create herds with a mix of *R. e. siamensis* and *R. e. thamin* ancestry. The Hainan population is now highly managed with most of the population in Datian NR, and other animals in Bangxi NR and Mihouling Reserve, and other captive facilities (Zeng et al. 2005). The species has been the subject of various avenues of research, for instance a population viability analysis was undertaken to examine demographic and environmental (drought) challenges to the two enclosed populations at the Hainan Datian Nature Reserve (Song 1996), with other research by Yuan et al. (1993) and Song (1993) (see also Zeng et al. 2005). But, the most important conservation measure needed appears to be strengthening of the protected area system so that deer can be re-established in a wild state without danger of poaching or habitat encroachment (see Zeng et al. 2005). India held 97 captive animals in 15 zoos in December 1992, a slight decline from over 100 in the mid-1980s (Decoux 1993); by 1994, 14 zoos held 107 animals (Sharma 1994). he population is probably inbred: all reportedly are descended from two founder pairs (Walker and Marimuthu 1991; see also Balakrishnan et al. 2003). The 40 km^2 Keibul Lamjao National Park was gazetted in 1977 specifically to protect R. e. eldii (Ranjitsinh 1978). Subsequently projects at the park have raised public awareness, and local support has been developed for conserving the endemic Manipur subspecies, and largely halted encroachment of the species habitat (Singh 2004). Status surveys of R. e. eldii have been carried out annually in Keibul Lamjao National Park since 1975 (Ranjitsinh 1996; Singh 2004), and research has been undertaken by Singh (1983; 2004). A Wildlife Institute of India proposal in the 1980s to establish a second free-ranging population in Assam (Khan et al. 1993) was not implemented. Conservation needs are essentially mitigating detrimental effects of the hydropower project on the park, strengthening the protected area management of the park (e.g. through even further improvement of local relations and capacity building), increasing the park's area to meet the needs of deer through appropriate research and potentially establishing further wild populations in other areas. In Myanmar, Chatthin Wildlife Sanctuary and Shwesettaw Wildlife Sanctuary were established specifically to protect *R. e. thamin* (Salter and Sayer 1986), but little management has been undertaken and despite promising initial efforts, management effectiveness appears to be decreasing (Myint Aung 1990; W.J. McShea pers. comm. 2008). Deer were found on the outskirts of only one other protected area, Alaungdaw Kathapa National Park, but protection efforts were never extended to cover the appropriate area, and the majority of the remnant area occupied by the species are likely to be outside protected areas (McShea et al. 2000; W. McShea pers. comm. 2008).

A number of reintroductions have taken place into enclosed sites such as Hlawga Wildlife Park, using animals from Rangoon Zoo and the wild. In 1995, an ecological study of radio-collared R. e. thamin was initiated in Chatthin Wildlife Sanctuary by the Smithsonian Institution, and a community education project commenced later that year to raise local awareness of the deer and the wildlife sanctuary (Wemmer 1995). The Wildlife Division surveys the population annually using a basic line transect method. Conservation measures needed were recently assessed to include extending the protected areas network (McShea et al. 2000) but in the interim the most pressing needs seem to be to prevent further deterioration at Chatthin Wildlife Sanctuary and to salvage the Shwesettaw Wildlife Sanctuary, if this is possible. McShea et al. (2005) argued that "only Cambodia, with 11% suitable forest protected, has placed sufficient dry dipterocarp forest under protected status. Other Southeast Asia countries have not recognized dry dipterocarp forest as a significant ecotype worthy of conservation status". It is certainly true that this forest type has been seen as a low priority in the conservation planning of all range states, but consideration that a country has placed 'sufficient' of any habitat type under protected status is subjective and a variety of measures can be used. In this case, 'sufficiency' was judged on the basis of proportion of a country's total suitable habitat that has been nominally protected. An alternative would be to consider factors like the actual amount of protected habitat, and whether this is enough to represent the range of biotic variation within the country; whether such habitat is in large enough blocks to remain ecologically functional into the long term, and whether protection is largely nominal or is genuinely effective. The last dichotomy is likely to be the chief determinant of whether Eld's deer will survive. Considering that no range states yet have existing protected areas which have secured the species on a long-term footing (although there are promising developments in several countries), and considering the speed with which populations can decrease, priority is necessary at present to concentrate most conservation actions for Eld's deer on securing existing protected areas for key populations which have real potential for success. Determining this potential will reflect situational factors as much or more than ?traditional? biological factors such as current Eld's deer status within them. Eld's deer could potentially persist in habitat mosaics of forest and agriculture, as they still do in some areas of Myanmar and Cambodia (McShea et al. 2000; T. D. Evans pers. comm. 2007), but cultural predisposition to hunting and competing land uses make it unlikely that poaching can be controlled, or habitat management developed, without the context of a protected area, in most range countries (India and perhaps Myanmar being the most likely exceptions).

There are large captive populations of the species, estimated in 2003 to number 180 *R. e. eldii,* 1,100 *R. e. thamin,* 23 *R. e. siamensis* (non-Hainan stock) and 280 *R. e. siamensis* (Hainan stock), but most captive groups are not managed with conservation in mind (Siriaroonrat 2004). There is some scope for using these animals in reintroduction programmes, particularly in **Thailand** where Eld's deer is nationally extinct yet the country is now probably in the best financial and bureaucratic position? of all range states to conserve landscape-scale field populations of the species (McShea et al. 2005).

Schomburgk's Deer *(Cervus schomburgki)*

Range Description:

Rucervus schomburgki was endemic to **Thailand** and is thought to have become extinct when the last captive individual was killed in 1938; the last known wild animals were killed in 1932 near Sai Yoke and Kwae Yai, although this date was not universally accepted at the time (Harper 1945; Lekagul and McNeely 1988). The recent range in **Thailand seems to have lain within 13°30'–18°N, 98°30'–102°E, comprising the central plain of Thailand (Giles 1937; Harper 1945; G.J. Galbreath in litt. 2008). Attempts to circumscribe the range have, however, been bedevilled by a surprisingly large number of statements of occurrence elsewhere.**

Statements of occurrence in the Sanda Valley, Yunnan, China by Sclater (1891: 180) and Bentham (1908) led to this locality being included without any caveat by Grubb (2005), and relate to both a frontlet and a skin. However, Sclater (1891) listed the set of antlers as of unknown locality, and the skin has a mark that makes its Sanda Valley locality no more than tentative (G.J. Galbreath in litt. 2008). Then, Bentham (1908) indicated that the antlers had been collected in the Sanda Valley by John Anderson in 1878, without giving explicit source for this statement. In fact, Anderson was in the Sanda Valley in 1868, not 1878, and moreover recorded no such antlers in his thorough write-up of zoological results of his Yunnan expeditions. This locality assignment for the antlers should be seen as an error of Bentham's, presumably stemming from the earlier, tentative, assignment of Sanda Valley to the deer skin Sclater also reported (G. J. Galbreath in litt. 2008). Pocock (1943) included under distribution of the species *"N. Siam* [**Thailand**] and, it has been alleged, Yunnan and Lao PDR", indicating his own concern with the Sanda valley 'evidence'. The Shan states (Myanmar) have also figured as part of the species's range, e.g. by Blanford (1891: 540). Reference to historical occurrence in these areas seems to stem from Brooke (1876), who wrote, based on written communication from a Dr Campbell of the Bangkok British Consulate) in reference to new specimens sent to him that ". . . all specimens were procured in northern Siam, probably even in the tributary states named Lao PDR and Shan".Campbell seems to have based his opinion at least partly on what professional indigenous hunters told him or others. Brooke gave no definition of "northern Siam" (conceivably it simply meant some way north of Bangkok), and Lao PDR in this context could well have included part or all of the Korat Plateau and other areas in present-day **Thailand** which are ethnically Lao and were often referred to in the past by terms such as 'Siamese Lao PDR'. It is not even clear that Lao PDR and Shan were not just speculation on the part of Campbell and/or Brooke (G.J. Galbreath in litt. 2008). Not surprisingly, Harper (1945) considered the Shan states report to be "highly indefinite" and occurrence in "Indo-China" [presumably = Lao PDR, Cambodia and Vietnam] to be "in error", although he took the 'Sanda valley' frontlet as "apparently authentic", perhaps being focussed on the identity (which is not in doubt) rather than the provenance (presumably assuming that the association with the highly-respected Anderson's name was enough).

A more modern reanalysis concluded that "the northern **Thailand,** Myanmar, Yunnan, and Lao PDR range extensions suggested in the literature are based on erroneous or quite inconclusive items of evidence", sometimes perhaps influenced by a switch of Thai names between Eld's and Schomburgk's deer by Flower (1900) (G.J. Galbreath in litt. 2008). Two loose antlers, presumed to be a pair, of Schomburgk's deer photographed in Phongsali province, far northern Lao PDR, in 1991 were a most surprising find. This has been taken to suggest the species might survive, at least into the 1980s, and that it did indeed occur in Lao PDR (Schroering 1995) and it was even taken as sufficient evidence that the species was not extinct by MacPhee and Flemming (1999). However, while these are indeed Schomburgk's deer antlers, it has proven impossible to recover a consistent story from their owner of the antlers' origin (e.g. Duckworth et al. 1999) and there is no compelling evidence that they had come from a recently killed animal, rather than being decades-old stock still in trade; Williams (1941) noted that even then that Schomburgk's deer antlers "keep coming down to Bangkok from Paknampo and Korat by railway amongst collections of ordinary deer-horns consigned to Bangkok Chinamen . . . the horns are probably quite old . . ." and Schomburgk's deer antlers are still in trade in Thailand (Srikosmatara et al. 1992). The owner of the antlers was, according to his daughter, a middleman trader in wildlife parts sourced from all over Indochina (W.G. Robichaud pers. comm. 2008, based on visit in 1996). Consequently, while remains of the species have been seen in Lao PDR in recent times, they cannot be taken as evidence that the species survives today, or ever occurred in Lao PDR.

Countries: Regionally extinct:

Thailand

Population:

Schomburgk's deer was apparently "not uncommon" in the late nineteenth century and herds still occurred around 1900–1910 (Harper 1945), but the species underwent a very rapid decline: Harper (1945) stated that it was never seen in the wild state by a European (although this may be better taken as a lack of such reports, because Europeans working, for example, on the Paknampo railway in the early days would almost surely have seen and hunted these deer; G.J. Galbreath in litt. 2008). The usually successful hunter Arthur S. Vernay made three trips to Thailand specifically for the species, the first in 1920, but failed to see the species. There are some hundreds of skulls, frontlet and antlers in collections (Harper 1945), but the species is now extinct.

Habitat and Ecology:

The species inhabited seasonally inundated swampy plains with long grass, cane, and shrubs; it apparently avoided forest (Giles 1937), although there seems to be no information on where animals went at the height of the wet season, when much of the dry-season range was under water.

Systems: Terrestrial; Freshwater

Major Threat(s):

Commercial production of rice for export began in the late nineteenth century in **Thailand's** central plains, leading to the loss of nearly all the grassland and swamp areas that this deer depended on, and greatly fragmented what remained. Intensive hunting pressure at the turn of the 19th–20th century restricted the species further and it disappeared in the 1930s. Schomburgk's deer was prominent in the antlers sought by the Chinese medicine trade (Harper 1945). During the wet season, animals marooned on higher ground were hunted readily with spears from boats (Harper 1945), no doubt hastening the species' decline.

Conservation Actions:

As an extinct species there are no appropriate conservation measures. The indication that a remnant population might still survive (Schroering 1995 and in litt. as reported in Grubb 2005; MacPhee and Flemming 1999) was followed up by investigation, independently by several people, of Phongsali province, Lao PDR, by field survey and interview. This failed to find any further evidence of the species there or to elicit a consistent story of the origin of the antlers (Duckworth et al. 1999; per J.W. Duckworth in litt. 2008). Given regional trends in Schomburgk's deer's sole known habitat, non-forest floodplains (its massive conversion to agriculture over the last 150 years) it is unlikely that the species could survive. However, any future indication of its existence should be followed up carefully. If by chance the species does survive, legislation giving it the highest levels of protection (in whatever country it was found), and an assessment of specific conservation needs would be imperative.

Sambar Deer *(Cervus unicolor)*

Description:

The appearance and the size of sambar vary widely across their range, which has led to considerable taxonomic confusion in the past; over forty different scientific synonyms have been used for the species. In general, they attain a height of 102 to 160 centimetres (40 to 63 in) at the shoulder and may weigh as much as 546 kg (1,204 lb), though more typically 100 to 350 kg (220 to 770 lb). Head and body length varies from 1.62 to 2.7 m (5.3 to 8.9 ft), with a 22 to 35 cm (8.7 to 13.8 in) tail. Individuals belonging to western subspecies tend to be larger than those from the east and females are smaller than males.Among all living cervid species, only the moose and the elk can attain larger sizes.

The large, rugged antlers are typically rusine, the brow tines being simple and the beams forked at the tip, so that they have only three tines. The antlers are typically up to 110 cm (43 in) long in fully adult individuals. As with most deer, only the males have antlers.

The shaggy coat can be anything from yellowish-brown to dark grey in colour and, while it is usually uniform in colour, some subspecies have chestnut marks on the rump and underparts. Sambar also have a small but dense mane, which tends to be more prominent in males. The tail is relatively long for deer, and is generally black above with a whitish underside.

Adult males and pregnant or lactating females possess an unusual hairless, blood-red spot located about halfway down the underside of their throats. This sometimes oozes a white liquid, and is apparently glandular in nature.

Range Description:

The Sambar extends from India and Sri Lanka east along the southern Himalayas (including Nepal and Bhutan) through much of south China (including Hainan Island) to Taiwan (where it occurs in the central and eastern parts; Lin, C.-Y. and Lee, L.-L. pers. comm. 2008). Further south it occurs in Bangladesh, throughout mainland South-east Asia (Myanmar, **Thailand**, Lao PDR, Cambodia, Vietnam, West Malaysia) and many of the main islands of the Greater Sundas (excepting Java): Sumatra, Siberut, Sipora, Pagi and Nias islands (all Indonesia), and Borneo (Malaysia, Indonesia, and Brunei) (Grubb 2005).

The current distribution is now highly fragmented in much of this range (see Population). Payne et al. (1985) also listed the Philippines, but the Sambar does not occur there. A record from Ujung Kulon, Java, in van Schaik and Griffiths (1996: 107) is presumably an error for Javan Rusa *R. timorensis*. The Sambar has been introduced widely outside its native range, e.g.: San Luis Obispo Country, California; the Gulf Prairies and Edwards Plateau regions of Texas (Ables and Ramsey 1974); the St. Vincent Islands, Franklin Country, Florida (Lewis et al. 1990); Australia (Slee 1984; Freeland 1990); New Zealand (Kelton and Skipworth 1987); and Western Cape Province, South Africa (Lever 1985). These introduced populations are not included in the distribution map.

Countries: Native:

Bangladesh; Bhutan; Brunei Darussalam; Cambodia; China (Guangxi, Guizhou, Hainan, Hunan, Jiangxi, Sichuan, Yunnan); India; Indonesia (Sumatera); Lao People's Democratic Republic; Malaysia; Myanmar; Nepal; Sri Lanka; Taiwan, Province of China; **Thailand;** Vietnam

Introduced:

Australia; New Zealand; Saint Vincent and the Grenadines; South Africa; United States (California, Florida, Texas)

Population:

Almost everywhere outside Taiwan, the Sambar is in rapid decline leading to a widespread distribution of very low numbers and many local-level extinctions, with occasional healthy populations reflecting special circumstances. Few surveys specifically target Sambar, but a reasonable picture of its current status is emerging through surveys for Tiger Panthera tigris and through general wildlife surveys. Tiger-oriented camera-trapping predictably yields photographs of Sambar when they are present, and most Tiger surveys afford priority to elucidating Sambar status as a key prey, and therefore population determinant, of Tiger. Understanding Sambar status through surveys of Tigers or, sometimes, of Leopard *P. pardus* (e.g. Ngoprasert et al. 2007) needs care, because such surveys have generally selected the best remaining areas for very large mammals within the country or region in question, so as to maximize chances of picking up Tigers. They therefore cannot be seen as indicative of Sambar status in the wider landscape. Most importantly, in the several cases (identified below) where such surveys have found few Sambar, these indicate cause for alarm, because there is little reason to believe that there ought to be larger populations in many other areas in the country or region in question, even though large parts of it may remain unsurveyed. Occasional herds can survive in unexpected places through atypical site-specific circumstances in some countries (e.g. Tungittiplakorn and Dearden 2002). Populations outside the native range are not treated here, as they are irrelevant to assigning the Red List category and criteria. In Sri Lanka, numbers have probably declined substantially through poaching which increased during the civil war. Poaching occurred widely across the country, including within protected areas, particularly where there are concentrations of resettled refugees (Santiapillai and Wijeyamohan 2003). In India, although the Sambar occurs widely and in many habitat types, and large populations occur in well-secured protected areas, nowhere is it now regionally abundant (Sankar and Acharya 2004). It has been recorded in 208 protected areas (National Wildlife Database, Wildlife Institute of India, cited in Sankar and Acharya 2004); its distribution outside protected areas is now highly scattered.The reported ecological densities of Sambar in India mostly fall within 1–10 animals per km² within the protected area network, and depending on the varying levels of protection efficacy: Bhadra Tiger Reserve, 0.89 +/- 0.23 (SE) per km² (Jathanna et al. 2003); Madhya Pradesh Pench National Park dry deciduous forest, 9.6 animals per km² (Karanth and Nichols 2000); Kanha moist deciduous forest, 1.5 animals per km² (Karanth and Nichols 2000); Nagarahole moist deciduous forest, 5.5 animals per km² (Karanth and Sunquist 1992); Bandipur dry deciduous forest, 5.6 animals per km² (Karanth and Nichols 2000); Tadoba-Andhari dry deciduous forest, 3.3 animals per km² (Karanth and Kumar 2005); Melghat dry deciduous forest, 2.7 animals per km² (Karanth and Kumar 2005); Maharashtra Pench dry deciduous forest, 5.9 animals per km² (Karanth and Kumar 2005); Ranthambore semi-arid dry deciduous forest, 10.7 animals per km² (Kumar 2000); and Gir semi-arid dry deciduous forest, two animals per km² (Khan et al. 1996). Similar surveys at Kaziranga found too few Sambar to estimate populations there (Karanth and Nichols 2000), this presumably representing habitat characters rather than defective protection, given the buoyant populations of other deer at that site (Hog Deer *Axis porcinus* and Barasingha *Rucervus duvaucelii).*

Outside protected areas, Sambar is present mostly in very low numbers, although larger numbers can still be found where its habitat is almost inaccessible to people. The recorded Bhadra density is low, reflecting poaching and forest-resource extraction (Jathanna et al. 2003), and the population density is steadily increasing following removal of these pressures in 2003 (K.U. Karanth and N.S. Kumar unpublished data). Nepal supports an important population in Bardia National Park (Dinerstein 1979). This remains healthy and there are also good populations in Parsa Wildlife Reserve, Chitwan National Park, and Sukla Phanta Wildlife Reserve. Across the country Sambar has declined so rapidly and so widely that it is now very rare outside protected areas, mostly recorded as isolated reports from forests adjoining these protected areas, and from south-east Nepal. Sambar has been victim of poachers, probably because of its large size (thus, more meat per animal and, for males, a more impressive trophy), perhaps compounded by nocturnal and relatively docile nature; recent declines are of the magnitude to meet IUCN criteria for Vulnerable on a national assessment (all information: Hem Sagar Baral pers. comm. 2008). It is unclear whether the demand for antler in velvet in eastern Asia is stimulating poaching in Nepal. No information concerning current status was traced from Bhutan but (simply on an area basis) numbers are unlikely to be large enough to influence the overall global population trend. In Bangladesh, Sambar was common in north-east, south-east and hill -tract forests in the 1960s (Khan 1985) but now only a very few persist in the south-east: "habitat loss and hunting pressure made them critically endangered in the country. The pressure became severe in the 1970s. I see little hope for this species in the next 30 years" (Md Anwarul Islam in litt. 2008). In China there were major declines in Sambar over much of the twentieth century, and populations probably decreased by more than 50% in the 30 years from 1978; records from Guangxi suggest that, there at least, the 1970s–1980s was the era of fastest decline (J. Fellowes pers. comm. 2008; B.P.L. Chan pers. comm. 2008). In some provinces populations seem now to have stabilized, and in some areas, particularly in Hainan (where several apparently stabilised populations are present, but all are small, reflecting habitat patch size: the largest may be of only a few hundred animals) and north Guangdong province (perhaps also the east of that province, where there is a nature reserve specifically for the species), populations may even have been increasing once again in the last few years. This is not so for Guizhou and the provinces between Guangdong and Yunnan (B.P.L. Chan and M.W.N. Lau pers. comm. 2008). Further north, in the provinces of Fujian, Jiangxi, Zhejiang, Hunan and Hubei, a multi-method Tiger survey in 2001 found that Sambar is very rare overall and scattered in present-day distribution (Tilson et al. 2004). For elsewhere in China, Smith and Xie Yan (2008) mapped wide presence in Sichuan, plus isolated populations in the provinces of Qinghai, Tibet, Chongqing, and Fujian. The original sources are not clear, and R. Harris (pers. comm.) doubts that Sambar has lived anywhere in Qinghai in recent times. In Taiwan, numbers and geographic distribution reduced rapidly through hunting before 1987, then expanded somewhat since the mid 1990s reflecting reductions in hunting. The average population density in the Yu-Li Wildlife Refuge was 10.6 animals per km^2 in 2005. In some areas at high altitude, densities exceed 40 animals per km^2, and the animals have a large impact on the coniferous saplings in this area. But at low to mid altitude, hunting pressure remains high enough to restrain their distribution and density. Based on the known geographical distribution, there are five populations: A-Yu Mountain, Nanhu, Shei-Pa, Jade Mountains –Central Mountains, and Tawu. The A-Yu population was re-introduced (about 70 animals) by local deer farmers in 1988–1991, when the velvet price was very low through the impact of deer TB in 1985. This population has not expanded reflecting continuous poaching; it may be genetically polluted. The other four populations are all natural, assumed to be isolated by cross-island roads, and may have very low gene flows with each other. Sambar are largely restricted to protected areas, only occurring outside them in a few nearby areas. The population sizes and distributions in these areas are smaller than in the protected areas. The poaching pressure is an important limitation to Sambar in these unprotected areas (all information: Lin, C.-Y. and Lee, L.-L. pers. comm. 2008). In Myanmar, Sambar was detected at 12 of 15 camera-trapped Tiger sites during 1999–2002, indicating that it is still widespread (Lynam 2003). The Myanmar Forest Department and Wildlife Conservation Society have jointly now (as of May 2008) camera-trapped 21 survey areas (including the earlier 15) and found Sambar in 13 (Saw Htun pers. comm. 2008).

This lower ratio of proven presence indicates how the areas selected for Tiger survey were above average for prey and may indicate increasing local-level extinctions. For example, in a multi-method study of one area not expected to hold Tiger, hunted animals showed that Sambar still occurred in Naungmung and Machanbaw (south of Hkakaborazi National Park, Kachin state), but it was so scarce that it was not detected by substantial amounts of either camera trapping or sign surveys (Rao et al. 2005). Even within these best areas, large declines are suspected and overall national population losses of about 50% over the last 25–30 years and in the following such period are likely (Than Zaw pers. comm. 2006; Saw Htun pers. comm. 2008).

In Thailand, several protected areas maintain big Sambar populations; these are mostly large with very few or no people living within them. Over much of the country Sambar is extinct or very nearly so, including in many protected areas, particularly in the north where most parks are small and heavily poached (Anak Patannavibool pers. comm. 2008). Sambar is rare in most areas of Thung Yai Naresuan, where commercial hunting has targeted large ungulates (Steinmetz et al. 2008). Sambar was recorded at over 40% of camera-trap locations during a Leopard survey in Kaeng Krachan National Park, and clearly remains widespread there. Abundance did not vary with proximity to villages, even though a variety of hunting methods occurs on site (Ngoprasert et al. 2007). Sambar is also locally abundant in Khao Yai National Park, Thap Lan National Park and Phu Khieo Wildlife Sanctuary (Lynam et al. 2006). In Huai Kha Khaeng Wildlife Sanctuary current Sambar density is about 2–3 animals per km^2 and it seems to be increasing (Anak Patannavibool pers. comm. 2008).

In Lao PDR, Sambar was described as very common throughout the wooded parts by Delacour (1940). Some decades later, the species was reported during 86% of 1988–1993 village interviews (n = 328) and was then still widespread throughout, except in the most heavily settled areas (Duckworth et al. 1999: Annex 5). However, sightings on surveys during 1992–1998 were very rare, and while calls heard by night and signs indicate a wide distribution, numbers are heavily depressed (Duckworth et al. 1999; Timmins and Ou 2001). Only in Nam Et–Phou Louey NPA did Sambar seem at least locally abundant (Davidson 1998), and reasonable populations were confirmed still to persist in 2003–2004 (Johnson et al. 2004); an exceptional area in an Indochinese context, that still even supports significant Tiger numbers. The Nakai Plateau and the mountainous interior of Nakai-Nam Theun NPA and a few other, localized, areas of Lao PDR, including the Dong Kalo tract of Xe Pian NPA, the Nam Hiang area of the Bolaven Plateau, and parts of Nam Kading NPA also seemed to support relatively high densities during the mid 1990s. The few sites resurveyed during 2004–2007 have in general shown major declines (e.g. Timmins and Robichaud 2005). There was not a single camera-trap photograph during fairly intensive camera-trapping in Nakai–Nam Theun NPA during 2006–2007, although the species has subsequently been found to persist in a few parts of the protected area (Johnson and Johnston 2007; W.G. Robichaud and Nam Theun Watershed Management Protection Authority pers. comm. 2008) and it is not clear whether quirks of camera-trap placement (e.g., hypothetically, a focus on slopes rather than valley bottoms) may have been a major contributor to the lack of photographs. The Nakai Plateau still supports noteworthy numbers in a Lao context, but even here there were signs of decline (Dersu 2007; R.J. Timmins pers. comm. 2008). Surprisingly, given the proximity to China and Vietnam, a good population was found to persist in Phou Dendin NPA in the far north-east of Lao PDR in 2004–2005 (W.G. Robichaud pers. comm. 2008). In Vietnam, in most of the northern highlands, in particular, and the northern and central Annamites to an increasing extent, Sambar is very rare or has been hunted out from many areas of otherwise suitable habitat (Timmins et al. 1999; Grieser Johns 2000; Timmins and Trinh Viet Cuong 2001; R.J. Timmins pers. comm. 2008; B. Long pers. comm. 2008). Camera-trapping studies in Pu Mat National Park, Thua Thien Hue province and Quang Nam province between 1998 and 2007 recorded only a handful of records and local communities in all these areas report a major decline in numbers (B. Long pers. comm. 2008). With so many higher-profile species long listed as threatened on the Red List, many of them with restricted ranges, the status of Sambar in Viet Nam has largely been ignored, and the paucity of evidence during many surveys has rarely been highlighted in reviews.

As with many large mammals, populations are thought to be best in southern parts of the country. It remains common in Cat Tien National Park (Polet and Ling 2004), but this is very much the exception, and it was not considered at all common in surveys of several areas of prime habitat in Dak Lak Province in 1997 (Le Xuan Canh et al. 1997). In Cambodia, Sambars "appear to be at least relatively common in some areas, though entirely absent from others, a feature which appears to have a strong correlation to local hunting pressure...At the current rate of persecution, this species is likely to decline sharply over the next few years" (Walston 2001). Sambar is still widespread, although heavily targeted by hunters and most populations are now very depressed (Timmins and Men 1999; Timmins et al. 2003; Tordoff et al. 2005: 171; Timmins 2006). In the eastern plain of Cambodia, where populations of many other ungulates are magnitudes more numerous than in other areas of Indochina, Sambar numbers are clearly low, almost certainly as a result of targeted hunting: numbers in some areas seem depressed even well below those of wild cattle (Timmins and Ou 2001). Protection activities in a few localised areas appear to have stabilised declines, but this is an exception to a general scenario of rapid decline. These declines are evidently faster and recent than in either Lao PDR or Viet Nam, reflecting the rapid pace of change of economic and logistical factors in Cambodia since the mid 1990s. In West Malaysia, Sambar was regularly camera-trapped in Taman Negara (Kawanishi and Sunquist 2004); population densities were taken to be 0.01–0.20 per km^2 (Kawanishi and Sunquist 2004). It is unclear how close this estimate is to a pre-exploitation density, because although these authors adjudged large mammal poaching to be negligible in Taman Negara, high levels were confirmed there by Wan Shahruddin (1998); whether Sambar was being targeted is unclear. Moreover, camera-trapping (suitable in methodology and intensity to find Sambar if any significant numbers were present) did not record Sambar in a fourth sector of Taman Negara in 1999 (Lynam et al. 2007). The great current rarity of Sambar in Peninsular Malaysia was evinced by a camera-trap survey for Tiger: it was photographed at only one of the nine areas surveyed during 1997–1999, Temenggor, where it was the most commonly photographed mammal (Lynam et al. 2007). Sambar was found in only one of 14 sites (seven virgin jungle reserves and seven adjacent logged areas) in a study assessing the importance of virgin jungle reserves, making it the most localized mammal species detected. The occupied site was a logged area (Laidlaw 2000). None was found in the Jerangau Forest Reserve, from which it was suspected to have been eradicated by hunting (Azlan 2006). Sambar is still regularly camera-trapped in Johor (M. Gumal pers. comm. 2008) but seems to be generally rare throughout other areas. In Indonesia, there are various recent records from Sumatra, e.g. Bukit Barisan Selatan National Park (O'Brien et al. 2003), Gunung Leuser (van Schaik and Griffiths (1996), and Way Kambas National Park (S. Hedges pers. comm. 2008), but in these areas it was recorded rather rarely by comparison with other ungulates. Sambar was nearly six times as abundant in areas of Bukit Barisan Selatan National Park with low than with high human population density within 10 km of the park boundary, suggesting low resilience to human presence, presumably the effects of hunting, and reduced populations overall (O'Brien et al. 2003). In Way Kambas National Park the population is reportedly stable (G. Semiadi pers. comm. 2008). Overall in Indonesia (Sumatra and Borneo), Sambar is not under major threat except where forest is being converted (i.e. most of the lowlands). Reflecting this, E. Pollard (pers. comm. 2008) considered the status of Sambar, in suitable habitat, to be much more healthy there than in Cambodia and Viet Nam, although there is a significant market for Sambar antlers in Sumatra (D. Martyr pers. comm. 2008). On Borneo, Sambar has been extirpated or persists only at extremely low densities in Lambir Hills National Park, Sarawak, Malaysia (Azlan and Engkamat Lading 2006). An intensive survey of Batang Ai National Park, Sarawak, found Sambar only once; this was stated to be a genuine reflection of actual rarity, and to result from intensive hunting at salt-licks (Meredith 1995). Across Sarawak it is seriously reduced by hunting (Bennett et al. 2000, see Threats; A.C. Sebastian pers. comm. 2008). Some better populations survive in Sabah, Malaysia, e.g. Danum Valley (Heydon 1994), Tabin Wildlife Reserve (Matsubayashi and Sukor 2005), and Deramakot Forest Reserve (Matsubayashi et al. 2007) but populations are in general also much reduced there (Bennett et al. 2000). In the Indonesian part of Borneo populations seem to be less depleted, particularly in East Kalimantan, which is believed to hold the largest Sambar populations among all Kalimantan provinces; in the Penajam Paser Utara district of East Kalimantan, approximately 60 hunted animals are sold monthly in several traditional markets (Semiadi et al. 2004; G. Semiadi pers. comm. 2008).

Population Trend: Decreasing

Habitat and Ecology:

No large Indian ungulate has adapted itself to a wider variety of forest types and environmental conditions than has Sambar (Schaller 1967). Within India, Sambar occurs in the thorn and arid forests of Gujarat and Rajasthan, in the moist and dry deciduous forests throughout peninsular India, in the pine and oak forests at the Himalayan foothills, and in the evergreen and semi-evergreen forests of northeastern India and the Western Ghats (Sankar and Acharya 2004; N.S. Kumar pers. comm. 2008). Outside India it extends into temperate-latitude and alpine-zone woodlands of Taiwan (Lin, C.-Y. and Lee, L.-L. pers. comm. 2008). This habitat flexibility is permitted by its broad diet: Sambar has been documented to eat 130–180 species of plants in India alone (Schaller 1967; Johnsingh and Sankar 1991; N.S. Kumar pers. comm. 2008), with food requirements less specialised than those of other deer (Schaller 1967). Sambar grazes or browses depending upon the forage available at that time (Schaller 1967; Richardson 1972; Martin 1977; Bentley 1978; Dinerstein 1979; Kelton and Skipworth 1987; Ngampongsai 1987; Sankar 1994; Semiadi et al. 1995). Across the very wide altitudinal range occupied in Taiwan, the diet varies reflecting major changes in plant communities (Lin, C.-Y. and Lee, L.-L. pers. comm. 2008). Sambar was found to live in much higher densities in moist than in dry deciduous forests of Nagarahole National Park (Karanth and Sunquist 1992) and there are no doubt also patterns of habitat selection across the rest of its range. Although the highest densities of Sambar so far recorded were in the semi-arid forests of Ranthambore (Kumar 2000), across most of its Indian range Sambar seems to thrive best in well-watered, moist deciduous hilly terrain (N.S. Kumar pers. comm. 2008). The Sambar occurs up to at least 3,825 m asl on Siouguluan Mountain, the highest peak of the Central Mountains in Taiwan; elsewhere on the island it ranges down to 150 m asl, mostly living at 2,000–3,500 m (Lin, C.-Y. and Lee, L.-L. pers. comm. 2008). It occurs up to 3,000 m asl on Gunung Kinabalu, Sabah, Borneo (Payne et al. 1985). In Myanmar, recent camera-trap photographs spanned the range of 0–2,150 m asl (Saw Htun pers. comm. 2008). Sambar is largely restricted to hilly terrain in the Terai Arc Landscape (Johnsingh et al. 2004), although how much this reflects real habitat selection and how much it is an artificial pattern produced by human effects (habitat conversion and hunting) is unclear. More widely in India, there does seem to be a marked preference for undulating terrain (N.S. Kumar pers. comm. 2008). Kushwaha et al. (2004) found that in Kumaon Himalaya (India), Sambar usage was greater of the higher than the lower altitude area. However, it makes wide use of plains areas elsewhere, where these have not been destroyed, e.g. the Hukaung Valley in Myanmar (J.W. Duckworth pers. comm. 2008).In Borneo, while Payne et al. (1985) considered Sambar "most common in secondary forests of gently-sloping terrain" they also knew of occurrence in "tall dipterocarp forests on steep terrain and in swamp forests". **In Thung Yai, Thailand,** Sambar signs were twice as abundant in lowland forest as in montane forest, although this difference was not statistically significant (Steinmetz et al. 2008). In Southeast Asian regions of dense evergreen closed-canopy forest, Sambar is highly tolerant of forest degradation: indeed, much higher numbers are found in encroached stands than in pristine forests, if hunting is under control (Rijksen 1978; Heydon 1994; Stuebing 1995; Davies et al. 2001). In Sabah, Sambar was camera-trapped in both mature and young forest stands (Matsubayashi and Sukor 2005).All the remaining Lao populations are centred around areas with extensive open, or at least broken, habitat amid forest (Duckworth et al. 1999), and the general paucity of records from the interiors of large blocks of closed-canopy evergreen and semi-evergreen forests, which generally support the least depleted large mammal populations, probably reflects natural patterns of habitat occupation (R.J. Timmins pers. comm. 2008). Similarly, in extensive tracts of deciduous dipterocarp forest the species occurs primarily around patches of denser habitats, and this seems to be a natural rather than hunting-led pattern (Timmins and Ou 2001). In this sense, in Indochina Northern Red Muntjac Muntiacus vaginalis is much more tolerant and versatile than is Sambar. In southern and central India where much of the forest is deciduous, Sambar is a true forest ungulate and conspicuously avoids disturbed and open forests, and is highly sensitive to any sort of forest resource extractive activities (N.S. Kumar pers. comm. 2008). Kushwaha et al. (2004) found that in Kumaon Himalaya (India), Sambar was primarily in areas of high tree and herb density with low shrub density. O'Brien et al. (2003) found no statistically significant difference in Sambar abundance between areas within 1 km inside the boundary of Bukit Barisan Selatan National Park, Sumatra, and those deep in the interior of the park.

Caution is needed in inferring that there is truly no association of Sumatran Sambar with forest edges (which would be astonishing by comparison with its habitat use elsewhere) because the study does not specify whether this 'non-significant' result might simply reflect methodological factors rather than a genuine lack of biological effect (see, e.g., Johnson 1999). Despite the population rises that occur in post-logging forest, there is no evidence that Sambar can survive landscape-level conversion to exotic plantations or other non-forest land-uses, although many live in coconut plantations (G. Semiadi pers. comm. 2008) and it is relatively common in immature Acacia mangium plantations within the matrix of plantation and natural forest of the Sarawak Planted Forests area, Bintulu (Belden Giman pers. comm. 2008). A study of coffee estates around Bhadra Wildlife Sanctuary, India, recorded Sambar only in those coffee areas within 1 km of the sanctuary's boundary (Bali et al. 2007). In forest protected areas of Thailand such as Khao Yai National Park and Phu Khieo Wildlife Sanctuary, Sambar populations are often concentrated around anthropogenic grass and scrub (e.g. sites of former villages) rather than in the forest itself (Trisurat et al. 1996; Lynam et al. 2001) and this also seems to be true on remote parts of the Bolaven Plateau of south Lao PDR (Evans et al. 2000). In the Annamite mountains of Lao PDR and Viet Nam, Sambar seems often to be associated with degraded valley bottom areas, largely the result of long human influence: although the effects of a permanent water source and differences in terrain in determining distribution are difficult to disentangle (R.J. Timmins pers. comm. 2008). Sambar regularly visits salt licks (e.g. Matsubayashi et al. 2007), perhaps especially when growing new antlers. This predictability exposes it to high levels of hunting, where this is not effectively controlled. It seems that within an area, not all sort of licks are visited. At the Seima Biodivesity Conservation Area, Cambodia, Sambar photographs were initially very few at mineral lick camera-traps. But in 2007, cameras placed on trails and at some permanent water sources (especially in semi-evergreen forest) recorded Sambar much more often. Checking tracks at more than 40 licks indicated that some were used by Sambar and Red Muntjac, but not Gaur Bos gaurus or Asian Elephant Elephas maximus, while the big ones with many cattle and elephant prints had few Sambar prints (E. Pollard pers. comm. 2008). Sambar was considered to be mostly nocturnal by Kawanishi and Sunquist (2004), to show 'no pattern' (meaning unclear; perhaps intended to imply no significant variation through the 24-hour cycle, although this is itself a pattern) by O'Brien et al. (2003), and to be cathemeral by van Schaik and Griffiths (1996). In fact, its activity pattern may vary across sites, but in general it is mostly crepuscular, with significant nocturnal activity as well (Schaller 1967; J.W. Duckworth pers. comm. 2008). Sambar is essentially non-social, stable groups being at most family associations (Schaller 1967; Karanth and Sunquist 1992). However, it is often in groups, presumably temporary, of 12–30 in dry deciduous and semi-arid forests of India (e.g. Bandipur, Pench, Melghat and Ranthambore) (N.S. Kumar pers. comm. 2008), and the same is true around the secondary grasslands within Khao Yai National Park, Thailand (J.W. Duckworth pers. comm. 2008). During peak summer, such associations of up to 80–100 near large waterbodies are not uncommon in Pench in central India (N.S. Kumar pers. comm. 2008). A detailed review of information on grouping in Sambar is in Sankar and Acharya (2004). Results from many areas show a strong bias in sex ratio to females, perhaps reflecting selective predation; alternatively, stags may be more vulnerable to stress (Sankar and Acharya 2004). Breeding is rather seasonal in most areas where studied, for example Schaller (1967) reported that in Kanha the rut spreads over a period of at least seven months with a peak in November–December. Stags during the rut sometimes can cover large distances, up to 10–20 km in one night. Further information for India is reviewed in Sankar and Acharya (2004). The bucks of Formosan Sambar have a seasonal antler cycle, and the mating season is from June to January with a peak in September–October. Males wallow and mark with scent glands in this season. Fawns are born (usually singly) from March to August, with a peak in May and June (Lin, C.-Y. and Lee, L.-L. pers. comm. 2008). Predation appears to be the major cause of Sambar mortality. In Bandipur, Sambar is one of the most important prey in terms of the biomass taken by Tiger Panthera tigris (31%), Leopard P. pardus (8%) and Dhole C. alpinus (13%) (Andheria et al. 2007). In Bandipur Sambar remains were found in about 22.3% of Tiger faeces, 6.1% of Leopard faeces and 7.3% of Dhole faeces (Andheria et al. 2007). In adjoining Nagarahole, proportion of Sambar in Tiger's diet was about 25–29%, derived from faeces and kill data respectively (Karanth and Sunquist 1995). In Tadoba-Andhari, Sambar together with Gaur contributed to nearly 70% of the prey biomass consumed by Tiger, whereas in Pench-Maharashtra it was 80% (Karanth and Kumar 2005). Sambar occurred in 50% of Tiger scats in Pench-Maharashtra (Karanth and Kumar 2005). In well protected forest reserves where different body-sized ungulates are abundantly available, Sambar together with Gaur are selectively predated by Tiger.

Major Threat(s):

Habitat encroachment and hunting are both widespread in the Sundaic region and in much of the rest of Sambar's range. Indeed, these two threats are often associated and all Borneo's ungulates suffer from the increased hunting that often accompanies logging (Bennett and Dahaban 1995; Bennett and Gumal 2001), such that hunting to local extinction is the chief detrimental effect of logging on ungulates in Sarawak (Bennett and Gumal 2001). This conclusion probably is applicable across the range of Sambar, within which it is heavily hunted (e.g. Duckworth et al. 1999; Walston 2001; Tungittiplakorn and Dearden 2002; Linkie et al. 2003; Santiapillai and Wijeyamohan 2003; Tilson et al. 2004; Rao et al. 2005; Steinmetz et al. 2008; Saw Htun pers. comm. 2008). It is among the most sought-after wild meats in Indonesia (Semiadi 2005), Sarawak (Belden Giman pers. comm. 2008), and Vietnam (B. Long pers. comm. 2008), and during 1988–1993 it was a common and preferred food in rural Lao PDR (Duckworth et al. 1999, Table 1), and is probably near the top of chosen wild meats throughout most of its range (GMA Indonesia workshop); however, it is less favoured in Taiwan compared with Formosan Serow *Capricornis crispus* and Reeves's Muntjac *Muntiacus reevesi* (Chang Shih-Wei pers. comm. 2008). However, it is now so rare in Nam Ha National Protected Area, Lao PDR, that a more recent hunting study found it to be eaten only rarely (Johnson et al. 2003). In Vietnam, muntjac meat is now often served in wildlife restaurants as Sambar, because real Sambar meat is now so difficult to procure (B. Long pers. comm. 2008). These declines in consumption are probably representative of all Lao PDR and Vietnam, and of increasing proportions of the rest its range, reflecting population losses There are major, ongoing, declines in at least Vietnam, Lao PDR, Cambodia, **Thailand,** Malaysia, Myanmar, Bangladesh and Nepal, and probably Sri Lanka and Indonesia (see Populations) which can plausibly only be driven by hunting, because suitable habitat for Sambar is abundant in these countries but is almost or actually bereft of the species. Even in India, poaching has seriously depleted the abundance of large mammals in most areas; even in high-profile areas such a Corbett Tiger Reserve and Rajaji National Park there are still some instances of poaching (Johnsingh et al. 2004). An intensive study of hunting on mammals in two Hmong villages of **northern Thailand** classified quarry species into tiers representing the sequence of loss through over hunting; Sambar fell in the second tier (aside, e.g., big cats Panthera) in the extinction sequence (Tungittiplakorn and Dearden 2002). This study found that Sambar persisted anomalously long in one place under an "exceptional village wildlife hunting regulation promulgated by the village headman until 1994", but was generally extinct in the landscape. A study at Nagarahole National Park compared an area which was only moderately hunted with a heavily hunted site, but found no significant difference in Sambar densities between the two areas (Madhusudan and Karanth 2002). This was speculated to reflect the greater difficulty of hunting Sambar (highly dispersed and usually in rather thick vegetation), despite it being a prized species, than, particularly, Chital Axis axis, a herding species often out in the open, and the benefit to poachers of getting in, making a kill, and getting back out again as fast as possible to reduce detection chances; therefore, this pattern seems to have arisen through hunter choice. In the context of Indochina, even the heavily-hunted site would, however, rank as lightly hunted (J.W. Duckworth pers. comm. 2008, based on visit and discussion). For these two reasons, their finding, of an apparent high resilience of Sambar to hunting, is not applicable to the rest of the species' range (particularly outside well-secured protected areas). However, somewhat reflecting this, Sambar's status in non-Sundaic South-east Asia (still widespread but in ongoing steep decline) relative to other large deer ecologically more similar to Chital (herding; often outside dense forest), Eld's Deer Rucervus eldii and Hog Deer Axis porcinus (extinct across most of their former range through former steep declines) does indeed suggest a broad applicability of a conclusion that Sambar is somewhat less rapidly reduced by hunting than are other sympatric deer, excepting muntjacs Munitacus spp.Deer are hunted in India using snares, dogs, and guns (Jathanna et al. 2003; Kumara and Singh 2004) and these methods are general across their range. Some hunting is for village consumption of meat, but most (at least in South-east Asia) is probably sold commercially, to the affluent urban classes (e.g. Duckworth et al. 1999; Walston 2001; Kumara and Singh 2004), and to itinerant labourers, such as logging crews in at least Sarawak (Bennett and Gunal 2001) and gold-diggers and rattan-cutters in Myanmar (J.W. Duckworth pers. comm. 2008).

Such commercially-driven hunting causes major declines (e.g. Steinmetz et al. 2008) because the market is limitless. Adult males suffer additionally because antlers are widely displayed as trophies and are used in traditional medicine (e.g. Martin 1992, Baird 1995). At least in Lao PDR, their market value means that, compared with muntjacs, a high proportion is sent to towns and traded internationally, and many are sold openly in tourist centres such as Louangphabang (Duckworth et al. 1999). In Taiwan, in the past live males were sought for farming velvet; this was far more valuable than the meat. Presently, velvet can be bought legally and cheaply from farms and hunting for velvet is now insignificant, as is, here, hunting for internal organs (Chang Shih-Wei pers. comm. 2008). Hunting pressures and patterns vary across Sambar's range: three areas (Indochina; Malaysian Borneo; Taiwan) are profiled below, and all are very different not just from each other, but from southern and central India, where (excepting Taiwan) the species has by far the healthiest conservation status. But even across India, hunting for Sambar meat is almost ubiquitous, even occurring within well-secured protected areas, and to greater extents in other protected areas and outside the protected area system (N.S. Kumar pers. comm. 2008). Vietnam, Lao PDR and to a lesser extent Cambodia and northern **Thailand** are apparently the areas of Sambar's range where generalised mammal hunting is heaviest. Sambar is hunted within a thriving hunting and wildlife trading culture in this area and in adjacent countries such as China and Thailand, involving many land vertebrate species, along with other forest products such as orchids and Aquilaria resin (e.g. Compton and Le Hai Quang 1998; Compton et al. 1999; Noreen and Claridge 2001; Smith and Xie Yan in press). Such hunting reaches all areas, although in large rugged mountain forest blocks trade-driven hunting of species valued only for their meat (rather than high-value, low-weight, generally medicinal products that can be carried out efficiently from even the remotest areas) is still limited by economics of accessibility. However, the ability of these areas to protect Sambar at the regional scale is questionable, because no areas of rugged evergreen forest are known with high Sambar densities: such habitat is rather marginal. The regional wild meat trade has little likelihood of abating as long as there are any of the most resilient species (pigs, muntjacs, and civets) to be hunted. The human population of Vietnam is more than 84 million, that of China more than 1,000 million. Together, they comprise an enormous market for wildlife products. For example, tens of millions of wild turtles are imported, legally and illegally, into China annually (van Dijk et al. 2000 and papers therein). There has been no comparable study of ungulate trade levels. Within the Northern and Central Annamites, every square kilometre of Vietnamese forest and of Lao forest that is within 5 km of the Vietnamese border probably has snares capable of capturing Sambar set in it every year (Timmins et al. 2007; R.J. Timmins pers. comm. 2008). Intensity in some areas probably reaches many thousands of snare-nights per km^2 per year (Timmins et al. 2007). Snaring is less intensive in Lao PDR at least away from the Vietnamese border, but is increasing dramatically (Timmins and Robichaud 2005; W.G. Robichaud pers. comm. 2007; R.J. Timmins pers. comm. 2008). Hunting intensity has in recent decades been generally lower in southern Vietnam than in Lao PDR and northern Vietnam, leaving many wildlife populations less depleted (e.g. Le Xuan Canh et al. 1997; Timmins and Duckworth 2000; Polet and Ling 2004), but the general increase in regional hunting intensity and economic wealth indicate that hunting intensity in this region must also be increasing. Rapid economic development and expanding wealth, particularly South-east Asia and China, are increasing the demand for wildlife meat and 'medicines'(e.g. Timmins et al. 2007). Timmins et al. (2007) highlighted a common misconception, especially in documentation of development projects in the region, that poverty is a principal cause of biodiversity loss: as they pointed out for Saola Pseudoryx nghetinhensis, the main driver of threats to wild ungulates in Indochina, at least for the mid-term, is not rural poverty but increasing urban wealth in Indochina and China. In Viet Nam "the free market economy has resulted in feverish periods of trade in wild species nationwide, with negative impacts on biodiversity" (Government of the Socialist Republic of Vietnam 2004). In Cambodia, the same factor has fostered a thriving bush meat market and hunting of species for international wildlife trade (Timmins and Ou 2001; Timmins 2006; Maxwell et al. 2007), and the intensity of hunting there for some species (Sambar included) is likely to exceed even that in Vietnam. This reflects the logistical ease of hunting and trading with few controls and the relative abundance of high-value quarry species (R.J. Timmins pers. comm. 2008). There is a rapid, ongoing, expansion of wealthy social strata in Lao PDR, Cambodia, and Vietnam, directly accelerated by illegal trade in timber, wild meat, and other forest resources, and by the economics of large infrastructure projects (e.g. Nan Theun 2 hydro-electric power project), creating a significantly greater in-country demand for luxury meats such as venison (W.G. Robichaud pers. comm. 2007).

Markets along major roads such as route 13 through Lao PDR (e.g. at Ban Namthon) have expanded, not contracted, in the last 15 years and have a huge (albeit unquantified) turnover of wild meat; at this specific market, multiple stalls are selling dried deer meat daily. The effects of hunting in Lao PDR, Vietnam, Cambodia and **Thailand** have been exacerbated during the last two decades by habitat loss and various socio-economic factors, of which the most biologically significant are discussed in the Red List account for Large-antlered Muntjac *M. vuquangensis*. The most significant constraint to Indochina's Sambar populations is the long-term uncertainty of success in protected areas, even in those currently effective in conservation management. Even the most successful protected areas face an uncertain future with the possibility of degazettement of conservation status of parts of them, the lack of, or possible future loss of, adequate external funding necessary to maintain high standards of management, the lack of, or possible future loss of, political support necessary to uphold high protection standards and the uncertainties of maintaining a motivated and well-trained staff. On Borneo, Sambar are also widely and heavily hunted, largely for meat consumed in-country (Bennett et al. 2000; Mohd Azlan J. pers. comm. 2008; Belden Giman pers. comm. 2008; G.M. Fredriksson pers. comm. 2008; A.C. Sebastian pers. comm. 2008; Siew Te Wong pers. comm. 2008). Traditionally, hunters used traps, dogs and spears, and blowpipes. Hunters in remote parts of the interior still use these methods, but most animals now die by gunfire. Sarawak held 60,000 legally registered shotguns while Sabah held almost 13,000 in the mid 1990s (Bennett et al. 2000). Methods used in Indonesia include guns and spotlights along logging roads, snaring, and dogs (G.M. Fredriksson pers. comm. 2008; S. Hedges pers. comm. 2008). Snaring is still very heavy outside well-secured areas, but shotguns remain the even bigger problem (Bennett et al. 2000; Mohd Azlan J. pers. comm. 2008; A.C. Sebastian pers. comm. 2008; Siew Te Wong pers. comm. 2008). Over most of (at least) Malaysian Borneo, "in general, everyone will hunt and eat anything" (Bennett et al. 2000). Much hunting is for the market rather than local use, so there is a limitless demand. Bennett et al. (2000) and Bennett and Gumal (2001) profiled the hunting of ungulates in Sarawak in the mid 1990s thus: Sambar (along with Bearded Hog Sus barbatus and muntjacs Muntiacus) is heavily shot for wild meat trade across Sarawak, and these ungulates are the species most sought by the restaurants; Sambar in particular is an important source of meat for logging company employees in Sarawak. Wild meat is widely sold in towns, villages, restaurants and logging camps throughout Sarawak; 250 sales outlets were estimated across the state in 1996, with an annual trade worth 3.75 million USD. Along the Rejang river alone, in the mid 1980s, 1,500 Sambars were sold per year (Caldecott 1988). The market for meat is great and probably expanding. Bennett et al. (2000) considered the effects of this onslaught upon Sambar through a lengthy hunting study in February 1993 to June 1995, in both Bornean states of Malaysia (Sarawak and Sabah), each with eight study areas. Sambar constituted only 6.7% of animals killed in Sabah, but 35% by weight of wild meat; negligible numbers were taken in Sarawak, reflecting the major declines that had already reduced the species to rarity. Index counts of signs show an inverse relationship between sign index and hunting pressure so strong that the latter over-rode the effects of habitat variables in determining their densities. Hunting was so ubiquitous, even in protected areas, that in the two states, only one site (part of Danum Valley, Sabah) could be found as a control, where there was primary forest but negligible hunting levels. They concluded that "the only single factor offering any effective protection for [quarry species of] wildlife is difficulty of access"; whether animals are protected or non-protected species, inside or outside protected areas, had no significant restraint on hunting levels. There were, then, no official controls on commercial hunting of non-protected species. The Master Plan for Wildlife in Sarawak (Wildlife Conservation Society and Sarawak Forest Department 1996) introduced a strict ban on all wildlife trade, shotgun ownership and cartridge sales in the late 1990s (Bennett and Gumal 2001). Substantial numbers of Sambars are, nevertheless, still killed in the state, as they are elsewhere in Borneo. Moreover, in the interim, much forest has been lost and Sambar populations in the smaller and more isolated areas that remain must now be even more vulnerable to local extinctions. Taiwan contrasts with the rest of Sambar range in that hunting has been so successfully reduced that populations are now increasing in large parts of the island. It was formerly a big problem, for venison and, for medicinal purposes, velvet and penis. The reduction during and since the 1990s reflects five factors. Hunting of Sambar was banned in 1989 through the Wildlife Conservation Act.

Velvet prices fell steeply during 1985–1990 with deer TB in 1985, reducing returns from poaching absolutely an effect exacerbated by rapid economic development in 1980s leading to higher daily wages. Former logging road systems are deteriorating within protected areas, making removal of deer meat less attractive. The general awareness of ecological conservation has much enhanced. And eco-tourism through community forestry plans has given alternative employment to former hunters (all information: Lin, C.-Y. and Lee, L.-L. pers. comm. 2008). Other sympatric ungulates are hunted broadly as avidly as Sambar, but there is suggestive evidence from two areas of **Thailand** (Kuiburi National Park and a 20 km^2 sector of Thung Yai Naresuan Wildlife Sanctuary) that when Sambar densities are very much reduced, recovery may be more difficult than with Gaur Bos gaurus and muntjacs. In both these areas the institution of effective protection fuelled a rise in sign density, and presumably animal populations, for Gaur and muntjacs, but not for Sambar (Steinmetz et al. 2007, in prep.; R. Steinmetz pers. comm. 2008). If this finding is widely applicable (and B. Long pers. comm. 2008 suspects a similar contrast between Sambar and other ungulates in a protected part of Cambodia), the implications for Sambar populations in Lao PDR, Viet Nam, northern Thailand, Sarawak and other areas where populations are mostly down to scattered tiny numbers, may be profound.

Over many centuries, the loss of wilderness to human settlement, cultivation and industry has resulted in massive loss of potential Sambar range, particularly in the densely-settled countries of India, China, Bangladesh, **Thailand,** and Vietnam. Throughout the range of Sambar, natural and semi-natural habitats (forest, scrubland and grassland) continue to be converted to anthropogenic habitats, degraded (e.g. stands are being changed in character through logging) and fragmented (broken into smaller blocks). The relative proportions of these activities vary across the range as does the percentage of former forest land so affected. The role of these habitat changes as a direct stimulant of the major declines in progress is unclear. Sambar is highly tolerant of forest degradation: indeed, much higher numbers are found in encroached stands than in pristine forests, if hunting is under control (Rijksen 1978; Heydon 1994; Stuebing 1995; Davies et al. 2001). Degradation per se is unlikely to be a significant threat. Many Indian forest areas are severely encroached by exotics such as *Lantana camara, Parthenium* spp. and *Chromolaena odorata*, and these are suspected to effect major changes to forest structure (Hiremath and Sundaram 2005): but their effects on Sambar populations warrant further study. There is no evidence that Sambar can survive complete conversion of landscapes to non-forest uses such as rice paddies, and its ability to use tree plantations in the absence of natural forest is unclear. Probably, like most deer, provided strips of semi-natural vegetation survive (e.g. along streams and on steep slopes) are present, it can use them to some degree. In most of its range (parts of India excepted) it is unlikely that habitat extent is the determining factor of current Sambar numbers, because large tracts of apparently suitable habitat now support no or only tiny numbers of Sambar. By far the most serious effect of current habitat trends is that fragmentation opens up wilderness areas to people, making it easier to hunt throughout them and cheaper to remove bulk wildlife products such as Sambar meat. In both Malaysian Borneo and Lao PDR the 'natural protection' afforded to an area through difficulty of access has been considered to be the chief factor allowing populations of large, hunted, mammals to survive (Timmins and Duckworth 1999; Bennett et al. 2000), and this is probably widely true for Sambar outside India, Nepal and the few other areas where active protective measures are effective. In sum, the proximate cause of the major ongoing Sambar declines in most of its range is hunting, for which habitat conversion and fragmentation exacerbate the effects; and these processes are resulting in an ever-smaller maximum possible population, were the hunting issues to be successfully addressed. Threats other than hunting and, locally, habitat loss, seem to be insignificant at the species level. Hybridization with Javan Rusa *Rusa timorensis* has been documented (Van Mourik and Schurig 1985), and is apparently a localized threat, where the latter has been introduced into Sambar range, in Pewajam District, East Kalimantan (G. Semiadi pers. comm. 2006). Hybridization has also been suspected in Sumatra, with the introduced *Chital Axis axis* (G. Semiadi pers. comm. 2006). However, Sambar and Chital co-occur extensively in India without hybridizing. Sambar is an occasional crop pest in mainland China, notably around the Ailoshan range (J. Fellowes pers. comm. 2008), India (N.S. Kumar pers. comm. 2008), Lao PDR (Duckworth et al. 1999), Myanmar (J.W. Duckworth pers. comm. 2008), Sumatra (D. Martyr pers. comm. 2008), and probably throughout its range.

There is no evidence that Sambar is a serious crop pest away from the fields directly abutting forests. Nonetheless, crop damage is often stated by affected farmers to lead to a requirement for lethal control. However, the ready market for meat means that this is most probably an excuse for killing that would enthusiastically be undertaken anyway (N.S. Kumar and J.W. Duckworth pers. comm. 2008): it is therefore unlikely that control of crop-degrading animals is a significant cause of the species' decline anywhere in Sambar's range. Grazing competition with domestic stock seems to be much less of a threat to Sambar than to sympatric Gaur and Chital: in fact the key study quantifying just how serious a threat such competition is in Indian protected areas (Madhusudan 2004) found that it had negligible effects on Sambar densities, presumably because Sambar is an efficient browser. Threats to southern India's forest ungulates by competition with domestic stock grazing within protected areas are exacerbated where dung is collected for export to adjacent coffee areas. Fuel wood removal may also be at levels sufficient to disrupt nutrient cycles of the habitat (Madhusudan 2005). In the Indian Terai Arc Landscape, lopping of tree branches for cattle fodder is a threat to Sambar as it effectively removes most browsable vegetation from the landscape (B. Long pers. comm. 2008). These are probably more serious concerns for Sambar than is grazing by domestic stock.

Conservation Actions:

The Sambar is found in many protected areas throughout its range, although in most of these areas this legal status has not stemmed declines and local extinctions from hunting. Similarly, although it is protected from hunting by legislation, even outside protected areas, in most or all range states, these laws are challenging to enforce, given the trade demand for meat and antlers (GMA Indonesia Workshop). Currently the law has a strong protective effect in Taiwan (Lin, C.-Y. and Lee, L.-L. pers. comm. 2008) and in various protected areas of India (N.S. Kumar pers. comm. 2008). In South-east Asia, Sambar declines have lagged behind those of other sympatric deer (excepting muntjacs). There is no doubt that within a decade or two Sambar will be, like Hog Deer and Eld's Deer are already, almost absent from South-east Asia unless effective protection from trade-driven hunting is instituted. In India the species's status is less grim, and this is dependent upon continuation of the current protected areas system, and the expansion of effective threat reduction into a greater proportion of protected areas: in many Indian protected areas, poaching is rampant and in such vulnerable areas Sambar numbers are still declining. Given the major challenges combating hunting of such a desired animal, long-term survival is most realistic in large protected areas (exceeding 1,000 km^2 where possible), lacking people living within them, with an adequate number of motivated and capable staff (Anak Patannavibool pers. comm. 2008; A.J. Lynam pers. comm. 2008). The precise ways for protected areas to function effectively in the conservation of large mammals depend upon their own specific circumstances, but field presence of staff, dispersed across the area, is likely to be important in almost every area. For example, Lynam et al. (2006) found in **Khao Yai National Park, Thailand,** that encounter rates of tracks and signs of Sambar (and other large mammals) decreased with distance from ranger sub-stations. Deep inside the forest to the south and east, far away from ranger substations, Sambars were locally extinct. Similar patterns are shown by current Sambar distributions in other **Thai protected areas such as Khao Sok, Phu Khieo and Kaeng Krachan** (A.J. Lynam pers. comm. 2008). In some protected areas (recent examples being Bhadra and Kudremukh in Karnataka), formerly forest-dwelling human communities have resettled closer to markets, clinics, school and other services (Karanth and Karanth 2007). This expands habitat available to Sambar (that which was formerly fields and houses) but more significant is the removal of a source of hunters and the cessation of grazing competition with domestic stock (see Karanth et al. 2006).Such conservation interventions have helped recovery of Sambar populations in this region. Increasing development aspirations mean that out migration is likely from further areas, e.g. Mudumalai. Most of the remaining forest within its Chinese range are already established as nature reserves (M.W.N. Lau pers. comm. 2008). As the majority of the Sambar population is in South Asia, conservation activities are important there. It is also important that populations are maintained at least in several areas in Southeast Asia, to maintain Sambar presence through its historical range, to preserve Sambar genetic diversity and through its importance to the ecosystems within which it lives. Given the regional pattern of threats and current successes to date, implementing effective conservation interventions is considerably more challenging, and therefore a higher priority in the Southeast Asian region. The most important conservation measure for Sambar in Cambodia is the continued strengthening and support of conservation efforts in Siema Biodiversity Conservation Area and the Srepok Wilderness Area, and to a lesser extent support of protected area conservation management in the northern plains and central Cardamoms.

Likewise in Vietnam, the most important conservation measure is the continued support of protection efforts within Cat Tien National Park, and in **Thailand** it is to continue the consolidation of the 'Western Forest Complex'. In Myanmar the Hukaung Valley is an exceptional area of lowland plains forest, grassland, and wetlands. Part lies within the Wildlife Sanctuary and some of the rest of the outstanding plains habitat is within an enormous proposed extension. Through its size and mostly little-encroached condition, this is the most outstanding remaining landscape-level floodplains habitat for very large mammals remaining in tropical Asia, and although wildlife populations are highly depleted, warrants the strongest effort to conserve it. It faces a number of daunting challenges (J.W. Duckworth pers. comm. 2008). In Lao PDR, any of a large number of areas could become key Sambar conservation sites, but there is as yet no precedent in the country for effective conservation of high-trade-value large mammals. Xe Pian, Nam Et–Phou Louey, and Nakai–Nam Theun NPAs could all be highly significant areas for the species, but so could almost any other NPA; the reality is that the most success with Sambar conservation is likely to come through identifying areas with positive underlying situations to achieve conservation, rather than identifying the area with the most intrinsic importance for Sambar as the focus for efforts. Captive breeding as it is being developed in East Kalimantan. A captive herd of 223 heads in 10 ha enclosure was established in 1998 in Penajam District, as a pilot project. Several private herds exist in the province, such as at Bearau and at Nunukan, both with more than 20 heads (G. Semiadi pers. comm. 2008). Captive Sambar herds are also widespread in mainland Asia.

Indian Muntjac *(Muntiacus muntjak)*

Description:

The Indian muntjac has a short but very soft, thick, dense coat, especially those living in cooler regions. Coloration of the coat changes from dark brown to yellowish and grayish brown depending on the season. The muntjacs' coat is golden tan on the dorsal side and white on the ventral side of the body, the limbs are dark brown to reddish brown, and the face is dark brown. However, the ears have very little hair which barely covers them. Male muntjacs have antlers that are very short, about 1–2 inches, usually consisting of only two or three points at the most and protrude from long body hair covered pedicels on the forehead. Females have tufts of fur and small bony knobs where the antlers are located in males. Males also have slightly elongated upper canines about an inch long that curve slightly outward from the lips and have the capability to inflict serious injury upon other animals or to other members of the population while exhibiting aggression. Males are generally larger than females. The body length of muntjacs varies from 35–53 in long and their height ranges from 15–26 in tall.

Range Description:

Southern Red Muntjac, as defined here, occupy part of the Thai–Malay peninsula and occurs on the main islands of the Greater Sundas (Borneo, Java, Bali and Sumatra) and on various small islands (Chasen 1940; Groves 2003; Meijaard 2003). On Sumatra, it is not geographically limited to southern and eastern parts of the island despite various statements that it is (R.J. Timmins pers. comm. 2008). Taking the northern limit on the peninsula as the Isthmus of Kra (see Taxonomy), M. muntjak (as here defined) is assumed to inhabit southern **Thailand** and might be found to occur in southernmost Myanmar. It is now extinct in Singapore (Baker and Lim 2008).

Countries: Native:

Brunei Darussalam; Indonesia; Malaysia; **Thailand**

Regionally extinct:

Singapore

Population:

Recent camera-trap studies on all main occupied landmasses show this to be a common species. The following is only a part collation of existing studies. In Sumatra, muntjacs (not identified to species but presumed, given the habitats and altitudes, to be largely or entirely *M. muntjak)* were camera-trapped very commonly in Bukit Barisan Selatan National Park and around Gunung Leuser (van Schaik and Griffiths 1996; O'Brien et al. 2003). In **Thailand,** Red Muntjacs were recorded at over 40% of camera-trap locations in Kaeng Krachan National Park, a site that may contain either M. vaginalis or M. muntjak or both.

Abundance did not vary with proximity to villages, even though a variety of hunting methods occurs on site (Ngoprasert et al. 2007). In West Malaysia, Red Muntjacs were commonly camera-trapped, at a high proportion of camera-sites, in Taman Negara, where densities were estimated at 3.2–4.05 per km^2 (Kawanishi and Sunquist 2004), were found to be ubiquitous and common at all nine potential Tiger Panthera tigris areas surveyed by Lynam et al. (2007), and were common in the Jerangau Forest Reserve, an area of secondary forest some of which was being logged at the time of study and where hunting was heavy enough to have eradicated (or reduced to very low densities) species such as Sambar Cervus unicolor (Azlan 2006). During a study of the mammals of West Malaysia's virgin jungle reserves, they were found in 13 of 14 sites (seven virgin jungle reserves and seven adjacent logged areas), a much greater spread of occurrence than for most other mammals (Laidlaw 2000). In Java muntjacs have "persisted in many areas where there is some forest cover" (Whitten et al. 1996), were camera-trapped commonly in Ujung Kulon National Park (van Schaik and Griffiths 1996) and Gunung Halimun National Park (Suyanto 2003), and were studied in Baluran National Park (East Java) in the 1990s by Tyson (2007); the latter study found densities of 23–25 animals per km^2 in the park's mosaic of open woodland, grassland, scrub at under 300 m asl during the dry season. These densities (considerably above those from Taman Negara) may be inflated by daily or seasonal movements of animals coming down to the coastal waterholes. The Baluran population may have benefited from a particular combination of seral processes including the growth of the exotic Acacia nilotica. Muntjacs remain locally very common on Bali (Oka 1998). Determining population status in Borneo is more difficult because of the difficulties of separation from *M. atherodes,* but it is clearly at least fairly common (Heydon 1994; for further discussion see Habitat and Ecology).

Population Trend: Decreasing

Habitat and Ecology:

Southern Red Muntjac are associated with forest, but occur widely even in heavily degraded forest and, in areas adjacent to forest, in plantations of coffee, rubber, sugarcane, cassava, coconut, and teak (Oka 1998; Laidlaw 2000; Azlan 2006; G. Semiadi pers. comm. 2008). However, its ability to survive in landscape-level stands of teak is probably dependent upon the retention of fringes of secondary growth along streams, gullies and other such features (M. Tyson pers. comm. 2008) and this may be so for the other plantation species, too. Peak densities are probably not in pristine forest: in east Java, secondary forest with gallery forest mosaic seems to hold higher densities, and this muntjac may also benefit even from agricultural conversion at forest edges (M. Tyson pers. comm. 2008). Muntjac (presumed to be largely or entirely *M. muntjak)* abundance was higher in areas within 1 km of the protected area boundary than in the interior of Bukit Barisan Selatan National Park, Sumatra (although this pattern was only weakly statistically significant), suggesting an association with edge, broken and secondary habitats (O'Brien et al. 2003). In contrast, an avoidance of the edge of the Taman Negara protected area in Malaysia was suspected by Kawanishi and Sunquist (2004). In Danum Valley (Sabah, Borneo), an area with negligible hunting, *M. muntjak* strongly increased in densities after logging (Heydon 1994; Davies et al. 2001); a weaker increase was found by Duff et al. (1984).

In the Sarawak Planted Forests, Bintulu Division, Sarawak, muntjacs (partly this species) are among the commonest species camera-trapped in young acacia plantations (under four years old) and have been seen browsing on young acacia shoots. Footprints are common in the adjacent newly planted areas and forest fragments (Belden Giman pers. comm. 2008). However, this area contains a substantial proportion of natural, if somewhat degraded, forest, which in theory might be found to be necessary for populations in the highly altered habitats. Information on the species in landscape-scale plantations, remote from even secondary forest, has not been traced. Most of its range is dominated by evergreen vegetation, but it readily uses deciduous forests and mosaics of grassland, scrub, and forest (e.g. on Java; Tyson 2007); on Bali and Java, *M. muntjak* routinely uses woodland savanna as a feeding ground (Oka 1998; S. Hedges pers. comm. 2008). This muntjac has a wide altitudinal range. In Java, S. Hedges (pers. comm. 2008) has seen them over 0–800 m. In Sumatra, it occurs widely in the lowlands and the second species there, *M. montanus,* seems to be montane; whether *M. muntjak* ascends to the highest forests is yet unclear (R.J. Timmins pers. comm. 2008, based on examination of specimens).

On Borneo, Red Muntjac lives up to at least 1,000 m asl on the Usun Apau plateau (Payne et al. 1985), who stated that "available data suggests that [M. atherodes] predominates over the Red Muntjac in low hill ranges and coastal regions", but Meijaard and Sheil (2008) pointed out that still "no robust quantitative data exist to support this [pattern]". Red Muntjac is scarcer than *M. atherodes* in Sungai Wain forest, Kalimantan, which spans 30–150 m asl (G.M. Fredriksson pers. comm. 2008), in the Sarawak Planted Forests, Bintulu Division, a mix of acacia plantation and natural forest (Belden et al. 2007; Belden Giman pers. comm. 2008), and (slightly so) in the Ulu Segama area of Danum Valley Conservation Area, Sabah, which lies mostly at about 300 m asl (Siew Te Wong pers. comm. 2008). In Borneo, muntjacs were frequently seen, suspected to comprise roughly equal numbers of Red and Yellow, in the Batang Ai National Park, Sarawak, which lies mostly at 100–760 m asl (Meredith 1995). The diet is mostly fruits, buds, tender leaves, flowers, herbs and young grass (Kitchener et al. 1990; Oka 1998). In Taman Negara, Malaysia, camera-trapping showed Red Muntjac to be mostly diurnal (Kawanishi and Sunquist 2004) whereas in Gunung Leuser, Sumatra, it was classed as cathemeral (i.e., sporadic and random intervals of activity during the day or night) (van Schaik and Griffiths 1996) and this is true for East Java (S. Hedges pers. comm. 2008); there may be some variation between localities in balance of day and night activity. It is a mostly solitary species that is capable of breeding through the year, and has been stated to be territorial (Kitchener et al. 1990; Oka 1998). However, Tyson (2007) found no evidence of territoriality in radio-collared female muntjacs of Baluran National Park, Java, although range overlap declined in the rainy season compared with the dry season. Nor was there evidence of males using exclusive ranges in any season, or any clear evidence of territorial defence. As in the case of *M. vaginalis*, occasional sightings of small groups probably refer to mating pairs, or to females with offspring (Oka 1998), or to loose aggregations at abundant food resources (M. Tyson pers. comm. 2008).

Systems: Terrestrial

Major Threat(s):

Habitat encroachment and hunting are both widespread in the Sundaic region. Indeed, they are often associated and all Borneo's ungulates suffer from the increased hunting that often accompanies logging (Bennett and Dahaban 1995, Bennett and Gumal 2001), such that hunting to local extinction is the chief detrimental effect of logging on ungulates in Sarawak (Bennett and Gumal 2001). This conclusion probably is applicable across the Sundaic region. Southern Red Muntjac are widely hunted across its range (e.g. Bennett et al. 2000; Linkie et al. 2003; S. Hedges pers. comm. 2008, based on unpubl. data, East Java, 1991–2002); it is the most sought-after wild meat in Peninsular Malaysia (Kawanishi and Sunquist 2004) and is among the most preferred in Indonesia, although Sambar is the preferred species at least in East Kalimantan (Semiadi et al. 2004; Semiadi 2005). However, there is no strong evidence that either hunting or habitat disruption are actually threats to the survival of populations except in the case of islands such as Singapore, where it is now extinct (Baker and Lim 2008). Peak densities are not in pristine forest (see Habitat and Ecology), and in Danum Valley (Sabah, Borneo), an area with negligible hunting, M. muntjak strongly increased in densities after logging (Heydon 1994; Davies et al. 2001); a weaker increase was found by Duff et al. (1984). Hence, even quite severe habitat disruption can increase ecological carrying capacity for this muntjac, at least temporarily. In an area with heavy hunting, recorded densities of muntjacs (presumed *M. muntjak)* were twice as high in areas with low than with high human population density within 10 km of the Bukit Barisan Selatan National Park, Sumatra, but this difference was not statistically significant (O'Brien et al. 2003).Poaching pressure in Kerinci-Seblat National Park, Sumatra, among snared ungulates was greatest for muntjac as assessed by the diameter of the cable. Snaring is locally very high in the protected area, e.g. 51 snares (mostly of size suitable for muntjac) were found in 1 km^2 (Linkie et al. 2003), although it is much less intense than in areas of Indochina supporting the related *M. vaginalis* (J. Holden pers. comm. 2008). Data are too sparse to determine whether this hunting is driving declines. Habitat loss is also severe including within parts of the park (Linkie et al. 2003), and is an increasing phenomenon in Java (S. Hedges pers. comm. 2008).

Muntjacs are widely and heavily hunted on Borneo for meat, skins, and as a source of traditional remedies (Bennett et al. 2000; Mohd Azlan J. pers. comm. 2008; Belden Giman pers. comm. 2008; G.M. Fredriksson pers. comm. 2008; A.C. Sebastian pers. comm. 2008; Siew Te Wong pers. comm. 2008). Traditionally, hunting used traps, dogs and spears, and blowpipes. Hunters in remote parts of the interior still use these methods, but most animals now die by gunfire. Sarawak held 60,000 legally registered shotguns while Sabah held almost 13,000 in the mid 1990s (Bennett et al. 2000).

Methods used in Indonesia include guns and spotlights along logging roads, snaring, and dogs (G.M. Fredriksson pers. comm. 2008; S. Hedges pers. comm. 2008). The combination of dogs and guns may be particularly efficient (Belden Giman pers. comm. 2008). Snaring is still very heavy outside well-secured areas, but shotguns remain the even bigger problem (Bennett et al. 2000; J. Mohd Azlan pers. comm. 2008; Siew Te Wong pers. comm. 2008). Unaccompanied packs of stray or even perhaps feral dogs are also likely to be a threat, particularly in forest isolates close to urban areas (e.g. Similajau National Park; J.W. Duckworth pers. comm. 2008). It is unlikely that hunters make any distinction between the two Bornean muntjac species directly, although it is plausible that *M. muntjak* is under lower hunting pressure than is *M. atherodes* reflecting its purported predominance in the hills and mountains (although there is little empirical evidence to support this), whereas most of the logging concessions currently lie in the lowlands. Some ethnic groups, at least the Punan, have traditional beliefs against eating muntjac meat (Belden Giman pers. comm. 2008), but over most of (at least) Malaysian Borneo, "in general, everyone will hunt and eat anything" (Bennett et al. 2000). Much hunting is for the market rather than local use, so there is a limitless demand. Bennett and Gumal (2001) profiled the hunting of ungulates in Sarawak in the mid 1990s thus: muntjacs (along with Bearded Hog *Sus barbatus* and *Sambar Rusa unicolor*) are heavily shot for wild meat trade across Sarawak, and these ungulates are the species most sought by the restaurants. Wild meat is widely sold in towns, villages, restaurants and logging camps throughout Sarawak; 250 sales outlets were estimated across the state in 1996, with an annual trade worth 3.75 million USD. Along the Rejang river alone, in the mid 1980s, 250 muntjacs were sold per year (Caldecott 1988). The market for meat is great and probably expanding. There were, then, no controls on commercial hunting of non-protected species (including muntjacs). The Master Plan for Wildlife in Sarawak (Wildlife Conservation Society and Sarawak Forest Department 1996) introduced a strict ban on all wildlife trade, shotgun ownership and cartridge sales in the late 1990s (Bennett and Gumal 2001). Substantial numbers of muntjacs are, nevertheless, still killed in the state, as they are elsewhere in Borneo.

Bennett et al. (2000) considered the effects of this onslaught upon muntjacs (not distinguished to species) through a lengthy hunting study in February 1993 to June 1995, in both Bornean states of Malaysia (Sarawak and Sabah), each with eight study areas. Muntjacs constituted only 3–10% of animals killed, but index counts of signs show an inverse relationship between sign index and hunting pressure so strong that the latter over-rode the effects of habitat variables in determining their densities. Hunting was so ubiquitous, even in protected areas, that in the two states, only one site (part of Danum Valley, Sabah) could be found as a control, where there was primary forest but negligible hunting levels. They concluded that "the only single factor offering any effective protection for [quarry species of] wildlife is difficulty of access"; whether animals are protected or non-protected species, inside or outside protected areas, had no significant restraint on hunting levels. In the interim, much forest has been lost and muntjac populations in the smaller and more isolated areas that remain must now be more vulnerable to local extinctions.

Conservation Actions: Southern Red Muntjac is found in many protected areas throughout its range (GMA Indonesia Workshop). It is a protected game species in West Malaysia under the Wildlife Protection Act of 1972, meaning that it can be hunted only with a license and only outside protected areas (Kawanishi and Sunquist 2004). Indonesian forestry law protects all species of muntjac. Species protection laws relating to ungulates in many areas of Sumatra have not been widely publicized. A significant percentage of Sumatra's forests are protected, but protected areas and the authorities responsible for conservation of catchment forests are often under-funded and almost all are grossly understaffed so that there is little ranger presence in the field. Where there are field ranger teams, these focus generally on flagship species such as Asian Elephant *Elephas maximus,* Sumatran Rhinoceros *Dicerorhinus sumatrensis* or Tiger *Panthera tigris*, although Tiger protection rangers do also conduct ungulate protection activities and in 2007 three hunters arrested in the national park with two muntjacs carcasses received custodial sentences at Kerinci district court (D. Martyr pers. comm. 2008). Much of Java's remaining forest is officially protected. Heavy hunting with shotguns was found to be a severe problem for larger mammals and birds across Sarawak and the need for its control (of guns and of sale of ammunition, with firm legal underpinnings) was recognised in the Master Plan for Wildlife in Sarawak (Wildlife Conservation Society and Sarawak Forest Department 1996). As everywhere, people devise ways for circumventing the controls (illegal trade in bullets and locally made shot-guns, called bekakok, which have no administrative or legal paper-trail) and strong enforcement is required for the foreseeable future. There seem to be no species-specific conservation measures in need of implementation, above turning existing laws on protected areas and protected wildlife into reality, with a particular focus on eradicating wildlife trade.

Gaur *(Bos frontalis)*

Description:

The Gaur is a strong and massively built species with a high convex ridge on the forehead between the horns, which bends forward, causing a deep hollow in the profile of the upper part of the head. There is a prominent ridge on the back. The ears are very large; the tail only just reaches the hocks, and in old bulls the hair becomes very thin on the back. In colour, the adult male gaur is dark brown, approaching black in very old individuals; the upper part of the head, from above the eyes to the nape of the neck, is, however, ashy grey, or occasionally dirty white; the muzzle is pale coloured, and the lower part of the legs are pure white or tan. The cows and young bulls are paler, and in some instances have a rufous tinge, which is most marked in groups inhabiting dry and open districts. The tail is shorter than in the typical oxen, reaching only to the hocks. They have a distinct ridge running from the shoulders to the middle of the back; the shoulders may be as much as 12 cm (4.7 in) higher than the rump. This ridge is caused by the great length of the spinous processes of the vertebrae of the fore-part of the trunk as compared with those of the loins. The hair is short, fine and glossy, and the hooves are narrow and pointed.

The Gaur has a head-and-body length of 250 to 330 cm (8 ft 2 in to 10 ft 10 in) with a 70 to 105 cm (28 to 41 in) long tail, and is 165 to 220 cm (5 ft 5 in to 7 ft 3 in) high at the shoulder. The average weight of adult gaur is 650 to 1,000 kg (1,430 to 2,200 lb), with an occasional large bull weighing up to 1,500 kg (3,300 lb). Males are about one-fourth larger and heavier than females. In general measurements are derived from gaurs surveyed in India and China. The Seladang, or Malayasian subspecies, may average larger but no scientifically published measurements are known.

Gaur do not have a distinct dewlap on the throat and chest. Both sexes carry horns, which grow from the sides of the head, curving upwards. Between the horns is a high convex ridge on the forehead. At their bases they present an elliptical cross-section, a characteristic that is more strongly marked in bulls than in cows. The horns are decidedly flattened at the base and regularly curved throughout their length, and are bent inward and slightly backward at their tips. The colour of the horns is some shade of pale green or yellow throughout the greater part of their length, but the tips are black.The horns, of medium size by large bovid standards, grow to a length of 60 to 115 cm (24 to 45 in).

The cow is considerably lighter in make and in colour than the bull. The horns are more slender and upright, with more inward curvature, and the frontal ridge is scarcely perceptible. In young animals the horns are smooth and polished. In old bulls they are rugged and dented at the base.

Gaur are among the largest living land animals. Only elephants, rhinos, the hippopotamus (*Hippopotamus amphibius*) and the giraffe (*Giraffa camelopardalis*) consistently grow heavier. Two species that naturally co-exist with the gaur are heavier: the Asian elephant (*Elephas maximus*) and Indian rhinoceros ((*Rhinoceros unicornis*),

Range Description:

Gaur historically occurred throughout mainland south and southeast Asia and Sri Lanka. It currently occurs in scattered areas in the following range states: Bhutan, Cambodia, China, India, Lao PDR, Malaysia (Peninsular Malaysia only), Myanmar, Nepal, Thailand, and Vietnam, but is extinct in Sri Lanka and also, as a resident, apparently in Bangladesh (Grubb 2005; Md Anwarul Islam in litt. 2008; Hedges in prep.). The species is now seriously fragmented within its range, and the mapped distribution is generalised, especially in India, Myanmar, China and Malaysia.

The domesticated form of Gaur, considered by IUCN a separate species *(Bos frontalis; Mythun, Mithan or Gayal)*, occurs in parts of India, China, and Myanmar as feral, semi-feral, and domestic animals. This animal is excluded from the red-listing considerations for Gaur.

Countries: Native:

Bangladesh; Bhutan; Cambodia; China; India; Lao People's Democratic Republic; Malaysia (Peninsular Malaysia); Myanmar; Nepal; **Thailand;** Vietnam

Regionally extinct:

Sri Lanka

Population:

The global population is estimated to lie within 13,000–30,000 animals. Field data suggest that the proportion of mature individuals in the population is likely to be 0.4–0.6, indicating a total of 5,200–18,000 mature individuals, with no population known to have over 1,000 individuals (S. Hedges pers. comm. 2000). Most population figures given for numbers of Gaur are little more than guesses. In Bangladesh, Gaur is probably extinct (Md Anwarul Islam in litt. 2008). A few were thought still to occur in the Chittagong, Chittagong Hill Tracts, Sylhet, and Mymensingh areas in the early 1980s (Sarker and Sarker 1984), but none had been seen in Pablakhali Wildlife Sanctuary in the Hill Tracts since the early 1970s (Khan 1985). Choudhury (2002) stated that Gaur individuals from Mizoram and Tripura (India) still cross into Bangladesh, although the records on which this was based were not given. In Bhutan, Gaur apparently persists all over the southern foot-hill zone, notably in Royal Manas National Park, Phipsoo Wildlife Sanctuary and Khaling Wildlife Sanctuary (Choudhury 2002). In Cambodia, Gaur probably declined by 90% or more in the period from the late 1960s to the early 1990s, at which time it still remained widespread, outside the lower Mekong and Tonle Sap lowlands, although in generally low numbers (Heng Kimchhay et al. 1998; Daltry and Momberg 2000; Timmins and Ou 2001; R. J. Timmins pers. comm. 2008). By far the most substantial Gaur population of the country remained in eastern Cambodia centred on Mondulkiri province, where in the late 1990s potentially several hundred to a thousand may have survived in a forested landscape of over 15,000 km² (Timmins and Ou 2001; Tordoff et al. 2005; R.J. Timmins pers. comm. 2008). Further declines, of about 50%, took place from the early 1990s to the present (R.J. Timmins pers. comm. 2008). As of 2007 the most significant population by far in Cambodia still remained within Mondulkiri Province (R. J. Timmins pers. comm. 2008 based on WCS and WWF unpublished data; T. Clements pers. comm. 2007; T.D. Evans pers. comm. 2007). Protection efforts in part of this landscape have probably stabilised the population there (R. J. Timmins pers. comm. 2008 based on WCS and WWF unpublished data; T. Clements pers. comm. 2007; T.D. Evans pers. comm. 2007).Protection outside these two areas is limited and for Cambodia as a whole the trend is still a significant decrease (R. J. Timmins pers. comm. 2008). In China, Gaur occurs in Yunnan and southeast Tibet (Ma Yiqing and Wang Yingxiang 1995). All sites in Yunnan (except Yingjiang) formed, until the 1950s–1960s, one large contiguous population, but this is now heavily fragmented. By the 1980s, the species was extinct in Lancang county, and the remaining animals were split into two populations: Xishuangbanna–Simao and Cangyuan (Wang Yingxiang verbally 1998 to Duckworth and Hedges 1998). Some sources suggest that a population of 600–800 existed in Yunnan Province in the mid-1990s, with the majority occurring in Xishuangbanna National Nature Reserve (Ma Yiqing and Wang Yingxiang 1995); other sources suggest that only small numbers of Gaur remain in Yunnan (along the border with Myanmar) and that hunting had led to the local extinction of Gaur in Xishuangbanna (Xiang and Santiapillai 1993). In India, three major (Western Ghats,

Protection outside these two areas is limited and for Cambodia as a whole the trend is still a significant decrease (R. J. Timmins pers. comm. 2008). In China, Gaur occurs in Yunnan and southeast Tibet (Ma Yiqing and Wang Yingxiang 1995). All sites in Yunnan (except Yingjiang) formed, until the 1950s–1960s, one large contiguous population, but this is now heavily fragmented. By the 1980s, the species was extinct in Lancang county, and the remaining animals were split into two populations: Xishuangbanna–Simao and Cangyuan (Wang Yingxiang verbally 1998 to Duckworth and Hedges 1998). Some sources suggest that a population of 600–800 existed in Yunnan Province in the mid-1990s, with the majority occurring in Xishuangbanna National Nature Reserve (Ma Yiqing and Wang Yingxiang 1995); other sources suggest that only small numbers of Gaur remain in Yunnan (along the border with Myanmar) and that hunting had led to the local extinction of Gaur in Xishuangbanna (Xiang and Santiapillai 1993). In India, three major (Western Ghats, Central India and North-East) and two minor (Bihar and West Bengal) "Gaur conservation areas" have been identified, reflecting the remaining distribution (Sankar et al. 2000; Choudhury 2002). The Western Ghats and their outflanking hills in south India constitute one of the most extensive extant strongholds of Gaur, with good numbers in Wynaad – Nagarahole – Mudumalai – Bandipur complex (Ranjitsinh 1997). Recent distributional assessment surveys in the Karnataka part of the Western Ghats showed the presence of Gaur in more than 60% of the 22,000 km^2 landscape (K.U. Karanth and N.S. Kumar unpublished data). Ranjitsinh (1997) estimated 12,000–22,000 in India, while Choudhury (2002) suggested that there were about 23,000–34,000 Gaur in India, Bhutan, and Bangladesh; but the true number is simply not known, as shown by the wide ranges. Major populations have been reported in Nagarahole National Park (probably over 2,000), Manas Wildlife Sanctuary (not known; may not be large), Bhadra Wildlife Sanctuary (over 800), Melghat Tiger Reserve (perhaps 500), Bandipur National Park (over 2,000), Kanha National Park (probably under 200), Radhanagari Wildlife Sanctuary (not known; may not be large), Tadoba–Andhari Tiger Reserve (over 1,000), Periyar Tiger Reserve and adjoining forest complex (500–1000), Annamalai range complex (Parambikulum, Valparai, and Grass Hills) (500–700), Silent Valley and adjoining forest complex (500–1000), Agastyamalai forest complex (about 500), Biligirirangangswamy Wildlife Sanctuary and Malemahadeswara Hill range (over 1000), Anshi–Dandeli Tiger Reserve and adjoining forest mosaic (perhaps about 400), Kudremukh National Park and Someshwara Wildlife Sanctuary (about 200–400), Brahmagiri–Pushpagiri–Talakaveri Sanctuary, Mukurti National Park, Pench Tiger Reserve, Jaldapara Wildlife Sanctuary ('a large population'), Chapramari Wildlife Sanctuary ('a large population'), and Gorumara Wildlife Sanctuary ('a large population'). Many other areas are known or suspected to hold small populations, but in many sites such small numbers are unlikely to be viable (Choudhury 2002). The foregoing figures were assigned by N.S. Kumar (pers. comm. 2008) and are based on a variety of sources and methods, and from most areas are for guidance only. An increasing rigour is being applied to large mammal population estimation in India and some of the figures here will warrant modification and future changes to the quoted populations should not necessarily be seen as indicating that a real change in numbers has occurred. In recent years Gaur has reportedly been exterminated from three protected areas, Thattekad Wildlife Sanctuary (Kerala), Bhandhavgarh (Madhya Pradesh) and Kanger Valley National Park (Madhya Pradesh) (Pasha et al. 2004). Information on population status from the states of Andhra Pradesh, Orissa and North-East is limited (Choudhury 2002); there are demonstrably very few Gaur in Kaziranga (Karanth and Nicholls 2000). Recent population trends appear to have been stable in well protected areas. The following densities have been estimated: Bhadra Tiger Reserve, 1.48 +/- 0.63 (SE) per km^2 (Jathanna et al. 2003); Pench (Madhya Pradesh) dry deciduous forest, 0.7 animals per km^2 (Karanth and Nichols 2000); Nagarahole moist deciduous forest, 9.6 animals per km^2, making it the second-most abundant ungulate there (Karanth and Sunquist 1992); Bandipur dry deciduous forest, 7.0 animals per km^2 (Karanth and Nichols 2000); Tadoba-Andhari dry deciduous forest, 1.8 animals per km^2 (Karanth and Kumar 2005); Melghat dry deciduous forest, 1.0 animals per km^2 (Karanth and Kumar 2005); Pench (Maharashtra) dry deciduous forest, 0.8 animals per km^2 (Karanth and Kumar 2005). No densities could be estimated, despite the use of suitable survey methodology, at Kanha or Namdapha, implying relatively low populations, and none was found during such surveys at Kaziranga or Ranthambore (Karanth and Nichols 2000); only small numbers persist in Kaziranga (N.S. Kumar pers. comm. 2008). The Bhadra density is low, reflecting poaching (using snares, dogs and shotguns) and livestock grazing.

However, the Gaur population there is now steadily increasing with successful conservation interventions (K.U. Karanth and N.S. Kumar unpublished data). A study at Nagarahole National Park compared the fauna of an area which was only moderately hunted with a heavily hunted site: this found respective densities of six and two Gaurs per km^2 (Madhusudan and Karanth 2002). In Lao PDR, Byers et al. (1995) stated that the national population was approximately 1,000. This number is perhaps possible (field records exist from many areas), but animals would have been widely dispersed and at very low density in the extensive forest tracts across the country, with perhaps fewer than six sites holding more than 50 animals (Duckworth and Hedges 1998; Duckworth et al. 1999; Steinmetz 2004; R.J. Timmins pers. comm. 2008). A large population of Gaur was found in Nakai–Nam Theun NPA and the adjoining Nakai Plateau in the mid 1990s, when signs were readily found and even the animals themselves seen directly (Duckworth 1998; Duckworth and Hedges 1998; Evans et al. 2000). Resurveys of both areas, although using comparable methodology only on the plateau, indicate massive declines in the subsequent decade, with, for example, fairly intensive camera-trapping not recording the species at all (Dersu 2007; Johnson and Johnston 2007). The last decade has seen major reductions (in general, probably of well over 50%), in many large mammal populations across Lao PDR and it is now likely that many of the populations listed in Duckworth and Hedges (1998) have become locally extinct (Timmins and Robichaud 2005; R.J. Timmins pers. comm. 2008). In Malaysia, in 1994 the population was stated to be less than 500 individuals and it was thought to be declining (J. Brennan pers. comm. to Read et al. 1994; a number of other stated numbers from around that time, all of the same general order of a few hundred, and none derived allowing a quantification of error margin, are presented in Hedges in prep.), with then confirmed records from various sites, e.g. the Sungai Singgor area of Temenggor Forest Reserve, Hulu Perak (Ratnam et al. 1995). In the subsequent few years, the population collapsed, despite optimism through relatively stronger control of guns in Malaysia than elsewhere on the southeast Asian mainland, and a perception (e.g. Nash 1993) of relatively strict control of wildlife trade in Malaysia. Densities in Taman Negara were crudely estimated from surveys during 1999–2001 as 0, 3 and 22 Gaurs per 100 km^2 in three study sites (Kawanishi and Sunquist 2004). Although these authors adjudged large mammal poaching to be negligible in Taman Negara, high levels were confirmed there by Wan Shahruddin (1998) and camera-trapping (suitable in methodology and intensity to find Gaur if any significant numbers were present) did not record Gaur at a fourth sector of Taman Negara in 1999 (Lynam et al. 2007). No Gaur was photographed at all in eight other potential Tiger areas surveyed by Lynam et al. (2007) during 1997–1999, including Temenggor (see Ratnam et al. 1995); because these were selected as the areas most likely to support Tigers and as Tiger-oriented camera-trapping elsewhere in South-east Asia usually yields photographs of Gaur, this study proves that only very low numbers of Gaur remain in Malaysia, no doubt way below the estimate in Read et al. (1994). Further supporting this, none was found in the Jerangau Forest Reserve, from which it was suspected to have been eradicated by hunting (Azlan 2006). Gaur was found in only two of 14 sites (seven virgin jungle reserves and seven adjacent logged areas), making it among the most localised mammal species which was detected in a study assessing the importance of virgin jungle reserves. The two occupied sites were a contiguous pair in each management class, and among the biggest sites surveyed (over 100 km^2); the virgin jungle reserves themselves rarely exceed 3 km^2 (Laidlaw 2000). In the 1960s, there were estimated to be approximately 40 Gaur in Krau Wildlife Reserve, a figure deemed optimistic at the time; in the 1980s there were thought to be 22–31; in 1995 the estimate was only seven. When surveying the reserve in 2000 R. Dunlop (pers. comm. 2008) covered the whole reserve (about 440 km^2) with camera-traps in 119 one km -grid squares. Out of a total of 6,918 periods of 24-hour trapping, 61 species were recorded in 1,804 photographs of wildlife, yet only one photograph was of a Gaur. In almost 100 days of field searching for signs, Gaur tracks were recorded only on eight days, on each occasion as prints of isolated animals. The photograph, the sole sighting, and the eight track records were all in the south-east lowland area of the reserve, moving between an area of salt-licks and grazing just outside the reserve. This survey corroborated the 1995 estimate of seven animals, and the 2000–2001 figure could be even lower. Sumatran Rhinoceros Dicerorhinus sumatrensis and Asian Elephant *Elephas maximus* have both disappeared from Krau, and Gaur may be the next species to go. R. Dunlop (pers. comm. 2008) considers there no reason to expect this recent history not to typify that of Gaur populations at all sites in Malaysia

There are very recent records from: Tanjong Mentong in Taman Negara, where tracks were found during elephant surveys in late 2006 (Nurul Huda, WCS Malaysia pers. comm. 2008); Belum, in 2007, where animals were camera-trapped at salt-licks several times, including a herd of seven (Ahmad Zafir, WWF Malaysia pers. comm. 2008); and Temenggor, where individual animals (no herds) were camera-trapped nine times (never at salt licks, always on logging roads) in 2,364 trap-nights in 25 locations (Mark Darmaraj, WWF Malaysia pers. comm. 2008). In Myanmar, the status of Gaur is poorly known (J.W. Duckworth and Than Zaw pers. comm. 2006). A March 1994 survey of Taminthi Wildlife Sanctuary, Myanmar's largest protected area, guessed that 100–200 Gaur remained, and that they were in danger of being extirpated from the area if current levels of poaching continued (Rabinowitz et al. 1995). A national Tiger survey camera-trapped Gaur in 11 of 15 camera-trapped sites, commonly in several (Lynam 2003); these sites were selected for their chances of holding Tigers, so, because the two species face similar threats, it may present the most positive picture of Gaur occurrence in the country.
In Nepal, the Gaur population was estimated to be 250–350 in the mid-1990s, with the majority in Chitwan National Park and Parsa Wildlife Reserve; population trends appeared to be relatively stable (J. Heinen pers. comm. to S. Hedges 1995). A few other areas hold small numbers (Choudhury 2002).

In Thailand, an estimated total of 920 Gaur remained in 1994. Only 6–7 populations were thought to contain 50 or more animals and there were no Gaur outside protected areas. The most important protected areas for Gaur were Khao Yai National Park, Tap Lan National Park and Pang Sida National Park, Huai Kha Khaeng Wildlife Sanctuary, Thung Yai Naresuan Wildlife Sanctuary, Kaeng Krachan National Park and Mae Nam Phachi Wildlife Sanctuary, the Khlong Nakha complex, and possibly Om Koi and Mae Tuen Wildlife Sanctuaries. Even in protected areas with much suitable habitat Gaur densities were very low, for example in Huai Kha Khaeng Wildlife Sanctuary the population density was less than 50% of the estimated carrying capacity of the area (Srikosamatara and Suteethorn 1995; S. Hedges pers. comm. 2000). Gaur populations are probably now increasing in several sites in Thailand, notably in Huai Kha Khaeng Wildlife Sanctuary and Thung Yai Naresuan Wildlife Sanctuary, with effective tackling of poaching. In Thung Yai Naresuan Wildlife Sanctuary, sign monitoring in one small area (20 km²) over six years since poaching absolutely ceased indicated an annual population growth rate there of 0.31, although the relative contributions of birth and immigration could not be distinguished. Breeding herds became much more common, compared with the situation 10–15 years previously at the height of poaching (Steinmetz et al. in prep.). In the Khao Paeng Ma reforestation site adjacent to Khao Yai National Park the population apparently expanded from two to 100 Gaur in less than 10 years, initially reflecting recolonization but apparently mostly from reproduction among animals at the site, all within a 10 km² area of secondary forest. The population stabilized at about 100 animals during 2004–2007 (based on counts from hilltops). As the forest grows and browse and grass diminish, reproduction is likely to decline (R. Steinmetz pers. comm. 2008). This herd has flourished only by dint of the provision of 24-hour security (A.J. Lynam pers. comm. 2008). Numbers also seem to be expanding rapidly in Kuiburi National Park in southern Thailand, again from an artificially lowered population and again within a secondary forest area adjacent to semi-evergreen forest. In early 2008, a (temporary) herd of almost 50 animals was spotted by elephant-watchers in a tree hide; only a few years previously, there were probably far fewer than this in the whole area (Steinmetz et al. 2007; R. Steinmetz pers. comm. 2008). By contrast, Gaur is largely gone from southern Thailand forest remnants, although it may be safe on the Malaysian border, reflecting tight control of illegal access and activities by the Border Patrol Police, and the danger of operations because of insurgency (A.J. Lynam pers. comm. 2008). In Vietnam, the current status of Gaur is poorly known. Several areas in Dak Lak Province (the Easo and Easup areas and Yok Don National Park) were known to contain Gaur in 1997 (Le Xuan Canh et al. 1997) but very little reliable information was traced for other parts of the country by Duckworth and Hedges (1998). Several herds of Gaur persist in Cat Tien National Park and in adjacent state forest enterprises (Polet and Ling 2004). There seems little room for doubt that the Gaur population is in serious decline, and many of the sites reported in Duckworth and Hedges (1998) may now have lost the species.

Population Trend: Decreasing

Habitat and Ecology:

The Gaur occurs from sea level up to at least 2,800 m asl (Wood 1937; Wharton 1968; Choudhury 2002). Despite the many reports that call it an animal of hill-country, low-lying areas seem to comprise optimal habitat (Choudhury 2002): in Conry's (1989) study area, elevations ranged from 46 to 1,079 m asl but the three radio-tracked Gaur only used areas below 381 m. Elevations below 61 m were used most intensively and all three animals selected these low-lying areas; elevations above 61 m were selected against or used in proportion to availability (Conry 1989). Similarly, in the Tenasserim–Dawna mountains, Thailand, signs of Gaur were more abundant in the lowlands than in the hills, noting that this was the opposite of the patterns that would be predicted if hunting (itself concentrated heavily in the lowlands) was the chief determinant of population densities, although solitary animals were found mainly in the hills (Steinmetz et al. 2008). Schaller (1967) compared vegetation maps of India with the (mid-20th century) distribution of Gaur and concluded that the species was largely confined to evergreen, semi-evergreen, and moist deciduous forest but it also occurred in dry deciduous forest areas at the periphery of its range. He considered Gaur habitat to be "characterized by large, relatively undisturbed forest tracts, hilly terrain below an altitude of 5,000 to 6,000 ft (1,500–1,800 m asl), availability of water, and an abundance of forage in the form of coarse grasses (including bamboo), shrubs, and trees". However, He cautioned that the apparent preference of Gaur for hilly terrain may be partly due to the earlier conversion of most of the plains and other low lying areas to croplands and pastures. Similarly, Wharton (1968) concluded that Gaur appeared to avoid evergreen rainforest, "preferring foothill tracts of sub-humid or deciduous forest adjacent to savanna forest, glades or other open terrain affected by man and fire thus co-existing with and exploiting low human populations in hill zones with moderate to heavy rainfall". Throughout the range of Gaur, habitat destruction and human presence is usually higher in lowlands than in hills, meaning that differential human pressures may have caused the apparent preference of Gaur for hills. Gaur occurs in many forest types in Lao PDR, Vietnam and Cambodia including truly evergreen and montane forest, but populations in closed canopy forests especially those at higher elevations might naturally be at much lower densities than those in lowland areas with a mosaic of open and closed forest types, and in closed canopy forests areas the species appears to frequent grassy openings (Duckworth et al 1999; Greiser Johns 2000; Duckworth and Hedges 1998; Timmins and Ou 2001; R.J. Timmins pers. comm. 2008). No significant difference was found in Gaur densities in moist compared with in dry deciduous forests in Nagarahole, India (Karanth and Sunquist 1992). Steinmetz et al. (in prep.) considered that Gaur herds in an expanding population were consistently concentrated in mixed deciduous forest rather than semi-evergreen forest, although the total area of study, that was safe from poaching, was only 20 km². Gaur can persist in fragmented areas with some habitat disturbance and hunting, including in landscapes dominated by plantations (such as on parts of the Valparai plateau, south India), but in Southeast Asia it has been too heavily hunted in such areas to survive in them. Gaur can better tolerate rugged terrain and denser forest than other wild and domestic cattle, provided that there are adequate water sources (N.S. Kumar pers. comm. 2008.; Smith and Xie Yan in press). In Thailand it uses former cultivation amid forest, and populations in such areas can recover well if not hunted (Anak Pattanavipool pers. comm. 2006; R. Steinmetz pers. comm. 2006). A study of coffee estates around Bhadra Wildlife Sanctuary, India, recorded Gaur only in those coffee areas within 1 km of the sanctuary's boundary (Bali et al. 2007). Gaur both grazes and browses, reportedly eating mostly young green grasses but also leaves, fruit, twigs, and bark of various woody species, as well as coarse dry grasses, and bamboo. It seems able to maintain good condition on relatively low quality feed. At least 180–190 species of plants have been recorded in the diet (N.S. Kumar pers. comm. 2008). More information is given about Gaur's diet in Dunbar Brander (1923), Hubback (1937), Schaller (1967), Wharton (1968), Krishnan (1972), Khan (1973), Conry (1981), Ebil Yusof (1982), Ramachandran et al. (1986), Prayurasiddhi and Smith (1993) and Sankar et al. (2000). Captive Gaur are reported to eat about 20 kg of green fodder per day (Ramachandran et al. 1986).

During the cool season (November–February) the Gaur that Schaller (1967) studied in central India ate various herbs, large quantities of the leaves and seeds of bamboo, various other grasses, and the leaves of a number of tree species; but by the beginning of the hot season (March–June) coarse dry and semi-dry grasses made up the bulk of their diet, averaging 85% by volume (range 66–100%; four rumen samples), with browse only making up about 10%. Gaur also feeds on crops including corn, cassava, and young rubber trees, sometimes causing great damage (Wharton 1968; National Research Council 1983). Schaller (1967) reported that Gaur in Kanha National Park drank at least once per day during the hot season, usually in the evening. Sahai (1972) stated that Gaur drink at least twice in a 24-hour period but there appears to be no fixed time for drinking. In undisturbed areas, Gaur is mainly diurnal, but in areas of high human disturbance, it is reported to become mainly nocturnal. The basic social group appears to be a female–juvenile unit with larger groups tending to be more-or-less temporary assemblages (as in the other large species of Asian Bovini). Maternal herds containing several adult cows, juveniles, and calves frequently occur but their composition is somewhat fluid, particularly during the rut; these herds can often contain several subadult and adult males. The number of bulls in a herd changes with the time of year. Mature males associate with cows primarily during the rut: during Schaller's study in Kanha National Park 65% of herds were accompanied by a mature male during April (the peak of the rut) but during the second half of the year the proportion dropped to 31% (Schaller 1967). Neither Schaller (1967) nor Sahai (1972) found any evidence to support Powell's (1964) statement that young bulls usually live apart from the maternal herds; nor did Schaller's observations support Dunbar Brander's (1923) statement that the 'master bull' drives all other males from the herd. Males not associating with mixed herds either live in all-male groups or are solitary. The bull herds are generally unstable, although sometimes two or more animals may frequently associate. The typical herd size appears to be in the range of 5–12 animals and herds are rarely larger than 20 (Dunbar Brander 1923; Hubback 1937; Hislop 1961a; Schaller 1967; Prater 1971; Sahai 1972; Pal and Guin 1986). Large groups of 20, 40, or even 100 animals are probably temporary aggregations of smaller groups (Pal and Guin 1986).

The majority of the 125–150 Gaur studied by Schaller (1967) in Kanha National Park (India) ranged over an area of at least 78 km²; and the herds he studied travelled on average 2–3 miles (3.2–4.8 km) a day, although on occasion the distance could be much less. In West Malaysia Gaur home ranges have been estimated to be: approximately 13 km² (the Kuala Tahan herd in Taman Negara National Park) (Weigum 1972); and 29.9 km² for a yearling male, 52.1 km² for a yearling female, and 137.3 km² for an adult male in Pahang (Conry 1989). The large differences between the two studies probably reflect the different methods employed by the two workers (see Conry 1989: 60). Home ranges that can be as large as 137 km² hinder the setting aside of conservation areas for Gaur, particularly as Gaur often occurs in those areas where the pressures for economic development are highest. In central India, most Gaur mate in December and January, and calves are born in August and September according to Dunbar Brander (1923); Stebbing (1911) and Sanderson (1912) generally agreed with Dunbar Brander but noted that calves may also be born in April, May, and June. In Kanha National Park (central India) Schaller (1967) recorded rutting bulls from December to June, with an apparent peak of sexual behaviour in March and April; however, the occurrence of a few new-born animals throughout the year indicated that some mating must occur in every month. In southern India the majority of matings reportedly take place between November and March (Morris 1937). In Cambodia the majority of Gaur births occur in December and January (Wharton 1957). In Myanmar, calves are born throughout the year (Evans 1912; Peacock 1933); and in West Malaysia young Gaur are seen in all months except October to December (Hubback 1937). The gestation period of Gaur is reported to be nine months by Hubback (1937) and Crandall (1964) and 270 days, 'a little shorter than for Banteng or domestic cattle and longer than for yak and Kouprey', by National Research Council (1983: 49). Only one calf is born to a cow per pregnancy; there are no records of twins (Hubback 1937; Schaller 1967). Females probably first mate when about two years old and thus have their first calf at three years old (Schaller 1967). If the food supply is favourable, most adult females probably calve every year. In Kanha National Park, approximately 90% of adult cows bore young during a year but calf mortality was high, with about 50% of the calves dying before reaching one year old; the main cause of death was predation by Tigers Panthera tigris (Schaller 1967).

Rarely, bulls kill each other when fighting, and occasionally they are severely injured while fighting. Predation is by far the major cause of Gaur mortality, perhaps followed by disease (N.S. Kumar pers. comm. 2008). For example in Bandipur, Gaur is the most important prey in terms of the biomass taken by Tiger (42%) and Leopard *Panthera pardus* (15%) (Andheria et al. 2007). In Bandipur, Gaur remains were found in about 24% of Tiger faeces and 9% of Leopard faeces (Andheria et al. 2007). In adjoining Nagarahole, the proportion of Gaur in Tiger's diet was about 45% in kills, 17% in scats (Karanth and Sunquist 1995). In Tadoba-Andhari, Gaur together with Sambar contributed to nearly 70% of the prey biomass consumed by Tiger, whereas in Pench-Maharashtra it was 80% (Karanth and Kumar 2005). In well protected forest reserves where different body-sized ungulates are abundantly available, Gaur is selectively predated by Tiger (N.S. Kumar pers. comm. 2008).

Systems: Terrestrial

Major Threat(s):

The set of threats and the gravity of each threat to Gaur differs greatly between south and central India, and Southeast Asia; China and most of northeast India are more similar to Southeast Asia in this respect Farshid Ahrestani pers. comm. 2008), and are so treated in the following discussion.

In Indochina, Myanmar, and northeast India, hunting is the major problem, compounded by loss of suitable habitat: large tracts of suitable habitat hold few or no Guar. Ongoing habitat degradation and conversion is continually reducing the potential population, should hunting be brought under control (Duckworth et al. 1999; Duckworth and Hedges 1998; Timmins and Ou 2001; Choudhury 2002; Steinmetz 2004; Than Zaw pers. comm. 2006; R.J Timmins pers. comm. 2008; Farshid Ahrestani pers. comm. 2008). Hunting is often for meat and a variety of methods is used (Choudhury 2002). There is also a flourishing international trade in Gaur parts, both horns (decorative) and internal organs (for medicinal purposes). Given the small size of the remaining Southeast Asian subpopulations and the number of trophies found for sale in Cambodia, Lao PDR, **Thailand,** and Myanmar during essentially opportunistic surveys, trade is clearly a major threat (Srikosamatara et al. 1992; Baird 1993; Srikosamatara and Suteethorn 1994, 1995; Martin and Phipps 1996; Srifa La-Ong et al. 1997; Duckworth et al. 1999). The most important population in Cambodia is scattered through a forest landscape that encompasses four provinces (Mondulkiri, Kratie, Stung Treng and Ratanakiri) and five conservation areas (Lomphat Wildlife Sanctuary, Phnom Prich Wildlife Sanctuary, Siema Biodiversity Conservation Area, Mondulkiri Protection Forest (including the Srepok Wilderness area) and Phnom Nam Lyr Wildlife Sanctuary). Hunting is still rife in much of this area, and forest fragmentation is rapidly accelerating with human population immigration, infrastructural developments (especially roads), commercial agricultural expansion, economic land speculation and mineral extraction. However, although perhaps less than 20% of this area is well protected from the aforementioned threats and protected area management is only close to effective in two areas, the most significant constraint to the area's Gaur is the long-term uncertainty of continuation of effective conservation management of the Srepok Wilderness Area and Siema Biodiversity Conservation Area. Although conservation efforts for Gaur and many other species are, in both these areas, extremely encouraging, both face an uncertain future with the possibility of degazettement of conservation status of parts of them, the possible loss of adequate external funding necessary to maintain high standards of management, the possible loss of political support necessary to uphold high protection standards and the uncertainties of maintaining a motivated well trained staff. Almost throughout Southeast Asia, protected areas are not yet stalling declines in mammals with commercial value, such as Gaur. Areas with stabilized populations (in addition to the two Cambodian areas, perhaps only a few areas in Thailand (see Population) and Cat Tien National Park in Viet Nam) are exceptional (R. Steimetz pers. comm. 2008; R.J. Timmins pers. comm. 2008). Surprisingly, given its retention of healthy populations of other high-value very large mammals (Tiger and Asian Elephant *Elephas maximus;* Lynam et al. 2007; S. Hedges pers. comm. 2008 for elephants), Malaysia has among the most rapidly declining populations of Gaur.

This reflects targeted hunting for large mammals of high commercial value, even within protected areas (Abdul Kadir 1998; Wan Shahruddin 1998; Lynam 1999; DWNP/Danced 2002; Lynam et al. 2007). Given the rapid decline within the 1990s and the lack of information traced since 2002, the exact current situation of Gaur in Malaysia is unclear. Kemasul Forest Reserve, where the species was found in Laidlaw's (2000) survey of virgin jungle reserves, probably no longer exists as such. Even at the time of survey, most of the Forest Reserve was due to be converted into an Acacia mangium plantation, although still nominally a Forest Reserve. Kemasul was, long ago, part of a massive block of lowland forest which included the famous Jengka Triangle and Krau "Game" Reserve. Only Krau remains, albeit much reduced (R. Dunlop pers. comm. 2008). Even Krau, which was set up (in 1923) principally for the protection of Gaur, numerous in the area then, faces multiple threats (R. Dunlop pers. comm. 2008). In the surveys for DWNP/DANCED (2002), evidence of hunting, including cartridges, was found within the reserve, particularly in the easily accessible lowland areas; there were plenty of hunting stories by and about all of the different local communities, without exception, although their preferred target species differed. Conversion from natural forest to other land-uses of adjacent state-land combined with new ease of access has undoubtedly made Krau more of a focal point for such activity. There is a new (metalled) road and land conversion right up to the boundaries, facilitating access and easing the use of hunting methods such as spotlighting, as well as reducing habitat. The salt licks (for which Krau used to be famous) increasingly fall outside the forest (reflecting shrinking reserve boundaries and farmland encroachment into the reserve) and are now mostly in zones of high human activity of one sort or another. Temporary grasslands generated through traditional indigenous farming, which were often fairly difficult to get to and thus relatively free from hunting, were previously important Gaur habitat. A forester who used to work in Malaysia in the 1940s said that Gaur herds used to follow the Orang Asli settlements, moving in when they moved out, and Weigum (1972) even recommended "...to maintain herds in the park it is recommended that 45 acres of land be clear-cut preferably in 15 acre plots ½ mile apart and be recut every three years". Gaur meat is, or at least was being, sold at restaurants in placed like Cameron and Grik. In south and central India, densities are likely to be widely below ecological carrying capacity, through hunting and competition with domestic livestock, although this conclusion is based on rather few empirical studies and there are some outstanding populations in well-secured protected areas. Substantial declines and local extinctions of Gaur were driven by hunting (for meat), competition with livestock, and habitat conversion, particularly during the first half of the twentieth century, resulting in a drastically reduced and fragmented distributional range, especially outside the protected area network (N.S. Kumar pers. comm. 2008). Hunting is much less prevalent than in Southeast Asia, reflecting the general enforcement of wildlife protection laws (Farshid Ahrestani pers. comm. 2008). Nonetheless, even in high-profile reserves, it is still suppressing densities significantly. For example, densities in Bhadra Tiger Reserve are very low, because of poaching, by comparison with what the habitat should support (Jathanna et al. 2003), and the measured density in a heavily hunted area of Nagarahole National Park was only a third of that in a well secured area (Madhusudan and Karanth 2002). Even though all hunting is illegal in this protected area, preventive measures are uneven across it, and Gaur is still poached, at least sporadically, wherever protection effectiveness is low (N.S. Kumar pers. comm. 2008). In recent years Gaur, has reportedly been exterminated from three Indian protected areas, Thattekad Wildlife Sanctuary (Kerala), Bhandhavgarh (Madhya Pradesh) and Kanger Valley National Park (Madhya Pradesh) (Pasha et al. 2004). Interaction with domestic stock is greatly under-appreciated as a threat to Gaur, but is probably the main factor which currently limits populations in south and central India: livestock are widely grazed even in otherwise well-managed protected areas. Mean densities of Gaur were fully 132 times higher in livestock-free areas than in adjacent livestock-grazed areas of Bandipur, and in shared grazing areas Gaur densities declined sharply with increasing livestock densities; in the grazed area studied, halving the livestock density allowed increases of Gaur by a factor of 57 (Madhusudan 2004). Threats to southern India's forest ungulates by competition with domestic stock grazing within protected areas are exacerbated where dung is collected for export to adjacent coffee areas. Fuel wood removal may also be at levels sufficient to disrupt nutrient cycles of the habitat (Madhusudan 2005). Many Indian forest areas are severely encroached by exotics such as Lantana camara, Parthenium spp. and Chromolaena odorata, and these are suspected to effect major changes to forest structure (Hiremath and Sundaram 2005): but the effects of these on Gaur populations warrant further study.

Killing of Gaur in retaliation to crop damage (through consumption and trampling) occurs, but is of limited significance to conservation (Choudhury 2002; Farshid Ahrestani pers. comm. 2008). Some animals also die from ingestion of agrochemicals, and human response when wandering into farmland and villages; these could be significant for already isolated and reduced populations (Choudhury 2002). Throughout the range, diseases, particularly rinderpest and foot-and-mouth disease, transmitted by domestic cattle are a potentially serious threat. Foot-and-mouth is the most frequent, but rinderpest has had a particularly dramatic impact in India and several sub -populations of Gaur were nearly destroyed as a result of the disease in 1968 (Krishnan 1972; Pal and Guin 1986; Choudhury 2002), while in Bhadra, it nearly wiped out Gaur populations in 1988–1989, when nearly 800 individuals died (Girish D.V. pers. comm. 2008). In north and central Myanmar anthrax was a major factor in the decline of the species (Salter 1983). Pasha et al. (2004) considered that "in fact no wild animal in India [is] so profoundly influenced by transmitted infections from domestic livestock as Gaur". There is little research or other evidence to determine impacts of introgression with domestic oxen on Gaur populations, although such factors might be expected to be most serious in India, given the wide overlap of domestic stock with Gaur. Farshid Ahrestani (pers. comm. 2008) saw a Gaur herd of 20 individuals far from human habitation with one domestic ox in the Palni Hills, so it seems likely that introgression occurs to at least some extent. Srikosamtara and Suteethorn (1995) drew attention to the pressing threats facing Gaur, placing it in the same league of trade-driven threat as better-known species such as Tiger and Asian Elephant, and emphasising the international dimension of trade. Subsequently, the population collapse in Malaysia seems to have been driven by hunting teams from Lao PDR and Thailand (as proven by language at hunting camps in Temenggor), co-opting local Malaysian associates (A.J. Lynam pers. comm. 2008). It is reasonable to expect that when returns in Malaysia drop (as they probably have done already) these teams may turn their attention to other countries' Gaur populations.

Conservation Actions:

Gaur is listed in CITES Appendix I, and is legally protected in all range states. It occurs in many protected areas, although in north-east India, Myanmar, China, Lao PDR, Vietnam and Malaysia many such areas retain only tiny populations or have already lost the species.

South and central India have a well established network of wildlife refuges and protection activities, to a large extent, are effective in many of them (Farshid Ahrestani pers. comm. 2008). However, conservation-oriented research and interventions on Gaur are limited and seem unlikely to expand much in the short-term, meaning that the species benefits largely through general protection activities (particularly those focused around Tiger and Asian Elephant), while species-specific negative factors or worrying long-term trends are likely to go unmanaged and even unnoticed (Farshid Ahrestani pers. comm. 2008). The chief of these seems to be competitive interaction with grazing stock: Gaur densities were the most depressed of all ungulates surveyed by Madhusudan (2004), perhaps reflecting higher niche overlap with domestic cattle than shown by the others (chiefly deer). There is long-term monitoring of some Gaur populations, for instance Mudumalai, Bandipur, Bhadra and Nagarahole (Farshid Ahrestani pers. comm. 2008). Rinderpest has reportedly been eradicated from India (Choudhury 2002). Choudhury (2002) listed proposed aspects of protected area expansion and consolidation to slow further fragmentation of Gaur populations in southern and central India. In some protected areas (recent examples being Bhadra and Kudremukh in Karnataka), formerly forest-dwelling human communities have resettled closer to markets, clinics, school and other services (Karanth and Karanth 2007).This expands habitat available to Gaur (that which was formerly fields and houses) but more significant is the removal of a source of hunters and the cessation of grazing competition with domestic stock (see Karanth et al. 2006). Such conservation interventions have helped recovery of Gaur populations in this region. Increasing development aspirations mean that out migration is likely from further areas, e.g. Mudumalai (Farshid Ahrestani pers. comm. 2008). As the majority of the Gaur population is in South Asia, conservation activities are important there. It is also important that populations are maintained at least in several areas in Southeast Asia, to maintain Gaur presence through its historical range, to preserve Gaur genetic diversity and through its importance to the ecosystems within which it lives.

Given the regional pattern of threats and current successes to date, implementing effective conservation I interventions is considerably more challenging, and therefore a higher priority in the Southeast Asian region. Gaur occurs mainly in protected areas in Southeast Asia, although Cambodia and Myanmar may both have large populations outside them and some may persist is such areas in Lao PDR (Duckworth 1996; Hansel 2004). In most protected areas of Lao PDR, Vietnam and Cambodia which still retain Gaur, protected status offers little, if any, real protection to the species, exceptions being Cat Tien National Park in Vietnam, the Srepok Wilderness Area and the Siema Biodiversity Conservation Area in Cambodia; in the latter two Cambodian areas, there are now even signs of population recovery (R.J. Timmins pers. comm. 2008), as there are at **Huai Kha Khaeng Wildlife Sanctuary and Thung Yai Naresuan Wildlife Sanctuary, Thailand, with effective tackling of poaching (R. Steinmetz pers. comm. 2008). Huai Kha Khaeng Wildlife Sanctuary and Thung Yai Naresuan Wildlife Sanctuary form part of Thailand's 'Western Forest Complex' which if these recent gains are consolidated will, in the long term, hold the largest functional Gaur population in Thailand and perhaps South-east Asia.** The most important conservation measure for the species in Cambodia is the continued strengthening and support of conservation efforts in Siema Biodiversity Conservation Area and the Srepok Wilderness Area, and to a lesser extent support of protected area conservation management in the northern plains and central Cardamoms. **Likewise in Vietnam, the most important conservation measure is the continued support of protection efforts within Cat Tien National Park, in Thailand it is to continue the consolidation of the 'Western Forest Complex' and to continue the high-intensity protection of the rebuilt herd in Khao Paeng Ma reforestation site, adjacent to Khao Yai National Park.** In Myanmar the Hukaun Valley is an exceptional area of lowland plains forest, grassland, and wetlands. Part lies within the Wildlife Sanctuary and some of the rest of the outstanding plains habitat is within an enormous proposed extension. Through its size and mostly little-encroached condition, this is the most outstanding remaining landscape -level floodplains habitat for very large mammals remaining in tropical Asia, and although wildlife populations are highly depleted, warrants the strongest effort to conserve it. It faces a number of daunting challenges (J.W. Duckworth pers. comm. 2008). In Lao PDR, any of a large number of areas could become key Gaur conservation sites, but there is as yet no precedent in the country for effective conservation of high-trade-value large mammals. Xe Pian, Nam Et–Phou Louey, and Nakai–Nam Theun NPAs could all be highly significant areas for the species (Duckworth and Hedges 1998; Steinmetz 2004; Johnson et al. 2006) but so could almost any other NPA; the reality is that the most success with Gaur conservation is likely to come through identifying areas with positive underlying situations to achieve conservation, rather than identifying the area with the most intrinsic importance for Gaur as the focus for efforts. Choudhury (2002) listed aspects of protected area expansion and consolidation to slow further fragmentation of Gaur populations in north-east India, Nepal, and Bhutan, and perhaps allow epopulation of Bangladesh. The most important actions in Malaysia are to determine if any viable population remains in the country and, if so, to implement effective anti-poaching measures and to work at the policy level to allow effective control of large mammal hunting and trading. Under the current circumstances, any site-based activities are likely to require the sort of 24-hour security employed at the reforestation site adjacent **to Khao Yai National Park, Thailand, which has allowed a rapid population growth of Gaur. It may also require reintroduction. The Department of Wildlife and National Parks have a breeding population of captive Gaur in enclosures in the Jenderak ranger station (within Krau Wildlife Reserve), for captive release.** A large captive population exists of wild-type Gaur (as well as of Mythun); given the complexity of field conservation of large ungulates, particularly in South-east Asia, it is clearly in resolving issues of hunting, trade, and protected area effectiveness that conservation interventions must focus.

Banteng *(Bos javanicus)*

Description:

The banteng is similar in size to domestic cattle, measuring 1.55 to 1.65 m (5 ft 1 in to 5 ft 5 in) tall at the shoulder and 2.45–3.5 m (8 ft 0 in–11 ft 6 in) in total length, including a 60 cm (2.0 ft) tail. Body weight can range from 400 to 900 kg (880 to 1,980 lb). It exhibits sexual dimorphism, allowing the sexes to be readily distinguished by colour and size. In mature males, the short-haired coat is blue-black or dark chestnut in colour, while in females and young it is chestnut with a dark dorsal stripe. Both males and females have white stockings on their lower legs, a white rump, a white muzzle, and white spots above the eyes. The build is similar to that of domestic cattle, but with a comparatively slender neck and small head, and a ridge on the back above the shoulders. The horns of females are short and tightly curved, pointing inward at the tips, while those of males arc upwards, growing 60 to 75 cm (24 to 30 in) long, and being connected by a horn-like bald patch on the forehead.

Range Description:

The species historically occurred from southern China (Yunnan) throughout mainland Southeast Asia, through Peninsular Malaysia to the islands of Borneo, Java, and probably Bali (see below).

No fossil evidence is known from Bali, and Grubb (2005) stated that Banteng was introduced to the island; Corbet and Hill (1992) did not include Bali in the range of wild Banteng and various other sources simply state that it occurs as domestic animals. Nevertheless, it does not seem unreasonable to suggest that Banteng should be included among the native fauna of Bali since several Indo-Malayan animal species are known to have penetrated eastwards as far as Flores and Timor during times of lowered sea-levels in the Pleistocene (Hooijer 1975). Ashby and Santiapillai (1988) treated the small number of free-living Banteng's which in occur in west Bali as wild animals as did Watling (1991); although Watling noted that interbreeding with domestic Bali cattle is a problem and the population is unlikely to consist solely of pure-bred animals. Wind and Amir (1977) had earlier raised similar fears. Domestic Banteng has been introduced to several of the islands of eastern Indonesia including Sulawesi, Sumbawa, and Sumba (Van der Maarel 1932; Rollinson 1984). Feral Banteng occur in Kalimantan (National Research Council 1983). Introduced Banteng (probably feral animals) occurs on the Indonesian islands of Enggano (off Sumatra) and Sangihe (off Sulawesi) (Corbet and Hill 1992; Grubb 2005). Domestic Banteng has also been introduced to New Guinea and Australia and there are now large feral herds in the Northern Territory (Kirby 1979; Bowman 1993; Bradshaw et al. 2006).Wild Banteng currently occurs on Java and possibly Bali, in Kalimantan [Indonesian Borneo], Sabah [part of Malaysian Borneo], Myanmar, **Thailand,** Lao PDR, Vietnam, and Cambodia.

A few may remain in Sarawak Malaysian Borneo. (D. Labang, cited in Caldecott 1988) but it is thought to be extinct in Brunei (Payne et al. 1985). Banteng is extinct in Bangladesh (Gittins and Akonda 1982; Md Anwarul Islam in litt. 2008), and in India (Prater 1971; IUCN 1978), if it ever occurred there: Groves (2003) stated that "there have been constant reports from Manipur, but there is as yet no confirmation". In West Malaysia it has probably been extinct since at least the 1950s (Hislop 1961). Medway (1983) thought that Banteng may have still occurred in parts of Kedah and Kelantan after that time, but there have been no recent records. In China, Banteng has been reported to occur in one area of southeast Yunnan around Tongbiguan Nature Reserve, along the border with Myanmar; however, the source for this is unclear; and presence in China should be considered tentative at best. In Cambodia the species is still likely to be widespread in the northern and eastern forests and as well as parts of the Cardamoms Mountain range, with the bulk of the population remaining in the eastern forests, centred on Mondulkiri Province (see Population). On Borneo, the species occurs in parts of eastern Sabah (Malaysia), including Tabin Wildlife Reserve, and in Kalimantan (Indonesia), it occurs along the border with Sarawak (Malaysia), as well as in the south central region and two isolates in the eastern parts of the province (in and around Kutai National Park and Hutan Kapur Sangkulirang Nature Reserve). Distribution is other countries is detailed under Population.

Countries: Native:

Cambodia; Indonesia (Bali, Jawa, Kalimantan); Lao People's Democratic Republic; Malaysia (Peninsular Malaysia - Regionally Extinct, Sabah); Myanmar; **Thailand;** Vietnam

Regionally extinct:

Bangladesh; Brunei Darussalam; India

Population:

The world population of Banteng is unlikely to be more than 8,000 and is quite possibly fewer than 5,000 animals. No subpopulations of more than 500, and only 6–8 subpopulations of more than 50 animals, are known, with 4–5 on Java and 1–2 in Thailand (S. Hedges pers. comm. 2000; Pudyatmoko 2004). A once fairly widely distributed species, it is now largely reduced to small isolated populations, most of which are still in decline.

In Cambodia, Banteng probably declined by 90% or more between the late 1960s and the early 1990s. At this latter time they still remained widespread, although in generally low numbers, in the lowland forests of the north and east, and also, probably somewhat more sporadically, in the south and west including the Cardamom Mountain range (Daltry and Momberg 2000; Heng Kimchhay et al. 1998; Timmins and Ou 2001; R. J. Timmins pers. comm. 2008). From at least this time onwards the most substantial Banteng population has been centred on Mondulkiri Province where in the late 1990s at least several hundred to a perhaps over a thousand Banteng survived in a forested landscape of over 15,000 km^2 (Timmins and Ou 2001; Tordoff et al. 2005; R. J. Timmins pers. comm. 2008).Further declines took place from the early 1990s to the present, with quite probably more than a 50% decline in this time period, and the resultant loss of Banteng populations from significant parts of the still forested landscape (Timmins and Ou 2001; Timmins et al. 2003; Timmins 2006; Bezuijen et al. in prep.; R. J. Timmins pers. comm. 2008). As of 2007 the most significant population by far in Cambodia remained that in Mondulkiri Province, still thought to be hundreds of animals (WCS unpublished data; WWF unpublished data; R. J. Timmins pers. comm. 2008). Protection efforts in parts of this landscape have probably stabilised declines locally, although for the area as a whole the trend is still likely to be a significant decrease (WCS unpublished data; WWF unpublished data; T. Clements pers. comm. 2007; T.D. Evans pers. comm. 2007; R. J. Timmins pers. comm. 2008). The Banteng's other present-day stronghold is Java, where six large subpopulations (those with more than 50 animals) occurred in 1990, but these declined to only 4–5 by 2004 (Pudyatmoko 2004). Pudyatmoko (2004) estimated: Ujung Kulon National Park (300–800 individuals in 2003), Cikepuh-Cibanteng Nature Reserve (25–65 individuals in 2003), Bonjonglarang-Jayanti Region (a small stable population of unknown size in 1988, with no recent data available), Cimapag Region (occurrence recorded until 1970, with current status unknown), Leuweng Sancang Nature Reserve (10 individuals in 2000, extinct in 2003), Cikamurang Region (occurrence recorded until 1970, with current status unknown), Pangandaran Nature Reserve (25–65 individuals in 2003), Kediri Region (occurrence recorded until 1970, with current status unknown), Coast of Blitar (10 individuals in 1988, with current status unknown), Coast of Malang (6 individuals in 1988, with current status unknown), Meru Betiri National Park (200 individuals in 2000, with at least 57 individuals in 2002), Alas Purwo National Park (with at least 80 individuals in 2002), and Baluran National Park (around 206 individuals in 2002). Number in all areas are declining (Hedges and Tyson 1996; S. Hedges unpub. data. 1991–2002; Pudyatmoko 2004).

Population numbers and even trends are difficult to quantify for Borneo but there have been declines, especially in Kalimantan (S. Pudyatmoko pers. comm. 2006; G. Semiadi pers. comm. 2006), and in Sabah (Han pers. comm.). The Bornean population is thought to have been particularly negatively affected by interbreeding with domestic and/or feral oxen of types other than pure-bred 'Bali cattle'. Banteng occurs in Kutai National Park and Hutan Kapur Sangkulirang Nature Reserve, Kalimantan (Indonesia). A small population was reported from Belantikan Hulu in Central Kalimantan in November 2007 (tracks found) and April 2008 (local reports of capture of cow and calf); the population is believed to be small (Togu Simorangkir, Chair of Yayorin pers. comm. 2008). Populations in Sabah are now confined in protected areas scattered mostly in the east coast of the state. Current reports include the Tabin Wildlife Reserve, Kulamba Wildlife Reserve, Danum Valley Conservation, Ulu Kalumpang Forest Reserve, Ulu Segama-Malua Forest Reserve, Kinabatangan Wildlife Sanctuary, and Deramakot Forest Reserve. Over 100 individuals live in Kulamba Wildlife Reserve, and the state-wide gross estimation of the population size is 300–500 individuals. However, this is only based on presence–non-detection surveys and the figure should be regarded as a crude approximation (Ahmad AH pers. comm. 2008). The biggest decline happened during the intensive and extensive timber operation and transformation of lowland forests large-scale plantations (palm oil, cocoa and other crops) late in the twentieth century (Ahmad AH pers. comm. 2008). The declines are over 50% in Sabah and Borneo as a whole (Belden Giman pers. comm.).

In Thailand, the only subpopulations with more than 50 animals are Huai Kha Khaeng Wildlife Sanctuary and perhaps Kaeng Krachan National Park; smaller numbers do or may persist in Mae Ping National Park, Phu Luang Wildlife Sanctuary, Thap Lan National Park, Khwo Ang Ru Nai Wildlife Sanctuary, Khao Soi Dao Wildlife Sanctuary, and Khao Sok National Park. Banteng range probably declined by approximately 85% in Thailand from 1980 to 2000 (S. Hedges pers. comm. 2000). The Banteng population in Huai Kha Khaeng Wildlife Sanctuary is reported to have increased over the past 10 years (per R. Steinmetz pers. comm. 2008).

Numbers in Lao PDR are now likely to be very low, with probably no area supporting more than 50 animals (Duckworth and Hedges 1998; Duckworth et al. 1999; R. J. Timmins pers. comm. 2008). Evidence of Banteng was found for a number of areas in the 1990s (Duckworth and Hedges 1998; Duckworth et al. 1999; Steinmetz 2004), but there has been very little survey effort since 1998. Confirmation of presence after the mid-1990s, when populations were in steep decline (Duckworth and Hedges 1998), is lacking. Most areas have received little if any protection: so it is quite likely that Banteng has been hunted out from several, perhaps most, conceivably all, of them (R.J. Timmins pers. comm. 2008). The most likely areas for continued presence are in the southern half of the country. A similar scenario is likely in Viet Nam, with little recent evidence of continued presence in most areas (M. Pedrono pers. comm. 2008), and perhaps only one region (the Yok Don National Park area along the Cambodian border) has a population that might exceed 50 animals; current presence is also possible elsewhere in the western border areas from Kontum Province to Dak Lak Province and in Krong Trai Nature Reserve, Rung Kho Phan Rang Nature Reserve, and Cat Tien National Park (Duckworth and Hedges 1998; Eames et al. 2004; Tordoff et al. 2005; M. Pedrono pers. comm. 2008; R. J. Timmins pers. comm. 2008). Little recent information has been traced from Myanmar: a national camera-trapping survey aimed at Tiger Panthera tigris, which will not necessarily have covered all, or even the best, areas for Banteng, recorded it in and around Alaungdaw Kathapa National Park, Mahamyaing Reserve Forest, and the central Bago (= Pegu) Yoma (Lynam 2003). Current numbers are unknown but are sure to be declining, given general trends in ungulates in better-studied parts of the country. There are no quantitative data available on remaining populations in China (Smith and Xie Yan 2008), but these are clearly small. A domesticated form of Banteng, often called 'Bali cattle', and considered to be the same species (contrasting with general taxonomic treatment for several other oxen) occurs widely in Indonesia and has been introduced to other areas of the world. However, few if any populations of this form are now likely to be pure-bred, due to interbreeding with domestic oxen derived from other Bos species (Bradshaw et al. 2006). One population of approximately 6000 animals, which appears now to be superficially indistinguishable from wild Banteng, thrives in Garig Gunak Barlu National Park, on the Cobourg Peninsula in Northern Territory, Australia.

These animals derive from approximately 20 domesticated 'Banteng' taken from Denpasar, Bali in 1849, and they have recently been shown to be genetically identical to wild Banteng, at least to the extent of the molecular sequences analysed. This suggests that they have not been cross-bred previously to any significant extent with other domestic cattle (Bradshaw et al. 2006). Some domestic oxen lineages in mainland Southeast Asia, for instance in Cambodia, may have at least partial Banteng ancestry (e.g. Timmins and Ou 2001).
Population Trend: Decreasing

Habitat and Ecology:

Wharton (1968) summarized information about Banteng habitat from throughout its range and he concluded that on the Asian mainland it avoids evergreen rainforest and is usually within more open dry deciduous forests; but within the more humid areas of Java and Borneo it occupies secondary forest formations resulting from logging and fires, and enters tracts of sub-humid forest on occasion. However, on mainland Southeast Asia, human use of these preferred areas has increased, may have forced surviving Banteng to retreat to relatively closed formations where consequently they now overlap more with Gaur Bos gaurus; although good evidence for such an assertion is lacking.
In Myanmar, Banteng is reported to prefer flat or undulating terrain with light deciduous (particularly indaing forest) or mixed deciduous and evergreen forest, with grassy glades which burn annually, and patches of bamboo; but it has retreated to denser hill forest in the face of advancing cultivation (Peacock 1933; Wharton 1968; Tun Yin 1967; Prater 1971). Similarly in **Thailand,** Cambodia, Vietnam, and Lao PDR Banteng occurred) in open mainly deciduous forest with glades, parklands, and dense forest patches; and 'it is claimed by some that they prefer rocky, slightly hilly country, but they are certainly at home wherever the Kouprey [Bos sauveli] is found. Their range in Cambodia [and elsewhere in South-east Asia] suggests that Banting [sic] are more adaptable than Kouprey and perhaps less exacting in their habitat requirements' (Wharton 1957: 46; Wharton 1968). Recently the attachment to open, more deciduous forest types was reported in Xe Pian National Protected Area, Lao PDR (Steinmetz 2004). In Borneo Banteng is 'locally common in logged forest on flatland. And in dipterocarp, swamp and beach forests' particularly along river courses (Payne et al. 1985). In Java Banteng occurs in semi-deciduous monsoon forest with small clearings, more open grassland-monsoon forest mosaics, and 'parkland' formations (Halder 1976; Hommel 1987; S. Hedges pers. comm., based on unpubl. data 1991–2002).
Banteng generally occurs from sea level up to at least 2,100 m asl (Hoogerwerf 1970; National Research Council 1983; Pudyatmoko 2004; S. Hedges pers. comm. 2008 base on unpubl. data 1991–2002).
Banteng reportedly drinks large quantities of water and prefers feeding grounds near a permanent water supply (Hoogerwerf 1970). When conditions are favourable it drinks daily, preferring standing water; but during droughts animals can survive several days without drinking (National Research Council 1983). Mineral licks are also an important feature of Banteng habitat; in coastal regions which lack mineral licks animals seem to meet their mineral requirements by occasionally drinking seawater (Halder 1976; Payne et al. 1985; Alikodra 1987; S. Hedges pers. obs.).Although probably a grazer by preference Banteng should perhaps be considered an intermediate feeder since it can and does consume a lot of browse and fruits depending on season and local food availability. It eats grasses, sedges, herbs, bamboo, as well as the leaves, fruits, flowers, bark, and young branches of woody shrubs and trees including palms. The young shoots of alang-alang grass (Imperata cylindrical) are apparently a favoured food source (Hoogerwerf 1970). Lists of food plants are provided by Wharton (1957), Hoogerwerf (1970), Halder (1976), Sumardja and Kartawinata (1977), Djaja et al. (1982), Alikodra (1983), Prayurasiddhi and Smith (1993), and Hedges (in prep.). However, very little is known about the Banteng's nutritional requirements, seasonal and annual variation in their dietary preferences, or the composition and quality of their diet. Hoogerwerf was of the opinion that Banteng were largely dependent on grasses and sedges and, with reference to Ujung Kulon in West Java, he stated that 'there were no proofs that browse formed a substantial part of the diet.'

However, Schenkel and Schenkel-Hulliger (1969) reported that Banteng and deer in Ujung Kulon are forced to travel widely in the dry season and 'feed on bamboo, young palm leaves, bush, and saplings as well.' And the stomachs of a number of bulls which were shot in the South Tjiandur region of Java were almost completely full of non-graminaceous species. In Pananjung-Pangandaran Reserve in West Java Banteng were observed to graze on a variety of plants but appeared to prefer the grasses *Ischaemum muticum, Axonopus compressus, Paspalum conjugatum, and Cynodon dactylon;* and a woody forest shrub ,i>*Psychotria malayana*. Grazing intensity was lower in the forest than in the open fields where it was inversely proportional to the amount of *Eupatorium odoratum and Imperata cylindrical* although they did eat these species (Sumardja and Kartawinata 1977).

When undisturbed, Banteng displays a more or less fixed diurnal pattern of behaviour with periods (usually two to three hours long) of feeding alternating with rest periods of a similar length during which the animals ruminate. In areas which are subject to frequent human disturbance it becomes rather nocturnal and generally enters open areas only at night, although in especially attractive localities it will tolerate human presence to a certain extent (Halder 1976). In undisturbed areas Banteng's 'do not differentiate much in their activities between day and night, but...really large assemblages on open plains occur almost exclusively by day...This round the clock activity (which is, however, interrupted countless times to rest and/or ruminate) may not be regarded as abnormal' (Hoogerwerf 1970: 211-212).

The basic social group appears to be female–juvenile unit (as in the other large Asian Bovini) with larger groups tending to be more-or-less temporary assemblages. Maternal herds containing several adult cows, juveniles, and calves occur; these groups can often contain one or more subadult and adult males. Groups of cows without calves are also seen. For much of the year (adult) Banteng's are largely sexually segregated and all-male groups are frequently encountered at this time. Solitary animals tend to be mature bulls or sometimes old cows. The composition of small groups of cows with calves or juveniles, and the solitary state of old individuals may remain the same for months or even years. The composition of other small groups, particularly the unisex groups, usually varies from day to day. During the mating season the male groups disband and dominant males compete for access to receptive females (Hoogerwerf 1970; Halder 1976; S. Hedges unpubl. data 1991-2002).

In open areas large herds of more than 100 animals have been reported but these herds appear to be temporary associations formed when smaller groups and solitary animals arrive in an area to feed and the animals tend to reform into smaller groups when they leave the feeding areas. The composition of these groups often varies on a daily basis although there may be a stable nucleus of animals which returns to the same clearing day after day. The large herds may have an established dominance hierarchy but it seems to vary depending on the habitat. In more densely vegetated areas solitary animals or small groups, generally with fewer than eight animals, are more typical (Halder 1973, 1976; S. Hedges unpubl. data 1991–2002).

In **Thailand** the mating season is in May and June (Lekagul and McNeely 1977); while breeding reportedly occurs the whole year round in Myanmar (Peacock 1933). Wharton (1957) collected information from Cambodians living within Banteng habitat which suggested that the majority of births occurred in December and January which, assuming a gestation period of about 10 months, suggests that most conceptions take place in March and April. However, Hermanns (cited in Hoogerwerf 1970) was of the opinion that Banteng in mainland Asia mate from September to October. Hoogerwerf (1970) recorded mating taking place all year round in West Java but there was a peak of activity in October; and he thought that those cows which mated outside the peak period were probably exceptional. He was of the opinion that Banteng's in both East and West Java generally calved in July and August which would suggest that mating generally took place from October to December; more recent observations in East Java have also revealed evidence of year-round matings (neonates have been seen in all months) but with an apparent peak of births between June and September (S. Hedges unpubl. data 1991–2002).

Systems: Terrestrial

Major Threat(s):

The major threats to Banteng are hunting and habitat loss (Duckworth and Hedges 1998; Pudyatmoko 2004; Steinmetz 2004; S. Hedges unpub. data 1991–2002). In Sabah habitat loss to permanent agriculture is a serious threat, although hunting is equally significant and the species has been rapidly exterminated from many areas there (Payne et al. 1985; Ahmad AH pers. comm. 2008). Habitat loss has also been serious in Java since 1998. Elsewhere, hunting is the most widespread and significant threat, and is exacerbated especially in mainland Southeast Asia by human repopulation of lowland forest areas and associated habitat fragmentation, that is, the very areas where most Banteng's occur (Duckworth and Hedges 1998; Timmins and Ou 2001; R.J. Timmins pers. comm. 2008). Although huge tracts of suitable habitat were lost in the twentieth century, and continue to be converted, this has probably largely occurred after Banteng have been hunted out (R.J. Timmins pers. comm. 2008). Thus, the hunting is the proximate cause of decline, but habitat loss is continually reducing the maximum population possible if hunting issues were to be controlled.

The magnitude of the threat posed to Banteng by international trade in trophy horns is difficult to quantify. Nevertheless, given the small size of the remaining Banteng population and the number of trophies found for sale in Cambodia, the Lao PDR, Thailand, and Viet Nam, during what were essentially opportunistic surveys, it is clearly a major threat on the Asian mainland (Srikosamatara et al. 1992; Baird 1993; Srikosamatara and Suteethorn 1994, 1995; Martin and Phipps 1996; Srifa La-Ong et al. 1997; Duckworth et al. 1999; Timmins and Ou 2001). The threat posed by use of traditional medicinal substances derived from wild oxen is even harder to determine in the case of Banteng and essentially remains unknown, although it is thought to be a source of significant threat to Gaur (see 2008 Red List account).

The most important population in Cambodia is scattered through a forest landscape that encompasses four provinces (Mondulkiri, Kratie, Stung Treng and Ratanakiri) and five conservation areas (Lomphat Wildlife Sanctuary, Phnom Prich Wildlife Sanctuary, Siema Biodiversity Conservation Area, Mondulkiri Protection Forest (including the Srepok Wilderness Area) and Phnom Nam Lyr Wildlife Sanctuary) (Timmins and Ou 2001; Tordoff et al. 2005; R. J. Timmins pers. comm. 2008). Hunting is still rife in much of this area, and forest fragmentation is rapidly accelerating with human population in-migration, infrastructural developments (especially roads), commercial agricultural expansion, economic land speculation and mineral extraction (R. J. Timmins pers. comm. 2008). However, although perhaps less than 20% of this area is well protected from the aforementioned threats and protected area management is only close to effective in two areas, the most significant issue concerning the area is the long-term uncertainty of continuation of effective conservation management of the Srepok Wilderness Area and Siema Biodiversity Conservation Area. Although conservation efforts for Banteng and many other species have been, in both areas, extremely encouraging for the last few years, both areas face an uncertain future with the possibility of de-gazettment of conservation status of parts of them, the possible loss of adequate external funding necessary to maintain high standards of management, the possible loss of political support necessary to uphold high protection standards and the uncertainties of maintaining a motivated and well-trained staff.

On Java some populations are potentially threatened by heavy predation from Dholes Cuon alphinus (Hedges and Tyson 1996; Pudyatmoko 2004). All populations are also threatened by poaching and some, perhaps most, are threatened by habitat loss and degradation (S. Hedges unpub. data 1991–2002). During the 1980s–1990s, when poaching and land conversion were relatively well under control in Javan national parks, the chief threat to the large population of Banteng in Baluran National Park was loss of grazing area to invasion by the introduced tree Acacia nilotica (Leguminosae) that converts open grassland to dense thorny scrub-forest. This plant was introduced (without adequate risk assessment) as part of an attempt to create a living fire-break around the park's grasslands, wild fire then being adjudged the major threat to the park's monsoon forests. Since that introduction, repeated cutting of the acacia has led to coppicing into very dense thickets that contain little or no grass or other herbs and are difficult for the cattle to penetrate. Thus habitat loss and poaching are now serious limiting factors in Baluran National Park, and habitat loss/degradation remains a severe long-term threat to be addressed (S. Hedges pers. comm. 2008). Lantana camara (Verbenaceae) is also a problem in Banteng habitat in Baluran National Park and elsewhere on Java (S. Hedges pers. comm. 2008).

Bali cattle have long been interbred with other cattle: Banteng and Bali cattle can interbreed with both common cattle and mithan (Bos frontalis) (Fischer 1980; National Research Council 1983). Hybrids between Banteng and common cattle (Bos Taurus) of the zebu type are fully fertile; in hybrids between Banteng and Bos Taurus of the European type the males are sterile (Jellinek et al. 1980; National Research Council 1983). Domestic and feral livestock are thus a potential threat to the genetic integrity of wild Banteng populations and a number of reports suggest that wild Banteng does interbreed with domestic cattle. For example, Hoogerwerf (1970) referred to several sources from the 1930s and 1940s which mention that many groups of Banteng in Kalimantan (particularly East Kalimantan) were no longer pure-bred having interbred with stray domestic cattle. Wharton (1957) also found evidence of interbreeding with domestic cattle in Cambodia; and reports from Myanmar mention that Banteng feed alongside village cattle and occasionally interbreed with them (Tun Yin 1967; Salter 1983). In addition to the genetic threat, domestic livestock are a potential source of diseases and parasites. This can have very serious consequences for Banteng which appear to be particularly susceptible to a number of cattle diseases; for example, Banteng populations in Myanmar have been very badly affected by diseases from domestic cattle (Hundley cited in Wharton 1968).

Introgression with domestic cattle is not presently an issue in Sabah; there have been imports of Bali cattle mostly by large cattle farms who house animals in feedlots away from wild populations. Ahmad AH (pers. comm. 2008) is unaware of any instances of deliberate introduction of Bali cattle or other domestic oxen into forest areas, or of any plantation holders that have deliberately introduced their cattle into the range of wild Banteng. Although integration of livestock into oil palm plantation has been discussed for many years, this has not yet been put into practice (Ahmad AH pers. comm. 2008).

Conservation Actions:

This species is legally protected in all range states. The species is not listed on CITES, although a proposal was submitted to include it on Appendix 1 (Hedges 1996).

The most significant known Banteng populations remaining in Southeast Asia are in Java, Cambodia and perhaps Thailand and Borneo, and it is in these countries and around these populations that conservation efforts should be focused.

Banteng is mainly confined to protected areas throughout Southeast Asia (see Population), with the probable exception of Cambodia (and the situation in Myanmar is unknown), where significant number may still occur outside them (J.W. Duckworth and Than Zaw pers. comm. 2006; R.J. Timmins pers. comm. 2008). Presence in Lao protected areas is currently uncertain but there are at least 10 large forested protected areas in the south which may still support the species as well as three to four areas in the north, but any populations are certainly not large. In Vietnam, the species probably now occurs in only a few protected areas .

In Cambodia, most of the large forested protected areas, of which there are more than 15, are likely to retain at least residual Banteng populations. Unfortunately, in most protected areas of Lao PDR, Vietnam and Cambodia which still retain Banteng, protected status offers little if any protection to the species (R. J. Timmins pers. comm. 2008). Exceptions are Cat Tien National Park in Vietnam (although Banteng is not confirmed still extant), and the Srepok Wilderness Area and Siema Biodiversity Conservation Area in Cambodia; in the latter two areas there are now even signs of population recovery (WCS unpublished data; WWF unpublished data). The immediate conservation needs for the large Mondulkiri-centred Cambodian population are to continue strengthening site-based conservation management, in particular of the Srepok Wilderness Area and Siema Biodiversity Conservation Area, and laying the necessary groundwork to ensure that such management can be continued indefinitely.

Java has highly fragmented natural habitats and has done for centuries, reflecting longstanding high human population densities. Many protected areas were established during the Dutch colonial period but from independence up until the 1970s they were largely under-funded and neglected. After hosting the World Parks Conference in 1982, the Indonesian government gazetted a swathe of national parks and more structured conservation planning began, funded by the World Bank and other donors. The focus was largely on the 'multi-function' national parks and much money was spent on infrastructure, some staff training and increased personnel. The 'lesser' protected areas such as "game reserves" and "nature reserves" still had few staff and resources, and that has continued to the present. There was some habitat loss from protected areas through illegal logging, agricultural encroachment and other offtake, but the national parks of Java (several of which contain ideal habitat for Banteng and, at least into the mid 1990s, large populations) remained remarkably intact for much of the period. During the 1980s and early to mid 1990s guns were tightly controlled and the military and police were feared and respected. The strong culture of caged bird keeping meant that hunting, including that within protected areas, was primarily for birds and some small game, through various forms of trapping, including snaring. Thus there was relatively little hunting of Banteng and populations were stable or even increasing. Socio-political changes from 1997 led to a reduction in the respect for the police and military and the rise of a viewpoint that protected areas were the peoples' resources and would therefore benefit from decentralised management (M. Tyson pers. comm. 2008). This policy change, which risked a 'tragedy of the commons', has indeed led to increased destruction and poaching in the past decade. Banteng, as a large ungulate, is particularly threatened by this rapid rise in illegal hunting (S. Hedges pers. comm. 2008) and also by expansion of agriculture on Java and other forms of encroachment (G. Semiadi pers. comm.; S. Hedges pers. comm. 2008.). The decentralisation of conservation management decisions and actions, macro-economic fluctuations and reduced authority of The Law impedes control of illegal hunting (S. Hedges pers. comm. 2008).

In **Thailand** there are two populations thought to contain over 50 individuals, in Huai Kha Khaeng Wildlife Sanctuary and Kaeng Krachan National Park. Both areas are under active conservation management.

A large captive population exists (in addition to the large global population of the domesticated form which is believed to have hybridized widely with other oxen species) and a large introduced feral population, derived from Bali cattle, occurs in the northern territories of Australia (Bradshaw et al. 2006).

Kouprey *(Bos sauveli)*

Description:

The kouprey is believed to be a close relative to the aurochs and the gaur. A very large ungulate, the kouprey can approach similar sizes to the wild Asian water buffalo. These bovids measure 2.1 to 2.3 m (6.9 to 7.5 ft) along the head and body, not counting a 1 m (3.3 ft) tail, and stand 1.7–1.9 m (5.6–6.2 ft) high at the shoulder. Their weight is reportedly from 680 to 910 kg (1,500 to 2,010 lb).Unverified reports of a body mass up to 1,700 kg (3,700 lb) from Vietnam are considered dubious, since they far exceed other recorded weights for the species.

Kouprey have tall, narrow, bodies, long legs and humped backs. They can be grey, dark brown or black. The horns of the female are lyre-shaped with antelope-like upward spirals. The horns of the male are wide and arch forward and upward; they begin to fray at the tips at about three years of age. Both sexes have notched nostrils and long tails.

Range Description:

The historical distribution of this species was Cambodia, southern Lao PDR, south east **Thailand,** and western Vietnam (Grubb 2005). However, due to significant declines, this species is now thought to be possibly extinct.
Countries: Possibly extinct:
Cambodia; Lao People's Democratic Republic; Thailand; Vietnam

Population:

At an international workshop, held in Vietnam in January 1988, reports were presented which suggested that there were about 27 Kouprey in Vietnam, possibly 40?100 in the Lao PDR, and up to 200 in Cambodia, with perhaps a seasonal presence of a few animals in **Thailand.** These figures, which were little more than guesses, suggested that a total of about 100/300 Kouprey still existed in the late-1980s. Unfortunately, it now seems that these figures were too optimistic, especially for Lao and Vietnam where surveys in the 1990s were unable to document even significant populations of other species of wild oxen (Duckworth et al. 1994, 1999; Duckworth and Hedges 1998; Evans et al. 2000; Cox et al. 1991, 1992; Le Xuan Canh et al. 1997). Numbers of wild oxen were, however, much higher in Cambodia, even though, within the vast extents of habitat, their densities were already low as documented during an aerial survey of a large part of eastern Cambodia in 1994 (Olivier and Woodford 1994). The last published records of Kouprey are those of C. Wharton, who lead an expedition to capture Kouprey in 1963/64 (Wharton 1966). In 1969, J. Mellon saw two female Kouprey in the Chhep/Melouprey area of Cambodia, and in 1993 he was of the opinion that there may still be Kouprey in the area (J. Mellon in litt to D. Ashwell 1993). Later reports were also received by Pfeffer (1974 in litt. to IUCN/CMC). In the 1964?1970 period, Pfeffer undertook five expeditions to Indochina (each three months long) during which he collected information about Kouprey and took the only reasonable photograph of a wild Kouprey,(Kemf 1988; see Pfeffer and Ou Kim-San 1967; Pfeffer 1969).

None of the evidence for Kouprey since Mellons time, based on second-hand reports, hearsay, equivocal track identification or trophy horns (found in villages or wildlife markets and which could plausibly date back to Wharton's era) (e.g. MacKinnon and Stuart 1989, Duckworth and Hedges 1998; Duckworth et al. 1999; Timmins and Ou 2001, Timmins et al. 2003), has been particularly convincing in suggesting that viable populations of Kouprey remain; rather it suggests that a very rapid demise occurred. Wharton (1957) observed Kouprey to occur, even in what was considered optimal habitat, in lower numbers than banteng (his figures and other observations suggest a ratio of somewhere between 1:2 /1:10), and suggested that, because of their restricted range and habitat specificity, they were at elevated risk of extinction compared with the other wild cattle. Extensive survey work has now documented where significant wild cattle populations remain within the historical range of Kouprey, and in no area other than eastern Cambodia are wild cattle numbers high (Timmins and Ou 2001; R.J. Timmins pers. comm. 2008). In most areas wild cattle numbers are so low (low dozens of individuals scattered through hundreds of square kilometres of habitat) that it is no longer conceivable that Kouprey could survive (Le Xuan Canh et al. 1997, Duckworth and Hedges 1998; Timmins and Ou 2001; Timmins et al. 2003; RJ. Timmins pers. comm. 2008). Although it will, at least for the foreseeable future, be impossible to identify every remaining individual wild ox to species in such areas, there are no defensible grounds for considering the population of Kouprey to be anything other than negligible in such areas, primarily because of the naturally lower densities of Kouprey compared with banteng, the habitat specificity of Kouprey and its exceptional value in trade (Wharton 1957; Timmins and Ou 2001; R.J. Timmins pers. comm. 2008). This trade value pre-dates the meteoric rise in volume and financial value of wildlife trade out of Lao PDR, Vietnam and Cambodia during the 1990s and 2000s. Salter et al. (1990) found three sets of male and two sets of female Kouprey horns (all reportedly from Cambodia) in a village in southern Champasak Province (southern Lao PDR). The male horns were valued by the owner at USD4,000 and the female horns at USD1,600 per set. During a trip to Amphoe Muang (Mukdahan Province, on the border with Lao) on 29 March 1991, Kouprey parts were found to be available at one vendor. Horns were not on display but customers could examine photographs in an album: female Kouprey horns were offered at USD6,000-8,000 per pair and male horns at USD2,000 per pair. The male horns were polished so that the shredded ends could not be seen. During a second visit (in July 1991) the same vendor's stall was less active and no Kouprey trophies were on offer (although gaur and banteng could still be ordered). Wild cattle trophies were on offer at other vendors but no Kouprey horns or skulls were for sale at these stalls (Srikosamatara et al. 1992). During a visit to Ban Mai (Thai/Lao border) by Srikosamatara and his colleagues in April 1993 a vendor of wildlife products claimed to have sold a pair of Kouprey horns to a Thai buyer for USD800 two years before. He also claimed to have two more sets of Kouprey horns (both old males) for sale at USD2,800 and USD12,000 respectively. Three months later (July 1993) another survey (by I. Baird) found no Kouprey trophies for sale at the same vendors (Srikosamatara and Suteethorn 1994).] In eastern Cambodia there has now been substantial survey work (including observation-based field surveys and camera-trapping) which has documented hundreds of both banteng and gaur and even small numbers of wild water buffalo (R.J. Timmins pers. comm. based on WCS and WWF unpublished data; T. D. Evans and T. Clements pers. comm. 2008). This, in addition to suggesting Kouprey really has been hunted out, gives good numerical grounds to be confident that Kouprey (historically the rarer species) no longer occurs in the ratio found by Wharton and that Kouprey declined significantly faster and was almost certainly less resilient to hunting than are the other species (R.J. Timmins pers. comm. 2008). Certainly fewer than 2,000 wild cattle now survive within the historical range of Kouprey, 90% or more of those within Cambodia and the majority of those within the eastern provinces (R.J. Timmins pers. comm. 2008). If Kouprey still survives, it is certainly only as individuals (not functional groups) in very low numbers; its extinction, if not yet upon us, is certainly sealed.

Population Trend: Unknown

Habitat and Ecology:

Little is known of its biology and ecology, the only significant work being that of Wharton (1957). The species travels in small herds, primarily of females and calves, of up to 20 animals, which commonly associate with banteng. Mature males form bachelor herds. The diet was grasses, sedges, and some browse.
The Kouprey was primarily an animal of open deciduous dipterocarp forests, especially those areas with extensive grasslands. Although deciduous dipterocarp forests are extensive in Cambodia, and also in parts of adjoining countries (especially Lao PDR and Vietnam), the preferred facies of Kouprey, are much more localised, and perhaps account for less than 30% of the total area of the lowland mosaic forests dominated by deciduous dipterocarp forest (Timmins and Ou 2001; Tordoff et al. 2005; R. J. Timmins pers. comm. 2008). The species appeared to use patches of mixed deciduous and semi-evergreen forest which also occur in such landscapes. Most of the kouprey's range lies in a highly seasonal area receiving less than 2,000 mm of precipitation per annum. The terrain in this area is generally flat or undulating lowlands. The presence of pools and mineral licks were certainly important.
Systems: Terrestrial

Major Threat(s):

Hunting, both for local consumption and for trade (meat and body parts, especially horns and skulls), is (or, if the species is extinct, was) the major threat throughout the kouprey's range (Duckworth and Hedges 1998; Timmins and Ou 2001; Tordoff et al. 2005). Diseases from domestic and/or free-ranging livestock could have disastrous consequences given the already severely reduced Kouprey population. Habitat loss as a result of the rapidly increasing land clearance for cultivation (local and commercial), mining and logging, as well as increasing levels of other human disturbances, are also threats, but are insignificant compared with hunting (Tordoff et al. 2005; R.J. Timmins pers. comm. 2008; see also 2008 account for Banteng). Many wildlife species of high monetary value still remain widespread and, relative to neighbouring countries, abundant in the extensive lowland forests of Cambodia, and with the fall of the Khmer Rouge and a rapidly growing free market economy there has been a surge of hunting to supply bush meat, trophy antler/horn and medicinal markets, which is leading in many cases to very rapid declines in large quarry species (e.g. macaques have declined in some areas by over 90% in as little as five years; Timmins 2006; Bezuijen et al. in prep.; R.J. Timmins pers. comm. 2008), thus, however, rare Kouprey becomes it will continually be at risk because, unlike in a single-quarry species system where at certain levels it becomes uneconomic to seek out the last few individuals, hunting levels will remain high, fuelled by returns from the more common species (Duckworth and Hedges 1998; Timmins and Ou 2001; Tordoff et al. 2005; R.J. Timmins pers. comm. 2008). Wild oxen in Cambodia are now low in number and in most areas now rare or already hunted out (see 2008 IUCN Red List accounts for banteng, gaur and wild water buffalo). Knowledge of the Kouprey is widespread among rural people in Cambodia, concerning its rarity and the value of trophies and perhaps other body parts; as such it is a more desirable target than most in Cambodia.

Conservation Actions:

This species is listed in CITES Appendix I, and is legally protected in all range states. If the species is still extant it is most likely to be in eastern Cambodia in one of four protected areas (Lomphat Wildlife Sanctuary, Phnom Prich Wildlife Sanctuary, Mondulkiri Protection Forest and or Siema Biodiversity Conservation Area) (Timmins and Ou 2001; R.J. Timmins pers. comm. 2008). There are no Kouprey in captivity. There have been suggestions that domesticated Kouprey may survive in Cambodia (Hassanin et al. 2006), but this seems very unlikely (R.J. Timmins pers. comm. 2008), although domestic oxen in Cambodia may well carry Kouprey genes (Hassanin et al. 2006).
It is unlikely that specific survey work for Kouprey would produce any better evidence than has already been documented, and the best conservation measures for the species now would be to concentrate on in situ protection activities for large mammal communities in eastern Cambodia, especially building upon and strengthening the existing projects within the Srepok Wilderness Area of the Mondulkiri Protection Forest and the Siema Biodiversity Conservation Area.

Wild Asian Water Buffalo *(Bubalus bubalis)*

Description:

The skin of river buffalo is black, but some specimens may have dark, slate-coloured skin. Swamp buffalo have a grey skin at birth, but become slate blue later. Albinoids are present in some populations. River buffalo have comparatively longer faces, smaller girths, and bigger limbs than swamp buffalo. Their dorsal ridges extend further back and taper off more gradually. Their horns grow downward and backward, then curve upward in a spiral. Swamp buffalo are heavy-bodied and stockily built; the body is short and the belly large. The forehead is flat, the eyes prominent, the face short, and the muzzle wide. The neck is comparatively long, and the withers and croup are prominent. A dorsal ridge extends backward and ends abruptly just before the end of the chest. Their horns grow outward, and curve in a semicircle, but always remain more or less on the plane of the forehead. The tail is short, reaching only to the hocks. Height at withers is 129–133 cm (51–52 in) for males, and 120–127 cm (47–50 in) for females. They range in weight from 300–550 kg (660–1,210 lb), but weights of over 1,000 kg (2,200 lb) have also been observed.

Tedong bonga is a black pied buffalo featuring a unique black and white colouration that is favoured by the Toraja of Sulawesi.[9]

The swamp buffalo has 48 chromosomes; the river buffalo has 50 chromosomes. The two types do not readily interbreed, but fertile offspring can occur. Buffalo-cattle hybrids have not been observed to occur, and the embryos of such hybrids do not reach maturity in laboratory experiments.

The rumen of the water buffalo has important differences from that of other ruminants.It contains a larger population of bacteria, particularly the cellulolytic bacteria, lower protozoa, and higher fungi zoospores. In addition, higher rumen ammonia nitrogen (NH_4-N) and higher pH have been found as compared to those in cattle.

Range Description:

During the Pleistocene epoch the genus Bubalus was widely distributed throughout Europe and southern Asia and contained forms conspecific with *B. arnee*. When the climate became drier the genus was restricted to the Indian subcontinent, mainland South-East Asia, and some of the South-East Asian islands. In historical times B. arnee ranged across South and South-East Asia, occurring from Mesopotamia to Indochina (Epstein 1971; Mason 1974; Cockrill 1984).

Remnant populations of Wild Water Buffalo are thought to occur at single sites in each of southern Nepal, southern Bhutan, western Thailand, eastern Cambodia, and northern Myanmar, and at several sites in India: in the Bastar region of Madhya Pradesh, in Assam, in Arunachal Pradesh, and possibly in Meghalaya, Orissa and Maharashtra. Wild Water Buffalo is believed to be extinct in Bangladesh, Peninsular Malaysia, and on the islands of Sumatra, Java, and Borneo. The domestic form (considered by IUCN as B. bubalis) occurs as feral and domesticated populations worldwide (Grubb 2005).

The situation in Indochina is less certain. Lao PDR, Cambodia, and Vietnam were not included within the range of Wild Water Buffalo given in Corbet and Hill (1992). Free-living buffalo of unknown pedigree occur throughout the region (e.g. Sayer 1983; Laurie et al. 1989; Salter et al. 1990; S. Hedges pers. comm. 2008) but Wild Water Buffalo is probably extinct in Viet Nam and almost certainly in Lao PDR (Groves 1996; Grubb 2005; Duckworth et al. 1999; Tordoff et al. 2005; R. J. Timmins pers. comm. 2008).

The origin and current genetic status of the herds of apparently wild buffaloes in Sri Lanka is uncertain but it is thought unlikely that any true wild buffaloes remain there today. Corbet and Hill (1992) included Sri Lanka within the historical range of wild buffalo, although Ellerman and Morrison-Scott (1951), Gee (1964), and Maia (1970) thought that the Sri Lankan buffaloes were descended from introduced domestic stock. Certain ancient texts seem to support this view (Ashby and Santiapillai 1983). The fact that no buffaloes occur south of the Godavari river in India has also been taken to suggest that Sri Lanka possesses only feral buffaloes descended from introduced animals. However, Deraniyagala (1953), considered that the occurrence of fossil buffalo teeth in the gem sands of the Ratnapura area disproved this view, although it is not clear how old these buffalo teeth are (and Gaur remains found at similar depths in the same area were less than 1,000 years old). Moreover, morphometrics suggest that there was an ancestral population of animals on that island closer to Wild Water Buffalo than to Domestic Water Buffalo (Groves and Jayantha Jayawardene unpublished). Nevertheless, even if the Water Buffalo is indigenous to Sri Lanka the question of whether the free-living herds found there today should be treated as wild B. arnee still arises. In the nineteenth-century, free-ranging herds were common over much of the island's dry low country but they were nearly eliminated by an outbreak of rinderpest at the end of the century, and for a time their survival was in doubt (Phillips 1935). Phillips reported that small populations might have survived in the hill country but the subsequent intensification of agriculture probably led to their demise (Ashby and Santiapillai 1983). After the rinderpest outbreak buffalo recolonised much of the dry zone but most of them had apparently interbred with domestic stock and in 1953 Deraniyagala wrote '[the] relatively purest herds are restricted to Yala Game Sanctuary, but much vigilance will be necessary if this remnant is to be kept free from domestic animals which are now encroaching upon this once inaccessible area'. Woodford (1979) also suggested that the genetic integrity of the wild form has already been lost in Ruhuna. To conclude, even if it is assumed that Wild Water Buffalo once occurred on Sri Lanka it seems unlikely that they survived the rinderpest outbreak and the subsequent genetic swamping by feral and domestic buffalo: consequently all free-living buffalo populations on Sri Lanka almost certainly contain genetic input from domestic or feral stock. Neither Java nor Sumatra are included within the original range of wild Bubalus arnee as presented in many accounts. Nevertheless Stremme (1911) thought that the occurrence of the fossil *B. palaeokerabau* in Java made it probable that the buffaloes there belonged to the original fauna of the island (as Cuvier believed). Merkens (1927) also doubted, on historical grounds, the domestic origin of all free-living buffaloes on the island as did Mason (1974) who stated that domestic buffalo were present on Sumatra and Java long before the Hindus arrived almost 2,000 years ago. Moreover Van der Maarel (1932) provisionally regarded the fossil specimens which he obtained from Java (and indeed *B. palaeokerabau*) as specifically indistinct from modern buffaloes, pointing to a Pleistocene presence of the species on the island (cf. Medway 1972). Corbet and Hill (1992) also thought it probable that wild buffaloes occur on Java and Sumatra. Despite the doubts raised by Van der Maarel and Dammerman there is in fact little doubt that all the apparently wild buffaloes now living on Java and Sumatra are descended from domestic animals, or from Wild Water Buffaloes that have interbred with domestic and/or feral buffaloes (S. Hedges pers. comm. 2008).

Opinion is divided over whether to include Borneo within the historic range of the species. Corbet and Hill (1992) did not list it, and neither Mason (1974) nor Payne et al. (1985) considered it likely that Water Buffalo was part of the indigenous fauna. Lydekker (1898), by contrast, described the small buffalo of Sarawak as a separate subspecies (B. b. hosei), although Mason (1974) thought that Lydekker was probably describing the feral animals which were common there. Cockrill (1968) suggested that traders from the Hindu empire in Sumatra may have introduced the buffalo in the twelfth and thirteenth centuries and he did not think that there was any convincing evidence that would suggest that it was an indigenous species

Nevertheless, animal remains from the Niah caves indicate a Stone Age presence of buffalo in Sarawak (Harrisson 1961). Van Strien (1986) also considered it probable that Bubalus bubalis was part of the original fauna of the island and gave north-west Borneo as its current distribution. Harrisson, however, thought that the wild form was extinct. Feral (and semi-feral) buffaloes were formerly numerous throughout Borneo but the current status of the island's feral population is poorly known. What does seem certain, however, is that even if the species is indigenous to the island (as seems to be the case) no true wild B. arnee occur there today since they would have been genetically swamped by the numerous feral animals some of which were descended from buffaloes introduced from outside Borneo (S. Hedges pers. comm. 2008).

Countries: Native:

Bhutan; Cambodia; India; Myanmar; Nepal; **Thailand**

Possibly extinct:

Vietnam

Regionally extinct:

Bangladesh; Indonesia; Lao People's Democratic Republic; Sri Lanka

Population:

The total world population of Wild Water Buffalo is almost certainly less than 4,000 animals and may well be less than 200, occupying an area of less than 20,000 km². Indeed it is possible that no pure-bred Wild Water Buffaloes remain. However, these figures are little more than informed guesses, since any assessment of buffalo numbers is hampered by the difficulty of distinguishing between free-ranging domestic buffaloes, feral buffaloes, truly wild buffaloes, and hybrids between wild and other buffaloes. Individuals of Wild Water Buffalo and Domestic Water Buffalo are difficult to distinguish in some areas, and some domestic populations may be very closely related to (perhaps identical to) Wild Water Buffalo, as in Cambodia, where traditional forms of buffalo husbandry allow herds to range freely in forest areas (Timmins and Ou 2001; R. J. Timmins pers. comm. 2008). There have been few detailed analyses of the purity of the presumed remaining Wild Water Buffalo populations, nor in many cases is it obvious how such an assessment would be made. The Domestic Water Buffalo occurs as feral and domesticated populations worldwide, including in sympatry with most remaining populations of Wild Water Buffalo (Grubb 2005; S. Hedges pers comm. 2008). Some feral and domestic populations may well have conservation significance, retaining some of the genetic stock of the wild populations for that particular region, this may be especially true in Indochina due to traditional methods of Water Buffalo husbandry (S. Hedges pers comm. 1994, 2008; Timmins and Ou 2001; R.J. Timmins pers. comm. 2008).

In India, Wild Water Buffalo is now largely restricted to Assam, Arunachal Pradesh, and Madhya Pradesh. In Assam, the species is found in and around Manas Sanctuary, Laokhowa Sanctuary, Kaziranga National Park, and Dibru Sanctuary. In Arunachal Pradesh, the species occurs in and around Namdapha Sanctuary. There are two populations in Madhya Pradesh (Bastar district), one in Indravati National Park, and another in Udanti Sanctuary, which might extend into adjacent parts of Orissa. However, most, if not all, surviving populations are believed to have interbred with domesticated and/or feral Domestic Water Buffalo. In the late 1980s, Divekar suggested that there were fewer than 100 truly wild buffaloes left in India (in Madhya Pradesh); and by 1992 the numbers had dropped even further to an estimated 50 animals. However, Choudhury (1994) reported that in the early 1990s there may still have been about 3,300–3,500 wild buffaloes in Assam and the adjacent states of northeast India, plus a small number in Madhya Pradesh. A 1997 CAMP workshop assessed the number of mature individuals in India to be less than 1,500 (S. Hedges pers. comm. 1995). The population data suggest a decline of about 80% between 1966 and 1992 in central India. In northeast India, insufficient data exist to enable trends to be calculated but there is little doubt that the number of true wild buffalo has declined and is continuing to decline as a result of interbreeding with domestic and/or feral buffalo, hunting, and habitat loss (S. Hedges and J. W. Duckworth pers. comm. 2000).An unknown number of Wild Water Buffaloes, believed to include truly wild individuals, occurs in and around Bhutan's Royal Manas National Park. This is the only subpopulation in Bhutan and is contiguous with that in Manas Tiger Reserve in India.

Kosi Tappu Wildlife Reserve contains the only population in Nepal, where there were 159 wild Asian Buffalo in 2004 but this sole Nepali population is currently seriously threatened (Heinen and Kandel 2006).

There are no population estimates for Myanmar, reflecting a paucity of recent surveys. The extent of suitable habitat means the species could be still extant, and a few wild-living animals independent of human husbandry live in the Hukaung Valley of Kachin state (Than Zaw pers. comm. 2006; Saw Htun pers. comm. 2006; A.W. Tordoff pers. comm. 2006 to J.W. Duckworth pers. comm. 2008). Pending further information, these should be seen as likely to be genuinely wild stock or at least if feral, of archaic origin, and so, in either case, of conservation significance.

In Thailand, 40–50 wild buffaloes are reported to occur in the Huai Kha Khaeng Wildlife Sanctuary; this is the only population remaining in Thailand (Rattanawat Chairat in litt. to Anak Pattanavibool pers. comm. 2006). This population may be interbreeding with Domestic Water Buffalo, although individuals can often be distinguished in the field (Anak Pattanavibool pers. comm. 2006). In Thailand, wild buffalo numbers have apparently been relatively stable since the mid-1980s (R. Steinmetz pers. comm. 2006).

The population in Cambodia, which is confined to a small area of easternmost Mondulkiri and possibly Ratanakiri Provinces is very small, with perhaps a few dozen individuals remaining (Timmins and Ou 2001; Tordoff et al. 2005). There has not been any thorough analysis of the purity of these animals (i.e. whether they are truly wild, rather than some mix or perhaps just feral domestic animals), as is the case with most other putative populations of the species (Timmins and Ou 2001; R.J. Timmins pers. comm. 2008).

Before 1999, there was reportedly a population in Mom Ray, Vietnam, but this population no longer exists (Dang Huy Huynh in litt.; Do Tuoc pers. comm. 2006). Rumours still persist of presence in other areas, for example the Satay region (Tordoff et al. 2005; R.J. Timmins pers. comm. 2008). Wild buffaloes of the Cambodian population may wander into western Dak Lak province, but no recent records in Viet Nam have been confirmed to represent wild Water Buffalo (Le Xuan Canh et al. 1997; Eames et al. 2004; Tordoff et al. 2005; R.J. Timmins pers. comm. 2008).

The species was thought to formerly occur in Lao PDR, but the date of extinction is not known (Duckworth et al. 1999).

Population Trend: Decreasing

Habitat and Ecology:

Wild buffaloes are tied to the availability of water: historically their preferred habitats were low-lying alluvial grasslands and their surrounds, with riparian forests and woodlands also used (Lydekker 1926; Prater 1971; Choudhury 1994). In Indochina the species frequented the lowlands dominated by deciduous forests and with a marked dry season, where it apparently used small pools and marshes, in addition to permanently flowing rivers (Wharton 1957). Free-ranging domestic animals continue to live like this today; in such areas buffalo retreat to the vicinities of larger more permanent waterbodies during the height of the dry season, but move widely through the forests at other times of the year (Timmins and Ou 2001; R.J. Timmins pers. comm. 2008).

Before their major population decline, wild buffaloes were known to make long-distance movements with season, such that when an undoubted wild buffalo turned up in Panna district, India, some 225 km northwest of its nearest known present-day occurrence in Raipur district, this was not an unprecedented movement (Hasan 1980; Arun Singh 1980).

The upper elevational limit is difficult to determine, although animals certainly occur down at sea level; it possibly relates to a need for level land rather than preference for lower elevations, thus allowing for plateaux, and wild-living feral buffaloes live up to at least 1,000 m asl in East Java (S. Hedges pers. comm. 2008).Little has been published on the diet of wild (or feral) Water Buffaloes. They are probably grazers by preference, feeding mainly on grasses when available, but they also eat herbs, fruits, and bark as well as browsing trees and shrubs. Daniel and Grubh (1966) listed Cynodon dactylon, Themeda quadrivalvis, and Coix sp., as grasses known to be eaten by wild buffaloes in India. They also saw them feeding on the sedge Cyperus corymbosus. Dahmer (1978), Shrestha (1981), and Hiralal Prasad Kushwaha (1986) provided a little information about the diet in Nepal. Wild Buffalo also feeds on crops, including rice, sugar cane, and jute, sometimes causing considerable damage (Lekagul and McNeely 1977; Hiralal Prasad Kushwaha 1986; Bauer 1987b). Wild Water Buffalo can be both diurnal and nocturnal. Typically, it forms maternal groups of loosely structured herds, typically containing 10–20, but up to 100, individuals, year round. Adult males form bachelor herds of up to 10 individuals, with older males often solitary. The species exhibits a polygynous mating system, with females typically giving birth to single offspring, although twins are possible. It is a seasonal breeder in most of its range, typically in October and November, however, some populations breed year round. Its gestation lasts 10-11 months, with an interbirth interval of one year. Age at sexual maturity is 18 months for males, and three years for females. The maximum known lifespan is 25 years in the wild (Nowak 1999).

Major Threat(s):

The most important threats to Wild Water Buffalo are interbreeding with feral and domestic buffalo, hunting, and habitat loss/degradation. Diseases and parasites (transmitted by domestic livestock) and interspecific competition for food and water between wild buffalo and domestic stock are also serious threats. The scale of the threat posed to Wild Water Buffalo by the trade in wildlife products is difficult to quantify, not least because it is unclear how many of the trophies from reportedly wild buffalo are in fact from wild buffalo rather than from feral or hybrid animals (Divekar and Bharat Bhusan 1988; Heinen 1993; Choudhury 1994; S. Hedges pers. comm. 2000).

Most of the species' former lowland habitat has been lost to agriculture, and what remains is highly fragmented. However, especially in countries such as Cambodia and Lao PDR, vast tracts of suitable lowland forest remain from which the species has long since been hunted out, and certainly there are several viable tracts of habitat in which the species could be re-introduced (Tordoff et al. 2005; R. J. Timmins pers. comm. 2008).

The Cambodian population is under severe threat from hunting for trophy horns both by Cambodians and Vietnamese hunters crossing the border (Timmins and Ou 2001; Tordoff et al. 2005; R.J. Timmins pers. comm. 2008; Do Tuoc pers. comm. 2006). This threat is currently somewhat alleviated through to an active protected area management project, but hunting is still rife in much of the surrounding area, as is forest fragmentation due to human population in migration, infrastructural developments (especially roads), commercial agricultural expansion, economic land speculation and mineral extraction (Tordoff et al. 2005; R. J. Timmins pers. comm. 2008). However, the most significant threat to the area is the long-term uncertainty of continuation of effective conservation management of the Srepok Wilderness Area. This area faces an uncertain future. with part already degazetted from conservation status and the possibility that more would be excised in the future, the lack of long-term security of external funding adequate to maintain high standards of management, the uncertainty of long-term political support to uphold high protection standards and the uncertainties of maintaining a motivated and well-trained staff (R. J. Timmins pers. comm. 2008).

In Myanmar, the population of wild-ranging buffaloes either truly wild or at least living outside human custody for a long time in and around the proposed Hukaung Valley Tiger Reserve faces a very uncertain future. The seasonally-flooded plains, a matrix of grass and scrub, lies largely outside the already-established Hukaung Valley Wildlife Sanctuary and the areas of highest conservation significance lie south of the boundary even for a vast proposed extension to form the Hukaung Valley Tiger Reserve. All level areas in this region are under very high threat for conversion to agriculture, both rice and industrial-scale plantations of cash-crops. There is also an active hunting system for wild meat to feed hundreds of thousands of itinerant labourers in this area (J.W. Duckworth pers. comm. 2008).

The invasive Mikania sp. also potentially threatens wetlands in the South Asian range of wild Water Buffaloes (Hem Sagar Baral pers. comm. 2008).

Hydropower development and resulting changes in water flow and level conditions downstream also threaten the ecological maintenance of floodplain areas (Odden et al. 2005).

Disease epidemics spreading from domestic livestock presumably pose a threat, especially given the close overlap of Wild Water Buffalo populations and domestic livestock in South Asia, the high densities especially of the latter, and the small and localised nature of Wild Water Buffalo populations.

Conservation Actions:

Wild Water Buffalo is included in CITES under Appendix III (Nepal). It is legally protected in Bhutan, India, Nepal, and **Thailand.** The Cambodia population was a major factor in the decision to place further protection on the area, now designated as Mondulkiri Protection Forest, and a significant factor in the decision to establish the Srepok Wilderness Area Project of the Cambodian Department of Forestry and WWF (Goodman et al. 2003; R. J. Timmins pers. comm. 2008).

Most of the known populations are within protected areas.In Myanmar, which retains true lowland floodplains in near-natural state in the Hukaung Valley, the protected area system originally almost entirely excluded floodplain grasslands. The inclusion of some areas is now under consideration, and would benefit this species. However, the outcome is far from certain (J.W. Duckworth pers. comm. 2006). Most remaining populations still require protection from hunting (especially those in **Thailand** and Cambodia, and Myanmar if still extant), but probably the greater threat in South Asia is protection from contact with other domestic bovines, especially Domestic Water Buffalo.

There is an urgent need to evaluate the integrity of wild-living buffalo populations, including those generally taken as being truly wild and those living as wild animals within the native range, using habitat typical of wild animals, and which have lived outside even occasional husbandry for a long time, in order to determine populations of conservation priority. This should involve the assessment of the relationship of such populations in the context of obvious domestic lineages, especially those in close proximity to wild populations. Such an approach should use multiple genetic markers in addition to an assessment of morphological characteristics.

Surveys to investigate current status of wild-living populations are needed in Myanmar and perhaps the Satay district of Vietnam.

Mainland Serow *(Nemorhaedus sumatraensis)*

Description:

Belonging to a group known as the goat-antelopes, the Sumatran serow is a rather small-bodied animal, with dark upperparts and whitish underparts. The hair of the coat is long and coarse, and a long mane of white, brown or black occurs on the neck . Male and female serows are similar in appearance, with both bearing stout, slightly curved horns which can be used to defend themselves to deadly effect. The long ears are narrow and pointed, the face bears large scent glands below the eyes, and the tail is fairly bushy. A number of subspecies of serow were previously recognised, but are now considered full species. However, the taxonomy of the serow is not resolved and further research is needed.

Also known as

mainland serow.

Size

Head-body length: 140 – 180 cm

Tail length: 8 – 16 cm

Shoulder height: 85 – 94 cm

Weight

50 – 140 kg

Range:

The Sumatran serow occurs in Indonesia (Sumatra), peninsular Malaysia and **southern Thailand.**

Biology and Ecology:

The Sumatran serow is generally a solitary animal, although it may sometimes move about in groups of several individuals. Each serow inhabits a small area which is well marked with trails, dung heaps, and scents. This small area of habitat is selected so it can provide all the needs of the serow, such as sufficient grass, shoots and leaves on which to feed on during the early morning and late evening, and suitable sheltered resting places in caves or under overhanging rocks and cliffs. This home range is defended against any intruding serows by using their dagger-like horns, which are also used by this rather aggressive goat-antelope to fight off predators.

Although less specialised for climbing rugged mountains than some of its relatives, and with a somewhat slow and clumsy gait, the serow is nevertheless adept at descending steep, rocky slopes, and is also even known to swim between small islands in Malaysia. The mainland serow is territorial and lives alone or in small groups. It usually stays in a small area of only a few square miles where it grazes on grass, shoots and leaves from along beaten paths. It marks its territory with droppings and markings. It is most active at dawn and dusk, and spends the rest of the day in thick vegetation. The mainland serow gives birth to a single young usually in September or October. The gestation period is about eight months.The Serow is most active at dawn and dusk. During the heat of the day, the Serow takes shelter among the rocks, in caves, under overhanging rocks or cliffs, or in dense underbrush. The Serow is not as agile as the Goral, nor does it move as quickly, but the Serow is sure-footed, clambering easily along well-defined trails on mountain slopes. Furthermore, the Serow's smell, vision, and hearing are acute. The Serow is also a good swimmer.

The Serow mates in late autumn. The gestation period of the Serow is about 200 to 230 days. Serow mothers give birth in the summer, each to a single kid weighing about 3.6 kilograms (8 pounds). The Serow kid reaches full size and leaves its mother's territory one year after birth. The Serow reaches sexual maturity at approximately three years old. The Serow's life span in the wild is unknown, but Serow in zoos have lived for over 10 years. The Serow's natural predators include bears, tigers, and wolves. The Serow may use its horns to defend itself.

Present Status:

The Mainland Serow is categorized as Vulnerable in the 2003 IUCN Red List of Threatened Species and is listed under Appendix I of CITES. The Serow is under second class protection in China and is included in Schedule II, Part I of the Indian Wild Life (Protection) Act of 1972.

Major Threats:

The main threat to the Serow's survival is habitat loss. The Serow's habitat has been lost as a result of logging and the conversion of forest into cropland and pastureland. The Serow is also hunted, for food, as game, and to obtain its body parts for sale. The Serow's body parts are used in traditional medicine; serow oil supposedly has bone and joint healing properties. Additionally, the Serow is sometimes trapped in snares set for other animals.

Conservation Status:

It is on Appendix I of CITES, and in Class II in China's list of protected species. This serow exists in many protected areas including: Sichuan-Baihe, Fentongzhai, Jiuzhaiguo, Labahe, Mabian Dafengdin, Meigu Dafengdin, Tangjiahe, Tiebu, Wanglag, Wolong and Xiaozhaiziguo; Shaanxi-Tabaishan, Fuping (Wu *et al.*, 1986) and Takin Reserve; Gansu - Baishuijiang and Tou'ersantan; Yunnan - Fenshuilin, Gaoligongshan, Tongbiguan, Xishuanbangnan and Yulonghabaxueshan (Lijiang and Zhongdian); Fujian - Daiyunshan and Wuyishan; Zhejiang - Fengyangshan and Jiulongshan; Jiangxi - Jiulianshan and Wuyishan; Guangxi – Congzuo Xiling, Huaping, Qianjiadong, Shouchen and Yindinshan; Gu&hou - Fenginshan; Guangdung - Chebaling. Conservation measures proposed for China: 1) Reclassify serow as Class I in the national wildlife protection list. 2) Ban hunting and 3) the State Commercial System should be prohibited from purchasing products of this animal. 4) Conduct basic population surveys to determine distributions and numbers for development of conservation actions. 5) Undertake studies of ecology, population and management problems of this serow as part of the conservation management program for the giant panda.

Serow is also protected in Thailand. It is known or expected to occur in the following protected areas: National Parks – Chaloem Rattankosin (Than Than Lot); Doi Inthanon; Doi Khuntan; Erawan; Khao Chamao-Khao Wong; Khao Luang; Khao Pu-Khao Ya; Khao Sam Lam; Khao Sam Roi Yot; Khao Yai; Nam Nao; Namtok Phliu (Khao Sabup); Phu Kradung; Rhamkhamhaeng; Si Nakarin; Si Satchanalai; Thaleban and possibly Ao Phangnga. WiZdZifi Sanctuaries – Doi Chiang Dao; Doi Pha Muang; Huai Kha Khaeng; Khao Banthat; Om-Koi/Mae Tun; Maenam Pha-chi; Phu Khieo; Phu Luang; Phu Miang-Phu Thong; Salawin; Thung Yai Naresuan; Tong Nga Chang. Conservation measures proposed for Thailand: Determine the serow's status to assess the effects that logging and poaching are having on its distribution and numbers.

The taxonomic validity of this species, and its relationship to other species in the genus *Capricornis* needs to be assessed.

Chinese Goral *(Nemorhaedus caudatus)*

Description:

The Chinese goral is intermediate between an antelope and a goat and grows to a shoulder height of 50 to 78 cm (20 to 31 in) and length of 82 to 130 cm (32 to 51 in). It is stockily built with long, stout limbs and broad hooves. The horns are short and conical and the ears fairly long and pointed. The four teats are in contrast to sheep and goats which normally have two. The coat consists of a short, dense under layer and an upper layer of longer, semi erect, coarse guard hairs. The colour is somewhat variable, ranging from pale grey to dark brown or reddish-brown. A dark stripe runs along the back and the throat and underparts are pale.

Range Description:

Long-tailed goral are found in eastern Russia (Primorsky and Khabarovsk Territories), northeastern China, the Republic of Korea, and the Democratic People's Republic (DPR) of Korea (Grubb, 2005). In Russia, long-tailed goral were previously distributed along almost the entire southern half of Sikhote-Alin range, (i.e. the entire Primorsky Territory and southern part of Khabarovsk Territory) and on the southern end of the Bureya range, although its occurrence in this last area is uncertain. In the 1960s and 1970s goral were repeatedly observed along the left tributaries of the Khor river, along the Kafen, Chuken and Sukpay rivers, on the western slopes of Central Sikhote-Alin (approximately 47°N, 137°E) (Dunishenko, 1983). The third part of this species range occupies mostly the eastern slope of Sikhote-Alin along the coast of the Sea of Japan, between 43°40'N and 45°N, and also all the southern end of Sikhote-Alin range (Myslenkov and Voloshina, 1989). The fourth part occurs along the Chinese-Russian border in the Khasan region. A fifth area reportedly exists along the same border south of Khanka lake. Its distribution within these parts is patchy because goral are confined to specific habitat—steep rocky slopes covered with sparse, montane broad-leaved forest, from sea level to about 1,000 m (Heptner et al., 1961).

In China the long-tailed goral is found in the northeast, and stretches along the Xiao Hinggan Ling mountains (Lesser Khingan range), along the lower reaches of Sungari and Amur rivers in eastern Jilin and Heilong Jiang, and eastern Liaoning, and includes the Changbaishan range (Jilin) on the border with North Korea (Wang 2002, Smith and Xie 2008).

In DPR Korea, little is known of the recent distribution of this species (Shackleton, 1997). It is likely that it occurs (or occurred) in the Hamgyong mountains which lie inland from the northeastern coastline, and in the Taebaek mountains in the southwest and which continue into the Republic of Korea. A third area where it may occur is the Nanghim mountains in the north-central part of North Korea. These are extensions of the Changhai mountains from Jilin (China). said to have been seen in the mountains of **northern Thailand** are probably erroneous In the Republic of Korea, it is restricted to the Seorak mountains at the northern end of the Taebaek range (Won 1997).

Countries: Native:

China; Korea, Democratic People's Republic of; Korea, Republic of; Russian Federation

Population:

In 1977, the total number in Russia was estimated at 600 to 750 animals (Bromley, 1977). Myslenkov and Voloshina (1989) indicated no decline in numbers or disappearance of any known local population, and that numbers along the Sea of Japan may have been slowly increasing. Lazo District has about 300 animals, Terneisky District around 250, Olginsky District 80, and Dalnegorsky District 15 to 20 goral. Population density in several places, for example in Sikhote-Alin Reserve, reaches 35 animals/km^2. In South Korea, less than 50 animals are estimated to remain on Konbong mountain in Konsong-gun, Kangwon Province, near the demilitarised zone (DMZ) (Won, 1988). There are no estimates for DPR Korea. Won (1997) estimated less than 50 animals, but the basis for this is unclear. There are no reliable populations estimates for long-tailed goral in China. Smith and Xie (2008) considered it to be much reduced in numbers.

Population Trend: Decreasing

Habitat and Ecology:

The species inhabits steep mountainous areas and will sometimes use evergreen forests near cliffs, but primarily stays within rugged, rocky terrain. It inhabits steep and rocky terrain in evergreen and deciduous forests, especially with exposed grassy ridges from about 500-2,000 m asl. It eats a wide range of plant material: grass, herbs and shoots, leaves of small trees, nuts, and even some fruit. Group home range size is typically around 40 hectares, with males occupying marked territories of 22-25 hectares during the mating season. They typically live in small groups of 4-12 individuals, with older males usually solitary. Gorals are diurnal, and are most active in the early morning and late evening, but can be active throughout on overcast days. It keeps to steeper slopes where it is very agile over rocky crags and cliffs. The gestation length is 250-260 days (Myslenkov and Voloshina, 1989). Males and females reach sexual maturity at approximately three years, with a lifespan of up to 15 years or so. Mating takes place in early winter, and one, or rarely two-three, kids born about six months later.

Systems: Terrestrial

Major Threat(s):

The threats in Russia are poaching, which may be increasing. In DPR Korea and the Republic of Korea, major threats to its survival include habitat loss due to forestry, agriculture, and poaching. Threats in China include hunting, snaring, habitat degradation, and competition from domestic livestock (Smith and Xie 2008). Hunting is both for traditional medicine, and meat.

Conservation Actions:

All gorals are listed on Appendix I of CITES.

In Russia, it is listed as Category I of the Russian Red Data Book (Borodin 1984). It is protected in five Nature Reserves: Lazo Reserve harbours some, Sikhote-Alin Reserve has about 200, Zheleznyakovsky Sanctuary about 50, Vasilkovsky Sanctuary about 50, and an unknown but small number in Ussuri Reserve (Myslenkov and Voloshina, 1989). Captive breeding programs are being carried out in Lazo and Sikhote-Alin Reserves. Conservation measures proposed for Russia: 1) Reintroduce goral in appropriate areas formerly inhabited by the species to help restore its range. 2) Enlarge the size of the Sikhote-Alin Reserve to include the area south along the coastline. 3) Prohibit boats from approaching protected shores, keeping a distance of at least 0.5 km.

In North Korea, two areas proposed as Biosphere Reserves (Poore 1986), Mount Myohyung Nature Reserve (37,500 ha) 180 km NE of Pyongyang and Mount Paektu (Paekdu) Biosphere Reserve (132,000 ha; 41°56'N, 128°10'E), are reported to have goral. These proposed reserves lie across the border from the Changbaishan Biosphere Reserve in Jilin (China). Its status within North Korea is Indeterminate. Conservation measures proposed for North Korea. Surveys of population status and distribution, followed by 2) development of conservation actions.

I

In South Korea, the Amur goral has been designated as a Natural Treasure by the Cultural Property Preservation Law in 1968, while hunting of all species was banned throughout the mainland between 1972 and 1981 (Won, 1979). Mount Seorak National Park, which contains goral, was approved as a Biosphere Reserve in 1982, and is protected by the National Monument Protection Law (No. 2233) of 1910, the law of Forestry (act 67.68) of 1908, and the National Park Law of 1962. Goral is found in two protected areas in South Korea; in Mount Seorak National Park, and rarely in Mount Odae National Park. Goral was designated as Natural Monument No. 217 on 14 November 1968 and theoretically receives full protection. Conservation measures proposed for South Korea. Fully re-evaluate the species' status. 2) Determine the feasibility of providing adequate protection through enforcing current protection measures, creating additional protected areas, or both. 3) Determine if captive breeding may also be required. However, without adequate habitat protection, this measure would be basically academic.

In China, long-tailed goral are categorized as a Class II species. It occurs in most of the nature reserves located within its range in the North China and the Northeast China Regions. These include Taoshan and Jinpuohu (Heilongjiang); Chanbaishan and Zuojia (Jilin); and Suzihe (Liaoning).

The taxonomic validity of this species, and its relationship to other species in the genus Naemorhedus needs to be assessed.

Biography

Obtained degrees in Ecology, Wildlife Conservation and Species Ecology from Manchester Open University.

Work on species protection with English Nature (Now Natural England) from 1989 to 2010. Working mainly with all species of British Bats

Founder member of East Lancashire Bat group from 1989 until 2008

Director of The Michael Birt Consultancy from 1987 until present

Chairman of East Lancashire Beekeepers Association from 2005 until 2013

Consultant at Big Bee Farm Thailand from 2015 to present day

Became interested in Thailands Mammals and bees since first travelling to Thailand in 2003 and then moving to Thailand in 2013.

The last serious work on the mammals of Thailand was written in 1977 which is 42 years ago, so I thought I would try and bring up to date the latest information and research on such an important topic.

www.ingramcontent.com/pod-product-compliance
Ingram Content Group UK Ltd.
Pitfield, Milton Keynes, MK11 3LW, UK
UKHW061957041026
15591UKWH00016B/1050

9 781912 271542